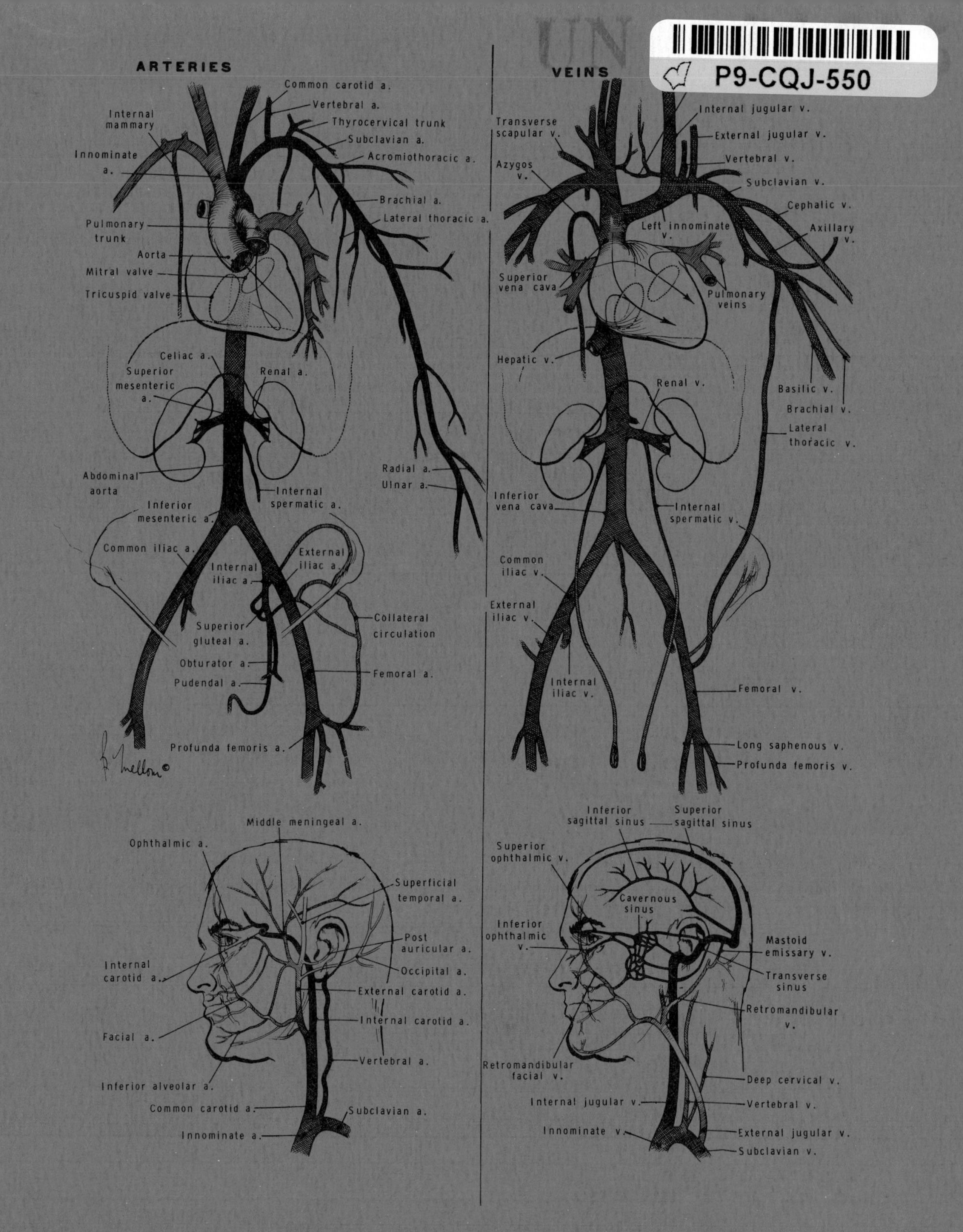
ARTERIES
Common carotid a.
Vertebral a.
Thyrocervical trunk
Internal mammary
Subclavian a.
Innominate a.
Acromiothoracic a.
Brachial a.
Lateral thoracic a.
Pulmonary trunk
Aorta
Mitral valve
Tricuspid valve
Celiac a.
Superior mesenteric a.
Renal a.
Abdominal aorta
Radial a.
Ulnar a.
Internal spermatic a.
Inferior mesenteric a.
Common iliac a.
External iliac a.
Internal iliac a.
Superior gluteal a.
Collateral circulation
Obturator a.
Pudendal a.
Femoral a.
Profunda femoris a.
Middle meningeal a.
Ophthalmic a.
Superficial temporal a.
Post auricular a.
Internal carotid a.
Occipital a.
External carotid a.
Internal carotid a.
Facial a.
Vertebral a.
Inferior alveolar a.
Common carotid a.
Subclavian a.
Innominate a.
VEINS
Internal jugular v.
Transverse scapular v.
External jugular v.
Vertebral v.
Azygos v.
Subclavian v.
Cephalic v.
Left innominate v.
Axillary v.
Superior vena cava
Pulmonary veins
Hepatic v.
Renal v.
Basilic v.
Brachial v.
Lateral thoracic v.
Inferior vena cava
Internal spermatic v.
Common iliac v.
External iliac v.
Internal iliac v.
Femoral v.
Long saphenous v.
Profunda femoris v.
Inferior sagittal sinus
Superior sagittal sinus
Superior ophthalmic v.
Cavernous sinus
Inferior ophthalmic v.
Mastoid emissary v.
Transverse sinus
Retromandibular v.
Retromandibular facial v.
Deep cervical v.
Internal jugular v.
Vertebral v.
Innominate v.
External jugular v.
Subclavian v.

The Human Body

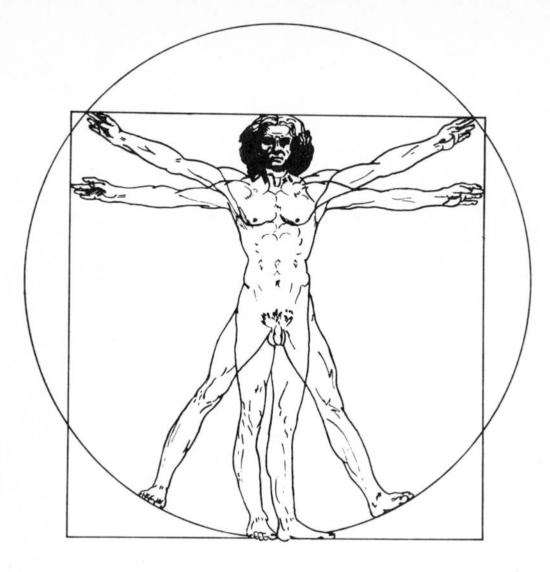

The Human Body

Its Structure and Physiology

Third Edition

Sigmund Grollman,
M.S., Ph.D., F.A.C.S.M.
Professor of Zoology
University of Maryland

Macmillan Publishing Co., Inc.
New York
Collier Macmillan Publishers
London

Macmillan Publishing Co., Inc.
866 Third Avenue, New York, New York 10022

Collier-Macmillan Canada, Ltd., Toronto, Ontario

Library of Congress Cataloging in Publication Data

Grollman, Sigmund.
The human body.

Includes bibliographies.
1. Physiology. 2. Anatomy, Human.
I. Title. [DNLM: 1. Anatomy, Regional.
2. Physiology. QT 104 G875h 1974]
QP34.5.G76 1974 612 72-91266
ISBN 0-02-348100-5

Printing: 1 2 3 4 5 6 7 8 Year: 4 5 6 7 8 9 0

Preface

The recent rapid expansion of experimental knowledge directed toward clarifying the relation between structure and function, and the desire of the author to clarify and strengthen the foundations established in the previous editions of the text and to deepen and broaden concepts were the guiding influences in this revision. The book is intended as an introduction to the subject of human structure and function. All the important terms, theories, and concepts of modern mammalian physiology are discussed. Pertinent aspects of inorganic, organic, and physical chemistry are discussed to bring about a better understanding of the physiological material. It is designed to suit the needs of the modern nursing and paramedical curriculum and undergraduate programs in physical education, behavioral and social sciences, agricultural sciences, and pre-law.

The fundamental unit of living systems is the cell, so our understanding of any living system must be based upon an understanding of the structure and function of cells. Attention is focused on the cellular, subcellular, and molecular levels as well as the organ level to clarify the relation between structure and function. In this way it is hoped that the student will see more clearly how the fundamental unit of structure and function affects and contributes to the over-all function and structure of the organ system and, finally, how the body acts as a single functional entity. Pervading emphasis on the interworkings of the various parts of the body helps the student grasp the over-all hierarchy of biological organization rather than viewing the various systems as totally separate entities.

Extensive changes have been made in the chapters on the cell, the nervous system, and the endocrine system. Typical of the new material is a section on the articulatory system. There are enlarged sections on cellular energetics, the thalamus, hypothalamus, reticular activating system, amygdala and synaptic transmission, and the actions of hormones and enzymes. In addition, many new tables, electron micrographs, and illustrations have been included to provide more detailed information and comprehension of both anatomical and physiological concepts as well as to reinforce those concepts.

To aid the student who is interested in studying a particular problem in depth, there are carefully selected up-to-date reference lists at the end of every chapter. In addition, the names of investigators and reference to their original research have been included to stimulate the interest and intellectual curiosity of the student.

S. G.

Contents

10 Respiration 311

11 The Digestive System 358

12 Metabolism 392

13 Nutrition 436

14 Kidney Function 488

15 *The Reproductive System* 519

16 *The Endocrine Glands* 568

Index 599

1

The Cell: Physical and Chemical Structure

In order to understand the complex systems that the higher living organism represents, the biologist must resort to methods in which the system is broken down gradually from the whole animal to organ systems, to organs, to tissues, and finally to single cells, making thorough observations at every step and correlating function with the level of organization.

The cellular physiologist, attracted by the simpler interactions, prefers the lowest level of organization, the cell, as material for study. This is also a good starting point for all who are interested in studying the function and structure of animals; for the cell theory, which was formulated in 1839 by Schleiden and Schwann, tells us that all plants and animals are made up of cells and that all biological phenomena are cellular. In order to understand all function, we must first understand function at the cellular level.

General Characteristics of the Cell

The term **cell** is rather difficult to define, because it represents an abstract generalization that attempts to cover a field that is too complex. It is possible, however, to give a general description that covers the majority of units of the cellular level of organization. First, we may make observations on the general characteristics of cells in regard to size and shape. In studying cells, the first thing we recognize is the great variation in size among different types of cells. Some cells are extremely small; in fact, there seem to be some bacteria that are invisible under the highest powers of the light microscope. Among organisms that have the smallest living mass are the microbes of the **pleuropneumonia group (PPLO),** which produce infectious diseases in various animals and man. They range in diameter from 0.25 micrometer (μm)—the limit in resolution of the optical light

Figure 1-1. Variation in size and shape of various types of cells drawn to scale. Liver, 30 μm; erythrocyte 2×, 7.2 μm; giant motor cell of cortex 100×, 500,000 μm; mature ovum, 300 μm; skeletal muscle fiber 40×, 100,000 μm.

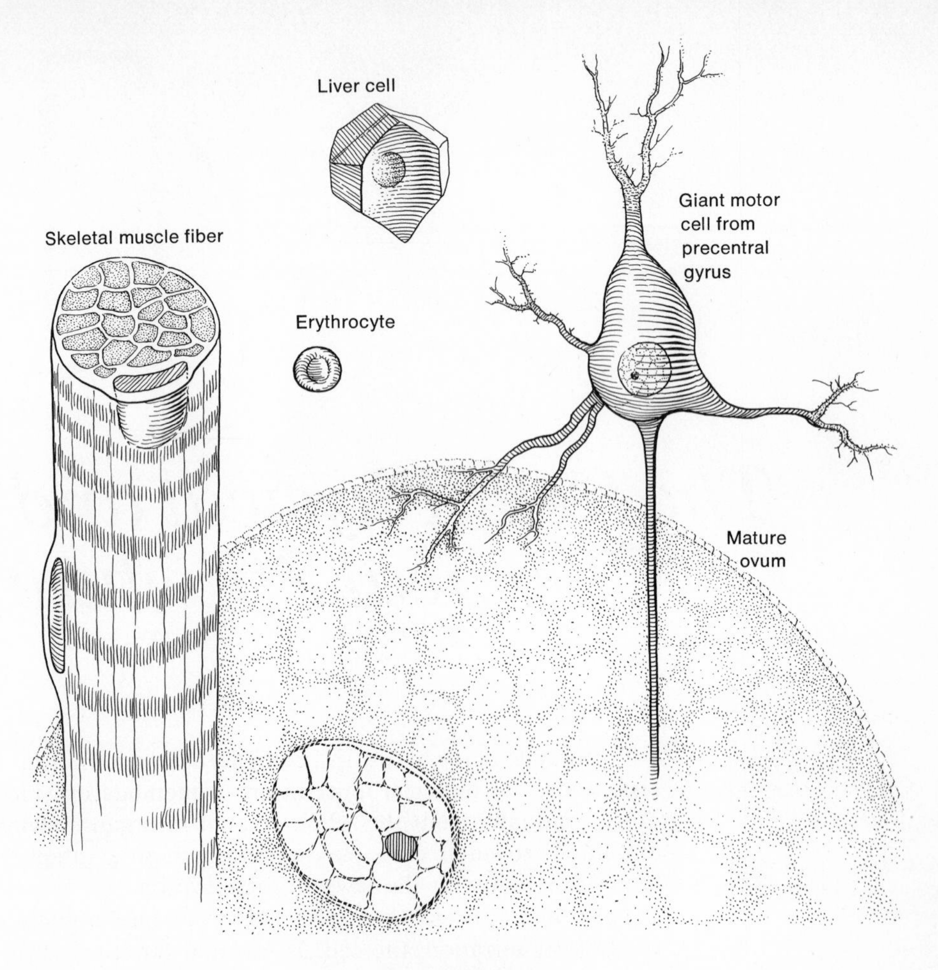

Figure 1-2. Typical architectural make-up of a cell as seen under the light microscope.

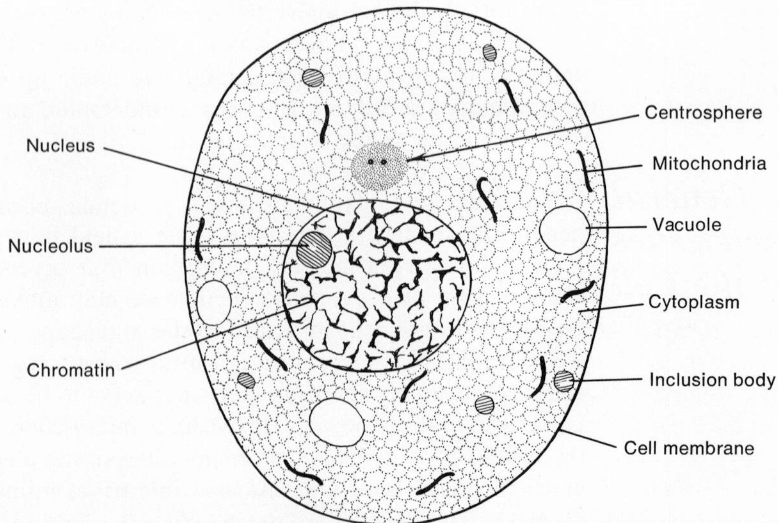

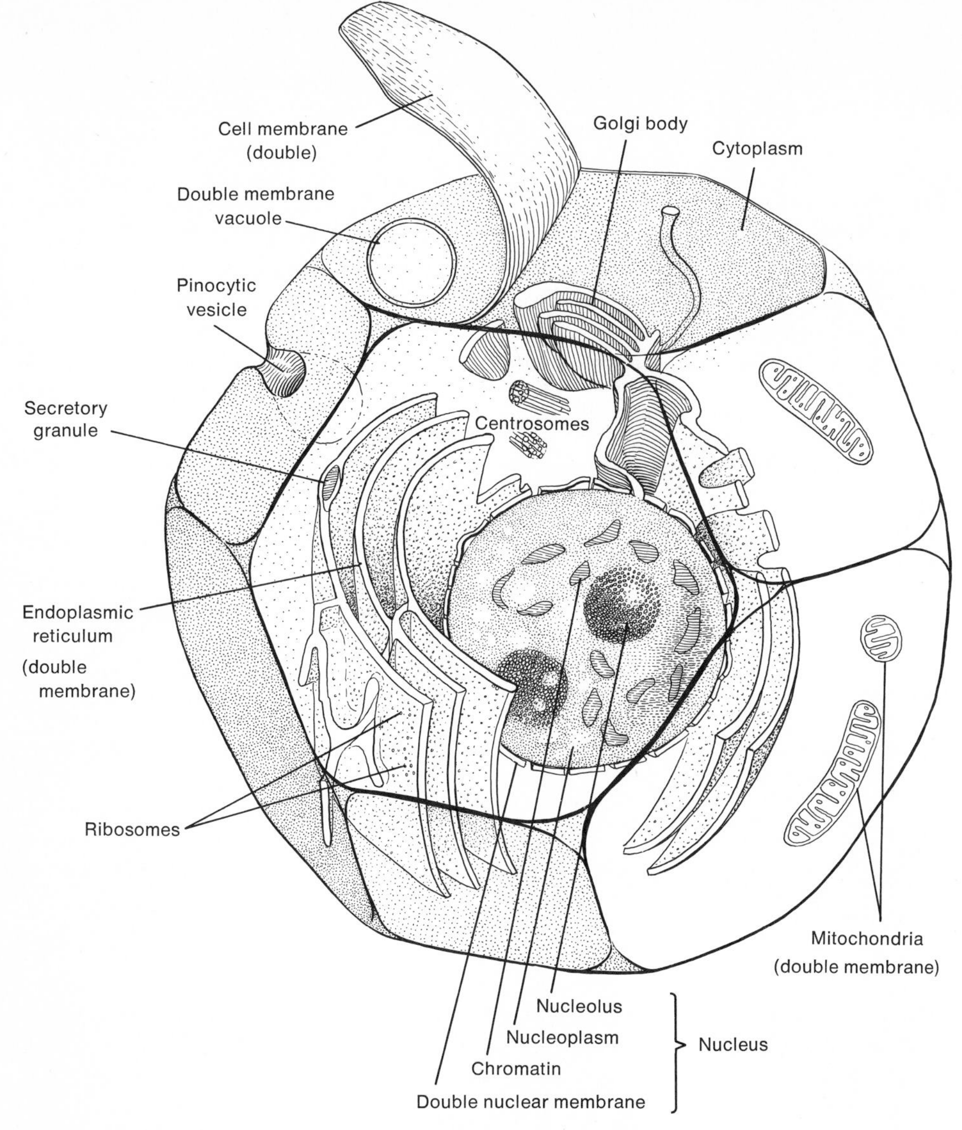

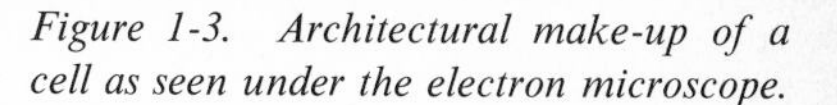
Figure 1-3. Architectural make-up of a cell as seen under the electron microscope.

microscope—to 0.1 μm. They correspond in size to some of the large viruses. The average bacterium is a thousand times larger (diameter 1 μm). PPLO are a million times smaller than animal cells. Although most cells are microscopic in size, some are large enough to be seen without the use of specialized magnifying equipment; these cells are considered macroscopic in size. The yolk of a hen's egg is a good example of such a cell, being about 1 inch in diameter.

The function that a cell must perform seems to have much to do with determining its size. Nerve cells acting as telegraph wires, picking up and sending electrical messages throughout the body, have to be extremely long. Female reproductive cells, which store food materials for sustaining a developing embryo for a long time, are relatively large and massive. Blood cells, which pass through minute tubules in the body, the capillaries, have to be extremely small. Even though we find this extreme variation in size of different kinds of cells, it appears to be a fundamental truth that the cells of any given tissue of a particular organism are nearly uniform in size and independent of the size of the mature individual. If we were to examine the cells that constitute the liver of a dog, we would find that all the cells of this organ are nearly uniform in size. If we compared these cells with the liver cells of some other organism, both specimens would be almost equal in size. The differences in the total mass of an organ are due to the number and not the volume of the cells (Driesch's law of constant cellular volume).

Table 1-1. ***Important Structural Components of a Cell, and Their Functions***

Structure	*Biochemical Composition*	*Function*
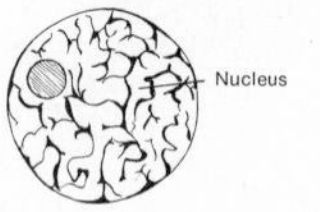 Nucleus	Protein, deoxyribonucleic acid, ribonucleic acid	Transfer of information; formation of tRNA, mRNA, DNA; regulation of metabolism
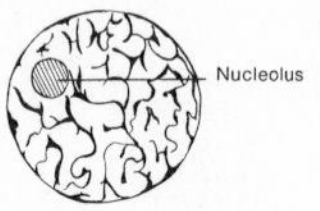 Nucleolus	Protein, ribonucleic acid	Important in metabolism of nucleus during stages of division; storage of tRNA
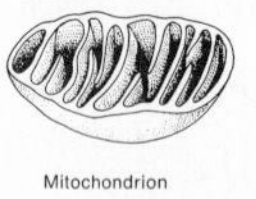 Mitochondrion	Double lipoprotein membranous sac; phospholipids, cytochromes, flavoproteins	Energy released from carbohydrate, fat, protein stored as ATP
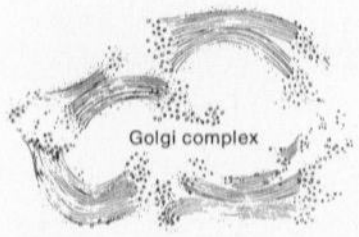 Golgi complex	Complex of many foldings of lipoprotein membrane	Storage and secretion of fats
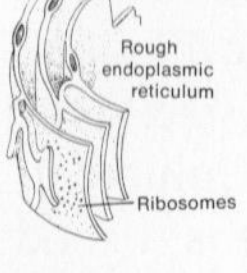 Rough endoplasmic reticulum; Ribosomes	Double lipoprotein membranous channel in cytoplasm containing granules of ribonucleic acid	Channels for transfer of materials; synthesis of steroid; ribosomes—area of protein synthesis
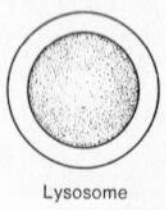 Lysosome	Double lipoptrotein membranous sac; enzymes	Contains enzymes for splitting of lipid, protein, and nucleic acids

As we have found this great variation in size of different kinds of cells, so we will also find even a greater variation in the shapes of cells. Cells are closely packed together in the multicellular organism to form sheets of cells, called tissues, and because these cells are relatively soft, jellylike bodies, they are subject to the laws of mechanics and thus will form many-sided figures of various types. The shape of a cell will also be influenced by its function. The typical cell is spherical; thus, cells that usually exist alone, uninfluenced by pressures of surrounding cells, will demonstrate this spherical shape. Egg cells, blood cells, and some protozoa are examples of the typical spherical shape that cells assume. The reason for this is that a fluid system, owing to the tension of the surface membrane, tends to assume the most compact form with the least possible surface exposed for a given mass. The sphere represents such a figure. Figure 1-1 (p. 2) shows some possibilities of variation in shape and relative size of cells.

Structure of the Cell

The typical architectural make-up of a cell includes a nucleus, surrounding cytoplasm, and a plasma or cell membrane (Fig. 1-2, p. 2). On closer examination and by using special techniques, other important structures may be identified (Fig. 1-3, p. 3 and Table 1-1). First let us consider the nucleus of a cell.

Nucleus

The **nucleus** appears as a more refractive intracellular sphere located somewhere near the center of the cell. Its protoplasm, usually referred to as the **nucleoplasm,** is separated from the protoplasm of the cell proper by the **karyotheca,** or **nuclear membrane** (Fig. 1-4). Scattered throughout the nucleoplasm of fixed preparations

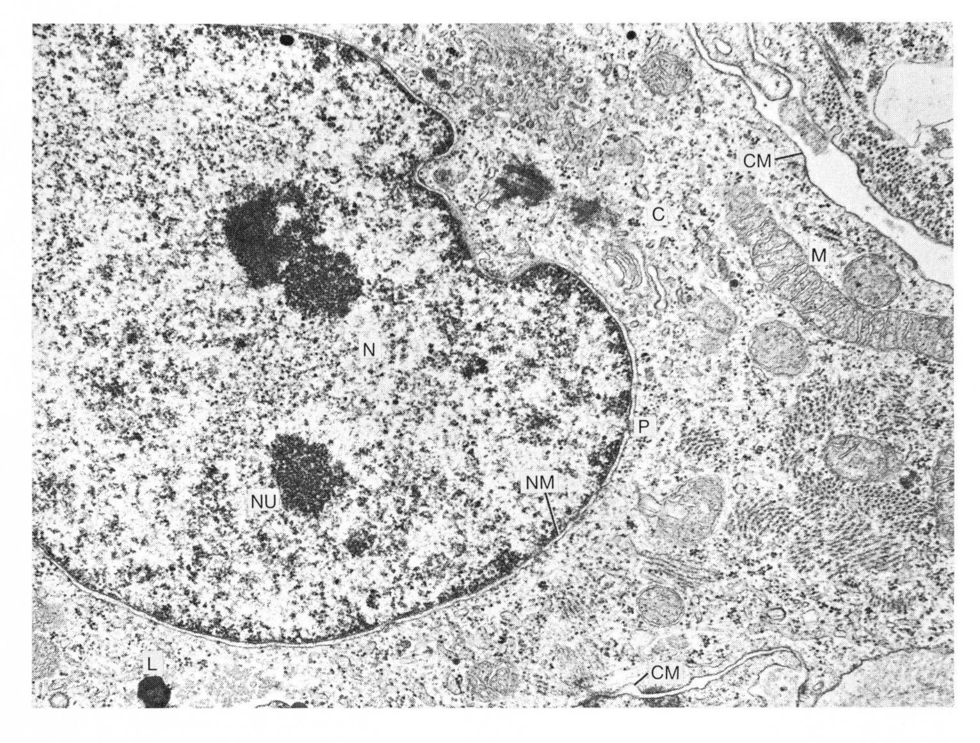

Figure 1-4. Electron micrograph of an animal cell revealing that the nuclear membrane is fenestrated and probably double. C, cytoplasm; CM, cell membrane; M, mitochondrion; N, nucleus; NU, nucleolus; NM, nuclear membrane; P, pore of nuclear membrane. (Courtesy George D. Pappas, College of Physicians and Surgeons, New York, and The Upjohn Co., Kalamazoo, Mich.)

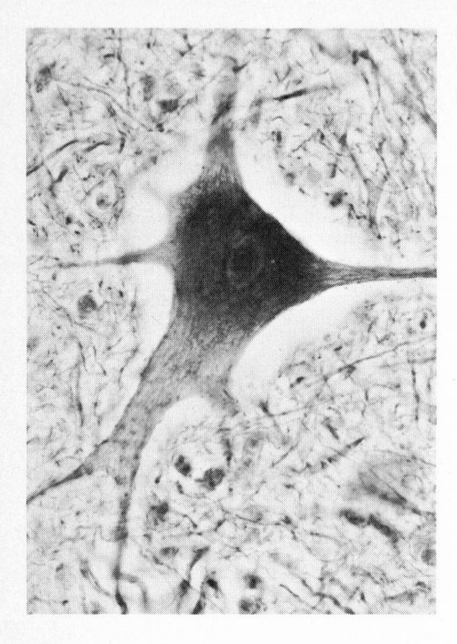

Figure 1-5. Section through spinal cord of mammal showing neurofibrillae in nerve cell. (Courtesy Ward's Natural Science Establishment, Inc., Rochester, N.Y.)

can be distinguished a series of twisted and interlaced filaments, the **chromonemata,** which contain a substance, **chromatin,** that stains intensely with certain dyes. Connecting the chromonemata are fine filaments, the **linin fibers.** One or more spherical bodies, the **nucleoli,** can be distinguished with appropriate methods of staining. We can see that the nucleus is not just a homogeneous mass of protoplasm but is a complex organization in itself, carrying out very specific functions for the cell and for the animal as a whole. In general, we can say the nucleus has some controlling influence on the activity of the cell, for all cells that engage in metabolic activity have a nucleus, and its absence is incompatible with a long and biochemically active life. The chromatin found in the nuclear material is concerned with the hereditary function carried out by the cell, for it is this material that forms the chromosomes, which, in turn, carry the genes.

Chemical analysis has shown that chromatin consists of four major macromolecules: a low-molecular-weight protein, a more complex protein, deoxyribonucleic acid (DNA), and ribonucleic acid (RNA). Studies have shown that DNA is the key molecule that enables chromatin to function as the hereditary unit within the cell. Caspersson and Schultz have demonstrated that the nucleolus contains ribonucleic acid, probably in the form of nucleoprotein. Ribonucleic acid plays an important role in the metabolism of the nucleus during the stages of division and of protoplasmic growth. However, the function of the nucleolus and, particularly, the significance of the cyclic changes that it undergoes during mitotic activity are still open to interpretation.

Cytoplasm

Commonly, the protoplasm found surrounding the nucleus and limited on its outer periphery by the cell membrane is referred to as the **cytosomic protoplasm,** or **cytoplasm,** to differentiate it from the nucleoplasm. Here again we find that, on more specific qualitative examination of the cytoplasm, many formed constituents can be found. These are divided into **organoids** (organelles) and **inclusions.** The organoids are capable of self-perpetuation, which would indicate they are specialized particles of living substance. They include such structures as fibrils, centrioles, mitochondria, lysosomes, and the Golgi apparatus. The inclusions consist of accumulations of lipids, carbohydrates, proteins, pigments, secretory granules, and crystals. **Fibrils** are thin protoplasmic threads running through and confined to the protoplasm. In many cells they are referred to as **tonofibrillae** and are presumed to offer stability to the cell. In muscle and nerve cells the fibrils are very highly developed and are referred to as **myofibrillae** in the former and **neurofibrillae** (Fig. 1-5) in the latter. They are concerned directly with contraction in muscle cells and conduction in nerve cells. **Centrioles** appear as small dots in a clear area of cytoplasm known as the **cell center,** or **attraction sphere,** which is very closely associated with the nucleus but not found in it. The centrioles are believed to initiate mitotic activity. The **mitochondria** are common to all animal cells and can be easily demonstrated when the stain Janus green B is applied to the tissue. They may be described as granular, rod-shaped, or filamentous; that is, they are pleomorphic bodies continually changing shape and position in the living cell. Lysosomes (Fig. 1-6) are oval-shaped bodies filled with powerful enzymes that can digest proteins, nucleic acids, and carbohydrates.

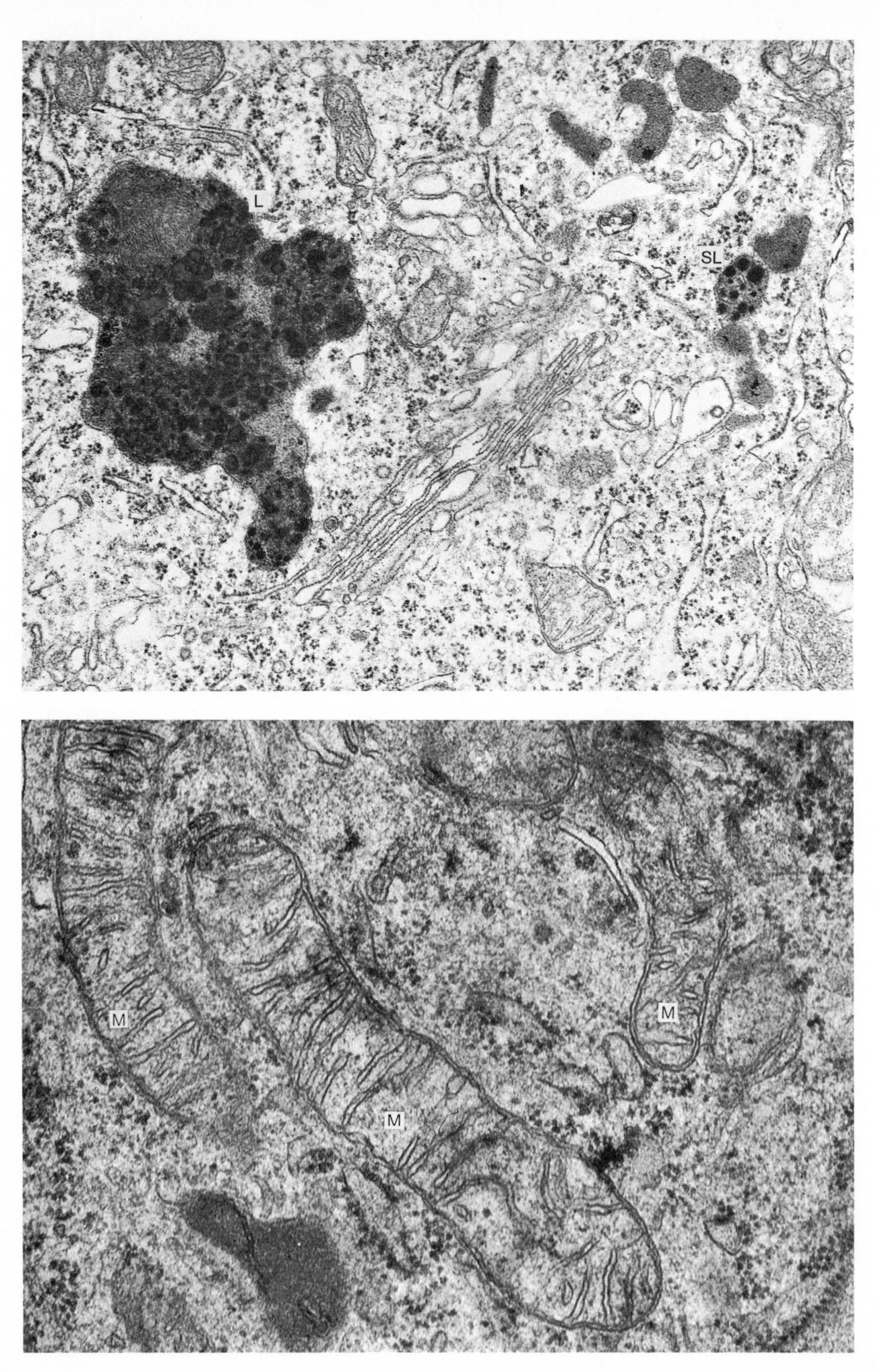

Figure 1-6. Lysosomes are present in most cells but are particularly abundant in white blood cells and macrophages. The powerful lytic enzymes contained within the lysosome are effectively separated from the cytoplasm by the special double membrane surrounding the structure: L, large phagocytic lysosome; SL, young lysosome. (Courtesy Timothy Mougel, Electron Microscopy Laboratory, Department of Zoology, University of Maryland, College Park, Md.)

Figure 1-7. Electron micrograph of pancreatic exocrine cell. The edge of the nucleus is at the bottom; several round or oval profiles of mitochondria are scattered across the picture from lower right to upper center. The parallel rows occupying the space between are elements of the endoplasmic reticulum. (Courtesy G. E. Palade, The Rockefeller Institute for Medical Research, New York, and The Upjohn Co., Kalamazoo, Mich.)

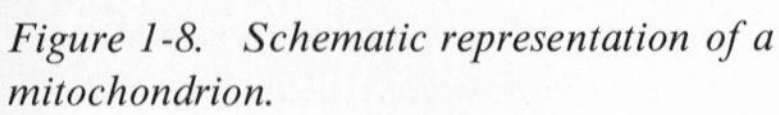

Figure 1-8. Schematic representation of a mitochondrion.

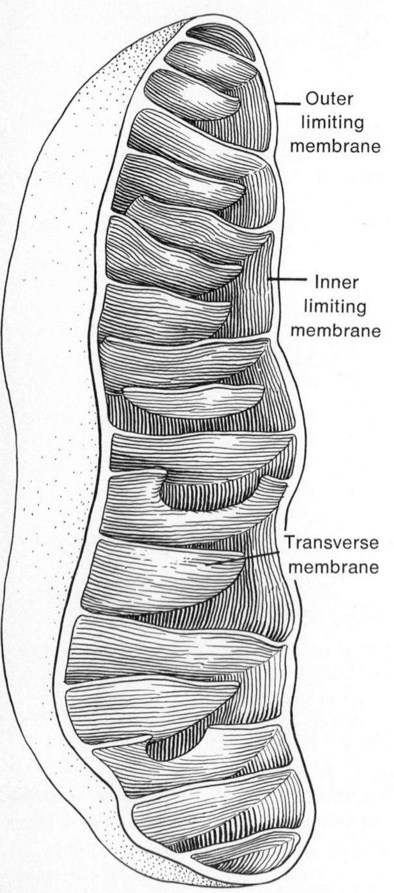

The structure of the mitochondria has been clearly revealed by the electron microscope (Fig. 1-7). A double lipoprotein membrane, similar to the nuclear and cytoplasmic membrane, surrounds the individual mitochondria. The inner membrane is thrown into many fingerlike projections toward the interior to provide a greater surface area (Fig. 1-8). The mitochondrion functions to oxidize the simple forms of proteins, fats, and carbohydrates that are delivered to the cell by the blood (that is, amino acids, fatty acids, glucose) into carbon dioxide and water or into a smaller acid or alcohol. The energy released in this process is transferred into the production of a high-energy phosphate compound known as adenosine triphosphate. This molecule is then secreted back into the cytoplasm and utilized in the cell whenever energy is needed.

The **Golgi apparatus** generally can be made visible by using an osmic acid and silver impregnation technique (Figs. 1-9 and 1-10). It appears as an irregular reticulum, situated around the centrosphere (central body). This structure has been thought by many authors to play an important role in cellular activities, particularly those dealing with the secretion of fats. **Cytoplasmic granules,** the inclusions referred to as **secretory granules,** are found in many types of cells, especially those having a secretory function. The exact nature of these granules is not known, although in some cases it has been correlated with the chemical composition of the secretion. Electron microscopy has revealed that the cytoplasm contains a series of channels coursing through its main mass. The membranes making up

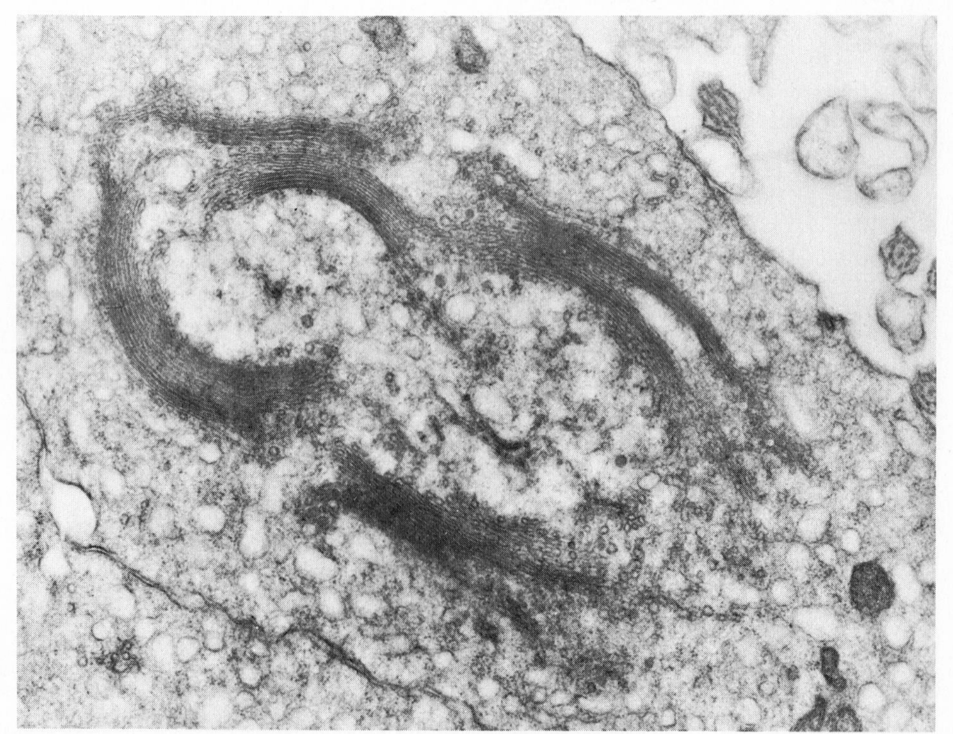

Figure 1-9. Electron micrograph of animal cell showing the Golgi complex. (Courtesy L. E. Roth, Iowa State University, Ames, Iowa, and The Upjohn Co., Kalamazoo, Mich.)

its walls may be variously arranged. These are referred to as the **endoplasmic reticulum** (Figs. 1-7 and 1-11). Lining the membrane are numerous particles called **ribosomes,** which have been shown to be rich in RNA. These particles are particularly active in the synthesis of proteins. The membrane itself is involved in the synthesis of steroids, a group of compounds that includes certain of the hormones.

Cell Membrane

The **cell membrane** has often been described as the outer living boundary of the cell. Like most membranes, the plasma membrane is selectively permeable, controlling the types of molecules which can enter the cell from the surrounding environment, or pass out of the cell into the environment. Studies with the electron microscope have shown the plasma membrane to be 0.01 μm or 100 angstroms Å thick, and double in appearance (Figs. 1-12 and 1-13). Because the resolution of the light microscope is only 0.2 μm, the cell membrane cannot be seen with this type of microscope. Thus, the actual demonstration of the membrane was dependent upon the advent of the electron microscope. It is interesting to note that the human eye has a resolving power of about 100 μm.

1 micron = 10,000 angstroms
1 millimeter = 1,000 microns

The contour of the cell membrane is very complex, with many infoldings and outpouchings (Fig. 1-13). The complex contour serves not only to dovetail cells together, thereby promoting tissue strength, but provides a large surface area for absorption. Recent experimental work indicates that the primary effect of insulin may be its influence on the integrity of this structure.

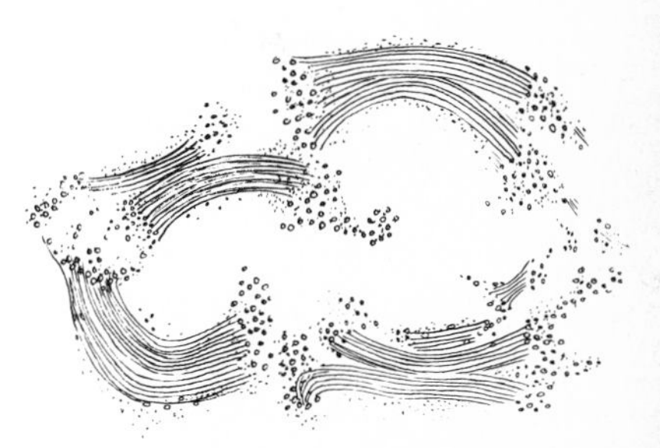

Figure 1-10. Schematic representation of Golgi complex.

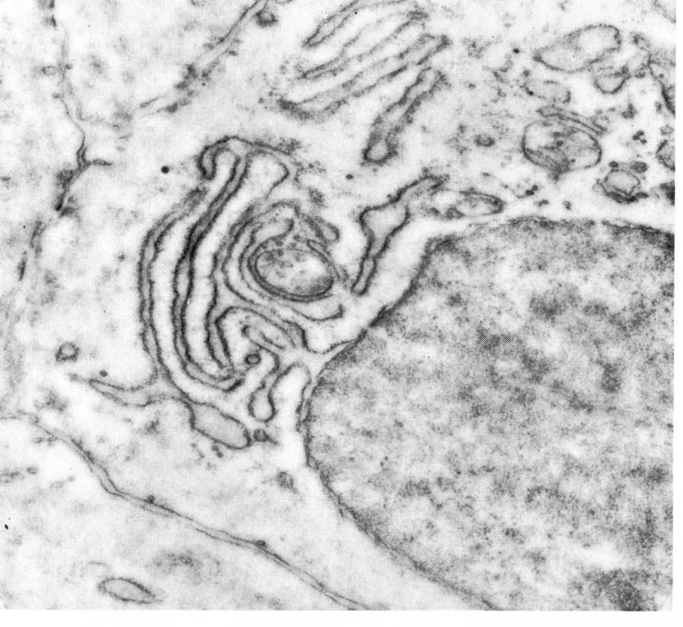

Figure 1-11. Electron micrograph of animal cell showing that the nuclear membrane running across the lower right quarter of the picture is continuous with the endoplasmic reticulum. (Courtesy Keith R. Porter, The Rockefeller Institute for Medical Research, New York, and The Upjohn Company, Kalamazoo, Mich.)

Figure 1-12. Hypothetical structure of the cell membrane. A double layer of lipid material is sandwiched between a double layer of protein.

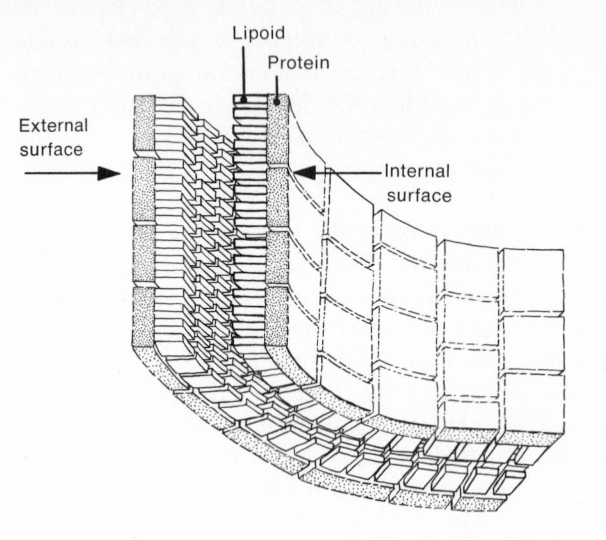

Figure 1-13. Electron micrograph showing the interdigitation of the plasma membranes (arrows) *of adjacent absorbing cells from a normal rat jejunum. The tissue was fixed in 2 per cent veronal acetate–buffered osmium tetroxide according to Palade. The magnification is 39,000×.* (*Courtesy M. Lee Vernon and R. C. Calvert, Hazelton Laboratories, Falls Church, Va.*)

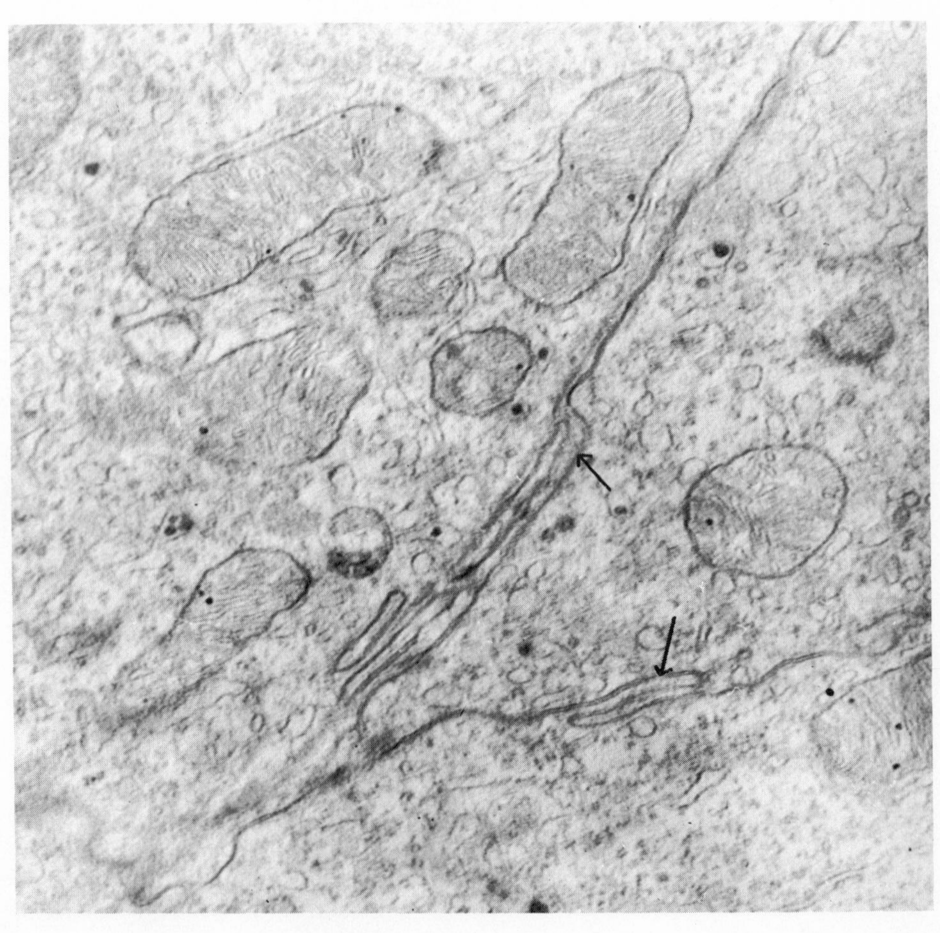

Chemical Composition. Permeability and surface-tension studies on various kinds of cells indicate that the membrane is mostly **lipoprotein** (fat plus protein). The **phospholipids** and **cholesterol** are the chief types of fatty materials found. **Protein** and small amounts of **carbohydrate** make up the remainder.

Fat solvents readily penetrate the cell, suggesting the lipid structure of the membrane while its low surface tension, which gives the membrane a characteristic wettability, suggests its protein make-up. It is known that protein molecules can fold or unfold, and the cell membrane also demonstrates this flexibility.

Lipid. Studies on plant cells and red blood cells of animals have led to the concept of a plasma membrane structure as a mosaic of lipid and protein and a small amount of carbohydrate. Lipids make up 20 to 40 per cent of the dry weight while the protein make up 60 to 80 per cent. The carbohydrates distributed between both the lipids and protein account for about 5 per cent. The phospholipids, which account for more than half of the total lipid content of most tissues, are located almost exclusively in the membranous components of the cell. There are five principal phospholipids of this type:

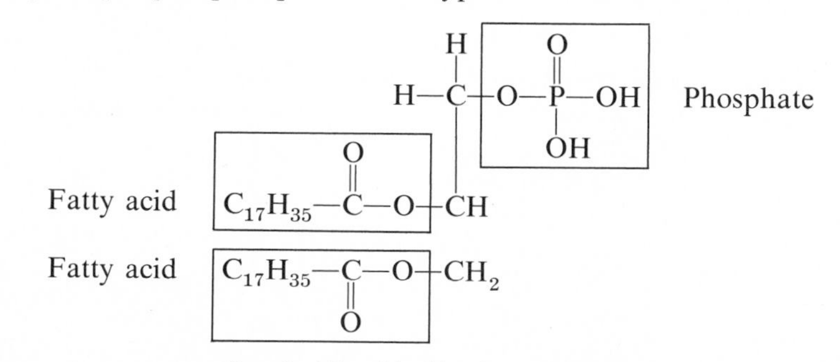

Phosphatidic acid with a base phosphate

Four other phospholipids are present, each of which has a different group—choline, inositol, ethanolamine, and serine—attached to the phosphate. These are named, respectively, as phosphatidyl choline, phosphatidyl inositol, and so on. Choline and inositol are B vitamins; ethanolamine is an alcohol, and serine is an amino acid. The choline version, usually called lecithin, is the most abundant:

$$CH_2{-}N^{+}(CH_3)_2{-}CH_2{-}CH_2{-}O{-}P(O)(O){-}O{-}CH_2$$
$$HC{-}OOCH_{35}C_{17}$$
$$H_2C{-}OOCH_{35}C_{17}$$

Phosphatidyl choline or lecithin

In the cell membrane these phospholipids comprise the middle section and are arranged with their water-insoluble fatty acid tails pointing toward each other and their water-soluble heads, glycerol and phosphate, facing outward and attached to two sheets of protein, which cover the lipids like two slices of bread in a sandwich.

Carbohydrate. Hexose (glucose, mannose, galactose), hexosamine (glucosamine, galactosamine, mannosamine), fucose, and sialic acid make up most of the carbohydrate, which is mainly bound to proteins. Only a small amount of carbohydrate exists in the form of glycolipid in the cell membrane. Gangliosides (glycolipid) are important constituents of the nerve membrane and are probably involved in ion transfer. Lapetina et al.[1] have demonstrated a specific concentration of gangliosides in the membranes of nerve endings, especially those that are rich in acetylcholinesterase.

Protein. The proteins, which constitute the bulk of the plasma membrane, have been characterized as high-molecular-weight proteins of unknown size. The amino acid composition and other properties of these proteins resemble those of actin in muscle. Some 30 enzymes have been detected in isolated plasma membranes. Those most constantly found are Mg^{2+} ATPase, sodium-activated Mg^{2+} ATPase, potassium-activated Mg^{2+} ATPase, alkaline phosphatase, acid phosphomonoesterase, ribonuclease, and 5-nucleotidase.

Water and some water-soluble materials readily pass through plasma membranes, which suggests that there are pores present which are approximately 7 Å in diameter. Thus, materials could enter and leave the cell by two routes: (1) through solution with the lipoid material, which would limit the materials to lipid-soluble substances, and (2) through the pores between the protein and lipid area, which could afford a means of entry for substances of low molecular volume such as water-soluble organic compounds and inorganic ions. It is also now known that the membrane contains a group of enzymes known as **permeases,** which are capable of pulling through certain molecules such as amino acids and glucose.

In addition to these two methods of transport, some cells are capable, by two other processes, of bringing large molecules incapable of passing through the membrane into the interior of the cell from the fluid environment. In one instance the plasma membrane invaginates, forming a pocket that allows materials to drop into it, and then pinches off, forming a vacuolelike structure that is incorporated into the cytoplasm to be digested. This process is referred to as **pinocytosis,** from the Greek words meaning "drinking cell." In another process arms of cytoplasm stream out, surround large particulate matter, and draw these materials into the interior of the cell, where they are digested by enzymes. This process, which resembles the act of eating, is referred to as **phagocytosis,** from the Greek words meaning an "eating cell."

Protoplasm

Up to this point we have considered only the visible physical structures of the cell, but still no clue has been given as to what life really is. Perhaps the answer will be found in a thorough study of the ground substance of which all cells are composed, the basic substance of life. This is the substance that early investigators

[1]E. G. Lapetina, E. F. Soto, and E. DeRobertis, "Gangliosides and Acetylcholinesterase in Isolated Membranes of the Rat-brain Cortex." *Biochim. Biophys. Acta,* **135**:33 (1967).

referred to as "gelatinous juice." It was not until the beginning of the nineteenth century that the soft substance of the interior of the cell came under the scrutiny of investigators. In 1835, Dujardin called this soft, gelatinous substance "sarcode." He was the first to describe the substance in great detail. Later, Purkinje and von Mohl called it **protoplasm,** a term that has persisted to our time. Max Schultze, in 1861, was the first to establish the similarity between protoplasm of plant and animal cells, which led to the development of the **protoplasmic concept.** This concept attempts to tell us what the scientist believes protoplasm really is. It states that only when a collection of certain chemicals has an organization so arranged as to have the attributes of life can it be considered protoplasm. We know from this concept the following about protoplasm. First, it is a collection of certain chemicals, not any or all types, but certain ones. Second, a definite organization is required of these chemicals. Finally, when the first two conditions are met, we should have a mass showing the attributes, or properties, of life. These properties of living material have been listed in a variety of ways. In general, they are contractility, conductivity, irritability, reproduction, and metabolism. All living things should respond to external stimuli (irritability). They may do this by a withdrawal or attraction reaction, which is realized by movement of protoplasm (contraction). The ability to pass this reaction along to the mass of protoplasm of neighboring cells is conductivity. All living things, from the single-celled organism to the most complexly organized animal, are capable of reproducing themselves, and, thus, we find reproduction as a property of life to differentiate such material from nonliving material. Finally, living substance must be able to carry out spontaneously a variety of biochemical reactions which are quite necessary for life and which, in general, require oxygen and liberate carbon dioxide as a universal end product. This over-all activity is grouped under the heading "metabolism." Now that we know what we mean by the term *protoplasm,* perhaps we will get a better understanding of this basic substance through a detailed study of its chemical and physical organization.

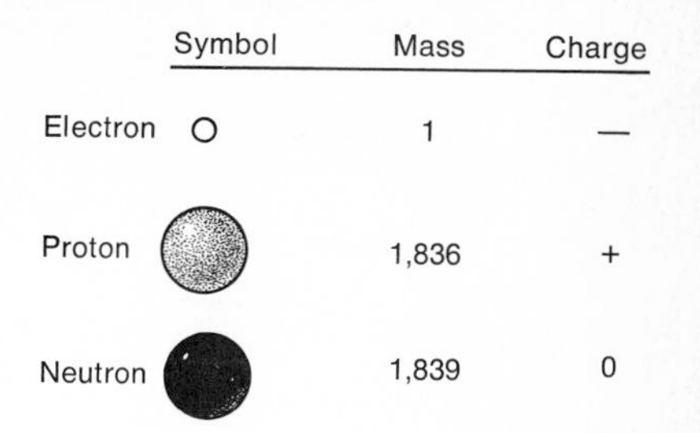

Figure 1-14. Types of particles found in an atom.

Chemical Organization

Science holds that all the realities of the physical world are expressions of two fundamental entities—matter and energy. **Matter** may be defined as something that occupies space and has mass (inertia) and weight. All matter is endowed with the power of energy and is always in action to a greater or lesser extent because of energy change; that is, the activities of matter are the result of energy changes.

Energy can be defined as the capacity to do work and may be in the form of either potential or kinetic energy. Any matter capable of doing work is said to contain **potential energy,** or stored energy. The current in an electric battery, a lump of coal, unexploded gunpowder—all contain potential energy. The ability to express this energy is called **kinetic energy.** Kinetic energy is, then, the energy of movement, whether it is in a falling body or an explosion.

It is important to remember that the various manifestations of energy are capable of being transformed from one to the other. Heat energy may be transformed into motion, into electric current, or, in living organisms, into growth and reproduction.

Units of Matter

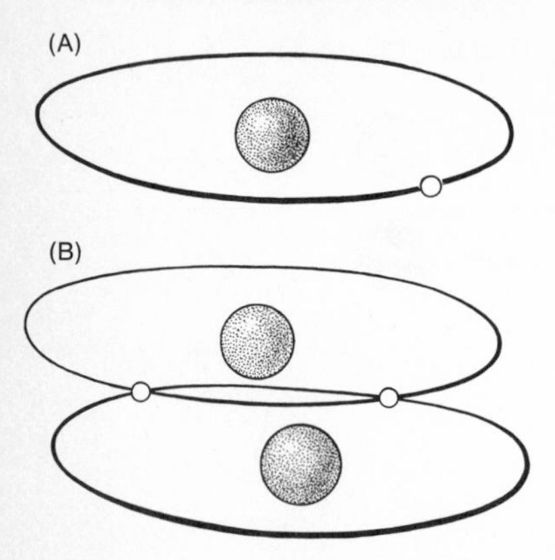

Figure 1-15. Organization of atoms showing the relationship of protons, electrons, and neutrons. A. *Hydrogen atom.* B. *Hydrogen molecule. Compare this with the more complex helium and carbon atoms.*

All material things are made up of minute, discrete particles called **molecules,** which are the smallest units into which a substance can be divided without destroying its identity. The molecule, however, is not the smallest unit of matter, for all molecules are composed of lesser particles called **atoms.** Each kind of molecule consists of a definite kind and number of atoms linked together by chemical bonds in a definite way. The simplest molecules consist of combinations of two atoms of the same kind, as when two atoms of hydrogen combine to form a molecule of hydrogen.

Even the atom is not an ultimate unit, for every atom is composed of minute particles, of which there are three main kinds: **protons, electrons,** and **neutrons** (Fig. 1-14). Protons are minute particles having a single positive electrical charge. Electrons are minute particles with a single negative electrical charge. A neutron is a particle without charge. Each kind of atom has a definite number of protons, electrons, and neutrons organized into a definite system (Figs. 1-15 and 1-16).

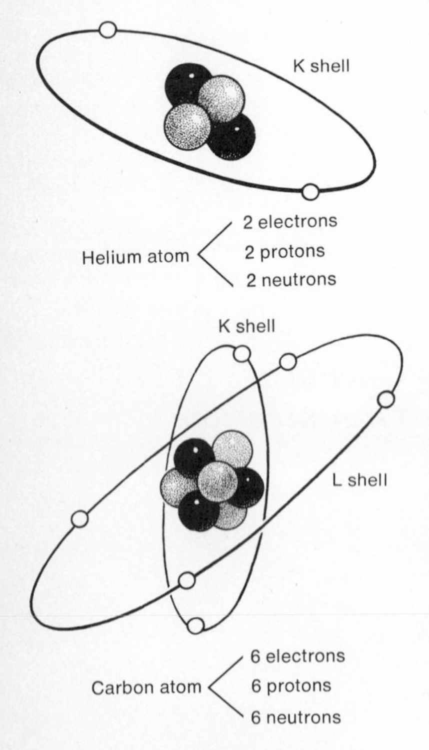

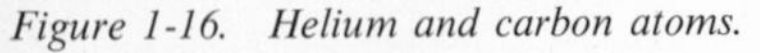

Figure 1-16. Helium and carbon atoms.

Atoms are normally neutrally charged so that there must be a positively charged region within the atom, called the nucleus of the atom, balancing the negative charges of the electrons. The electrons are arrayed in "shells" around the positively charged core of the atom. The innermost shell is called the *K* shell. The next innermost shell is called the *L* shell and the outermost one the *M* shell. Evidence indicates that the *K* shell is limited to two electrons, the *L* shell to eight electrons, and the *M* shell to eighteen. Every atom with an unfilled outer shell has a tendency to enter into combination with other atoms to satisfy its outer shell. For instance, the hydrogen atom consists of one unit of positive charge (nucleus) and one electron; therefore, its *K* shell is unfilled. Every atom with an unsatisfied outer shell has a tendency to enter into combination with other atoms in a manner that leaves it with a filled outer shell. The hydrogen atom's tendency is to fill the *K* shell with another electron, and it can do this in several ways. First, two hydrogen atoms can pool their single electrons and, by sharing the two electrons, mutually fill their *K* shells (Fig. 1-15), forming a hydrogen molecule. Hydrogen gas almost always exists in the form of a pair of atoms. To separate the two atoms and free them as atomic hydrogen takes a good deal of energy. The carbon atom (Fig. 1-16) with six electrons, two on the *K* shell and four on the *L* shell, will share each of these four electrons with a different hydrogen atom, thereby filling the *K* shells of the hydrogen atoms and in turn filling its own *L* shell by sharing their electrons, thus forming the stable methane molecule, CH_4 (Fig. 1-17).

The electron of the hydrogen atom has become partially incorporated into the electron structure of the carbon atom, forming a chemical bond. The electrons are shared by both atoms, and electric interaction holds the atoms together. The number of bonds that can be formed with a given atom depends upon the number of electrons on its outer shell. Hydrogen is capable of forming only a single bond; carbon can form four bonds. A chemical bond is generally represented as a line joining two atoms. Sometimes a double bond may be formed between two atoms, as in the combination of the carbon atom and two oxygen atoms to form carbon dioxide, CO_2; two lines are used to indicate such double bonding, O=C=O. In this case each oxygen atom with six electrons on its outer shell (needs two to

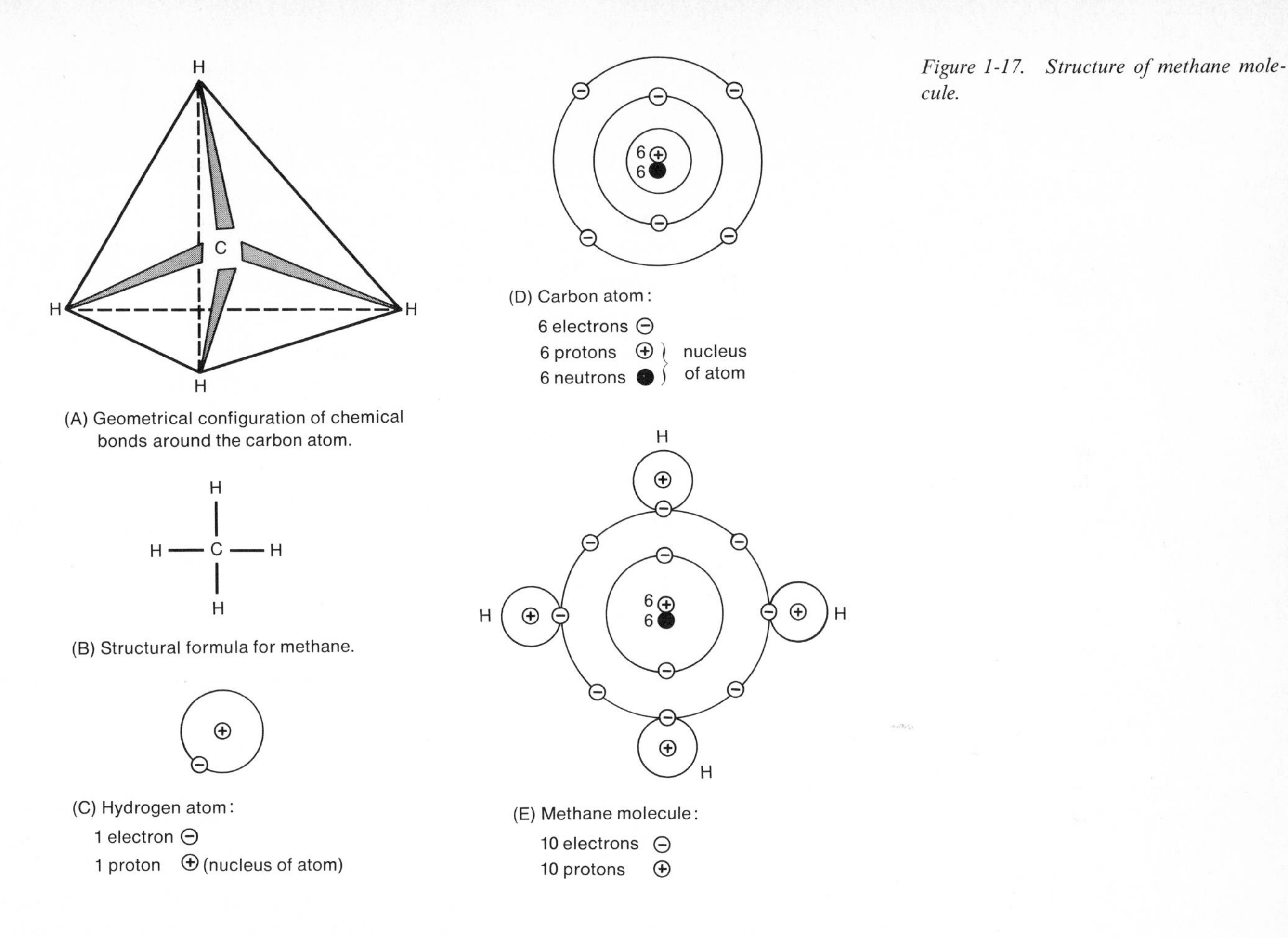

Figure 1-17. Structure of methane molecule.

complete shell) will share two of the four electrons of the carbon atom. Carbon still forms four chemical bonds and oxygen only two.

Just as atoms are electrically neutral, the molecules formed by the bonding of the same or different atoms are electrically neutral. However, because of the unequal distribution of electrons between two atoms, one portion of the molecule may be relatively more positive and another relatively more negative. Such molecules are said to be polarized. For example, the water molecule is a polar molecule. The negative electron of each hydrogen atom is more strongly attracted to the positive oxygen nucleus than to its own positive nucleus since the oxygen nucleus contains six protons while the hydrogen nucleus contains only one. The two hydrogen atoms are slightly positive, whereas the oxygen atom becomes slightly negative:

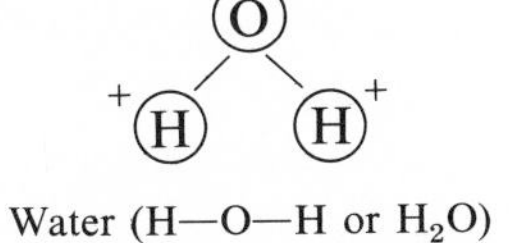

Water (H—O—H or H_2O)

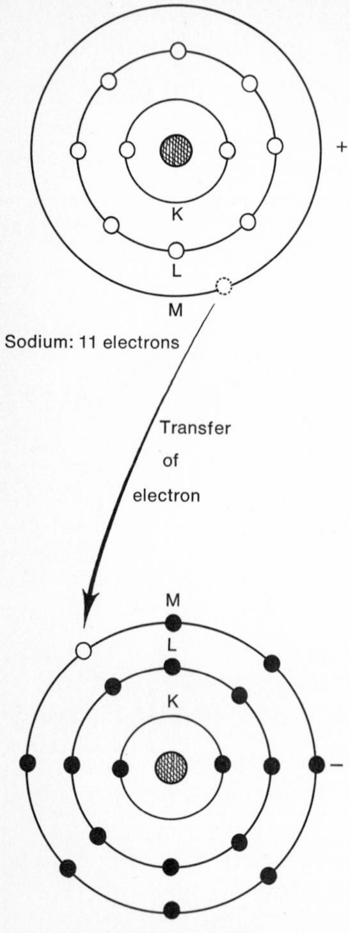

Figure 1-18. Union of oppositely charged atoms to form a molecule. Sodium loses its one electron in its outer shell to chlorine, making the sodium atom positively charged. The chlorine atom capturing this extra electron satisfies its outer M *shell (eight electrons) but in turn becomes negatively charged. The combination of the two oppositely charged ions results in the formation of a molecule of sodium chloride.*

The formation of such polar bonds in a molecule provide modes for variable interaction with other molecules. The slightly negative oxygen atom of one water molecule is electrically attracted to the slightly positive hydrogen atom of a second water molecule, forming an electrostatic bond between the two water molecules that does not involve an electron exchange between the two. Each water molecule is able to form four such electrostatic bonds, one with each of its two hydrogen atoms, and two with its oxygen atom. In the liquid state water exists as a three-dimensional lattice network of water molecules electrostatically bonded. At low temperatures practically all the water molecules form closely packed crystal lattices or ice. At high temperatures the weak electrostatic bonds are broken and the freedom of movement of individual water molecules is increased, allowing the molecules to escape into the gaseous state.

Atoms that have their outer shells completely filled do not combine with other atoms and are therefore stable. Helium, which has only two electrons, is an example of such an atom. It is stable and will not combine with other atoms. The atomic numbers of the elements as they appear in the periodic table represent the number of electrons in their atoms. Hydrogen, with one electron, is the first element in the periodic table; the second element is helium, with two electrons; the third, lithium, with three, and so on, all the way to lawrencium, with 103 electrons.

Another way in which an atom with an unsatisfied outer shell may fill its shell is to capture an electron from another atom or donate an electron to another atom. For example, the sodium atom, with eleven electrons—two on the *K* shell, eight on the *L* shell, and one on the *M* shell—readily surrenders its one *M* shell electron so that its outer *L* shell is filled, while the chlorine atom tends to capture an electron to add to the seven electrons on its outermost *M* shell and thus satisfy it (Fig. 1-18). Therefore, sodium and chlorine have an affinity for each other; when they combine, sodium donates its *M* electron to chlorine to fill the latter's *M* shell. Sodium with an electron subtracted becomes positively charged because the positive charges or protons in the nucleus of an atom do not change and are always equal in number to the electrons originally possessed by the atom. Chlorine, on the other hand, with one extra electron, carries a net negative charge. The mutual attraction of the opposite charges holds the two ions together. The compound is called sodium chloride. Organization in atoms, as well as in living units of life such as the cells, lends character to each kind of unit and confers upon it its own peculiar properties.

From a chemical viewpoint, we can define protoplasm as a complex mixture of specialized substances and compounds that are capable of carrying on the activities of life. An analysis of protoplasm shows the following elements to be present: carbon, hydrogen, oxygen, phosphorus, potassium, iodine, nitrogen, sulfur, calcium, iron, magnesium, sodium, and chlorine. Ninety-six per cent of the whole body is composed of four of these elements: carbon, oxygen, hydrogen, and nitrogen. Of these four, carbon is the most important, for it has chemical properties that favor the formation of complex molecules so abundant in living matter. Although we find these four elements in protoplasm, elemental analysis reveals that the free elements very seldom exist in their free form in living material, but they do appear in combinations with one another to form substances. The properties of a substance are very different from the properties of the free elements that combine to form that substance. In comparing the properties of certain elements and the substances formed by these elements, we can see these differences a little better:

Element	*Symbol*	*Properties*
Hydrogen	H	Gas, compressible, explodes, lighter than air
Oxygen	O	Gas, supports combustion, same density as air
Sodium	Na	Solid, reacts violently with water, poisonous to living material
Chlorine	Cl	Gas, heavier than air, poison
Carbon	C	Solid, black, high kindling temperature

A substance or compound is generally a definite combination of two or more elements; we find that water is composed of two atoms of hydrogen and one atom of oxygen, forming a molecule of water. Salt is a combination of an atom of sodium and an atom of chlorine:

Substance	*Symbol*	*Properties*
Water	H_2O	Liquid, heavier than air, puts out fire, can be frozen or changed into steam and vapor
Salt	NaCl	Solid, goes into solution very easily, dissociates into ions in water
Sodium bicarbonate	$NaHCO_3$	Solid, soluble, neutralizes acids
Carbon dioxide	CO_2	Gas, does not support combustion

There are five kinds of substances always present in living matter that are never found naturally in lifeless matter. These characteristic substances are carbohydrates, lipids (fatlike substances), proteins, enzyme proteins, and nucleic acids.

Carbohydrates

The **carbohydrates,** often termed **starches** or **sugars,** are widely distributed—both in animal and in plant tissues. Although they are part of the structure of the inclusions in the cell and play an important role as part of the normal environment of the cell, their chief function is furnishing the most available source of energy for the organism. When combined with oxygen, carbohydrates burn to form carbon dioxide and water, and in this process **potential (stored) energy** is converted into heat, motion, growth, or any other manifestation of **kinetic**

(released) energy that may be needed in a particular cell. Carbohydrates in the form of starch (plant) and glycogen (animal) are stored in certain cells and constitute a reserve fuel supply for emergency uses.

In general, carbohydrates are made up exclusively of carbon, hydrogen, and oxygen. There are always twice as many hydrogen as oxygen atoms in the molecule. For example, **glucose,** the simplest sugar, is composed of six atoms of carbon, twelve atoms of hydrogen, and six atoms of oxygen, having the formula $C_6H_{12}O_6$.

Besides glucose the sugars **fructose, galactose,** and **mannose** can also be represented by the same formula, $C_6H_{12}O_6$. They differ only in the position of the OH groups, which endows them with different chemical properties even though they have the same chemical composition. Such substances are called *isomers.* For example, the difference between the two sugars glucose and galactose depends on the position of the OH group on the fourth carbon atom:

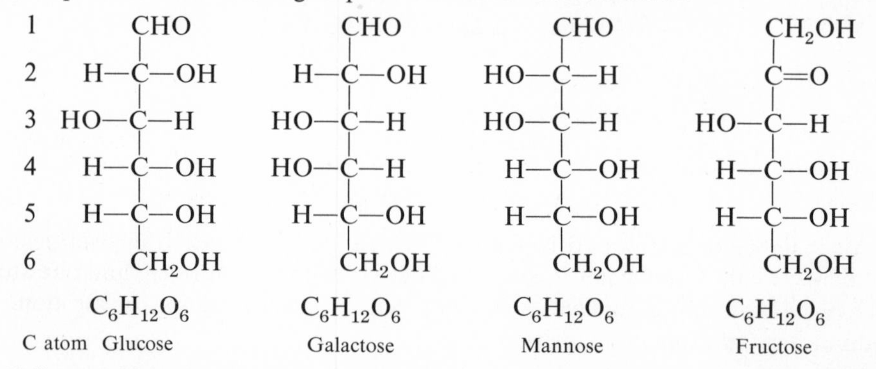

Cells are able to discriminate between these isomers because of the high degree of specificity of reaction shown by the enzymes. Most cells of the body cannot use any hexose other than glucose. Table sugar, or cane sugar, has the formula $C_{12}H_{22}O_{11}$. Carbohydrates are divided into three major groups: monosaccharides, disaccharides, and polysaccharides. The **monosaccharides** are sugars that cannot be hydrolyzed (split with the addition of water) into a simpler form. These are often designated as *simple sugars.* The general formula for these sugars is $C_nH_{2n}O_n$. The monosaccharides may be further designated as trioses (3-carbon sugars), tetroses (4-carbon), pentoses, hexoses, or heptoses, in accordance with the number of carbon atoms that the molecule possesses. The most important simple sugars physiologically are the pentoses and hexoses:

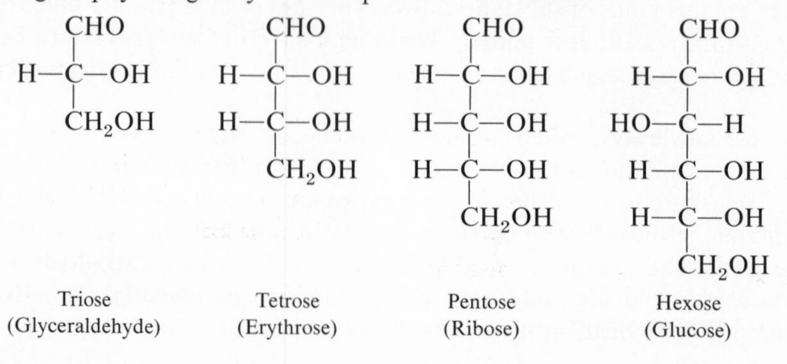

Among the pentoses, ribose and deoxyribose are important constituents of nucleic acids and nucleotides. Glucose is the hexose mainly involved in the energetic changes of the cell.

The structure of the carbohydrate molecule may be represented in several ways. One is the conventional representation, which may be the open-chain form, which occurs in aqueous solution but is in equilibrium with the ring form. The ring form is more common with the longer-chain carbohydrates. The rings are formed by the reaction of the **hydroxyl group (OH)** in the 4 or 5 position with the **carbonyl group** (CHO or $\underset{\diagdown H}{\mathrm{C{=}O}}$). To show the correct spatial relations of the various groups another representation is the three-dimensional structure, in which the ring is imagined as being looked at from above, and the three thickened edges of the bonds connecting carbon atoms 1, 2, 3, and 4 are imagined as being nearest the observer:

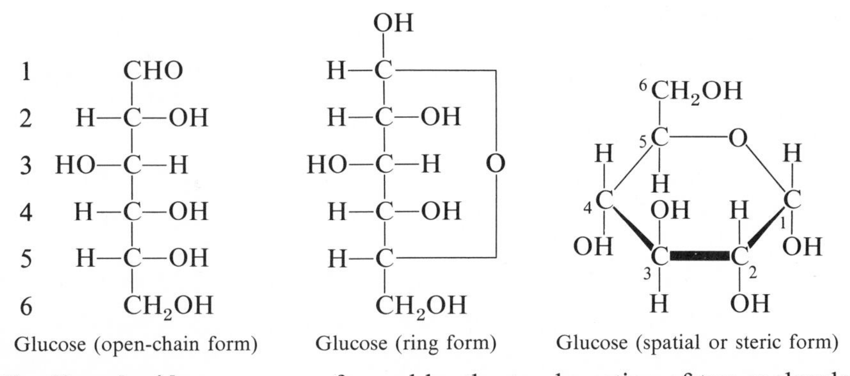
Glucose (open-chain form) Glucose (ring form) Glucose (spatial or steric form)

The **disaccharides** are sugars formed by the condensation of two molecules of monosaccharides with the loss of one molecule of water, or, in other words, they are those sugars that yield two molecules of the same or of different monosaccharides when hydrolyzed. The general formula is $C_nH_{2n-2}O_{n-1}$. The most important substances of this group are sucrose, maltose, and lactose. Sucrose, for example, is regarded as being formed from the condensation of glucose and fructose by the elimination of water:

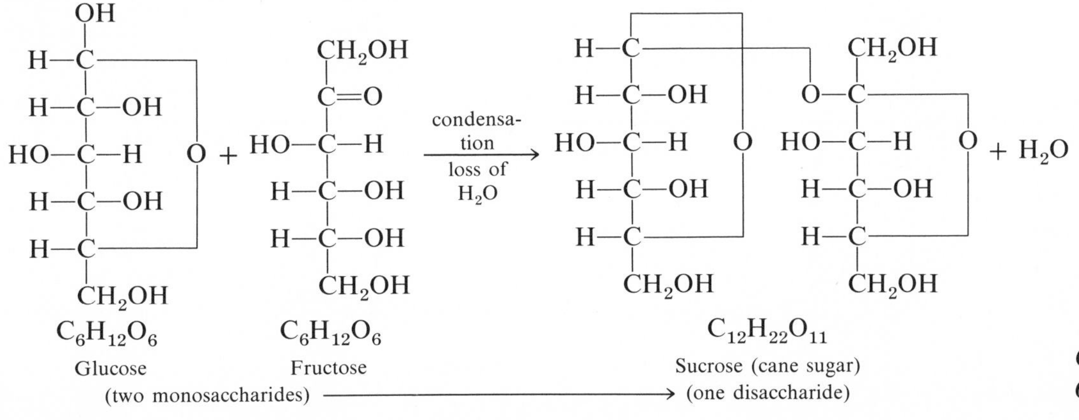

$C_6H_{12}O_6$ Glucose; $C_6H_{12}O_6$ Fructose (two monosaccharides) → $C_{12}H_{22}O_{11}$ Sucrose (cane sugar) (one disaccharide)

The **polysaccharides** are the result of the condensation of many molecules of monosaccharides with a corresponding loss of water molecules. Their empirical formula is $(C_6H_{10}O_5)_n$, and they will yield by hydrolysis n molecules of simple sugars. The most important polysaccharides for the animal are starch and glycogen. Glycogen is a substance consisting of long chains of glucose molecules. It is found in the largest proportion in liver and muscle. Starch is usually found in the plant cell in the form of cell inclusions, whereas glycogen, being to a certain degree soluble in water, may be dissolved in protoplasm. In the living cell glycogen is invisible because of its solubility but can be demonstrated histochemically by means of the iodine reaction, which gives a reddish-brown color.

The carbohydrates are important to living material in three ways:

1. They furnish the most readily available energy source.
2. They are important articles of food storage (glycogen).
3. They play a role in furnishing part of the structural material or, at least, are part of the essential environment of living cells.

Lipids and Fats

The **lipids and fats** are found abundantly, in the form of droplets, scattered throughout the protoplasm. They are substances that, like the carbohydrates, contain only carbon, hydrogen, and oxygen in their molecule. The fats differ from the carbohydrates in that there are many more carbon and hydrogen atoms in the fat molecule than there are in the carbohydrate molecule. Fats are able to provide much more heat energy per unit weight than can carbohydrates, because fats can combine with many more atoms of oxygen in the oxidative process. Fats are compounds of fatty acids and other substances. The main properties of fats are that (1) they have fatty acids in their molecule; (2) they are insoluble in water and soluble in organic solvents, such as ether, chloroform, and benzene; and (3) they are utilized as foodstuffs by living organisms.

Natural fats contain many different fatty acids, which may vary from one species to another. The fatty acids that occur in natural fats usually contain an even number of carbon atoms and are straight-chain derivatives. The chain may be saturated or unsaturated. **Saturated fatty acids** are those containing no double bonds. The general formula is $C_nH_{2n+1}COOH$. Examples of some of the acids in this series are listed in Table 1-2. **Unsaturated fatty acids** are those containing one or more double bonds. The double bonds invest these acids with a greater chemical reactivity, and they are more easily oxidized and combine with a larger quantity of iodine than saturated fatty acids. The **iodine number,** or **iodine value,** of a fat is the amount of iodine that can be combined with 100 grams (g) of fat, and it indicates the number of double bonds in a fatty acid.

Lipids can be divided into (1) simple lipids, (2) compound lipids, (3) sterols, and (4) hydrocarbons.

Simple Lipids

Simple lipids are esters of glycerol or another alcohol and fatty acid. They can be divided into

1. *Fats and oils,* usually called natural fats or glycerides. These are esters of glycerol and fatty acids.
2. *Waxes,* which are esters of fatty acids and an alcohol other than glycerol. They are of high molecular weight.

Table 1-2. ***Major Saturated and Unsaturated Fatty Acids***

Name	Number of C Atoms	Structural Formula
		Saturated Fatty Acids
Butyric	C_4	C_3H_7COOH
Caproic	C_6	$C_5H_{11}COOH$
Caprylic	C_8	$C_7H_{15}COOH$
Lauric	C_{12}	$C_{11}H_{23}COOH$
Myristic	C_{14}	$C_{13}H_{27}COOH$
Palmitic	C_{16}	$C_{15}H_{31}COOH$
Stearic	C_{18}	$C_{17}H_{35}COOH$
Arachidic	C_{20}	$C_{19}H_{39}COOH$
		Unsaturated Fatty Acids
Palmitoleic	C_{16} (1 double bond)	$CH_3(CH_2)_5CH{=}CH(CH_2)_7COOH$
Oleic	C_{18} (1 double bond)	$CH_3(CH_2)_7CH{=}CH(CH_2)_7COOH$
Linoleic	C_{18} (2 double bonds)	$CH_3(CH_2)_4CH{=}CH{-}CH_2{-}CH{=}CH(CH_2)_7COOH$
Linolenic	C_{18} (3 double bonds)	$CH_3{-}CH_2{-}CH{=}CH{-}CH_2{-}CH{=}CH{-}CH_2{-}CH{=}CH(CH_2)_7COOH$
Arachidonic	C_2O (4 double bonds)	$CH_3{-}(CH_2)_4CH{=}CH{-}CH_2{-}CH{=}CH{-}CH_2{-}CH{=}CH{-}CH_2{-}CH{=}CH(CH_2)_3COOH$

Compound Lipids

Compound lipids are esters of fatty acids and alcohols combined with other substances. These can be divided into

1. *Phospholipids,* made up of two fatty acids (therefore they are diglycerides), glycerol, a nitrogenous base, and a phosphate molecule in place of the third fatty acid. The phospholipids, which account for more than half of the total lipid content of most tissues, are located almost exclusively in the membranous components of the cell (see the section on the cell membrane).
2. *Cerebrosides or glycolipids,* which are fats made up of the fatty acids, phrenosinic or lignoceric acid, galactose or another sugar, and sphyngosine. The two most important cerebrosides are phrenosin and kerasin.

Sterols or Steroids

The **sterols or steroids** are not true fats, but alcohols with a cyclic nucleus; that is, they are ring structures in which the atoms of the molecule are arranged in the form of a ring. It is customary to abbreviate the ring structure in such compounds by using the figure of a hexagon if six carbon atoms are present in the ring or a pentagon if five carbon atoms are present, implying the presence of a carbon atom at each corner of a ring and hydrogen atoms added until the valence number reaches 4. Steroids are derivatives of **phenanthrene,** a hydrocarbon skeleton which may be represented as follows:

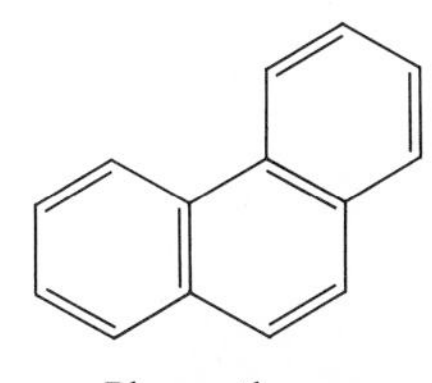

Phenanthrene

and are found free or in esters with fatty acids known as cholesterides. They are extracted by organic solvents together with the fats. Important steroids are cholesterol, coprosterol, ergosterol, the sex hormones, adrenocortical hormones, and D vitamins.

The best-known steroid is **cholesterol,** since it has been studied extensively because of its suspected relationship to heart disease and atherosclerosis. It is widely distributed in the tissues and is particularly abundant in brain and other nervous tissue, in the skin, and in the adrenal glands. In the tissues, it exists partly in the free state and partly in the form of esters, both types uniting with protein to form water-soluble complexes found in blood plasma. It has a complex structure:

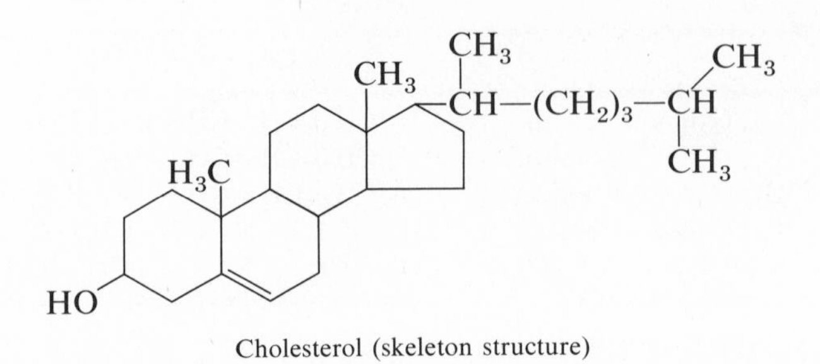

Cholesterol (skeleton structure)

Hydrocarbons

Hydrocarbons are substances that bear no obvious structural relationship to fatty acids and are classed as lipids merely because they exhibit similar solubility properties. They have considerable biological importance and include such compounds as squalene, the plant pigments known as carotenes, and chemically related compounds such as vitamins A, E, and K.

Neutral Fats

The **neutral fats** are the most abundant group of lipids in nature. These are esters of fatty acids with glycerol, $CH_2OHCHOHCH_2OH$. The most important chemical reaction of the neutral fats is their process of hydrolysis to yield three molecules of fatty acid and one of glycerol:

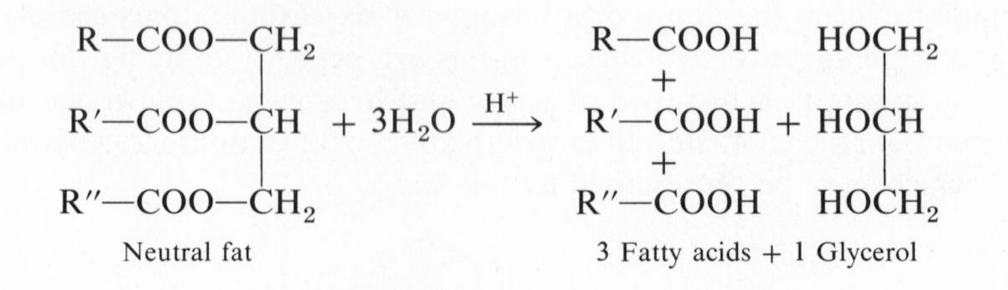

Neutral fat 3 Fatty acids + 1 Glycerol

RCOOH, R′COOH, and R″COOH represent three molecules of either the same or different fatty acids.

The fats function importantly in several ways (1) they provide a food with a high caloric value; 1 g of fat yields 9.3 kilocalories (kcal), twice the energy yield of 1 g of carbohydrate or protein; (2) fats are the most important reserves of energy in the organism, and they can be stored with a much smaller portion of water than carbohydrates; (3) fats are the vehicle of the natural fat-soluble vitamins, especially of vitamins A, D, and E; and (4) the fats are important structural components of the cell, especially the cell membrane.

Proteins

Proteins differ in their composition from carbohydrates and lipids in that, in addition to carbon, hydrogen, and oxygen, they always contain nitrogen and, with a few exceptions, sulfur also. The average elementary composition of protein is C, 45 per cent; H, 7 per cent; O, 22 per cent; N, 16 per cent; and S, 1 per cent. Proteins are made up of numerous units known as **amino acids.** Natural amino acids are fatty acids in which one of the alpha hydrogen atoms has been replaced by an amino (NH_2) group. Therefore, an amino acid has a basic (NH_2) group and an acid carboxyl (COOH) group. The simplest amino acid is glycine, which is aminoacetic acid:

$$\underset{\text{Amino group}}{NH_2} + \underset{\text{Acetic acid}}{CH_3COOH} \longrightarrow \underset{\text{Glycine}}{\underset{NH_2}{\overset{}{CH_2COOH}}}$$

The next simplest is alanine:

$$CH_3\underset{|\atop NH_2}{C}HCOOH$$

Amino acids may be represented by formulas of the following type, in which R represents the rest of the amino acid molecule:

$$R\underset{|\atop NH_2}{C}HCOOH$$

The basic group of one amino acid may combine with the acid group of another, resulting in the linkage of the two amino acids and the loss of water:

$$\underset{\text{Alanine}}{CH_3CHNH_2CO}\underbrace{:\!O\!:\!H\!:\; + \;:\!H\!:}_{H_2O}\underset{\text{Glycine}}{NHCH_2COOH} \xrightarrow[\text{(loss of } H_2O)]{\text{dehydration}}$$

$$\underset{\text{Alanylglycine}}{CH_3CHNH_2CO \cdot NHCH_2COOH} + H_2O$$

The union CO · NH is known as a peptide linkage. The reverse process, which also takes place in the peptide linkage, causes the separation of amino acids, and this reaction is a process of hydrolysis:

$$\underset{\text{Alanylglycine}}{CH_3CHNH_2CONHCH_2COOH} + H_2O \xrightarrow[\text{(addition of } H_2O)]{\text{hydrolysis}}$$

$$\underset{\text{Alanine}}{CH_3CHNH_2COOH} + \underset{\text{Glycine}}{CH_2NH_2COOH}$$

In theory, the 23 naturally occurring amino acids could form an enormous number of proteins, but there is evidence that this is not so. Amino acids seem to be grouped into units, and proteins are formed by the combination of these. Proteins differ from one another in their molecular size, the sequence of the various constituent amino acids, the nature and relative proportions of these acids in the chain, and the way the chain itself is arranged. Svedberg was the first to classify proteins according to molecular weight. From his studies on various proteins using the technique of ultracentrifugation, he suggested that the molecular weight of all proteins was a multiple of 17,600. In light of current methods of study using accurate measurements either of osmotic pressure or of sedimentation velocity and sedimentation equilibrium in the ultracentrifuge, Svedberg's classification proved to be oversimplified. There are many proteins having molecular weights less than 17,000, for example ribonuclease (12,700), and many more that are not multiples of 17,600. The best thing we can say is that the molecular weight (that is, the particle size) of most proteins is very high.

The architecture of the protein molecule has been studied by means of X-ray diffraction photographs. By shining X-rays on proteins congealed into crystals, the scientist can obtain a photographic plate which is spotted by the X-ray beam passing through a crystal. The location of every atom in the crystal and therefore of a single protein molecule can be discerned from this photographic plate after mathematical calculations are made involving the location and brightness of the spots. Fibrous proteins, such as keratin, collagen, and myosin, have long, threadlike molecules that can be stretched and afterward recover their initial length. It appears that the molecule, by a process of folding and unfolding, changes its length. Globular proteins, such as blood plasma proteins and egg albumen, are formed by chains or filaments wound in a compact mass.

Proteins form colloidal solutions because of the large size of their molecules. Since amino acids are amphoteric (substances which act as both acid and base), proteins show this characteristic also. The acid group (COOH) can set free hydrogen ions (H^+), and the basic group (NH_2) can accept H^+ or set free hydroxyl ions (OH^-). In an acid medium proteins behave as bases, and in an alkaline medium, they behave as acids, forming salts called **proteinates,** such as sodium or potassium proteinate. At a certain hydrogen-ion concentration of the solvent, the protein molecule is electrically neutral; its dissociation is at a minimum, and the concentration of acid ion is equal to that of the basic ion. This is known as the **isoelectric point.** Some proteins precipitate out at the isoelectric point; thus we find that isoelectric precipitation is used to separate and purify proteins.

The proteins occupy a most important position in the architecture and functioning of living matter. They are intimately connected with all phases of chemical and physical activity that constitute the life of the cell. Proteins participate in all important physiological functions because they are part of enzymes and some hormone structure. They are concerned with muscular contraction, oxygen carrying, immunological mechanisms, and hereditary factors. The proteins are quantitatively the main material of animal tissue, for they constitute approximately 75 per cent of the solid substance. The proteins are important for two main reasons:

(1) they act as growth material for the organism, and (2) most organized structures of living cells are composed largely of proteins of different types.

It is of interest to note that every different species of animal or plant has its peculiar protein content. It is probably largely as a result of the high specificity of proteins that species differ from one another as they do. Even individuals of the same species differ slightly from one another in their protein make-up.

Enzymes

Enzymes may be described as complex catalysts of biological origin. They are protein in nature and function to increase the speed of a reaction without being used in the reaction itself. They are not only found in all cells but are secreted by cells into the blood and into the digestive tract, where they take part in chemical reactions in which complex materials are broken down into simple substances, or they may act in chemical activity in which simple substances are synthesized into more complex compounds. Many enzymes can be shown to consist of two portions: One part, which is heat labile and nondialyzable, is presumably a protein. It is referred to as the **apoenzyme.** The other portion of the enzyme, which is heat stable and dialyzable, is nonprotein in nature and is attached as a **prosthetic** group to the enzyme protein. This is the **coenzyme.** Unless both components of enzymes are present, an enzyme usually remains inactive. The combination of an enzyme and a coenzyme is called a **holoenzyme.** Many of the B vitamins have been shown to function as constituents of coenzymes.

In addition to these organic cofactors there are a number of inorganic cofactors, mostly certain metal ions, such as iron, copper, zinc, cobalt, and magnesium, necessary for electron transport or to bind the substrate molecule to the enzyme active site. Many enzymes, especially those concerned with digestion, do not require coenzyme for their activity.

It is necessary to activate a substance that is to undergo chemical activity preliminary to the reaction, and enzymes act as these activators. The energy required for this activation is reduced considerably by an enzyme or other catalyst. It is assumed that a combination of enzyme and substrate (substance undergoing activity) is required before the action of the catalyst can be exerted. The reaction is represented as

$$\text{Enzyme (E)} + \text{substrate (s)} \rightleftharpoons \text{E–S complex} \longrightarrow \text{Products of the reaction} + \text{enzyme.}$$

Mechanisms of Enzyme Action. It has been postulated that the close association of specific enzyme and substrate (E–S complex) imposes a temporary reduction in bond stability or a bond strain in such a way that electron-orbital shifts are more likely to occur in the substrate molecule, thus bringing about a chemical transformation or chemical reaction (Fig. 1-19). The combination of a substrate with an enzyme is very specific; that is, each enzyme has a unique area or surface that interacts with the substrate molecule. This area, which may involve only a few of the hundreds of amino acids present in the molecule, is known as the enzyme's **active site.**

Because proteins have three-dimensional structure, owing to the coiling of the

Figure 1-19. Reaction of enzyme with substrate to form an intermediate, which induces a bond strain that permits easier attachment by colliding H_2O molecules; this action produces a second reaction, which gives the final product and active enzyme. A. *Possible arrangement of hydrogen-bonding groups on active site of enzyme molecule.* B. *Bond strain produced on substrate molecule complexing with enzyme, permitting easier attachment by colliding water molecules.* C. *Formation of new product from substrate and return of enzyme to its original state.*

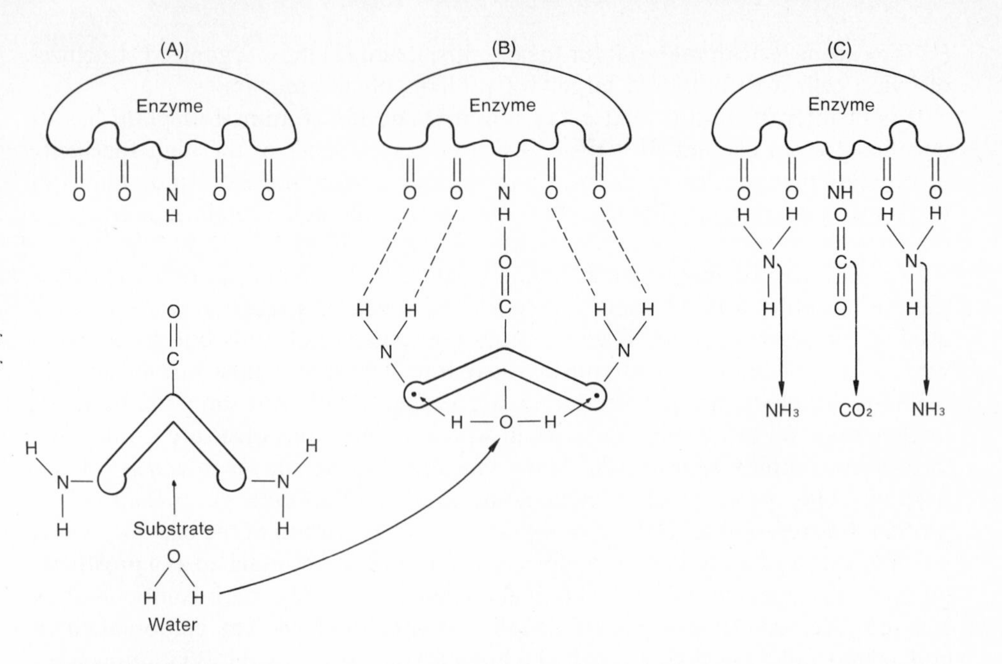

amino acid chain, they present unusual folded surfaces and only specifically shaped molecules can gain access. For this reason the enzyme–substrate complex is often compared to a lock and key, as shown in Fig. 1-20. The name first applied to enzymes gave no indication of their activity. The present convention is based on a system in which the name of the enzyme is derived from the substrate on which it acts with the suffix **-ase.** Enzymes that split fat (lipose) are called *lipases,* those that act on the sugar maltose are called *maltases,* and those that split starch (amylum), *amylases.* It has been estimated that a given cell could contain 1,000 enzymes. Enzymes exist that are capable of reacting with every organic compound that occurs in nature and with many inorganic compounds. These enzymes bring about synthesis, digestion, oxidation, hydrolysis, and other chemical changes that occur in metabolic activity of the cells of which an animal is composed.

Nucleic Acids

Although proteins have always been considered the basic substance of living material, more recent views assign proteins to a more restricted role and emphasize the importance of nucleic acids. **Nucleic acids** are necessary for protein synthesis in the body and also for the transmission of genetic information in the cell. Like proteins, nucleic acids are quite complex in their structure, but unlike the protein, the basic structure of which is the amino acid, nucleic acids consist of long chains of **nucleotides.** Nucleotides are phosphorylated nucleosides, which may be represented by the formula

$$\text{Nucleoside} + H_2PO_4 \xrightarrow{-OH^-} \text{Nucleotide.}$$

A **nucleoside** is a combination of a 5-carbon sugar, such as **D-ribose or D-2-deoxy-**

ribose, with an organic **purine base,** such as **adenine** or **guanine,** or with a **pyrimidine base,** such as **cytosine, thymine,** or **uracil:**

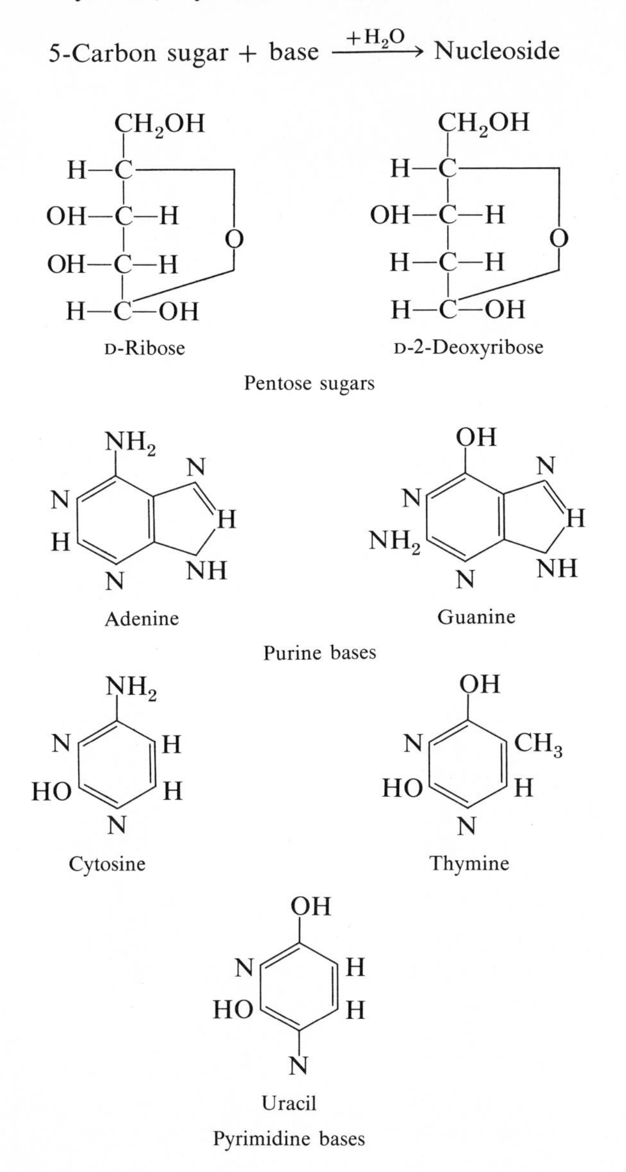

A great number of nucleotides link together to form nucleic acid. If the sugar present in the molecule is D-ribose, the resulting nucleic acid is called **ribonucleic acid (RNA),** and the chief bases present within the molecule are adenine (A), guanine (G), cytosine (C), and uracil (U). If the sugar present is D-2-deoxyribose, it is known as **deoxyribonucleic acid (DNA),** and the chief bases present are adenine, guanine, cytosine, and thymine (T). DNA, which is located only in the cell nucleus (specifically in the chromosomes, where it forms the fundamental

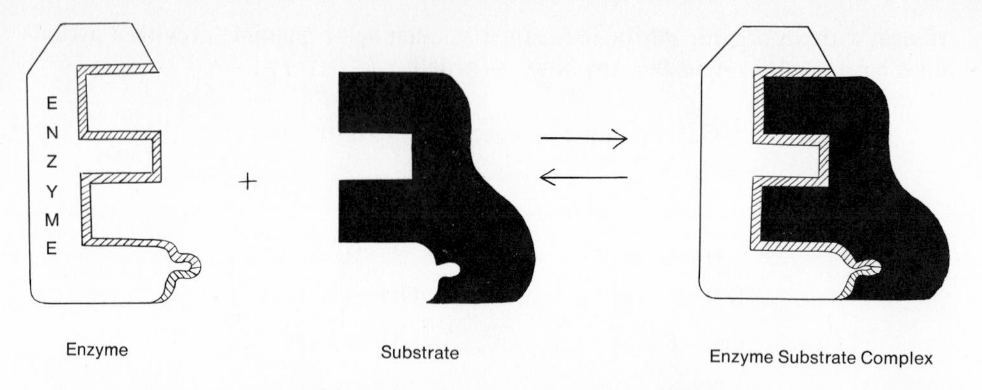

Figure 1-20. Formation of an enzyme–subtrate complex. The shaded area on the enzyme represents the active site, consisting of specific chemical groups, such as SH, anions, OH, and metals, of which only specifically shaped molecules can gain access.

genetic material of the chromosomes) is pictured as a structure of two nucleotide chains twisted into a double helix (Fig. 1-17). The two helixes are held together by hydrogen bonding between the bases adenine on one strand and thymine on the other; similarly, a quanine base bonds with a cytosine. This type of structure was first postulated for DNA by two biologists, J. D. Watson and F. H. Crick. DNA is looked upon as the key to heredity since all the genetic information carried by the cell is coded on the DNA molecule. The code of instructions for the synthesis of specific enzymes and other proteins is apparently in the arrangement of the four bases, adenine, guanine, thymine, and cytosine (A, G, T, and C). From these four bases, usually occurring in triplet, code words are formed which direct the placement of amino acids in a specific order to form a specific protein. It has been predicted that with a three-letter code word there are 64 possible triplets. There are only 23 amino acids, so there must be more than one code word for placement of each amino acid. This genetic code, which is embodied in DNA, is transcribed as a message in the formation of messenger RNA, as indicated in Fig. 1-21. It appears that all substances of living matter, such as enzymes, proteins, and all others whose production is catalyzed by enzymes, depend on this nucleic acid DNA.

RNA, although found in the nucleus, is chiefly found in the cytoplasm. It is known to exist in several forms: (1) **soluble** or **transfer RNA,** which is responsible for transferring specific amino acids to protein-synthesizing sites and is symbolized as **tRNA.** It has a molecular weight ranging from 20,000 to 30,000 and is confined to the cytoplasm; (2), as **ribosomal RNA** of high molecular weight (2,000,000) found in the minute ribosomes of the cytoplasm, which are believed to be the protein-synthesizing centers (p. 9); and (3) as **messenger RNA (mRNA),** which is responsible for carrying information from the nucleus to the protein-synthesizing centers in the cytoplasm.

Recent work,[2,3] including X-ray diffraction studies, on the physical properties of crystalline tRNA has thrown some light on the structure and function of this

[2] M. W. Nirenberg and J. H. Matthaei, "The Dependence of Cell-free Protein Synthesis in *E. coli* upon Naturally Occurring or Synthetic Polyribonucleotides." *Proc. Natl. Acad. Sci.* (*U.S.*) **47:**1588 (1961).

[3] M. Spencer, W. Fuller, M. H. F. Wilkins, and G. L. Brown, "Determination of Helical Configuration RNA Molecules by X-ray Diffraction." *Nature* (*London*), **194:**1014 (1962).

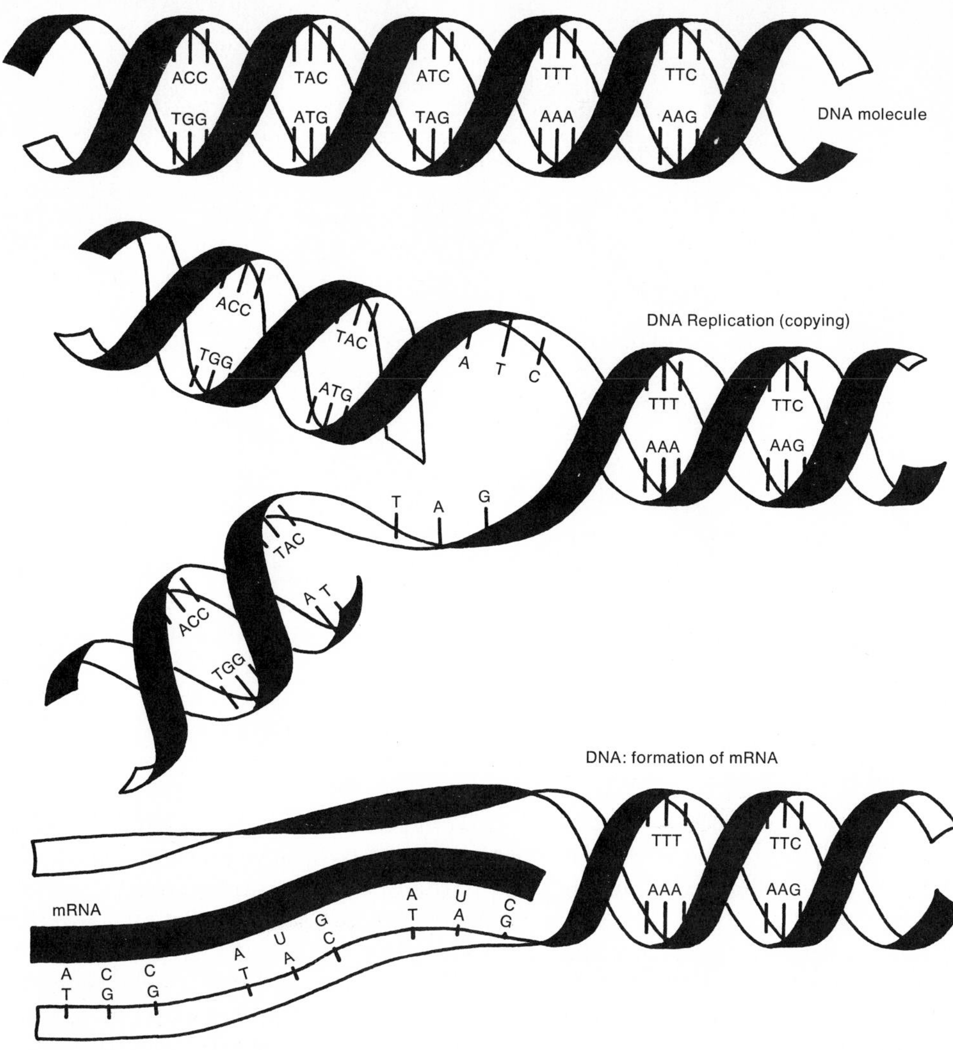

Figure 1-21. Structure of the DNA molecule, showing two nucleic acid chains twisted into a double helix. The two helixes are held together by hydrogen bonding (not shown) between the bases adenine and thymine and cytosine and guanine. In cell division two double strands are made (DNA replication), each containing one old nucleic acid chain and one new chain. In this way two complete sets of hereditary instructions are produced at every cell division. The lower illustration shows the synthesis of mRNA on one strand of a DNA molecule. Notice that uracil, not thymine, is coded by adenine.

complex molecule. Transfer RNA is made up of a carbon chain of 80 nucleotides. The chain is folded back on itself, forming a double helix (winding like a spiral staircase) with the opposing bases on each of the two halves being linked together by hydrogen bonding. The constituent nucleotides forming the chain are joined together by phosphate bonds connecting carbon atom 5 in the ribose of one nucleotide to carbon atom 3 in ribose of the next nucleotide in sequence. The present concept of the structure of tRNA is shown in Fig. 1-22, which also includes current evidence for the RNA code for the amino acid phenylalanine.

The closed end of this special helical form of transfer RNA containing the unpaired bases may provide the code for the proper placement of the amino acid it is specific for transporting when it reaches the site of protein formation. The amino acid is attached to the free end of the molecule, and there is at least one

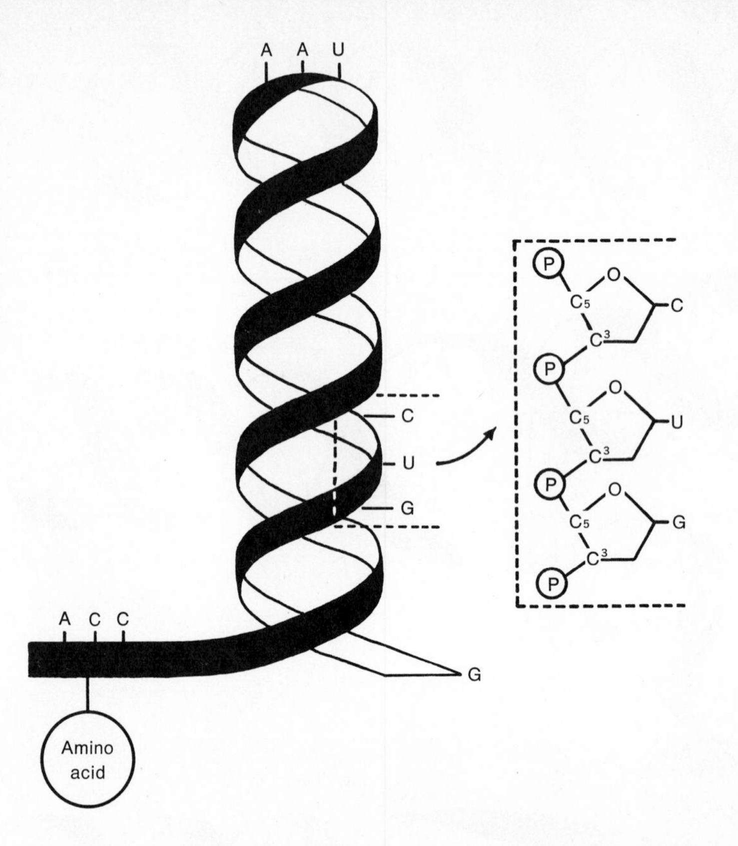

Figure 1-22. Model depicting the structure of transfer RNA. Notice the closed end of this special helical form of nucleic acid, containing the triplet of bases AAU that will provide for the proper placement of the amino acid molecule that it is transferring to the ribosomal surface. The amino acid is attached to the free end of the molecule. The insert depicts the structure of a segment of the nucleic acid chain. Phosphates connect ribose molecules between the 3- and 5-carbon atoms of different ribose units.

transfer RNA for each of the 20 amino acids. All, however, seem to carry the bases ACC where the amino acid is attached and G at the opposite end. The amino acids must be activated prior to being attached and do so by reacting with ATP in the presence of a specific amino acid–activating enzyme. Transfer RNA appears to be produced within the cell under the influence of specific genes.

Messenger RNA differs from transfer RNA in its internal structure. mRNA is a single stretched-out filament, unlike tRNA, which is bent back on itself like a twisted hairpin. This single-filament structure, because of the lack of hydrogen bonding between bases which occurs in tRNA, enables this molecule to act as the agent that transcribes the genetic code from DNA (genes) and carries it to the ribosomes in the cell, where protein synthesis takes place. It is now known that the synthesis of strands of mRNA is patterned on strands of DNA within the nucleus. Messenger RNA is similar to DNA except that the former contains the sugar ribose instead of deoxyribose and the base uracil instead of thymine. That is, when mRNA is being formed on a DNA strand, uracil appears instead of thymine in the RNA chain wherever adenine is located at the complementary site on the DNA chain. Messenger RNA so formed leaves the nucleus of the cell and attaches to the ribosome, where it directs the synthesis of protein. The sequence (which presumably occurs in triplet) of bases in the messenger RNA specifies the amino acid sequence in the protein to be synthesized. This sequence

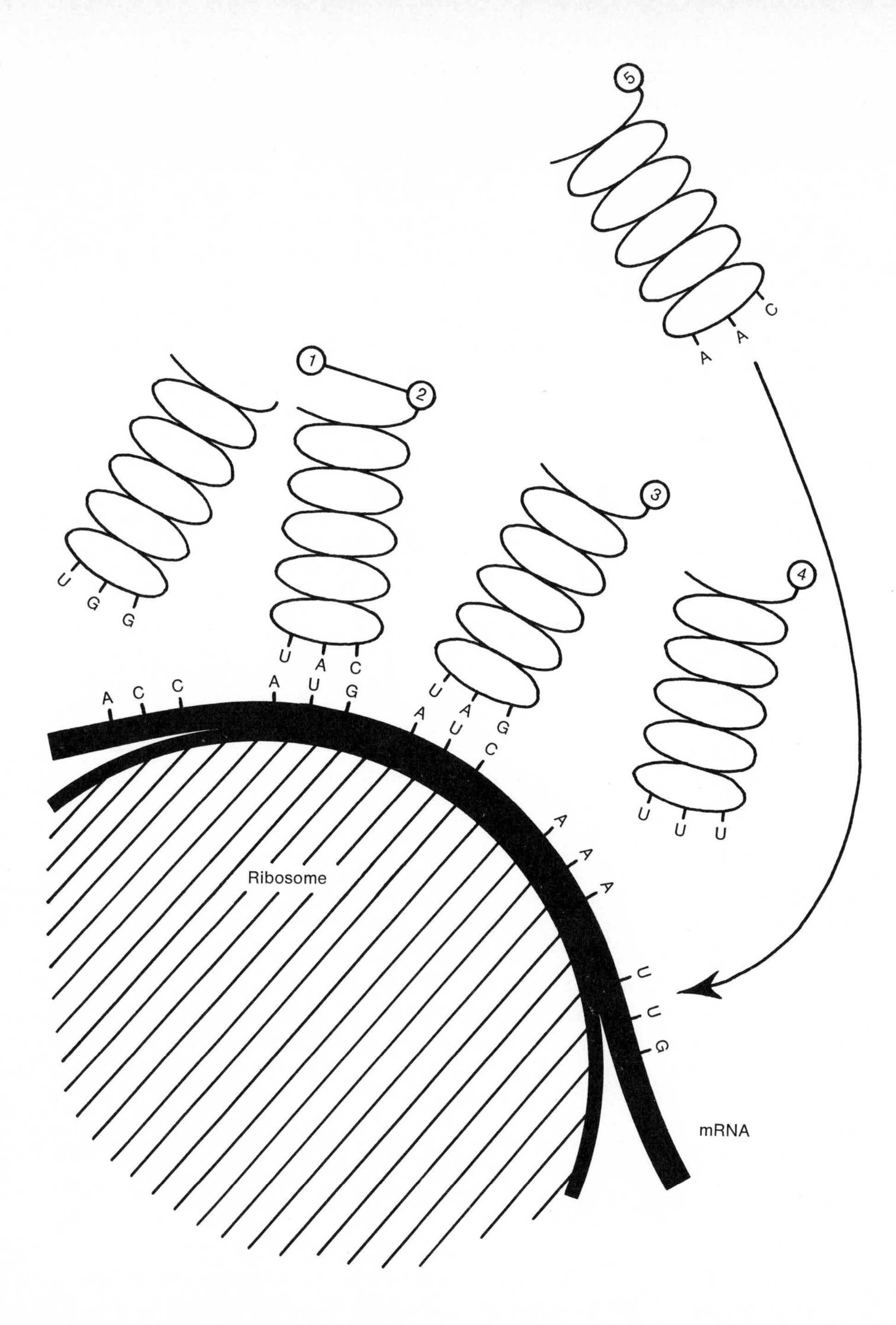

Figure 1-23. Synthesis of protein on a ribosome. The mRNA has reached the ribosome and has stretched out on its surface. Specific tRNA molecules pass successively along the mRNA molecule, each seeking out the proper placement of its amino acid by the code it carries on its closed end, which matches the code on mRNA (C bonds only with G, and A bonds only with U); thus, it contributes its amino acid molecules in the proper sequence to the developing protein molecule. The tRNA then returns to the amino acid pool in the cytoplasm for another load.

is determined by the tRNA, which recognizes the code for the amino acid that it is carrying at a particular spot on the ribosome where the mRNA has already attached itself. This recognition involves the ability of nucleic acid bases to hydrogen bond specifically with other base residues. Adenine pairs only with uracil or thymine and guanine with cytosine. Nirenberg and Matthaei (1961) and Mattaei et al.[4] have made spectacular progress in identifying what combination of bases corresponds to a particular amino acid, by use of a system in which synthetic polynucleotides will act as templates. The replication of DNA, its transcription into messenger RNA, and the method of recognition between transfer RNA and messenger RNA is schematically represented in Figs. 1-21 and 1-23 (p. 31).

Other Substances in Protoplasm

While the substances just discussed are largely peculiar to protoplasm, there are always present many very common substances, the chief of which are **water** and **inorganic salts.** The main function of water is to act as a solvent for the crystalloids and a medium for the colloidal material. This is of great importance in the body because substances in solution separate into small particles, frequently of atomic or molecular dimensions. Certain substances on dissolving in water give rise to another kind of particle, ions. An ion is an electrically charged atom or group of atoms. The charge on the ion is negative in some cases, positive in others. Sodium chloride in water is charged:

$$NaCl \longrightarrow Na^+ + Cl^-$$

$$\text{Salt} \xrightarrow{\text{(dissociation)}} \underset{\text{(positive ion)}}{\text{Sodium ion}} + \underset{\text{(negative ion)}}{\text{chloride ion.}}$$

Any substance capable of dissociation is considered an electrolyte. Water is an electrolyte because it dissociates or ionizes itself:

$$H_2O \longrightarrow H^+ + OH^-.$$

The total body water is equal to about 40 to 65 per cent of the body weight and is distributed throughout two main parts of the body. The extracellular part (plasma and interstitial fluid) contains about one fourth of the total body water, and the intracellular part (water inside the cell) contains about three fourths of the total body water. Water has many properties that make it an important solvent in biological systems:

1. It has a high **dielectric constant,** that is, the ability to ionize a salt or the capacity of a medium to affect the force exerted between electric charges immersed in the medium. The higher the dielectric constant, the greater the dissociation of ions. The dielectric constants of some common compounds are as follows:

H_2O	79
Air (vacuum)	1
Alcohol	26
Ether	4
Hydrocyanic acid	96

[4] S. H. Matthaei, O. W. Jones, R. G. Martin, and M. W. Nirenberg, "Characteristics and Composition of RNA Coding Units." *Proc. Natl. Acad. Sci.* (*U.S.*), **48:**666 (1962).

When we say that the dielectric constant of water is 79, what is meant is that the force required to separate two oppositely charged particles is one seventy-ninth of that required to effect the same separation in a vacuum, or air.
2. Water itself is an electrolyte; that is, it dissociates itself. In 10 million g of water are 1 g of hydrogen ions (H^+) and 17 g of hydroxyl ions (OH^-).
3. Water has a high specific heat. The specific heat is the amount of heat necessary to raise 1 g of water 1° C of heat. The excess heat of the animal body is taken up by water and transferred to the surface by the blood, where it is dissipated.
4. Water has a high heat of evaporation. The heat of evaporation is the amount of heat necessary to evaporate 1 g of water. This has a tendency to help cool the body, as in the evaporation of sweat from the surface of the body.
5. Water has a low viscosity. This is important in the passage of blood through the small blood vessels, because water lowers viscosity of blood.

In addition to proteins, fat, carbohydrates, and other organic compounds, mineral substances, or **inorganic compounds,** are very important to protoplasmic activity. The inorganic constituents are found in the cell in the form of salts or in combination with proteins, carbohydrates, and lipids. Salts are electrolytes and thus exist principally in protoplasm dissociated into electrically charged particles. For example, sodium chloride exists dissociated as positive (cations) sodium ions (Na^+) and negative (anions) chloride ions (Cl^-). The principal ions found in protoplasm are **sodium, chloride, potassium, calcium, magnesium, phosphate, carbonate,** and **sulfur.** The inorganic constituents are not distributed evenly throughout the tissues but are more concentrated in some portions of the body than in others. **Calcium** ions are found in the blood, in parts of the cell, especially the cortex, and abundantly in bone and calcified cartilage. **Phosphate** occurs in the blood and tissue fluids as the free ion but more abundantly in the bound form with lipids, proteins, and carbohydrates. The phosphate ion in the bound form, especially in adenosine triphosphate, is an important source of energy, and such bonds are known as high-energy bonds. Phosphate is needed for normal bone and tooth structure.

Chloride ion circulates in the blood and intercellular fluid in free ionic form as well as bound form. It is found not only in the cell but also in various secretions, for example, gastric juice—where it is in the form of hydrochloric acid and sodium chloride. Chloride is important in osmotic regulation of cells and in maintaining globulins in solution. **Potassium** is present within the protoplasm of the cell in relatively high concentration. Potassium is of importance in the red cell, in the transport of oxygen and carbon dioxide, and also in regulation of pH. It is believed to have an important physiological part in nerve conduction and muscle contraction. **Sodium ion** is found, in contrast to potassium ion, mainly outside the cell in the intercellular fluid, lymph, and plasma. Sodium is important in the regulation of blood pH and in maintenance of osmotic pressure. **Sulfur** is found in several amino acids that form important metabolites of active cells. Sulfur in the form of a sulfhydryl group is found in the contractile protein of muscle, myosin, and in certain essential enzymes, thus making it a vitally important ion in the activity of contraction. Sulfur is also found in the sulfate ion in intercellular fluids.

In general, the inorganic compounds maintain the acid–base equilibrium and

Table 1-3. *Size Range for Different States of Matter*

State of Matter	*Division of Matter*	*Size (mm)*
Crystalloidal	Atoms	$10^{-7} = \frac{1}{10,000,000}$
	Molecules	$10^{-6} = \frac{1}{1,000,000}$
Colloidal	Aggregate of molecules	$10^{-5} = \frac{1}{100,000}$
Coarse suspensions and emulsion	Particles	$10^{-4} = \frac{1}{10,000}$
		$10^{-3} = \frac{1}{1000}$
		$10^{-2} = \frac{1}{100}$
		$10^{-1} = \frac{1}{10}$
		1

regulate the osmotic pressure. They are also important in forming compounds that help regulate cellular activity, such as hormones, and help in the release of energy necessary for normal cell activity.

Physical Organization of Matter

Protoplasm is not a true solution, but a heterogeneous mixture of many different substances and compounds that are in colloidal suspension or solution. Since protoplasm is a **colloid,** some knowledge of colloids and their properties is indispensable in our quest for an understanding of the activity of living substance. A colloid can be simply defined as a suspension of minute particles in a solution. Particles in the colloidal state range in size from 1 nanometer (nm) (1/1,000,000 mm) up to 100 nm (Table 1-3). Smaller particles are said to be in true solution. Larger particles than those of colloidal dimensions are in suspension or emulsion. Colloids exist in two phases: one is known as the **sol phase,** and the other is known as the **gel phase.** A sol occurs when the particles in suspension (dispersed) are widely separated from one another so that the entire medium (particles plus suspending medium) has a rather fluid consistency. The white, or albuminous, material of a raw egg is a good example of colloidal sol. Gelatin dissolved in warm water before it sets is also an example of a sol phase of a colloid. The gel phase exists when the particles in suspension are closely packed together so that the entire medium takes on a rather firm or solidlike consistency, such as that seen in the white of a hard-boiled egg and in firm gelatin. A **gel** may be defined as a semirigid mass of a lyophilic sol in which the dispersion medium has been completely absorbed by the particles so as to leave a sponge structure of micelles. A **micelle** can be defined as a particle of substance together with the absorbed solvent or other material. A **lyophilic sol** is a sol formed by dispersion of particles in water in which the particles in suspension have an unusual attraction for the water molecules and form a gelatinous mass. Under suitable conditions, almost any lyophilic sol can be converted into a gel. Change in temperature, addition of chemicals, or physical agents can cause gelation of colloids.

Because protoplasm is a colloid, we find it exhibiting the characteristics of colloids. Solation and gelation are possible in protoplasm because of the action of enzymes. Thus, we find that not only can the sol phase be converted to the gel phase, but also the opposite can occur. This phenomenon has been used as an explanation for movement in the ameba.

If a strong beam of light is passed through a colloidal solution, its path is made visible by the light reflected from the particles. This phenomenon was studied by Tyndall in 1869 and became known as the **Tyndall effect.** True solutions show no such effect because the molecules in solution are not sufficiently large to reflect the light. This furnished a simple method for distinguishing between a colloidal suspension and a true solution.

Other properties exhibited by colloids in addition to the Tyndall effect are as follows: The boiling and freezing points of colloidal systems are practically the same as those of the dispersion media. The particles of a colloidal system are constantly in motion, owing to their bombardment by the solvent (usually water) molecules. This disordered movement of the particles is referred to as **brownian motion.** Colloidal particles pass through ordinary filter paper but not through an animal or vegetable membrane. Substances in true solutions do pass through a membrane, and so they can be separated from colloidal material by this process, called **dialysis.** Colloidal particles, owing to their fine state of division (which means they provide a large surface area), have high absorptive capacity. In addition, one of the most striking characteristics of colloids is their tendency to form films at their surfaces; that is, the particles in solution are more closely packed together at the surface of the colloid than below the surface. Plasma membranes are nothing more than colloidal materials in which the particles dispersed through the suspending medium are more closely packed at the surface. These films, or membranes, are semipermeable. Certain materials are capable of passing through membranes, whereas others are excluded.

Osmosis

The movement of solvent molecules (H_2O in biological systems) across a boundary (for example, membrane) into a phase in which there is a higher concentration of a solute (substances in solution: salt, sugar, protein, and the like) to which the boundary is impermeable is called **osmosis.** Perhaps osmosis may be defined more clearly as a process which takes place when a solution of a given solute is separated from another solution of lesser solute concentration by a semipermeable membrane (that is, a membrane permeable only to water and not to solutes), and water passes through the membrane in the direction that tends to equalize the solute concentrations on both sides of the membrane. The plasma membrane of a cell is often pictured as a combination of a lipid skeleton and other radially aligned molecules sandwiched between layers of protein molecules that are penetrated by a network of channels or pores; it can certainly be considered as some barrier (that is, semipermeable) to the movement of ionic and nonionic substances.

The movement of the solvent molecules across the membrane has been considered to be due to the fundamental physical mechanism of diffusion; that is, the direction and equilibrium of osmotic processes can be predicted accurately by the laws of diffusion. However, it is incorrect to say that osmosis is due only to the diffusion of solvent (water) through a semipermeable membrane since this interpretation has been recognized to be inconsistent with more recent observations. Investigations[5-7] of the rates of osmosis and diffusion have revealed that movements of water by osmosis occur more rapidly than could possibly be predicted by the theoretical maximum rate of diffusion and that an alternative mode of transport, which can be referred to as a *nondiffusional process,* must be considered.

A theoretical treatment of the *diffusion process,* which involves the inherent motion of molecules was given by Einstein in 1905, (Molecules are in ceaseless rapid motion and, although molecules of a liquid or gas cannot be seen, the random zigzag, or brownian, motion of larger suspended particles observed through the microscope is believed to be evidence of their motion.) Einstein postulated that if a molecule or particle executes a "random walk," then a group of similar particles, acting without mutual interference, will distribute themselves in space and time according to the basic diffusion equation of Fick:

$$\frac{\delta c}{\delta t} = D\frac{\delta^2 c}{\delta x^2},$$

where $\frac{\delta c}{\delta t}$ = change in concentration per unit time

D = constant

$\frac{\delta^2 c}{\delta x^2}$ = second derivative of concentration with respect to distance.

Pappenheimer in 1953 suggested that flow of water through membranes resulted primarily from a bulk (nondiffusional) flow of water in the membrane. Mauro in 1957, using a collodion membrane because the complexity of living membranes would make the results inconclusive, determined that the diffusional component of flow of isotopically labeled $H_2{}^{18}O$ in his membrane was only 1/730 of the total flow during osmosis. Late in 1960 Ray published a new hypothesis which showed how the driving force behind osmosis could be a hydrostatic pressure difference within the membrane. Dainty (1963) supported and further elucidated this hypothesis. The tremendous amount of experimental evidence now at hand substantiates the theory that osmosis is a bulk-flow phenomenon resulting from hydrostatic pressure differences within the membrane rather than a process resulting from random molecular movement (diffusion). Recent experiments by

[5] J. R. Pappenheimer, "Passage of Molecules through Capillary Walls." *Physiol. Rev.,* **33:**387 (1953).

[6] V. Koefold-Johnson and H. H. Ussing, "The Contribution of Diffusion and Flow to the Passage of H_2O Through Living Membranes." *Acta Physiol. Scand.,* **28:**60 (1953).

[7] H. H. Ussing and B. Andersen, "The Relation between Solvent Drag and Active Transport of Ions." *Proc. Intern. Congr. Biochem.,* pp. 434–440 (1955).

Mauro[8] (1965) in an elementary osmotic system showed that the flow of solvent through the membrane was caused by a hydrostatic pressure gradient that was generated by a drop in pressure just within the membrane at the membrane–solution interface.

The development of this hydrostatic pressure difference within the membrane may be schematically represented and perhaps a little better understood by referring to Figs. 1-24 to 1-28. In each case a membrane is arranged between two compartments and means are provided for viewing the movement of water across the membrane by use of a capillary tube. In Fig. 1-24 both compartments contain pure water (solvent).

When the system contains pure solvent on both sides of the membrane and the temperature and pressure are constant, the solvents on both sides of the membranes have the same chemical potential and therefore the system is at equilibrium. A system is said to be at equilibrium when the chemical potential of each substance is equal at every point in the system. The **chemical potential** of a substance x in a solution is its partial molar free energy; that is, it represents the maximum work at constant pressure and temperature that can be obtained from 1 mole of x in the solution. The chemical potential for water in an aqueous solution may be expressed by

$$u = \bar{V}P + RT \ln N, \tag{1-1}$$

where u = chemical potential of the solvent (water)
$\bar{V}$ = partial molar volume of water
$\left[\text{which equals } \left(\dfrac{\text{milliliters of water}}{\text{moles of water}}\right) \text{ about 18 ml}\right]$
P = pressure component (for example, atmospheric)
R = gas constant
T = temperature, °K
ln = natural logarithm
N = mole fraction of water

$$\left(\text{which equals } \frac{\text{number of water molecules}}{\text{number of solute molecules + number of water molecules}}\right).$$

$\mu_1 = \mu_2$ = Chemical potential
$P_1 = P_2$ = Atmospheric pressure

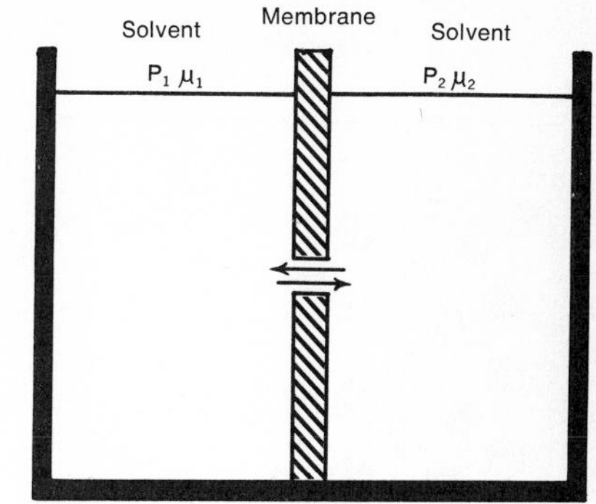

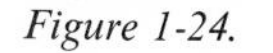
Figure 1-24.

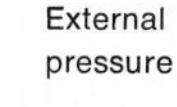

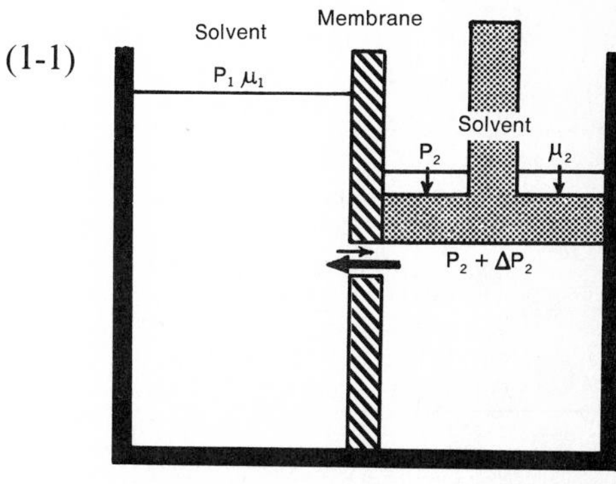

$P_1 = P_2$ = Atmospheric pressure
ΔP_2 = Applied pressure

Figure 1-25.

Considering again the situation in Fig. 1-24, the chemical potential of the water in the left-hand chamber (u_1) is equal to the chemical potential of water in the right-hand chamber (u_2), since both chambers are under the same atmospheric pressure and the mole fractions of both are equal to unity.

Thus

$$\bar{V}P_1 + RT \ln N_1 = \bar{V}P_2 + RT \ln N_2$$

or

$$u_1 = u_2 \quad \text{equilibrium.}$$

In Fig. 1-25 both compartments contain pure solvent, as in Fig. 1-24. However, if we apply an external pressure (in addition to the atmospheric pressure, P_2)

[8]A. Mauro, "Osmotic Flow in a Rigid Porous Membrane." *Science,* **149:**867 (1965).

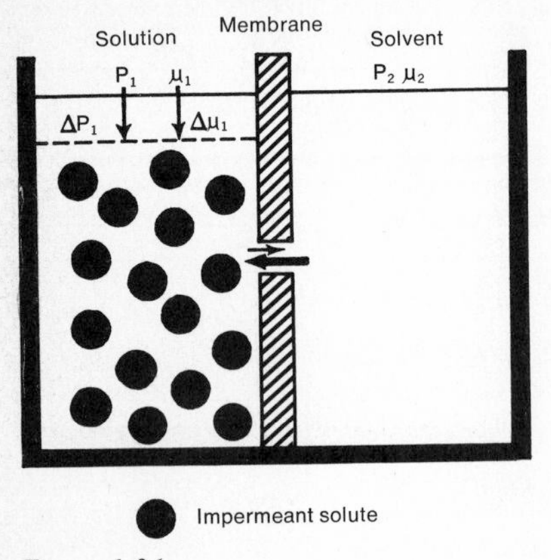

Figure 1-26.

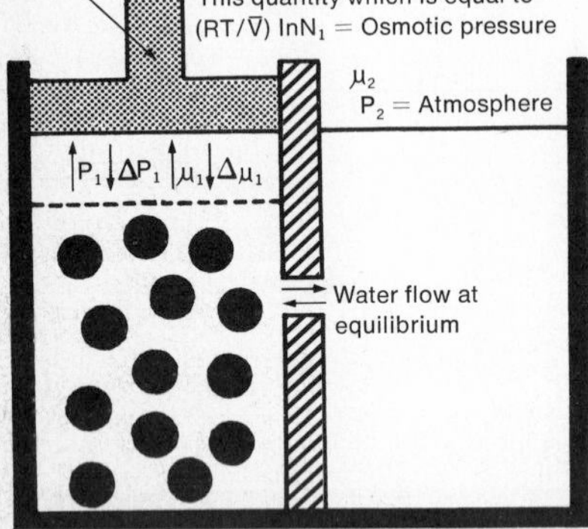

Figure 1-27.

on one side, we get a flow of water to the other compartment which we can consider as being due to increased pressure.

The intuitive expectation depicted in Fig. 1-25 can be predicted by the chemical potential changes in the solvent induced by the external pressure. Since

$$u_1 = \bar{V}P_1 + RT\ln N_1$$

and $$u_2 = \bar{V}(P_2 + \Delta P_2) + RT\ln N_2 = \bar{V}P_2 + \bar{V}\Delta P + RT\ln N_2,$$

then $$\Delta u_2 = \bar{V}\Delta P_2;$$

that is, the increase in pressure in the right-hand chamber increases the chemical potential of the water in that chamber in direct proportion to the increase in pressure (ΔP_2), and the constant of proportionality is the partial molar volume ($\bar{V}$). There, $u_2 > u_1$, and water would be expected to flow from the right chamber into the left one.

In Fig. 1-26 the compartment on the right contains pure solvent, the one on the left contains a solution, and the membrane separating the two compartments is absolutely impermeable to the solute. Under these conditions water will flow from the pure solvent compartment to the solution (solute and water) compartment, and it can be further observed that the flow of water increases as the mole fraction of water decreases in the solution compartment. The addition of solute causes a drop in the chemical potential of water in the solution compartment below that of the pure solvent compartment, as indicated.

The addition of the impermeant solute to the solvent in the left-hand chamber produced a mole fraction of water (in that compartment) less than unity, and therefore the chemical potential of the water in that compartment is lower; that is, $u_1 > u_2$, and water would be expected to flow from right to left as a consequence.

As can be calculated from Equation (1-1), the change in chemical potential

$$\Delta u_1 = RT\ln \frac{\text{number of water molecules}}{\text{number of water molecules} + \text{number of solute molecules}}.$$

Since the mole fraction of water (N_1) is a fraction whose logarithm is negative, the change in chemical potential caused by the addition of solute is also negative. If, however, hydrostatic pressure in excess of atmospheric pressure (present in both compartments) is applied to the solution side to bring the rate of water movement to zero (Fig. 1-27), it is found that the magnitude of this pressure is equal to the quantity

$$-\frac{RT}{\bar{V}}\ln N_1. \tag{1-2}$$

This hydrostatic pressure produces a positive increment in the chemical potential of water, $\Delta u_1 = \bar{V}\Delta P_1$, which is equal and opposite to the decrement in chemical potential caused by the addition of solute to the left-hand compartment. Thus, by the addition of hydrostatic pressure to the left-hand chamber, it was possible to establish an equilibrium state, that is, $u_1 = u_2$, with no net flow of water.

This hydrostatic pressure is also known as the **osmotic pressure**[9] of the solution and can be related to the drop in pressure as a result of the decrement in chemical potential in the solution compartment upon the addition of solute. This drop in pressure in the solution side associated with the drop in chemical potential of the solvent in the solution phase serves to act as the driving force for the osmotic flow. If one considers the third phase, which occurs in the membrane itself, that is, what is happening to the pressure and mole-fraction concentration of water within the membrane or barrier, it becomes apparent that there is only pure solvent inside the membrane and the change in pressure ΔP in that membrane must drop equal to the change in chemical potential Δu. The profile of the pressure drop within the membrane may be represented as depicted in Fig. 1-28.

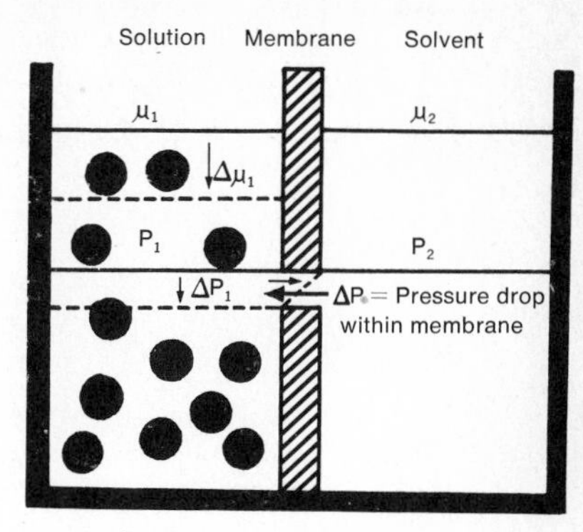

Figure 1-28.

It follows, then, that a drop of pressure ΔP must exist at the interface of the membrane and the solution and thus serve as the driving force for the osmotic flow.

This description of osmosis appears adequate for semipermeable membranes whose pores are larger than the diameter of the water molecule (about 3Å). For membranes with pore sizes approaching the dimensions of a single water molecule, it is not possible to distinguish whether osmotic or diffusion processes produce the flow of water across the membrane. Most living membranes, however, have relatively large pores and exhibit true osmotic (or nondiffusional) flow.

Osmosis may be studied more simply by using a cell (Fig. 1-29) prepared in the following way. A porous earthenware cylinder is filled with cupric sulfate solution and immersed in a solution of potassium ferrocyanide. These two compounds will react to form a precipitate of cupric ferrocyanide, which forms a gelatinous film across each pore of the cell and thus builds up a semipermeable membrane. If the cell is filled with a sugar solution, and a stopper with a long

[9]The usual equation to express the osmotic pressure of a solution is

$$\pi = \frac{n_2}{V} RT,$$

where π = osmotic pressure of solution
V = volume of solution
n_1 = number of moles of solvent
n_2 = number of moles of solute
R = gas constant
T = temperature, °K

Equation (1-2) is converted to this form by assuming a dilute solution, that is, one in which $n_2 >> n_1$. Then the ln N_1 reduces to $-n_2/n_1$. So, by substituting into Equation (1-2),

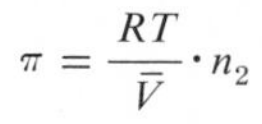

$$\pi = \frac{RT}{\overline{V}} \cdot n_2$$

and since
$$n_1\overline{V} = V,$$

therefore
$$\pi = \frac{RTn_2}{V}.$$

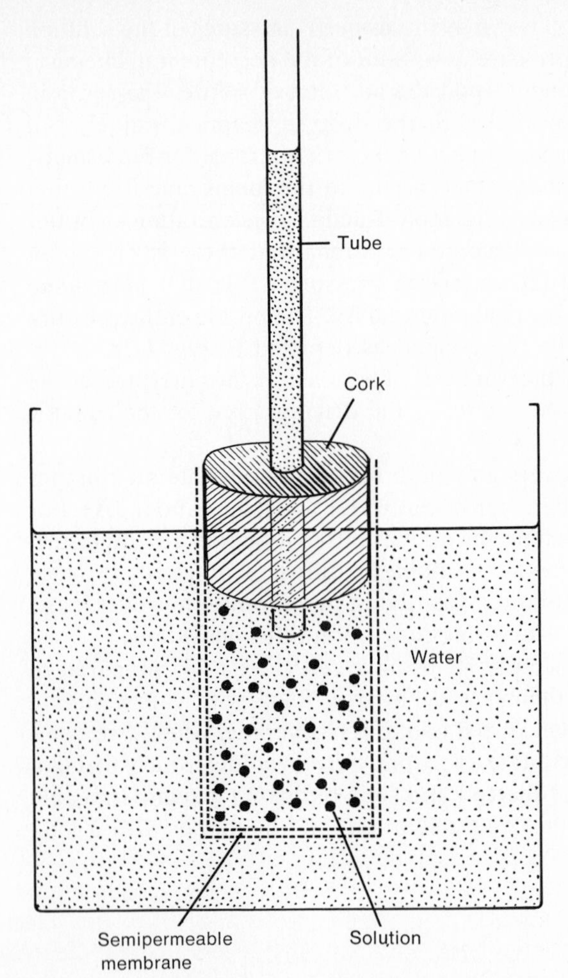

Figure 1-29. Simple osmometer for the study of osmosis across a semipermeable membrane.

glass tube of small bore attached to it is applied to the top of the cell, and the cell is then immersed in water, osmosis occurs through the semipermeable membrane into the cell. The level of the solution in the glass tube will rise continuously, perhaps to a height of several feet. The same effect can be observed using a simpler cell. A carrot or other root vegetable is bored out and filled with sugar solution. A rubber stopper with attached glass tube is fitted into the top of the carrot. When the carrot is immersed in water, osmosis takes place, and the level of the solution in the tube rises to a height of 10 ft or more in the course of 24 hours. The "head" of solution within the tube constitutes a hydrostatic pressure, and the maximum height of the aqueous column is a measure of the osmotic pressure of the solution in the cell. A more accurate measure of the osmotic pressure of the original solution is, however, the pressure that just prevents the passage of water through the membrane into the solution.

Owing to osmosis and osmotic pressure, cells may swell up and burst, or they may lose their liquids to the outside and shrivel, depending on conditions around

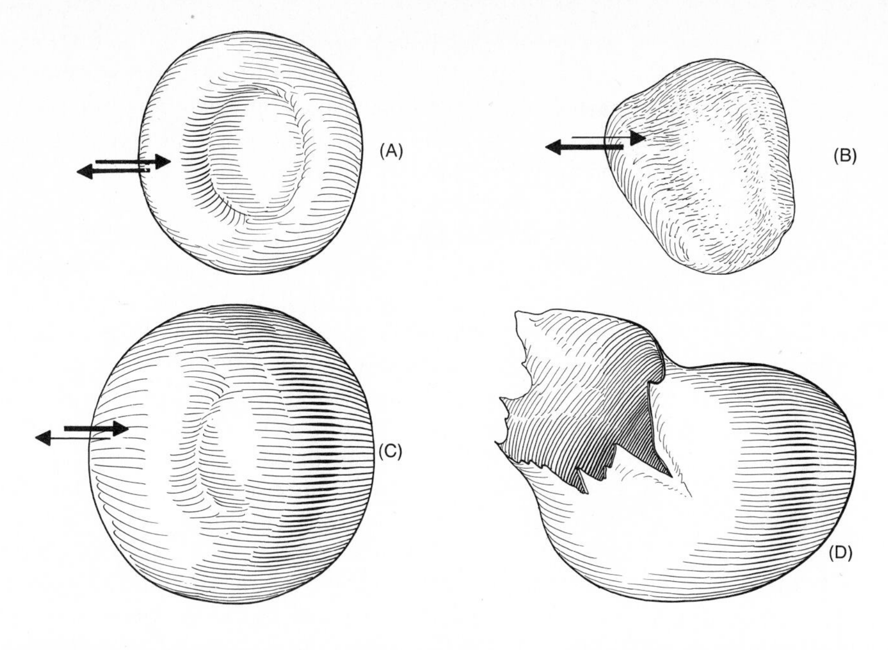

Figure 1-30. Effect of different concentrations of liquids on red blood cells. A. *Cell surrounded by isotonic solution.* B. *Cell surrounded by strong salt solution.* C *and* D. *Cell surrounded by distilled water.*

the cell. This can be easily demonstrated in the laboratory by using red blood cells and seeing what effect various concentrations of solutions have on the cell (Fig. 1-30). For example, if we place some red blood cells in distilled water, more water passes into the cell than out, and eventually the cell breaks open. In a strong salt solution, the opposite effect takes place; that is, more water passes out of the cell than in, and the cell shrinks. If you take a physiological salt solution, which is a solution containing 8.6 g of salt for every 1,000 ml of water, and place red blood cells in this solution, there will be no bursting or shrinking, since the amount of water passing into the cell will exactly equal that passing out of the cell. Such a solution is called an **isotonic solution,** one that has the same molecular concentration as protoplasm and in which osmosis into the cell is equal to osmosis out of the cell. To distinguish this from distilled water, the latter or any solution containing a molecular concentration less than that of protoplasm is referred to as a **hypotonic solution.** Cells in such a solution will swell up and burst. The strong salt solution has a greater molecular concentration that protoplasm. Cells in such an environment will lose large quantities of materials to the outside and, therefore, collapse or shrink. Such a solution is spoken of as a **hypertonic solution.** The tonicity of a solution is very important when bathing living tissue with the solution or injecting it into the body.

Diffusion

Another physical phenomenon very similar to osmosis is **diffusion.** Substances in a cell are continually moving around. From a physical standpoint we can describe diffusion as a passage of molecules of one substance between the molecules of a second substance to form a mixture of the two substances. For example, if a concentrated solution of cupric chloride is run into a beaker of water so as to form a layer at the bottom, the diffusion is shown by the upward spreading of the blue color. Weeks may elapse before the blue color reaches the top of the cylinder. Diffusion involves not only the migration of dissolved substances, but also the migration of particles in suspension. Osmosis involves migration of solvent molecules only. Diffusion is responsible for the distribution of materials throughout the mass of protoplasm within the cell; otherwise, there would be a tendency for materials to accumulate around the periphery of cells.

Osmosis, osmotic pressure, and diffusion play a very important role in the physiology of living organisms. Many interesting facts and relationships have been established by study of living cells even though they were cells and organs removed from living organisms. Animals are composed of millions of individual cells, each surrounded by a cell membrane, which functions as a more or less perfect semipermeable membrane. Water and solutes of low complexity pass through such membranes into the cell where they may be resynthesized into more complex substances. The surroundings of cells have a great effect on the normal functioning and even the shape of cells, as demonstrated by the immersion of red blood cells in various concentrations of salt solution.

Energy Extraction and Transformation by Living Cells

In order for living cells to preserve the integrity of their organization, they must have a constant supply of **energy** to convert the nutrients which pass through the cell membrane to new carbon skeletons that are needed to synthesize vital compounds. In addition to the chemical work required to preserve their integrity of organization, cells must transform energy to do the varieties of work that constitute the life processes of organisms (mechanical, electrical, chemical, osmotic). Cells extract energy from the chemical bonds in fuels (lipid, carbohydrate, protein) and use this energy at a fairly constant and low temperature. They do this by burning or oxidizing these complex food materials in the process called **respiration,** using molecular oxygen (O_2) from the atmosphere. As the energy is employed by the cell to carry out its biological work, carbon dioxide is given off to the environment as an end product. The process of respiration is carried on by **mitochondria,** large numbers of which are found in almost all cells. In the mitochondria the energy extracted from foodstuffs in the process of respiration is converted into the chemical-bond energy of **adenosine triphosphate (ATP).** Adenosine triphosphate contains three phosphate groups linked together (see p. 86). The chemical energy of the bond is made available to the cell for **biosynthesis, transfer of materials across membranes,** or **muscle contraction** by detaching the terminal phosphate of the ATP molecule through hydrolysis to yield **adenosine diphosphate (ADP)** and simple phosphate. In the energy balance of the cell, ATP may be considered as the energy-rich form of the energy carrier and ADP as the energy-poor form. Energy is constantly being cycled through ATP molecules, so its function is not so much the storage of energy as the transfer of energy. The advantage to cells of having an intermediate energy carrier is that the vital activities carried out by the cell are no longer dependent upon a single supply of fuel such as glucose, because the energy stored in ATP may have been obtained from other sources, such as fats or proteins.

The energy cycle of the cell (Fig. 1-31) and therefore of life is based on the sun as the ultimate source of energy. Solar radiation drives photosynthesis, which

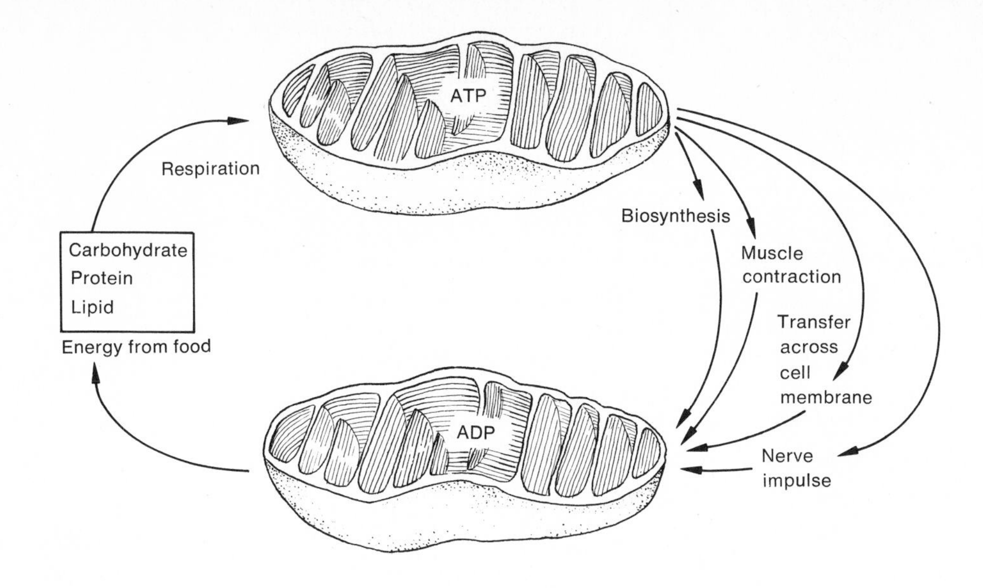

Figure 1-31. Energy cycle of cell. Glucose and other fuels are broken down by the cell and the energy transferred to ATP in the mitochondria. ATP supplies energy for muscle contraction, synthesis, transfer of materials across membranes, and propagation of nerve impulses. Energy-poor ADP thus formed is converted back to energy-rich ATP by food energy.

allows plants to build energy-rich glucose from carbon dioxide and water and other fuels, such as lipid and protein, from it. These fuels are then broken down to carbon dioxide and water by animal cells, which use the energy extracted in the process of respiration and conserved in the phosphate-bond energy of ATP to do their work. The breakdown of polysaccharides (glycogen) and monosaccharides (glucose) by the cell may be taken as typical of the process by which most known aerobic cells obtain energy. Essentially the energy stored in the carbohydrate molecule during photosynthesis (689 Kcal/mole) is released during catabolism and carbon dioxide and water are generated. The over-all reaction can be written

$$\underset{\text{Glucose}}{C_6H_{12}O_6} + 6O_2 \longrightarrow 6H_2O + 6CO_2 + \text{Energy (689 kcal/mole).}$$

The total energy released by glucose is released not in one reaction as indicated above, but in small quantities in a series of many reactions collectively referred to as the **metabolic pathway for glucose degradation.** This is divided into three major interrelated pathways:

1. The **glycolytic,** or **Embden-Meyerhof pathway.**
2. The **citric acid,** or **Krebs cycle pathway.**
3. The **pentose phosphate shunt** or **phosphogluconate pathway.**

Glycolytic Cycle

The apparently simple formation of lactic acid from glycogen **(glycolytic cycle)** in the cell in reality consists of a number of reactions involving many types of molecules, and the energy relationships primarily involve phosphate-bond energy and the oxidation–reduction potentialities of coenzymes. We have already seen that splitting of a high-energy phosphate bond is a typical energy-releasing reaction, which in the body yields energy for vital processes. For example, each

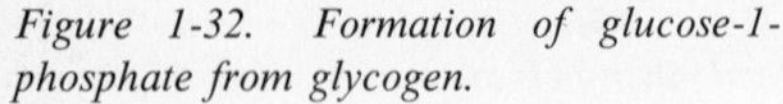

Figure 1-32. Formation of glucose-1-phosphate from glycogen.

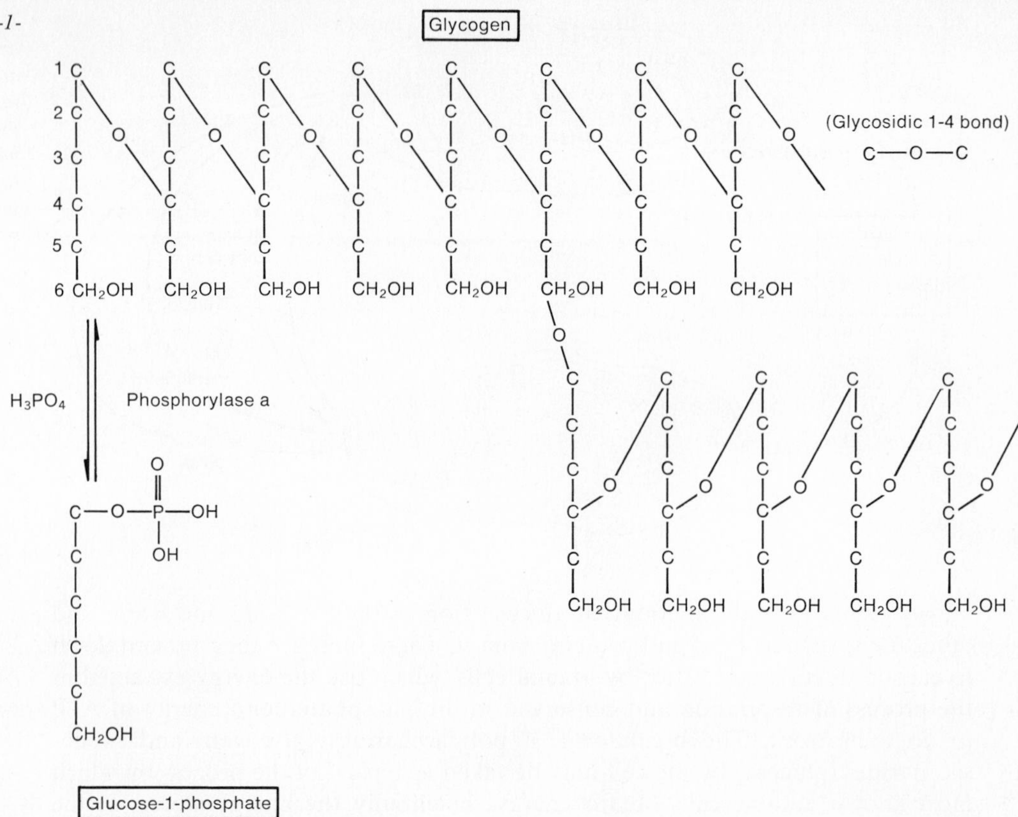

of the two terminal phosphate bonds of ATP will liberate on hydrolysis 11,000 to 12,000 small calories of energy per gram molecule. The high-energy bond of phosphocreatine will liberate a similar amount of energy on hydrolysis. The oxidation–reduction reactions also liberate energy, but to a smaller degree. Transfer of electrons is involved in all oxidation and reduction reactions. Oxidation is defined as removal of electrons with an increase in valence; reduction is defined as a gain in electrons with a decrease in valence. Cellular enzyme systems accomplish oxidation without oxygen by removal of hydrogen from an oxidizable substrate. Such a method of oxidation is extremely common in biological systems. In the following example, $A \cdot H_2$ is the reduced form of an oxidizable substrate, and B is the oxidized form of another metabolite of higher oxidation–reduction potential:

$$A \cdot H_2 + B \xrightarrow{\text{specific dehydrogenase}} A + B \cdot H_2.$$

The reductant $A \cdot H_2$ is termed a **hydrogen donor,** and the oxidant B is the **hydrogen acceptor.** In order for the reaction to proceed readily, a catalyst is necessary. This is supplied by the specific enzymatic dehydrogenases of the tissues. The energy transfers involved in oxidation–reduction systems are measured by the difference in the electrical potential of the various systems. The system with the higher potential oxidizes the one with the lower potential, with a consequent liberation of energy for vital processes.

The formation of lactic acid from glycogen (glycolysis) represents both types of energy relationships discussed. A simplified account of glycolysis follows. **Glycogen** is a branched complex polysaccharide with a molecular weight as high as 4 million (Fig. 1-32). The glucose molecules are linked together in a straight-chain fashion for as many as 18 units by a bond between the **number 1 carbon** of one glucose and the **number 4 carbon** of the next glucose. Branching from this straight chain are a number of 1-6 carbon bonds that then continue as 1-4 linkages. Enzymes attack glycogen at either the 1-4 or the 1-6 carbon bond. In muscle when glycogen is broken down to individual glucose units, it reacts first with inorganic phosphate at the 1-4 carbon bond to form glucose-1-phosphate. The reaction requires an enzyme (phosphorylase) and the process is designated a **phosphorolysis,** not a hydrolysis. Muscle phosphorylase exists in two forms, a and b. **Phosphorylase a** is the active form and is made up of four identical polypeptide chains, each chain containing one phosphate. **Phosphorylase b,** the inactive form of phosphorylase a, is brought about by a specific enzyme in muscle (phosphatase) which removes the phosphates from phosphorylase a, which causes it to dissociate into two double-stranded inactive polypeptide chains. The conversion of the inactive phosphorylase b to active phosphorylase a can be brought about by the utilization of ATP and the presence of another enzyme, **phosphorylase b kinase,** as shown in the reaction

$$\underset{\text{(inactive)}}{2\text{ phosphorylase b}} + 4\text{ATP} \xrightarrow{\text{phosphorylase b kinase}} \underset{\text{(active)}}{\text{Phosphorylase a}} + 4\text{ADP}.$$

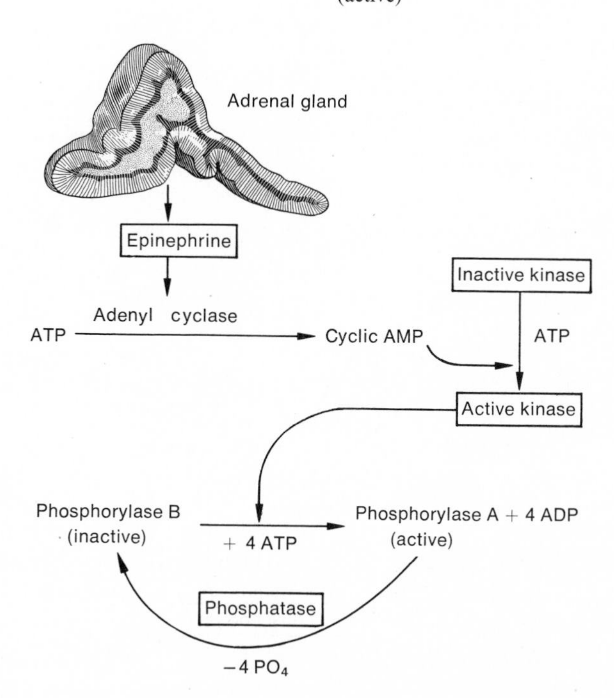

Figure 1-33. Conversion of inactive phosphorylase b into active phosphorylase a, which controls glycogen breakdown. As epinephrine is released, more glycogen is broken down into glucose-1-phosphate.

The activation of phosphorylase b kinase requires **cyclic AMP,** which is formed from ATP under the influence of the hormone epinephrine released from the adrenal glands. These mechanisms regulate glycogen breakdown in muscle, as shown in Fig. 1-33.

Glycolysis is a sequence of enzyme-mediated reactions which can operate without oxygen (anaerobically), leading from a molecule of **glucose** to two molecules of **pyruvic acid** or **lactic acid.** Whether pyruvic or lactic acid is formed is dependent upon the supply of oxygen to the cell. When there is an insufficient supply of oxygen, lactic acid is formed. The glycolytic pathway occurs in the cytoplasm of the cell. The reactions are shown in detail in Fig. 1-34. Glucose must first be phosphorylated before degradation, and ATP acts as a phosphate donor. The net yield of phosphate-bond energy (moles of ATP) derived from glycolysis of 1 mole of glucose can be easily determined (Fig. 1-34). Two moles

Figure 1-34. Glycolytic (Embden-Meyerhof) pathway, which results in a net yield of 8 moles of ATP from 1 mole of glucose. In anaerobic glycolysis there is a net yield of only 2 moles of ATP. In aerobic glycolysis the $NADH_2$ in reaction (5) gives rise to 6 moles of ATP by oxidative phosphorylation.

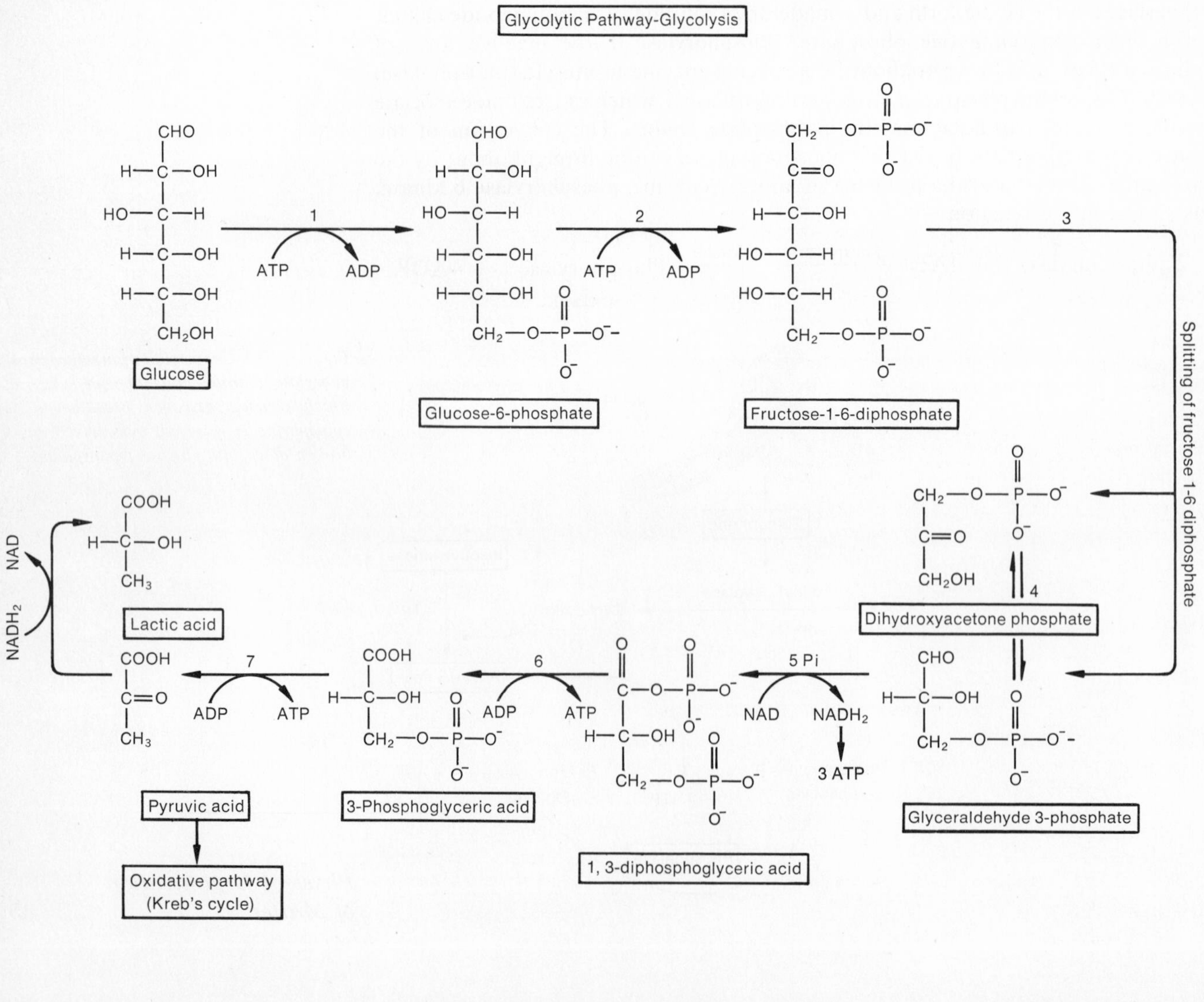

Glycolytic Pathway-Glycolysis

Glucose → (1; ATP → ADP) Glucose-6-phosphate → (2; ATP → ADP) Fructose-1-6-diphosphate → (3; Splitting of fructose 1-6 diphosphate) Dihydroxyacetone phosphate ⇌ (4) Glyceraldehyde 3-phosphate → (5 Pi; NAD → $NADH_2$ → 3 ATP) 1, 3-diphosphoglyceric acid → (6; ADP → ATP) 3-Phosphoglyceric acid → (7; ADP → ATP) Pyruvic acid → Oxidative pathway (Kreb's cycle)

Pyruvic acid → ($NADH_2$ → NAD) Lactic acid

of ATP are initially consumed (reactions 1 and 2). Four moles are generated (reactions 6 and 7), two from each of the three carbon molecules formed in **reaction 3,** which resulted from the splitting of the six carbon glucose molecules, thus yielding a net gain of 2 moles of ATP. ATP may also be synthesized from the $NADH_2$ formed in reaction 5 by the process of oxidative phosphorylation. This process results in 3 moles of ATP from each mole of $NADH_2$, for a total of 6 moles of ATP for each mole of glucose. The process of **oxidative phosphorylation** results in the transfer of the two hydrogen atoms to oxygen-generating NAD, which can again be used in reaction 5. This process is referred to as **aerobic glycolysis** and results in a net gain of 8 moles of ATP. In the absence of oxygen, $NADH_2$ can still be converted to NAD by transferring the two hydrogen atoms to pyruvic acid, resulting in the formation of lactic acid, as indicated in Fig. 1-34. In anaerobic glycolysis the net amount of ATP synthesized per glucose molecule is less than during aerobic glycolysis (2 moles in contrast to 8).

Krebs Cycle–Citric Acid Pathway

The oxidative metabolism of the three carbon molecules of lactic or pyruvic acids to carbon dioxide and water which takes place in the mitochondria provides the most efficient method of generating phosphate-bond energy in aerobic organisms. In contrast to the glycolytic pathway, which provides a net yield of eight high-energy phosphate bonds, the Krebs cycle provides a net yield of 30 high-energy bonds (15 moles of ATP for each mole of pyruvic or lactic acid entering this pathway). The over-all reaction can be written

$$\underset{\text{Pyruvic acid}}{2CH_3\text{—}\overset{\overset{\displaystyle O}{\|}}{C}\text{—}COOH} + 5O_2 \longrightarrow 6CO_2 + 4H_2O + 30ATP.$$

Thus a total of 38 high-energy phosphate bonds (molecules of ATP) can be synthesized from each molecule of glucose metabolized. Since each mole of glucose has an energy content of 690,000 cal (690 kcal) and each mole of ATP represents a transfer of 10,000 cal (10 kcal) of this energy to phosphate-bond energy, then 38 moles of ATP represents a transfer of 380,000 cal (380 kcal). Thus 55 per cent of the total potential energy in glucose has been transferred to ATP. The remaining energy escapes as heat.

Figure 1-35 summarizes the reactions of the **Krebs cycle.** The first reaction involves the decarboxylation (loss of CO_2) of **pyruvate,** which in turn combines with coenzyme A (CoA) to form an "active" two-carbon-unit **acetyl CoA.** This unit then enters the cycle by combining with a 4-carbon dicarboxylic acid, **oxaloacetic acid,** to form a 6-carbon acid, **citric acid.** In the several reactions that follow, citric acid is rearranged by successive loss and recapture of water to form isocitric. Isocitric is then oxidized to form **oxalsuccinic acid,** which loses CO_2 to form **α-ketoglutaric acid. Succinic acid** is formed from the oxidative decarboxylation of α-ketoglutaric acid. Succinic acid is then oxidized to **fumaric acid,** which is immediately hydrated to **malic acid.** Oxalacetic acid is regenerated by the oxidation of malic acid. The cycle is closed by the condensation of oxalacetic with **acetyl CoA** to form citric acid. To keep the cycle going, however, acetyl CoA must be supplied.

Figure 1-35. Krebs, citric acid, cycle: 15 molecules of ATP are synthesized per molecule of pyruvic acid. Since 2 molecules of pyruvate are formed from each molecule of glucose, a total of 30 molecules of ATP is formed.

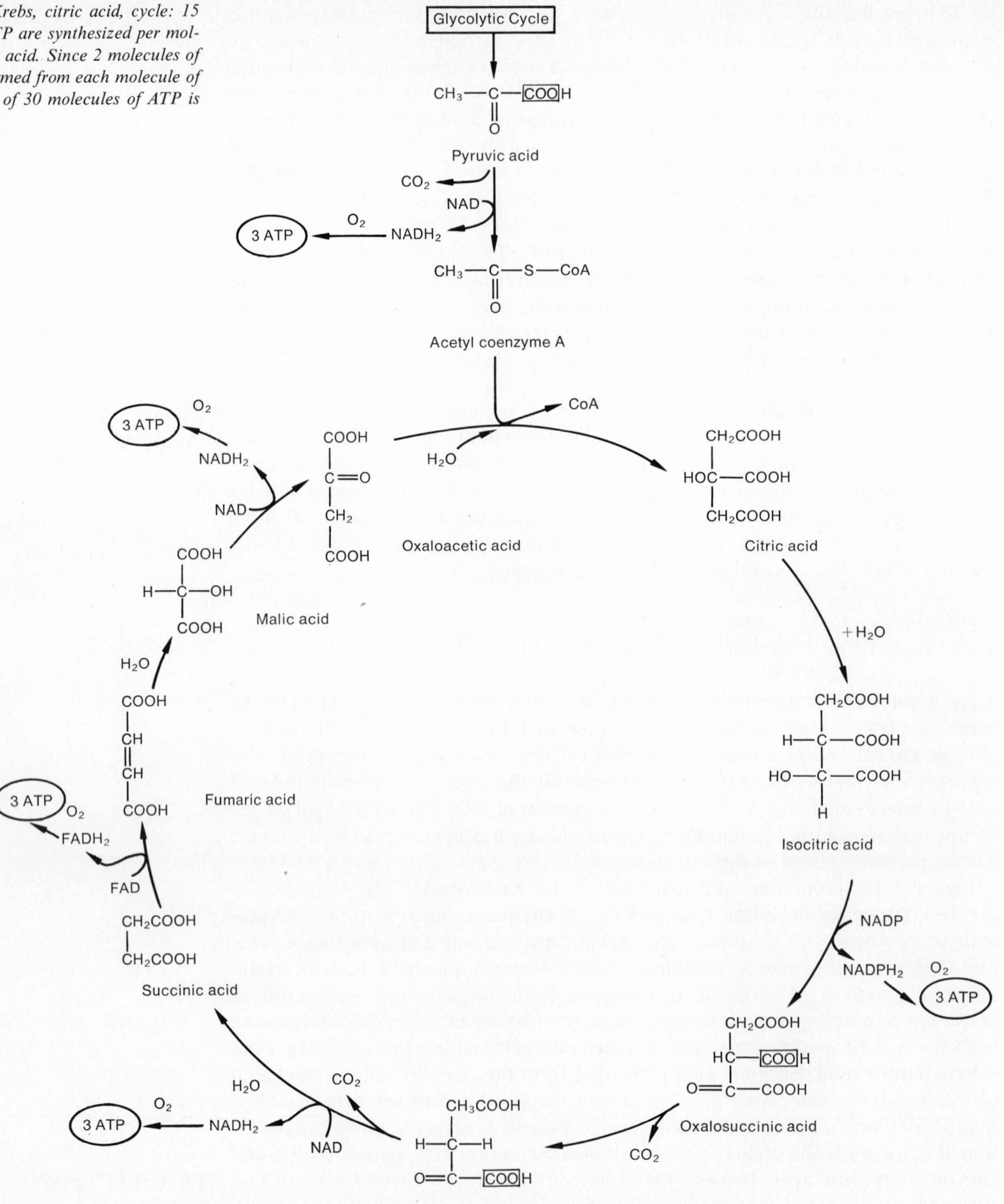

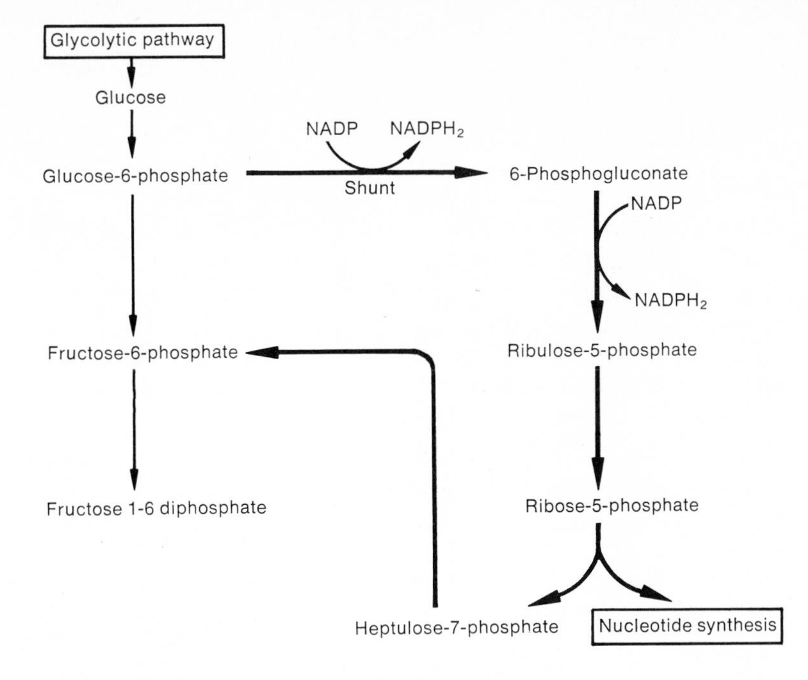

Figure 1-36. Pentose phosphate shunt.

The glycolytic and citric-acid-cycle pathways are the major sources of ATP energy. In addition, they are necessary for glycogen breakdown and provide intermediates for the synthesis of lipids and amino acids.

Pentose Phosphate Pathway (Phosphogluconate Shunt)

The **pentose phosphate shunt** provides pentoses, which are essential components of nucleotides and therefore nucleic acids. It also provides $NADPH_2$ (reduced nicotinamide adenine dinucleotide phosphate), which is essential in many syntheses reactions within the cell, such as in steroid and fatty acid synthesis. The term **shunt** refers to the alternate pathway this represents as an offshoot of the glycolytic pathway. The reactions in this pathway are shown in Fig. 1-36. The initial reaction is the oxidative conversion of glucose-6-phosphate from the glycolytic pathway to 6-phosphogluconate. This is further oxidized to ribulose-5-phosphate. Ribulose-5-phosphate is converted to ribose-5-phosphate, which in turn is converted to fructose-6-phosphate, thus entering back into the glycolytic pathway.

Production of ATP by Oxidative Phosphorylation (Biological Oxidation–Enzyme Systems)

Oxygen is the final acceptor of hydrogen in cellular respiration, and enzymes known as **dehydrogenases,** which can transfer hydrogen from a metabolite directly to oxygen, are known as **aerobic dehydrogenases.** Another class of enzymes, the **oxidases,** is often grouped with the aerobic dehydrogenases, but the oxidases can use only oxygen as hydrogen acceptor, whereas the aerobic dehydrogenases can utilize methylene blue or other artificial acceptors in addition to oxygen. Since most dehydrogenases cannot catalyze the transfer of hydrogen from a substrate directly to oxygen, other systems that will couple with the anaerobic dehydrogenase systems as well as with molecular oxygen must exist in the tissues. The iron-containing catalysts, or **cytochromes,** which are widely distributed in natural cells, are capable of performing this function. Three main cytochromes are known. They are designated cytochromes *a*, *b*, and *c*.

Many dehydrogenases require **activators,** or **coenzymes,** which are present in the intact cell. Three such coenzymes are *coenzyme I, or nicotinamide-adenine dinucleotide* (*NAD*); *coenzyme II, or nicotinamide-adenine dinucleotide phosphate* (*NADP*), and a new electron transport coenzyme, related to vitamin K, called *coenzyme Q*. In addition to these enzyme systems, there are **flavoproteins,** which evidence strongly suggests act as a link between the coenzymes I, II, Q and the cytochromes. The flavoproteins can act as hydrogen transfer agents in a manner similar to the action of coenzymes I and II. The *yellow enzyme of Warburg,* which can catalyze the reoxidation of coenzyme II, is an example.

In summary, the transfer of hydrogen electrons given off in the citric acid cycle to molecular oxygen to form water involves a series of enzyme systems, much like the transfer of a commodity through various "middlemen" before the ultimate consumer (oxygen) is reached. The flow of hydrogen electrons from the high energy level of reduced DPN to the lower energy level of the other coenzymes releases energy to form ATP. Thus, the oxidation of 1 mole of DPNH, in which a pair of electrons (2H) is transferred to oxygen, gives rise to the utilization of three inorganic phosphates to form three molecules of ATP. These three reactions can be indicated as follows, where the arrows indicate the direction of hydrogen electron flow:

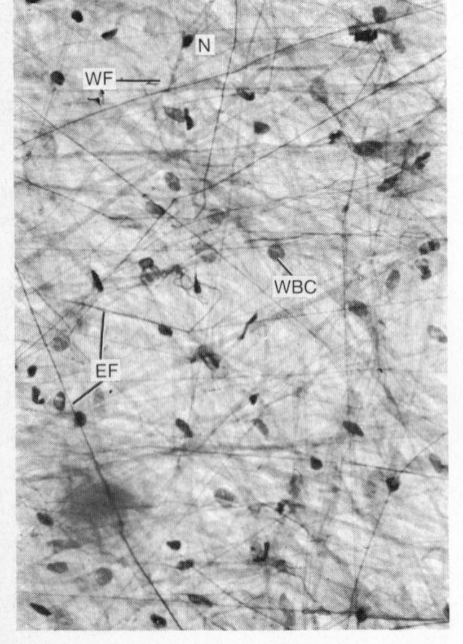

Figure 1-37. Areolar connective tissue showing white collagenous fiber (WF), yellow elastic fibers (EF), fibrous tissue nucleus (N), and various types of white blood cells (WBC).

$$\text{Substrate} \cdot H_2 + \text{DPN} \xrightarrow{\text{dehydrogenase}} \text{Substrate} + \text{DPNH}_2.$$

1. DPNH $\longrightarrow$ flavoprotein $\longrightarrow$ coenzyme Q = 1 ATP formed.
2. Coenzyme Q · H $\longrightarrow$ cytochrome *a* $\longrightarrow$ cytochrome *b* = 1 ATP formed.
3. Cytochrome *b* · H $\longrightarrow$ cytochrome oxidase $\longrightarrow$ $\frac{1}{2}O_2$ = 1 ATP formed.

Since one oxygen atom is used ($2H + O \longrightarrow H_2O$), the ratio of phosphate consumed to oxygen used is 3 (P/O = 3). Note that, in the citric acid cycle, the oxidation of one pyruvic acid (Fig. 1-35) causes formation of four molecules of reduced pyridine nucleotide, ($NADH_2$) and the oxidation of these molecules leads to the synthesis of twelve ATP molecules. Three more ATP molecules are synthesized from the oxidation of $FADH_2$, accounting for a net total of fifteen ATP molecules synthesized for each pyruvate molecule. The entire oxidation process is initiated by the action of dehydrogenase enzymes specific to the metabolite (isocitric dehydrogenase, α-ketoglutaric acid dehydrogenase, succinic acid dehydrogenase, malic acid dehydrogenase). These enzymes cannot transfer the hydrogen electrons directly to oxygen to form water but require intermediary coenzymes, which are also found in the mitochondria of the cell (see reactions 1, 2, and 3 above).

Although the citric acid cycle is considered most important in terms of the synthesis of ATP, it is also important in the production and utilization of the carbon skeletons of many other compounds important in biological activity.

Tissues

Very few cells exist as lone entities in the animal body; rather, they are associated into sheets of cells called tissues. **Tissues** are aggregates of similar cells performing a similar function. A group of liver cells is referred to as liver tissue; a group of bone cells is bone tissue. The types of tissue found in the animal body are limited to vascular, reproductive, epithelial, nervous, muscular, and connective tissue. **Epithelial tissues** cover the surface of the body or form the delicate linings of body cavities that open directly or indirectly to the surface. **Connective tissues** are found throughout the body—they help to form the framework of organs, and they pervade tissue spaces, filling in and connecting organs and various other structures. The intercellular substance, which varies considerably in different kinds of connective tissue, enables us to classify these tissues as fibrous, areolar, cartilaginous, body, adipose, and reticuloendothelial tissue.

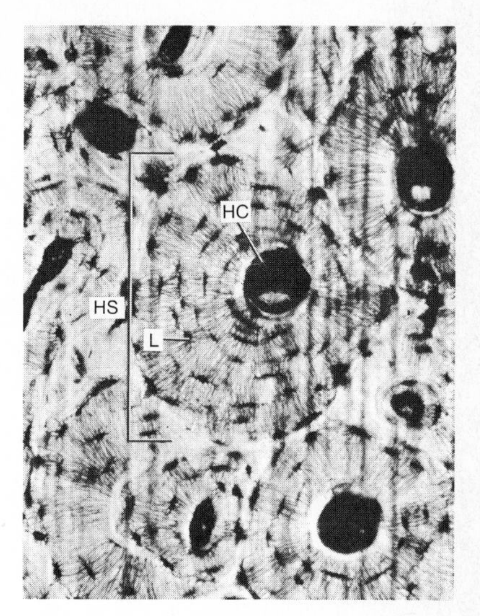

Figure 1-38. Cross section of human bone showing haversian system (HS), haversian canal (HC), and osteocytes in lacuna (L). Tiny radiations from lacunae are canaliculi.

Fibrous Connective Tissue. **Fibrous connective tissue** may be divided into **white fibrous** and **dense elastic,** or **yellow fibrous,** tissue. The former appears microscopically as dense parallel rows of white fibers in between which are scattered a few cells. White fibrous tissue appears as heavy, glistening pearly white cords or sheets, as in tendons, ligaments, and aponeuroses. Yellow fibrous tissue, as seen under the microscope, consists of coarse elastic fibers with little ground substances and few cells.

Areolar Tissue. Sometimes referred to as **loose connective tissue, areolar tissue** (Fig. 1-37) appears as a relatively thin and delicate gossamer weblike membrane. It consists of various types of connective tissue cells scattered between numerous white fibers and yellow elastic fibers. **Cartilage** is found at the end of bones and forms the pliable framework of the ear and the tip of the nose. Macroscopically cartilage appears as a bluish white, firm but not rigid substance and is quite transparent. Like osseous tissue, the cartilage cells are widely scattered throughout a homogeneous intercellular matrix and are situated in tiny spaces called **lacunae.**

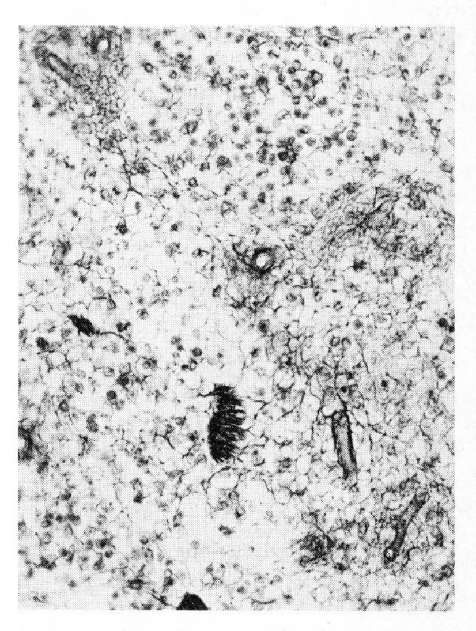

Figure 1-39. Section of spleen showing reticular type of connective tissue. (Courtesy Ward's Natural Science Establishment, Inc., Rochester, N.Y.)

Bony Tissue. **Bony tissue** (Fig. 1-38) is made up of a series of cells widely separated from each other by a large amount of solid intercellular substance, the **matrix.** The matrix is harder than that in cartilage, largely owing to the deposition of calcium salts. A more detailed discussion of osseous tissue will be found in Chapter 2.

Adipose Tissue. **Adipose tissue** is the fat tissue of the body. Microscopically, fat cells have the shape of signet rings, with large globules of fat within the cells surrounded by a thin margin of cytoplasm. The nucleus is squeezed to the periphery of the cell by the accumulation of fat within. Fat cells are more closely packed with less intercellular material than is in the other types of connective tissue.

Reticuloendothelial Tissue. **Reticuloendothelial tissue** (Fig. 1-39) consists of various types of connective tissue cells widely scattered throughout the body. Most of the cells of the reticuloendothelial system are phagocytic cells located mainly in the bone marrow, the liver capillaries, the splenic sinusoids, the lymph glands, and the serous membranes. These cells have great importance for the defense of the body against foreign particulate matter and microorganisms, since they are capable of surrounding local infections, ingesting invading materials, and probably synthesizing immune bodies. In addition, under conditions of stress the reticuloendothelial system can form blood cells if the marrow cannot do so as well as cells that are not blood cells (fibroblasts).

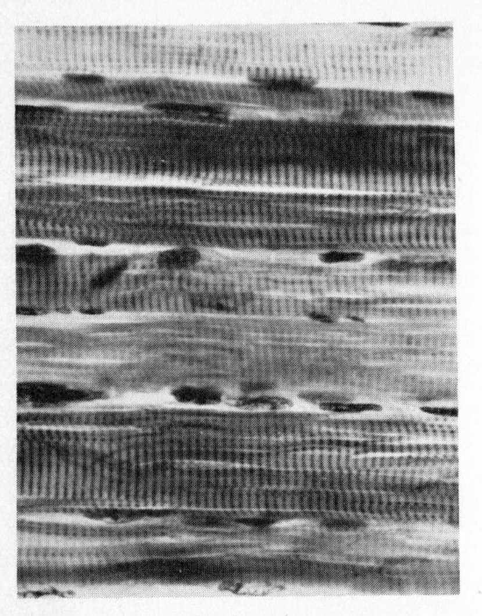

Figure 1-40. Skeletal muscle section showing multinucleated muscle fibers and cross striations. (Courtesy Ward's Natural Science Establishment, Inc., Rochester, N.Y.)

Vascular Tissue. Unlike most other tissues of the body, **vascular tissue,** another connective tissue type, consists of cells, not closely packed together in sheets, but individually suspended in a fluid environment called **plasma.** The cells are differentiated into erythrocytes, leukocytes, and blood platelets.

Muscle Tissue. Three types of **muscle tissue** are found in the body. These are classified as smooth muscle, skeletal muscle, and cardiac muscle. Smooth muscle is more closely associated with connective tissue structures of the body, being found in the walls of blood vessels, the walls of the digestive tract, and most ducts. It is the least differentiated type of all muscle tissue, the units of structure being the smooth muscle cell, which is spindle-shaped and contains a single oval nucleus. Skeletal muscle tissue, as the name implies, is associated with the skeletal parts. Its basic unit of structure may be described as an unbranching, multinucleated cylinder of varying length. The skeletal muscle cell is further differentiated from the smooth muscle cell in that it is cross-striated; that is, alternate dark and light

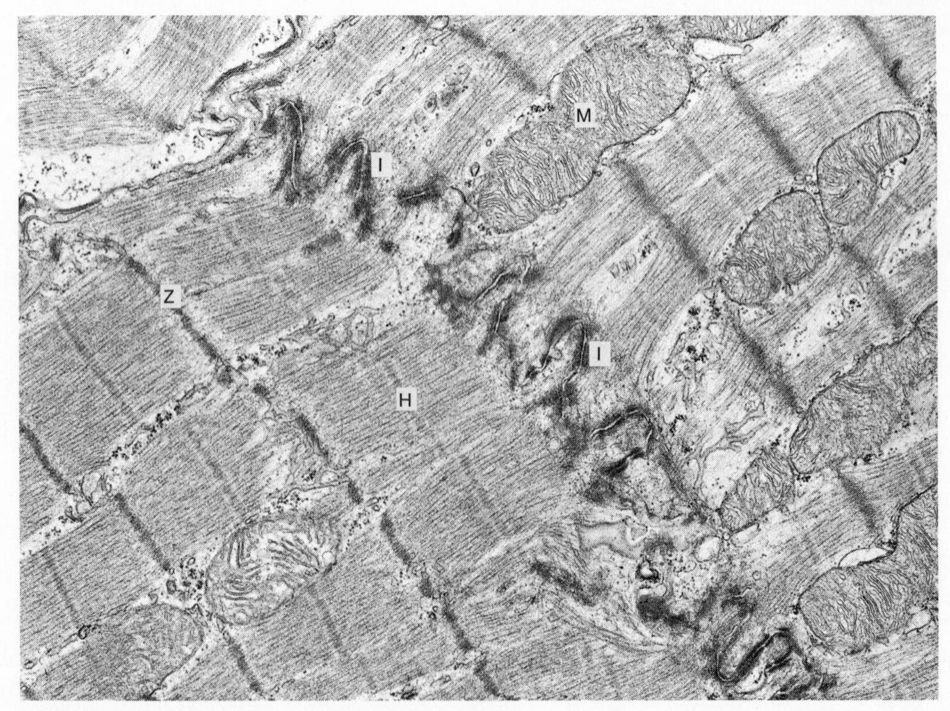

Figure 1-41. Cardiac muscle (dog heart) showing area of the intercalated disk. Magnified 30,450 ×. I, intercalated disk; Z, Z band; H, H band; M, mitochondria. (Courtesy Timothy Mougel, Electron Microscopy Laboratory, Department of Zoology, University of Maryland, College Park, Md.)

bands of material are found along its axis, giving the cell a striped appearance (Fig. 1-40). Cardiac tissue consists of units that are roughly rectangular, with one or both ends having short, stubby branches. A single oval nucleus is present, and peculiar bandlike structures called **intercalated disks** run across the cell, giving a cardiac tissue a characteristic appearance. The individual units of structure do not appear to have definite cell boundaries when viewed under the light microscope since adjacent units are closely fused with one another (Fig. 1-41).

Nerve Tissue. The basic unit of structure of **nerve tissue** (Fig. 1-42) is the **neuron,** or **nerve cell,** which differs from other cells of the body in that each one possesses two or more processes extending from the cell body. These processes are called **dendrites** or **axons.** Dendrites are shorter structures, and usually more than one is present on one cell, functioning in the transmission of nerve impulses into the cell body. Axons are usually much longer processes then dendrites, and only one is present on each nerve cell. They carry impulses away from the cell body. Grossly, nerve tissue appears as a creamy white or gray, soft, gelatinous mass.

All the tissues listed are associated with one another to form organs. An organ may be defined as an association of different tissues so constituted as to perform a specific function. As such, these tissues will be discussed more completely in the following chapters as each organ system of the body is considered separately.

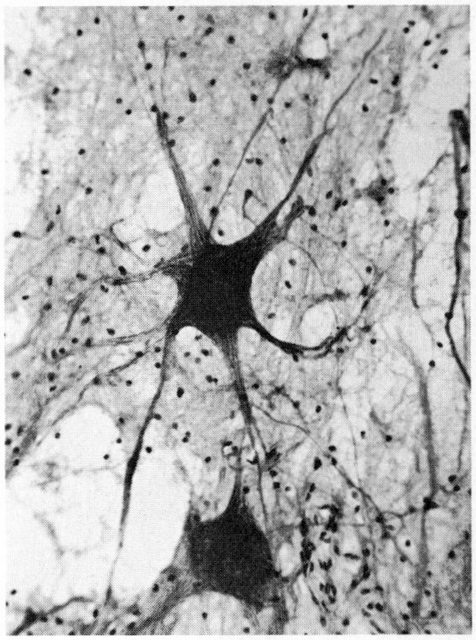

Figure 1-42. Giant multipolar motor neutons from ventral horn of spinal cord of ox. (Courtesy Ward's Natural Science Establishment, Inc., Rochester, N.Y.)

References

Bloom, W., and D. W. Fawcett. *A Textbook of Histology*. Philadelphia: Saunders, 1968.

Bolis, L., A. Katchalsky, R. D. Keynes, W. R. Loewenstein, and B. A. Pethica (eds.). *Permeability and Function of Biological Membranes*. New York: Am. Elsevier, 1970.

Boyer, P. D. (ed.). *The Enzymes,* Vol. 1: *Structure and Control,* 1970; Vol. 2: *Kinetics and Mechanism,* 1970. New York: Academic.

Edwards, N. *Cellular Biochemistry and Physiology*. New York: McGraw-Hill, 1972.

Green, D. E., and H. Baum. *Energy and the Mitochondrion*. New York: Academic, 1970.

Kotyk, A., and K. Janacek. *Cell Membrane Transport*. New York: Plenum, 1970.

Plowman, K. *Enzyme Kinetics*. New York: McGraw-Hill, 1972.

Plummer, D. *An Introduction to Practical Biochemistry*. New York: McGraw-Hill, 1972.

Temasheff, S. N. *Molecules and Life. An Introduction to Molecular Biology*. New York: Plenum, 1972.

Tosteson, D. C. (ed.). *The Molecular Basis of Membrane Function*. Englewood Cliffs, N.J.: Prentice-Hall, 1969.

Yost, H. T. *Cellular Physiology*. Englewood Cliffs, N.J.: Prentice-Hall, 1972.

Osteology: Physical and Chemical Structure of Bone, Bone Formation, and Function

The bones of the body with their blood and nerve supply constitute the organs of the skeletal system (Fig. 2-1). Bone itself is a highly specialized form of connective tissue, differing from connective tissue proper in that it is hard and contains cells peculiar to it, whereas for the most part the cells common to connective tissue are lacking. The hardness of bone is caused by the deposition of a complex of mineral salts within the soft organic matrix that makes up the framework of the bone. In general, the interstitial (between cells) substance, in addition to containing water, which varies greatly in amount and which is abundant in the bones of young animals, consists of two main components: (1) the organic framework, and (2) the inorganic mineral salts (bone ash).

The organic portion of bone makes up about 35 per cent of bone and consists chiefly of a protein called **bone collagen,** or **ossein.** In ordinary histological sections the intercellular portion of bone appears to be homogeneous, but with special staining methods the collagen can be demonstrated to be present in the form of fibers. Collagen is a soft gelatinous material found in white fibers. Most studies on collagen have been made on material from other sources, but enough information concerning collagen from bone is now available to make it clear that it shares common properties with that from other forms of connective tissue. Analysis has demonstrated that collagen is either a pure protein or a mixture of proteins

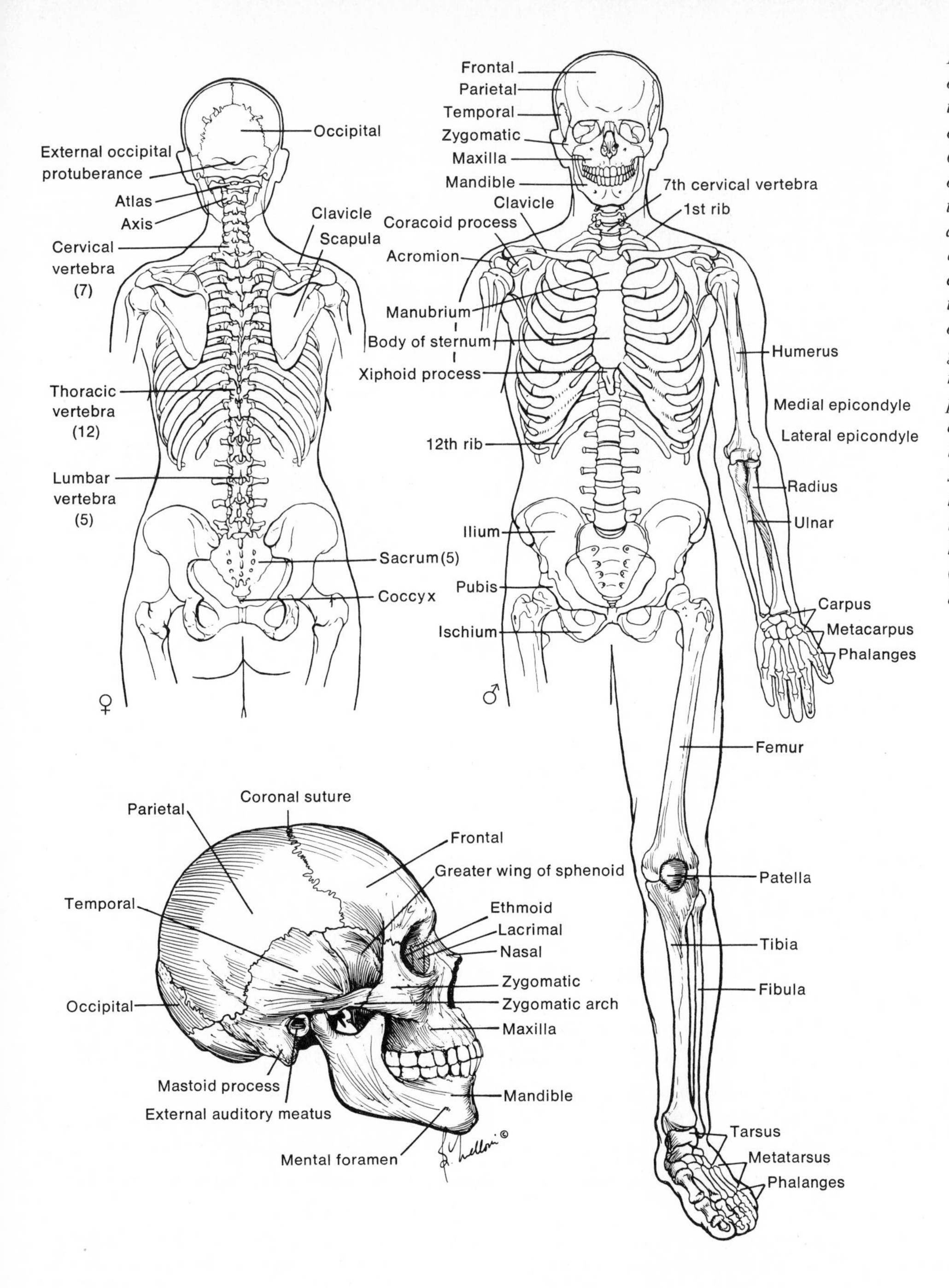

Figure 2-1. Anterior and posterior views of human skeleton. Axial skeleton includes the following: skull (29 bones), consisting of cranium, 8 bones; face, 14; ear ossicles 6; and hyoid, 1. Vertebral column, made up of 26 bones when the 10 terminal segments that fuse to form the sacrum and coccyx are considered as 2 bones. Thorax (sternum and ribs), 25 bones. The appendicular skeleton is divided into upper and lower extremities. The upper extremity is comprised of 64 bones, which includes the shoulder girdle (scapula and clavicle), arm (humerus), forearm (ulna, radius), hand [carpus (wrist), 8 bones; metacarpus, 5 bones; and phalanges (fingers), 14 bones]. The lower extremity contains 62 bones and consists of the hip [pelvic girdle (innominate bone)], thigh (femur), kneecap (patella), leg (tibia, fibula), foot [tarsus (ankle), 7 bones; metatarsus, 5 bones; and phalanges (toes), 14 bones]. Total number of bones in adult skeleton, 206.

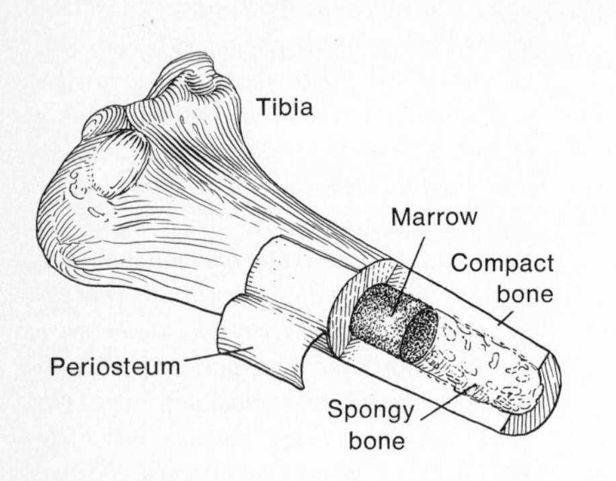

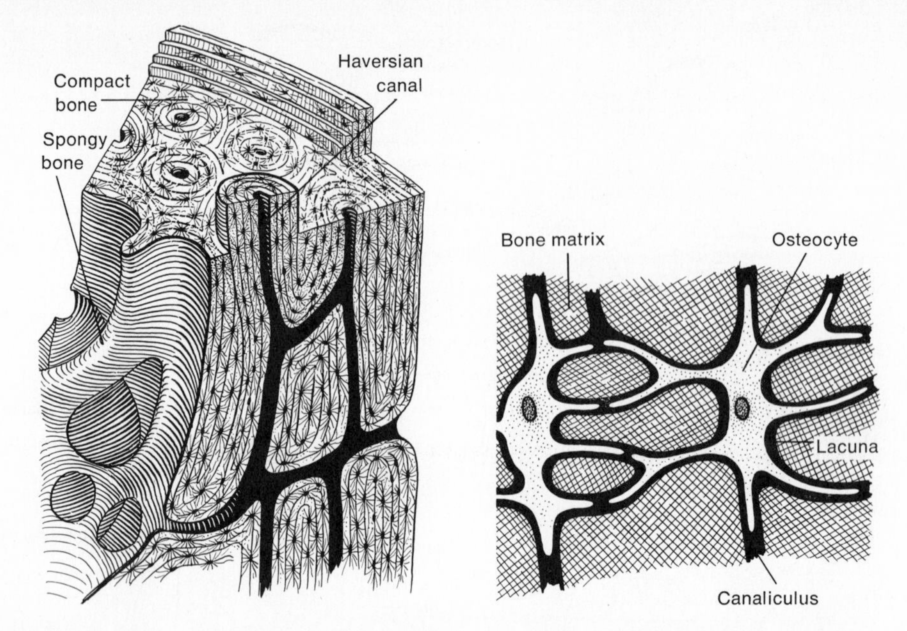

Figure 2-2. Structure of bone showing the relationship of the haversian system to compact and spongy bone.

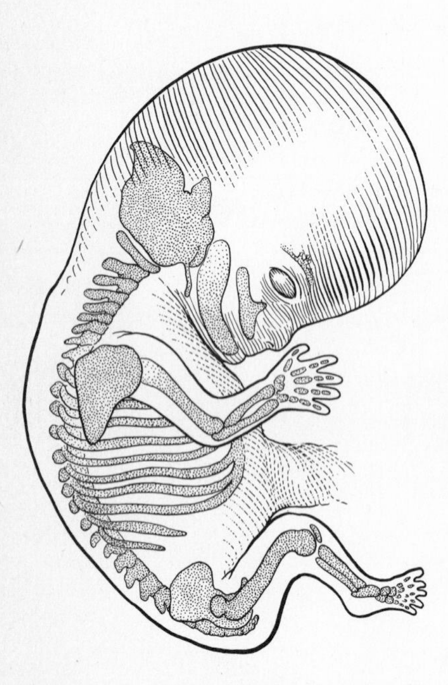

Figure 2-3. Human embryo 7 weeks old. The skeleton consists entirely of cartilage and membrane.

with perhaps other substances. This organic framework yields gelatin when boiled. Between the collagenous fibers is found a fluid, resembling tissue fluid, an amorphous ground substance whose main component is a mucopolysaccharide, chondroitin sulfate.

The inorganic portion makes up about 65 per cent of bone and is responsible for the hardness of bone. The mineral, commonly designated as the bone salt, together with the organic matrix in which it is deposited, makes up the interstitial substance of bone. Neglecting the small admixture of other elements, such as sodium, magnesium, potassium, chlorine, and fluorine, the bone salt consists of a complex of calcium carbonate and calcium phosphate, having the empirical formula $CaCO_3 \cdot nCa_3(PO_4)_2$, in which n has a value of approximately 2.5. A study of the X-ray pattern of the bone salts indicates a similarity to the naturally occurring mineral hydroxyapatite, the formula of which is said to be $Ca(OH)_2 \cdot 3Ca(PO_4)_2$ or $Ca_{10}(OH)_2(PO_4)_6$. It is believed that the crystal lattices of bone are similar to those of these apatite minerals, but that the elements may be substituted in bone without disturbing the structure. Despite its passive character, low metabolic rate, and great content of inorganic material, bone is a very plastic tissue and is highly sensitive to alterations of its normal mechanical function. Thus, disuse is followed by atrophy, in this case associated with loss of substance, and increased use is accompanied by hypertrophy, with an increase in the mass of the bone.

The organization of bone as a tissue is seen in the structure of the organic matrix and the relationship of this matrix to the bone cells, or **osteocytes.** The osteocytes are embedded in the matrix in small pockets called **lacunae,** which actually means "small lake." The osteocytes are separated from one another by the matrix but maintain indirect connections with one another by the small canals, called **canaliculi,** that course from lacunae. The unit of compact bone is the haversian system, which consists of various layers of osteocytes, their lacunae, and interconnecting canaliculi and matrix, circularly arranged around a narrow central lumen, the **haversian canal.** The canal carries one or more blood vessels, mainly capillaries and venules, and a nerve. The lacunae are arranged in concentric layers around the canal (Fig. 2-2). The canaliculi not only connect adjacent lacunae but also penetrate the central canal. The osteocytes fill its lacuna and send fine threads of living protoplasm some distance along the adjoining canaliculi. The canaliculi permit substances to pass from one cell to another and to and from the blood vessels in the haversian canals. In this way the living cells get rid of their waste products and receive the nourishment they must have to maintain normal function.

Bone Formation

The skeleton of the young embryo is composed of fibrous membranes and hyaline cartilage (Fig. 2-3). The formation of bone begins in these two tissues in the eighth week of embryonic life (human). Bone that forms in the fibrous membranes is called **intramembranous bone,** and bone formed in cartilaginous structures is called **intracartilaginous,** or endochondral, bone. These two terms do not imply any difference in structure once the bone is formed, but only indicate the method by which the bone starts to develop.

Intramembranous Ossification

Intramembranous ossification is the simpler, more direct type of bone formation. The flat bones of the face and skull and a part of the clavicle form in membrane. The continuous growth, internal reconstruction, and remodeling of every bone depends on intramembranous ossification. In the area in which bone formation is about to begin, mesenchymal cells congregate with an invasion of the area by many small blood vessels. The mesenchymal cells increase in size, clustering together to form long strands of cells, which run in all directions. The mesenchymal cells assume the form of osteoblasts and secrete collagenous fibrils. These fibrils form an axis for each elongated group of cells. The tissue formed is rather loose, and between the cells and collagenous fibrils is a semifluid substance called **osseomucoid.** The entire soft framework is called **osteoid.** Calcium salts are then deposited in the organic framework, and osteoid becomes true bone. This concept, postulated by Pommer, has come into dispute in the light of more recent work. McLean and Urist state that they do not recognize physiological osteoid; instead, they regard newly formed bone matrix as calcifiable, and for the most part, the matrix is calcified as it is laid down.

Intracartilaginous Ossification

Intracartilaginous ossification (Fig. 2-4) involves the same processes as intramembranous bone formation, but these processes are preceded by an initial period of cartilage erosion and calcification. The intracartilaginous, or **endochondral,** course of development occurs in most of the bones of the body, namely, the bones of the thorax, the limbs, most of the bones of the skull, and the hyoid bone. The entire process of ossification in cartilage models begins in a relatively small number of foci called **ossification centers.** Replacement of cartilage by bone extends centrifugally from these centers until the marrow cavity is free of cartilage cells. Some of the cartilage cells survive and become osteoblasts. The widespread replacement of cartilage is slowed down at the end of the cartilage model, which is destined to form the rounded end of the bone and the cartilaginous plate, which forms a boundary between the body, or shaft, of the bone and the rounded end. This plate, called the **epiphyseal cartilage plate,** represents a growth area by which an increase in length of bone is possible up to the twenty-fifth year of life. Growth in length of the long bones is thus a direct continuation of the intracartilaginous ossification within the embryonic cartilage model. Bone formation (osteogenesis) is brought about by osteoblasts. The cartilage model undergoes changes usually described as degenerative. The cartilage cells enlarge (that is, they hypertrophy) accumulating glycogen and the appropriate glycolytic enzymes and phosphatase. The matrix between the cartilage cells becomes calcified by the deposition of lime salts. Osteoblasts then surround and penetrate into the calcified cartilage matrix of the model, which soon undergoes erosion. The osteoblasts surrounding the cartilage model deposit a ring, or collar, of bony tissue around the calcified cartilage, becoming the compact bone of the shaft. Those osteoblasts that penetrate into the calcified cartilage form **trabeculae,** or bars, of bony tissue as the inner calcified cartilage erodes, thus forming primitive spongy bone with primary marrow spaces. These primitive bars of bone are resorbed by special cells called **osteoclasts,** and the primary marrow spaces unite to form the marrow cavity. The same changes occur at the ends of the bone during postnatal life with the exception that the trabeculae, or bars, of bone remain, forming spongy bone, and the cartilage forming the articular surfaces persists as the articular cartilages of the adult.

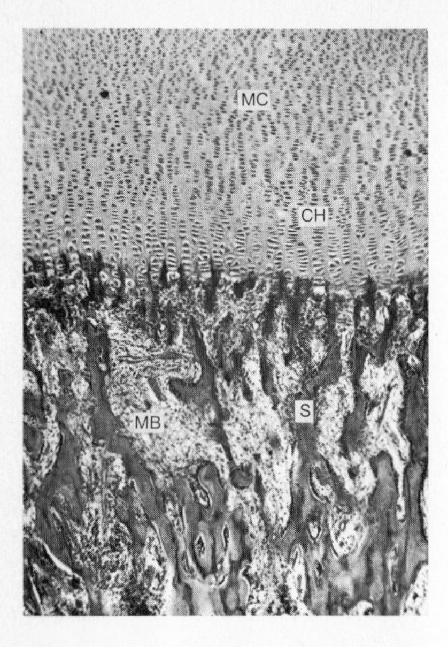

Figure 2-4. Intracartilaginous bone development, head of femur, human fetus. MC, matrix of cartilage; CH, *chondrocytes;* MB, *matrix of bone;* S, *"Spicules" of cartilage.* (*Courtesy General Biological Supply House, Chicago.*)

It must be emphasized that, regardless of the specific site at which bone formation is initiated, the actual formation of osseous tissue is in both cases exactly the same. The process of ossification as it occurs in long bones is illustrated in Fig. 2-5.

Macroscopically, mammalian bone is either **spongy** (cancellous) or **compact** in structure. Spongy bone consists of intercrossing and connecting bony bars of various thicknesses and shapes (Fig. 2-6). The arrangement of these interconnecting osseous bars gives the skeleton a maximum rigidity and resistance to changes in shape. Compact bone appears as a continuous hard mass in which spaces can be distinguished only with the aid of the microscope. Spongy bone and compact bone are merely different arrangements of the same histological elements and are not to be considered as having a different chemical structure. Practically every bone contains both types of osseous tissue.

The structure of a typical long bone, such as the femur or humerus, is as follows (Figs. 2-5 and 2-7). The shaft, or **diaphysis,** consists of compact bone and contains in its center a **bone marrow cavity.** The rounded end of the bone, called the **epiphysis,** consists of spongy bone with a thin peripheral layer of compact bone. In the growing animal the epiphysis and the diaphysis are separated by the **epiphyseal cartilage plate,** which is united to the shaft by columns of spongy bone called the **metaphysis.** The epiphyseal cartilage together with the metaphysis forms a **growth apparatus** from which growth in length of the long bones occurs. The growth apparatus is based on the epiphyseal cartilage but is not limited to this structure. The cartilage cells in the upper border of the plate, which end toward

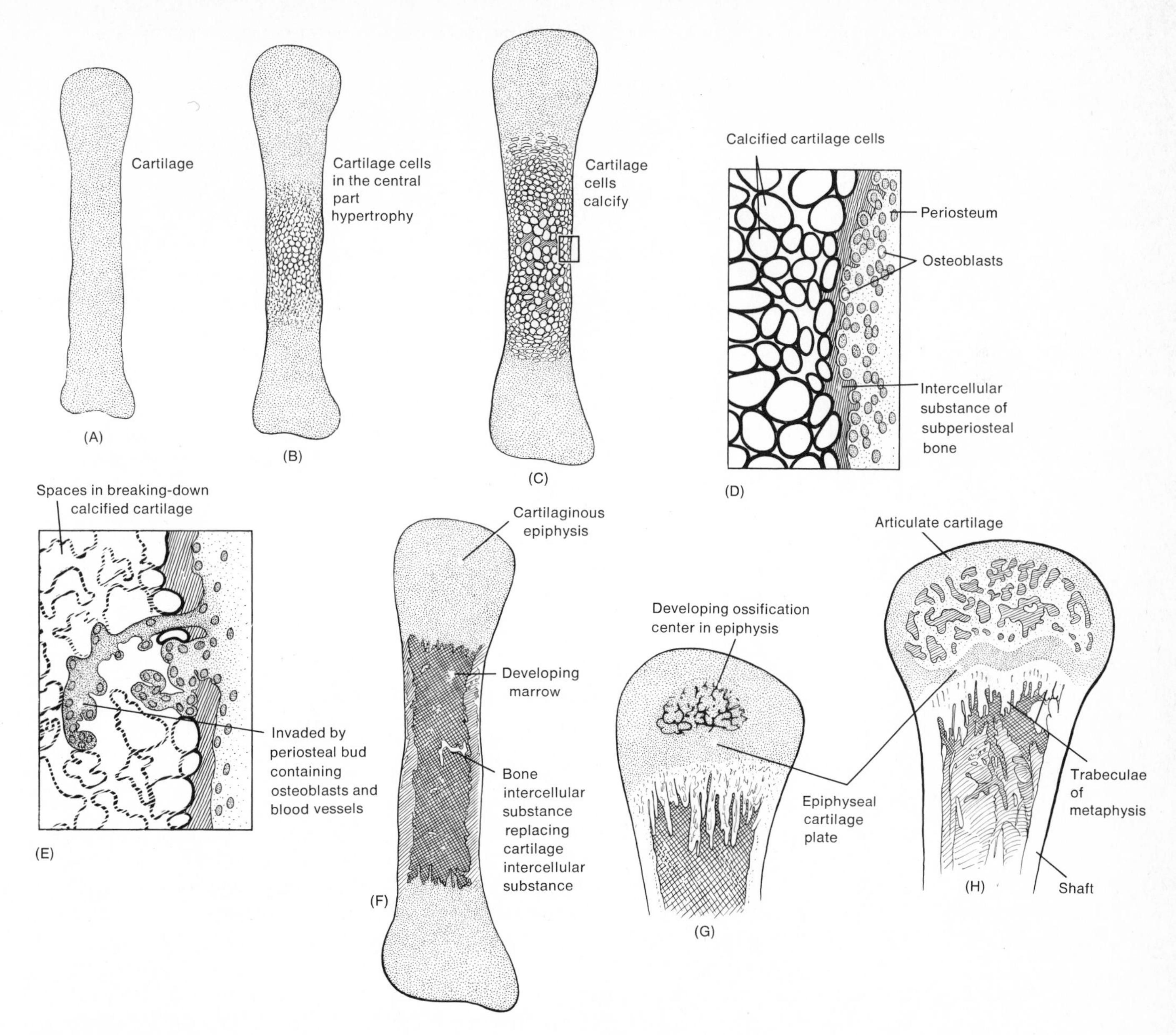

the epiphysis, continue to divide, giving rise to new cells commonly arranged in rows. For this reason, the dividing cells are known as mother cells. Replacement of the cartilage cells by bone occurs at the other face of the cartilage, the end that faces the shaft, or diaphysis. Ossification is complete when the cells of the cartilage cease to divide and the disks are entirely replaced with bone. The epiphyses are then united with the diaphyses, and growth is no longer possible. At approximately the seventeenth year of life this union occurs in the distal epiphysis of the tibia, and about the twentieth year, in the proximal one.

Figure 2-5. Process of ossification as it occurs in long bones. D *and* E *represent a magnified section out of* C.

Figure 2-6. Spongy bone from shaft of tibia; a very small section greatly enlarged.

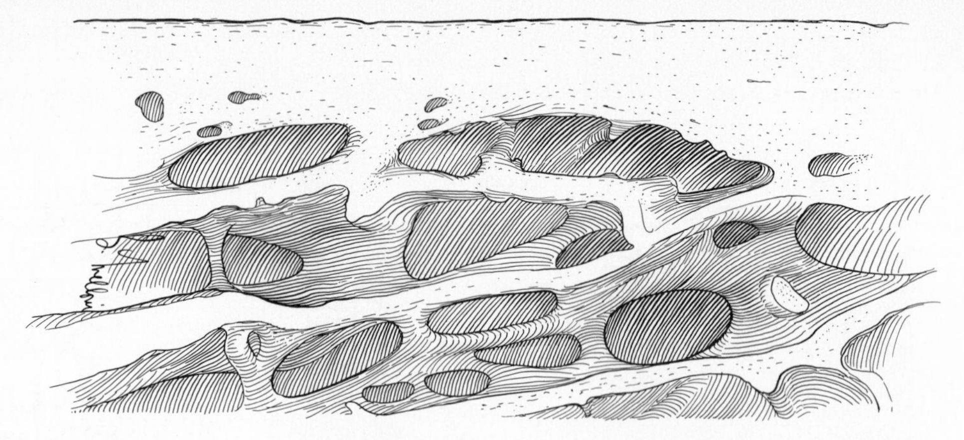

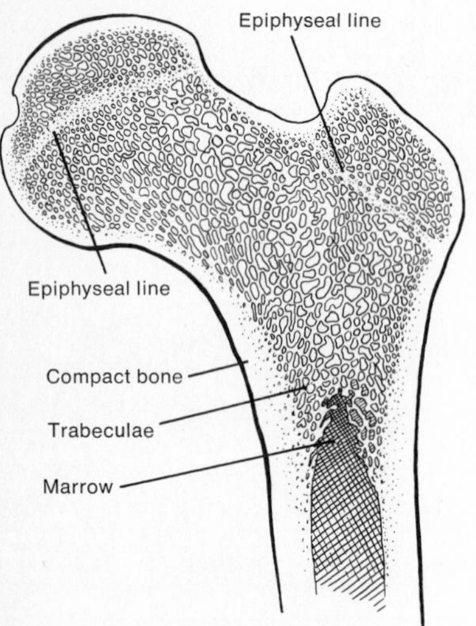

Figure 2-7. Structure of a typical long bone (section of femur).

All bones are covered with a modified connective tissue called **periosteum;** the articular surfaces may retain hyaline cartilage and its investing **perichondrium.** The outer layer of the periosteum is a network of dense connective tissue fibers and fibroblasts. Osteogenetic activity has not been demonstrated for this layer. The inner layer of the periosteum, sometimes called the **cambium layer,** is capable of bone formation during growth. Growth in circumference of the bone is accomplished by addition of layers of bone formed by the cambium layer. Ossification is not complete in all the bones of the body until about the twenty-fifth year. The growth of bone is markedly influenced by the growth hormone of the anterior pituitary. Hypophysectomy (removal of the pituitary) results in the cessation of intracartilaginous ossification. The administration of anterior pituitary extract reinitiates growth and, if continued for a sufficiently long time, may result in gigantism.

Bone Marrow

The marrow cavity is lined with a somewhat similar tissue corresponding to the periosteum, called the **endosteum.** The endosteum is a condensation of the stroma of the bone marrow. Its cells are reticular cells, identifiable as endosteal cells by their location. It has both **osteogenetic** and **hemopoietic** (blood-forming) potencies and, like the periosteum, takes an active part in the healing of fractures.

The cavity of a long bone is filled with **bone marrow;** this may be fatty or red, depending on the species of animal, its age, and the particular bone being studied. Marrow begins to form in the cartilage model soon after it is invaded by the osteoblasts, and it is obvious that these cells, which can form bone, can also form marrow. That is, the osteoblasts become differentiated in the marrow to provide for the production of reticuloendothelial phagocytic cells and of various kinds of blood cells. The mother cells from which these types arise retain their potentiality to form osteogenetic cells as well. We find that the bone marrow, although more commonly thought of in connection with its hemopoietic functions, also participates actively in formation of bone. Many of the cellular elements commonly seen in loose connective tissue are absent from bone tissue, but their counterparts are seen in numbers in the stroma of the bone marrow.

Fractures and the Repair of Bone

If bone is of normal structure, only forces many times greater than physiological forces or forces that are exerted in an unphysiological direction can produce a fracture. Such fractures are termed **traumatic.** There is a considerable quantity of data on the strength of bone in the literature. Although the classical analysis of beam mechanics began with Galileo in 1638, it was not applied to bone until the latter half of the nineteenth century. Hermann von Meyer demonstrated and described the barlike structure of spongy bone, and Culmann, a Swiss mathematician and engineer, saw the similarity of this structure to efficiently designed mechanical structures, such as the Fairbairn crane, and recognized the relation between structure and stress lines in bone. This resulted in the development of the **Culmann–Meyer theory,** which states that there is a functional relation between the stress lines in bone and its anatomical shape. This concept was further developed by Julius Wolff, who formally stated his **law of transformation of bone** in 1892. This law states that wherever stresses occur in bone, whether they are tensile or compressive stresses, there is formation of bone so that the strength of bone in that area is increased. It is believed that osteogenesis is stimulated by pressure or tension, and therefore, the greater the stress, the greater is the amount of bone formation by the osteocytes. Breaking-strength tests on whole femurs have shown that they can withstand compression tests of 1,547 to 1,990 lb before breaking. Torsion tests have indicated that whole femurs fail at 449 to 1,350 lb per inch, and bending-strength tests for rat femur showed a value of 10.6 lb before failure.

In contrast to traumatic fractures, there is also a pathological fracture, termed **spontaneous fracture,** in which the bone structure changed by disease may be broken even by physiological stresses. A fracture is termed **incomplete** or **complete** according to whether it leads to partial or complete discontinuity of bone. If the bone is splintered at the site of impact and smaller or larger fragments of bone are found between the two main fragments, this is called a **comminuted fracture.** If the site of fracture communicates with the outside through the traumatized area, the fracture is termed **compound,** or **open.** When the fracture is not open to the outside but is protected by the skin, it is termed **simple,** or **closed.**

During the healing of fractures, three stages can be observed (Fig. 2-8):

1. The formation and organization of the blood clot into the procallus.
2. The formation of the fibrocartilaginous callus.
3. The formation of the bony callus.

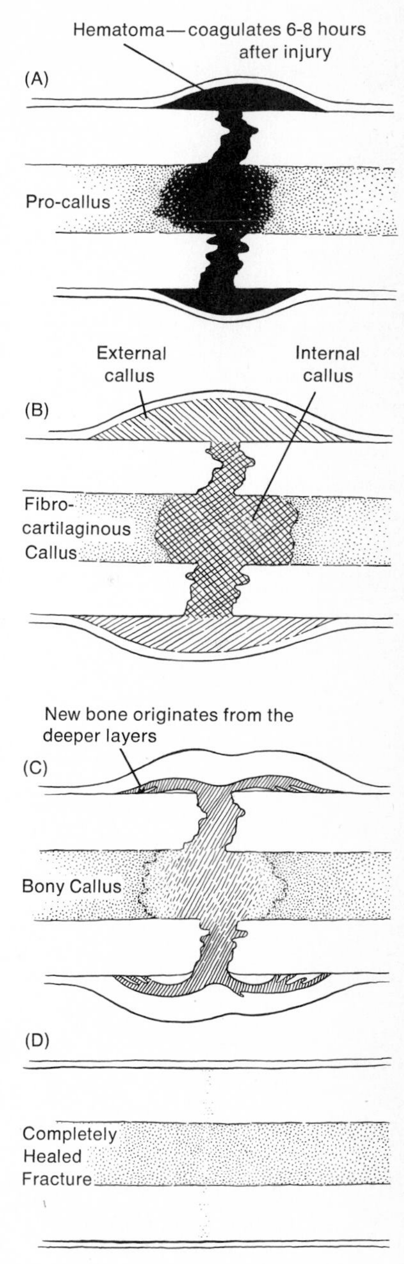

Figure 2-8. Stages in the healing of a bone fracture.

Following a fracture, there are first the usual reactions of any tissue to severe injury. The rupture of blood vessels in the bone marrow and in the periosteum causes development of a large hematoma, or layering of blood, around the fracture, with the bleeding extending into the bone marrow and into the surrounding soft tissues. The blood clots from 6 to 8 hours after the accident. Replacement of the blood clot by young connective tissue (granulation tissue) results in the organization of the blood clot into what is known as the **procallus.** The time needed for organization of the hematoma and its replacement by granulation tissue is variable, depending largely on its size. Organization of the procallus may be complete as early as the end of the first week, but may require from 30 to 60 days. The activity, resulting in the formation of the procallus, is initiated by chemical stimuli arising from the breakdown of proteins. It has been demonstrated that an aseptic wound will not heal if the blood clot and debris are entirely removed and the wound is completely protected from irritation. The primary function of granulation tissue is removal and replacement of necrotic (dead) tissues. Macrophages and polymorphonuclear leukocytes perform this function by their phagocytic activity and remove the engulfed particles by way of the lymphatics. Bone fragments are removed by osteoclasts, which differentiate from invading cells of the connective tissue. The granulation tissue becomes dense connective tissue as soon as its primary function has ended. The fibroblasts now produce numerous collagenous fibers, forming the **fibrocartilaginous callus.** This callus forms a spindle-shaped cuff around the fracture area, filling the gap between the fragments and sealing the marrow cavity of both fragments more or less completely. The new bone originates from the deeper layers of the periosteum, the fibrocartilaginous callus being replaced by new bone, the **bony callus.**

The healing time of each bone in the body is predetermined and related to regional conditions. It is known that the bones of the upper extremity, in general, heal more rapidly than those of the lower extremity. In man, the humerus and forearm bones unite in 3 months. The femur and tibia normally require 6 months. Separation of the fragments in fractures through compact bone will greatly increase the healing time. Young persons generally produce more callus and heal fractures more quickly. During callus formation, bone is produced in excess as a protective measure. After solid union, the surplus bone surrounding the site of fracture loses its functional importance and is gradually reduced and finally removed by resorption until the original shape and outline of the fractured bone have been re-established.

When calcification fails in growing animals or young children, the osseous tissue continues to grow, but the new, uncalcified tissue is known as osteoid. This failure of calcification to keep pace with growth causes a condition known as **rickets** and is usually associated with a diminished concentration of phosphate in the blood plasma. The preventive and curative action of sunlight and of vitamin D in rickets is generally attributed to increased absorption of calcium and phosphate from the gastrointestinal tract.

Functions of Bone

Bone has several functions, which may be listed as follows:

1. Bone performs a mechanical function in forming the skeletal support of the body and in forming a leverage system whereby work and movement are possible.
2. Bones serve as the basis for the attachment of muscles.
3. Bone protects the vital organs of the cranial and thoracic cavities.
4. The bones serve as a reservoir of minerals supplying calcium and phosphorus to the blood.
5. Bone lodges the bone marrow, which is important in the formation of blood cells (hemopoietic function).

A view of the skeleton will reveal a variety of types of bones. There are long, thin bones, which, being hollow, are light and are designed for locomotion and manipulation. Flat, circular bones are found in the vertebrae of the spine and are designed for flexibility and to dissipate shocks transmitted to it by the extremities. The longitudinal arrangement of the bones of the extremities allows the weight to be borne along the line of the strong, rigid framework. There are circular bones, as found in the ribs, forming the thoracic cage. These are designed to work with the diaphragm as an effective bellows in drawing air into the lungs and forcing it out again. Owing to the high tensile strength and the arrangement of the ribs, a great elastic quality is exhibited by the thoracic cage. Bierman found in his work on the protection of the human body from impact forces that most young men used as test subjects could withstand an impact force against the chest of 2,800 lb without injury. The thoracic cage not only serves a physiological function but also affords maximum protection for the vital organs that lie within the chest cavity. The bones of the shoulder and hip are shallow, flat pieces, and the skull is made up of numerous thin plates of bone forming a rounded case to protect the brain and containing deep caverns for the eyes. The pelvic girdle, formed by the bones known as the sacrum and innominate, functions as a base for the movement of the femur and at the same time forms a stable platform for the flexible shaft of the upper segments of the vertebrae. The innominate bones form an arch in which the sacrum, being wider at the top than at the bottom, enters as a wedge and binds the pelvic arch as a keystone. The weight of the trunk is effectively distributed along the vertical axis of the hip. Many protuberances are found on the various bones of the skeleton, which afford convenient and effective places for attachment of the muscles.

Lever Systems

Perhaps the most important mechanical function that the skeleton serves is its ability to provide a leverage system whereby the human body is able to accomplish work. Machines are devices for doing work, and so we find that the body, too, is a machine. In general, work is accomplished or performed when some object is moved through some distance against some kind of resistance. A source of energy is required, and in the body the energy source is the *prime movers,* or muscles, which are capable of converting chemical energy into mechanical energy. We shall see how the muscles are capable of doing this in Chapter 3. Work is accomplished when levers are moved, and the purpose of the body engine is to perform essentially the following types of work: walking, running, lifting, striking, grasping, and climbing.

A lever is a rigid bar turning about an axis, and the body levers are formed by the bones. Systems of levers that are studied in elementary physics are made up of three components: the **fulcrum,** a point on which the lever turns, represented in the body by the articulating surfaces of adjoining bones; the **resistance,** the weight to be moved, which is the product of the load of the lever segment plus the load of the object to be moved; and finally, the **effort,** the sources of energy that move the lever arms, represented in the body by the muscles. Three classes of levers are recognized and are classified as first-, second-, and third-class levers, depending on the relative locations of the fulcurm, resistance, and effort:

First-Class Lever. The fulcrum is between the resistance and the effort (Fig. 2-9).

Figure 2-9. *First-class lever.*

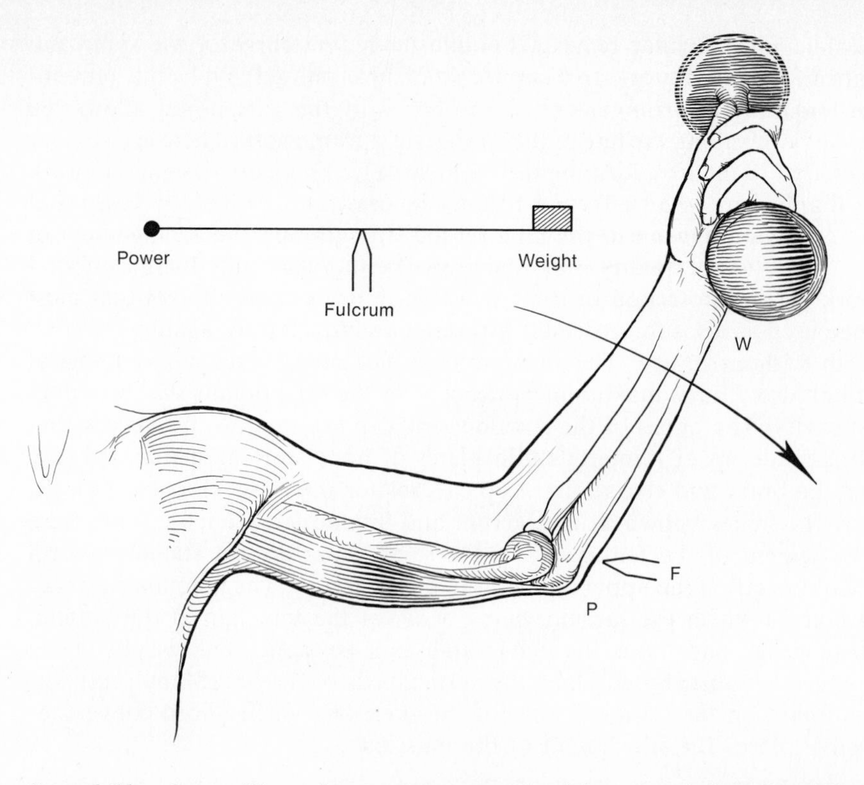

Second-Class Lever. The resistance is between the fulcrum and the effort (Fig. 2-10).
Third-Class Lever. The effort is between the resistance and the fulcrum (Fig. 2-11).

All three classes of levers are found in the human body, and many of the **joints,** representing the fulcrum, can be used as levers of more than one class. An example of a first-class lever is represented by the head and its joint with the neck. The atlanto-occipital joint, the joint between the skull and the first cerival vertebra, the atlas, may be considered the **fulcrum;** the weight of the head in front of the joint, the **resistance;** and the force applied by the splenius muscle to keep the head stable or tilt it up, the **effort.** This lever system becomes a third-class lever when the head is tilted down toward the chest, as the effort then resides between the fulcrum and the resistance. An example often cited of a second-class lever in the body is the system by which one rises to stand on his toes. The fulcrum in this system is represented by the ball of the foot; the resistance is the weight of the body, which is applied at the joint between the tibia and the ankle bones; and the effort is represented by the force applied upward by the gastrocnemius muscle at its insertion on the calcaneus. An example of a third-class lever is the system involving the elbow joint and the pull of the biceps muscle when the forearm is flexed while the hand is holding a weight. The joint of the elbow represents the fulcrum; the weight held in the hand, the resistance; the pull of the biceps to flex the forearm, the effort. Figures 2-9 to 2-11 diagrammatically represent examples of the three classes of lever systems found in the body. The leverage system extends man's ability to perform work and aids in making work easy and comfortable; that is, it gives man mechanical advantage. It is possible to determine the mechanical advantages of a lever system by the formula

$$\text{Mechanical advantage} = \frac{\text{Effort arm}}{\text{Resistance arm}}.$$

The **effort arm** is the distance between the fulcrum and the insertion of the muscle that will supply the effort or force. The **resistance arm** is the distance between the fulcrum and the center of the weight to be lifted. In addition to determining the mechanical advantage of a lever system, we can also determine the amount of effort or force the muscle must generate to overcome the resistance or weight to be lifted:

$$\text{Effort} = \frac{\text{Resistance (weight)}}{\text{Mechanical advantage}}$$

or

$$\text{Effort} = \frac{\text{Resistance} \times \text{resistance arm}}{\text{Effort arm}}.$$

Persons whose muscles are inserted farther from the joints about which levers move have greater mechanical advantage than those whose muscles insert closer to the fulcrum. These persons appear to possess superior muscular power. To

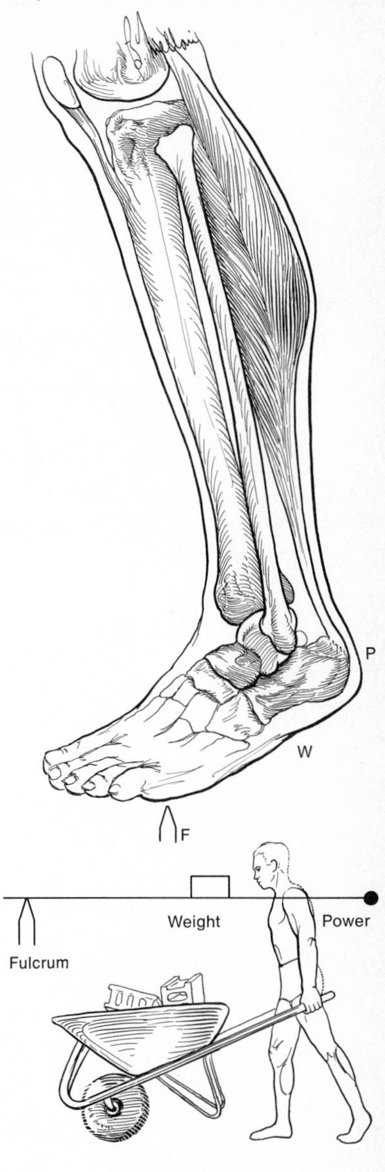

Figure 2-10. Second-class lever.

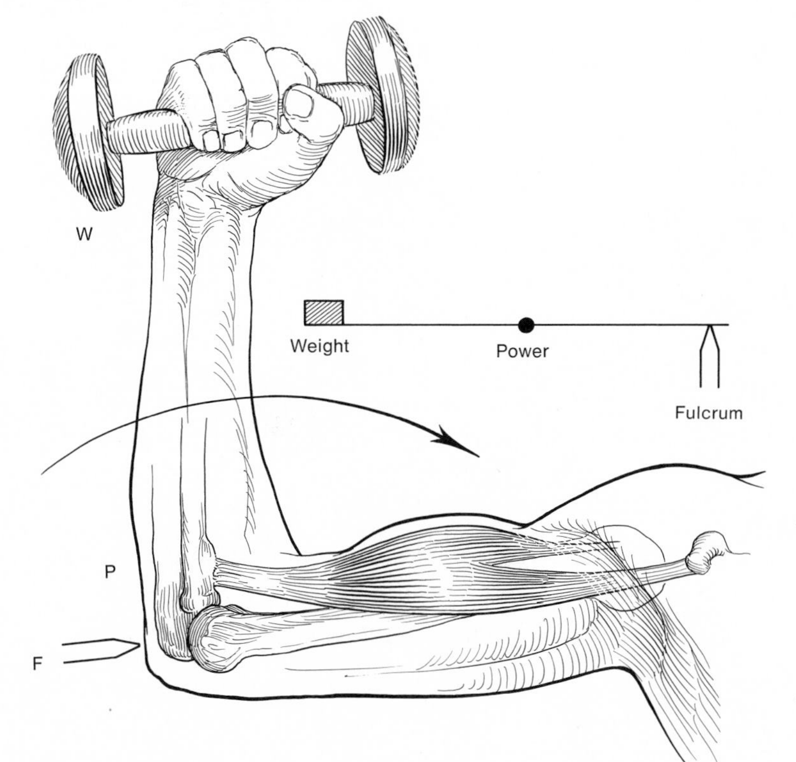

Figure 2-11. Third-class lever.

demonstrate this advantage, we can use as an example two persons of similar weight, build, and age. The person with the longer insertion can curl a 35-lb weight with less effort exerted and thus much more easily than the person with the shorter insertion.

Man A—long insertion of biceps muscle on bicipital tuberosity of the radius (Fig. 2-12):

Forearm	14 inches long
Effort arm	3 inches long
Resistance arm	16 inches long
Resistance or weight	35-lb dumbbell

$$\text{Mechanical advantage} = \frac{\text{Effort arm}}{\text{Resistance arm}}$$

$$MA = \tfrac{3}{16} = 0.187$$

$$\text{Effort} = \frac{\text{Resistance}}{\text{Mechanical advantage}}$$

$$E = \frac{35}{0.187}\ \text{lb} = 187\ \text{lb}$$

or

$$\text{Effort} = \frac{\text{Resistance} \times \text{resistance arm}}{\text{Effort arm}}$$

$$E = \frac{35 \times 16}{3} = \frac{560}{3}$$

$$= 186.6\text{, or } 187\text{, lb.}$$

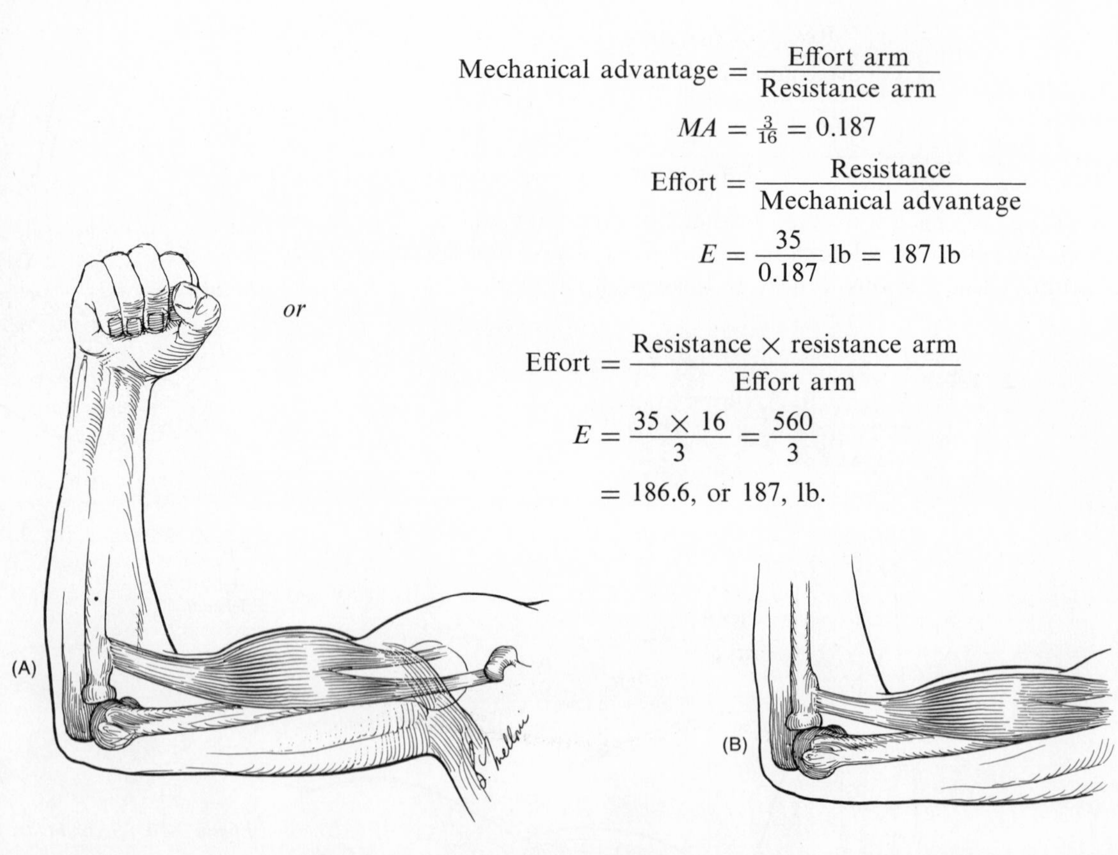

Figure 2-12. Insertion of biceps muscle on radial tuberosity. A *represents man with long insertion.* B *represents man with short insertion.*

Man B—same conditions except insertion is only 1 inch from fulcrum; that is, effort arm is equal to 1 inch:

$$\text{Mechanical advantage} = \tfrac{1}{16} = 0.062$$

$$\text{Effort} = \frac{35}{0.062} = 564.5\ \text{lb.}$$

We can see that the effort required to lift 35 lb by man A is 377.5 lb less than for man B. The mechanical advantage supplied by a difference of 2 inches in the distance where the biceps is inserted on the radius enhances the effective strength of a muscle because of improved application of effort.

Articulatory System

The **articulatory system** is comprised of the various simple and complex **joints** found associated with the skeletal system together with their blood and nerve supply. Bones are joined to one another by connective tissue structures, which permit varying degrees of movement between adjoining bones. A joint or articulation is a junction between two or more bones or cartilages. The joints present extreme variation in character, which depend primarily upon the type of tissue uniting the bones, the type of bones that are joined, and the varying degrees of motion permitted by the articulation. In some cases, the joints are immovable, as in the skull, and the connected bones are separated by a very thin layer of connective tissue. Other joints are slightly movable, as is found in the articulations between the vertebrae. The bones, in this case, are united by dense fibrous tissue and intervening cartilage. Other bones are freely movable upon one another, as the adjoining ends are completely separated by a short wide tube of strong fibrous tissue. Joints are classified according to the way the bones are united and also according to the degree of movement they permit.

Bones may be united in three ways (Figs. 2-13 and 2-14):

1. By **intervening fibrous tissue**—a *fibrous joint* or *syndesmotic joint.*
2. By **intervening cartilage**—a *cartilagenous joint* or *synchondrotic joint.*
3. By a **short wide tube of strong fibrous tissue** which contains an oily liquid, *synovial fluid,* secreted by a membrane that lines the tube—a *synovial joint.*

Figure 2-13. Three different classes of joints based on type of tissue uniting adjoining bones. Syndesmosis: fibrous tissue, such as interosseous ligament. Synchondrosis: cartilagenous disk, as seen in the amphiarthrotic joint. Synovial joint: fibrous tube lined with synovial membrane as seen in the diarthrotic joint.

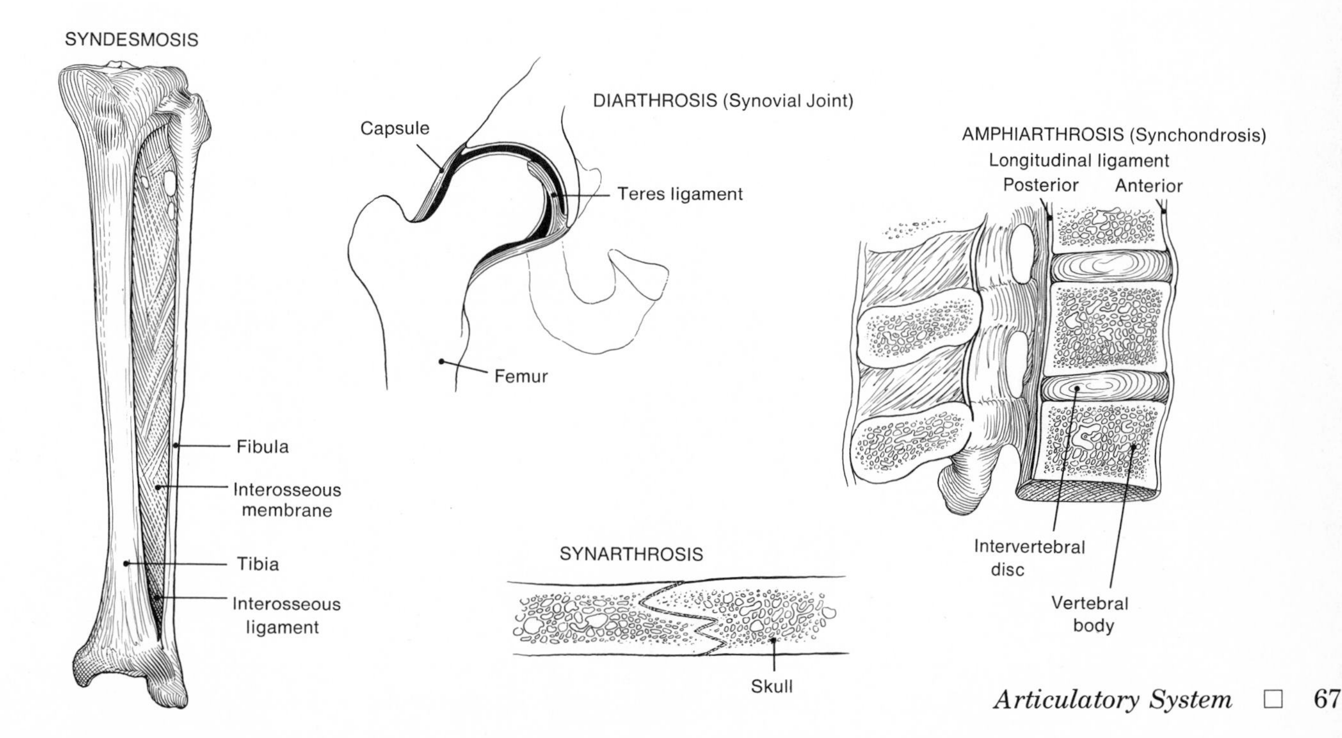

Depending upon the degree of motion allowed between adjoining bones the joints are further classified in the following ways:

1. Synarthroses. Joints in which there is practically no movement. These are found only in the skull. The articulating margins of the skull bones are united by an intervening thin ribbon of fibrous tissue (fibrous joint) called **sutural ligaments.** The lines of union are called **sutures.** In most instances the union of the bones is secured through the interlocking of projections on the serrated or denticulated margins (Fig. 2-13).

2. Amphiarthroses. Joints that allow limited movement and are either of the **fibrous joint type** or **cartilagenous joint type. Cartilagenous** or **synchondrotic joints** are found in the articulations between the bodies of the vertebrae, between the pubic bones at the **symphysis pubis,** and between the sacrum and ilium in the sacroiliac joints (Fig. 2-13). In this type of joint the articular surfaces of the opposing bones are covered or coated with hyaline cartilage and are united by an intervening disk of fibrocartilage surrounded by a ring of strong fibrous tissue which is continuous with the periosteum of the two bones. The center of the intervening cartilagenous disk is filled with a fibrogelatinous pulp that acts as a cushion or shock absorber. The **fibrous,** or **syndesmotic, joint** is found in articulation between the lower end of the fibula and tibia by means of the **interosseous ligament,** and between the scapula and clavicle by means of the **coracoclavicular ligament.** In this type of joint the bones are a little distance apart and are united by an intervening **ligament** or **membrane** of some thickness and width, which by twisting and slight stretching allows a little movement (Fig. 2-13).

3. Diarthroses. Freely movable joints, including most joints in the adult. They are of the synovial joint type and have a more elaborate structure than the immovable and slightly movable joints (Fig. 2-14). The two or more bones are united by an encircling band of fibrous tissue called the **articular** or **fibrous**

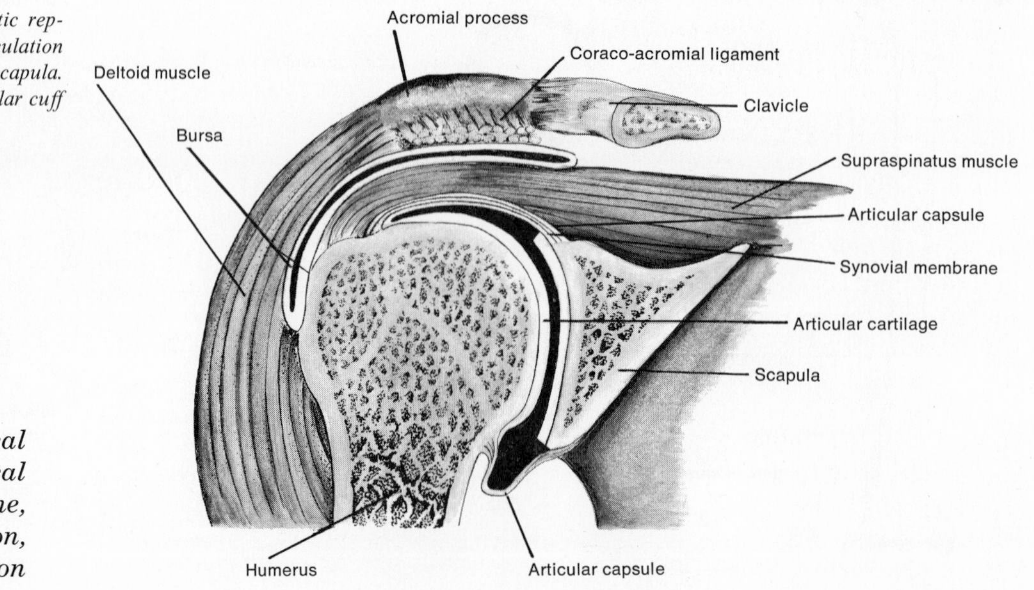

Figure 2-14. Diarthroses: schematic representation of shoulder joint or articulation between head of humerus and scapula. Synovial membrane lines the articular cuff or capsule.

capsule. The parts of the bones in contact with each other are coated with smooth hyaline cartilage. The fibrous capsule is composed of two layers: (1) a **capsular ligament,** and (2) its lining of **synovial membrane.** The capsular ligament is a short wide tube attached by its ends to the bones at a varying distance from the margins of the articular cartilages. It is made of strong fibrous tissue, and in many joints it is strengthened by thickenings of its own substance, by expansions from tendons and deep fascia, or by accessory ligaments. The synovial membrane lines the capsule and is reflected onto the bone, covering its nonarticular parts up to the margin of the articular capsule. The membrane is made of very fine fibrous tissue which contains numerous connective tissue cells. It secretes **synovial fluid** into the cavity, which lubricates the surfaces, nourishes the articular cartilage, and contains phagocytic cells, which remove microorganisms and the detritus from the articular surfaces.

Some diarthroidal joints are more freely movable than others. This variation in the degree of movement is determined by the shape of the articulating surfaces forming the joint, and according to structure the diarthroidal joint can be classified into the following types:

a. **Ball-and-socket joints**—movement is freest in this type of joint, in which a rounded head on one bone moves in a cuplike cavity of another. The shoulder joint and hip joint are examples of ball-and-socket joints (Fig. 2-14).

b. **Hinge joints**—movement in a hinge joint is around one horizontal axis and action is limited to movement in one plane, usually forward or backward. Examples are elbow and knee joints (Fig. 2-15).

c. **Pivot joints**—another kind of hinge joint, but the movement is around a longitudinal axis and exhibits a rotary movement in which a ring moves around a pivot. Examples: articulation between the first and second cervical vertebrae, where a ring, the atlas, rotates around a pivot; the odontoid process of the axis; and the head of the radius around the ulna.

d. **Gliding joints**—the two articulating surfaces are alike and very little movement takes place. There is a slight gliding or rotation of one surface on the other. Examples are the joints between the articular processes of the vertebrae uniting vertebral arches; and the joints of the carpal bones and the tarsal bones.

Movements at Joints. The following varieties of movement may be distinguished in true joints (diarthroses): (1) flexion, (2) extension, (3) abduction, (4) adduction, (5) gliding, (6) rotation, and (7) circumduction. **Flexion** is a bending movement in which the angle between the acting parts becomes smaller, as in bending the forearm toward the arm at the elbow joint. **Extension** is the reverse movement; it is a straightening out in which the angle between acting parts increases. Extension occurs at the elbow joint when the forearm is straightened on the arm. **Abduction** is a movement of a part away from the median plane of the body or some other plane agreed upon. Abduction occurs in the shoulder joint when the arm is raised to the horizontal position from the side of the body. **Adduction** is the reverse movement of abduction or movement toward the median plane of the body, as in lowering the arm from the raised horizontal position to the side of the body.

Gliding is a movement in which the surface of one part glides or moves over the surface of another part as in the movement of the wrist or carpal bones upon one another, and the gliding of the tarsal or ankle bones upon one another. **Rotation** is a movement in which one part revolves or turns on another part without displacement. Examples are rotation of the atlas on the axis and the radius about the ulna. Rotary movement of the radius, combined with rotation of the humerus, is called **supination** (as in the movement of turning the hand palm up) and **pronation** (as in the movement of turning the hand palm down). Rotation is never a complete movement but is limited by the ligaments of the joint and the muscles which cross it. **Circumduction** is a movement in which the moving segment describes or moves around a conical figure, of which the base is never quite circular. Circumduction combines the movements of flexion, extension, abduction, and adduction, as in swinging the arms in a circle.

Prominent **bursae** are found at the elbow, the hip, the knee, and the heel. These are synovial membranous sacs found outside joint cavities in places where friction occurs. The membrane secretes the characteristic fluid, synovia, which fills these sacs and serves to prevent friction. In some cases, bursae in the immediate area of a joint communicate with the joint cavity (Fig. 2-15).

Security of Joints. Joints depend for their security upon three factors: (1) closely fitting bony parts, (2) strong ligaments, and (3) the tension of overlying muscles. Since the shoulder joint has mainly the tension of overlying muscle to protect or secure it, it is the most easily dislocated joint. On the other hand, the elbow joint is rarely dislocated, because of the fact that it depends mostly, for its security, on closely fitting bony parts. It has less mobility than the shoulder joint, but it is much more secure.

Storage of Minerals in Bone

Bone is often described as a storehouse for calcium and phosphorus since calcium and phosphorus can be mobilized by resorption of bone. Calcium serves a number of important functions. In the form of carbonate and phosphate it forms the principal content of bone, dentine, and enamel. It is found in all the other tissues and in all glandular secretions. It is necessary for the clotting of blood, and it keeps the excitability of nerve endings at a normal level. A deficiency of calcium leads to undue excitability, especially of the motor nerve endings, and thus to tetany.

The blood calcium and phosphate level is maintained mainly by the action of the hormone secreted by the parathyroid gland and vitamin D. The parathyroid hormone increases or maintains the blood calcium level by withdrawing calcium from the bones of the skeleton through stimulating osteoclastic activity in bone resorption. Vitamin D causes a rise of blood calcium level or maintains it by increasing the absorption of calcium from the intestinal tract. It also influences the deposition of calcium in bone and cartilage. It appears that vitamin D and parathyroid hormone help maintain blood calcium level by antagonistic action. In fact, vitamin D, in aiding in the mobilization of external resources of calcium, restricts the withdrawal from the skeleton of the emergency supplies of calcium by inhibiting the action of the parathyroid glands.

Phosphorus as a constituent of bone and an important intermediary compound of carbohydrate metabolism is, like calcium, indispensable for normal function of the animal body. Phosphorus is liberated by bone into the circulatory system by osteoclastic activity. We find that liberation of calcium from bone necessarily involves liberation of phosphorus also. Numerous other mineral elements may be deposited in bone. Some of these serve no particular physiological function because they remain in the bone throughout the life of the individual. Others, such as sodium and magnesium, are mobilized when needed by other tissues of the body. The release of sodium and magnesium from bone into the circulatory system occurs by solution of surface-bound ions. Resorption of bone by osteoclastic activity, as in the release of calcium and phosphorus, is not required. Elements that are considered foreign to the body, such as radium, lead, and fluorine, are stored in bone once they have been introduced into the general circulation. Accumulation of such elements will alter the normal metabolism of bone, which, in turn, can affect all part of the animal's body, leading to various pathological conditions.

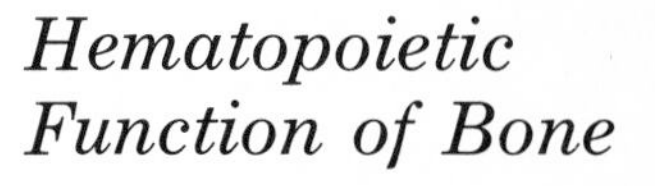

Hematopoietic Function of Bone

The bone marrow can be considered an organ in itself composed of many sections scattered throughout the bony framework of the body. It is one of the most important organs because it is the source of all red cells, hemoglobin,

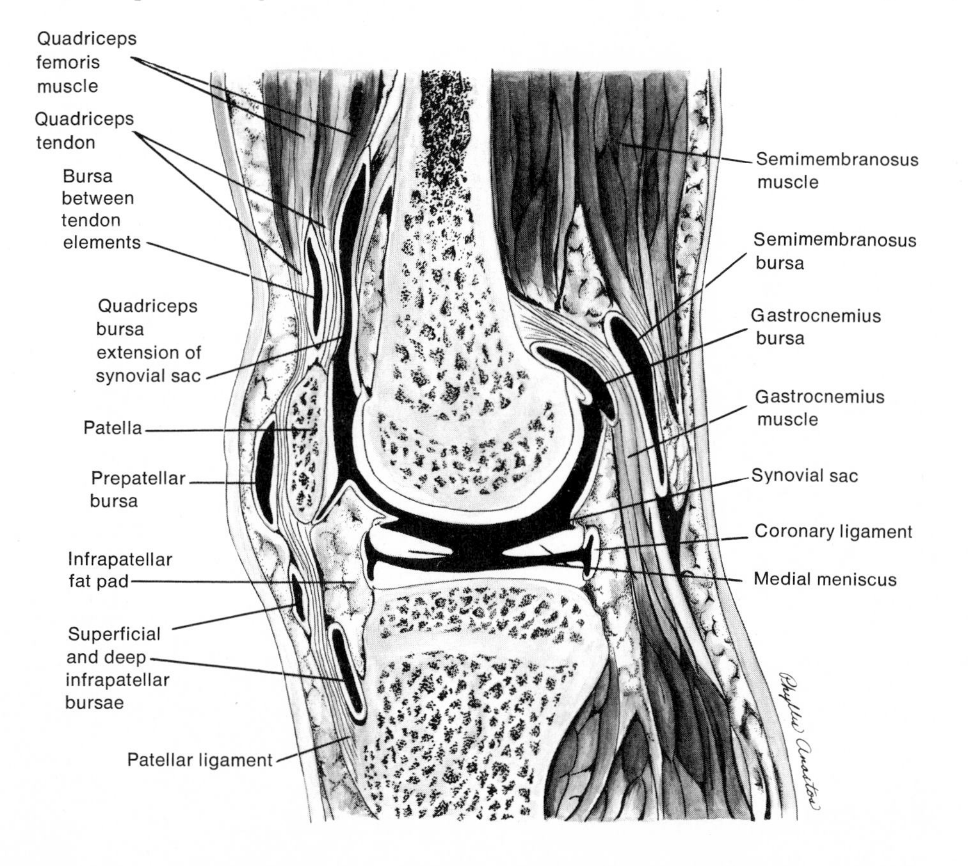

Figure 2-15. Hinge joint: schematic sagittal section through knee medial to midline.

granulocytes, and platelets. As we have already seen, long bones contain a medullary cavity, or marrow cavity, which is filled with marrow. In addition, the irregular spaces occurring between the interlacing bony bars of spongy bone are also filled with marrow. The marrow tissue of all bones has an over-all volume of approximately 1400 cc. The marrow contains only three groups of cells: endothelial, reticular, and fat cells. Owing to the accumulation of fat cells, marrow takes on a yellow color and is therefore called yellow marrow; this is the type of marrow normally found filling most of the shafts of the long bones. In flat, short bones, the production of red blood cells and granulocytes is normally sufficient to replace fat cells, and thus we find that the normal marrow of such bones is always red. Bones containing red marrow include the bones of the skull, the ribs, the sternum, the clavicle, the vertebrae, and the pelvic bones. Very few cells are normally formed in the long bones, and thus the marrow is yellow; however, it should be noted that yellow marrow has the capacity of rapid transformation into active production of erythrocytes and granulocytes and thus constitutes an immense emergency reserve for blood cell formation.

Investigators express widely differing views concerning the origin of blood cells. All agree that both red and white cells come from a common ancestor, stem cell, which on proliferation differentiates and develops to form mature cells. In the embryo, the blood cells are formed in the **mesenchyme,** which is derived from the **mesoderm.** At first, formation occurs outside the embryo in the yolk sac and abdominal pedicle, but the site of formation later shifts to within the embryo, chiefly the liver and spleen. Blood cell formation finally becomes localized in the bone marrow and lymphatic tissue, and this process of development is already complete at birth.

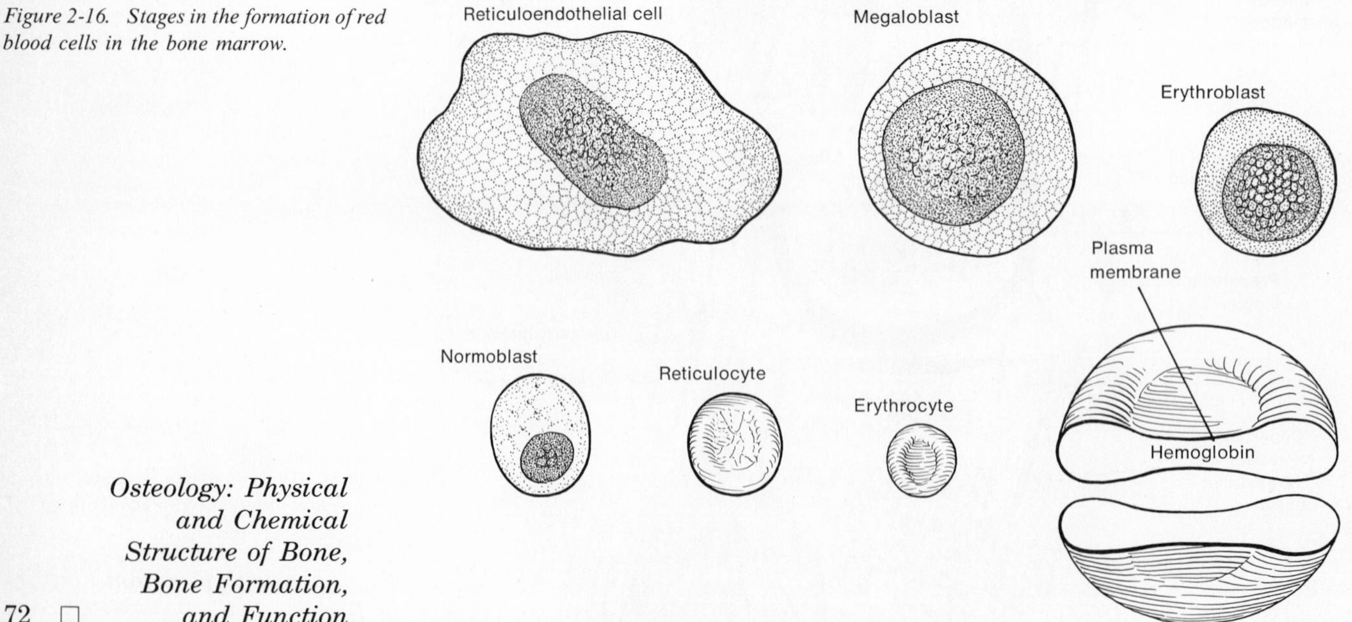

Figure 2-16. Stages in the formation of red blood cells in the bone marrow.

All immature blood cells reproduce by mitosis. The number of consecutive mitoses is limited, for when the cells mature they eventually lose their power of division. No longer capable of division, they enter the blood stream to fulfill their function and are finally destroyed. In healthy persons, it is in only extremely rare instances that cells are found in the blood in their immature form. The parent cell, which arises from reticuloendothelial cells, has been designated by different names but is most commonly called the **megaloblast.** According to the monophyletic theory[1] of blood cell formation, the megaloblast differentiates to give rise to a cell called the **pronormoblast,** which in turn gives rise to the **polychromatic** and **orthochromatic** normoblast, which in turn gives rise to the **reticulocyte,** which finally results in the mature blood cell, the **normocyte,** or erythrocyte (Fig. 2-16).

References

Bourne, G. H. (ed.). *The Biochemistry and Physiology of Bone,* Vols. 1, 2, and 3. New York: Academic, 1972.

Brunnstrom, S. *Clinical Kinesiology,* Philadelphia: F. A. Davis, 1966.

Copenhaver, W. M., R. P. Bunge, and M. B. Bunge. *Bailey's Textbook of Histology.* Baltimore: Williams & Wilkins, 1971.

Hall, M. C. *The Architecture of the Bone.* Springfield, Ill.: Thomas, 1966.

Menczel, J., and A. Harell (eds.). *Calcified Tissue: Structural, Functional and Metabolic Aspects.* New York: Academic, 1971.

Metcalf, D., and M. A. S. Moore. *Haemopoietic Cells.* New York: Am. Elsevier, 1971.

Milch, H., and R. Milch. *A Textbook of Common Fractures.* New York: Harper, 1959.

Nichols, G., and R. H. Wasserman (eds.). *Cellular Mechanisms for Calcium Transfer and Homeostasis.* New York: Academic, 1971.

Rodahl, K., J. T. Nicholson, and E. M. Brown. *Bone As a Tissue.* New York: McGraw-Hill, 1960.

[1] According to the monophyletic theory of blood cell formation, the first recognizable cell is capable of giving rise to red cells, granulocytes, lymphocytes, and monocytes. The polyphyletic theory holds that the earliest cells are already differentiated and cannot be changed into any other type of cell.

3

Myology: Microscopic Anatomy, Physical and Chemical Composition of Muscle

The study of the muscular system, including development, structure, physiology, and general anatomy, is referred to as **myology.** There are three classes of muscles in the body, which differ histologically, anatomically, and physiologically. They are designated as **skeletal** (striated, voluntary) (Fig. 3-1), **visceral** (smooth, involuntary), and **cardiac** (involuntary, imperfectly cross-striated) muscle. Although this chapter is concerned mainly with the physiology of skeletal muscle tissue, all three of these types will be considered briefly.

Classes of Muscles

Smooth Muscle

Smooth muscle (Fig. 3-2) is so called because it does not exhibit cross striations when examined under the microscope. It is also called visceral muscle because it shows a very close association with connective tissue, being widely distributed through the vertebrate body, occurring in the walls of the alimentary tract, in the arteries and veins, and in numerous other ducts. The isolated smooth muscle cell is spindle-shaped with an elongated oval nucleus occupying a central position. The cells vary considerably in length in different regions of the body, being found as short as 0.02 mm in small blood vessels and as long as 0.5 mm in the uterus. The cells are connected with each other by means of fine collagenous and elastic connective tissue fibers, which act as a cementing material. Although this cement-

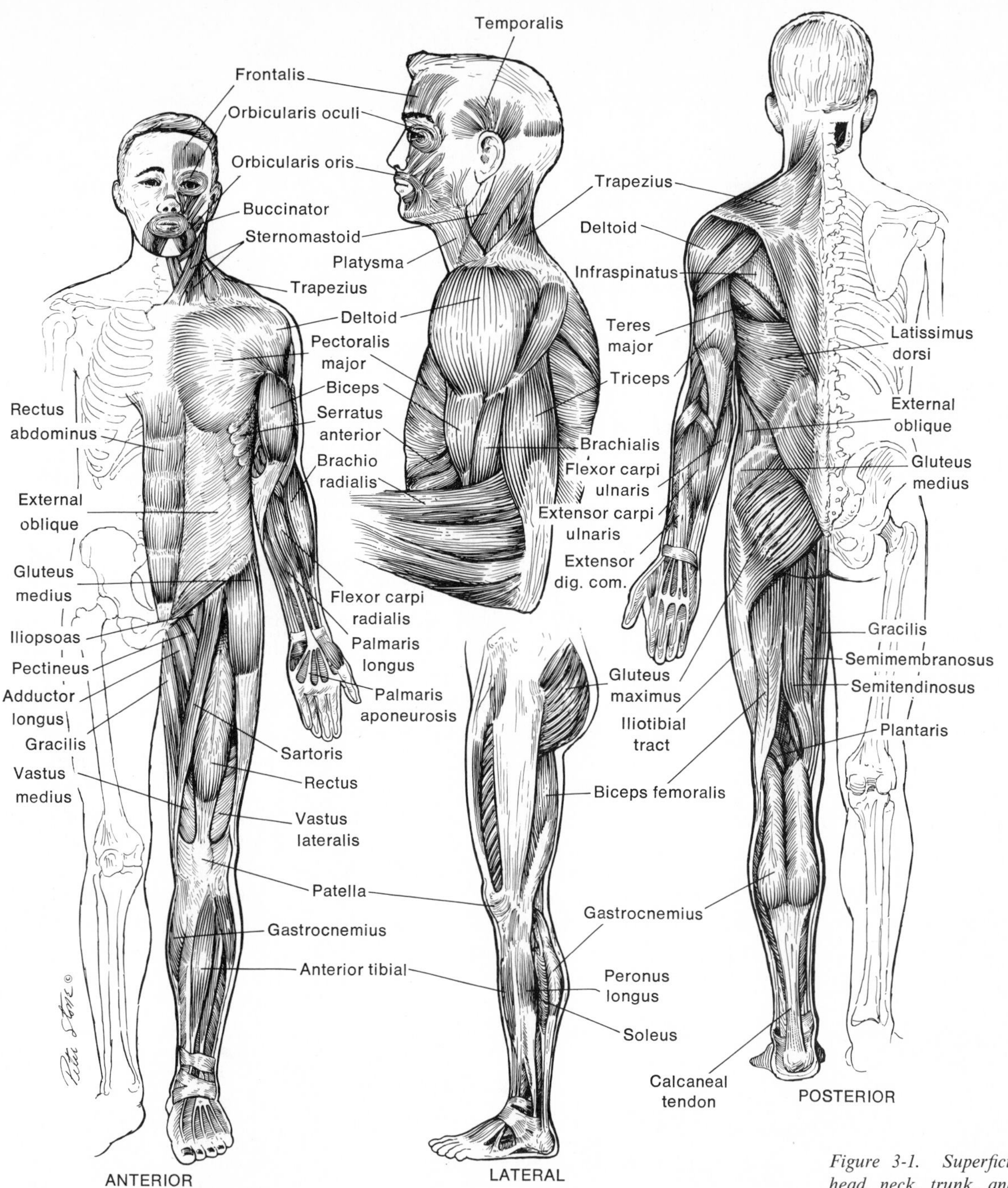

Figure 3-1. Superficial muscles of the head, neck, trunk, and upper and lower extremities. Anterior, posterior, and lateral views are shown.

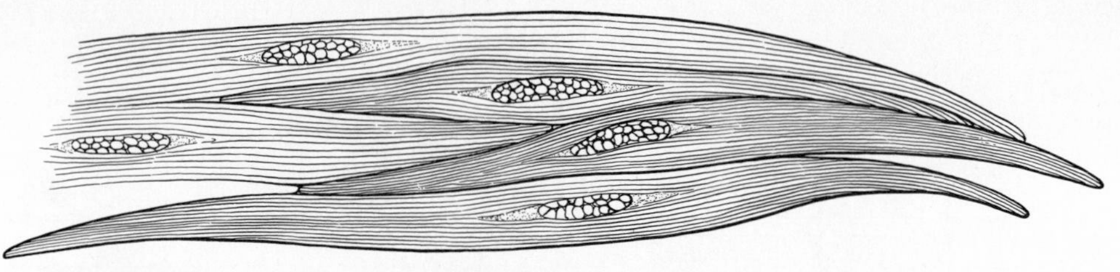

Figure 3-2. Smooth muscle fibers from human intestine.

ing substance is not evident in ordinary routine preparations, silver-staining techniques demonstrate the reticulum of fibers very nicely.

Contraction Quality. Physiologically, smooth muscle fibers are characterized by their **slow contraction** and **relaxation** periods, in contrast to the quick action of striated muscle. In addition, smooth muscle is **rhythmical** in its action, producing a series of alternate contractions and relaxations. Finally, it differs from striated muscle by the fact that the individual cannot control its action, and therefore, it is **involuntary.** It is involuntary in its action because its nervous innervation is from that part of the nervous system referred to as the **autonomic division.**

Innervation. Motor nerves can be traced to either smooth or striated muscle fibers. In the involuntary, or smooth, muscles, the nerves are composed mainly of **nonmyelinated fibers.** As the nerves near their terminal points, they divide into numerous branches, which communicate with one another, forming **plexi.** From these plexi, minute branches are given off, which divide and break up into ultimate **neurofibrillae,** of which the axon of the nerve is composed. These fibrillae course between the involuntary muscle cells and, according to Elischer, terminate on the surface of the cells opposite the nuclei, in minute swellings (Fig. 3-3).

Striated Muscle

Striated muscle (Fig. 3-4) is so called because its fibers exhibit alternating light and dark bands, called **striations.** It is also called skeletal muscle because of its close association with the skeletal parts. This type of muscle makes up the greatest bulk of the body and accounts for most general motion. The muscle fiber may be described best as an elongated, multinucleated cylinder of varying length (Fig. 3-5). The individual fiber is surrounded by a thin, transparent membrane known

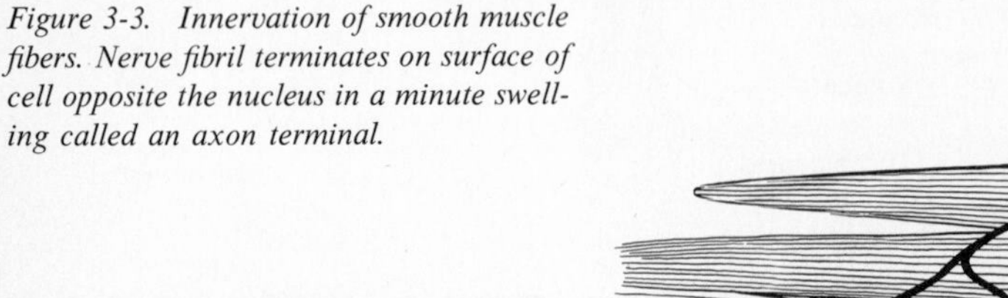

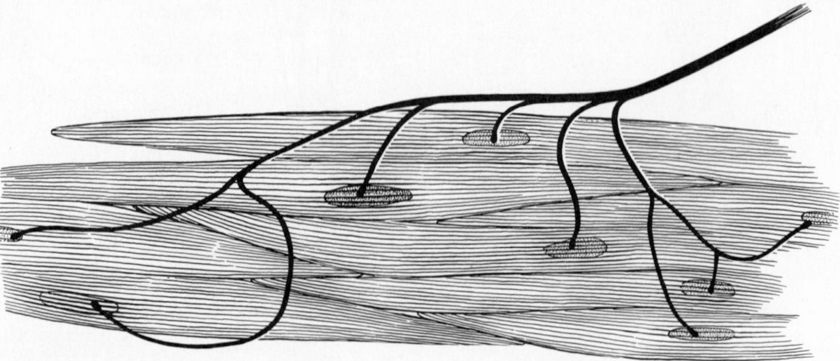

Figure 3-3. Innervation of smooth muscle fibers. Nerve fibril terminates on surface of cell opposite the nucleus in a minute swelling called an axon terminal.

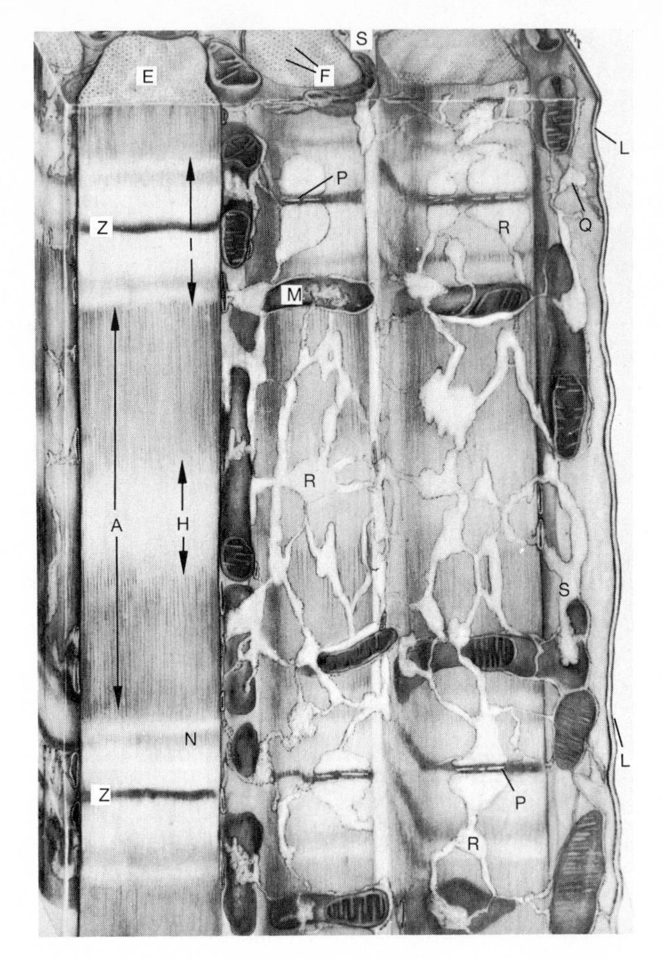

Figure 3-4. Block of striated muscle magnified 70,000 ×. Cross bands of muscle are designated by the letters A, H, I, Z, *and* N; *other identification is as follows:* S, *sarcoplasm;* L, *sarcolemma;* M, *mitochondria;* R, *sarcoplasmic reticulum;* P, *attachment of reticulum to myofibrils;* Q, *attachment of reticulum to sarcolemma;* E, *myofibril;* F, *myofilaments. [Courtesy Encyclopedia Britannica, Chicago. After Bennett,* Neurol., **8:**66 (*1958*).]

Figure 3-5. Striated muscle fiber as seen under high magnification. (Courtesy General Biological Supply House, Chicago.)

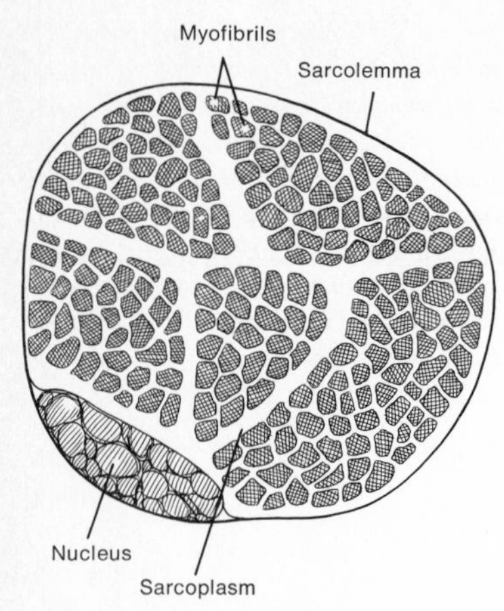

Figure 3-6. Cross section of a single muscle fiber.

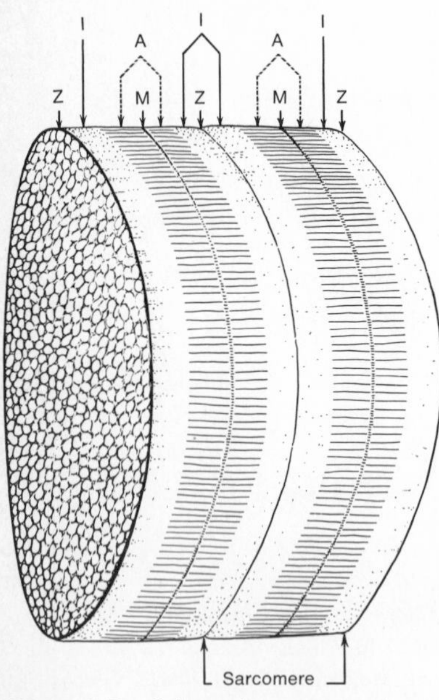

Figure 3-7. Schematic representation of muscle fiber magnified many times showing the differentiation of the cross striations. Compare with Figs. 3-4 and 3-5. The M band is comparable to the H band in Fig. 3-4.

as the **sarcolemma.** Within the sarcolemma are the fluid protoplasm, called the **sarcoplasm,** and numerous highly developed **myofibrils,** which run the length of the fiber parallel to each other. Nuclei are scattered in and limited to the sarcoplasm below the sarcolemma (Fig. 3-6). The myofibrils exhibit alternating light and dark bands, or disks, which, when regularly arranged with those of adjacent myofibrils, give to the fiber its striated appearance. These striations are quite definite and under the microscope can be differentiated into various types. The cylindrical muscle fiber enclosed by the sarcolemma is divided into definite compartments by the thin Z membranes (Fig. 3-7). It has been demonstrated that in mammalian muscle these Z membranes are 0.002 mm apart, separating the fiber into partitions known as **sarcomeres** (Fig. 3-7). The Z membrane may be involved in the transmission of the impulse from the surface of the fiber to the contractile material. On both sides of the Z membrane the material is less dense and appears lighter—this has been termed the I band.

Toward the middle of the sarcomere the protoplasmic material is more dense and thus appears darker. This area has been termed the Q or A band. On closer examination, it can be seen that this band has a very thin, membranelike formation in the middle, which has been termed the M membrane since it is in the middle of the sarcomere. It is of interest to note that controversy has arisen whether myofibrils actually exist as elongated protoplasmic threads or are artifacts. Szent-Györgyi, in his monograph on the chemistry of muscular contraction, presents evidence on which he bases his statement that "the muscle fiber contains no preformed fibrils and the contractile matter forms one continuous mass. Fibrils are artifacts, owing their origin to the treatment—chemical or mechanical injury outside the body."[1]

In skeletal muscles, individual fibers are surrounded by a thin sheath of connective tissue, called the **endomysium.** A sheath of similar connective tissue, but more abundant, surrounds bundles of fibers. This sheath is called the **internal perimysium,** and the bundle of fibers is referred to as a **fasciculus.** A heavier external sheath of connective tissue, the **external perimysium,** or **epimysium,** encloses large numbers of fasciculi to form a muscle (Figs. 3-8 and 3-10).

Skeletal muscle is quick-acting muscle requiring from 0.1 to 0.2 second in contraction in contrast to several seconds for smooth muscle. It also differs functionally from smooth muscle in that it is nonrhythmical and voluntary in its action. The nerve supply from the cerebrospinal nerves places this type of muscle under voluntary control; it is composed mainly of **myelinated fibers.** After entering the epimysium, the nerve breaks up into fibers of or bundles of fibers, which form plexi and gradually divide until a single nerve fiber terminates on a number of muscle fibers. If the muscle fiber is long, more than one nerve fiber may enter it. Within the muscle fiber, the nerve fiber terminates in a special expansion, called by Kühne, who first accurately described it, a **motor end plate** (Fig. 3-9). Each motor nerve fiber and the muscle fibers supplied by it constitute a **motor unit.**

For a long period, most histologists believed that the axis cylinder of the nerve

[1]A. Szent-Györgyi, *Chemistry of Muscular Contraction.* (New York: Academic, 1951), p. 19.

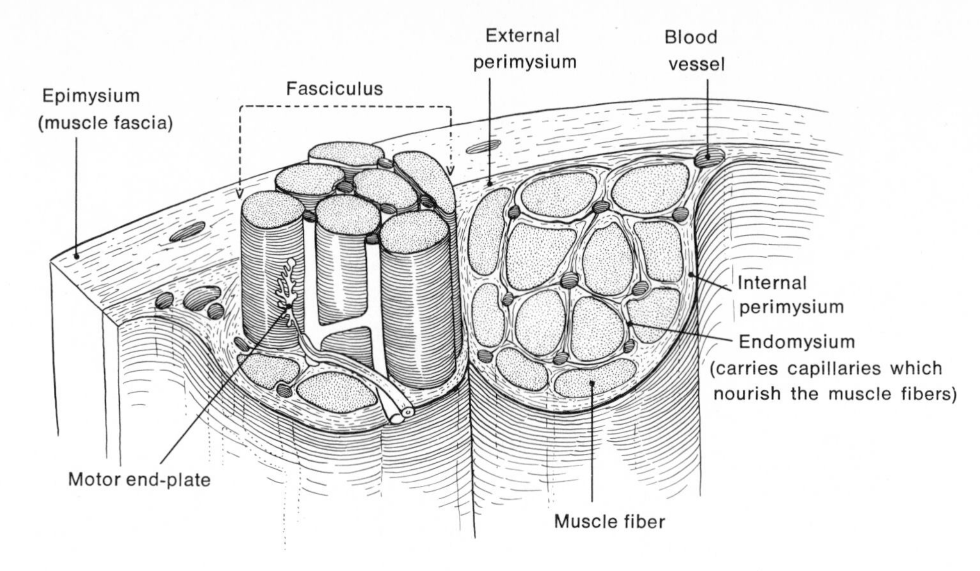

Figure 3-8. Cross section of muscle showing that it is constituted of bundles (fasciculi) of fibers. Each bundle has its own connective tissue sheath (internal perimysium), and all the bundles are surrounded by another sheath, the external perimysium.

fiber terminating on the muscle fiber was naked after leaving the most distal myelin and then pierced and lay under the sarcolemma, providing direct continuity between neuroplasm and sarcoplasm. Cajal showed that a layer of neuroplasm separates the motor end plate from the sarcoplasm. Kühne's work also showed that the end plate is on the surface of the muscle fiber and that a clear space covers the nerve fiber ending and the sarcoplasm at all points (Fig. 3-11).

The enlargment of a muscle, which takes place during growth or from exercise, is due to an increase in the diameter of its fibers rather than an increase in the number of its fibers. The number of fibers in a muscle does not increase after birth. Change in size is due to an increase in the constituent amount of protoplasm

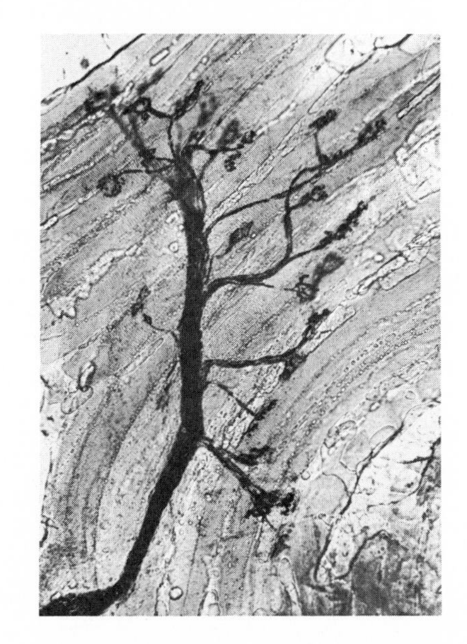

Figure 3-9. Motor nerve endings (end plates) in striated muscle. (Courtesy General Biological Supply House, Chicago.)

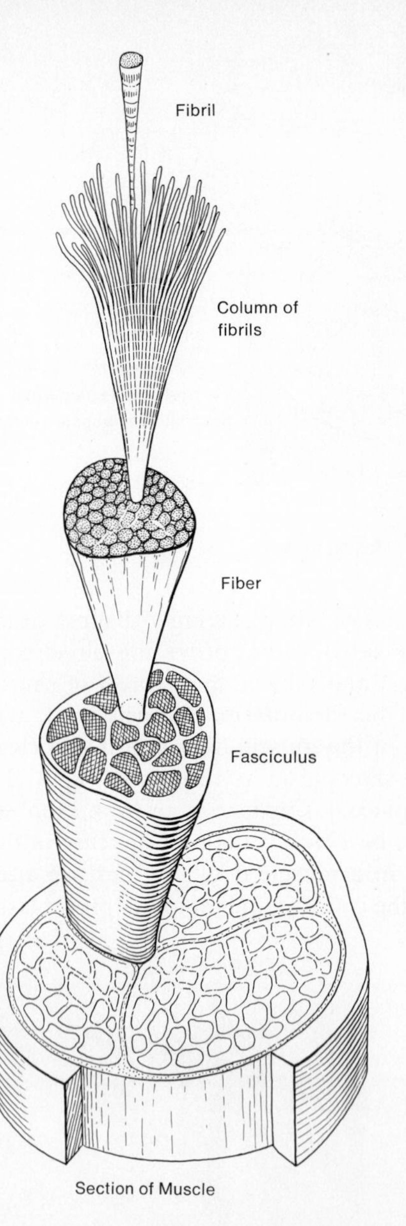

Figure 3-10. Construction of a section of muscle to show the relationship of its individual parts to the whole. An individual muscle fiber contains many fibrils, and a fasciculus contains many muscle fibers. A muscle is a composite of many fasciculi.

contained within the muscle fiber. In addition, we find that striated fibers lack the capacity to produce new fibers, so that regeneration of muscle after an injury is slight. An injured fiber can repair itself, but any large defect is replaced by connective tissue.

Cardiac Muscle

Cardiac muscle (Fig. 3-12) is so called because it is the muscle tissue of the heart. Histologically, it appears to be intermediate between smooth and skeletal muscle in that its fibers are striated but contain a single centrally located nucleus. If cardiac tissue is placed in dissociating fluid, it can be shaken into separate units

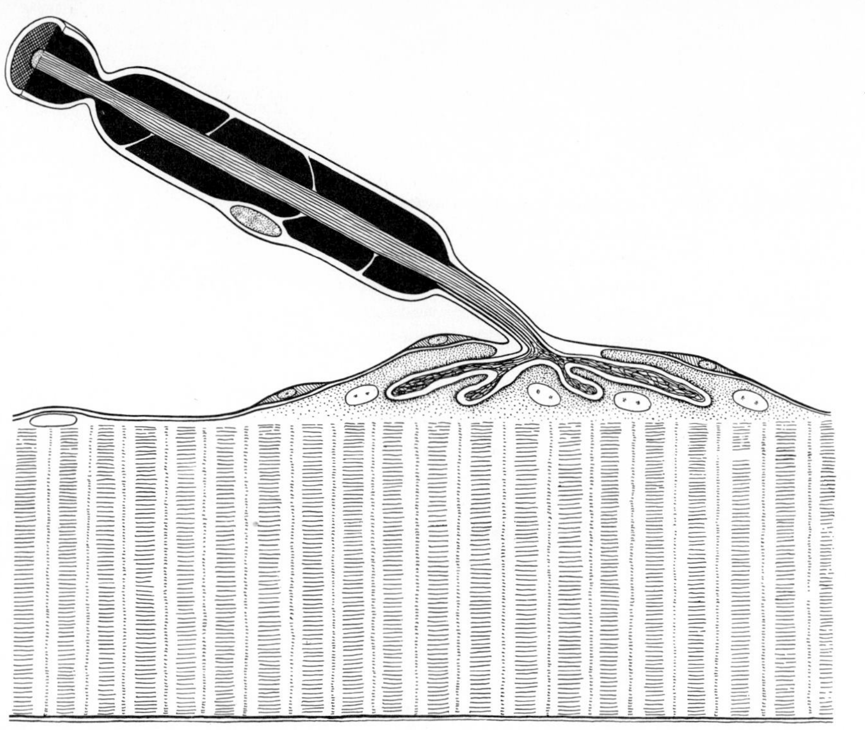

Figure 3-11. Schematic representation of an electron micrograph of the myoneural junction (magnification many thousand times). The end plate terminates on the surface of the muscle fiber, and it appears that a clear space covers the nerve fiber ending and the sarcoplasm at all points.

resembling cells under the light microscope. These units may be described as being rectangular with parallel sides having a definite sarcolemma that does not completely enclose the unit. The ends are uneven and may possess short, stubby branches, which freely unite with other units, forming a continuous network of cells, or **syncytium.** Since it appears that there is no definite completely enclosing cell membrane, and since myofibrils can be seen to extend continuously through several units, the isolated cardiac unit corresponds to a cell but was not considered a definite cell. Cardiac tissue was referred to as a *syncytium.*

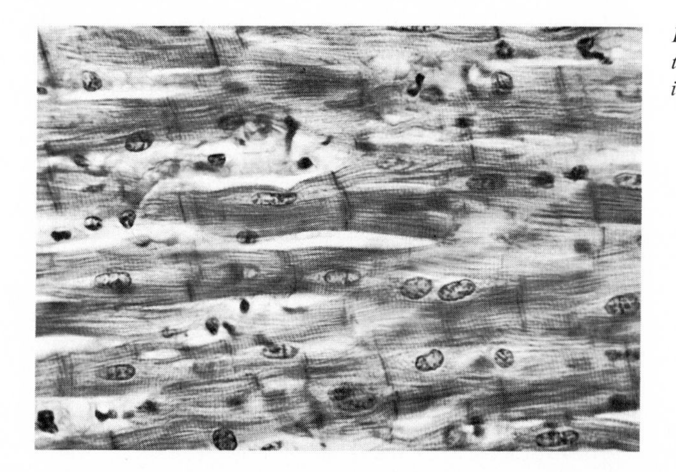

Figure 3-12 Heart muscle stained for intercalated disks. (Courtesy General Biological Supply House, Chicago.)

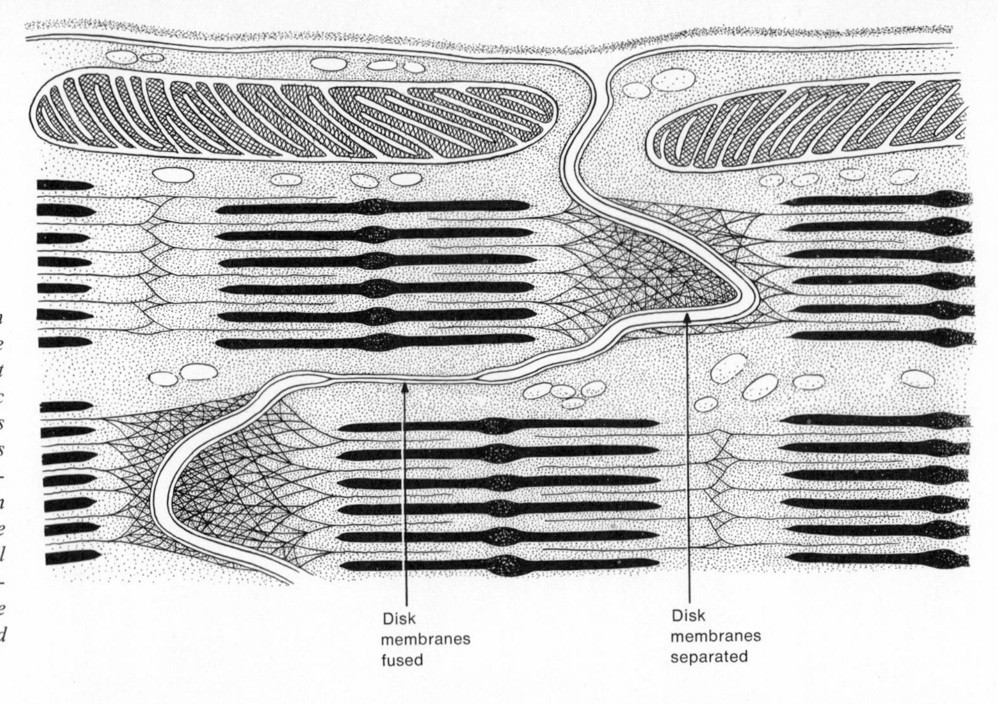

Figure 3-13A. Diagrammatic illustration of an electron micrograph of cardiac tissue showing the intercalated disk as a distinct membrane separating adjacent cardiac cells. The area of membrane fusion is known as the nexus. The ovoid structures at the top are mitochondria. The thick parallel bars are myosin filaments, between which course the thin actin filaments. The actin filaments terminate on the cell membrane in a dense meshwork which appears under the light microscope as the dark band characteristic of intercalated disks.

Studies of cardiac tissue under the electron microscope have shown cardiac muscle fibers to be separated into individual cells by extensions of the plasma membrane which are modified in places, becoming quite thick and forming the **intercalated disks** (Figs. 3-13*A* and *B*). It is no longer feasible to consider cardiac tissue to be a syncytium anatomically. In certain areas of the intercalated disks the membranes of individual fibers become fused to the membranes of neighboring fibers, called a **nexus.** These nexal regions, where the plasma membranes of adjacent cells actually fuse, are probably low-resistance bridges facilitating current flow between the interior of cells. From a functional viewpoint, cardiac tissue appears to be acting as a syncytium.

Functionally, the action of cardiac fibers can be described as strong and rhythmical. Like smooth muscle, the cardiac muscle exhibits a high state of tonus, possesses automaticity, and is innervated by the autonomic system. Innervation of cardiac muscle will be discussed more thoroughly on page 219.

Characteristics of Skeletal Muscles

Myology: Microscopic Anatomy, Physical and Chemical Composition of Muscle

The skeletal muscles make up from 40 to 50 per cent of the total body weight, and the special function of skeletal muscles—the production of voluntary movements—is responsible for one of the most outstanding characteristics of animal, as contrasted with plant, life. That characteristic is the ability of an animal to move some parts of its body or to change its own position with respect to its environment. The production of voluntary movements by skeletal muscle is one of the most essential activities of the body, enabling an animal to secure its food and protect itself in its environment, and, in its close association and interrelation with other organ systems, helps the animal maintain its normal status quo.

Muscles vary considerably in form, some being broad and sheetlike and others, cylindrical or spindle-shaped. Some muscles receive their names from their shape; many are named after the bones to which they are attached; others are named according to the action they perform. In conjunction with muscle study, the student should familiarize himself with such terms as *tendons, aponeuroses, fasciae, origin,* and *insertion.*

Tendons are white, glistening fibrous cords, varying in length and thickness and devoid of elasticity. The collagenous fibers of which a tendon is composed are in continuity with those of the muscle, the epimysium of the muscle forming a common covering for both. Tendons are connected to the one hand with the muscles and on the other with the movable structure, for example, the bones, cartilage, ligaments, and fibrous membranes.

Aponeuroses are flattened, ribbon-shaped tendons of a pearly white color.

The site where a muscle is attached to an immovable or less movable bone is spoken of as the **origin** of the muscle. This is usually the end nearest the body.

The attachment of a muscle to the more readily movable bone is known as the **insertion** of that particular muscle. This is usually the end farthest from the body.

Fasciae may be described as layers of white fibers and yellow elastic fibers united by a matrix, or aponeurotic laminae of variable thickness covering the softer and more delicate organs of the body as well as the muscles. Fascia is also found as connective tissue between muscles.

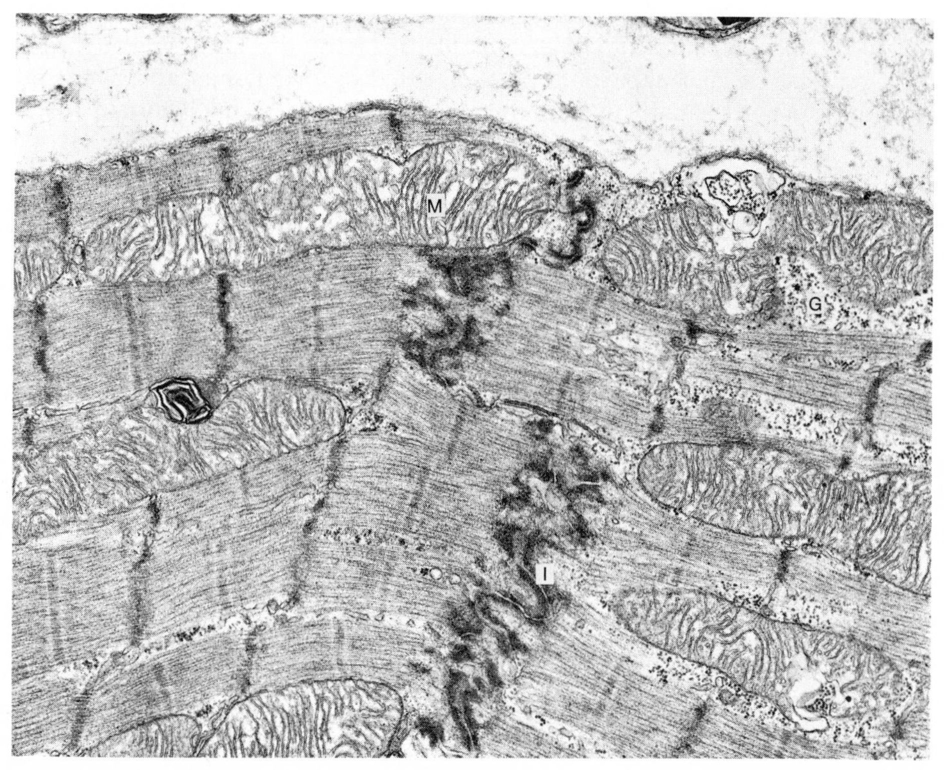

*Figure 3-13*B. *Electron micrograph of dog cardiac tissue, magnified 30,450* ×. *Compare with Fig. 3-13*A. I, *area of intercalated disk;* M, *mitochondrion;* G, *glycogen granules.* (*Courtesy Timothy Mougel, Electron Microscopy Laboratory, Department of Zoology, University of Maryland, College Park, Md.*)

The thickest part of a muscle, which is the intermediate part, is often called the **belly,** or **body,** of a muscle.

Chemistry of Muscle Tissue

The chemical constitution of the skeletal muscle has been most completely studied. Seventy-five per cent of skeletal muscle is water, and 25 per cent is solid material. Of the 25 per cent solid material, 20 per cent is protein, and the remaining 5 per cent is composed of inorganic material, certain organic extractives, and carbohydrate.

Proteins in Muscle

From the chemical point of view the contractile structure of muscle (myofibrils) consists almost entirely of protein. Perhaps 90 per cent of this substance is represented by three proteins, myosin, actin, and tropomyosin.

Myosin. **Myosin** is the most abundant protein in the muscle; about half the dry weight of the contractile part of the muscle consists of myosin. Straub, in 1942, working in Szent-Györgyi's laboratory, found that the contractile protein of muscle, the myosin of earlier investigators, consisted of at least two basic proteins, not one as previously supposed. For one of these proteins the name *myosin* was retained and for the other the term *actin* was given. Under certain conditions these two proteins are loosely bound in a complex called **actomyosin,** which is thought to be the structural protein of the resting myofibril. The myosin of the new system was found to be a complex protein consisting of long molecules, which by radiological methods can be demonstrated to form fibrils. Myosin has the following characteristics: It is water-soluble, markedly hydrophilic, and carries a positive charge that appears to be neutralized by magnesium and potassium ions, for which it has particular affinities. The addition of a small amount of salt, such as potassium chloride, will precipitate it from aqueous solution. It has numerous links to enable it to combine with a specific substance, adenosine triphosphate.

Myosin constitutes the thick filaments of the myofibrillae (see p. 85), and each such filament is made up of several hundred separate tadpole-shaped myosin molecules. One portion of the head end of the molecule can act enzymatically to split ATP and has been demonstrated to be adenosine triphosphatase. The other portion of the head is the site for binding to actin and forms the cross bridges proposed by Huxley and Hanson in the sliding filament theory of contraction (p. 89 and Fig. 3-14).

Actin. In contrast, the second protein of the muscle fibril, **actin,** is composed of relatively small globular molecules. It possesses the remarkable property of being able to assume either a dispersed globular form, G actin, or a fiberlike form, F actin. The F-to-G and G-to-F transformations, which probably occur by the process of polymerization, can be effected in vitro in the presence of small amounts of magnesium ions by varying the concentration of potassium or sodium.

When myosin and F actin are mixed in solution, a new complex, called F actomyosin, is formed. Addition of 0.1 per cent solution of ATP results in an immediate reduction in volume by as much as 44 per cent. This change in length

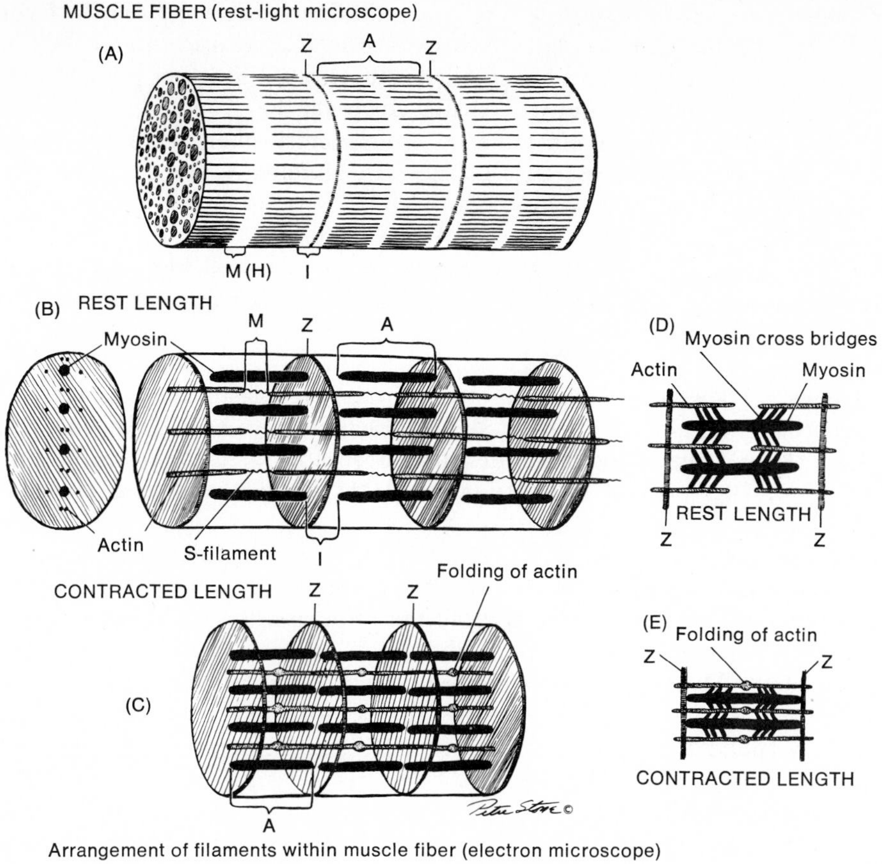

Figure 3-14. Arrangement of muscle fibrils in muscle showing thick and thin filaments.

of the threads of actomyosin can be observed under the microscope, and it may represent the fundamental change that is responsible for contraction of the myofibrils. In resting muscle the ATP is strongly adsorbed to the F-actomyosin complex. As a result of the ionic upset (depolarization) that occurs when a volley of motor nerve impulses reaches the muscle, the ATP–F actomyosin linkage is broken and G actomyosin–ADP is immediately formed. With this liberation of activation of the ATP, a process of dephosphorylation begins under the enzymatic direction of a substance called adenosine triphosphatase (ATPase). In this way ATP is broken down and provides the necessary energy for muscular contraction. The resting myofibril may be represented as

F actomyosin–ATP.

When contraction occurs, the myofibril may be represented as

G actomyosin–ADP.

Relaxation requires the conversion of G atomyosin–ADP back to F actomyosin–ATP.

Tropomyosin. **Tropomyosin** is a protein with interesting properties. It has been described by Bailey. It is prepared by extracting alcohol-dried muscle with a salt solution. In its amino acid composition and X-ray diffraction pattern it shows resemblance to myosin, but no function for it can be proposed.

In addition to the protein found in the myofibril, there are several proteins in the sarcoplasm of the muscle cell. These are globulin X, myogen, myoalbumin, and myoglobin.

Globulin X. **Globulin X** remains after myosin is removed from a saline extract of muscle. It coagulates at 50° C and has a molecular weight of about 160,000. Its function is unknown.

Myogen. **Myogen** is another protein of the sarcoplasm and has the general properties of an albumin. Two crystalline fractions have been obtained, myogens A and B, and enzymatic properties have been attributed to crystalline myogen A. Myogens have a molecular weight of 80,000 to 90,000 and coagulate at about 52° C.

Myoalbumin. **Myoalbumin** is a second albumin found in sarcoplasm and is present in the smallest amounts of all the proteins.

Myoglobin. **Myoglobin** is a conjugated protein, often called muscle hemoglobin. It has been isolated in crystalline form, and its molecular weight, isoelectric point, and absorption bands all differ from those of blood hemoglobin. The iron content of both is the same.

Muscle Extractives

Many of the nonprotein constituents of muscle are soluble in water, alcohol, or ether; they are usually called **muscle extractives.** They include those compounds containing nitrogen and are referred to as the **nonprotein nitrogen compounds (NPN).** These include creatine, phosphocreatine, creatinine, inosinic acid, adenylic acid, adenosine triphosphate, glutathione, purines, carnosine, anserine, choline, and acetylcholine. The other extractions of muscle, which do not contain nitrogen, are mainly the muscle carbohydrate glycogen and all its derivatives formed in glycolysis, such as glucose, inositol, hexosephosphate, and lactic acid.

Some of the individual extractions deserve mention at this point. One of the most important compounds in muscle is **adenosine triphosphate (ATP),** in which the terminal phosphate is linked by an energy-rich bond. It is customary to designate energy-rich phosphate bonds by the symbol ~ and the energy-rich phosphate as Ⓟ:

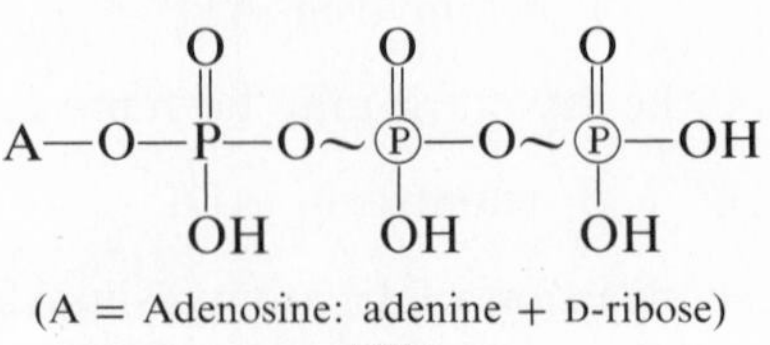

(A = Adenosine: adenine + D-ribose)
ATP

The rupture of the terminal phosphate of ATP with the liberation of inorganic phosphate and the formation of adenosine diphosphate (ADP) is associated with the release of a large amount of energy (11,500 cal per mole):

$$\text{ATP} \xrightarrow{\text{adenosine triphosphatase}} \text{ADP} + H_3PO_4 + 11{,}500 \text{ cal.}$$

It is this energy released by the breakdown of ATP that is used for the contraction of muscle, that is, the conversion of F actomyosin to G actomyosin.

Inositol is another extract of muscle that has recently been recognized as important in the metabolism of muscle tissue. There are a number of isomers of inositol in nature, the most important one being mesoinositol:

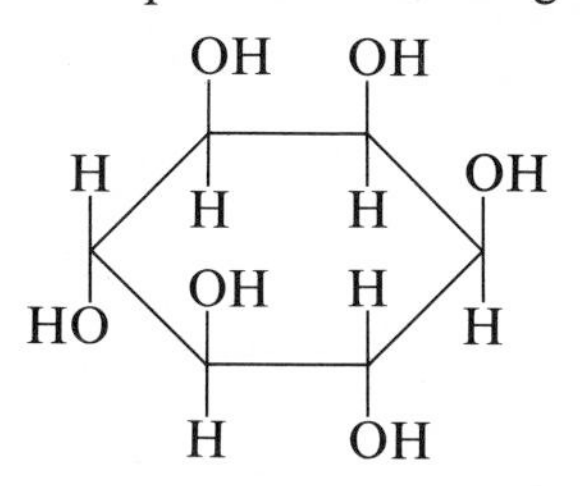

The other isomers differ from this in the arrangement of the OH's and H's in space. Inositol is widely distributed in the plant and animal kingdoms and is now recognized as part of the vitamin B complex. In sharks and certain other fish, inositol is stored up instead of glycogen. It is thus possible that this substance plays some role in carbohydrate metabolism, either as a substitute for glycogen or as an intermediate in the transformation of one monosaccharide into another.

Creatine and **phosphocreatine** are also involved in muscle metabolism. Their formulas and that of the structurally related creatinine are

Creatine	Phosphocreatine	Creatinine
NH_2	NH℗O(OH)$_2$	NH
|	|	|
C=NH	C=NH	C=NH
|	|	|
N—CH_3	N—CH_3	N—CH_3
|	|	|
CH_2	CH_2	CH_2
|	|	|
COOH	COOH	C=O

During muscular contraction phosphocreatine breaks down to creatine and phosphoric acid. On recovery, phosphocreatine is resynthesized. There is evidence that phosphocreatine acts as a reserve, or emergency, source of phosphate for the synthesis of adenosine triphosphate:

$$\text{Phosphocreatine} \longrightarrow \text{Creatine} + \text{phosphoric acid} + \text{energy},$$

$$\text{ADP} + \text{phosphoric acid} \xrightarrow{\text{energy}} \text{ATP}.$$

Salts of **lactic acid** are present because they result from the breakdown of carbohydrate in muscle metabolism:

$$\text{Glycogen} \longrightarrow \text{Lactic acid} + \text{energy}.$$

Lundsgaard demonstrated that muscles poisoned with sodium iodoacetate will not produce lactic acid on contraction. Thus the energy liberated in the breakdown of glycogen is not responsible for the contraction energy, since muscle will still contract as long as phosphocreatine and ATP are present. The energy released in the breakdown of glycogen is used for the synthesis of phosphocreatine from creatine and phosphoric acid:

$$\text{Creatine} + \text{phosphoric acid} \xrightarrow{\text{energy}} \text{Phosphocreatine}.$$

Part of the lactic acid formed in glycolysis (approximately one fifth) is oxidized to carbon dioxide and water with a release of energy, and the rest is converted back to glycogen by the liver, utilizing the energy released in the oxidation process:

$$\tfrac{1}{5}\ \text{Lactic acid} + O_2 \longrightarrow CO_2 + H_2O + \text{energy},$$

$$\tfrac{4}{5}\ \text{Lactic acid} \xrightarrow{\text{energy}} \text{Glycogen}.$$

Mechanism of Muscle Contraction

Chemical Basis of Contraction. The factors just discussed allow the following tentative summary of the mechanism of muscle contraction. In the resting muscle cell the myofibril is composed of F actomyosin with ATP strongly adsorbed to it to form an ATP–F actomyosin complex. As a result of the ionic upset that occurs when a volley of motor nerve impulses reaches the muscle fiber, the ATP–F actomyosin linkage is broken and G actomyosin is immediately formed, the energy liberated by the dephosphorylation of the ATP under the influence of the enzyme adenosine triphosphatase providing the fuel for the reaction. The chemical reactions in muscle contraction could be summarized as follows:

$$\text{(1) ATP–F actomyosin} \xrightarrow{\text{nerve impulses}} \text{G actomyosin} + \text{ATP}.$$

$$\text{(2) ATP} \rightleftharpoons \text{ADP} + H_3PO_4 + \text{energy}.$$

As the muscle fiber undergoes contraction, the following reactions occur and, it is believed, in the order listed.

$$\text{(3) Phosphocreatine} \rightleftharpoons \text{Creatine} + H_3PO_4 + \text{energy}.$$
$$\text{(4) Glycogen} \rightleftharpoons \text{Lactic acid} + \text{energy (glycolytic cycle)}.$$
$$\text{(5) } \tfrac{1}{5}\ \text{Lactic acid} + O_2 \longrightarrow CO_2 + H_2O + \text{energy (Krebs cycle)}.$$
$$\text{(6) } \tfrac{4}{5}\ \text{Lactic acid} + \text{energy} \longrightarrow \text{Glycogen}.$$

The energy liberated in reaction (2) is used for the actual shortening of the muscle fibril (contraction). The energy liberated in reaction (3) is used for the resynthesis of ATP. The energy released in the glycolytic cycle is used mainly for the resynthesis of phosphocreatine, and the energy released in the oxidation of lactic acid is used to drive the remaining lactic acid back to glycogen. However, it is to be noted that all energy released in the above reactions is first cycled through ATP (see the detailed discussion of energy extraction and transformation by living cells, p. 42). The excess energy liberated in all the reactions is used for heating the body, growth, thought, and other metabolic processes carried on within the living organism.

Physical Basis of Contraction. Recent observations by Hugh Huxley and Jean Hanson[2] using the X-ray diffraction technique and the electron microscope have shown that the myofibril is made up of two types of filaments which are arranged parallel to one another and overlap in certain areas, as shown in Fig. 3-14. These filaments have been shown to be the proteins actin and myosin. The thicker, shorter myosin filaments are lined up to form the A band of the myofibril, whereas the longer, thinner actin filaments form the less dense I band. It has been shown that the actin filaments are connected by a very thin filament called the S filament. The A band has a narrow, lighter area in the center which is quite noticeable in the relaxed muscle fiber and is the area of the filament where the actin filaments do not overlap the myosin filament. This area has been referred to as the H zone or M band (see Figs. 3-4 to 3-7). The myosin filament has been shown to have fingerlike projections at regular intervals along its length which form **cross bridges** with the actin filaments. These cross bridges are considered to have an important role in the **sliding filament theory of contraction** proposed by Huxley and Hanson. The theory proposes that the cross bridges, which are a permanent part of the myosin filament, hook onto an actin filament at an active site. The cross bridges are activated to oscillate back and forth by the energy released in the breakdown of ATP. This oscillation enables the thick filaments to pull the thin filaments closer together by a kind of ratchetlike action, thus causing the H zone, on the basis of the sliding in of the actin filaments toward the center, to disappear. This is exactly what happens when a muscle fiber is stimulated to contract. The H zone is seen to disappear as a change in the length of the muscle fiber changes the arrangement of the filaments (Fig. 3-14). As the thin filaments slide into the center to meet, they pull the *Z lines* (membranes that transect the fibrils and connect to the actin filament) with them. This also causes thick myosin filaments to meet adjacent thick filaments (as shown), producing new types of bands that appear as the ends of the filaments overlap one another.

The energy for contraction comes from the breakdown of ATP, the reaction being initiated by the myosin–ATPase. However, before ATP complexed to actomyosin may be split, the ATPase must first be activated. This has been demonstrated to be caused by the release of bound calcium from the sarcolemma and sarcotubular system, mainly in the region of the Z lines, by depolarization of the cell membrane by whatever means. The physical and chemical events leading to muscle contraction and relaxation are summarized in Table 3-1.

The considerable metabolic activity of muscle, particularly in carbohydrate metabolism, requires the presence of a number of enzymes. Some have been thoroughly studied, but many more have not. Some of the more important enzymes in muscle tissue catalyse the reactions occurring in the glycolytic and Krebs cycles. Examples of these enzymes and their actions are as follows.

[2] H. Huxley and J. Hanson, "Muscle Structure and Theories of Contraction." *Progr. Biophys.,* **7**:255 (1957).

Table 3-1. ***Sequence of Events in the Physical (Molecular) Mechanism of Muscle Contraction and Relaxation***

1. Nerve impulse reaches axon terminals at myoneural junction and releases acetylcholine from vesicles.
2. Acetylcholine drips onto receptor sites located on the muscle cell membrane at myoneural junction, causing depolarization of sarcolemma.
3. Depolarization spreads to sarcotubular system, causing a release of free calcium ion into the sarcoplasm.
4. The increased concentration of free calcium ion activates myosin–ATPase, causing a split of ATP $\longrightarrow$ ADP + PO_4 + energy.
5. Energy released activates cross bridges of myosin, producing sliding filament action (actin threads slide into center of sarcomere over myosin filaments—magnesium ions are needed for actin and myosin to unite), and the resulting state of the myofibril becomes actomyosin–ADP.
6. Relaxation (return of muscle to original length) follows when nerve impulse is cut off and muscle membrane is repolarized.
7. Concentration of calcium ions in sarcoplasm falls as calcium is pumped back into the sarcotubular sacs by a calcium-specific ATP pump.
8. Inactivation of the myosin ATPase allows the myosin–ADP to become bound to ATP either by direct exchange with sarcoplasmic ATP or by the action of ATP formation coupled with the phosphocreatine reaction resulting in the relaxed state of the myofibril as F actomyosin–ATP.

Glycolytic Cycle

(1) Glycogen $\xrightleftharpoons{\text{phosphorylase}}$ Glucose-1-phosphate.
(phosphorylation)

(2) Glucose-1-phosphate $\xrightleftharpoons{\text{phosphoglucomutase}}$ Glucose-6-phosphate.
(transition)

(3) Glucose-6-phosphate $\xrightleftharpoons{\text{isomerase}}$ Fructose-6-phosphate.
(transition)

(4) Fructose-6-phosphate $\xrightleftharpoons{\text{phosphohexokinase}}$ Fructose-1-6-diphosphate.
(phosphorylation)

The diphosphofructose molecule now splits under the influence of the enzyme aldolase into two smaller molecules, each containing three carbon atoms and one phosphorus atom, as indicated below:

(5) Fructose-1-6-diphosphate $\xrightleftharpoons{\text{aldolase}}$
3-Phosphoglyceraldehyde + dihydroxyacetone phosphate.

In the next stage 3-phosphoglyceraldehyde is phosphorylated:

(6) 3-Phosphoglyceraldehyde $\xrightleftharpoons[\text{dehydrogenase}]{\text{phosphoglyceraldehyde}}$
(phosphorylation) 1-3-Diphosphoglyceraldehyde.

(7) 1-3-Diphosphoglyceraldehyde $\xrightleftharpoons{\text{dehydrogenase + CoI(NAD)*}}$
(reduction) 1-3-Diphosphoglyceric acid + HCoI.

(8) 1-3-Diphosphoglyceric acid $\xrightleftharpoons{\text{phosphoglyceric kinase}}$
(splitting) 3-Phosphoglyceric acid + H_3PO_4.

(9) 3-Phosphoglyceric acid $\xrightleftharpoons{\text{phosphoglyceromutase}}$
(transition) 2-Phosphoglyceric acid + energy.

(10) 2-Phosphoglyceric acid $\xrightleftharpoons{\text{enolase}}$ Enol phosphopyruvic.
(reduction)

(11) Enol phosphopyruvic acid $\xrightleftharpoons{\text{pyruvic phosphokinase}}$
(splitting) Pyruvic acid + H_3PO_4 + energy.

(12) Pyruvic acid + HCoI $\xrightleftharpoons{\text{lactic dehydrogenase}}$ Lactic acid + COI.
(reduction)

The triose dihydroxyacetone phosphate formed in reaction (5) is converted into more 3-phosphoglyceraldehyde, as the compound is consumed, as shown in reaction (6):

Dihydroxyacetone phosphate $\xrightleftharpoons{\text{phosphotrioseisomerase}}$ 3-Phosphoglyceraldehyde.

One fifth of the lactic acid produced is oxidized (aerobically) to form water and carbon dioxide with a release of energy. This process takes place by the **citric acid cycle,** or **tricarboxylic acid cycle,** of Krebs, which operates in the mitochondria.

Citric Acid Cycle

(1) Lactic acid + CoI $\xrightleftharpoons[\text{}]{\text{−2H lactic acid dehydrogenase}}$ Pyruvic + HCoI.

(2) Pyruvic acid $\xrightarrow[\text{Co carboxylase}]{+H_2O-CO_2-2H}$ Acetic acid.

(3) Acetic acid $\xrightarrow{\text{Co acetate}}$ Citric acid.

(4) Citric acid $\xrightarrow{-H_2O\text{ aconitase}}$ *cis*-Aconitic acid.

(5) *cis*-Aconitic acid $\xrightarrow{\text{aconitase + }H_2O}$ Isocitric acid.

(6) Isocitric acid $\xrightarrow[-2H]{\text{isocitric dehydrogenase}}$ Oxalosuccinic acid.

(7) Oxalosuccinic acid $\xrightarrow[-CO_2]{\text{decarboxylase}}$ Ketoglutaric acid.

(8) Ketoglutaric acid $\xrightarrow[+H_2O-2H-CO_2]{\text{decarboxylase}}$ Succinic acid.

$$\text{(9) Succinic acid} \xrightarrow[-2H]{\text{dehydrogenase}} \text{Fumaric acid.}$$

$$\text{(10) Fumaric acid} \xrightarrow[+H_2O]{\text{fumarase}} \text{Malic acid.}$$

$$\text{(11) Malic acid} \xrightarrow[-2H\ \text{CoI} \longrightarrow \text{CoIH}_2]{\text{dehydrogenase}} \text{Oxaloacetic acid.}$$

$$\text{(12) Oxaloacetic acid} \xrightarrow{\text{acetylation}} \text{Acetic acid.}$$

The empirical formula is usually used to show the oxidation of lactic acid by way of the citric acid cycle:

$$\text{Lactic acid} + O_2 \longrightarrow CO_2 + H_2O,$$
$$C_3H_6O_3 + 3O_2 \longrightarrow 3CO_2 + 3H_2O.$$

Carbon dioxide is given off in reactions (2), (7), and (8), showing that three molecules of carbon dioxide are produced in the citric acid cycle.

Six pairs of hydrogen atoms are transferred to hydrogen acceptors in reactions (1), (2), (6), (8), (9), and (11), and they ultimately combine with oxygen to form water. Of the six molecules of water thus formed, four can be accounted for as used within the cycle, reactions (2), (5), (8), and (10), and the remaining two are given off along with the water formed in reaction (4) to give three molecules of H_2O, as noted in the empirical formula summarizing oxidation of lactic acid.

Inorganic Constituents of Muscle

Muscle tissue contains the same inorganic constituents as extracellular fluid. The cations of muscle are potassium, sodium, magnesium, and calcium. Potassium predominates in muscle. The anions include phosphate, chloride, and a small amount of sulfate. The high inorganic phosphate content may actually be an artifact produced by the breakdown of the organic phosphates of ATP and phosphocreatine. Intracellular potassium plays a very important role in muscle metabolism. When glycogen is deposited in muscle and when protein is being synthesized, a considerable amount of potasssium is also incorporated into the tissue. The calcium and magnesium of muscle appear to function as activators or inhibitors of intramuscular enzyme systems. Calcium ion is needed specifically for the activation of myosin–ATPase. Magnesium ion is necessary to bind actin to myosin.

Muscle Stimulation and Contraction

A basic characteristic of all animals is their ability to move in a purposeful fashion. Animals move by contracting their muscles; so muscle contraction is one of the key processes of animal life. In order for a muscle to contract, it must be stimulated by an outside disturbance. All skeletal muscles receive motor nerves from the central nervous system, the fibers of which make intimate contact with the muscle fiber (Fig. 3-15). When the nerve fibers are stimulated in the central nervous system, the impulse created is transmitted to the muscle fiber and the muscle decreases in length (shortens) and thickens (contraction). Since muscular contraction is initiated by a reduction in the magnitude of the resting membrane potential of the muscle fiber, the immediate function of the neuromuscular transmission process may be expected to be the alteration of the membrane-

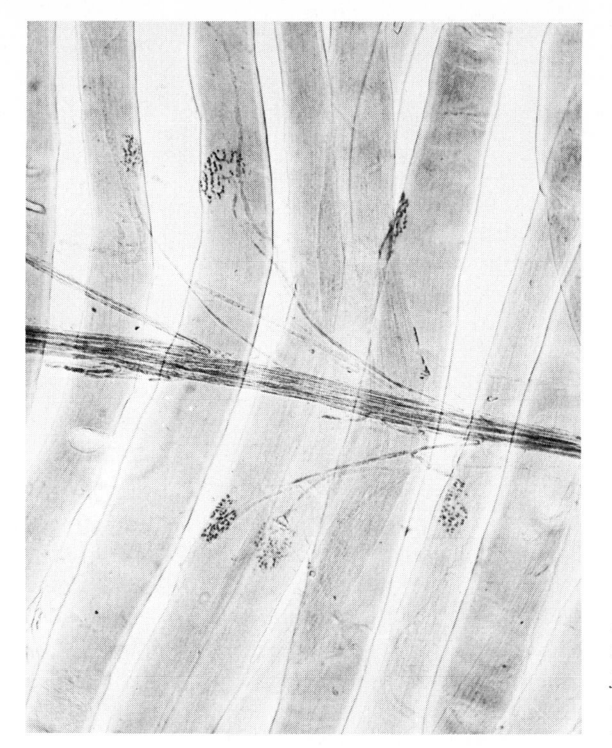

Figure 3-15. Motor nerve endings. Preparation shows both the branching nerve fibers and the end plates. (Courtesy Ward's Natural Science Establishment, Inc., Rochester, N.Y.)

potential level. Modern knowledge of the mechanism of the transmission of the nerve impulse into the muscle started to accumulate after Dale, Feldberg, and Vogt, in 1936, showed that when motor nerves of the perfused tongue of the cat or the calf muscle of the dog, the cat, and the frog are stimulated electrically, **acetylcholine** is released at the myoneural junction. It was also demonstrated that when very small amounts of acetylcholine are injected intra-arterially into the gastrocnemius muscle, impulses are evoked in the muscle and contraction ensues. The response to acetylcholine can be inhibited by curare, and the action of acetylcholine can be intensified and prolonged by the inhibition of the enzyme cholinesterase. On the basis of these observations, a chemical hypothesis of neuromuscular transmission was developed—that acetylcholine is released at the motor end plate of skeletal muscle following the arrival of a nerve impulse, which in turn initiates contraction of a muscle. The chemical reaction at the motor end plate initiated by the arrival of the impulse occurs in the following way. Choline, acetic acid, coenzyme A, and the enzymes cholineacetylase and cholinesterase are all normally present at the motor end plate. The arrival of the impulse supplies sufficient energy to cause choline and acetyl coenzyme A to combine to form acetylcholine. The enzyme cholineacetylase catalyzes the reaction:

$$\text{Choline} + \text{acetyl CoA} \xrightarrow{\text{cholineacetylase}} \text{Acetylcholine} + \text{CoA.}$$

The liberation of acetylcholine from the vesicles of the axon terminals raises the permeability of muscle cell membrane, so that ions that do not diffuse through

Figure 3-16. Liberation of acetylcholine from end plate, which leads to depolarization of muscle cell membrane.

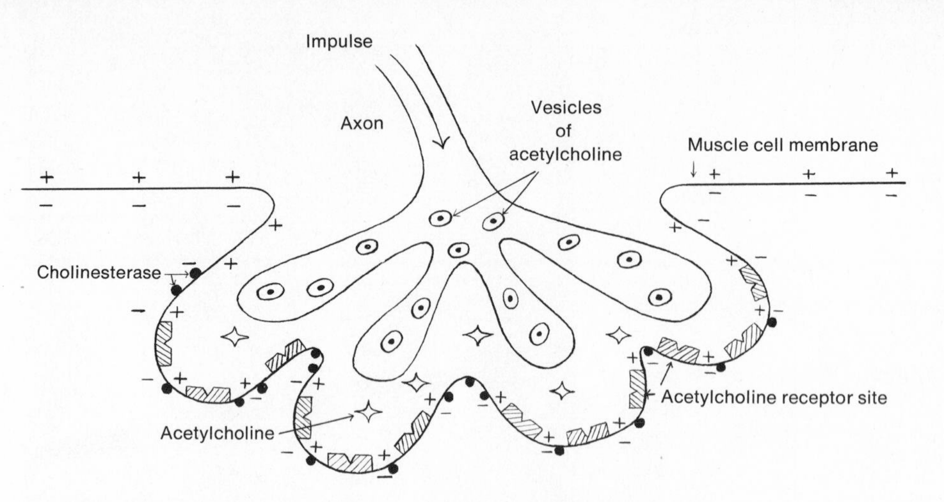

the resting membrane (polarized) can now pass across it (depolarized effect) (Fig. 3-16). This ionic disequilibrium sets off a series of chemical reactions which result in the contraction of the muscle with a release of energy (see p. 88). In order for the muscle thus stimulated to relax, the acetylcholine must be removed. The enzyme cholinesterase which is present in the muscle membrane at the myoneural junction acts on the acetylcholine, breaking it back down into choline and acetic acid:

$$\text{Acetylcholine} \xrightarrow[\text{relaxation}]{\text{cholinesterase}} \text{Choline} + \text{acetic acid.}$$

That acetylcholine is necessary for contraction and its removal necessary for relaxation can be demonstrated by the use of the drugs curare and physostigmine (eserine).

When **curare** is injected into the animal, stimulation of the nerve leading into the muscle will not cause a contraction. Stimulation of the muscle directly will produce a contraction. The muscle itself is not effected by the curare, nor is the nerve, since stimulation of the nerve will invoke an impulse, but this will not be transmitted through the myoneural junction. After the release of acetylcholine at the motor end plate it is apparently bound to some type of receptor in the muscle membrane. This complex is believed to be responsible for the change in membrane potential. Curare is believed to compete with acetylcholine for these receptor sites, forming a complex which does not change the membrane potential, thus inhibiting depolarization. A muscle poisoned with curare will not contract unless directly stimulated because of a lack of sufficient receptor sites for formation of acetylcholine complex.

Another drug, **physostigmine,** will inhibit the enzyme cholinesterase. An animal injected with physostigmine will undergo contraction of muscle when stimulated, but since there is no enzyme to break down the acetylcholine produced, the muscle will not relax. From these and related experiments we may conclude the following:

1. Muscle fiber cells are independently irritable.

2. The myoneural junction is a specialized structure, as demonstrated by its sensitivity to curare, but the nerve or muscle tissue is not.
3. Some drugs are extremely specific in their action, affecting only certain structures without acting on other parts of the body.
4. Some chemicals counteract the activity of other chemicals, such as curare (counteracts cholineacetylase) and physostigmine (counteracts cholinesterase). This effect is known as *physiological antagonism.*

Simple Muscle Contraction

A single nerve impulse of sufficient strength evokes a muscle twitch, or simple muscle contraction, followed by relaxation. In order to analyze the complex phenomena that take place, muscles have been separated from the organism and stimulated directly by electric shocks. The frog gastrocnemius muscle has been extensively used in the laboratory to demonstrate and to study muscle contraction experimentally. It is possible to record graphically such a contraction (Fig. 3-17). (See laboratory manual of mammalian physiology[3] for the experimental arrangement whereby the contraction of an exercised skeletal muscle can be recorded.) The muscle is stimulated by single or multiple electric shocks sent into it through platinum electrodes. The shocks are generated by an **induction apparatus.** This consists of two coils of wire, which can be set at various distances from each other. By the closing of a key, an electrical current driven by a 2- to 4-V dry-cell battery is caused to flow through the primary coil. As the current rises from zero to a maximum value, a brief electrical current is "induced" in the secondary coil. This *induced* or *faradic current* is led into the muscle by the platinum electrodes. The contraction occurring from such a stimulation can be recorded on a revolving drum. The tracing that is developed is referred to as a **myogram.** An examination of the myogram of the contraction of the frog gastrocnemius muscle (Fig. 3-18) will show that a simple muscle contraction, or muscle twitch, can be divided into three periods: (1) the latent period, (2) the contraction period, and (3) the relaxation period.

Contraction does not occur immediately with the presentation of an adequate stimulus. A finite time is necessary for the chemical processes to take place which result in the contraction. This time, brief as it is, is called the **latent period.** For the frog gastrocnemius this period lasts about 0.01 second. The latent period differs in different muscles. Cooling prolongs the latent period, and warming shortens it. At the end of the latent period, the muscle begins to shorten and thicken; that is, it changes its shape but not its size. During this **period of contraction** tension is developed. This contraction period lasts about 0.04 second. The muscle will then pass into the **relaxation period,** during which the muscle will return to its normal length. In the frog gastrocnemius the relaxation period takes about 0.05 second. Figure 3-18 shows a typical myogram of a simple muscle twitch.

Such contractions are produced experimentally in the laboratory to study their nature and to study what and how various conditions will affect them. Normal contractions of our muscles in everyday activity are not simple muscle contractions (muscle twitches) but are **tetanic contractions. Tetanus** is defined as a sustained

[3] Sigmund Grollman, *Laboratory Manual of Mammalian Anatomy and Physiology,* 3rd ed. (New York: Macmillan, Inc., 1974).

Figure 3-17. Arrangement of apparatus which records the contraction of an excised skeletal muscle.

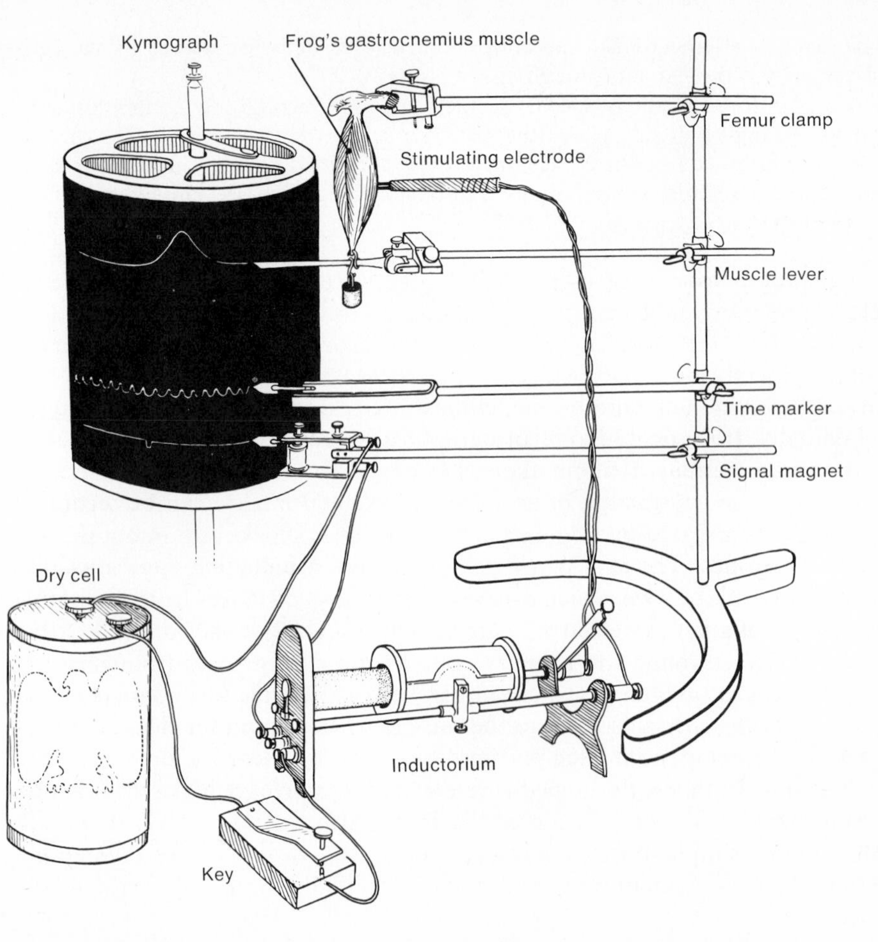

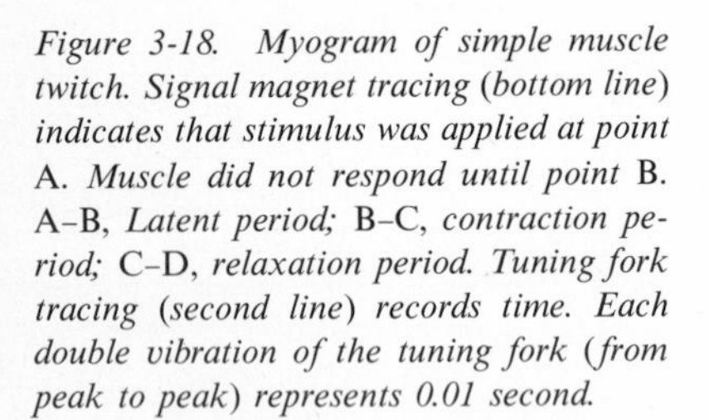

Figure 3-18. Myogram of simple muscle twitch. Signal magnet tracing (bottom line) indicates that stimulus was applied at point A. *Muscle did not respond until point* B. A–B, *Latent period;* B–C, *contraction period;* C–D, *relaxation period. Tuning fork tracing (second line) records time. Each double vibration of the tuning fork (from peak to peak) represents 0.01 second.*

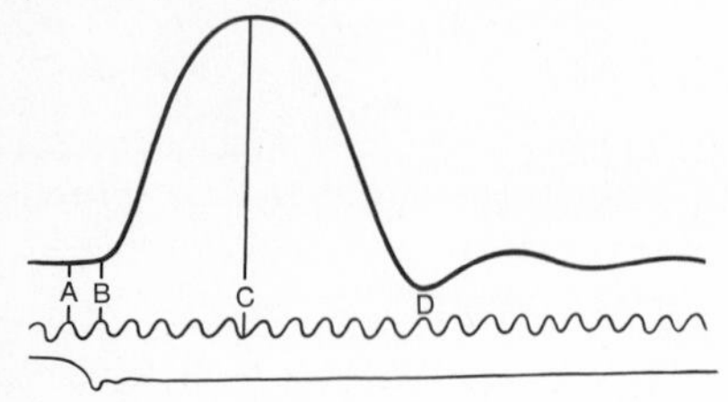

contraction due to the fusion of many twitches following one another in rapid succession. In order to understand our definition better, the genesis of tetanus, or normal contraction, of muscle can be demonstrated in the laboratory in the following way.

One stimulus of sufficient strength when applied to a muscle will produce a simple contraction or twitch. When two stimuli, each capable of causing a simple contraction, follow each other rapidly so that the second stimulus occurs before the first contraction is completed, the second twitch is superimposed on the first and is usually greater in extent. This is referred to as *summation of twitches* or *wave summation* (Fig. 3-19). The second stimulus, being applied before the muscle can completely relax, brings the active state again to its full intensity. The fibrillae shorten and begin to pull on the fibrillae that are already stretched, and a higher tension will be developed than can be achieved by a single shock. There is no algebraic summation of the stimuli following one another, but the tension developed by each consecutive stimulus will depend on the amount of stretch already existing. This is actually the sum of two waves of contraction in the same muscle cell.

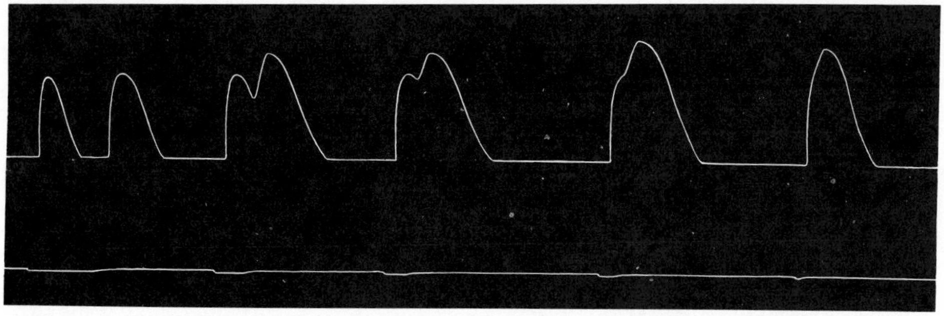

Figure 3-19. Myogram of "wave summation." The first two curves represent simple muscle twitches following the application of independent single stimuli. The third curve represents the summation of twitches when the second stimulus is applied before the first contraction is completed. The fourth and fifth curves represent the same thing except for the fact that the second stimulus has been applied more swiftly. The last curve represents complete fusion of two independent twitches.

When a large number of stimuli are applied to a muscle in rapid succession so that relaxation commences before the succeeding stimulus arrives, a sustained contraction curve is observed with the top of the curve showing undulations. This type of contraction is called *incomplete tetanus* (Fig. 3-20). With further increase in the rapidity of the stimuli delivered to the muscle, so that the interval between stimuli is less than the contraction time, the curve written by the muscle is a smooth line and is referred to as a *complete tetanus* or *tetanus* (Fig. 3-21). Normal contraction is of this type, owing to waves of impulses going over the nerve to the muscle. In man a rate of 50 to 200 per second, depending on the muscle involved, causes complete tetanus.

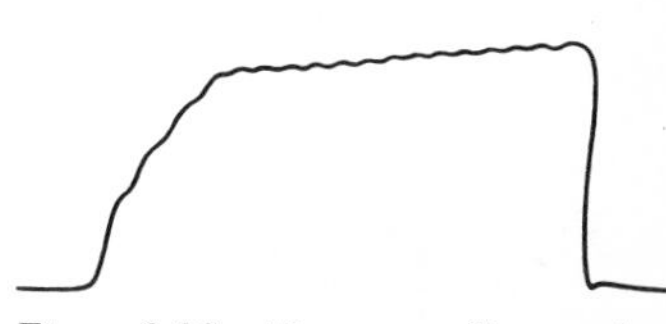

Figure 3-20. Myogram of incomplete tetanus.

Resting muscles are never completely relaxed. There is a steady stream of nerve impulses going to the muscles, keeping them in a partial state of contraction. This is spoken of as *muscle tone,* or **tonus.** It is a reflex phenomenon caused by the stretching of receptors, presumably muscle spindles, in the muscle itself. Tone serves the purpose of keeping the muscle always taut and ready for action. The isometric tonus of the postural muscles, such as the leg, trunk, and neck muscles, maintains posture. The tone of our jaw muscles helps maintain the mandible in a closed position. Thus, the mouth remains closed when we are relaxed instead of falling open. Muscles can lose their tone in several ways. A muscle to which the nerve is cut, injured, or diseased does not receive a constant flow of impulses, and thereby loses its tone. The muscle soon becomes flaccid and atrophies. This is called the **atrophy of denervation.** Tone may also be lost by immobilization of a part or inactivity of a part, illustrated by bedridden people and by limbs that have been in casts for long periods of time. The normal flow of nerve impulses into such muscles is reduced, and atrophy will occur. This is referred to as **atrophy of disuse.** Atrophy of disuse is different from atrophy of denervation since the nerve connection remains normal and the muscle can recover when it is again normally active. In general, we can say that muscles in contraction "pull," in contrast to muscles in tonus, which "hold."

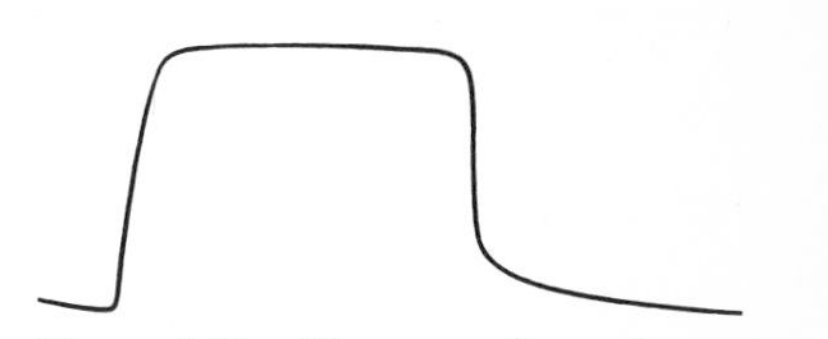

Figure 3-21. Myogram of complete tetanus.

Mechanical Work and Muscle Action

An engine is a device for transforming energy. As an electrical motor converts electrical energy into mechanical energy and a steam engine converts heat energy into mechanical energy, the muscles act as engines converting chemical energy into mechanical energy. The amount of work that can be performed by a muscle is determined by multiplying the weight of a load lifted by the height to which the load is lifted.

$$\text{Work} = \text{Height} \times \text{weight}.$$

Work is expressed in units of foot-pounds, kilogram-meters, and gram-millimeters.

When a muscle contracts without there being a load to be lifted or when the load is too heavy to lift, no work is done and all the energy produced is given off as heat energy. Up to a certain point, as we gradually increase the weight lifted by a contracting muscle, there is an increase of the amount of work performed. Beyond this point, less work is performed until the weight becomes so heavy the muscle cannot develop sufficient tension to overcome the load and will not shorten, thus accomplishing no work. The load at which the most work is done is called the **optimum load.** We may make the general statement that, with the same strength of stimulus, a fairly well-loaded muscle does more work than one underloaded or overloaded.

As previously stated, the conversion of potential energy (ATP) into muscle energy (contraction) is accomplished by the liberation of some energy in the form of heat. In fact, only a small percentage of the energy set free is utilized for mechanical work; the largest part escapes in the form of heat. When a muscle contracts isometrically, all the energy escapes as heat, since no mechanical work is performed. An **isometric contraction** is one in which the length of the muscle may be regarded as constant. As in instances where the muscle is unable to develop sufficient tension to overcome the external load, no work is performed under these conditions, even though the muscle is exerting its maximum tension. Thus all heat energy is liberated as heat since none is converted into mechanical energy. When the tension developed by the contracting muscle is greater than the load, the muscle shortens and performs mechanical work. This is referred to as an **isotonic contraction,** and much less heat is evolved since part of this energy is converted to mechanical energy. But in either isometric or isotonic contraction, all energy is evolved as heat if no work is done (no weight lifted).

The total heat produced by a contracting muscle is divided into **initial heat** and **recovery heat.** The initial heat makes up 45 per cent of the total heat produced and is developed during the anaerobic chemical processes of contraction. The initial heat is divided into **contraction heat** and **relaxation heat.** The contraction heat makes up 65 per cent of the initial heat and is incident to the breakdown of phosphocreatine. The relaxation heat makes up 35 per cent of the initial heat and is the result of the tension developed by the contracting muscle. The recovery heat makes up 55 per cent of the total heat and is the heat developed during the aerobic chemical processes occurring in the muscle. This heat appears after relaxation and is due to oxidation of lactic acid; it will not be formed in the absence of oxygen.

Factors Determining the Extent of Contraction. A muscle's response to a stimulus is determined (1) by the strength of the stimulus and (2) by the state of the protoplasmic structure.

In the laboratory, by varying the strength of the stimulus applied to a muscle by the use of an inductorium, we can demonstrate the relationship that exists between the extent of contraction of a muscle and changes in the strength of stimulus. When the current is just strong enough to produce a very feeble break contraction in the muscle, that is, a contraction produced by the stimulus of the breaking of an applied current, this strength of stimulus is spoken of as the **minimal, threshold, liminal,** or **rheobase** stimulus. It is really a measure of the irritability of the tissue, and with it we can compare the irritability of two different muscles or the same muscle under different conditions. As we gradually increase the strength of our stimulus from this point, the muscle will contract more and more until a strength of stimulus has been reached in which no further increase in contraction takes place because all the motor units of the muscle have been thrown into action. This strength of stimulus is spoken of as the **maximal stimulus.** A stimulus of a strength between the liminal and maximal is known as a **submaximal stimulus.** A stimulus below the liminal in strength is termed the **subliminal stimulus.** Although a single subliminal stimulus cannot make a muscle respond, a series of such subliminal stimuli applied to the muscle can elicit a response. This is spoken of as the *summation of inadequate stimuli.*

An individual muscle fiber or muscle cell, when stimulated with a liminal strength of stimulus, responds to the stimulus to its fullest extent, or complete contraction takes place. This phenomenon is known as the *law of all or none.* Although the individual fiber follows this law, the muscle as a unit does not, and therefore it is possible for muscle to execute contractions of graded extent or force. The reason for this is that all the individual fibers in a muscle do not have the same threshold of irritability, and many may not be within the physical range of the stimulus applied.

Fatigue. When an excised muscle is stimulated with a single induction shock of sufficient strength, a contraction of a certain height is obtained. If another stimulus of the same strength is applied after the first has had an opportunity to complete its effect, the second contraction is a little higher than the first. By stimulation of the muscle with a series of such stimuli and under the condition that each contraction is allowed to be completed before the next stimulus is applied, a series of contractions is obtained in which each twitch is higher than the preceding one. This is referred to as *treppe,* or *staircase.* This phenomenon has been explained on the basis of an increased hydrogen-ion concentration and an increase in temperature within the muscle acting as a subsidiary stimulus to increase the threshold of irritability of the muscle cells, thereby creating a more favorable condition for further work (a warming-up process both chemically and physically). After the contractions have reached their maximum height, this height may be maintained for a short time, but soon, continued stimulation causes a gradual decrease in contractions until further stimulation brings forth no response of the muscle at all, no matter how strong the stimulus may become. The muscle has gradually lost its irritability; the term **fatigue** has been applied to this condition. Fatigue of the muscle is caused by (1) the exhaustion of the energy sources of the muscle and (2) the accumulation of waste products, such as lactic acid, carbon dioxide, and ketone bodies. In the intact animal, fatigue in most instances is due to events at the synapses in the central nervous system and is not the experimental fatigue of muscle discussed above. This has been demonstrated in the laboratory in the following way. By noxious stimulation of the foot of a spinal frog, a reflex withdrawl of the foot is obtained; by continual stimulation, the reflex contraction of muscle is lost, and the foot no longer withdraws. If at this stage the nerve of the flexor is stimulated, a contraction again occurs. The cessation of this contraction must not be due to fatigue of the muscle or of the nerves, but it must be due to changes in the spinal cord. This same condition may be demonstrated in man, by making a subject contract his leg or arm muscles repeatedly with a weight attached to the part. When he is no longer able to lift the weight voluntarily, electrical stimulation of the motor nerve through the skin produces a powerful contraction. This may be regarded as a protective mechanism, since central fatigue of the nervous system occurs before there is any block at the neuromuscular junction or before the muscle itself has lost its irritability. It should be remembered that ther term "fatigue" has also a psychological meaning, and the term as used here is purely physiologically defined.

Oxygen Debt, Steady State, and Recovery. If muscular activity is moderate and uniform, the oxygen intake rises gradually and then in a minute or two levels off and remains at this level for the duration of the exercise. Since the other bodily functions, such as respiration, heartbeat, and lactic acid production, also maintain a steady level, this is called a **steady state.** Oxygen intake is equal to oxygen expenditure. When the intensity of work rises to such an extent that a steady state cannot be maintained because more oxygen is required than can be supplied, it is obvious that additional work has to depend entirely on the anaerobic chemical processes in the muscles. Lactic acid will accumulate during this period to be oxidized when the body is able to supply sufficient oxygen for its oxidation. This accumulation of lactic acid with other products of metabolism has been referred to as the **oxygen debt.** The amount of oxygen used, in excess of the resting level, during the period beginning at the moment muscular activity ends and terminating when recovery is complete, for example, when oxygen intake has returned to its resting level, may be considered to be the oxygen debt incurred during the period of exercise. Thus the extent of debt is determined by measuring the total amount of oxygen consumed during the period of recovery and subtracting from it the amount of oxygen that would have been normally consumed during the same period if the subject had remained at rest. After exercise, oxygen consumption remains increased from some time until the oxidation of accumulated products of the anaerobic chemical reactions of the muscle has been completed and the muscle has been recharged. This period of recharging is called the **recovery period,** and, as A. V. Hill has said, "the subject is paying his oxygen debt."

As we can see from the foregoing discussion, it is possible for a person to take muscular exercise that requires far more oxygen than can conceivably be supplied by respiration and circulation during the period of exercise. In the severest forms of exercise, something over 22 liters of oxygen may be required to provide the energy used in 1 minute. To deliver to the tissue this amount of oxygen is an impossible task for the respiratory and circulatory systems, which even in a well-developed athlete may supply only 4 or 5 liters a minute. By contracting an oxygen debt of 17 to 18 liters, the demand for 22 liters a minute can be met.

The amount of oxygen debt the individual is capable of accumulating will be limited by the degree of body tolerance for the accumulation of products of anaerobic decomposition, chief among which is lactic acid.

When the concentration of lactic acid in muscle tissue reaches 2.8 per cent, the muscle cannot contract. This level, however, is seldom reached.

Formerly it was believed that there was a quantitative relationship between oxygen debt and lactic acid production. This belief led to the statement that lactic acid is the security given for the payment of the oxygen debt. Now it is known that this theory is only partially correct and that the oxygen debt consists of two parts: an alactacid and a lactacid. The alactacid debt is paid back at a much faster rate than the lactacid debt. Experiments have indicated that the alactacid portion of the oxygen debt is due to the accumulation of ketone bodies as a result of the increased metabolism of fat during exercise. These ketone bodies, which include acetoacetic acid, beta hydroxybutyric acid, and acetone, accumulate as does lactic acid in the condition of exhaustion, waiting for more favorable circumstances (an ample oxygen supply) so that they may be oxidized, releasing carbon dioxide, water, and energy, the energy being used for resynthesis of more phosphocreatine.

The maximum oxygen debt to which the body can submit depends on various factors. It is really a measure of the maximum amount of lactic acid and ketone bodies that the tissues can tolerate, and the factors involved are

1. The amount of alkali available in the buffers of the tissue and blood to neutralize the acids formed during the metabolic processes.
2. The rate at which lactic acid and ketone bodies are liberated. If this liberation is great, the hydrogen-ion concentration in the blood will not rise immediately as high as in the muscles, and consequently respiratory distress will not occur for some time after the acid has been liberated because the distress is caused largely by acid acting on the nervous system.
3. The resolution of the individual in pressing his body to exhaustion.

It has been reported that in one unusual case, a trained athlete was capable of incurring an oxygen debt of 18.6 liters with a lactic acid concentration in the muscles of 0.30 per cent. In less-able-bodied persons, the greatest oxygen debt tends to be about 12.5 liters and the concentration of lactic acid in his muscles, about 0.21 per cent.

References

Bard, Phillip. *Medical Physiology*. St. Louis: Mosby, 1966.

Best, C. H., and N. B. Taylor. *The Physiological Basis of Medical Practice*. Baltimore: Williams & Wilkins, 1966.

Bloom, W., and D. W. Fawcett. *A Textbook of Histology*. Philadelphia: Saunders, 1968.

Dreyfus, J. C., and Georges Schapira. *Biochemistry of Hereditary Myopathies*. Springfield, Ill.: Thomas, 1962.

Dubuisson, M. *Muscular Contraction*. Springfield, Ill.: Thomas, 1954.

Evans, F. G. *Biomechanical Studies of the Musculo-skeletal System*. Springfield, Ill.: Thomas, 1961.

Gergely, J. (ed.). *Biochemistry of Muscle Contraction*. Boston: Little, Brown, 1964.

Gutmann, E. *The Deinnervated Muscle*. New York: Plenum, 1962.

Hoyle, Graham. *The Nervous Control of Muscular Contraction*. New York: Cambridge U.P., 1958.
Kaplan, E. B. *Duchenne's Physiology of Motion*. Philadelphia: Saunders, 1959.
Karpovich, P. V. *Physiology of Muscular Activity*. Philadelphia: Saunders, 1959.
Mommaerts, W. F. *Muscular Contraction*. New York: Wiley-Interscience, 1950.
———, A. J. Brady, and B. C. Abbott. "Major Problems in Muscle Physiology," in: Annual Review of Physiology, Vol. 23. Palo Alto, Calif.: Annual Reviews, 1961, p. 529.
Morehouse, L. E., and A. T. Miller. *Physiology of Exercise*. St. Louis: Mosby, 1971.
Paul, W. M., E. E. Daniel, C. M. Kay, and G. Monckton (eds.). *Muscle*. Elmsford, N.Y.: Pergamon, 1965.
Pernow, B., and B. Saltin. *Muscle Metabolism during Exercise*. New York: Plenum, 1971.
Poglazov, B. F. *Structure and Function of Contractile Proteins*. New York: Academic, 1966.
Rodahl, Kaare, and Steven Horvath. *Muscle as a Tissue*. New York: McGraw-Hill, 1962.
Sandow, A. "Skeletal Muscle," in: *Annual Review of Physiology*, Vol. 32. Palo Alto, Calif.: Annual Reviews, 1970, p. 87.
Simonson, E. *Physiology of Work Capacity and Fatigue*. Springfield, Ill.: Thomas, 1971.
Szent-Györgyi, A. *Chemical Physiology of Contraction in Body and Heart Muscle*. New York: Academic, 1953.
———. *Chemistry of Muscular Contraction*. New York: Academic, 1951.

4

The Nervous System

Bodies of all animals are formed of cells and their products, specialized in various ways to perform specific functions and organized into tissues, for example, epithelial tissue for protective covering, connective tissue for support, muscle tissue for contraction, and nervous tissue for conduction. Although the organization of the nervous system reaches an unimaginable complexity, its functional part consists of only one type of cell, the neuron.

The nervous system in general is divided into **central** and **peripheral** portions. The central portion includes the **brain** and the **spinal cord.** The peripheral portion includes those structures, referred to as **nerves,** that connect the central portion to all other parts of the body (Fig. 4-1). The nerves attached to the brain are called **cranial nerves,** and those that are attached to the spinal cord, the **spinal nerves.** The peripheral portion also includes a special division, the **autonomic.** The autonomic is further divided into a **parasympathetic** and **sympathetic portion,** sometimes referred to as the **vegetative nervous system.**

The Brain

The **brain** consists of the **cerebrum,** the **cerebellum,** and the **brain stem** (Figs. 4-2 and 4-3).

The cerebrum consists of two cerebral hemispheres and is divided into frontal, temporal, parietal, and occipital regions (Fig. 4-4). On closer examination of the cerebral hemispheres, certain characteristics become apparent. A deep depression, the **longitudinal fissure,** divides the cerebral hemispheres into right and left halves. The surface of this structure is extensively wrinkled (Fig. 4-5). The deeper depressions are called **fissures.** The shallow ones, called **sulci,** give rise to the wrinkles, or **convolutions. Gyri** are portions of convolutions. The pattern of cortical convolutions is constant enough for a terminology to have arisen. Though most of

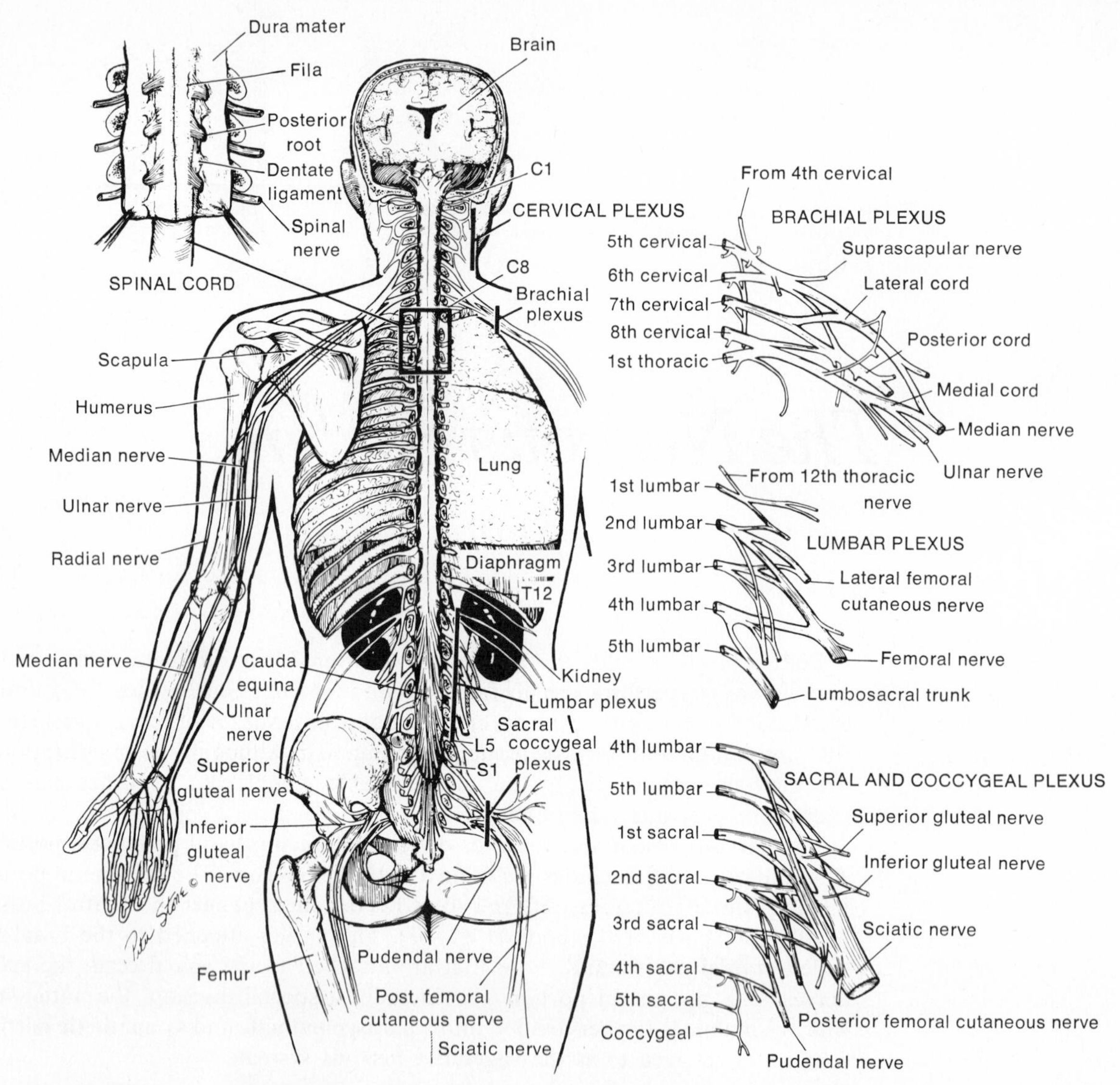

Figure 4-1. Superficial distribution of spinal nerves. The main plexi formed by the spinal nerves are shown also.

the fissures, sulci, and gyri are named, this does not imply that they are always definite in size, shape, and location (Fig. 4-6).

A cross section of the cerebral hemispheres (Figs. 4-7 and 4-8) shows the following structures. The layer of material forming the surface, or rind, of the cerebral hemispheres is grayish in color and is called the **cortex.** Masses of gray matter located internally in the cerebrum are called the **basal ganglia** and consist of the following structures: the *caudate nucleus, putamen, globus pallidus, thalamus, hypothalamus,* and *amygdala.* The putamen and globus pallidus are known together as the **lentiform nucleus.** The caudate nucleus, the lentiform nucleus, and special tracts of white matter that connect these two structures, the *internal capsule,* form what is known as the **corpus striatum.**

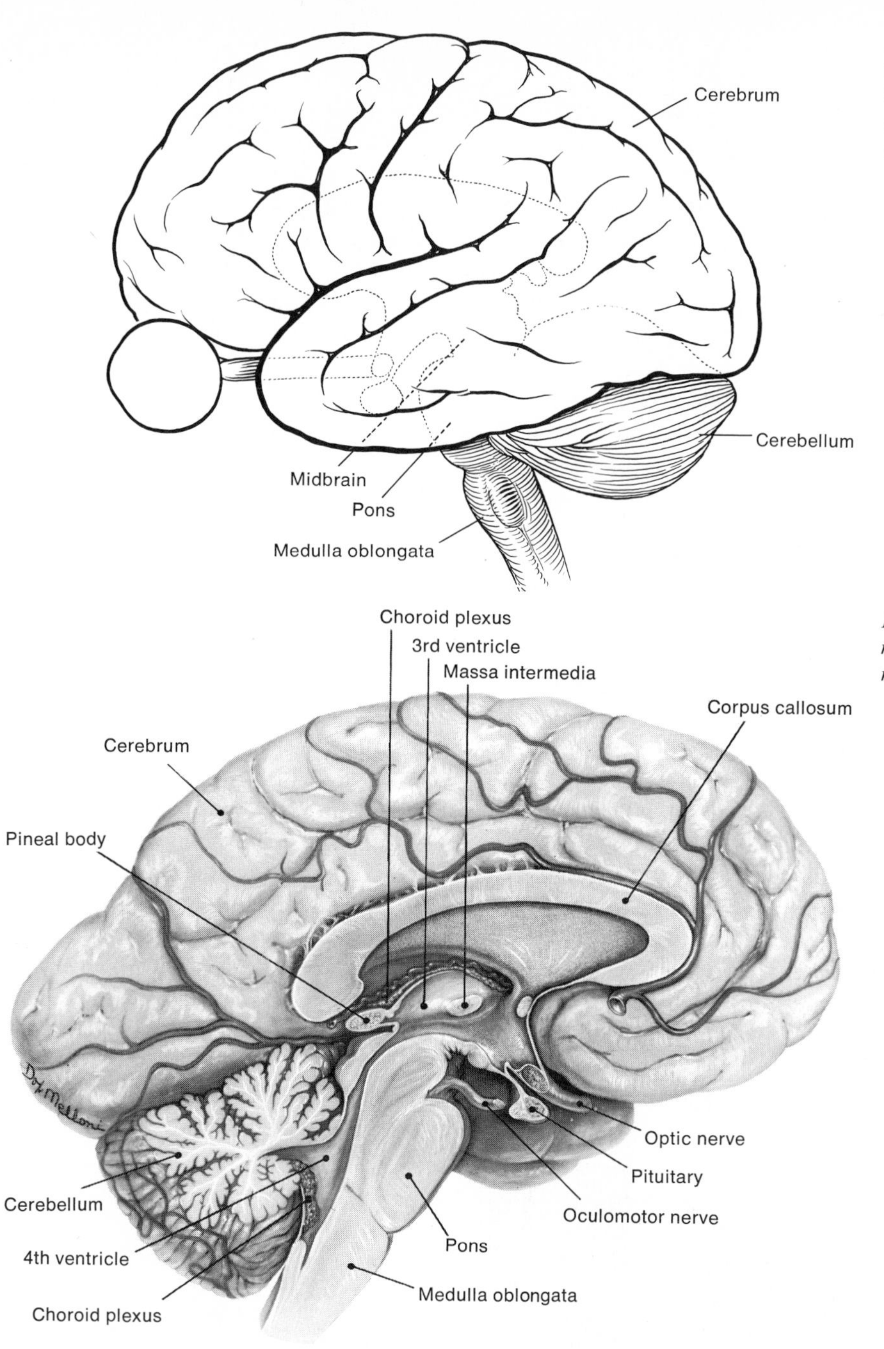

Figure 4-2. Lateral view of brain showing its three major divisions: cerebrum, cerebellum, and brain stem. The brain stem is comprised of the midbrain, pons, and medulla.

Figure 4-3. Sagittal section through the midline of the brain showing detail of the medial surface of the left half of the brain.

The cerebral hemispheres also contain the lateral and third ventricles, which are part of the cavities found within the central nervous system.

Between the cortex and the basal ganglia is material referred to as the **white matter. Gray matter** throughout the nervous system is partly composed of the cell bodies of nerve cells, and the white matter is to a larger extent formed by the process of nerve cells.

Cerebral Cortex

The gray matter of the **cerebral cortex** is divided into **motor, sensory,** and **association areas.** These areas have been studied by direct stimulation of the cortex electrically and by the observation of the responses elicited from experimental subjects. Much information has been gathered by observation of subjects with various injuries or lesions of the cortical regions and by removal of parts of the cortex in experimental animals. Such studies, referred to as *mapping out of the brain,* have led to the numbering of the cortex in areas from 1 to 52.

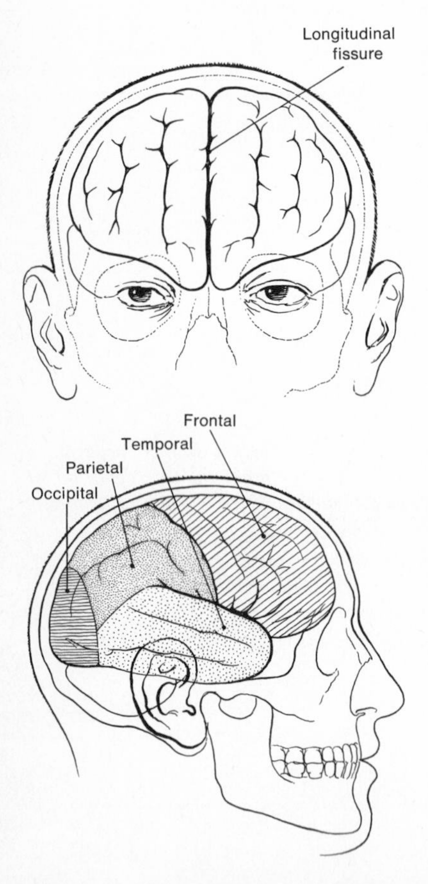

Figure 4-4. (Above) *Cerebral hemispheres in situ showing the division of the various lobes.*

Motor Cortex. The **motor cortex** has been divided into subdivisions, of which areas 4, 6, and 8 have been most extensively investigated (Fig. 4-9). If area 4 is explored with a stimulating electrode, it is found that the upper or dorsal part is associated with movements of the lower extremities of the body. The middle portion of area 4 is associated with movements of the upper extremity, and stimulation of the lower, or ventral, part of the area results in movements of cranial muscles and occasional vocalization. The motor cortex contains cells whose large axons connect with those motor neurons of cranial and spinal nerves that innervate skeletal muscle, and electrical stimulation of this area produces discrete muscular response. Stimulation of area 6 gives rise to more complex coordinated movements, which are frequently similar to those seen in supporting and postural mechanisms. Area 8 is primarily concerned with the control of eye movements.

Sensory Cortex. The **sensory cortex** is concerned with the interpretation of sensory impulses and is located mainly in the postcentral convolution (Fig. 4-10). This area is divided into a large number of sensory foci, each of which when stimulated produces sensory impressions on a special segment of the limbs or a special part of the body. The sensory areas of the postcentral convolution are designated areas 1, 2, and 3. The receptive areas for vision are in the occipital lobes and are numbered areas 17, 18, and 19. The receptive areas for hearing are in the temporal lobes and are numbered areas 22 and 41.

Association Cortex. The **association cortex** occupies the greater part of the lateral surfaces of the occipital, parietal, and temporal lobes and of the frontal lobes

Figure 4-5. (Right) *Lateral surface of cerebrum showing the fissures, sulci, convolutions, and gyri.*

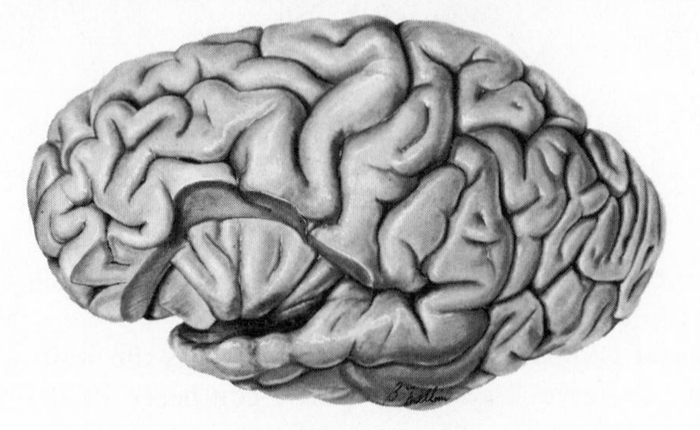

anterior to the motor areas (Fig. 4-11). These areas are concerned with the emotional and intellectual processes involving memory, reasoning, and judgment. The **temporal association areas** are numbered 35, 36, and 37, and they are believed to have the integrating function whereby we recognize body image, individuality, and continuity of the personality and self in relation to the environment. The

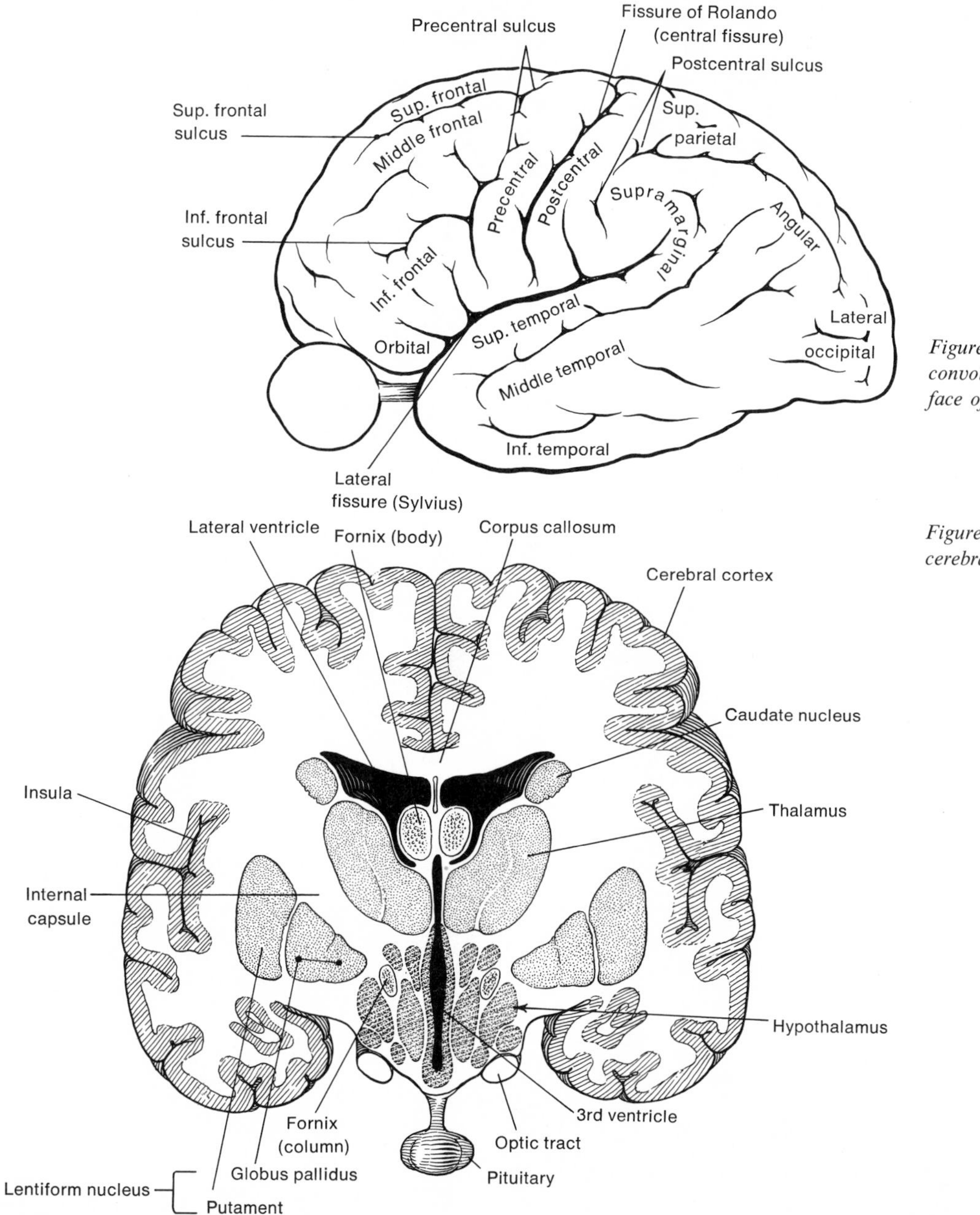

Figure 4-6. Principal fissures, sulci, and convolutions of the cerebrum. Lateral surface of left cerebral hemisphere.

Figure 4-7. Frontal section through the cerebral hemispheres (coronal slice).

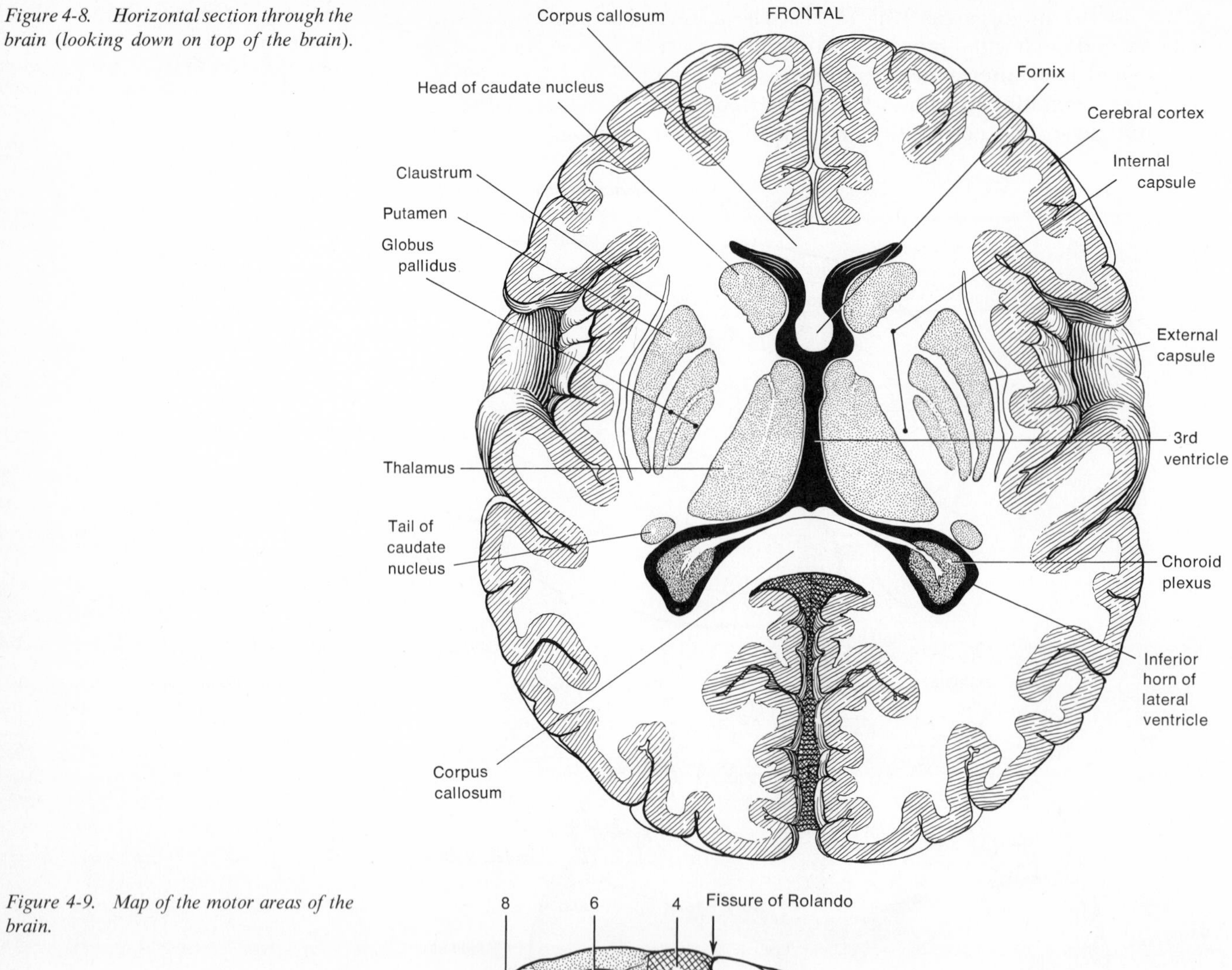

Figure 4-8. Horizontal section through the brain (looking down on top of the brain).

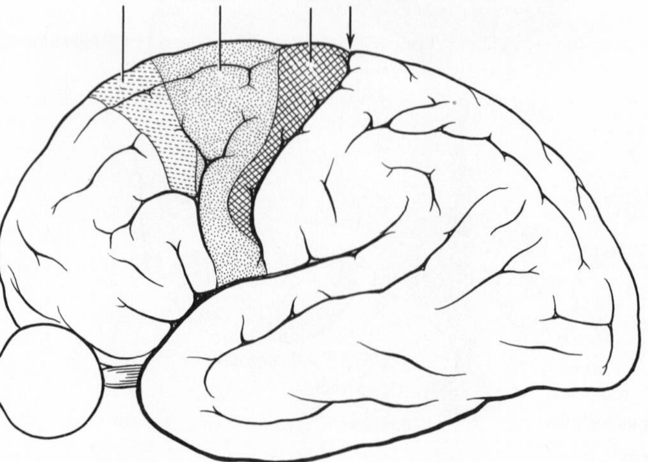

Figure 4-9. Map of the motor areas of the brain.

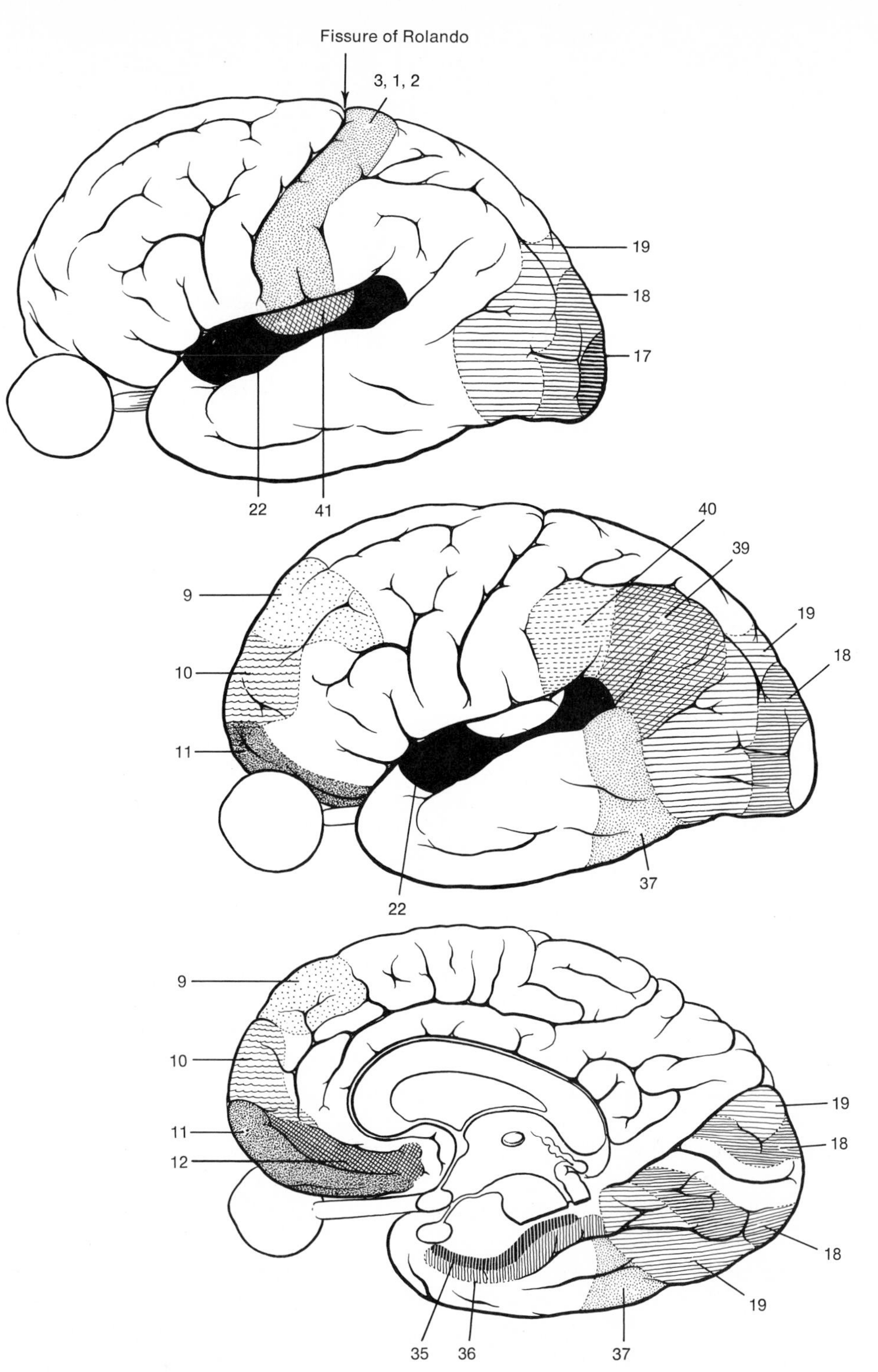

Figure 4-10. Lateral view of left cerebral hemisphere showing the location of the various sensory areas. Receptive areas for sensory impressions on special segments of the body are located in areas 1, 2, and 3. Receptive areas for vision are 17, 18, and 19 for hearing, 22 and 41.

Figure 4-11. Association areas of the cortex. (Top) *Lateral view of left cerebral hemisphere.* (Bottom) *Medial view of right cerebral hemisphere.*

parietal association cortex is divided into area 40, which is concerned with organization of thought, and areas 22 and 39. Area 22 is functionally complex. It is, in part, the auditory psychic area where spoken speech is understood; another part is a musical memory area, where, on stimulation, the subject is caused to hear tunes or songs, often forgotten ones. A third part of area 22 is concerned with speech. Injuries in this area prevent the subject from forming grammatically correct sentences. In mild forms, prepositions and connectives are omitted. Area 39 is in the posterior half of the inferior parietal lobule. Lesions in this area will result in the subject's inability to call common objects by name. Common symbols, like denominations of coins, numbers, maps, and musical scores, are meaningless, and although thoughts and feelings are unaffected, the subject cannot express them due to a loss of words. The **occipital association areas** include areas 18 and 19 and are known as the visual association area and the psychic area, respectively. Stimulation of area 18 produces complex visual hallucinations, and if it is destroyed, the subject does not recognize individuals, objects, or food. Area 19 receives and combines recognized visual images from area 18 and sends extensive associations forward to be combined with images from other senses.

The **frontal association areas** include areas 9, 10, 11, and 12. These areas function to conceive and will the motor acts; they permit man to deliberate on sensory data, intellectualize, and rationalize. The regions of the cortex already discussed, parietal, temporal, occipital, and frontal, are best known because they are most accessible and most significant. Other regions of the cortex, such as the insula, the cingulate region, and the olfactory, are less accessible for study and, therefore, less understood.

Basal Ganglia

The interior of the cerebral hemispheres is composed partly of white matter and partly of well-defined areas of gray matter known as the **basal ganglia.** These ganglia are arranged medially, forming the lateral walls and floor of the ventricles (Figs. 4-7 and 4-12). They include the following structures.

Corpus Striatum. The **corpus striatum** structure includes the caudate nucleus, which forms the side walls of the first and second ventricles, and a lateral group of gray material composed of the **putamen** and **globus pallidus.** These two groups are connected by a portion of white matter known as the **internal capsule.** The putamen and globus pallidus together form a lens-shaped structure and are referred to as the **lentiform nucleus.** The main function of this entire unit is to exert a steadying effect on voluntary movement.

Thalamus. The pair of oblong-shaped masses of gray matter that bounds the third ventricle laterally is the **thalamus** (Figs. 4-13 and 4-16). The thalamus acts as the central relay station in a complex network of interconnections between the cortex and the spinal cord. All sensory input, except for olfaction, converges on the thalamus, where the sensory input is analyzed before being relayed to the appropriate areas of the cortex. All connections between thalamus and cortex are reciprocal, that is, all areas of the cortex which receive fibers from the thalamus send fibers back to it.

Cortical excitation is always accompanied by spontaneous activity of the thalamus; thus the cortex and thalamus function together. In man the thalamus is also a center of primitive, uncritical sensation. Electrical stimulation of the thalamus gives rise to an exaggerated sensation of pain or pleasure. An individual whose sensory cortex had been destroyed would find that the thalamus would make mildly unpleasant sensations unbearable while pleasurable sensations would become ecstatic. Laboratory experiments have shown that the thalamus apparently also has something to do with creating subjective feelings.

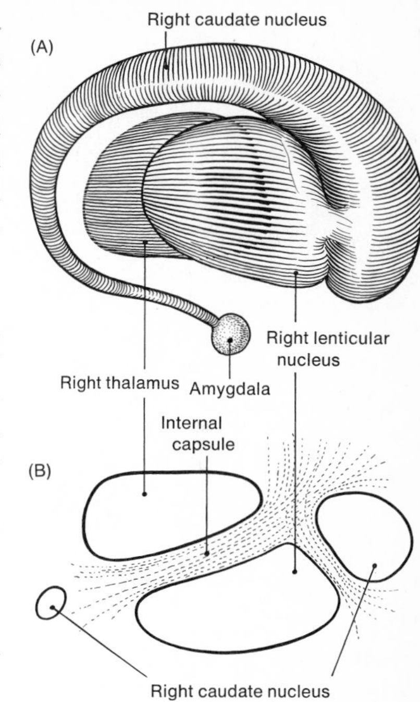

Figure 4-12. Basal ganglia isolated from the cerebral hemispheres. A. *Lateral view of the right basal ganglia.* B. *Horizontal section of the right basal ganglia. The tail of the caudate nucleus terminates at the amygdaloid nucleus.*

Hypothalamus. Below each thalamus, and forming the floor of the third ventricle, is the **hypothalamus** (Fig. 4-14). This area of gray matter is supplied by sensory fibers from the cerebral cortex, the thalamus, and the brain stem and also sends motor fibers to the thalamus, the brain stem, and the spinal cord. Electrical stimulation of this area in man is said to cause a decrease in heartbeat and to produce sleep, suggesting that this area, as well as the postulated sleep center, is a cardioinhibitory center. It has been demonstrated that the hypothalamus plays an important part in the regulation of water, fat, and carbohydrate metabolism. The temperature-increasing portion of the heat-regulating center is located here, as shown when ablation of this area in man produces hypothermia. The hypothalamus is the chief guardian of the body's basic well being. It continually monitors such vital functions as body temperature, heart rate, and blood pressure. In times of stress it prepares us for action by accelerating the heartbeat, increasing blood sugar and coagulant levels of the blood, and inhibiting the digestive process to conserve blood for the muscles. The hypothalamus also may be a source of many of our emotional reactions to the world around us. Electrical stimulation of the hypothalamus produces a whole range of emotional responses from serenity to anxiety and terror.

It should be pointed out that although the thalamus and hypothalamus have

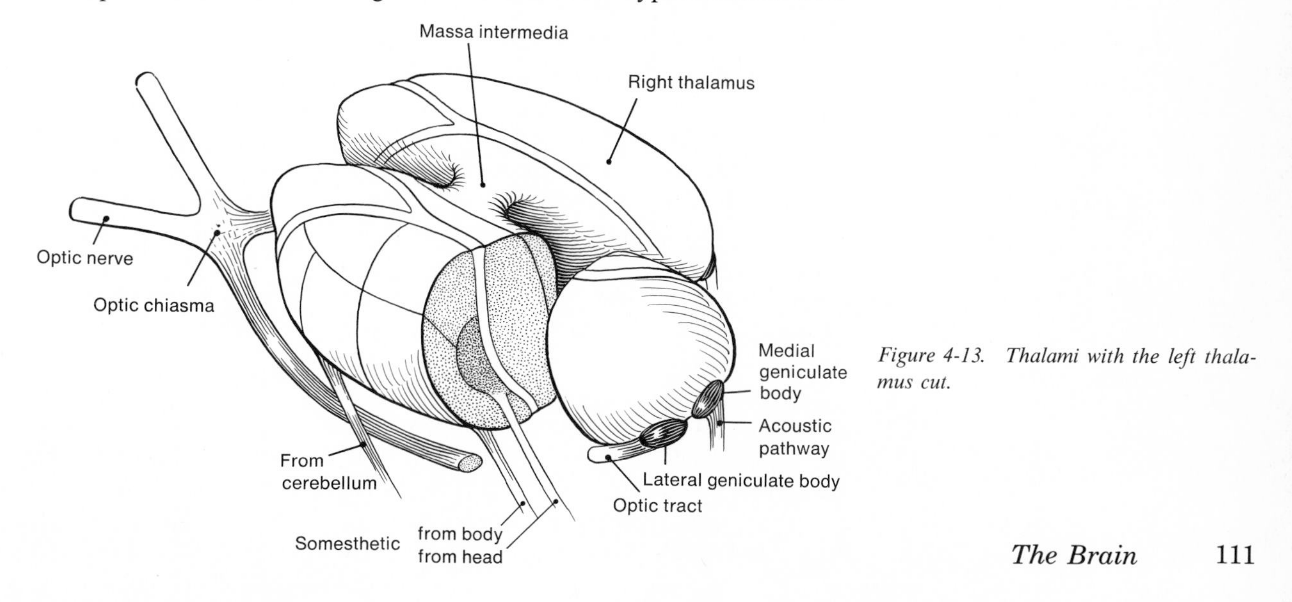

Figure 4-13. Thalami with the left thalamus cut.

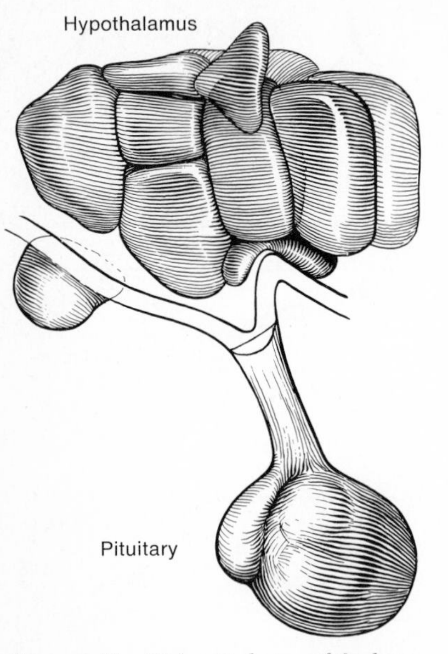

Figure 4-14. Midsagittal view of the hypothalamus in relation to the pituitary.

been included in the group referred to as the *basal ganglia,* they are often excluded because of marked functional differences between these two and the other masses of gray matter.

Amygdala. The gray material positioned at the tip and the rostromedial and rostrodorsal walls of the inferior horn of the lateral ventricle and which is more or less fused with the tail of the caudate nucleus is the **amygdala** (Figs. 4-12 and 4-15). Stimulation of this structure produces face and jaw movements and various visceral effects. These visceral effects are represented by a wide range of responses, such as pupillary changes, piloerection, changes in temperature and in blood pressure, increase in motility of the gastrointestinal tract, salivation and defecation, changes in respiration, and a gagging response. Fear and rage reactions have also resulted from stimulation of the amygdala. Bilateral destruction has produced drowsiness, indifference to surroundings, psychic blindness, and loss of appetite. It has been suggested that one function of the amygdaloid complex would be to reinforce cortical activity, to keep the individual awake, and to associate him with his environment. The amygdala becomes very active when one encounters anything new or unexpected and can also amplify the intensity of our emotional expression.

Other Important Structures of the Cerebral Hemispheres. In addition to the basal ganglia just discussed, there are other cerebral regions which share with these structures functional relationships out of which come patterns of behavior that are modifiable by situations and are accompanied by autonomic adjustments which adapt the individual to changes in both external and internal environments.

An area on each side of the cortex consisting of a band of a special type of cortex resembling a sea horse in general configuration is the **hippocampus** (Fig. 4-15). The function of the hippocampus is closely related to those of the temporal cortex and the amygdala as a result of their rich interconnections. There is evidence of an "arousal" effect of the hippocampus, which can be elicited experimentally by olfactory, auditory, or somesthetic stimulation. A monkey under anesthesia can be aroused by stimulation of the hippocampus but will go back to sleep after the stimulus ceases. Various investigators have regarded the hippocampus as an area that is essential for recent memory. Patients with large bilateral lesions in the area remember clearly events in their lives that occurred many years ago but show an inability to remember anything recent for more than a few moments. The hippocampus is also regarded as a regulator of hypothalamic centers concerned with emotional responses (anger or fear), as the behavior of patients having this structure removed because of disease demonstrate a lack of control of these emotions.

The **epiphysis** or **pineal gland** is a small oval mass of nerve cells located deep in the cerebral hemispheres just above the brain stem (Fig. 4-3). It is now considered to be the area of the brain which represents the *biological clock* that synchronizes the body's rhythmic changes with day and night cycles. It is responsive to changes in light, getting its information about light from the eyes. One of its hormones, **melatonin,** is known to cause concentration of the pigment

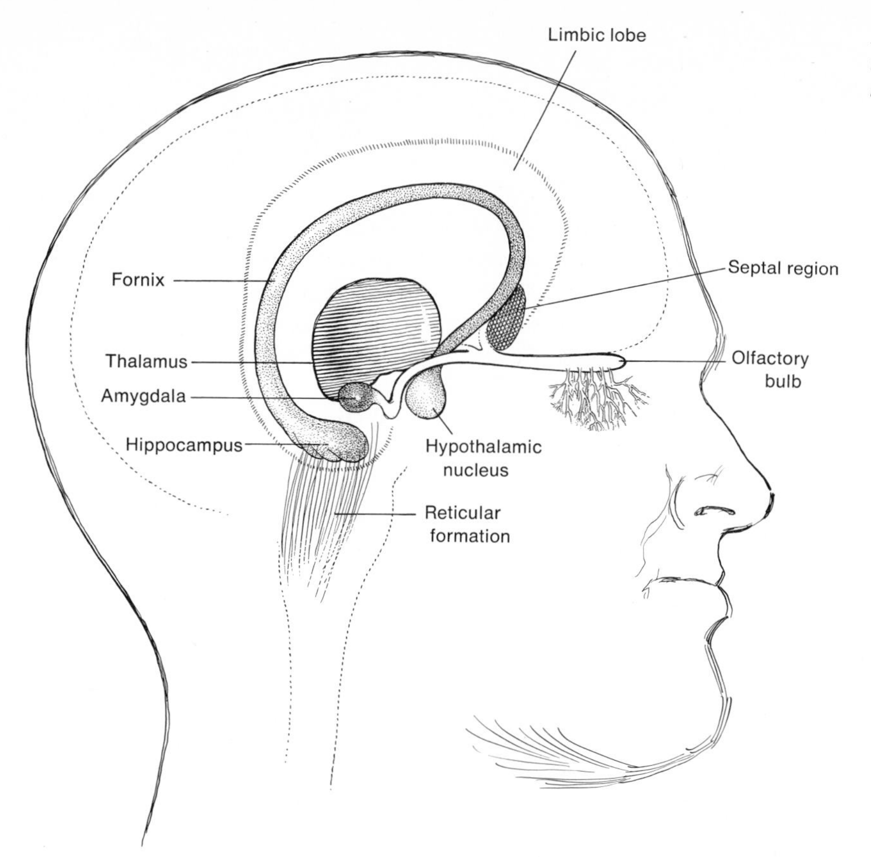

Figure 4-15. Cerebral structures that share important functional relationships with the basal ganglia.

granules in the skin of amphibians. Its function in mammals is unknown, but it has been suggested that melatonin may be a chemotransmitter, just as are acetylcholine and norepinephrine. Another hormone, **seratonin,** has been found to be present in relatively large quantities in the pineal gland. Because of the presence of this compound, it is thought that the gland may function as an antistressor organ, opposing the effect of ACTH produced by the pituitary gland.

Endocrinological experiments indicate that the function of the pineal gland is also in some way related to gonadal function. It also seems to control sexual cycles in animals. Rats kept in a lighted environment of constant intensity remain in heat continuously. In humans it probably assists the pituitary in controlling gonadal function, either by inhibiting the secretion of pituitary gonadotropin or it may lessen the sensitivity of the gonads to gonadotropin.

On the posterior surface of the midbrain just above the brain stem are four rounded eminencies referred to as **colliculi** (little hillocks). The upper two are the superior colliculi, and the lower are the inferior colliculi. The superior colliculi are all that remains of the optic lobe, which in more primitive animals is the most prominent part of the brain. In humans this area controls the visual reflexes—blinking, the aperture of the pupil, and the focus of the lens. The inferior colliculi adjust the ear to the volume of sound and probably reflexly make us jump at a loud noise.

The most medial portions of the frontal and temporal lobes of the brain involving the inferior frontal portion of the cortex and the associated deep nuclei in these regions (olfactory bulb, basolateral nuclei of the amygdalae, hippocampal structures), have been classically called the **rhinencephalon** or **smell brain** because of their association with olfaction; the olfactory tracts terminate in this area. These areas also have direct and indirect connections with the hypothalamus and thalamus and are intimately concerned with the emotional responses, instincts, and complex neuroendocrine regulatory functions in mammals and man. For this reason Broca introduced the term **limbic lobe** or **limbic system** for this part of the brain (Fig. 4-15). The major role of the limbic system is to recognize any upset in the state of physical and psychological equilibrium which the individual may encounter and aids in bringing about reactions that will help re-establish this state of equilibrium. The limbic system generates feelings that strongly influence our behavior. By modifying our thoughts with such emotions as pleasure, anger, and fear, it guides us to do what is most likely to aid us in survival, whether it be to run from what is dangerous or to seek out that which is pleasurable.

White Matter of the Cerebral Hemispheres

Generally, nerve fibers with similar connections and functions are grouped together in bundles. When appearing among nonneural structures of the body, these bundles of nerve fibers are called **nerves,** but when they appear within the brain and spinal cord, they are called **tracts.** Since nerve fibers are usually surrounded by a sheath of myelin, these tracts have a whitish appearance and are termed *white matter*. The white matter of the cerebrum is divided into three main tracts, which are referred to as the association, commissural, and projection tracts.

Association Tract. The **association tract** is made up of nerve fibers that connect adjacent convolutions and distant convolutions of cortex of the same hemisphere.

Commissural Tract. The **commissural tract** connects the two hemispheres with one another. The principal tract is known as the **corpus callosum** (Fig. 4-7).

Projection Tract. The **projection tract** is made up of nerve fibers that connect the cerebral cortex to other parts of the central nervous system.

Cerebellum

The **cerebellum** appears to be a miniature of the cerebral hemispheres attached to the brain stem by three pairs of massive bundles of nerve tissue called **peduncles** (Fig. 4-16). Like the cerebrum, it has a cortex of gray matter and an interior of both gray and white matter (Fig. 4-17). The cerebellum is grossly divisible into two lateral hemispheres and a midline connecting area, the **vermis.** Buried within the white matter of the cerebellum is the **dendate nucleus,** the most important nuclear mass of the cerebellum. The peduncles are connecting pathways between the cerebellum and the rest of the central nervous system. The inferior peduncle acts as a connecting link between the cerebellum and the cells of the brain stem and spinal cord. The middle cerebellar peduncle carries impulses from the gray matter of the pons to the cerebellum, and the superior peduncle carries impulses from the dendate nucleus to the thalamus, where they are relayed to the motor cortex of the cerebrum. Thus, there are intricate connections between the cerebellum and the cerebral cortex. The function of this structure is to act as a reflex center, through which coordination and refinement of muscular movement are brought about. The cerebellum appears to coordinate impulses from sensory organs concerned in equilibrium with accurate muscle contractions to maintain equilibrium. Lesions of this area cause disturbances of rate and force of voluntary movements, producing such effects as volitional tremor, abnormal grasping, inaccurate movement of the arm, and difficulty in making rapidly alternating movements. In Figs. 4-16 and 4-18 are identified the areas of the cerebellum discussed above.

Brain Stem

Issuing from the base of the cerebral hemispheres is a massive bundle of nerve tissue referred to as the **brain stem.** It is common practice to divide the brain stem into three areas: The **midbrain,** or **mesencephalon,** is that portion that is directly connected to the base of the brain and continues inferiorly into a massive, rounded structure known as the **pons.** This, in turn, gives way to a portion known as the **medulla oblongata,** which is directly continuous with the spinal cord at the foramen magnum.

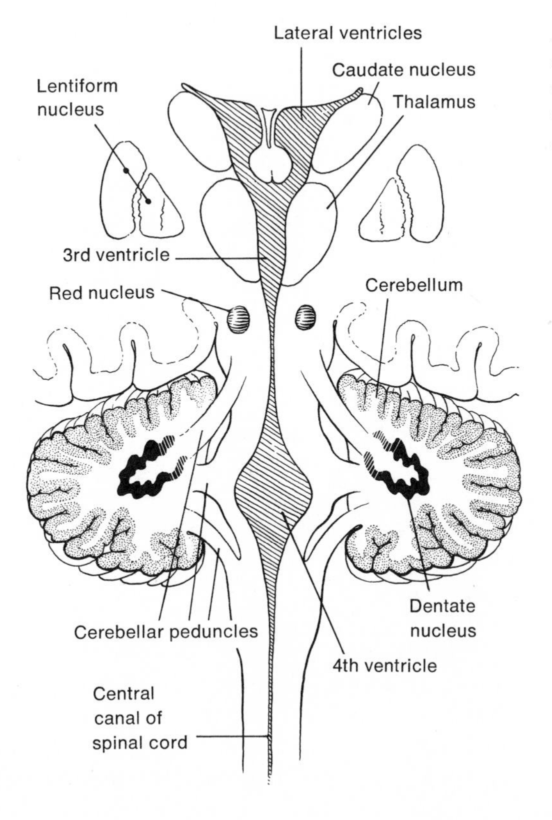

Figure 4-16. Schematic representation of cerebellum oriented to ventricles and various nuclei. The cerebellar peduncles, which attach the cerebellum to the brain stem, are shown also.

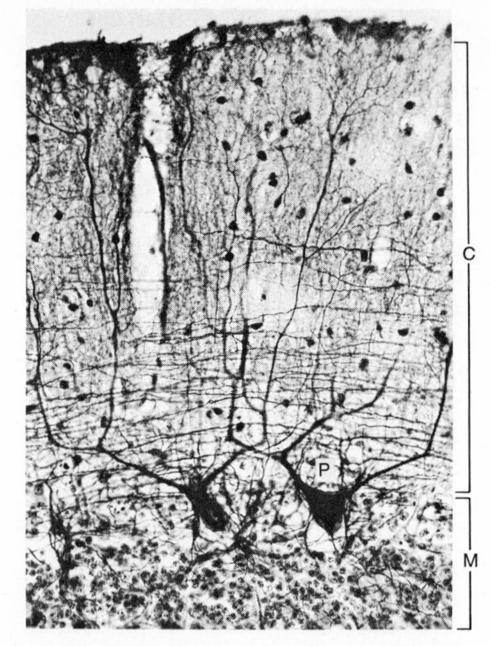

Figure 4-17. Detailed structure of mammalian cerebellum (sagittal section). C, *cortex;* M, *medulla;* P, *Purkinje cell.* (*Courtesy Ward's Natural Science Establishment, Inc., Rochester, N.Y.*)

The brain stem may be considered an appartus concerned with special senses, vital processes, and other visceral and somatic functions, all of which may be modified by impulses entering from the cerebellum and the cerebral cortex. Regulating all this sensory information coming into the brain is a cone-shaped diffuse network of neurons centered in the brain stem called the **reticular activating system (RAS).** The RAS (Fig. 4-15) decides what sensory information, if any, to pass on to our conscious minds, thereby allowing us to focus our thoughts amid the constant barrage of impulses from our senses. It is a filtering system for afferent information and a modulator of generalized ongoing activity. For example, the reticular system receives sensory information from the thalamus which comes in from the eyes, ears, and gustatory areas and which it, in turn,

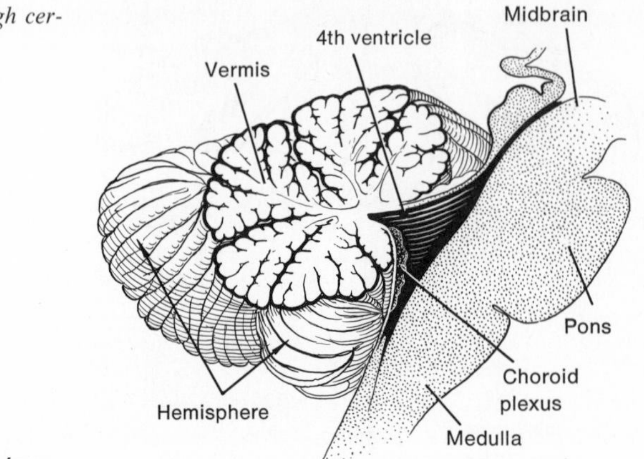

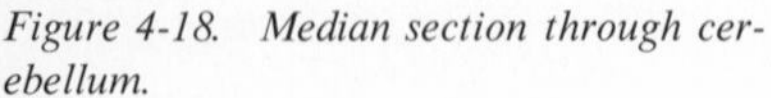
Figure 4-18. Median section through cerebellum.

may modify by increasing or decreasing its intensity before passing it on to the proper cortical areas. When the cortex receives sufficient sensory impulses, it is said to be aroused or activated or a state of consciousness and alertness is produced. In addition to screening sense information the reticular system with its interconnections with the cortex can produce cortical wakefulness even if the activity of the sensory systems is not in itself strong enough to do so. If neural activity in the RAS drops below a certain level, sleep is produced. This indicates the importance of the brain stem as a regulator of states of consciousness. The pons and the medulla contain the fourth ventricle, which is continuous with the third ventricle by a canal, the **aqueduct of Sylvius,** which is positioned in the substance of the midbrain. In the gray substance of the pons and medulla, below ventricle IV, is the center for control of respiration. Also located in the medulla are the defecation, deglutition, emetic, vasomotor, and cardiac centers. Table 4-1 summarizes all the major structures of the brain and their functions.

Table 4-1. ***Summary of Brain Function***

Cerebrum	*Component*	*Function*
Cerebral cortex	Motor area	Initiates voluntary muscular movement
	Sensory area	Interpretation of sensory input giving rise to sense of smell, taste, sight, hearing, equilibrium, and touch
	Association area	Concerned with the emotional and intellectual processes involving memory, reasoning, and judgement
Basal ganglia	Corpus striatum Caudate nucleus Putamen Globus pallidus	Exerts steadying effect on voluntary movement
	Thalamus	Central relay station for sensory input to the cortex; area responsible for subjective feelings and primitive uncritical sensation
	Hypothalamus	Monitors such vital functions as body temperature, heart rate, and blood pressure; regulates water, fat, and carbohydrate metabolism; also, a source of many of our emotional reactions
	Amygdala	Reinforces cortical activity; regulates visceral activity and amplifies the intensity of our emotional expression
Deep nuclei	Hippocampus	Arousal or activation of the cortex; area for recent memory and regulator of hypothalamic emotional (anger, fear) responses

(continued on next page)

Table 4-1. (*Continued*)

Cerebrum	*Component*	*Function*
	Pineal gland	Synchronizes body's rhythmic changes with light cycles; secretes melatonin and seratonin; controls sexual cycles in animals and probably gonadal function in man
	Colliculi Superior colliculi	Controls visual reflexes—blinking, the aperture of the pupil, and the focus of the lens
	Inferior colliculi	Adjusts ear to volume of sound
	Limbic system	Recognizes upsets in physical and psychological equilibrium; acts to re-establish equilibrium
Brain stem	Reticular activating system (RAS)	Filtering system for sensory information and a modulator for ongoing activity; arousal center
	Pons	Contains pneumotaxic center, which inhibits activity of inspiratory center; also, contains motor nuclei of facial, abducens, and trigeminal nerves as well as sensory nuclei of trigeminal nerves; it is structurally and functionally and upward continuation of the medullary RAS
	Medulla	Center for control of respiration; also, centers of control for cardiac activity, defecation, deglutition, and vasomotor control
Cerebellum	Neocerebellum (bulk of cerebellum), so called because it is the latest portion to have developed in the evolutionary scale	Important in proper timing of impulses from cortical motor areas during voluntary activity; coordination of muscular activity; aids in maintaining proper muscle tone
	Paleocerebellum (most of vermis and a part of the cerebellar hemispheres)	Modified and regulates postural tone; capable of inhibiting any or all stretch reflexes

Electroencephalogram

Since all active tissue gives rise to fluctuations in electrical potential, in the continuous activity of the cerebral cortex these variations of electrical potential can be picked up and recorded, as first shown by Hans Berger in 1929, by placing electrodes on the scalp and properly amplifying the electrical potential developed. The usual record **(electroencephalogram)** shows waves that appear at different frequencies and amplitudes, and these waves are classified according to their rate,

Figure 4-19. Recording electroencephalograms. Technician sits at the console of the electroencephalograph recording the electroencephalogram of a young patient. Both normal and abnormal tracings are shown.

which varies from 6 to 50 per second. These waves are arbitrarily divided into three groups, depending on this rate, and are referred to as *alpha, beta,* and *delta waves.* The alpha waves occur at the rate of 10 to 12 per second. The beta waves occur at a faster rate and are usually superimposed on the alpha. The beta waves have a rate of 16 to 24 per second. Waves slower than 5 per second are delta waves and are found in the normal person only during sleep. Figure 4-19 illustrates electroencephalogram tracings obtained from normal subjects.

Spinal Cord

The **spinal cord** is a massive nerve column continuous with the brain stem at the magnitude foramen and terminating at the upper border of the second lumbar vertebra. It is much shorter than the vertebral canal in which it rests. A cross section of the cord at any level will reveal a characteristic structure. Unlike the brain, the gray matter of the spinal cord is internally located—arranged in the form of a letter H, surrounded by white matter (Figs. 4-20 and 4-21).

The white substance of the cord is composed of parallel longitudinal bundles of myelinated fibers, which are grouped together into functional units called tracts (Fig. 4-22). The **descending tracts** carry impulses downward from the cerebral cortex to the motor cells in the anterior horn of the gray matter. This is the voluntary motor pathway by means of which we have conscious control over

Figure 4-20. Cross sections of the spinal cord at various levels show considerable variations in size and shape. The proportion of gray matter also varies and is much greater in the cervical and lumbar regions and greatest in the conus medullaris.

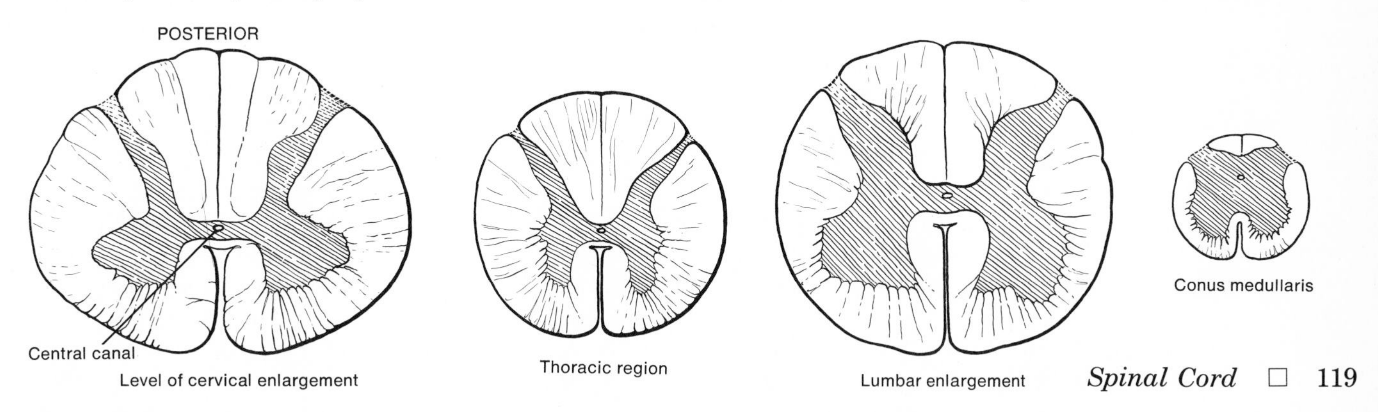

movements of the body that is, skeletal muscle. **Ascending tracts** carry sensory impulses up toward the brain. These fibers usually have their origin in the dorsal root ganglion of each spinal nerve. Some fibers whose cell of origin is in a dorsal root ganglion do not go all the way up to the brain but make connection with at least one neuron whose cell of origin is in the gray matter of the cord. This **intermediate,** or interneuron also makes connections with other neurons within the cord itself, thus making many spinal reflexes possible. A large part of the white matter is composed of **spinospinal tracts.** These are nerve fibers arising and terminating within the spinal cord, thereby linking various levels and providing coordinated activity.

The central bar of gray matter of the spinal cord is pierced by the central canal, which extends the entire length of the cord, entering into the fourth ventricle in the medulla. In the adult, the caliber of the canal is extremely small. The gray substance of the cord somewhat resembles the letter H. The upper arms of the letter H are referred to as the *dorsal horns,* and the lower arms are called the *ventral horns.* The nerve cells in the dorsal horn receive sensory impulses, which come in over the dorsal roots of the spinal nerves and relay these impulses to other parts of the central nervous system. The nerve cells in the ventral horn are motor in function, and their fibers form the ventral roots of the spinal nerves (Fig. 4-23).

Glial Cells

Many cells within the central nervous system, which are derived from the primitive spongioblast cells of the embryonic neural tube, function as supporting cells of the adult nervous system. Collectively these **glial cells** are known as **neuroglia,** and although they may differ in size and shape, they all have processes that weave around nerve cells and fibers (Fig. 4-24) and frequently attach to the walls of the blood vessels supplying nerve tissue. Special stains are necessary to bring out the structure of glial cells; by such techniques, it has been demonstrated that some neuroglia are star-shaped; these are, therefore, called **astrocytes.** Others are usually associated with the axons and cell bodies of neurons and are, therefore, called **oligodendroglia.** A third type of neuroglia, because of its small size in contrast to the other types, is referred to as **microglial cells.** Microglial cells, in addition to the function of support, can enlarge and become phagocytic, capable of engulfing nervous tissue after injury or destruction by disease. From a pathological viewpoint, neuroglial cells are important because they are the most common source of tumors of the nervous system.

Coverings of the Central Nervous System

The central nervous system is protected by a bony framework composed of the skull and vertebrae. If the skull and vertebrae are removed, we find that the brain and spinal cord are covered by a tough, fibrous membrane called the **dura mater** (Fig. 4-25). In certain areas the membrane contains blood channels, the **dural sinuses,** through which the venous blood flows on its way from the brain back to the heart. Below the dura mater is another membrane of connective tissue called the **arachnoid.** Internal to the arachnoid and very closely applied to the surface of the brain and spinal cord is a thin layer of connective tissue, the **pia mater.** These three membranes (Fig. 4-26) are collectively termed **meninges.** The

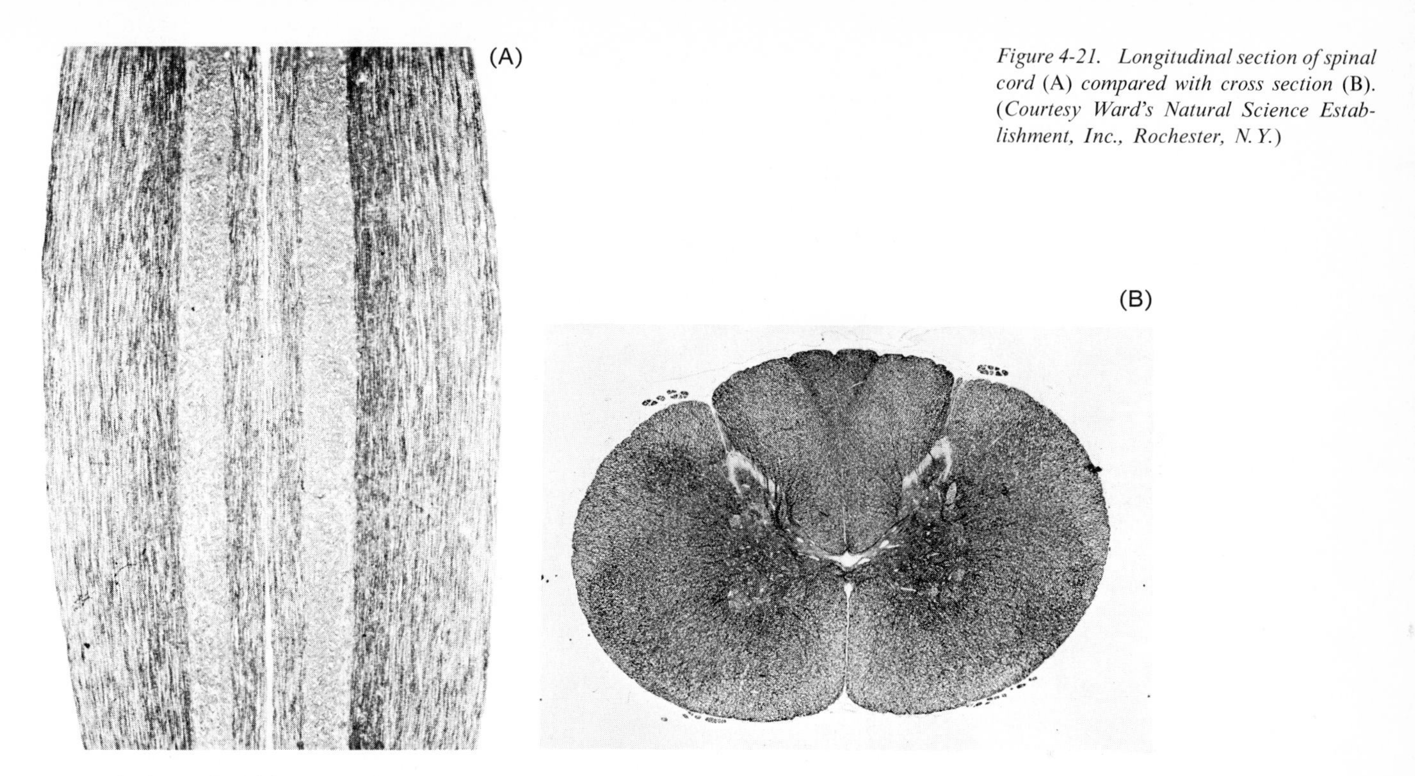

Figure 4-21. Longitudinal section of spinal cord (A) *compared with cross section* (B). *(Courtesy Ward's Natural Science Establishment, Inc., Rochester, N.Y.)*

space (subarachnoid space) between the arachnoid and the pia mater is filled with a clear, watery **cerebrospinal fluid.**

Ventricular System

The **ventricles** in the interior of the brain are filled with cerebrospinal fluid. The lateral ventricles communicate with the third ventricle by way of the **interventricular foramen** (foramen of Monro). The third ventricle communicates with the fourth ventricle, located in the pons and medulla, by a narrow channel, the **cerebral aqueduct,** or **aqueduct of Sylvius,** located in the midbrain. The fourth ventricle is directly continuous with a small channel, the **central canal,** which extends downward into the spinal cord. In the posterior wall of the fourth ventricle are openings, or foramina, that communicate with the subarachnoid space (Fig. 4-27).

Formation and Circulation of Cerebrospinal Fluid

All the ventricles of the brain are lined by a complex network of capillaries, the **choroid plexus.** Blood circulates through these vessels, and from this network is formed an almost protein-free fluid, which passes into the ventricles. This cerebrospinal fluid circulates from the lateral ventricles into the third ventricle and then into the fourth, where it is joined by more fluid formed in the latter two ventricles. The fluid then escapes into the subarachnoid space by way of the foramina, located in the posterior wall of the fourth ventricle, bathing the external surface of the brain and the spinal cord. The fluid is eventually absorbed into the blood stream through small villi, or arachnoidal tissue, which projects into the dural sinuses (Fig. 4-28).

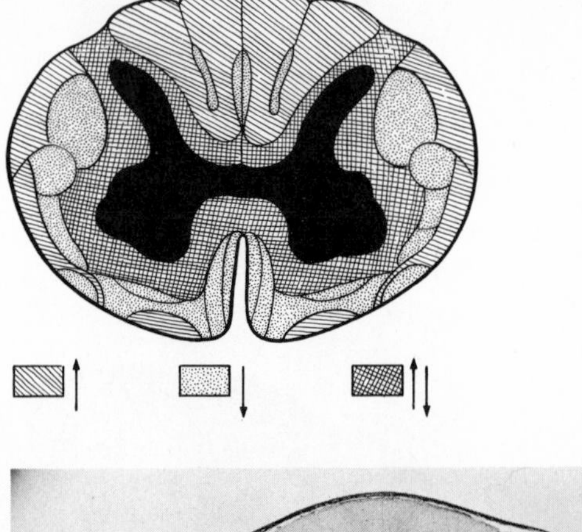

Figure 4-22. Ascending and descending pathways of the spinal cord. Intersegmental tracts (contain both ascending and descending nerve fibers) lie between the gray matter and the descending fibers.

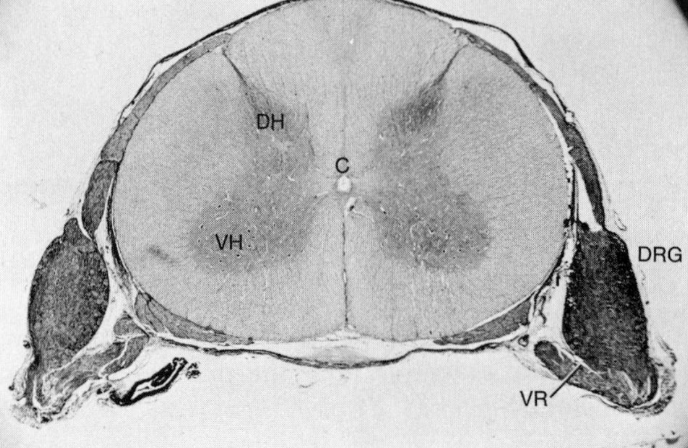

Figure 4-23. Cross section of spinal cord. C, *Central canal;* DRG, *dorsal root ganglion;* VR, *ventral root;* VH, *ventral horn;* DH, *dorsal horn.* (*Courtesy Ward's Natural Science Establishment, Inc., Rochester, N.Y.*)

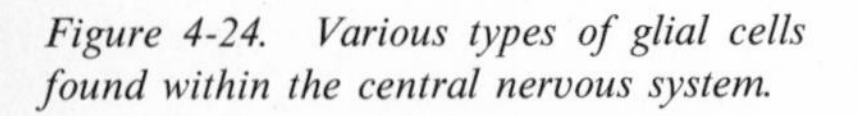

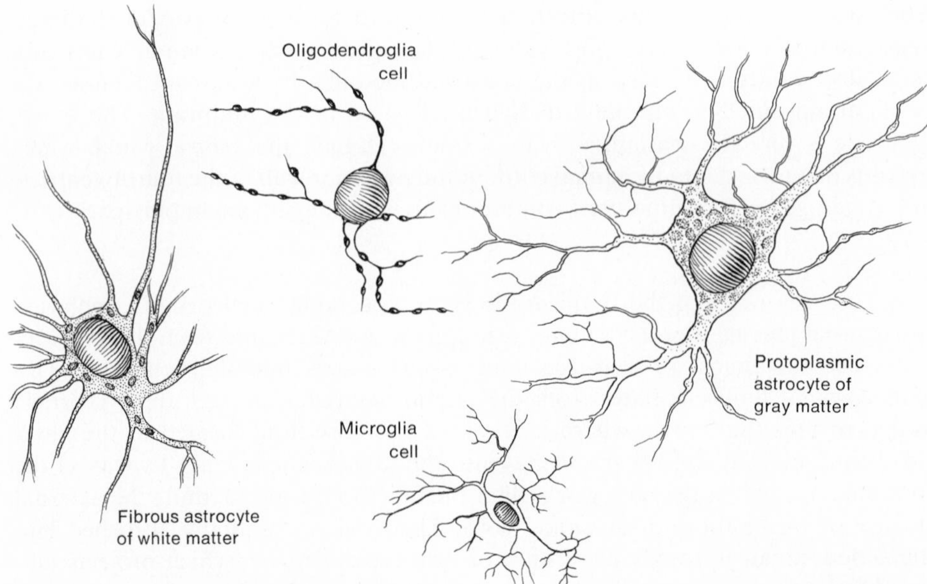

Figure 4-24. Various types of glial cells found within the central nervous system.

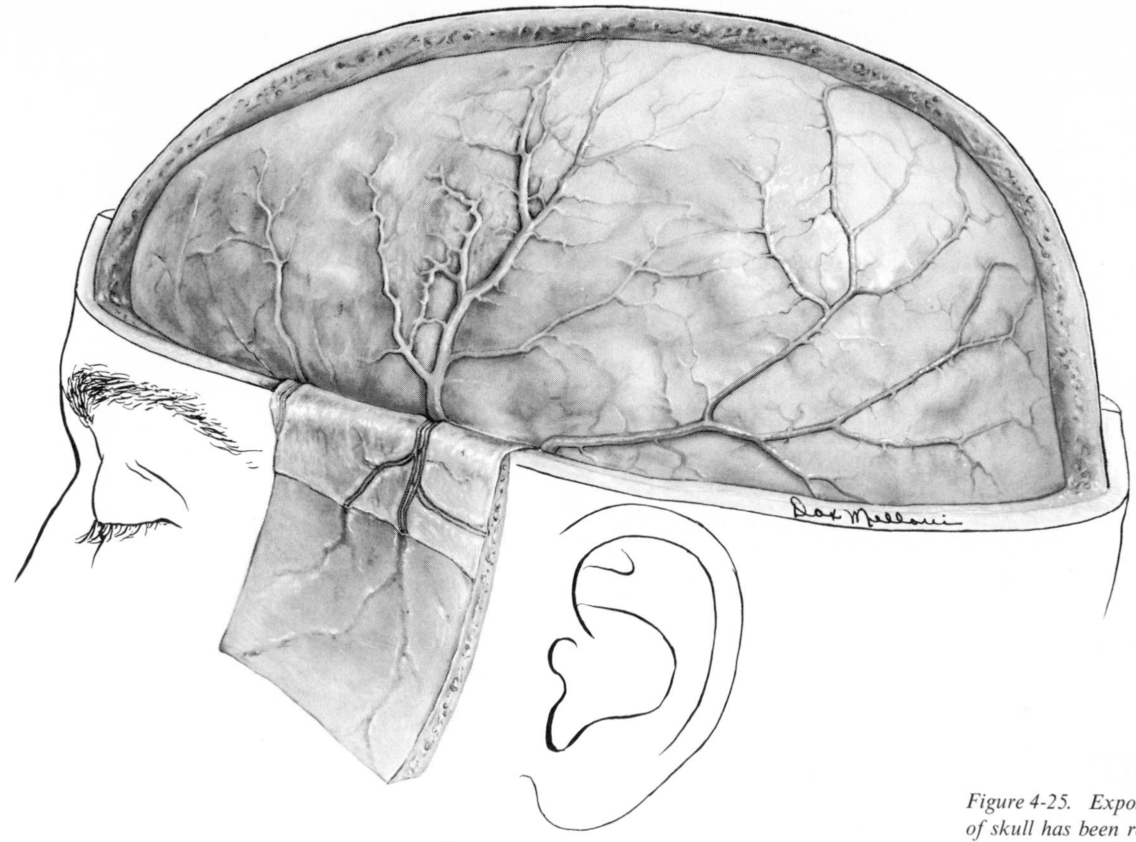

Figure 4-25. Exposed dura in situ. Section of skull has been removed.

The fluid is kept circulating in the direction indicated above because the blood pressure in the capillaries at the site of formation is higher than that at the site of absorption. Changes in the position of the body and changes in pressure with respiration also aid in driving the fluid from the ventricles into the subarachnoid space. The pressure of the fluid ordinarily amounts to 10 to 20 mm of mercury when the subject is in the horizontal position. Not all the functions of the cerebrospinal fluid are known, but since it completely surrounds the brain and spinal cord so that the central nervous system is literally floating in this liquid, it certainly serves a protective function. It also serves to nourish and remove metabolic wastes resulting from the activity of the brain and spinal cord.

Peripheral Portion of the Nervous System

The peripheral system is made up of those structures called nerves, which relate the central to all other parts of the body. Those that are connected to the brain are **cranial nerves;** those that are connected to the spinal cord are known as **spinal nerves.** Branches of both cranial and spinal nerves that supply muscles and skin make up the **somatic portion** of the peripheral system. The peripheral nervous system, in addition, includes a special division, the **autonomic nervous system,** which is composed of those branches of cranial and spinal nerves that supply visceral structures.

A **nerve** may be defined as a grossly visible collection of nerve fibers (neurons) with generally similar connections and function. Nerve fibers are very delicate, and therefore they must be supported by other tissues. Within a nerve, each fiber is wrapped with fine but tough connective tissue fibers called the **endoneurium.** Groups of these nerve fibers are surrounded by a distinct membrane similar to the endoneurium, but much thicker, the **perineurium.** Many of these bundles encased in their perineural sheaths are bound together by a still thicker membrane, the **epineurium** (Figs. 4-29 and 4-30). A nerve may be extremely large or quite minute, depending on the number of constituent nerve fibers.

Neuron

The **neuron** is the true structural and functional unit of the nervous system. All neurons consist of a cell body, which is the vital center of the cell, and one or more threadlike extensions, which may reach a length of several feet (Fig. 4-31). The extensions are terminal nerve fibers. The extensions that receive impulses and conduct them toward the cell body are **dendrites,** and those that carry them away from the nerve cell are **axons.** The dendrites in nearly every case are quite numerous and short, whereas the axons occur singly and are quite long. The term *nerve fiber* usually refers to these axons.

Such cells just described are said to be **multipolar,** and most nerve cells in the nervous system (motor neurons) are of this type. The exceptions to this type are the unipolar cells, found mainly in cranial and spinal (dorsal root) ganglia, whose

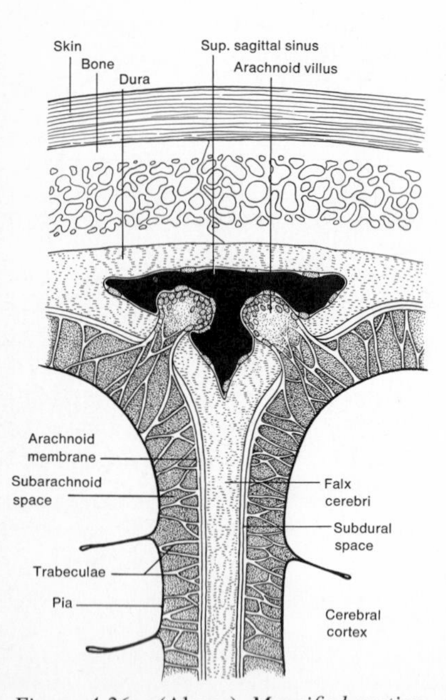

Figure 4-26. (Above) *Magnified section through a small portion of the skull and brain showing the location of the three membranes collectively termed the meninges. Detail of the arachnoidal villi projecting into the dural sinus is shown also.*

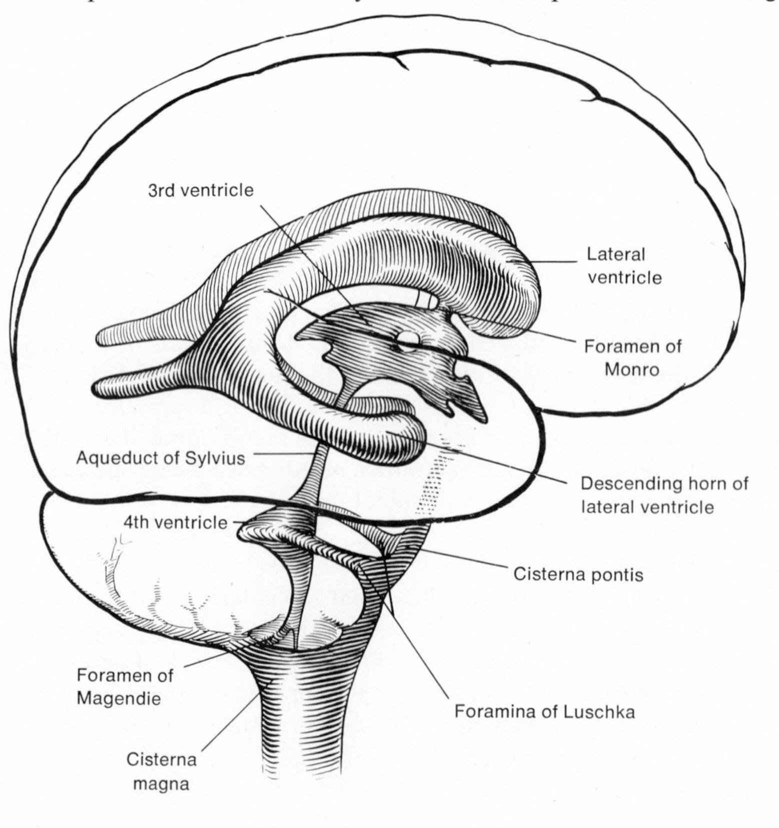

Figure 4-27. (Right) *Ventricles of the brain. Lateral view of cerebral hemisphere and cerebellum.*

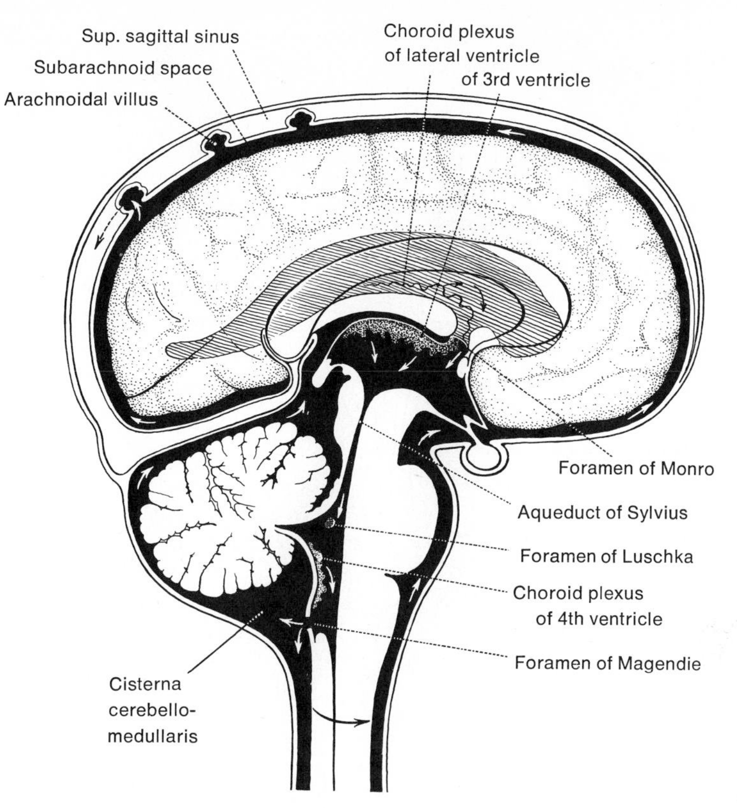

Figure 4-28. Sagittal section of the brain in situ showing the circulation of the cerebrospinal fluid from the ventricular system to the subarachnoid space. The fluid is eventually absorbed into the blood stream through the arachnoidal villi, which project into the dural sinuses. Also see Fig. 4-26 for detail of an arachnoidal villus.

single process which projects from the cell body divides into two branches. One of the branches projects toward the central nervous system, becoming the axon, the other toward the periphery, forming the dendrite (Fig. 4-32). Sensory neurons are of this type, having an extremely long dendrite and a relatively short axon. Figure 4-33 shows the pathways of such a neuron with a receptor in the skin and its termination in the spinal cord.

The nerve cell body and its extensions are limited by a double-contoured membrane about 100 to 150 Å thick. Few data are available on the structure of the nerve cell membrane, which is commonly called the neurilemma. Schmitt has postulated, by analogy with other cells investigated with the polarizing microscope, that the nerve cell membrane is composed of parallel leaflets of proteins, separated by a double layer of lipids.

Most peripheral nerve fibers are, in addition, surrounded by a fatty sheath of myelin arranged segmentally. Such a nerve fiber is referred to as being medullated. Brante, on analysis of the **myelin sheath,** found it to be a complex of protein and fat. He found that the phospholipid, cholesterol, and cerebroside components of myelin exist in a molecular ratio 2:2:1. The phospholipids have been assumed to interact with the proteins through their phosphate groups and to exist in a curled configuration. The cholesterol molecule appears to be a stabilizing factor, forming a complex with the phospholipid molecule. The cerebroside molecule

Figure 4-29. *Cross section of a large peripheral nerve. Individual nerve fibers encased in their endoneural sheaths are bound into bundles (fasciculi) by a connective tissue covering called the perineurium. Many bundles are bound together by the epineurium to form a nerve.*

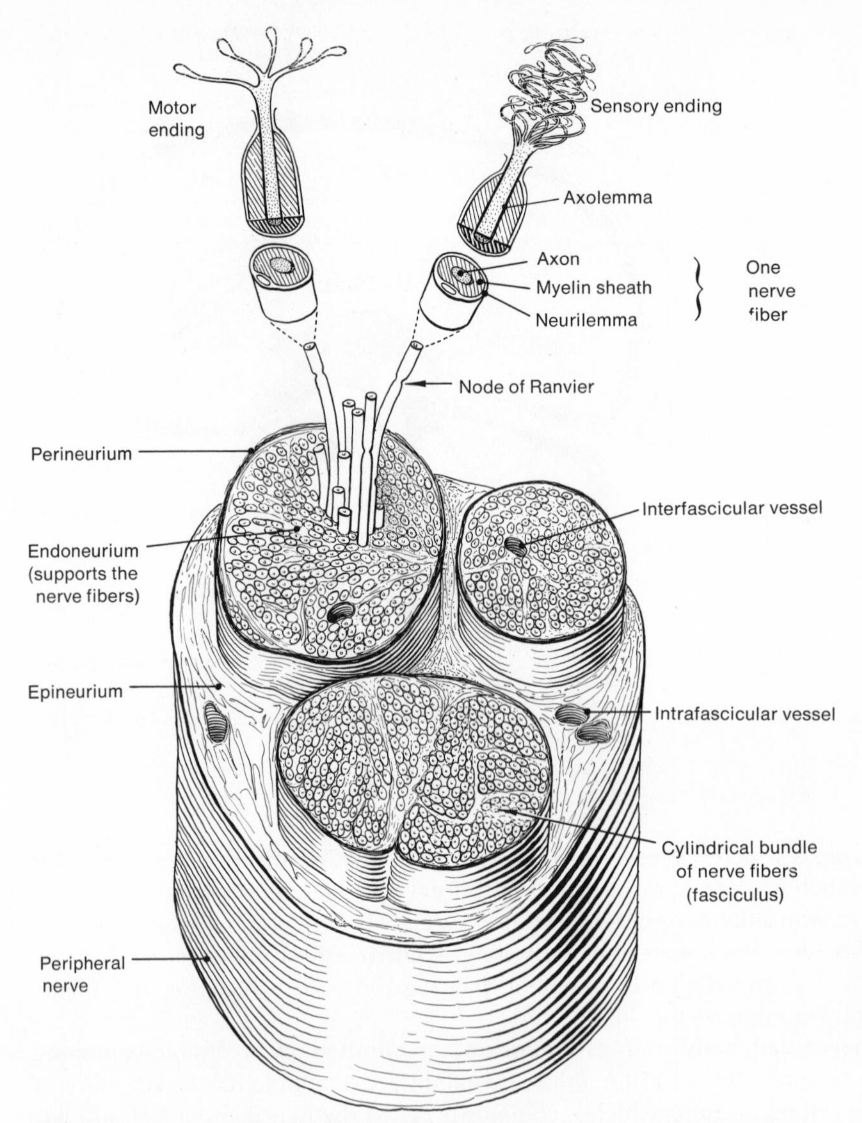

appears, from X-ray diffraction studies, to be located between two phospholipid–cholesterol complexes. The myelin sheath is not a continuous structure along the axon but is interrupted at intervals, showing a fundamental repeat length of about 171 Å. There seems to be a fundamental difference between myelin of central fibers and that of peripheral fibers. The myelin from central fibers seems to have a different lipoprotein association, in that central nerves contain considerably more proteolipid proteins than do peripheral nerves. In the electron micrograph the myelin of the central fibers also shows a repeating length of about 120 Å.

The myelin is surrounded by a thin sheet of cells called the **Schwann cells** (Fig. 4-34). It is believed that these thin cells are endowed with the capacity to give

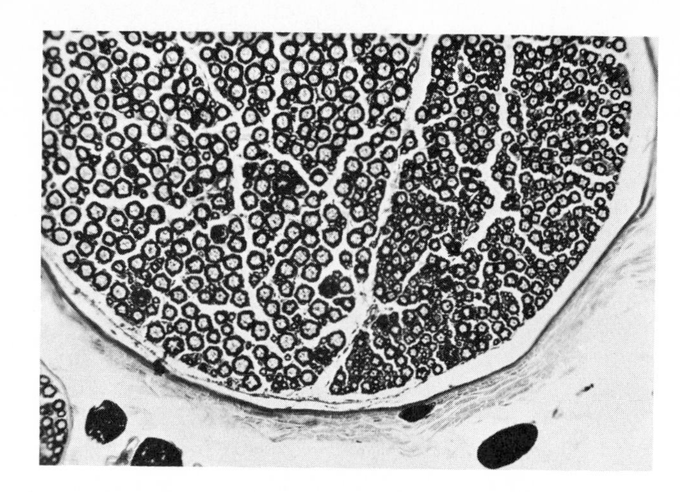

Figure 4-30. Cross section of peripheral nerve showing the myelin sheath (blackened rings) and the variation in size of single fibers. (Courtesy Ward's Natural Science Establishment, Inc., Rochester, N.Y.)

rise to the myelin sheath and may be the site of high energy turnover. The Schwann cells dip down to meet the axon where the myelin sheath is interrupted. These indentations are referred to as the **nodes of Ranvier.** The nodes are crucial structures in the bioelectrical activity of the axon, for it is only in these areas of the axon that ions, on which the transmission of the nerve impulse depends, can pass through the inner and outer surfaces of Schwann cells.

Axons of nerve cells terminate by branching into many fine filaments, called **terminals,** or **telodendrites.** Ramón y Cajal enumerates four anatomical types of axon terminals: small buds, or *boutons,* irregular enlargements containing a network of neurofibrils, terminals with compact groups of neurofibrils, and terminals in the form of small rings. The endings may establish contact with the tissue of an organ (neuroeffector junction) or with a dendrite or cell body (soma) or axon of another neuron. They are referred to as **axodendritic, axosomatic,** or **axoaxonic** contacts and the junction is referred to as a **synapse** (Fig. 4-35), a term contributed by the great English physiologist Sherrington. The function of the synapse is to transfer electrical activity from one neuron to another. If the first neuron can depolarize the second neuron to its threshold, an "all or none" impulse is then generated in the second neuron.

Synapses and Synaptic Transmission

Electrical Transmission

Transmission of impulses through synapses may be mediated either *electrically* or *chemically*. Electrical coupling between cells occurs in only a few instances in mammalian systems. For example, in the heart where the cells, although anatomically separate, form "tight" junctions as a result of the fusion of adjacent cell membranes at the nexal region (see discussion of cardiac tissue, p. 82). These **tight junctions** or similar regions of specialized membrane contact must be present to facilitate electrical coupling between the cells. The membranes of such cells act as low-resistance bridges, resulting in an electrical continuity between the cells and the same process by which potential spreads passively along the axon provides for spread into the postsynaptic cell. A few junctions of this type have been identified in the central nervous system and in this case are referred to as **electrical synapses.**

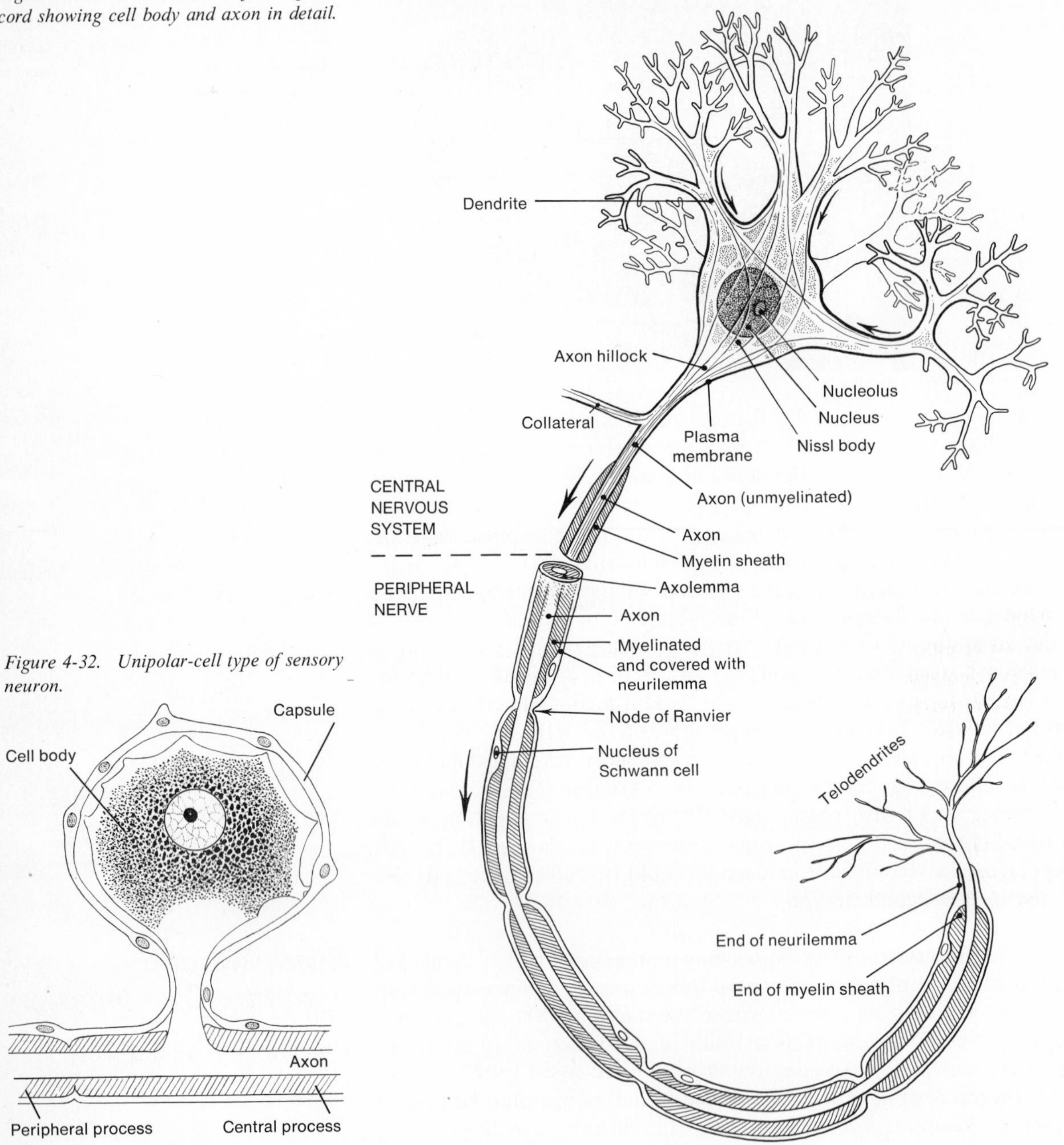

Figure 4-31. Motor neuron from spinal cord showing cell body and axon in detail.

Figure 4-32. Unipolar-cell type of sensory neuron.

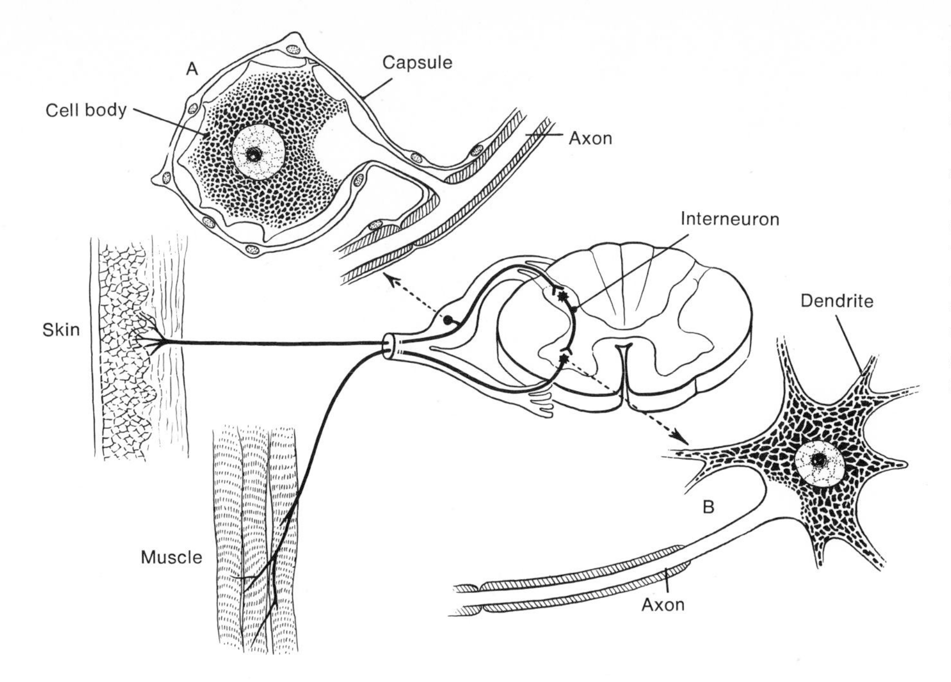

Figure 4-33. Relations of the spinal nerve to the spinal cord. Sensory neurons from the skin pass into the dorsal horn of gray matter by way of the dorsal root. The cell body of the sensory neuron is located in the dorsal root ganglion. A *represents a detailed study of a sensory neuron cell body in the dorsal root ganglion. Motor neurons, which have their cell bodies originating in the ventral horn of gray matter, pass to the effector organ (in this case, muscle) by way of the ventral root.* B *represents a detailed enlargement of the cell body of a motor neuron.*

Chemical Transmission

Most neurons in the mammalian nervous system do not form electrical synapses but form **chemical synapses.** The chemical theory of synaptic transmission implies that a chemical agent is released into the space (synaptic cleft) formed by the synaptic junction as a result of the nerve impulse. The chemical agent is stored in small membranous sacs in the terminal knobs of the telodendrites. The passage of a nerve impulse releases small quantities of the chemical transmitter into the synaptic cleft, which immediately combines with the reactive sites on the neuronal membrane of the postsynaptic cell lying just below the terminal knobs of the presynaptic neuron (Figs. 4-36 and 4-37). The combination of the transmitter with the reactive sites causes changes in the permeability characteristics of the postsynaptic cell. The time lapse between the arrival of the impulse at the presynaptic terminal and the membrane potential changes in the postsynaptic cell is called the **synaptic delay** and lasts less than 1 millisecond. Its duration is a function of the release mechanism, which frees the chemical transmitter from the membranous sacs within the terminal knobs of the telodendrites. The local potential caused by the release of the transmitter agent is called the **end plate potential**. Synaptic activity stops when the chemical transmitter is enzymatically split (Fig. 4-38) into ineffective substance or simply diffuses away from the reactive sites.

Acetylcholine is the transmitting agent at all synaptic junctions of the somatic peripheral nerves (myoneural junctions), at all peripheral ganglia of both parasympathetic and sympathetic systems, and at some synapses within the central nervous system. The synthesis of acetylcholine is shown in Fig. 4-38. In addition, other chemicals, such as serotonin, norepinephrine, glutamine, and gamma-aminobutyric acid (GABA), serve as transmitters within the central nervous system. Impulses arriving over sympathetic nerves at neuroeffector junctions are

Figure 4-34. Detail of axon showing the arrangement of myelin, Schwann cells, and glia cells.

Figure 4-35. Types of synapses.

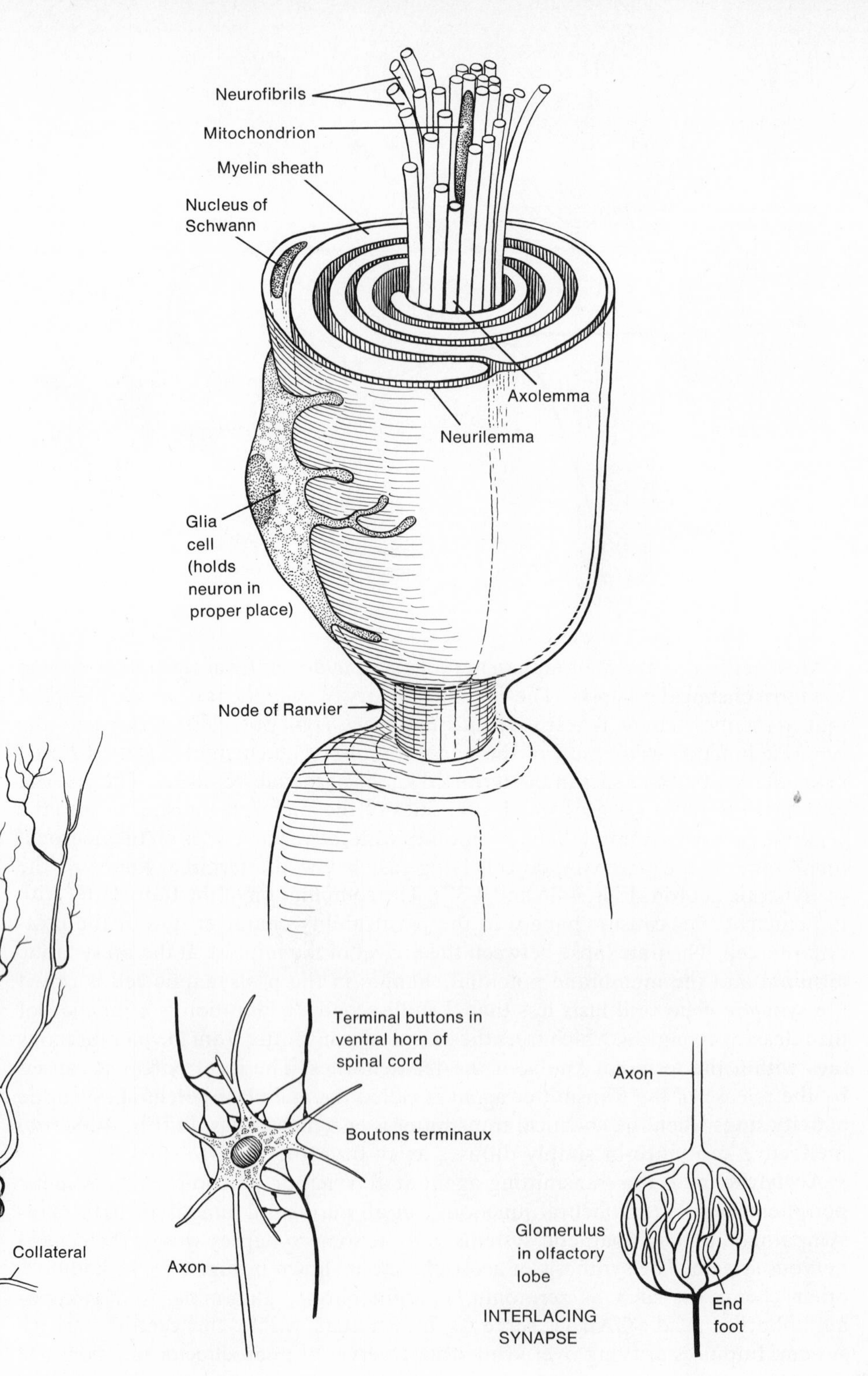

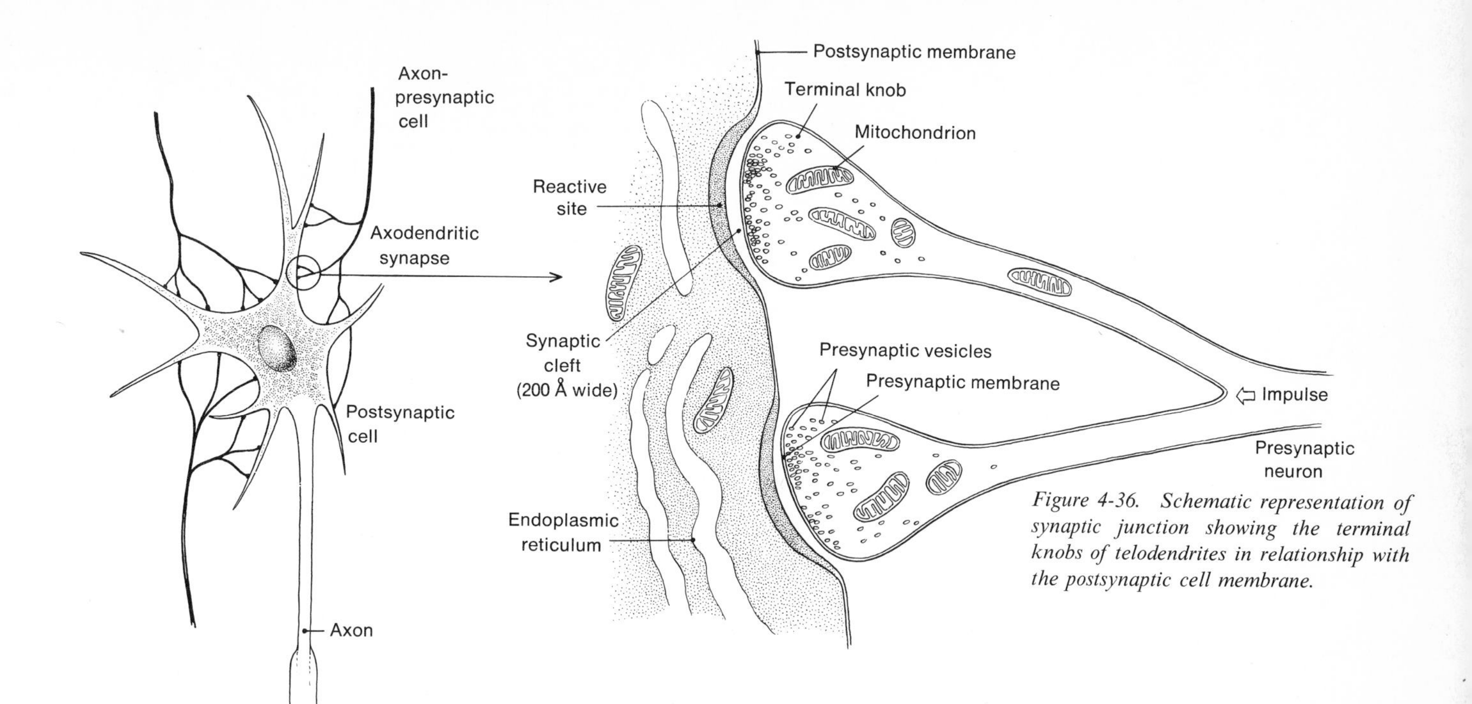

Figure 4-36. Schematic representation of synaptic junction showing the terminal knobs of telodendrites in relationship with the postsynaptic cell membrane.

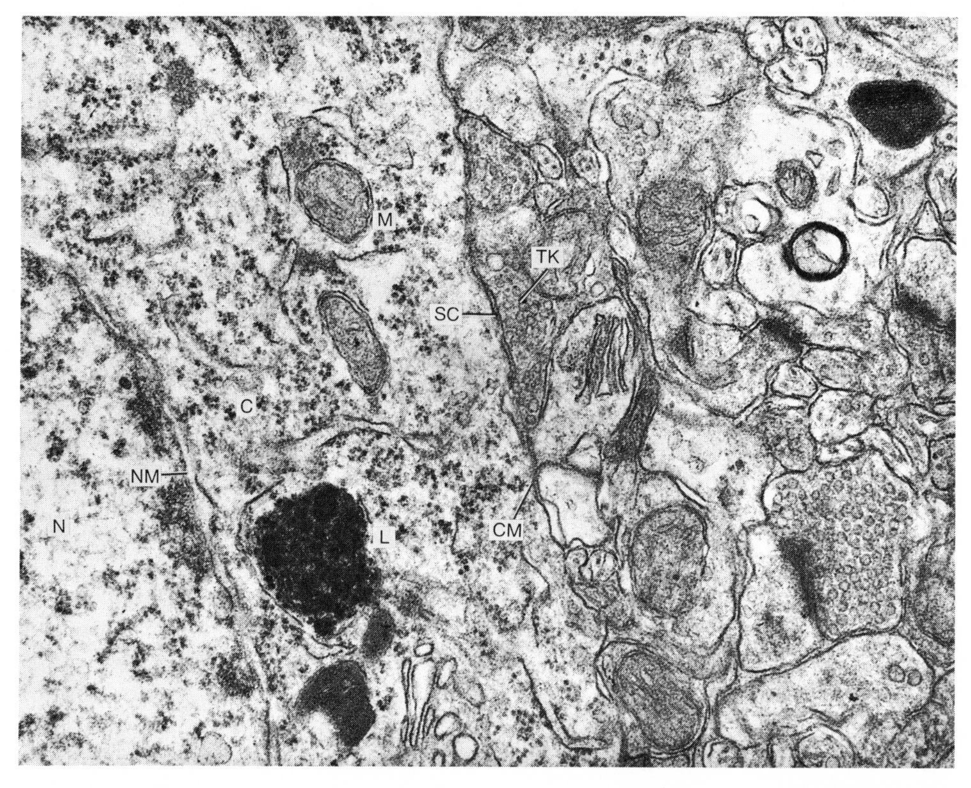

Figure 4-37. Electron micrograph of rat cerebral cortex (magnification 66,000 ×) showing terminal knobs (TK) *of presynaptic neuron terminating on cell membrane* (CM) *of postsynaptic cell.* SC, *synaptic cleft;* C, *cytoplasm;* L, *lysosome;* N, *nucleus;* NM, *nuclear membrane;* M, *mitochondrion.* (*Courtesy Timothy Mougal, Electron Microscopy Laboratory, Department of Zoology, University of Maryland, College Park, Md.*)

Figure 4-38. Pathway for acetylcholine synthesis and its degradation by enzyme activity.

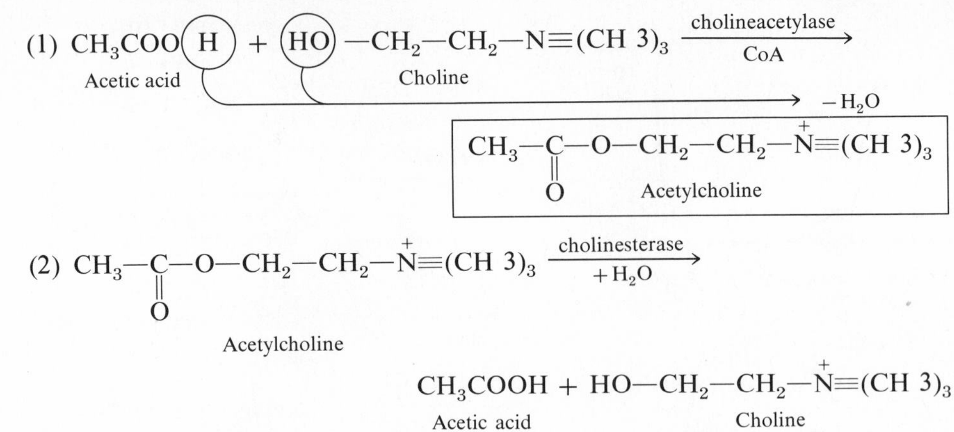

transmitted to the tissues of various organs by the formation of *epinephrine* and *norepinephrine*. Norepinephrine is probably a precursor of epinephrine, into which it is converted by a process of methylation. The exact mechanism of its formation at the synaptic junction is not known. The mixture of epinephrine is known as sympathin:

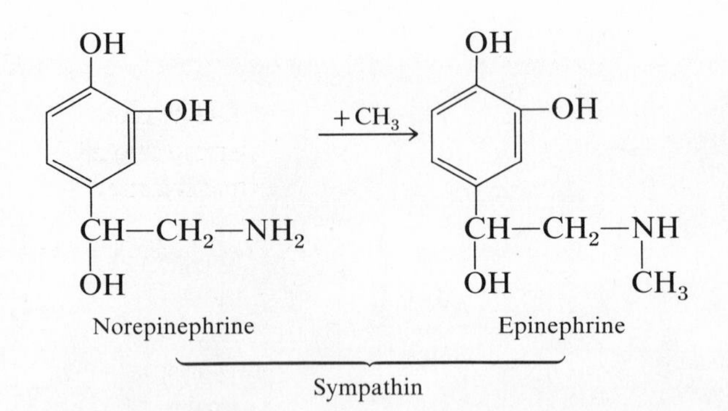

Excitatory and Inhibitory Synapses

Synapses are classified as excitatory synapses or inhibitory synapses. When the electrical activity in the presynaptic neuron influences the excitability of the postsynaptic neuron in such a way as to increase the possibility that its membrane potential will reach threshold and that an action potential will be developed, the synapse is called **excitatory.** In this case the postsynaptic membrane becomes more permeable to sodium ions, causing a slight depolarization of the postsynaptic cell, bringing it closer to threshold. Its only function is to help trigger an action potential. An **inhibitory synapse** is one in which the electrical activity in the presynaptic neuron lessens the likelihood that the postsynaptic neuron will develop an action potential. In this case the release of the chemical transmitter causes an increase in the permeability of the postsynaptic membrane to potassium and chloride ions and not to sodium ions. As a result there is a net movement of potassium and chloride ions out of the cell, producing an increased negativity or hyperpolarization of the cell membrane. The postsynaptic membrane threshold level is increased or moved farther away from threshold level. The type of synaptic inhibition just described is referred to a **postsynaptic inhibition.** A second type acts by reducing the amount of transmitter substance released from the presynaptic neuronal terminals and is called **presynaptic inhibition.** Experimental evidence indicates that the nerve impulse releases the chemical transmitter in packets or quanta rather than in a continuous stream. It has been hypothesized that a quantum corresponds to a single vesicle or sac as seen in the nerve terminal under the electron microscope. The normal motor end plate potential on nerve discharge appears to result from a simultaneous release of about 300 quanta. If the chemical transmitter is stored in quanta of a standard size and the total end plate potential developed on nerve discharge was 150 mV, each quantum or package represents a depolarization potential of 0.5 mV. Studies on spontaneous leakage of chemical transmitter (acetylcholine) from the nerve terminal at the neuromuscular synapse (motor end plate) indicates that very small depolarizations of the muscle membrane occur, even in the absence of an action potential. These are known as miniature end plate potentials (m.e.p.p.) and have been found to be for a particular muscle of an elementary size. For example, if it were found to be 0.5 mV, there would be no m.e.p.p. smaller than this value and would correspond to the leakage of chemical transmitter from one vesicle.

Spinal Nerves

In man there are 31 pairs of **spinal nerves,** which are divided into 8 cervical, 12 thoracic, 5 lumbar, 5 sacral, and 1 coccygeal (Fig. 4-39). Each pair of spinal nerves emerges into the body cavity by way of the intervertebral foramina (Fig. 4-40). In man, although the spinal cord does not extend through the entire length of the vertebral column, the emerging spinal nerves maintain their proper intervertebral exists. This necessitates the continuation of the posterior pair of spinal nerves within the vertebral column for increasingly longer distances before they finally emerge, with the result that a brush of spinal nerves is formed at the end of the cord called the **cauda equina,** or the "horsetail" (Fig. 4-38). In the embryo the nerve roots pass transversely outward from the spinal cord to exist through their respective intervertebral foramina. As a consequence of the unequal growth of the spinal cord and the vertebral column, the nerve roots become more and more elongated, so that the lumbar and sacral nerves descend almost vertically to reach their points of exit (Fig. 4-41). Figure 4-42 shows the spinal cord and its attached spinal nerves in the very young.

The connection between the spinal nerves and the spinal cord is effected through the dorsal and ventral roots of the nerves (Fig. 4-42). Dorsal roots are composed primarily of neurons that carry impulses into the central nervous system and, therefore, are termed **sensory roots.** The ventral roots, on the other hand, are made up of neurons that carry impulses away from the central nervous system and are commonly called **motor roots** because the impulses leaving by way of this route usually bring about motion or some activity on the part of an organ system. In general, peripheral nerves are classified as sensory, motor, or mixed, depending on the direction in which they carry impulses. Nerves carrying impulses toward the central nervous system are **sensory,** and those carrying impulses away

Figure 4-39. *Dorsal view of spinal cord showing the ganglia, spinal nerves, and cauda equina.*

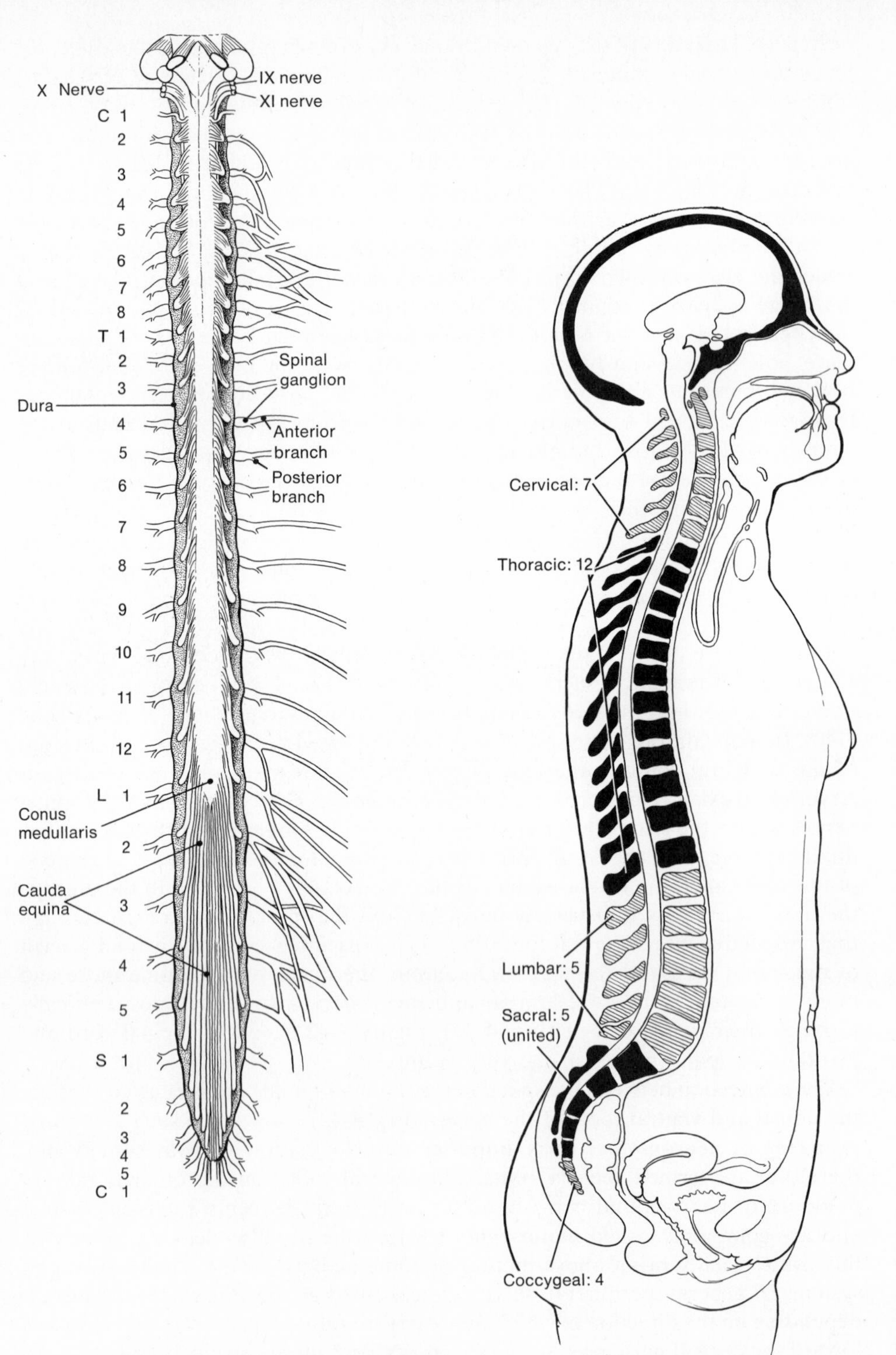

Figure 4-40. (Far right) *Longitudinal section of adult vertebral column. Between the vertebral bodies there are openings (intervertebral foramina) through which the spinal nerves emerge into the body cavity.*

from the central nervous system are termed **motor.** The spinal nerves contain both types of neurons, and those that carry impulses toward and away from the central nervous system are termed **mixed.** The cell bodies of the sensory neurons of the spinal nerves are located in an enlargement of the sensory root (dorsal root), called the **dorsal root ganglion.** The motor neurons of spinal nerves, on the other hand, have their cell bodies originating in the central horn of gray matter of the spinal cord (Fig. 4-33).

After leaving the intervertebral foramen, each spinal nerve gives off four branches (Fig. 4-43).: (1) a dorsal branch, called the **posterior primary branch,** which supplies the muscles and the skin of the back; (2) a ventral branch, the **anterior primary branch,** which supplies the muscles and skin of the sides, front,

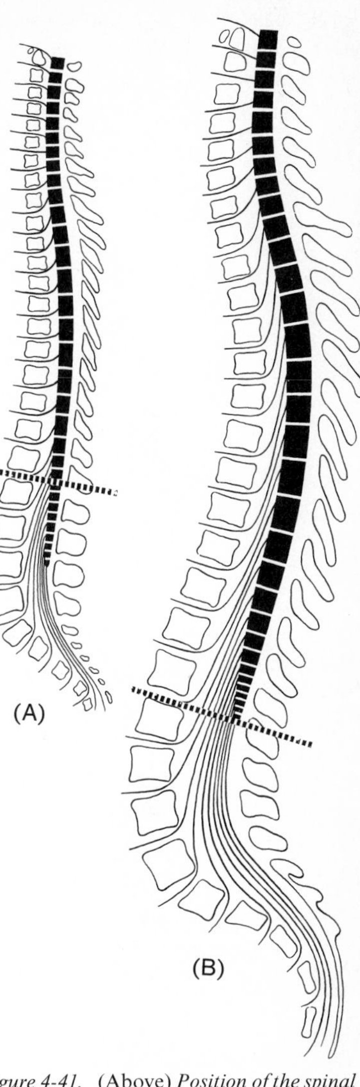

Figure 4-41. (Above) *Position of the spinal cord with reference to the bodies and spinous processes of the vertebrae. Tape is between the first and second lumbar vertebrae.* A *shows vertebrae of newborn infant.* B *represents adult vertebrae.*

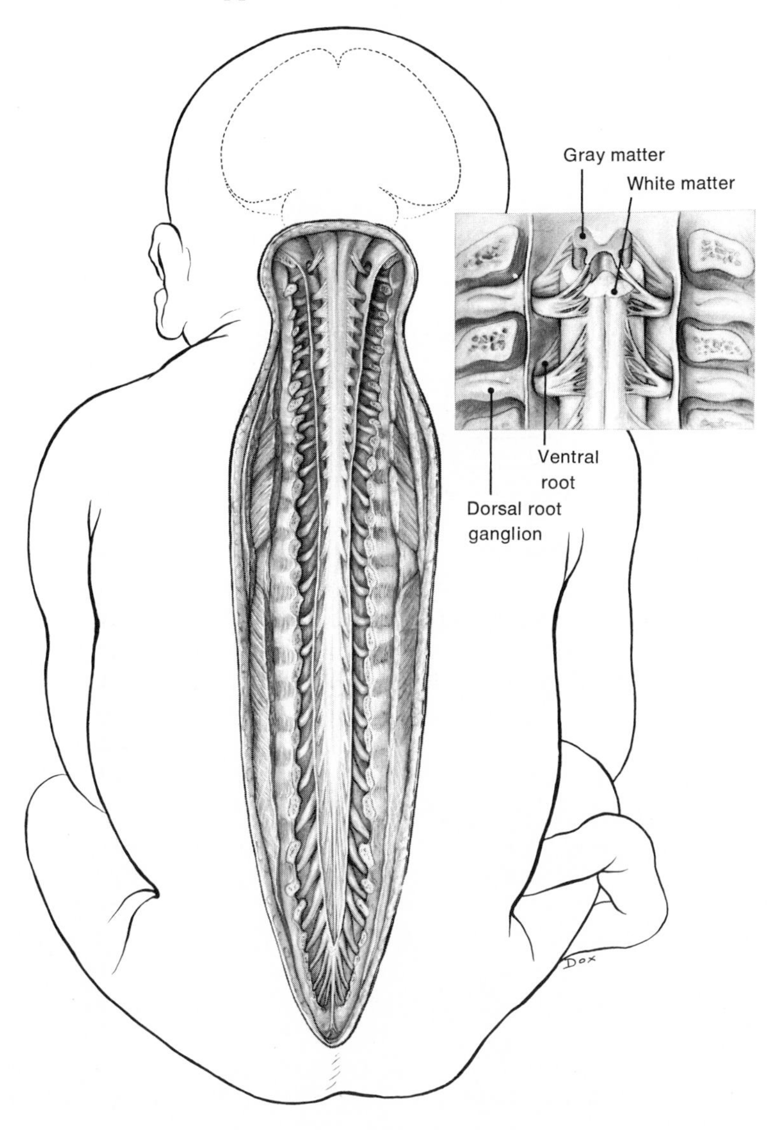

Figure 4-42. (Left) *Spinal cord in situ. Insert contains an enlargement of a section of the spinal cord showing the relationship of the dorsal and ventral roots to the spinal cord.*

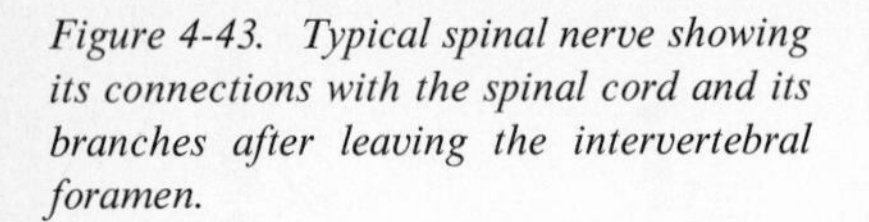

Figure 4-43. Typical spinal nerve showing its connections with the spinal cord and its branches after leaving the intervertebral foramen.

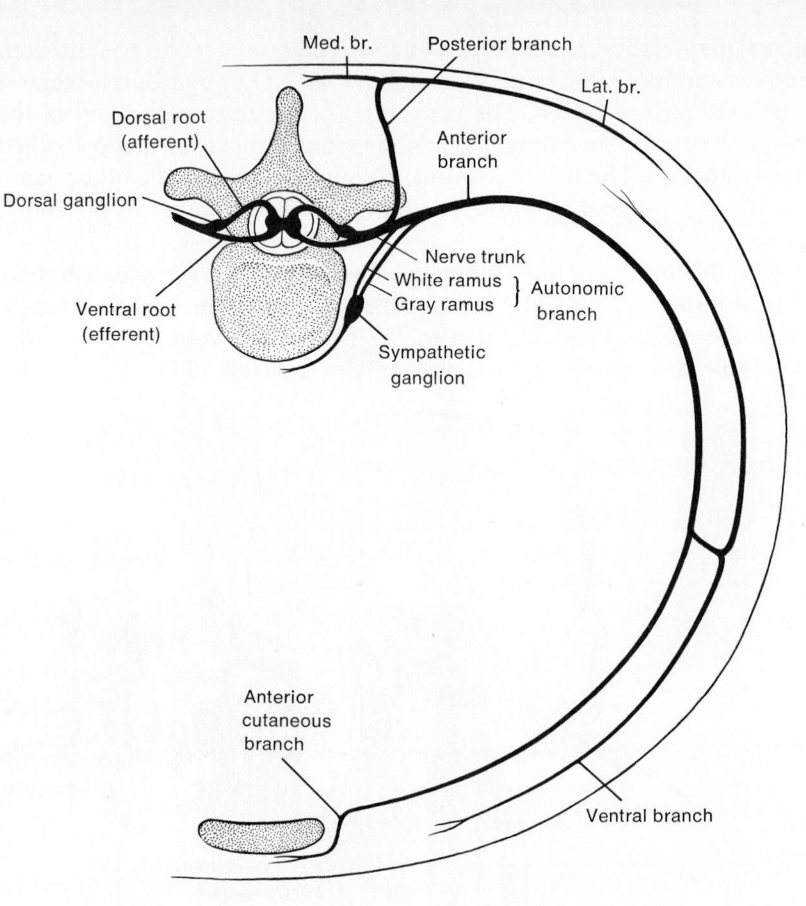

and extremities of the body; (3) a small **meningeal branch,** going back to the meninges of the cord; and (4) an **autonomic branch,** which connects with ganglia located along the vertebral column. The posterior and anterior primary branches supplying the skeletal muscles form the **somatic peripheral** portion of the nervous system.

Cranial Nerves

Cranial nerves are grouped together into twelve pairs (Fig. 4-44). They all arise from the brain stem except the first pair, which is located on the olefactory lobes of the cerebrum. The cranial nerves are classified into purely sensory, purely motor, or mixed. The sensory cell bodies of these nerves, as in spinal nerves, are associated with ganglia located outside the brain. The cell bodies of the motor neurons have their headquarters in the gray substance of the brain.

The twelve pairs of cranial nerves are named with reference to the parts they supply and are commonly designated by Roman numerals I to XII.

I. Olfactory Nerve. The fibers composing the **olfactory nerve** arise in the olfactory mucous membrane of the upper part of the nasal cavity and end in the olfactory bulb. This nerve is without a ganglion and is entirely sensory. It is associated with the sense of smell (Chap. 5).

II. Optic Nerve. The **optic nerve** has its point of origin on the retina of the eye and terminates in the optic lobe of the brain. It is a purely sensory nerve, carrying impulses from the eye to the brain, where they are interpreted as vision.

III. Oculomotor Nerve. The **oculomotor nerve** is purely motor, having lost its sensory component, and innervates a number of muscles attached to the eyeball, moving it in various directions. A portion of the nerve is distributed to certain muscles within the eye.

IV. Trochlear Nerve. The **trochlear nerve** runs from the midbrain to a single muscle attached to the eyeball. It is purely motor.

V. Trigeminal Nerve. The **trigeminal cranial nerve** is a mixed nerve and supplies the muscles of mastication, the skin of the face and part of the scalp, and the mucous membrane of the mouth and nasal cavity (Fig. 4-45). This nerve arises from the lateral surface of the pons.

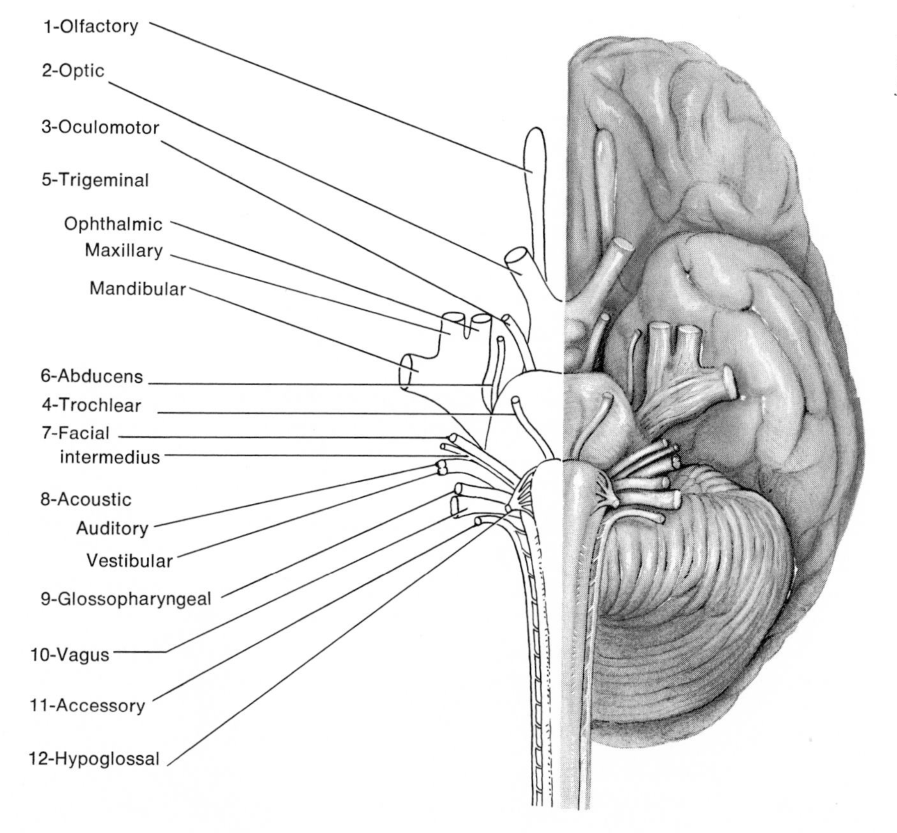

Figure 4-44. Base of the brain, the emergence of the cranial nerves, and their identification.

Figure 4-45. Sensory areas of the head showing the general distribution of the three divisions of the trigeminal (V) nerve.

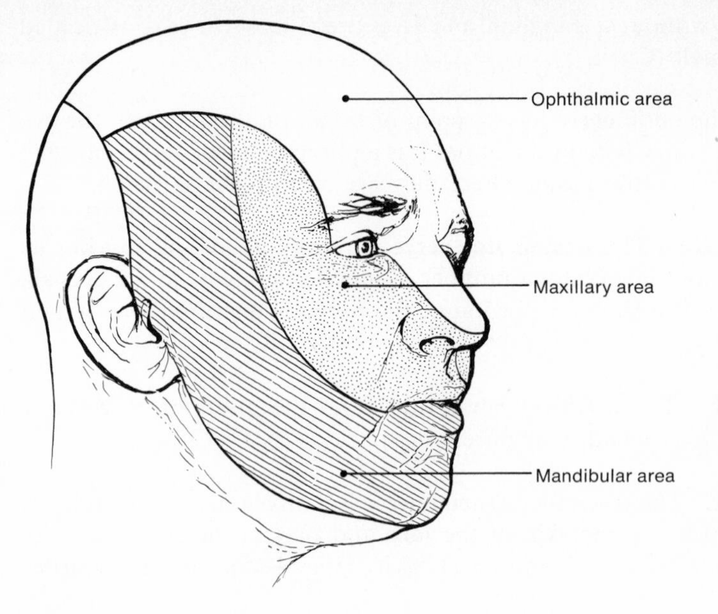

VI. Abducens Nerve. The **abducens nerve** is attached to the ventral surface of the pons and innervates a single muscle attached to the eyeball. It is entirely motor.

VII. Facial Nerve. The **facial nerve** is attached to the pons and is distributed to the facial muscles (muscles of expression). Some fibers are also distributed to the mucous membrane of the tongue and some to certain of the salivary glands. This is considered a mixed nerve.

VIII. Acoustic Nerve. The **acoustic nerve** is a sensory nerve and is associated with the senses of hearing and equilibrium.

IX. Glossopharyngeal Nerve. The **glossopharyngeal nerve** supplies the throat, a salivary gland, and the mucous membrane of the tongue. It is a mixed nerve.

X. Vagus Nerve. The **vagus nerve** supplies the mucous membrane and muscles of the larynx and pharynx. It also innervates all the visceral organs in the thorax and abdomen and is considered a mixed nerve.

XI. Spinal Accessory Nerve. The **spinal accessory nerve** joins the vagus in supplying the muscles of the larynx. It also supplies portions of the trapezius and sternocleidomastoid muscles and is considered a mixed nerve.

XII. Hypoglossal Nerve. The **hypoglossal nerve** is attached to the medullary portion of the brain stem and supplies the muscles of the tongue. It is entirely motor.

Autonomic System

The autonomic part of the peripheral nervous system is composed of branches of certain cranial nerves and spinal nerves that innervate organs in the thorax, abdomen, and pelvis and most of the blood vessels and glands in the body. It is anatomically divided into parasympathetic and sympathetic portions and is usually an efferent or motor system.

Parasympathetic Division. The **parasympathetic division** includes the branches from the following cranial nerves supplying visceral structures: the oculomotor, facial, glossopharyngeal, and vagal. There is also a spinal portion composed of branches of the second and third sacral nerves, which supply viscera in the pelvis. Figure 4-46 shows the components of the parasympathetic division. Numerous ganglia occur in this division but do not appear in a definite trunk, as in the sympathetic division.

Sympathetic Division. The **sympathetic system** consists of a pair of long nerve trunks located on either side of the backbone, called the **paravertebral sympathetic trunks.** A number of ganglia are located along its length. The three ganglia in the neck region are the cervical ganglia; there are ten to twelve pairs in the thoracic region, four pairs each in the lumbar and sacral regions, and one pair in the coccygeal area (Fig. 4-47). Nerve branches from these trunks are given to all the spinal nerves as they emerge from the vertebral column into the body cavity. These are the **gray rami communicantes** and consist of postganglionic fibers that radiate to the effector organs. In addition, the trunks receive branches from the thoracic spinal nerves and the first three lumbar spinal nerves, which form the **white rami communicantes.** These branches form the **preganglionic fibers,** extending from the central nervous system to the ganglia. The sympathetic trunks, their connections with the rest of the nervous system, and their branches to the visceral organs constitute the sympathetic division of the autonomic system.

The autonomic system is essentially a two-cell and two-fiber system; the cell of origin arises in the gray matter of the brain or spinal cord, and then the preganglionic fiber extends from the central to the peripheral ganglia. From the peripheral ganglion, where the second cell body makes its synapse with the preganglionic fiber, the postganglionic fiber, which has no medullary sheath (therefore is gray in color), extends from the ganglion to the visceral structures it innervates (Fig. 4-48).

Transmission of impulses through the synapses of the autonomic system is mediated by two chemicals, acetylcholine and sympathin. The parasympathetic and sympathetic divisions are differentiated in this respect also. The parasympathetic is purely **cholinergic;** that is, transmission through the synapses at both the ganglion and neuroeffector junctions is mediated by the production of acetylcholine. The sympathetic division is considered to be **adrenergic,** since mediation through the synapses involves both acetylcholine and sympathin (epinephrine). Acetylcholine is the chemical mediator at the ganglion, and sympathin is responsible for the transmission of the impulse at the neuro-effector junction.

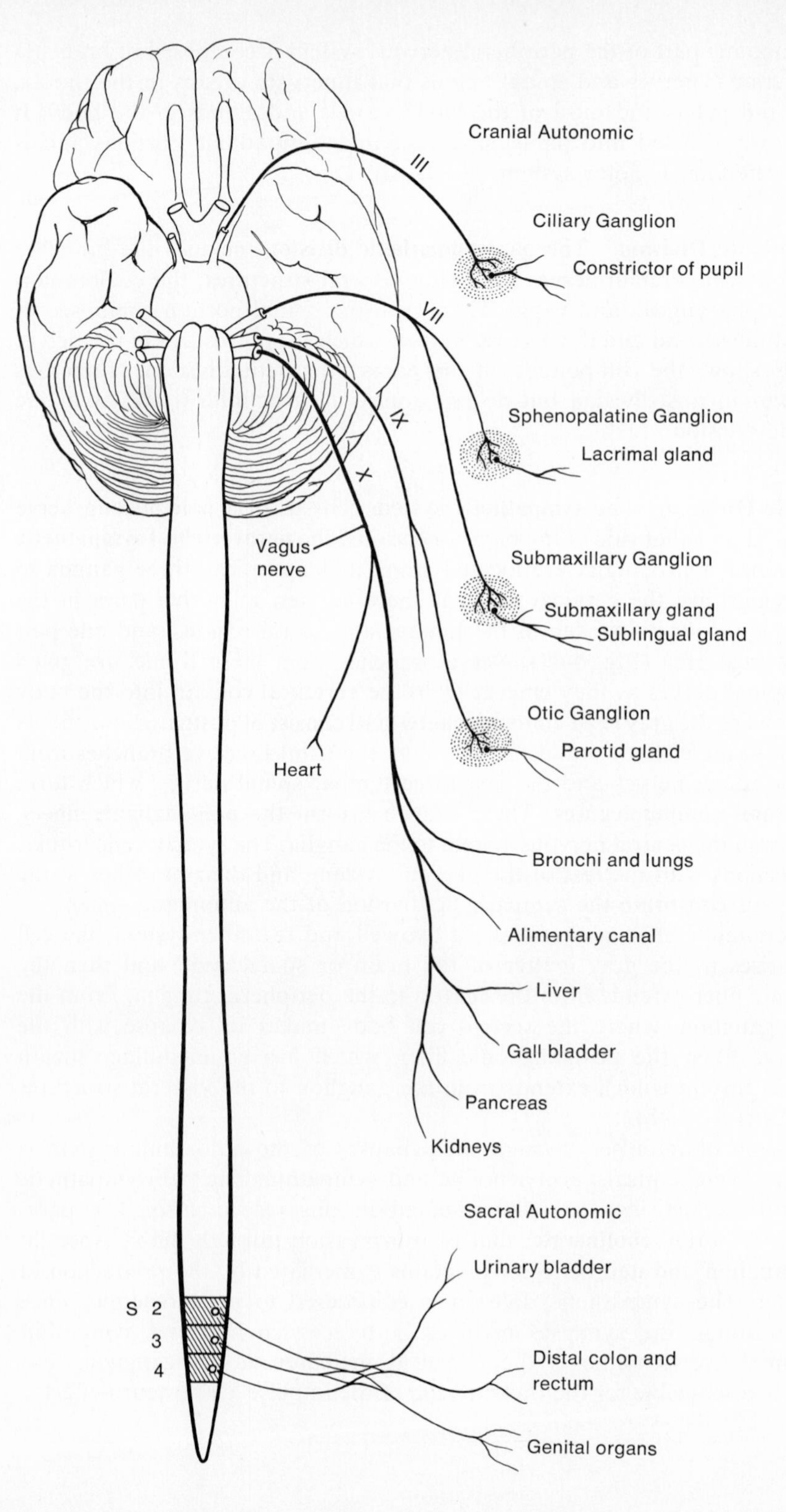

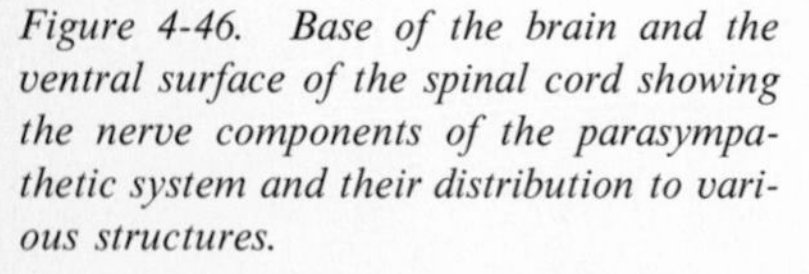

Figure 4-46. Base of the brain and the ventral surface of the spinal cord showing the nerve components of the parasympathetic system and their distribution to various structures.

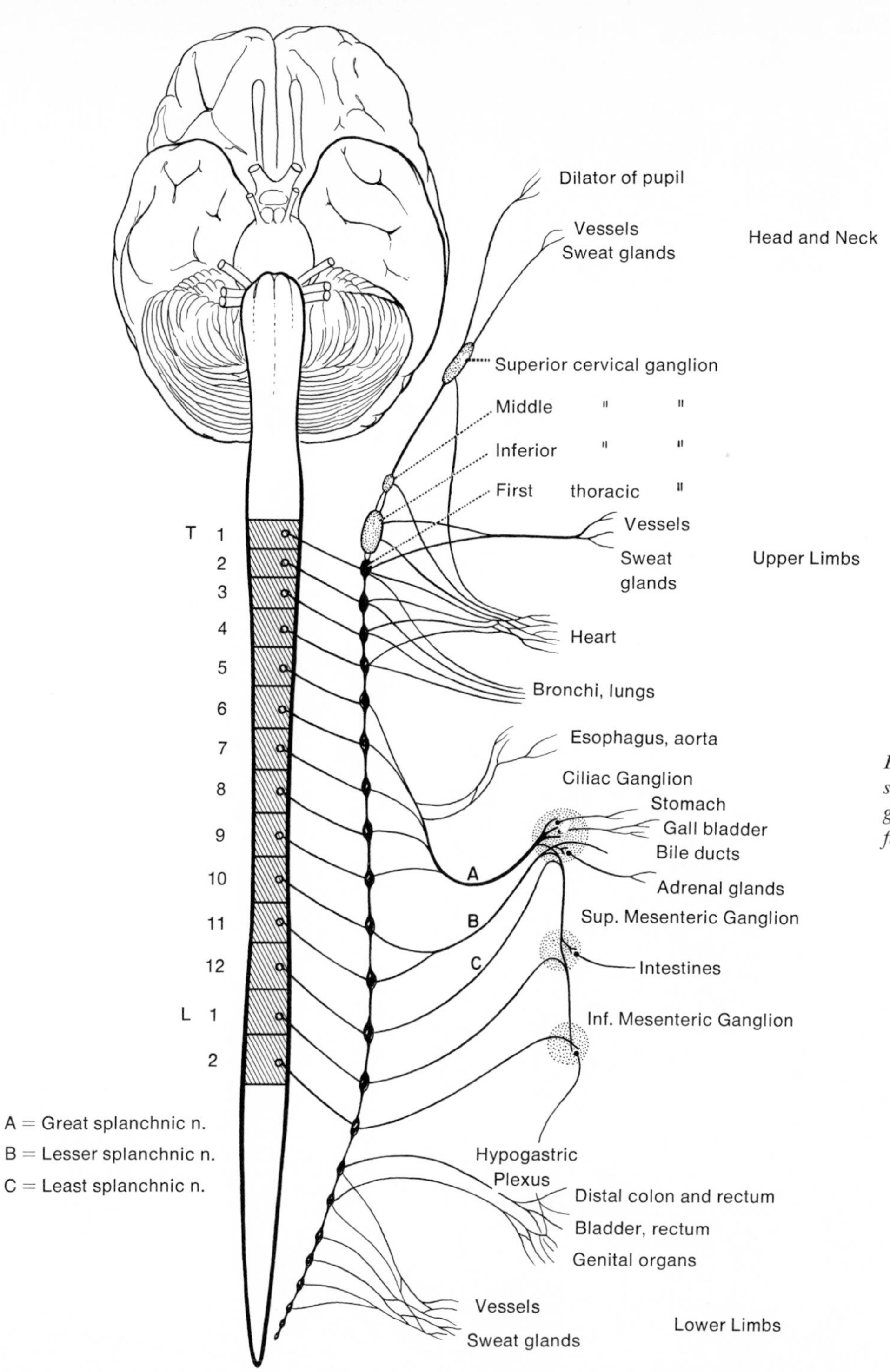

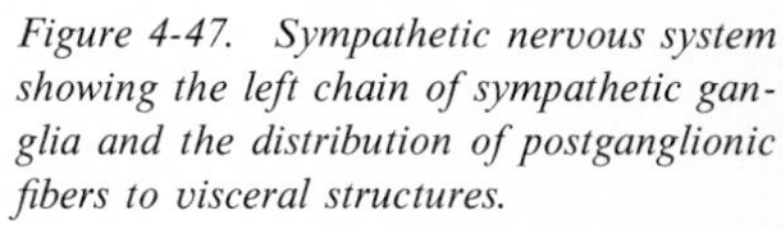
Figure 4-47. Sympathetic nervous system showing the left chain of sympathetic ganglia and the distribution of postganglionic fibers to visceral structures.

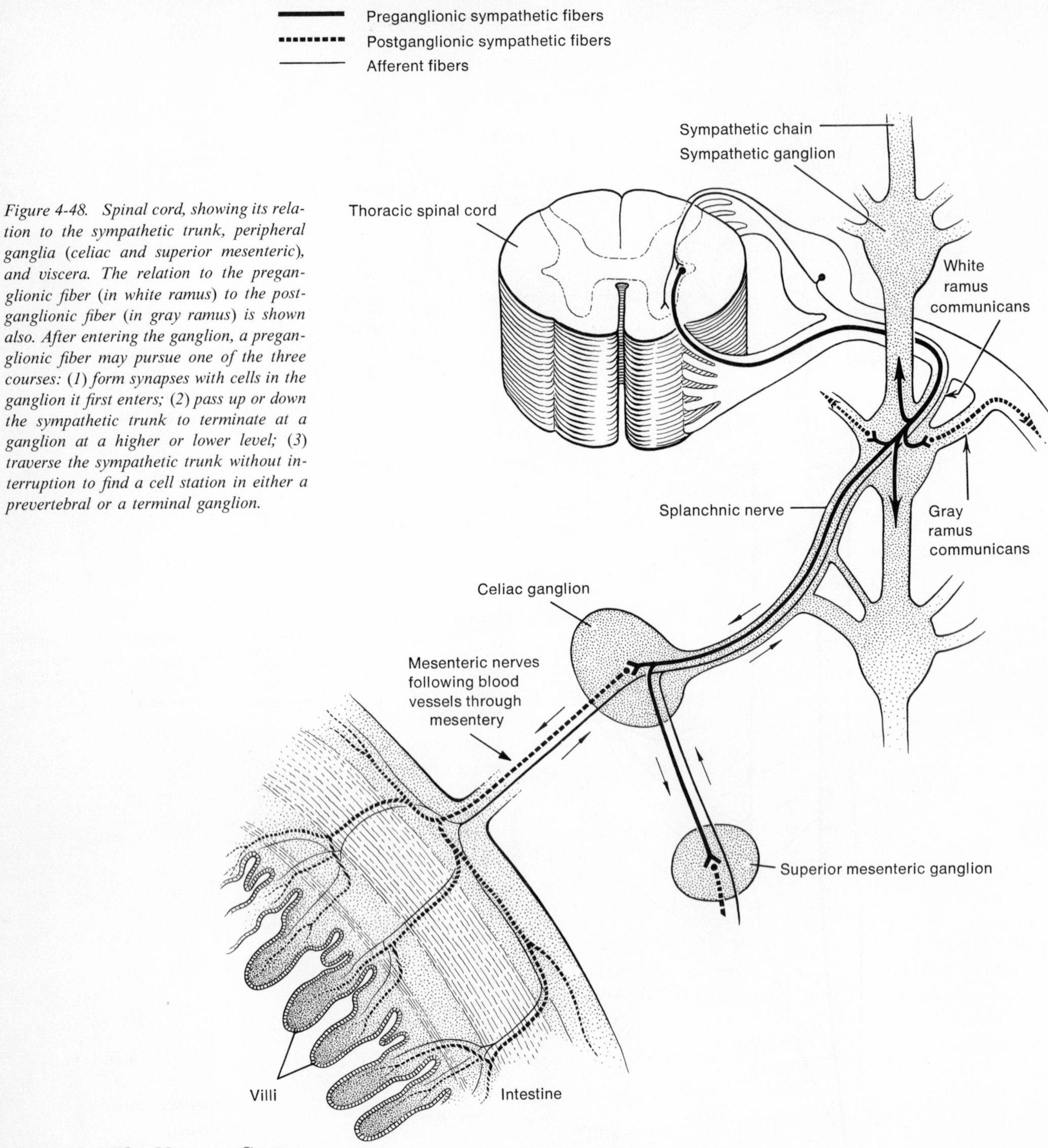

Figure 4-48. Spinal cord, showing its relation to the sympathetic trunk, peripheral ganglia (celiac and superior mesenteric), and viscera. The relation to the preganglionic fiber (in white ramus) to the postganglionic fiber (in gray ramus) is shown also. After entering the ganglion, a preganglionic fiber may pursue one of the three courses: (1) form synapses with cells in the ganglion it first enters; (2) pass up or down the sympathetic trunk to terminate at a ganglion at a higher or lower level; (3) traverse the sympathetic trunk without interruption to find a cell station in either a prevertebral or a terminal ganglion.

Function of the Sympathetic System. In general the actions of the sympathetic system are directed to strengthening an animal's ability to adapt itself to changes in its external environment, secure its food, and protect itself. Cannon, by removal of the sympathetic trunks, has demonstrated that an animal can survive without its sympathetic system. Such animals remain in good health as long as all their needs are taken care of by their laboratory handlers. However, such animals are incapable of hard work; glucose is not mobilized in the blood on demand; the spleen does not pour out more red blood cells during excitement or exercise; the usual reactions to cold, such as construction of vessels in the skin and elevation of the hair, fail; and epinephrine is not secreted in an emergency. It is evident that the sympathetic system is involved with processes concerning the expenditure of energy and that, without its sympathetic system, the animal could not fend for itself and, in the struggle for existence, would soon lose out to the various hazards of the environment.

Function of the Parasympathetic System. The parasympathetic system is concerned with processes that restore and conserve energy for the animal. In most instances the parasympathetic generally inhibits or slows down activities of organ systems (conserving energy), in contrast with the sympathetic, which generally stimulates or increases activities. The only stimulating activity of the parasympathetic is on the digestive tract through which the energy stores of the body are replenished. The sympathetic has the opposite effect on this system by inhibiting its activity. The action of the autonomic system on various structures is listed in Table 4-2.

Reflexes

A **reflex act** is commonly defined as a response that is the same for any given stimulus. Tapping the tendon of the quadriceps femoris muscle just below the kneecap causes extension of the leg. A part of the body inadvertently set on a hot object is immediately pulled away. The stimulus of food in the intestine results in the secretion of digestive juices. All these actions are quite automatic and predictable and occur independently of the higher centers.

Table 4-2. ***Action of the Autonomic System on Visceral Organs***

Organ	*Sympathetic*	*Parasympathetic*
Iris of eye	Dilates	Constricts
Ciliary muscles	Relaxes	Contracts
Salivary glands	Inhibits	Stimulates
Bronchi	Dilates	Constricts
Sweat glands	Stimulates	No action
Blood vessels	Mostly constricts	Mostly dilates (by inhibition of sympathetic fiber)
Adrenal gland	Stimulates	No action
Intestinal tract	Inhibits action	Stimulates action
Heart	Stimulates	Inhibits
Bladder	Inhibits	Stimulates

Reflex Arc

The pathway for reflexes consists of sensory nerves by which impulses initiated by the stimulus reach the spinal cord. Synaptic connections are made within the cord, so that impulses leave by way of motor nerves that transmit them on the proper muscles. The **reflex arc,** or **pathway,** is considered the functional unit of the nervous system, as contrasted with the structural unit, the neuron. Reflex arcs involve two or more neurons and, in addition, a nonnervous element in the form of a muscle or a gland that acts in consequence of the stimulus received (Fig. 4-49).

More specifically, a reflex arc consists in each case of at least five fundamental parts, each with a different function: (1) a **receptor**—this is a specialized termination of a sensory nerve fiber or nerve endings associated with special sense organs; (2) a **sensory transmitter**—a single neuron that passes the impulse, picked up by the receptor, to the spinal cord; (3) a **motor transmitter**—a single neuron that transmits the impulse from the spinal cord to the effector organ; (4) a **neuroeffector junction**—a specialized ending of motor nerves; and (5) an **effector**—a structure, such as a muscle or gland, that carries out the actual response. Some reflexes require a third neuron that is interposed between the sensory and motor neurons. This neuron, found in the gray matter of the cord, is called the **intercalated neuron,** or **interneuron.**

Reflexes may be classified with respect to function, as postural reflexes, blood pressure–regulating reflexes, extensor reflexes, flexor reflexes, and vasopressor and

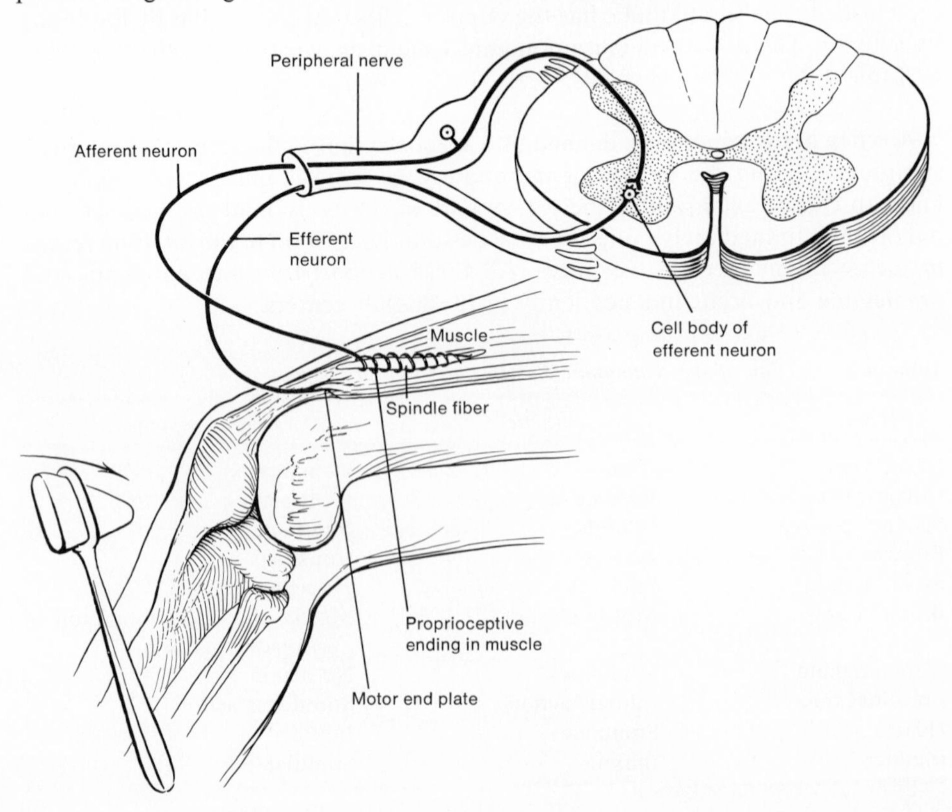

Figure 4-49. Simple reflex arc (knee jerk). The five fundamental parts of a reflex arc are shown: (1) a receptor, a proprioceptive ending in tendon; (2) a sensory transmitter, the afferent neuron; (3) a motor transmitter, the efferent neuron; (4) a neuroeffector junction, the myoneural junction; (5) the effector, the muscle.

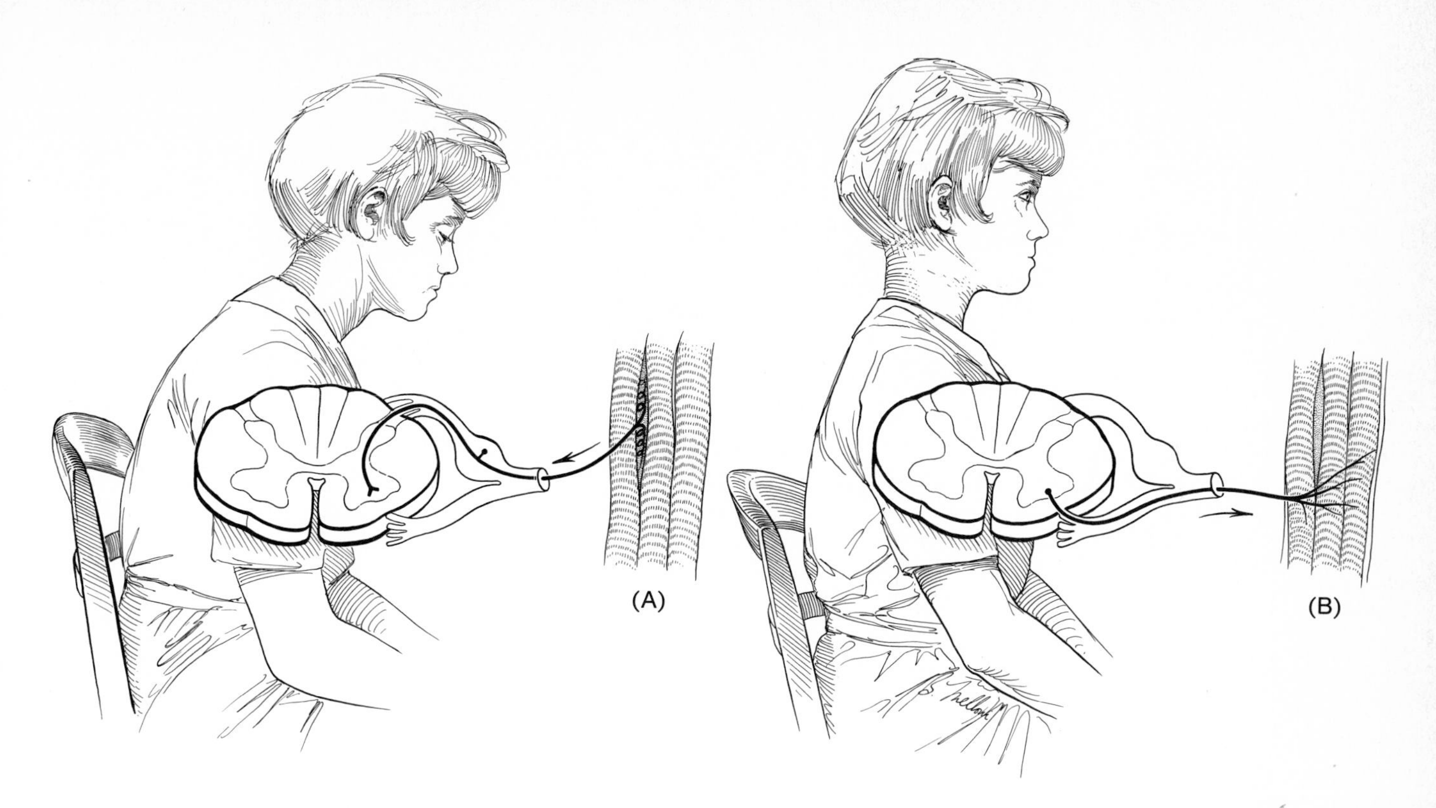

Figure 4-50. Postural reflexes. A *represents relaxation of trunk muscles, causing an increase in length of muscle fibers, which acts to stimulate the neuromuscular spindles (sensory receptor in muscles). The resulting impulse is transmitted to the spinal cord by way of a sensory neuron.* B *represents a motor impulse being transmitted to the muscle originally lengthened by relaxation. The muscle contracts and regains its former length so that the posture is also regained.*

vasodepressor reflexes. Certain of these reflexes may be represented by a pathway of two neurons, the simplest type, and others require a more complex pathway involving more than two neurons. The familiar knee jerk is an example of the two-neuron reflex arc and is considered an extensor reflex. Postural reflexes (Fig. 4-50) also involve two neurons and therefore one area of synaptic junction. Under most postural conditions there are many muscles that by their contractions serve to hold the body upright. Relaxation of any of these muscles causes an increase in length, which acts as a stimulus on the neuromuscular spindles (sensory receptor in muscles). The resulting impulse is transmitted to the spinal cord by a sensory neuron, which synapses with a motor neuron in the ventral horn of gray matter. The axon of the motor neuron transmits the impulse to the muscle originally lengthened by its relaxation. The muscle then contracts and regains its former length. This reflexly induced contraction, or state of tension, is commonly called **tone.** The antigravity muscles just cited are usually physiological extensors and are involved with extensor reflexes.

Flexor reflexes are those that bring about such movement as withdrawals. Painful stimuli received by receptors in the skin or in deeper tissue will activate flexor reflexes, as, for example, the withdrawal of the hand from a hot object or the removal of the foot from an irritating stimulus. The impulses in these cases are conducted over pathways whose simplest circuit involves at least three neurons: sensory, internuncial, and motor (Fig. 4-51). Many of these types of reflexes usually involve more than three neurons.

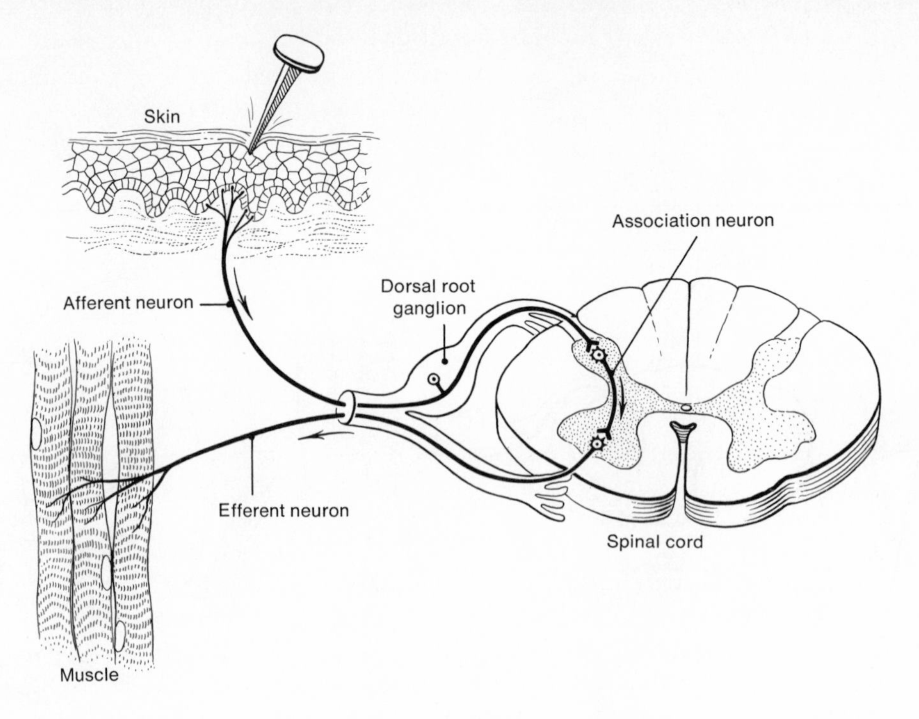

Figure 4-51. Three-neuron reflex arc. Painful stimuli initiate flexor reflexes, which bring about withdrawal movements. The association neuron (internuncial neuron) is found within the gray matter of the spinal cord.

Reflex arcs are not limited to the spinal cord (spinal reflexes) but occur over the brain stem areas as well (cranial reflexes). The sensory impulse also can be carried over fibers that reach the sensory area of the cerebral cortex, where they may be interpreted as sensations, but reflex action is not dependent on this interpretation. A withdrawal response to a painful stimulus can occur before the subject experiences pain. Experimental animals whose cerebral cortex has been removed and who therefore cannot experience pain will demonstrate very effectively withdrawal reactions from painful stimuli.

A frog may be decapitated by cutting with scissors between the back of the mouth and the neck, thus depriving the animal of many of the higher centers of sensory perception. Such an animal will hang limply with extended hind limbs and remain so indefinitely unless disturbed. Such an animal is referred to as a **spinal animal.** If an irritating or painful stimulus is applied to the skin of such an animal, reflex movement will be produced that will remove the part from the offending stimulus. Such movements are coordinated movement of the limbs rather than independent contractions of individual muscles. If, for example, acid is applied to the foot, the hind limb is flexed; if applied to the back, wiping movements are induced. Such reactions appear to be purposeful and enable even the animal deprived of its higher centers to show a measure of adaptation to the environment. These reflexes in the spinal frog are abolished by destroying the spinal cord, and it may be inferred, therefore, that the reflex arcs concern synapses within the cord.

Reaction Time

The interval occurring between the application of the stimulus and the beginning of the response is called the **reaction time.** It is evident that this time will vary, depending on the length of the nerve pathway, the number of synaptic junctions the impulse must pass over, the latency period of the receptors, and the tissue innervated. Increase in strength of stimulus may shorten this time, probably because of more rapid summation of effects of sensory impulses at the synapses. The time taken for the nerve impulses to pass along the lengths of the sensory and motor nerve trunks can be measured, as can the latent period (time it takes a stimulus to produce an impulse or reaction) of the receptor and effector. When this time is deducted from the total reflex time, the delay in the central nervous system can be deduced. This central delay is called the **reduced reflex time.**

Reflex Coordination

On casual examination, reflex activity appears to be an apparently simple mechanism, as in the extension of the leg in the knee jerk. Closer examination will reveal a more complex mechanism. For this response to occur, the opposing muscles that flex the leg must relax to a degree corresponding to the extension. The production of this relaxation is a function of the gray matter of the spinal cord. Sensory impulses reaching the cord to excite motor neurons that produce the extension will also reach motor neurons that supply flexor muscles, but in this case, inhibition of the activity of these neurons occurs. The inhibitory field of a reflex commonly embraces the motor neurons of muscles whose action is directly antagonistic to that of the excited muscles. This is the principle of **reciprocal innervation.** Although little is known about the mechanism of inhibition, it is nevertheless an important factor in muscular activity, for when muscles contract, those that oppose them must be coordinated in some way. If both flexors and extensors contract, a fixed position results, or the opposing muscles relax and a movement occurs. The causative inhibition in the relaxation of the opposing muscle group is frequently termed **Sherrington's inhibition,** or **reciprocal inhibition.**

The central nervous system has no peripheral inhibitory nerves such as those in the autonomic. Inhibition occurs by preventing or diminishing central excitation. Two theories have been developed to explain inhibition in the central nervous system: (1) Impulses reaching the synapses in the reflex arc may come in so rapidly as to keep the cell body in a prolonged refractory state, and (2) impulses from the antagonistic reflex or the cortex may produce inhibitory substance at the synapses of the reflex, thus blocking it.

Sensory impulses reaching the spinal cord in reflex activity may eventually cause more than one muscle to react, even though the motor neurons of these muscles are in different levels of the cord from the area of incoming sensory impulses.

Withdrawal movements usually involve the contraction of many muscles, which become involved in the reaction because the sensory fibers transmitting the impulse into the spinal cord by way of the dorsal root not only synapse with motor cells at the same level, but also send internuncial neurons up and down the cord to motor neurons at other levels (Fig. 4-52). These connections provide for diffuse responses and, if they are mitigated on the same side of the cord, are referred to as **ipsilateral responses.** If the connecting internuncial neuron crosses

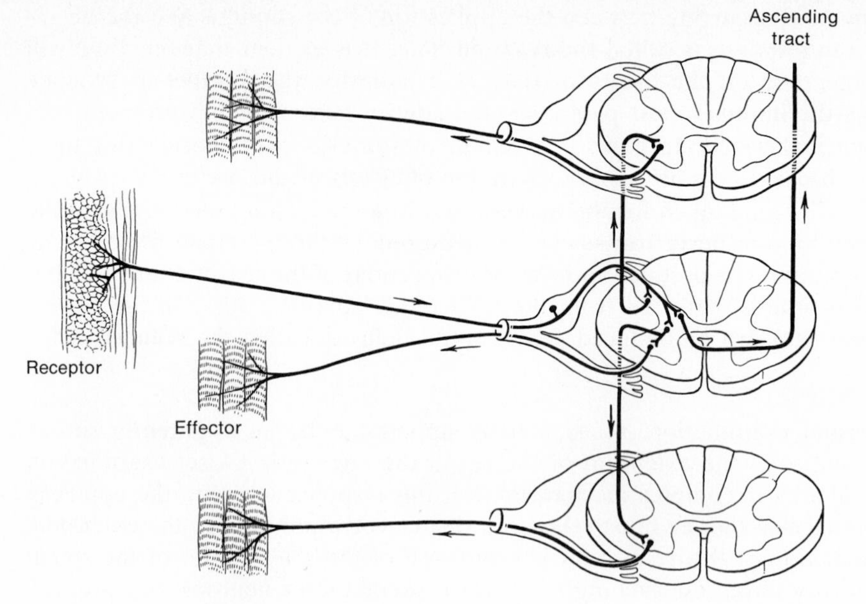

Figure 4-52. How a diffuse motor response on the same side of the body may be mitigated by sensory impulses passing into one segment of the cord (ipsilateral response). A stimulus entering the spinal cord through one neuron may be distributed to a large number of motor neurons in different segments of the cord by a number of internuncial neurons.

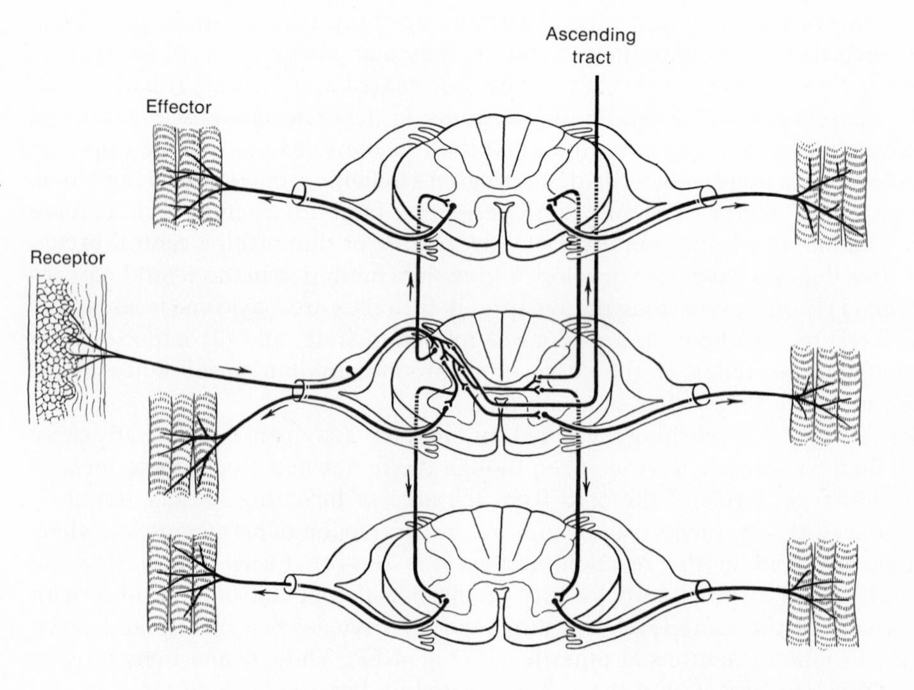

Figure 4-53. How a diffuse response may be mitigated on the opposite side of the body (contralateral response) from that of the incoming sensory impulse. Connecting internuncial neurons cross cover (second segment of spinal cord) to make connections with motor neurons on the opposite side of the spinal cord.

over (Fig. 4-53) to make connection with motor neurons on the opposite side of the cord, thereby causing responses on the side opposite that of the incoming sensory impulse, this response is spoken of as a **contralateral response.** Crossed extensor reflexes are of this type. For example, an animal may be painfully stimulated if it steps on a sharp object. The response will be to withdraw the foot (withdrawal reflex), and in doing so the animal must shift its weight on the opposite side in order to maintain its balance. After weight shifting, extension of the opposite limb occurs. Generally, strong stimulation of a sensory nerve of a limb leads to generalized flexion of that limb and generalized extension of the opposite limb. In contrast to the restricted two-neuron arc pathways, which limit the reflex action of stretch reflexes, the multiple internuncial pathway in flexor reflexes with their axons running up and down the spinal cord provides the structural mechanism for ensuring the spread of the flexor reflex to many flexor muscles of the limb on the same side and extensor muscles of the limb on the opposite side. Internuncial neurons vary in the number of branched fibers attached to them. Depending on the number of such processes, internuncial neurons may connect a varying number of sensory and motor neurons to each other or to higher brain centers. A stimulus entering the spinal cord through one neuron may be distributed to a large number of motor neurons and thus cause activity in a large number of motor neurons. This spreading effect is called **divergence.** On the other hand, more than one axon may be in synaptic connection with a single neuron. This converging effect of axons is called **convergence** (Fig. 4-54).

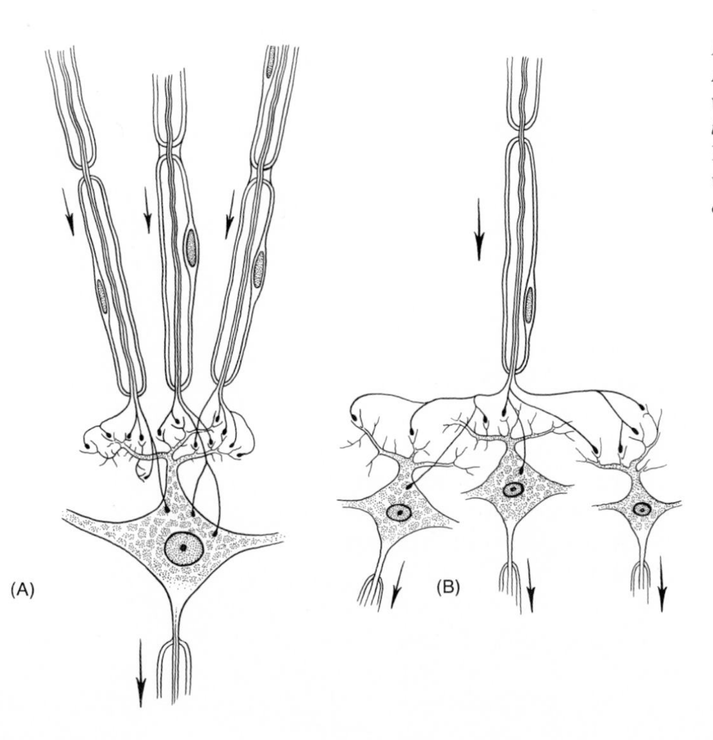

Figure 4-54. Divergence and convergence. A. *Three axons are in synaptic connection with a single nerve cell body. This converging effect of axons is called convergence.* B. *A single axon makes synaptic connection with three nerve cell bodies. This spreading effect is called divergence.*

Fatigue

Nerve axons are practically impossible to fatigue if the circulation is adequate, but a reflex arc is very susceptible. This may be due to the much higher metabolic rate of nerve cell bodies as compared with axons. The actual cause of fatigue is unknown, but it may be due to the exhaustion of the acetylcholine mechanism for synaptic conduction. Complete fatigue in man results from central fatigue and not from the inability of the motor nerves to carry impulses to produce muscular contraction.

Reflexes in the Human Being

Reflexes may be classified as deep, superficial, special, and abnormal.

Tendon, Myotatic, or Stretch Reflexes (Deep Reflexes). *1. Patellar Reflex.* The **patellar reflex** is more commonly called the *knee jerk* and is elicited by tapping the tendon of the quadriceps femoris muscle just below the patella. This reflex is best produced when the leg being tested is allowed to hang free. The muscle is slightly stretched by the tapping, and the receptors located around the muscle cells (neuromuscular spindles) are stimulated. The impulse set up in the receptors is transmitted to the spinal cord, the center of this reflex, by a sensory nerve fiber, which synapses with cells in the ventral horn of gray matter in the spinal cord. These large, multipolar cells are the cell bodies of motor neurons, which transmit impulses to the muscle originally stretched, which then contracts to regain its original length, causing an extension of the leg. Figure 4-49 illustrates the pathway of this reflex.

2. Achilles Reflex. **Achilles reflex,** or **ankle jerk,** is obtained by striking the Achilles tendon, just above its insertion into the heel bone (calcaneus). This causes a reflex contraction of the triceps surae muscle, resulting in plantar flexion, or dorsiflexion of the foot.

3. Biceps Jerk. A flexion of the forearm, the **biceps jerk,** is obtained by tapping the biceps brachii tendon. This is best elicited by placing the thumb of the left hand firmly on the tendon of the muscle and striking the thumb with a rubber reflex hammer.

4. Triceps Jerk. A **triceps reflex** is best obtained by the examiner's supporting the wrist of the flexed arm with his left hand and tapping the tendon of the triceps brachii muscle close to the tip of the elbow. This causes a reflex extension of the forearm.

The responses of these reflexes vary enormously with normal persons. Although it is true that exaggerated or depressed reflexes are frequently indicative of disease of the nervous system, some persons normally respond quite briskly on being reflexly stimulated. Whether the reflexes generally are sluggish, moderately active, or very active is not as important as the fact that they are equal on the two sides. If the corresponding reflexes on the two sides are unequal and this inequality is consistent in repeated tests, neurological disorder may be indicated.

Relaxation of the subject undergoing a test for reflex activity is most important in order to elicit satisfactory tendon reflexes, for the magnitude of the reflex varies directly with the degree of tone in the muscles. Most persons will tense their muscles the moment contact is made or are nervous and anxious, which causes an increase in the extent of the reflex as a result of increased muscle tonus and

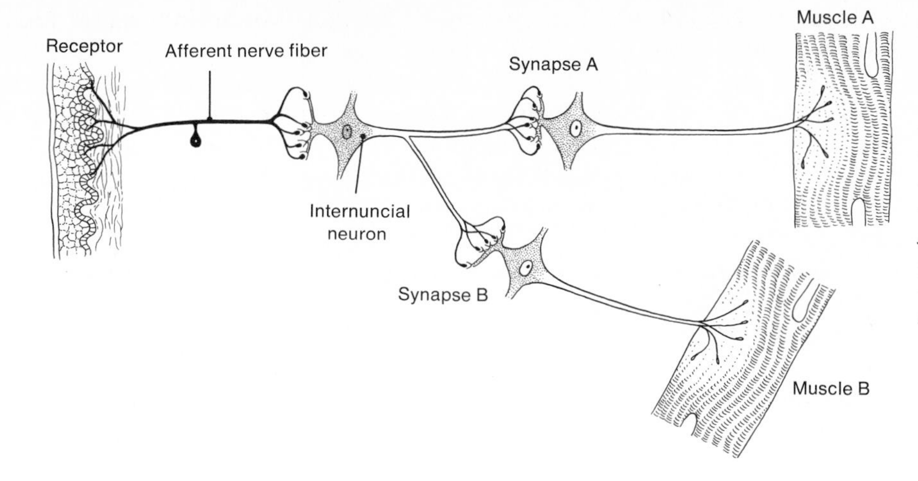

Figure 4-55. Synaptic resistance. Weak inpulses transmitted from the receptor to the internuncial neuron could possibly pass through both synapses A *and* B *to cause contraction of muscles* A *and* B. *However, since* B *is more resistant to the passage of the impulse than* A, *only muscle* A *will respond to the weak stimulus. More powerful impulses could cause both muscles to contract, since the more resistant synapse* B *will be overcome by the stronger impulses.*

decreased synaptic resistance. Increased strength of stimulus will also cause an increase in response. Another condition that will cause an increase in response is the **reinforcement phenomenon,** which results from simultaneous stimulation of a reflex arc other than the one under excitation. For example, if the subject of a patellar-reflex test clenches the fist or makes any other voluntary muscular movement at the same time that the stimulus to the knee tendon is applied, the extension of the leg is greater than it would normally be. This is explained on the basis that activity in one part of the nervous system tends to reduce **synaptic** resistant than others. Weak impulses are usually shunted through pathways that are less resistant and strong impulses are conducted through the more resistant paths. Also, activity in general has a tendency to reduce resistance (Fig. 4-55). This is substantiated by the fact that whatever increases or diminishes the activity of the central nervous system as a whole will also increase or decrease reflex activity.

Reflex activity may be decreased with any decrease in the irritability of the nervous system, or it may be entirely absent in persons with disease conditions that involve the sensory fibers or the motor fibers of the spinal cord. Conditions that usually decrease irritability of the nervous system are excessive alcohol content of the blood, general fatigue, and sleep. Destruction of the dorsal roots by disease or injury causes, in addition to complete loss of all forms of sensation, loss of spinal reflexes since the ingoing part of the reflex arc is interrupted; muscle tone is also reduced. The exclusion of proprioceptive impulses from the spinal cord prevents the patient from knowing, unless visually, the position of the muscles and joints in the affected area. As a result, voluntary movements are carried out most inaccurately. Disease produced by the invasion of the nervous system by syphilitic germ, as occurs in taboparesis and tabes dorsalis, usually involves the dorsal root fibers. Such patients have lost the patellar reflex but can extend the leg when asked to do so. Disease of the ventral roots causes paralysis of the muscles supplied by these roots with complete loss of voluntary and reflex movements. Diseases, such as poliomyelitis (infantile paralysis), that destroy the motor cells in the ventral horn of the gray matter or cause lesions in this area produce motor paralysis.

Superficial, or Cutaneous, Reflexes. **Superficial reflexes** are those elicited by cutaneous stimulation and are of the three-neuron flexor type. The most important of the cutaneous reflexes are the abdominal, cremasteric, and plantar responses. The **abdominal reflexes** are divided into two upper abdominals and two lower abdominals. The upper ones are elicited by making a stroke beginning from the side and extending to the sternum. The lower ones are obtained by making a light stroke beginning from the side and extending obliquely downward toward the symphysis pubis. Reflex contraction of the abdominal muscles will result if the reflex is present. The absence of abdominal reflexes on both sides may be indicative of pyramidal tract involvement. Excessive upper abdominal reflex and absent lower abdominals may indicate a spinal cord involvement at the level of the tenth thoracic spinal nerve.

Stroking the inside of the thigh of men produces the **cremasteric reflex,** which is a contraction of the cremaster muscle with a drawing up of the testicle on the side tested. Absence of this reflex may confirm pyramidal tract disease if it occurs in conjunction with other indications.

The most important cutaneous reflex is the response of the toes to stimulation of the sole of the foot, known as the **plantar reflex.** This is elicited when the sole of the foot from the heel to the base of the small toe is scratched. The normal response is a flexion of the toes. If the big toe extends upward and the smaller toes spread, this indicates involvement of the pyramidal tract on the side tested. This **abnormal,** or **pathological, reflex** is known as the **Babinski response.**

The **corneal reflex** is elicited by touching the cornea of the eye, which causes a reflex closing of the eyelids. The sensory pathway is the ophthalmic branch of the fifth cranial nerve and the trigeminal, which also innervates the conjunctiva and the upper lid. The center of the reflex is in the pons from which impulses then pass to the superior corpora quadrigemina and leave by the seventh cranial (facial) nerve, which supplies the obicularis oculi muscle. This blinking reflex has a protective function.

Special or Organic Reflexes. **Special reflexes** are those that involve visceral structures rather than skeletal muscles. Some of the more important organic reflexes are the following. The **photopupil,** or **light, reflex** is elicited by shining a light in the eye; the response is the narrowing of the pupils in both eyes, restricting the amount of light that can enter. The sensory pathway is by way of the optic nerve to the pretectal area of the midbrain. Impulses then pass to the Edinger–Wesphal nuclei of the oculomotor nerves on both sides, thus accounting for the fact that when light shines into one eye, both pupils contract. In certain diseases of the nervous system the light reflex is abolished, and this is probably due to a lesion in the pretectal area.

When the gaze is moved from a distant to a near object, the retinal images are blurred since they do not fall on corresponding points on the retina. Accommodation (see p. 196) abolishes the blurring by bringing the divergent rays of light to a focus on the retina, and fusion of the separate images is brought about by convergence. In addition, the pupils contract reflexly, and this is known as the **accommodation-pupil reflex.** It is an advantage to have the pupil contract during accommodation for near vision because the lens is quite curved at this time and its central portion has a different focusing power than its peripheral zone. The constriction of the pupil cuts off light from the peripheral zone so that blurring of the image does not occur, owing to the difference in focusing power of the two areas of the lens. When objects are viewed at close hand, the eyes must converge (that is, a medial movement of each eye must take place); the lens must become more curved to maintain sharp focus, and the pupil must contract to give greater sharpness to the image. Convergence is a voluntary act because it is dependent on the contraction of the internal rectus muscle, which is under the direct control of central motor neurons. Nevertheless, the movement of the eyes is usually an involuntary reflex act; that is, the eye automatically follows the shifting of the retinal images of moving objects. In the case of the movement of the eyes inward, the reflex is spoken of as the **convergence reflex** and is controlled by the nucleus of Perlia, situated between the two oculomotor nuclei. The oculomotor nerve supplies the voluntary recti muscles of the eye. In the nearsighted, accommodation and convergence are not necessary for seeing near objects since the distance between the cornea and the retina in these persons is abnormally great. Therefore, these persons can see clearly only objects that are comparatively near, since the image of near objects will fall correctly on the retina. The myopic person with suitable spectacles must accommodate for near objects in the normal way.

Another organic reflex which is of interest and is associated with the size of the pupil is the **ciliospinal reflex.** This is elicited by stimulation of the skin of the neck, which causes a dilation of the pupils. In fact, stimulation of any sensory nerve (except the trigeminal) causes a dilation of the pupil as a result of the fact that all sensory impulses reach cells in the spinal cord that give rise to sympathetic fibers and cause this dilation. Conditions such as fear, pain, or anger stimulate the pouring out of epinephrine from the adrenal glands, causing dilation as a result of stimulation of the sympathetic nerves.

Physiological Properties of Nerves

The nerve fiber may be stimulated electrically, thermally, chemically, or mechanically. In experimental work on living tissue, the electric current is usually employed as the stimulus because of (1) its convenience, (2) the accuracy with which it can be measured, and (3) the relatively undamaged state in which it leaves the tissue.

A stimulus may be defined as any change in the environment of a tissue that will cause it to react. Any of the above types of stimuli, if applied to a nerve fiber, cause a physical and chemical change in the fiber at the point of stimulation. This change has been called by Keith Lucas the **local excitatory state (l.e.s.).** If this change, or l.e.s., attains a certain value, a wave of excitation is transmitted along the nerve fiber. The developed disturbance is referred to as the nerve **impulse** and its passage from point to point along the fiber, as **conduction.**

In order for a stimulus to induce in a nerve fiber a local excitatory state of sufficient value to set up an impulse, it must meet the requirements of strength and duration.

Intensity or Strength

When electrical stimulation is employed, the intensity corresponds practically to the voltage. A stimulus just capable of producing an impulse is said to have an intensity of **threshold,** or **liminal,** value. Stimuli of less strength are called **subthreshold,** or **subliminal.** A subthreshold stimulus, although not of sufficient strength to generate an impulse, does not, however, leave the nerve unaffected. If a second stimulus of subthreshold strength or a series of such stimuli is applied to the nerve fibers, an impulse can be developed. The first stimulus is believed to cause a local excitatory state, which, although of too low a value to set up an impulse, can be built up to the required level by a second subthreshold stimulus or a series of them applied at regular intervals. This phenomenon is spoken of as the **summation of inadequate stimuli.**

Individual neurons follow the law of "all or none," as do muscle cells. The nerve, as a bundle of many nerve fibers, does not. Nerves stimulated by a strong-intensity current will carry more powerful impulses than nerves stimulated by a current of low intensity. This is due to the fact that a weak stimulus creates impulses in only a few of the nerve fibers making up the nerve under stimulation. The reason only a few fibers may respond to a weak stimulus is that (1) the fibers more centrally located in the nerve may be out of the physical range of the stimulus, and (2) all the nerve fibers making up a nerve do not have the same threshold of irritability; that is, whereas a stimulus of 1 V may be strong enough to create an impulse in some of the fibers, others will require stronger stimuli to make them respond. It is important to remember in this connection that the term *more powerful impulses* merely means an increased number of stimuli, produced either by the excitation of a larger number of sensory nerve fibers or by an increased frequency of stimuli in the same number of fibers.

Duration

A second requirement a stimulus must meet in order to set up an impulse in a nerve is that it must be applied for a very brief but definite length of time. Although it is true that no physiological effect is produced in the nerve by prolonging the period of current flow indefinitely, the current must flow for a certain period in order to be effective. Presumably, then, a certain duration is required for the current to bring about these physical and chemical changes in the nerve on which the local excitatory state depends. The brief effective period after the commencement of the current flow is called the **serviceable,** or **utilization, time.** This is the same as the latent period in a muscle contraction. The time is measured in milliseconds, or sigmas. One thousandth of a second is equal to one sigma. The length of utilization time varies in different tissues and can be determined very easily in the laboratory.

If a current of a certain duration and voltage is just capable of producing an impulse (liminal strength), reducing either the strength or the duration renders the stimulus ineffective. Beyond the utilization time no relationship exists between the strength and the duration of the current, so that when the current flow is of indefinite duration, its effectiveness depends entirely on its strength. Neurophysiologists prefer to call the liminal strength of stimulus for the nerve the **rheobase.** Thus, the rheobase may be defined as that intensity of a current which, when allowed to flow for an indefinitely long period, is just capable of creating an impulse. The relationship between strength and duration of current is illustrated in Fig. 4-56.

Determining Excitability of Nerve Tissue

The irritability, or excitability, of a tissue may be determined in one of two ways:

1. By ascertaining the rheobase, or the minimal strength of current that, when allowed to flow for an indefinite period of time, will create an impulse or below which no propagated response may be elicited, however long the application; this type of determination has been termed an intensity-threshold measurement.
2. By ascertaining the utilization time of a standard strength of current; this has been called the duration-threshold measurement, or **chronaxie.** A current having an intensity of twice rheobase is used, and the length of time it takes a current of this strength to create an impulse is determined. Chronaxie may be defined as the shortest duration of a current necessary for excitation when its strength is twice rheobase.

Electrical Change Associated With the Nerve Impulse

The nerve impulse is not an electric current, it is a wave of physiochemical activity in the nerve which moves rapidly along the fiber (conduction) and is accompanied by a potential change. The wave of activity travels from the cell body to the axon and along this structure to another nerve cell or to nonnervous tissue. As the wave of activity moves, each point it reaches along the axon becomes electrically negative to the inactive regions on each side of it. As a result of this change in electrical charge in the area occupied by the impulse, a difference in potential will exist between the two areas, and a current will flow.[1] This electrical activity can be demonstrated in the laboratory by placing two recording electrodes on an isolated nerve suspended in air, or in any other nonconducting medium, and attaching the electrodes to any adequate recording system, such as a cathode-ray oscilloscope or galvanometer (Fig. 4-57). Because the electrodes are placed a short distance apart on the surface of a resting nerve, the instrument shows no deflection since the entire surface of the nerve is positively charged. We can

[1] Everything physical is made up of **atoms,** but the atom in turn consists of several kinds of still smaller particles. One is the **electron,** the smallest quantity of **electricity** that can exist and which is considered to be **negative electricity.** An atom also consists of a central core, called the **nucleus,** around which one or more electrons revolve. The nucleus has an electric charge of the kind of electricity called **positive,** the amount being equal to the negative electrons circulating around the nucleus. These two opposite charges are strongly attracted to each other, whereas similar charges repulse one another.

A normal atom has an exact balance between positive and negative kinds of electricity. When an atom loses one of its electrons, it has more positive electricity than negative and is considered to be ionized; it is called a **positive ion.** If a normal atom picks up an extra electron, it becomes a **negative ion,** as it now has an excess of negative electricity. A positive ion will attract any stray electron in its area, including the extra electron attached to a negative ion. In this way electrons travel from atom to atom. The movement of ions or electrons constitutes the electric current. The intensity or strength of the current is determined by the rate at which electric charge (an accumulation of electrons or ions of the same kind) moves past a point in a circuit.

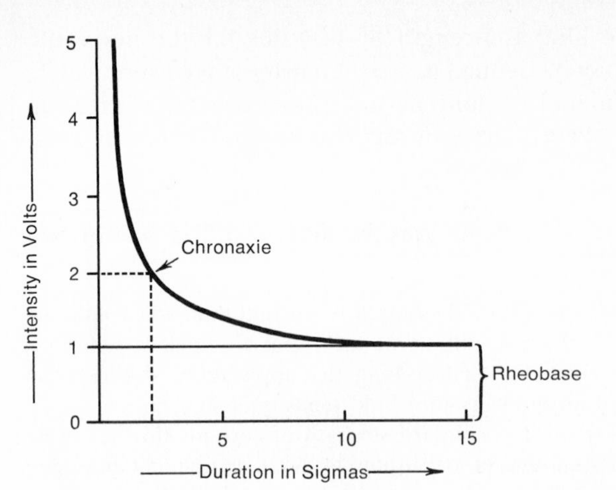

Figure 4-56. Relationship between strength and duration of current. The horizontal line represents strength values; 1 V represents threshold, or rheobase. The vertical line represents duration. The time it takes for a standard strength of current of twice rheobase (2 V on the graph) to flow to produce an impulse (approximately 2.5 sigma on the graph) is termed the chronaxie.

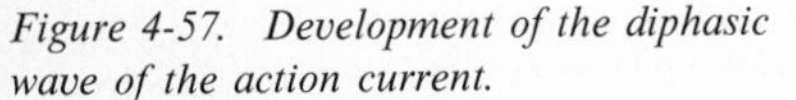

Figure 4-57. Development of the diphasic wave of the action current.

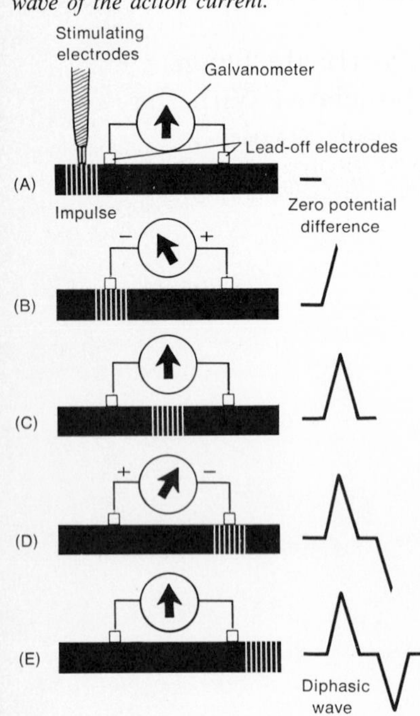

say, therefore, that all points of a resting nerve are isopotential (have similar charge). Stimulation of the nerve at one end will cause the instrument to register current flow as the wave of activity passes the first electrode (I), since this is now negative, to the second electrode (II). As the wave of activity proceeds to electrode II, this will now in turn become negative to the first, and the current will be reversed through the instrument. The apparent reversal of polarity is due to the electrodes' being connected to opposite inputs of the recording instrument.

The movement of the galvanometer needle first in one direction and then in the other may be recorded as a curve in which the needle resting at zero would record a straight base line (Fig. 4-57*A*). As the impulse reaches the first electrode, the needle moves from the zero position to the left, which can be compared to an upward deflection of the base line (*B*). As the impulse moves along the fiber and the area under the first electrode becomes repolarized, the needle falls back to the zero position, indicated on our record by the line's falling back to its base position (*C*). As the impulse passes under the second electrode, the needle shows a deflection to the right, which would be recorded as a deflection of the base line in a downward direction (*D*). As the impulse passes beyond the second electrode, the needle of the galvanometer falls back to the zero position, which would be reflected by the return of our line to the base position (*E*). The record obtained is, therefore, that of the upward deflection of the negative potential change at the first electrode followed by a downward deflection as the second electrode becomes negative to the first, in other words, a wave followed by its inverted mirror image. The interval between the first and second waves is determined by the speed of conduction and the distance between the recording electrodes. The potential change under the first electrode is the electrical sign of the nerve impulse at a single region in the nerve, and the record of this electrical activity is known as the **action potential** of the nerve. Study of the electrical activity of a nerve usually involves the analysis of one wave, since the second wave is absolutely the same, only occurring in a different area of the axon occupied by the impulse. Such a study of the action potential with more sensitive recording instruments reveals it to have several components. The recording instrument of

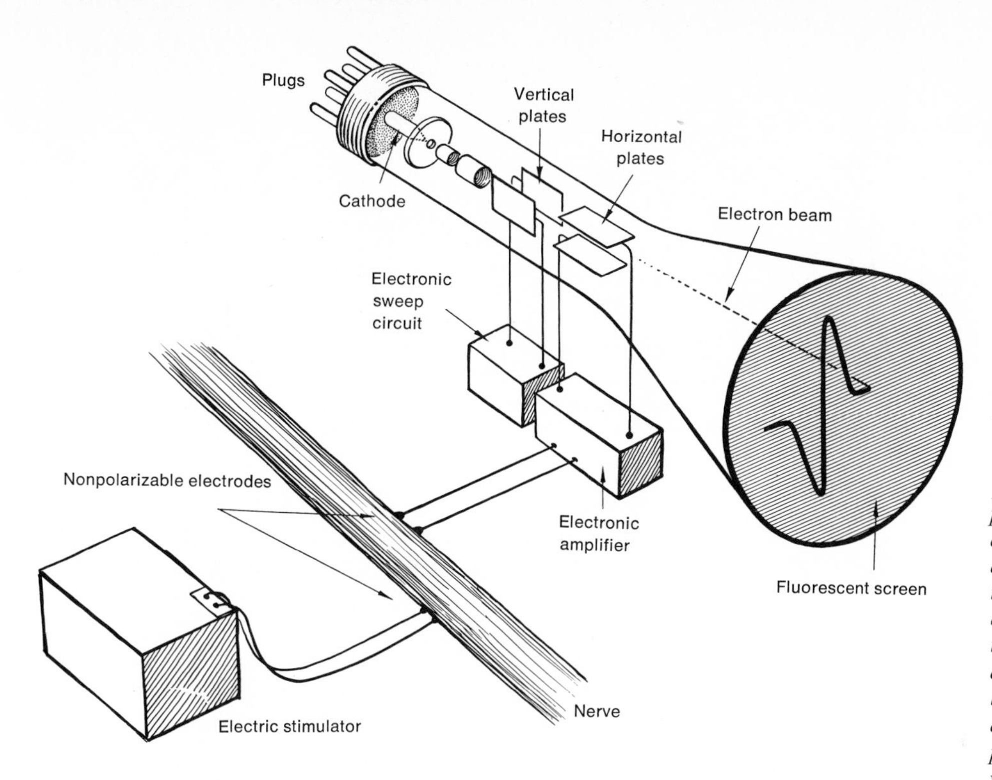

Figure 4-58. The instrument of choice for recording potentials in nerves is the cathode-ray oscilloscope. A cathode produces an electron beam, which is brought to a focus at a point on the fluorescent screen and may be viewed or photographed as a spot of light. A potential difference between the vertical plates causes the spot to shift horizontally from the left of the screen to the right at a certain velocity. Knowing the velocity, one can use this movement as a time measurement. A potential difference between the horizontal plates causes the spot to shift vertically. These plates are connected to an amplifier, which in turn is connected to the nerve by pickup electrodes. When the nerve impulse passes the proximal lead electrode, the electrical disturbance is amplified, causing a potential difference in the upper horizontal plate, which causes an upward deflection of the spot, which rises to a maximum and then declines. The impulse reaching the distal lead electrode causes a charge on the lower horizontal plate, which causes a deflection of the spot downward. The action potential, which produces a diphasic wave when two electrodes are used, may be recorded.

choice is a cathode-ray oscilloscope (Fig. 4-58). Permanent records are obtained by photographing the screen as the spot formed by the electron beam's striking the fluorescent material traces out the electrical changes impressed on the deflecting plates X and Y. Potential changes in the nerve, consequent to activity, are led to a vacuum-tube amplifier by means of recording electrodes and then to the Y (horizontal) plate to cause a vertical displacement of the electron beam. The sweep, operating through the X (vertical) plate, moves the electron beam in the horizontal axis so that the nerve activity is traced as a function of time as well as space.

Action Potential

Gasser and Erlanger have given us much of our detailed knowledge of the action potential of nerve, but the techniques by which it can be studied may be credited to Adrian and Bronk. It has been shown that the action potential is always made up of three components: the **spike,** the **negative after-potential,** and the **positive after-potential** (Fig. 4-59). The **spike** makes up the greatest portion of the action-potential wave and has the shortest duration, about $\frac{1}{2}$ to 1 millisecond. At constant body temperature the spike is of constant duration in all mammalian fibers, the rising potential taking up about one third of the total time; and the falling potential, two thirds of the time. The magnitude of the spike is not dependent on the strength of the stimulus but varies with the diameter of the fiber and the condition of the fiber when it receives the stimulus. Large fibers develop a greater magnitude of spike than smaller diameter fibers and average

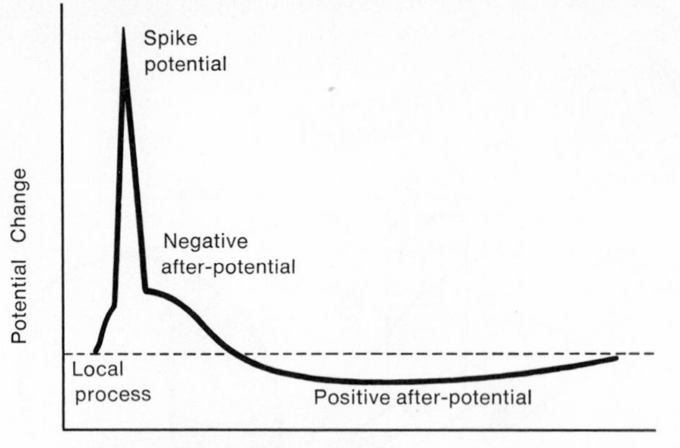

Figure 4-59. Action potential showing its three components: the spike, the negative after-potential, and the positive after-potential. The hatched line represents the base line or zero potential.

about 45 mV. The spike is the basic unit of the action potential and is that part of it which is associated with the passage of the nerve impulse. The physiological event contributing to the development of the action spike is believed to be the depolarization of the fiber membrane initiated by the applied stimulus, the depolarization being brought about by, first, an increased permeability of the fiber membrane to sodium ions. The rapid entry of sodium ions into the cell causes the outside of the membrane involved with the activity to become negative because of the movement of positive ions inward, and the inner surface to become positive. This exchange establishes a potential difference between the active and inactive regions of the neuron and a current flows between the two regions, resulting in the rising phase of the spike. After 1 or 2 milliseconds the increased permeability to sodium ion is switched off and converted into an increased selective permeability to potassium ion. The accelerated outflow of potassium ions, equal in quantity to sodium influx, allows the membrane potential to return rapidly to the original resting level, giving rise to the falling phase of the spike.

The **negative after-potential** component of the action-potential wave begins before the falling-potential part of the spike has a chance to return to zero potential (base line). It is believed to be caused by a change in the rapidity with which potassium ions are flowing through the membrane, and its beginning is seen as a break in the smooth decline of the spike potential, which represents a decrease in the speed of fall to zero. The duration of the negative after-potential is about 15 milliseconds.

The **positive after-potential** follows the negative phase and is represented in the action-potential wave as a slight fall below the base line. It is conventional for neurophysiologists to mark any movement above the line as negative and any movement below the line as positive. This is purely an arbitrary convention. The positive phase is of very low magnitude but is comparatively long lasting, persisting for about 70 milliseconds. It involves the interchange of sodium and potassium ions previously displaced, by the resumption of the normal functioning of the sodium and potassium pumps, which is a metabolic process.

Refractory Period

A very close correlation exists between the potentials of a nerve fiber and its ability to be excited. The three phases of potential change discussed previously (the spike, the negative after-potential, and the positive after-potential) are accompanied by three different levels of excitability.

If a second threshold stimulus to a nerve follows the original stimulus too closely, no response will be elicited in the nerve. It seems that the nerve fiber needs about one-thousandth of a second to recover from the passage of the *first* spike potential before it can transmit another. This period of time during which the nerve will not respond to another stimulus has been referred to as the **refractory period** of the nerve. More thorough study of this period has shown that it is made up of two parts:

1. An **absolute refractory period,** in which no stimulus, however strong, will excite the nerve fiber; this period coincides with the rise of the spike potential and with its fall to the point where the negative after-potential begins (Fig. 4-59); in large fibers the absolute refractory period lasts about 0.5 millisecond and probably corresponds to most of the time a given region is occupied by the active process.
2. The **relative refractory period,** in which the nerve fiber will respond to another stimulus, but the stimulus must be stronger than normal; this phase corresponds to the action-potential wave where the spike potential passes into the negative after-potential; the threshold returns to normal in about 0.5 to 2 milliseconds, depending again on fiber size.

After the relative refractory period in which the threshold of excitability of the nerve is raised, the nerve fiber passes into a **supernormal phase,** in which stimuli lower than threshold will excite the fiber. This period coincides with the negative after-potential phase of the action wave. The nerve fiber then passes into a subnormal phase in which again it takes stronger stimuli than normal to make the nerve fiber respond. This subnormal phase corresponds to the time in the action-potential wave occupied by the positive after-potential. All threshold variations cease after about 80 milliseconds.

The fact that during the passage of a spike potential a fiber is refractory to stimuli limits the number of impulses per unit time in any one fiber. We can calculate the maximum number of impulses per second that a nerve fiber is capable of transmitting by determining the refractory period. In large fibers the absolute refractory period is about 0.5 millisecond; thus a stimulus, no matter how strong, cannot develop a response more often than 2,000 times in a second. In natural conditions in the body, the frequency of impulses in any fiber is no more than 10 to 500 per second. Thus a large margin is available to deal with unusually intense stimuli.

Mechanism of Impulse Transmission: Conduction

Most neurophysiologists maintain that electrochemical changes are responsible for the passage of the nerve impulse. This is the most widely accepted theory, and it assumes that nervous conduction and excitation are the result of altered permeability of the nerve membrane. Originally proposed by Ostwald, it has received its most complete elaboration by Lillie and Bernstein.

According to the theory, the neuron is surrounded by a semipermeable membrane, probably a very thin surface layer made up of fatty material only one or two molecules thick, which is normally polarized, that is, has positive sodium ions on the outside and organic negative ions on the inside. This state of affairs exists as a result of the selective permeability exhibited by cell membranes to sodium and potassium ions. Measurement shows that about 20 times as much K^+ is concentrated within the cell as outside the cell. Concerning Na^+, there is approximately 10 times as much sodium outside the cell as inside. Negatively charged ions of both inorganic chloride and organic forms distribute themselves as best they can to achieve an electrically neutral state on the two sides of the membrane. However, the complex system of ions involved simply cannot overcome the selective property of the membrane in concentrating K^+ inside the cell and Na^+ outside, and therefore the ions cannot distribute themselves so that the same number of both ions and electrical charges appear inside and outside the cell. As a result, in the resting cell, the degree to which the ions involved are incapable of equally distributing themselves represents the 70-mV potential difference across the membrane; that is, the inside of the cell is almost always 70 mV negative to the outside, and this is called the **resting potential.** By convention, this resting potential is written with a minus sign to signify that the inside of the cell is negative to the exterior. This potential difference can be measured by microelectrodes, one inside the nerve cell and one on its surface. Connecting these electrodes to a voltmeter will illustrate how unequal distribution of ions can produce a voltage difference of 70 mV across the membrane (Fig. 4-60).

The membrane becomes more permeable (depolarizes) locally under the influence of a stimulus, so that there is a violent exchange of positive and negative ions previously separated by the membrane. This exchange establishes a potential difference between the active and inactive regions of the neuron, and a current

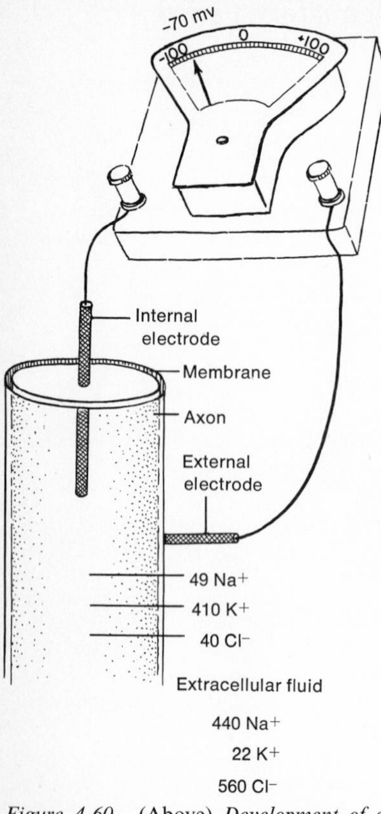

Figure 4-60. (Above) *Development of a potential difference of 70 mV across the nerve membrane because of unequal distribution of ions on either side of the membrane. A microelectrode must be placed on the inside of the nerve cell and one on its surface to measure this potential difference.*

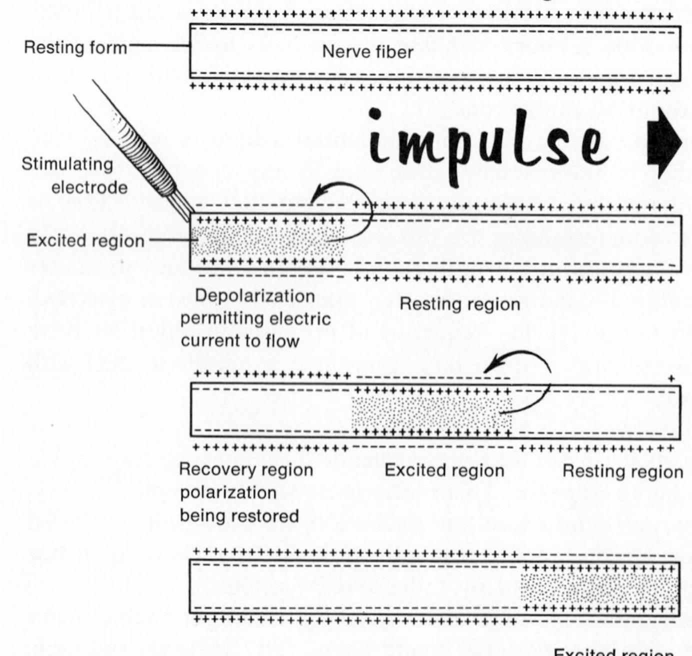

Figure 4-61. (Right) *Membrane theory of nerve conduction.*

flows between the two regions. The electric-potential difference across one point of the fiber membrane serves to excite the region ahead, with the result that this region now contributes a greatly amplified electric signal, capable of spreading to and exciting the next region. The nerve fiber is, in effect, a chain of relay stations. Each point along the fiber receives an electric signal from the preceding point, boosts it to full strength, and thus enables it to travel a little farther (Fig. 4-61). Experiments have fully confirmed this concept of how a nerve fiber transmits a signal. Hodgkins, at Cambridge, and Lorente de No, at the Rockefeller University, have demonstrated that excitation of a point on the fiber actually generates enough current to excite the next area. In fact, they found that a little more than 10 per cent of the current generated by the nerve is sufficient to excite the normal nerve fiber. In nonmedullated fibers the signal is relayed from point to adjacent point. The type of fiber that is surrounded by an insulating sheath of the fatty material myelin has its signal relay stations distributed at relatively few points, namely, at the nodes of Ranvier. Such medullated fibers pass impulses some ten times as quickly as nonmedullated fibers of the same diameter. The insulating layer restricts the time-consuming process of relaying electric signals to a few points along the lines.

Rates of Conduction

Helmholtz in 1850 was the first to study and measure conduction velocity in the nerve. Since then, many observations have followed, mostly based on measurements of motor or sensory reaction times. Erlanger and Gasser, and Bishop and Heinbecker, contributed much to the understanding of conduction velocities with their recognition of the three different types of fibers of which nerves are composed. The three types recognized have been referred to as A, B, and C fibers. The A fibers are the largest, ranging in size from 20 to 1 μm, and are the myelinated fibers of somatic nerves. The B fibers are the small myelinated fibers found in autonomic nerves, and the C fibers, which convey pain and warmth, are the smallest and have little apparent myelin. Rates of conduction vary according to fiber diameter. Large fibers conduct at a faster rate than small ones. The ratio is such that, for myelinated fibers in warm-blooded animals, multiplying the diameter by 8 gives the approximate conduction rate in meters per second. An axon 20 μm in diameter conducts about 160 m per second. The formula $V = KD$ describes the relationship of velocity to diameter, where V = velocity, K = a constant, and D = diameter. Wherever an axon arborizes into terminal branches, these branches conduct at a reduced velocity. Nonmyelinated fibers conduct more slowly than myelinated fibers.

Chemical and Thermal Changes

The metabolism of resting nerve is like any other living tissue in that oxygen is consumed and carbon dioxide is given off. The principal energy material in maintaining the resting state is glucose and also phospholipids. Excitation and conduction involve an increase in oxygen uptake and carbon dioxide production, but the passage of the nerve impulse takes energy from the fiber for its propagation and this is thought to come, not from glucose, but from the breakdown of phosphocreatine, which contains energy-rich phosphate bonds.

The role of acetylcholine in nerve metabolism is the subject of great controversy.

With the data available, the hypothesis that acetylcholine release is a fundamental step in the process of axon conduction has been open to question. Although the role of both acetylcholine and epinephrine is of primary importance for the transmission at the synapse, the presence of one or the other of these substances in some nerve fibers is believed to be unimportant for the propagation of impulses.

A. V. Hill has shown from a series of studies that the passage of a nerve impulse is accompanied by a heat production of approximately 10^{-7} cal per gram of nerve fiber. The major part of the heat produced is related to the restoration of the polarized state since it is maximal during this period, and only a small fraction is given off during the development of the spike potential. The amount of heat produced is directly related to the observed consumption of oxygen and the liberation of carbon dioxide. The active nerve consumes about one third more oxygen than does the resting nerve, and most of this oxygen is utilized during the phase of maximal heat production in the process of restoring resting conditions and not during the development of the spike potential.

References

Austin, G. *The Spinal Cord.* Springfield, Ill.: Thomas, 1971.

Bourne, H. G. (ed.). *The Structure and Function of Nervous Tissue.* Vols. 4 and 5. New York: Academic, 1972.

Bowman, R. E., and S. P. Datta (eds.). *Biochemistry of Brain and Behavior.* New York: Plenum, 1970.

Carson, D. (ed.). *Physiological and Biochemical Aspects of Nervous Integration.* Englewood Cliffs, N.J.: Prentice-Hall, 1968.

Eccles, Sir John. *The Inhibitory Pathways of the Central Nervous System.* Springfield, Ill.: Thomas, 1969.

Eleftheriou, B. E. (ed.). *Neurobiology of the Amygdala.* New York: Plenum, 1971.

Fiorentino, M. R. *Reflex Testing Methods for Evaluating C.N.S. Development.* Springfield, Ill.: Thomas, 1970.

Hockman, C. H. *Limbic System Mechanisms and Autonomic Function.* Springfield, Ill.: Thomas, 1971.

Lajtha, A. (ed.). *Handbook of Neurochemistry,* Vols. 1–8. New York: Plenum, 1972.

Lasansky, A. "Nervous function at the Cellular Level: Glia," in: *Annual Review of Physiology,* Vol. 33. Palo Alto, Calif.: Annual Reviews, 1971, p. 241.

Martini, L., M. Motta, and F. Fraschini. *The Hypothalamus.* New York: Academic, 1971.

Reynolds, D. V., and A. E. Sjoberg. *Neuroelectric Research.* Springfield, Ill.: Thomas, 1971.

Rodieck, R. W. "Central Nervous System: Afferent Mechanisms," in: *Annual Review of Physiology,* Vol. 33. Palo Alto, Calif.: Annual Reviews, 1971, p. 203.

Rusinov, V. S. (ed.). *Electrophysiology of the Central Nervous System.* New York: Plenum, 1970.

Russell, S. M., and J. W. Maran. "Brain-Adenohypophysial Communication in Mammals," in: *Annual Review of Physiology,* Vol. 33. Palo Alto, Calif.: Annual Reviews, 1971, p. 393.

Sidman, R. L., and M. Sidman. *Neuroanatomy: A Programmed Text.* Boston: Little, Brown, 1972.

5

Sense Receptors

An animal is under continual influence from stimuli in its environment. These stimuli, which are represented by physical and chemical changes, are capable of producing physiological reactions through which an animal can obtain important information about its surroundings. This information is needed by the animal to find food, mates, and dwellings and to protect itself from its enemies and other hazards that may beset it in its environment.

Stimuli may arise from the outside in the form of environmental changes in temperature, pressure, light, and vibrations, or they may arise from processes going on within the organism that cause change in the internal environment. The essential thing about any stimulus is the element of change, or the fact that something different in kind or degree is taking place in the surroundings of an animal. These changes excite specialized endings of sensory nerves, called **receptors,** which pass the impulse by way of the sensory nerve fibers to the brain for proper interpretation and also to motor neurons for proper action (reflex). The brain is the headquarters for all sensory perception, and the receptor and its sensory pathway are avenues of approach to the central nervous system. Thus we find that environmental changes are detected by the receptors, the sensory cortex of the **brain** interprets these changes as sensations, and the responses to these changes are carried out by **effectors.**

The fact that the interpretation of sensory impulses is carried out within certain areas of the sensory cortex does not mean we will feel the sensation in the brain, but, on the contrary, the sensation is felt in the area where the stimulus originates, that is, at the receptor. This is known as **sensory localization,** or **projection.** If the hand is placed on a hot object, the subject knows the point of stimulation. He is able to localize the stimulus by projection. Sensory neurons from each part of the body maintain definite pathways throughout the spinal cord and are projected onto a specific area of the sensory cortex. The individual learns to associate stimuli coming into a specific area of the cortex with particular parts of the body.

Receptors

Receptors are usually modified sensory nerve endings that are specialized to respond to stimuli of a single kind. For example, the sensory receptors of the eye will respond only to light waves; the taste buds of the mouth, only to chemicals in solution; the receptors of the ear, only to sound waves; and the receptors of the nose, to chemicals in the form of gas. This exclusive ability to respond to only one type of stimulus is a very great advantage, since in this way protection is gained against unnecessary confusion arising from stimulation other than the one to which the receptor in question has become adjusted.

Receptors were formerly classified according to the sensations produced through them. Thus, we had receptors for touch, taste, smell, sight, hearing, and the thermal sense. Today the classification is more complex and involves not only the subjective interpretations occurring in the cerebral hemispheres, but also the types of receptors and their adequacy for particular stimuli.

Receptors can be classified according to their location in the body. Those receptors that are superficially located, that is, in the skin, or receive stimuli directly from the environment are referred to as **exteroceptors.** Those receptors that are located around joints, in tendons, and in muscles are called **proprioceptors,** and those that are more internally located and associated with the visceral structure are referred to as **interoceptors.**

Exteroceptors

Most of the **exteroceptive receptors** occur in the skin and are differentiated by the type of sensation to which they give rise. We have receptors for touch, pressure, pain, cold, and heat. More complex exteroceptive receptors are the ear and the eye, which are responsible for our senses of hearing and sight. In the mouth, where the taste buds are located, are the receptors giving rise to our sense of taste, and in the nose are the receptors responsible for our sense of smell. The exteroceptors are classified into the following types:

Cutaneous receptors	1. Tactile receptors	Sense of light touch
	2. Pressure receptors	Sense of pressure
	3. Pain receptors	Sense of pain
	4. Thermal receptors	Sense of heat and cold
Special receptors	5. Auditory receptors	Sense of hearing and equilibrium
	6. Photoreceptors	Sense of sight
	7. Gustatory receptors	Sense of taste
	8. Olfactory receptors	Sense of smell

The receptors for the **sense of touch** are of three specialized anatomical types and are recognized as **Meissner's corpuscles, tactile disks,** and **free nerve endings** associated with hair follicles (Fig. 5-1). Deformation of any of these endings initiates nerve impulses, which are transmitted to specific areas of the cerebral cortex by sensory nerves and are there interpreted as light touch. Tactile receptors are distributed in groups throughout the skin and are known as *touch spots.*

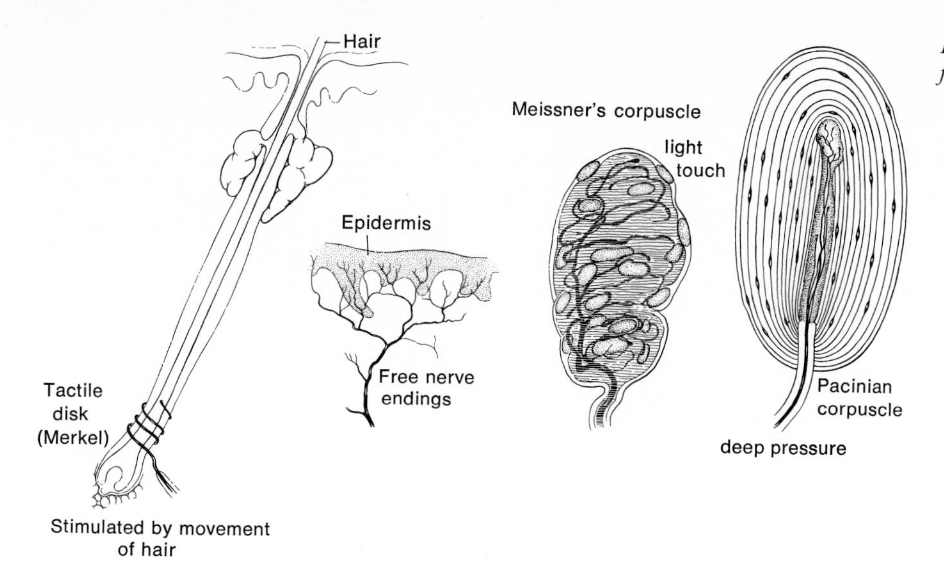

Figure 5-1. Various types of exteroceptors found in the skin.

Stimulation of areas between the spots does not give rise to tactile sensation. The fingertips are the most sensitive areas for touch because the tactile receptors are more numerous here than in any other area (Fig. 5-2).

Pressure receptors are represented in the skin by encapsulated nerve endings; that is, the terminal branches of the sensory neuron are encased by successive layers of connective tissue (Fig. 5-1). This morphological type of receptor is called the **pacinian corpuscle** and is stimulated when touch is sustained beyond transient contact. Although in many instances the same receptors give the sensations of both touch and pressure, the two sensations are differentiated. The sense of touch is more definitely localized in both time and space and varies in intensity from that of pressure. Another encapsulated end organ concerned with pressure is the **Golgi-Mazzoni corpuscle** (Fig. 5-3).

Pain receptors are the least differentiated of all the receptors. They are found as **bare nerve endings** and are widely distributed throughout the skin and mucous membranes of the body. The activity set up in this type of receptor by even a single stimulus tends to be repetitive, and there is little adaptation. This means that under prolonged stimulation the receptors continue to discharge at their initial frequency. Pain receptors differ in this respect from the more differentiated receptors, whose rate of discharge tends to fall off rapidly as the stimulus persists.

Another, more highly differentiated type of ending (Fig. 5-3) is the **end bulbs of Krause,** which respond to skin temperatures that fall below that of normal skin and thus give rise to sensations of cold. The **corpuscles of Ruffini** are excited when the temperature of the skin is raised above normal, and on reaching the brain, these impulses are interpreted as warmth.

It is assumed that five primary sensations can be derived from the skin (touch, pressure, pain, heat, and cold). Each of these five sensations has specific receptors, adapted to transmit only one kind of sensory impulse. All other cutaneous sensa-

Figure 5-2. Enlarged section of the skin of the fingertip, showing the location of the cutaneous receptors.

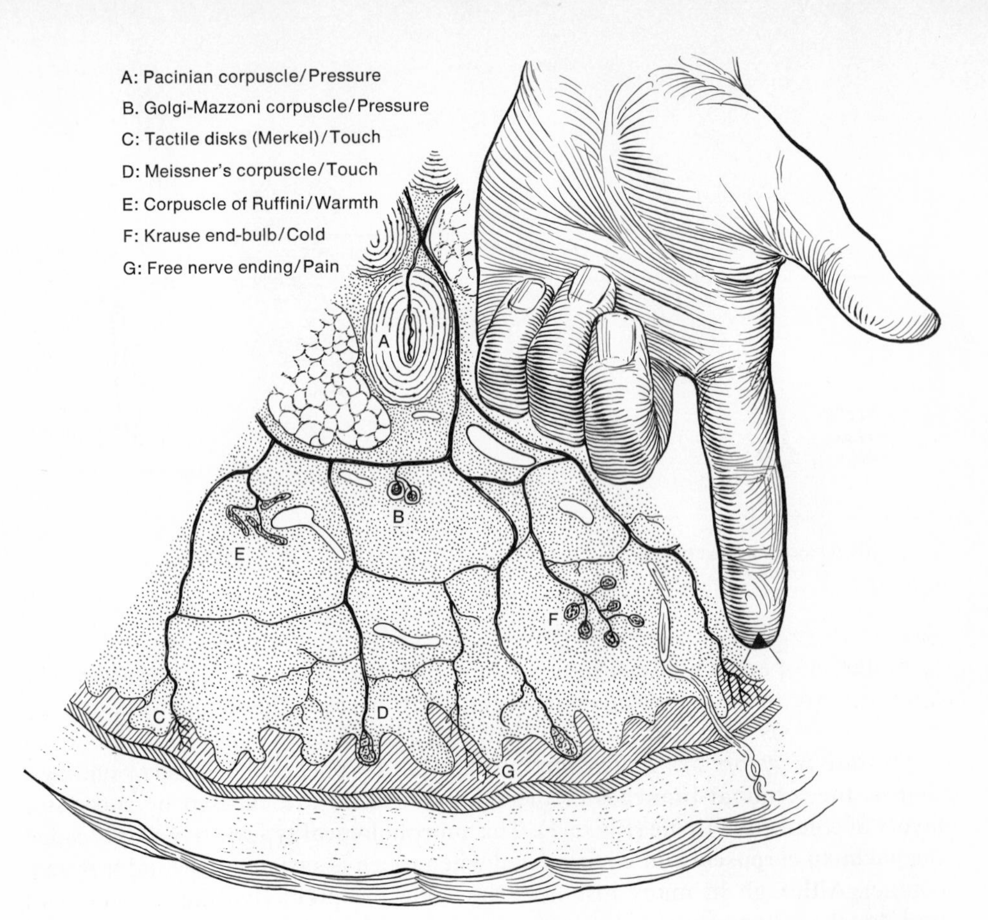

Figure 5-3. Exteroceptors found in the skin concerned with pressure, cold, and warmth.

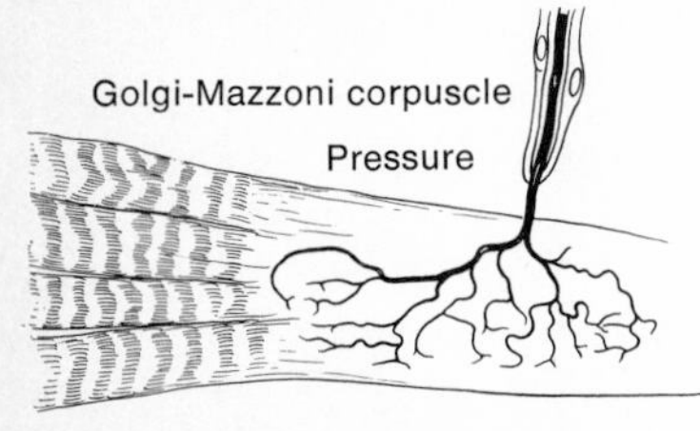

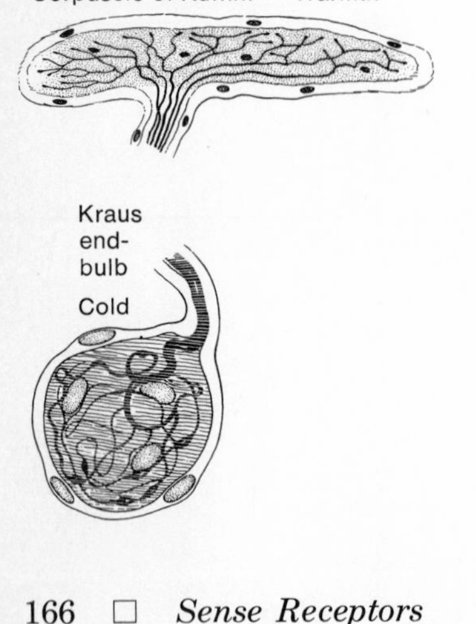

tions, such as tickling, itching, burning, and crawling, are presumed to be derived from some combination of these five primary modalities of cutaneous sensibility.

The thalamus of the brain is the receiving center for all types of sensory impulses (Fig. 4-13). It is also a center for awareness and apparently represents the seat of primitive emotion. The thalamus and, to a greater extent, the sensory cortex combine sensations into a pattern of spatial organization on the basis of an extremely accurate representation of the body as a whole. The thalamus sends projections to all parts of the cortex (Fig. 5-4), but the pathways subserving cutaneous sensation are concentrated in the postcentral convolution lying just behind the fissure of Rolando (Fig. 5-5). While the sensory cortex is the principal area for sensory perception, sensation involves broader cerebral activity. The sensory cortex is in intimate functional relationship with other parts of the cortex, so that all sorts of associations based on past experience can be called into play in evaluating a particular sensation. The individual can evaluate various levels of sensory impressions to localize accurately the source of stimulation (projection) and to make the discriminations that characterize the cerebral function of sensory experience.

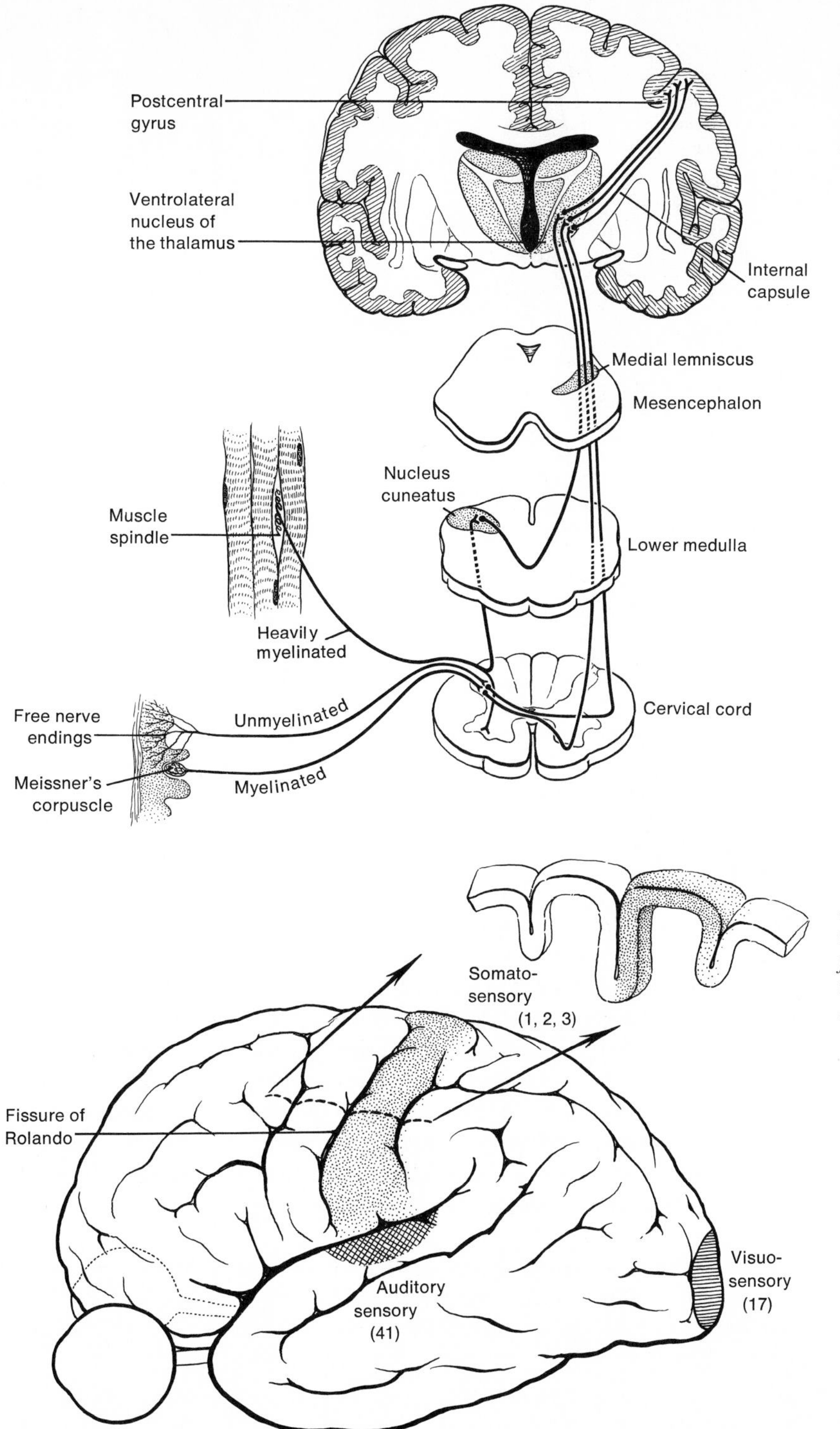

Figure 5-4. Sensory paths of the cerebral cortex. The thalamus sends projections to all parts of the cortex, but those subserving cutaneous sensations are concentrated in the postcentral convolution lying just behind the fissure of Rolando.

Figure 5-5. Extent of cutaneous projection to the cortex. The insert demonstrates the fact that the strongest projection reaches the buried part of the posterior wall of the central sulcus.

Figure 5-6. Type of proprioceptor found in muscle. Specialized spindle-shaped fibers found in skeletal muscles receive the sensory receptors, which are stimulated by any increase in muscle tension caused by lengthening or stretch. The endings of these sensory nerves are of two types: (1) annulospiral and (2) flower spray.

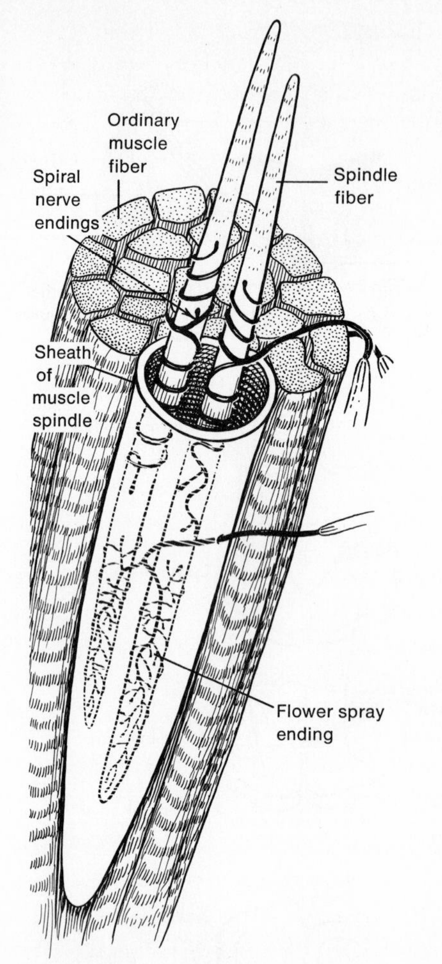

Proprioceptors

Proprioceptors are found in the fibrous cuff surrounding joints, in muscles, and in tendons. This receptor is represented by two morphological types, consisting of a nerve fiber end that winds around the cells of the above-mentioned structures to form a spiral or as a flower-spray ending (Fig. 5-6). Such a terminal is called a **neuromuscular spindle** in muscle and is stimulated by any increase in muscle tension due to lengthening or stretching.[1] Impulses initiated by this type of activity will set up a reflex act in that the muscle stretched will be stimulated by way of a motor neuron to contract to its original length. Those muscles that support

[1]Muscle spindles are highly complex structures that are found at the edge of a fasciculus or in the perimysium of the fleshy portions of most muscles. As the name suggests, the spindle consists of a spindle-shaped connective tissue capsule within which a specialized region of one or more muscle fibers is embraced by a complex sensory receptor. These specialized muscle fibers, sometimes called *intrafusal fibers,* are of very small diameter. It is the intrafusal fiber that receives the sensory innervation for proprioception. The endings of these sensory nerves on the intrafusal fiber may be of two types: (1) annulospiral and (2) flower spray.

the body against gravity are richly supplied with this type of receptor. Those nerve fibers ending in a spiral in tendinous structures are called **neurotendinous spindles.** They are stimulated by the tension produced in a tendon when its attached muscle is either stretched or contracted. The spindles found in the connective tissue capsule surrounding joints are stimulated by changes in position caused by movement at the joints. This gives rise to our position sense, which is the quality enabling a person whose eyes are closed to know just where his body and limbs are in space.

The proprioceptive receptors are very important in reflex activity. A clearer insight into the function of these receptors may be derived from a review of reflex activity discussed on pages 143 to 153.

Interoceptors

The structure and function of **interoceptive receptors** are not as well known as those of the other types of receptors discussed. Many appear to be free nerve endings terminating in the walls of the various organs, glands, and vessels that constitute the viscera of the body. The nerve endings appear to be stimulated on distention of the walls of these structures, bringing about a specific reflex activity. For example, nerve endings are found in the muscular wall of the heart, the aorta, and the large veins entering the heart, which appear to be particularly sensitive to change in blood pressure. There are nerve fibers which branch profusely throughout the lungs that are stimulated when the lungs are expanded during inhalation. Other types of interoceptors exist, but they have been less extensively studied.

The Special Senses

The **special senses** include the senses of taste, smell, hearing, sight, and equilibrium. They are classified as special senses because all the receptors for these modalities are located in the cranial region, and their sensory paths are over cranial nerves. In addition, the receptors of many of these are associated with, or are part of, more highly developed complex organs, which serve to facilitate the reception of specific stimuli.

The Gustatory Sense

Taste is a complex perception that usually involves not only the special chemical sense of taste, but also the senses of odor, temperature, and touch. By closing the nose and tasting aqueous solutions at the temperature of the body, the individual can distinguish four tastes: sweet, sour, salty, and bitter and, possibly, two more, alkaline and metallic.

Taste Buds

The receptors for taste are complicated end organs embedded in structures called **taste buds,** which are found in the tongue, soft palate, and the beginning of the throat (Figs. 5-7 to 5-9). Most of the taste buds are located in the tongue and are distributed in a definite pattern over the tongue. Those that are located toward the back of the tongue on its upper surface are responsible for the bitter taste. The taste buds located on the lateral surface of the tongue are excited by stimuli that are interpreted as salty and sour tastes, and those located near the tip of the tongue are responsible for the sweet taste (Fig. 5-10). Although four primary taste qualities are recognized in man, histological examination of the

Figure 5-7. Upper surface of the tongue. The tongue is made rough by numerous elevations called papillae. These papillae are of three kinds: (1) the high and narrow filiform papillae, (2) the low and wide fungiform papillae, and (3) the circumvallate papillae. The organs of taste, called taste buds, are most easily found in the side walls of the circumvallate papillae.

Figure 5-8. Mucous membrane covering the tongue greatly enlarged to show the location of taste buds in the papillae.

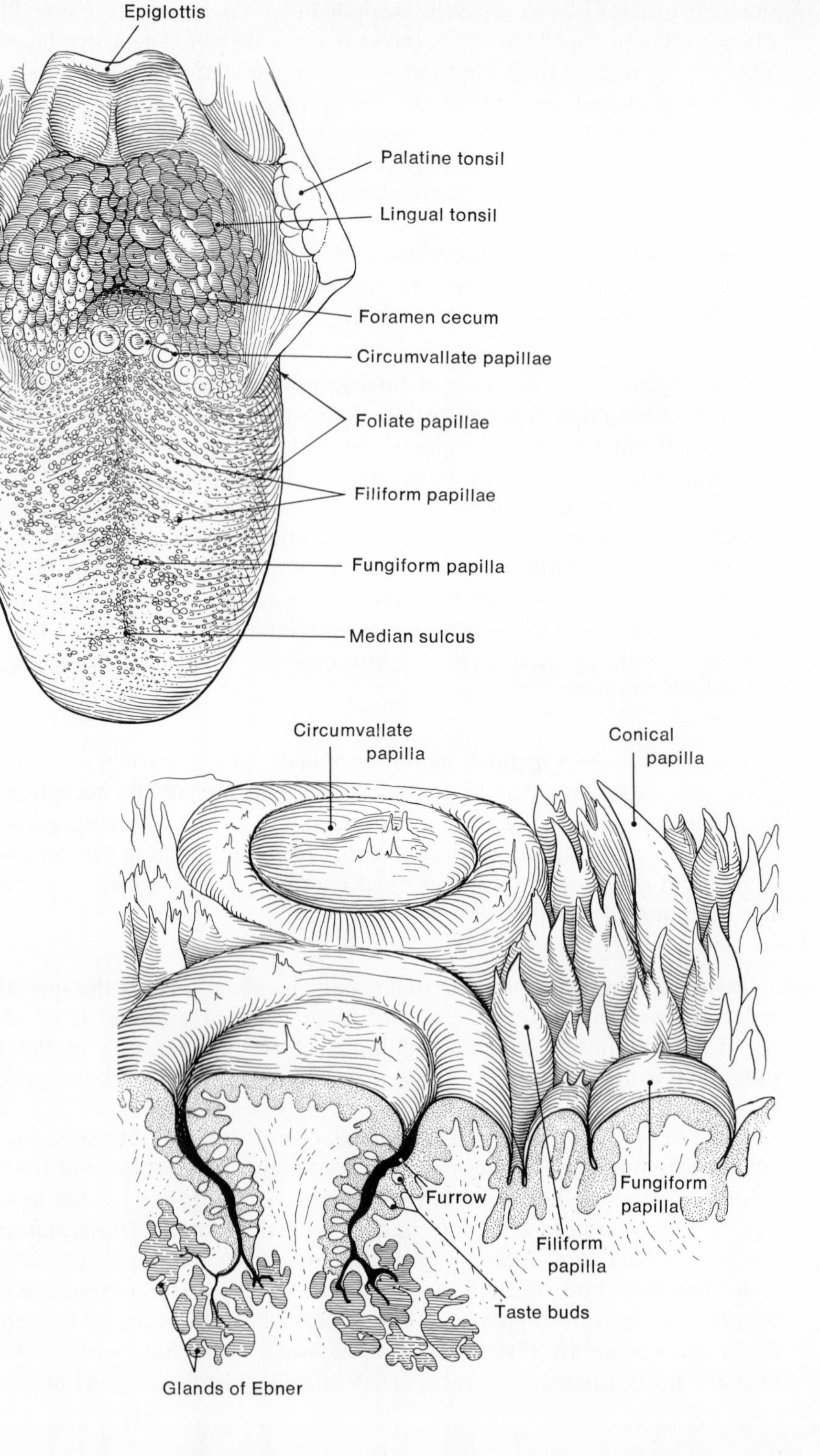

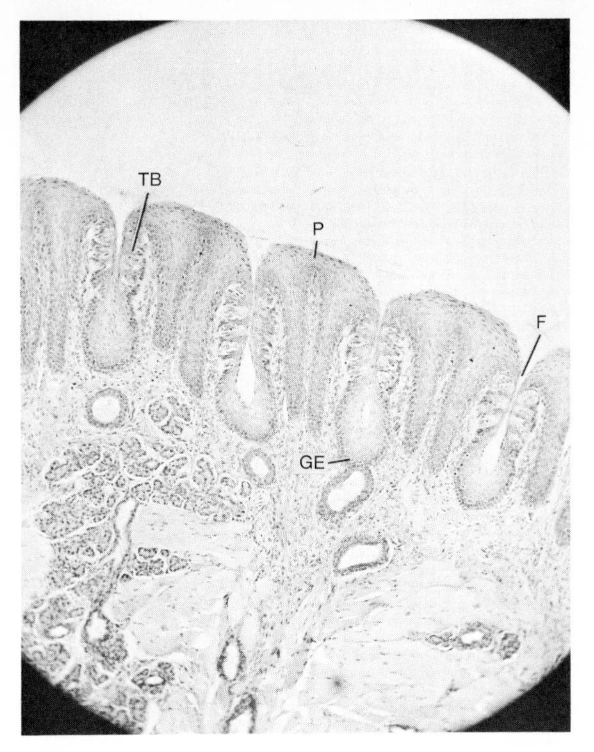

Figure 5-9. (Left) *Taste buds of the mammal.* TB, *taste bud;* P, *papilla;* F, *furrow;* GE, *glands of Ebner.* (*Courtesy Ward's Natural Science Establishment, Inc., Rochester, N.Y.*)

Figure 5-10. (Below) *Taste areas of the tongue. Impulses from taste receptors are carried by cranial nerves V, VII, and IX. V* (*trigeminal*) *and VII* (*facial*) *serve the sides and tip of the tongue. The back and sides are served by IX* (*glossopharyngeal*).

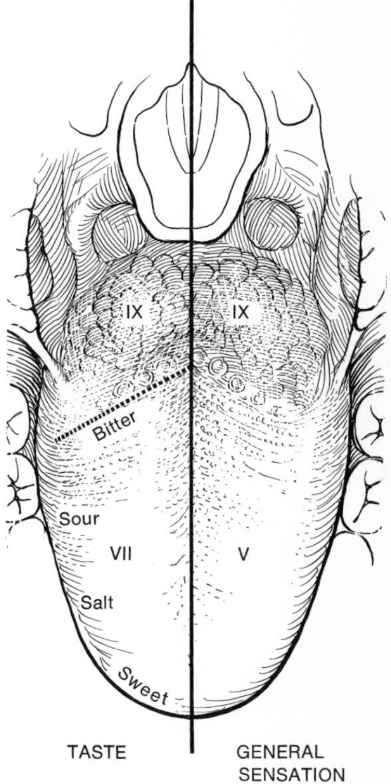

taste receptors reveals no difference in structure and composition. The structure of a typical taste bud is shown in Fig. 5-11.

Other tastes experienced by man are believed to be blendings of the four primary sensations: sweet, sour, salty, and bitter. Variations that are distinguished in a single taste are produced by interpretation of composite stimuli from taste, smell, and thermal receptors. Substances that have characteristic taste become either tasteless or very different in quality if held in the mouth while the nose is plugged or the temperature of the mouth cavity is changed.

Four of the cranial nerves are involved in the transmission of impulses from taste receptors to the sensory cortex of the brain. The back and sides of the tongue are served by the glossopharyngeal (IX cranial), and the sides and tip of the tongue are served by the lingual branch of the trigeminal (V cranial) and the chorda tympani of the facial (VII cranial). The laryngeal branch of the vagus (X cranial) serves the pharyngeal surface of the tongue (Fig. 5-12).

Stimulus for the Four Primary Qualities of Taste.

The process of excitation of the taste receptors appears to occur in the hairlike processes that protrude from the free surfaces of the taste cell (Fig. 5-11). It is assumed that these hair cells are so differentiated from one another as to be highly selective in their response to the chemical composition of substances in solution. It has, for example, been recognized that the **sour taste** is common to all acids. It is also a fact that all acids ionize in aqueous solution into negative ions and

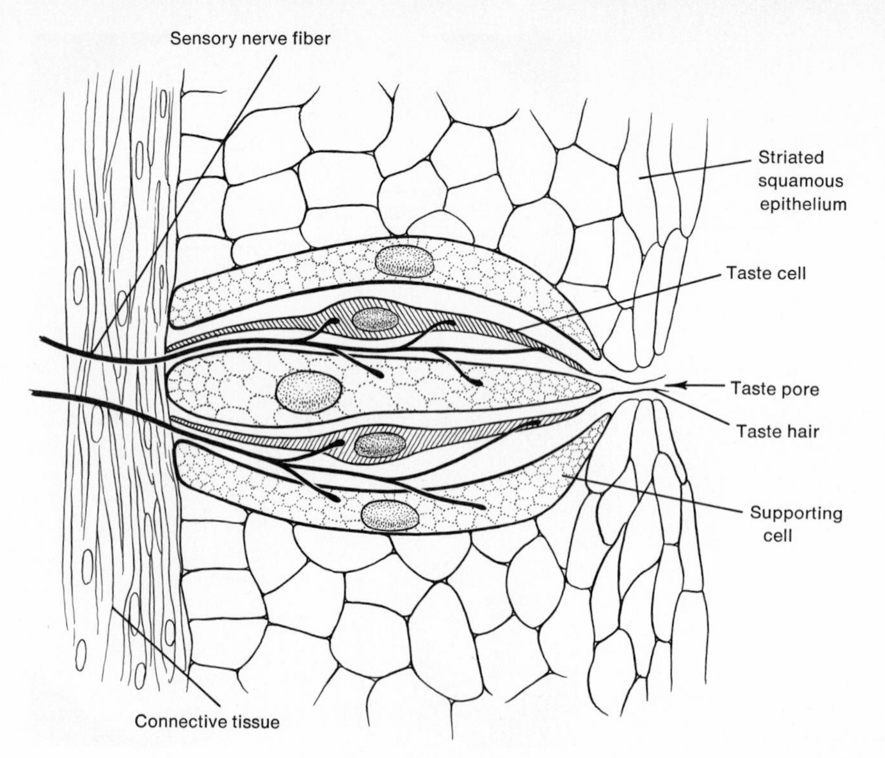

Figure 5-11. Section of taste bud showing the taste cell, the taste pore, and the hair-like processes that protrude from the free surfaces of the taste cell.

positive hydrogen ions. The higher the degree of dissociation of the acid, the greater is the concentration of hydrogen ions and the more sour the taste. The hydrogen ion appears to be the exciting agent in creating sour taste, and the degree of sourness is proportional to the concentration of these ions. In weak acids that do not dissociate to a great extent, such as acetic acid, the sour taste depends on both the hydrogen-ion concentration and the total undissociated acid.

The **salty taste** is commonly associated with the various inorganic salts, which may be divided into three groups with respect to the degree of saltiness elicited by their presence in the mouth at normal temperature. Those that produce a dominant salty taste include the common salt, sodium chloride, and the chlorides of potassium, ammonium, lithium, and calcium. Inorganic salts that produce a combination of both salty and bitter taste include potassium bromide and ammonium iodide. Those salts that produce a slightly salty taste include the bromides and iodides of cesium and rubidium and the iodide of potassium. Experimental evidence indicates that it is the negative ion of the salt that produces the taste and not the positive ion, but the positive ion has a specific effect since chlorides do not taste the same, nor do bromides or iodides. The chloride ion has a greater exciting ability to produce salty taste than either the bromide or iodide ion. Temperature has an effect on the degree of saltiness, as a warm salt solution tastes less salty than a cold salt solution.

The receptors for the **bitter taste** are concentrated in an area toward the back of the tongue. A variety of substances are able to elicit the bitter taste, but it

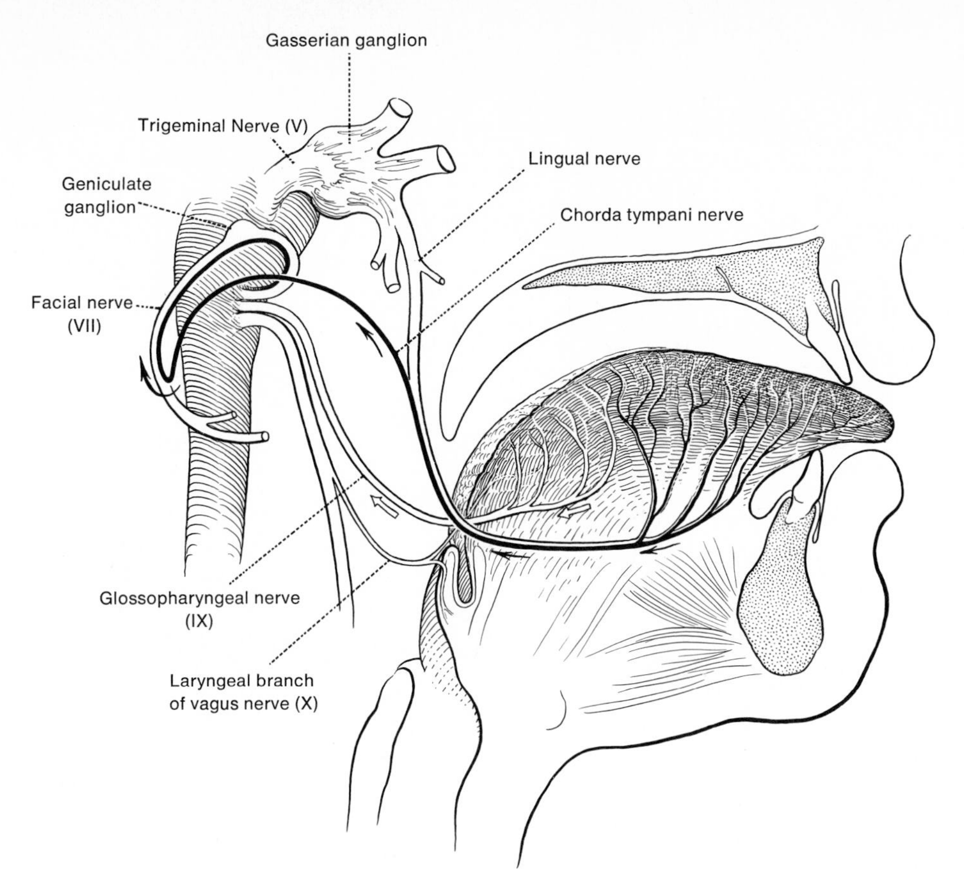

Figure 5-12. Sagittal section of oral cavity showing nerve pathways for the sense of taste. The sensation of taste does not pass through the gasserian ganglion or its sensory root. The glossopharyngeal nerve carries most of the taste fibers in man.

appears that all compounds that have three NO_2 groups in the molecule produce the bitter response more strongly than do others. The exact chemical structure that is able to excite the bitter receptors is not known.

The **sweet taste,** like the bitter, is produced by a variety of organic compounds, which as a rule are not ionized. Oertly and Myers postulated that a substance in order to be sweet must contain two different kinds of groups in the same molecule. One group they called a *glocuphore* and the other an *auxogluc.* These observations, however, lead to no more definite conclusion than do corresponding studies of bitter taste, as there are many compounds that produce the sweet taste but are not related in any way to the postulated structures.

The Olfactory Sense

The receptors for the sense of **smell** are situated very high in the interior of the nose between the median septum and the superior turbinate bone. This area is referred to as the **olfactory cleft** (Figs. 5-13 and 5-14) and is further differentiated from the other areas of the nasal cavity by its yellow-brown color. The pigment appears to be connected with the mechanism of the olfactory apparatus. The olfactory receptors consist of long, narrow pigmented cells, which have six to eight protoplasmic filaments extending from their free borders projecting into

Figure 5-13. Frontal section through the nose showing the relationship of the olfactory bulbs and nerves to the nasal cavity. The receptors for the sense of smell are situated very high in the interior of the nose between the median septum and the superior turbinate bones.

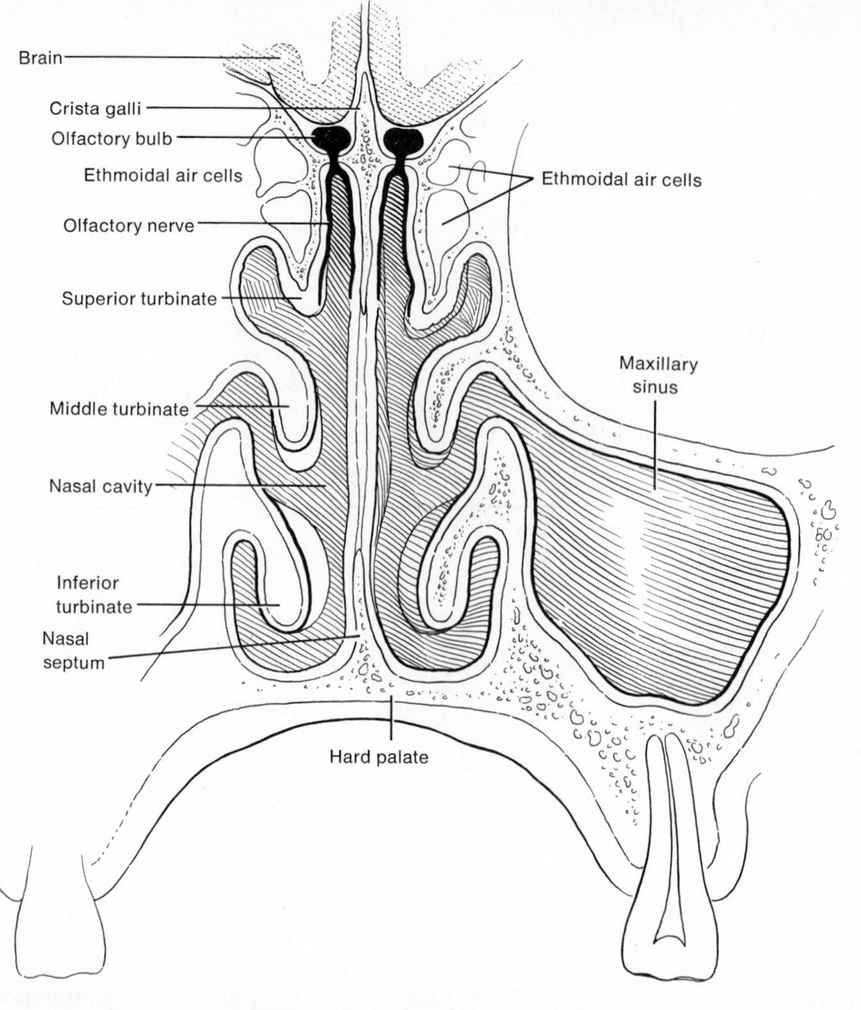

Figure 5-14. Section of mammalian olfactory epithelium, showing general structure. (Courtesy Ward's Natural Science Establishment, Inc., Rochester, N.Y.)

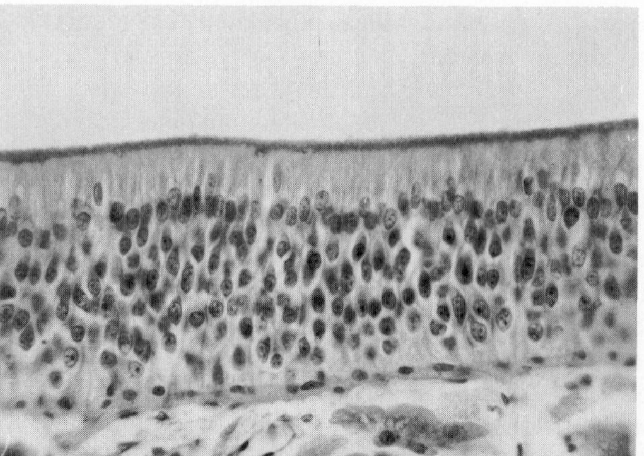

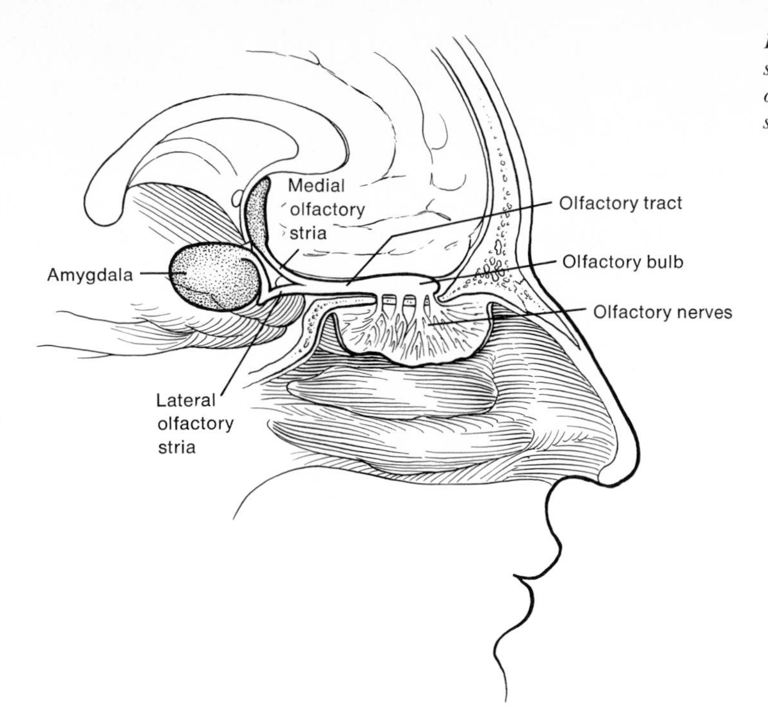

Figure 5-15. Olfactory receptive area showing the gross relationship of peripheral olfactory nerves and their terminations in specific brain centers.

the open olfactory cleft (Fig. 5-15). These olfactory cells are surrounded by epithelial cells for support and goblet cells, which secrete mucus that keeps the hairlike projections moist. It is generally believed the filaments attached to the olfactory cells are affected by molecules given off by gaseous materials. The impulses that are initiated in the olfactory cell travel to the brain by way of the olfactory nerve, where they are interpreted as various types of odors.

Many attempts have been made to classify the various types of odors. The best-known classification is that of Zwaardemaker, which he set up as far back as 1895 in his monograph *Physiology of Smell.* He grouped odors into nine classes, each of which may contain two or more subdivisions.

Odors

Class	*Subdivisions*
I. Ethereal	(a) Ether, (b) fruit, (c) beeswax
II. Aromatic	(a) Bitter almond, (b) lemon, (c) lavender, (d) cloves, (e) camphor
III. Fragrant	(a) Flowers, (b) vanilla
IV. Ambrosial	(a) Amber, (b) musk
V. Alliaceous	(a) Hydrogen sulfide, (b) arsine, (c) chlorine
VI. Empyreumatic	(a) Coffee, (b) benzene
VII. Caprillic	(a) Cheese, (b) rancid fat
VIII. Repulsive	(a) Belladonna, (b) bedbug
IX. Nauseating	(a) Carrion, (b) feces

It has been suggested that there are nine groups of receptor cells in the olfactory area, one cell corresponding to each of the nine classes of odors listed above. Thus, molecules given off by materials in one class would excite one type of receptor cell, and those given off by materials in another class would excite a different set.

However the classification, none has been very successful, because most systems are purely associative and subjective. In addition, the number of types of odors is so large that different judges do not agree as to their distinctions. What is nauseating to one may be fragrant to another.

The Auditory Sense

The ear is one of the more complex sense organs and serves a dual function, since it is both a receptor for **sound** and the organ for maintenance of equilibrium.

The **ear** can be divided into three areas: external, middle, and internal, practically all of which is enclosed by the temporal bone (Fig. 5-16). The **external ear** is that portion composed of the **auricle,** an oblong, flat fold of skin that is obviously situated on the lateral aspects of the head, and an **external auditory canal,** which leads from the auricle to the middle ear. The auditory canal is directed upward and backward in the skull and is about $1\frac{1}{4}$ inches in length. It terminates at the **tympanic membrane** (eardrum), which separates the auditory canal from the middle ear chamber.

The **middle ear** contains the mechanism for the conduction of airborne sound waves to the internal ear, but it does more than this since it acts similarly to a hydraulic press. Three small bones contained within the middle ear chamber, the **malleus,** the **incus,** and the **stapes,** which are united by true joints, form a lever system that transforms the small pressure of sound waves on the surface of the eardrum into a pressure 22 times greater on the fluid of the inner ear. The energy that usually would be reflected away when sound hits a solid surface is absorbed by the three small bones, which transmit it to the inner ear, acting as a mechanical transformer converting the large amplitude of the sound-pressure waves in the air into more forceful vibrations of lower amplitude.

The malleus (hammer) is attached to one end of the tympanic membrane and articulates with the incus (anvil) with its other end. The incus articulates with the stapes (stirrup), which in turn completes the bridge of the middle ear by attaching to the margin of the oval window of the vestibule by ligamentous fibers (Fig. 5-17). The eustachian tube, which is associated with the middle ear, is a communication between the nasal portion of the pharnyx and the middle ear. The eustachian tube allows for the equalization of pressure in the middle ear, since its entrance from the pharynx is guarded by a valve that is usually closed but can be opened when pressure changes in the environment necessitate such action. The act of swallowing or yawning will temporarily open the pharyngeal end of the eustachian tube. However, because this tube is lined with a mucosal membrane common to both the middle ear and the pharynx, an easy access is unfortunately provided for the passage of infectious material between these two areas.

The internal ear is the area of sensory reception for hearing and equilibrium. It contains the **cochlea** for the auditory sense and a series of intercommunicating

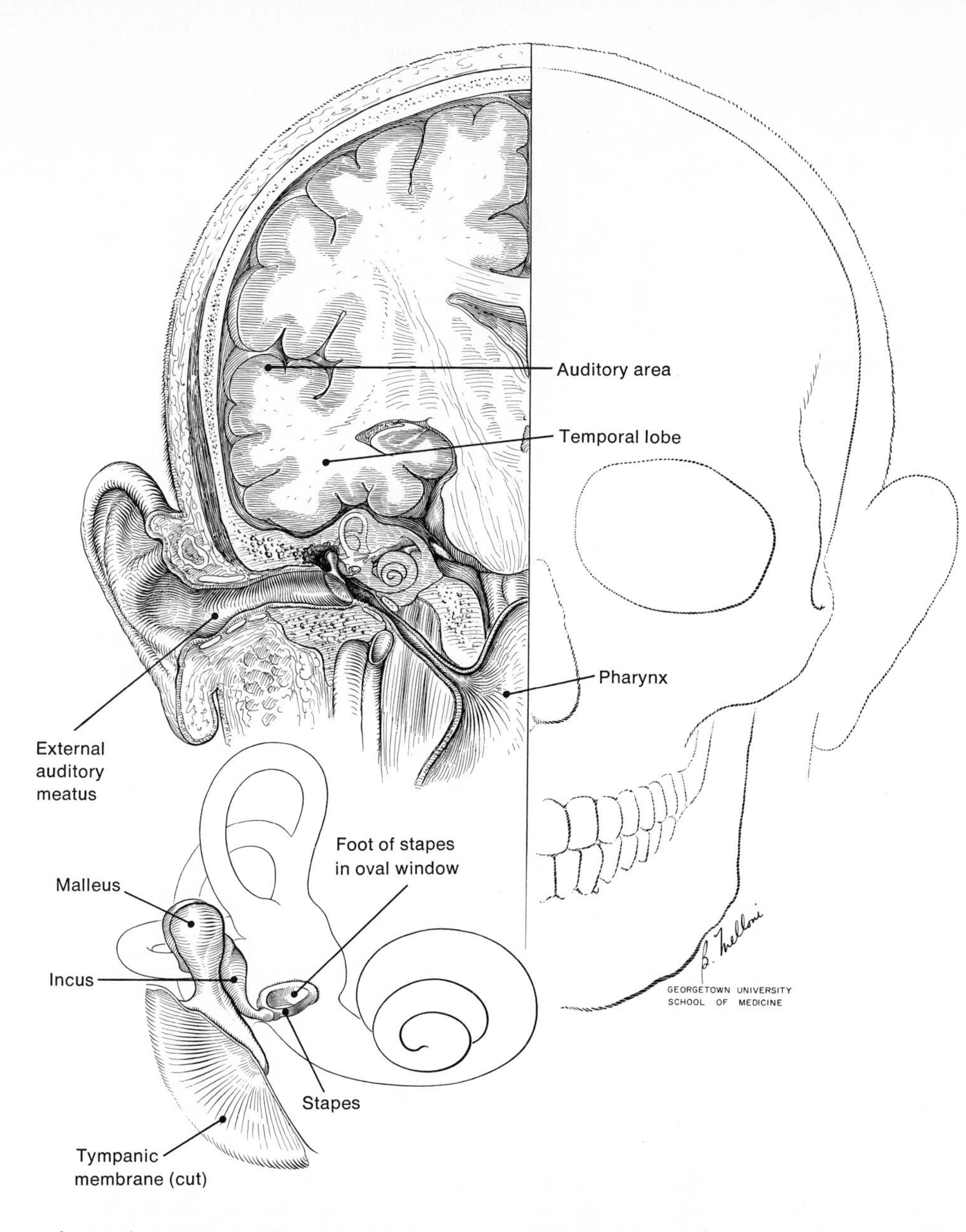

Figure 5-16. Principal features of the human ear. The external ear consists of the auricle and external meatus. The tympanic membrane is situated at the medial end of the meatus. The middle ear is situated within the temporal bone between the tympanic membrane and the lateral bony wall of the internal ear. It communicates with the pharynx through the eustachian tube (see Fig. 5-17). The internal ear is embedded deeply in the petrous portion of the temporal bone. The insert shows the three ossicles that bridge the middle ear in relation to each other and to the internal ear.

channels, the **semicircular canals,** the **utricle,** and the **saccule,** which are the essential organs for the sense of balance and position (equilibrium). The cochlea and semicircular canals are tubular systems composed of membranous tissue and are called the **membranous labyrinth** (Fig. 5-18). These tubes contain a fluid, the **endolymph,** which when set into motion by sound waves stimulates the sensory receptor cells located in their epithelium. The membranous labyrinth is contained in a bony canal, the **bony labyrinth,** which is also filled with fluid, called the **perilymph** (Fig. 5-19).

Figure 5-17. Parts of the ear are illustrated in cross section. Between the eardrum (tympanic membrane) and the fluid-filled inner ear are the three small bones (ossicles) of the middle ear. The elastic round window of the inner ear, which is the point at which the scala tympani terminates, is also shown.

Figure 5-18. The membranous labyrinth is divided into (1) the semicircular canals, (2) the utricle and saccule, and (3) the cochlear duct. The membranous labyrinth contains a fluid known as endolymph. The utricle and saccule are enlarged sacs of the membranous labyrinth. They also contain sensory endings of the vestibular nerve.

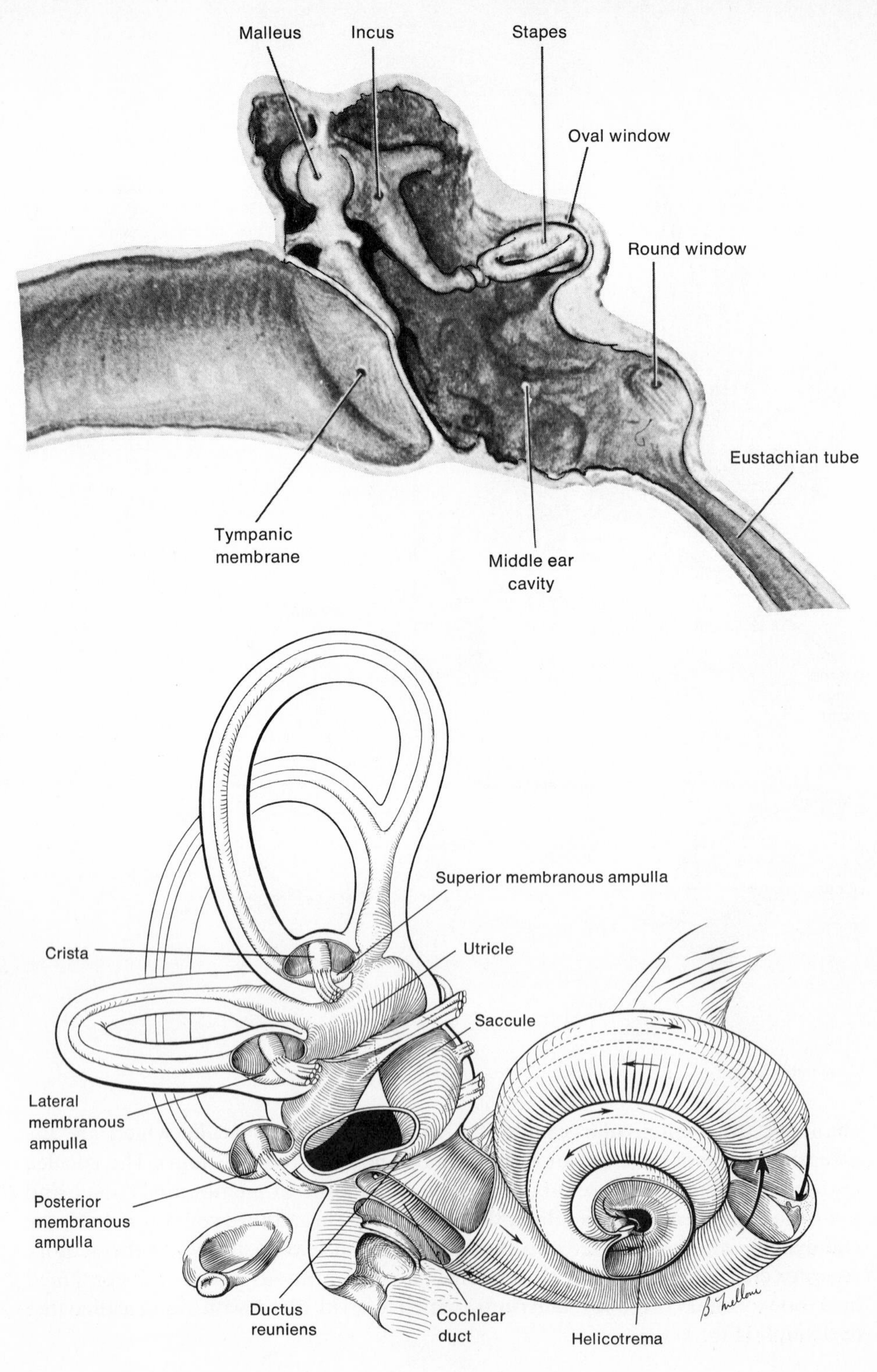

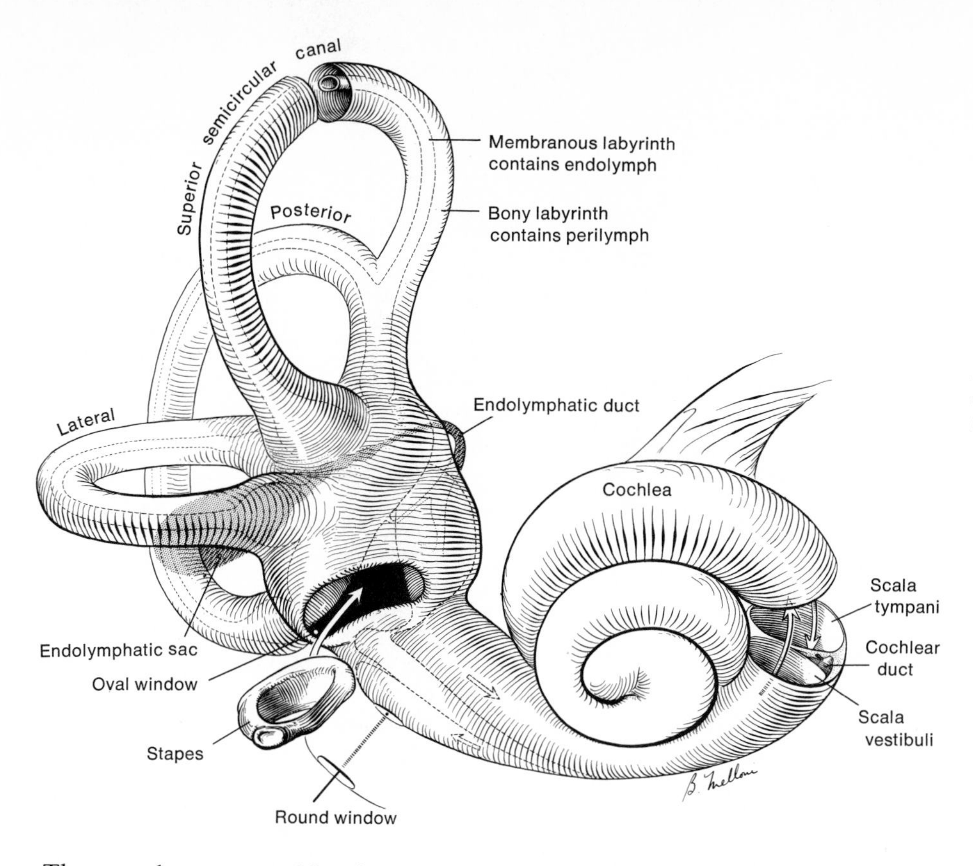

Figure 5-19. The internal ear is composed of a membranous labyrinth and its surrounding bony labyrinth. They are separated by perilymphatic fluid. The osseous labyrinth is composed of the vestibule, the semicircular canals, and the cochlea. The vestibule is directly medial to the oval window, which opens into it. The stapes has intentionally been displaced from the oval window in order to expose the vestibule. The cochlear duct (see Fig. 5-18) is a coiled canal, triangular in cross section, which runs through the entire length of the spiral canal of the cochlea, separating the scala vestibuli from the scala tympani.

Cochlea

The membranous **cochlea** is a coiled duct corresponding to the turns of the bony cohlear canal and is triangular in cross section. The cochlear ligament attaches the membranous cochlea to the wall of the bony labyrinth. The membranous cochlea, with its attaching membranes, separates the bony labyrinth into two portions, called the **scala vestibuli** and the **scala tympani.** The membranous cochlear duct is referred to as the **scala media** (Figs. 5-18 and 5-19).

The **organ of Corti,** a cochlear structure, is an epithelial ridge of considerable thickness and of extremely complex structure situated on the **basilar membrane** (Fig. 5-20). It contains the terminal nerve fiber of the cochlear branch of the eighth cranial nerve and extends spirally throughout the length of the cochlear duct. Two types of cells are associated with the organ of Corti: (1) the supporting and (2) the hair cells, which act as receptors of the stimuli produced by the sound waves. The **tectorial membrane** is a gelatinous mass that is connected to the bony spiral on one end and forms a canopylike covering over the hair cells with its free end (Fig. 5-21).

Physical Basis of Hearing

Sound production is dependent on two essential components: The first is a vibrating source and the second is an elastic medium by which the vibrations produced can be conducted some distance away from the original source in order that they can reach the ear of the receiver. The conducting medium ordinarily

Figure 5-20. (Right) *Longitudinal section of cochlea showing details of organ of Corti.* (*Courtesy Ward's Natural Science Establishment, Inc., Rochester, N.Y.*)

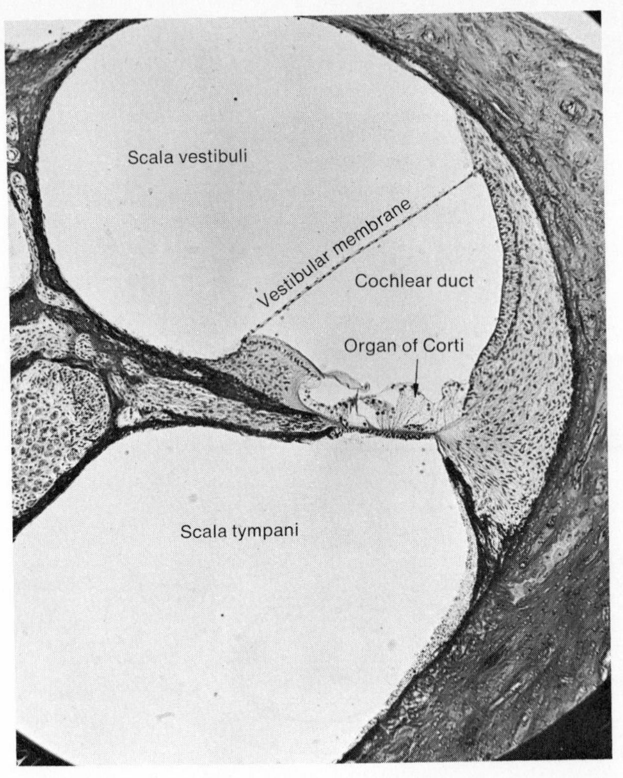

Figure 5-21. (Below) *The organ of Corti extends spirally throughout the length of the cochlear duct. It lies between the basilar and tectorial membranes. Within it are the hair cells, which are attached to a branch of the cochlear nerve and act as receptors of the stimuli produced by sound waves.*

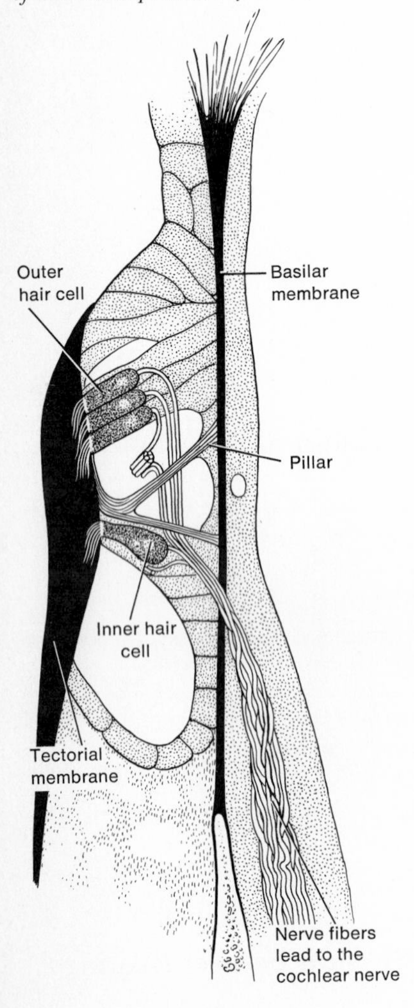

serving this purpose is air, although other gases, liquids, and even solids can transmit sound.

The vibrations imparted to the conducting medium by the sound-producing mechanism are commonly spoken of as sound waves, which consist of a series of alternate condensations and rarefactions propagated in a pendular manner in all directions. In order to get a visual image of the meaning of condensation and rarefaction, let us closely analyze a vibrating body, such as a steel rod, one end of which is tightly held in a block, while the other end is vibrating back and forth. As the free end of the rod moves toward the right, it pushes the air molecules in front of it somewhat closer together than normal, creating a compressed region, or **condensation,** in the air. As the rod goes back toward the left, a partial vacuum is created behind it, which causes the molecules of air to be pulled farther apart than normal. Such a region is called a **rarefaction** (Fig. 5-22). When these alternating compressed and rarefied parts of the atmosphere strike our eardrums, they cause them to vibrate at the same rate as the iron rod, and we hear sound. A sound wave in air consists of an abnormal distribution of air molecules in which compressed regions alternate with regions of partial vacuum. When a normal distribution of air molecules exists, no sound waves are traveling through the air.

The sound wave, which forms the units of hearing, has certain properties, such as pitch, amplitude (loudness), and quality, or timbre. The **pitch** is dependent on the frequency of the wave motion. Frequency may be defined as the rate of

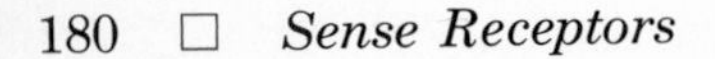

vibration of the source. The higher the frequency, the higher the tone, or pitch. Low tones, or pitch, correspond to sound waves of slower frequency. The range of hearing for the human ear lies between 16 and 28,000 double vibrations per second.

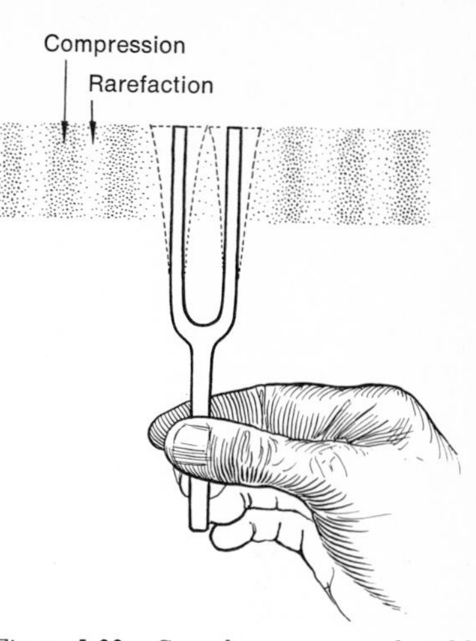

Figure 5-22. Sound waves as produced by a tuning fork or vibrating rod showing the production of condensations (compressions) and rarefactions.

Another important factor characterizing the sound wave is its **amplitude,** which is defined as the maximum displacement of a molecule of the medium from its normal position. This may be graphically represented by the distance between the crest and trough of the wave. In sound, variations in amplitude register on our hearing as variations in loudness, with greater amplitudes producing louder sounds. The third attribute of sound is quality, or **timbre,** which characterizes sounds produced by different instruments, such as the saxophone, piano, or violin, even though the pitch and intensity may be the same. Timbre is determined by what are known as overtones, which are produced when the vibrating source is vibrating in more than one piece at the same time.

Sounds are practically always combinations of several vibrations of different frequencies and amplitudes. The auditory mechanism functions not only in registering the sound patterns but also in interpreting them in the light of past auditory experience. Two types of sound exist and are designated as tones and noises. A **tone** is a sound produced by a number of equal and regular sound waves, and **noises** are sound waves irregular in length and tone. Speech is a composite of more or less pure tones and of noise, changing in composition and intensity from one moment to the next.

Sound Conduction

The auricle of the ear may be regarded as the expanded portion of a funnel formed by the auditory canal, functioning to gather sound waves coming from various directions in the environment. The auditory canal directs these sound waves toward the tympanic membrane. Because of its obliqueness and slight concavity, the tympanic membrane is admirably adapted for the transmission of the sound waves to the ossicular chain of the middle ear. The ossicles act as a lever system to convert the vibrations imparted by the sound waves to the tympanic membrane into vibrations of sufficient force to produce wave motion in the fluid of the internal ear. The stapes acts as a piston of the lever system, since it is inserted in the oval window of the vestibular apparatus on the other side of which is perilymph.

Sound Perception

Sound, which has its origin outside the body, reaches the cochlea in the manner described above. The force transmitted to the perilymph through the oval window is propagated as a wave through that part of the bony cochlea known as the scala vestibuli, and it continues into and through the scala tympani, ending at the entrance of the **round window.** Between the scala vestibuli and the scala tympani lies the middle compartment, the membranous cochlear duct, containing the organ of Corti, which forms the sensory receptor of the cochlear nerve. As the sound wave, now in the form of a liquid wave, passes through the bony cochlea, it impinges on the **membrana vestibuli** and the **basilar membrane** of the cochlear duct, setting into motion the endolymph. The organ of Corti, which rests on the basilar membrane, is set into motion by these waves so that the sound waves are communicated to these structures. The basilar membrane, running from the base of the cochlea in spiral fashion to the tip, is analogous to a system of stretched strings, each string being connected to a given nerve fiber and having a different length, increasing from the base to the tip. At the base of the cochlea the basilar membrane fibers are 64 to 128 μm long; at the apex (tip) they are 352 to 480 μm long. Different tones, ranging from 28,000 double vibrations per second at the base to 16 double vibrations per second at the tip, are made possible by this type of arrangement. Between these two ranges there are, of course, many tones in a range of about 11 octaves.

The actual mechanism of auditory excitation is quite complicated and controversial. The most widely accepted theory is the **resonance theory** proposed by the physicist Helmholtz. It assumes that the organ of Corti is analogous to the harp or piano, in which the basilar membrane fibers, as is well illustrated by the movement of piano strings, may be excited by sympathetic vibrations of the air to reproduce certain musical sounds originating elsewhere. Responding by sympathetic vibrations to external notes, the piano strings analyze the sound by picking out the different components of the complex air-pressure wave reaching the strings. In a corresponding way, the fibers of the basilar membrane are assumed to resonate to the vibratory motions of the endolymph and thus to excite specific neurons of the auditory apparatus in response to a given sound. The impulses that are generated by displacement of the hair cells of the organ of Corti are carried along the corresponding nerve fibers to the brain, where they are resynthesized to produce in consciousness the original sound or group of sounds carried to the ear (Fig. 5-23).

Regarding the other theories of hearing, mention should be made of the **place theory,** which postulates that frequency discrimination is a function of specific areas of the basilar membrane. According to this concept, the basilar membrane receives, as a result of endolymph disturbances, certain vibrations along given portions that are transmitted by means of the tectorial membrane to the hair cells of the organ of Corti. In turn, the impulses generated are carried by the auditory nerve to the brain, which sorts out the vibrations.

The **dual theory** of Wever, somewhat a composite of the above two theories, postulates that discernment of high frequencies of sound is a function of cochlear (basilar membrane) localization, but with low sounds it is a function of frequency of discharge of individual basilar fibers. This dual theory fits existing experimental data and currently seems quite acceptable.

The most recent theory of hearing, involving **cochlear microphonics,** had its impetus when Wever and Bray showed that, when a sound wave impinges on the eardrum, two types of electrical potential are created in the cochlea and auditory nerve. They also showed that if an electrode were correctly applied to the cochlea or the auditory nerve, these potentials could be conducted to a distant room, suitably amplified and fed into a loudspeaker, where sounds applied to the ear could be reproduced through the loudspeaker. It has been demonstrated that one type of electrical response is related to the movement of the stapes in and out of the oval window and is actually due to the changing of physical energy into electrical energy, a piezoelectric effect of distortion of the structures of the

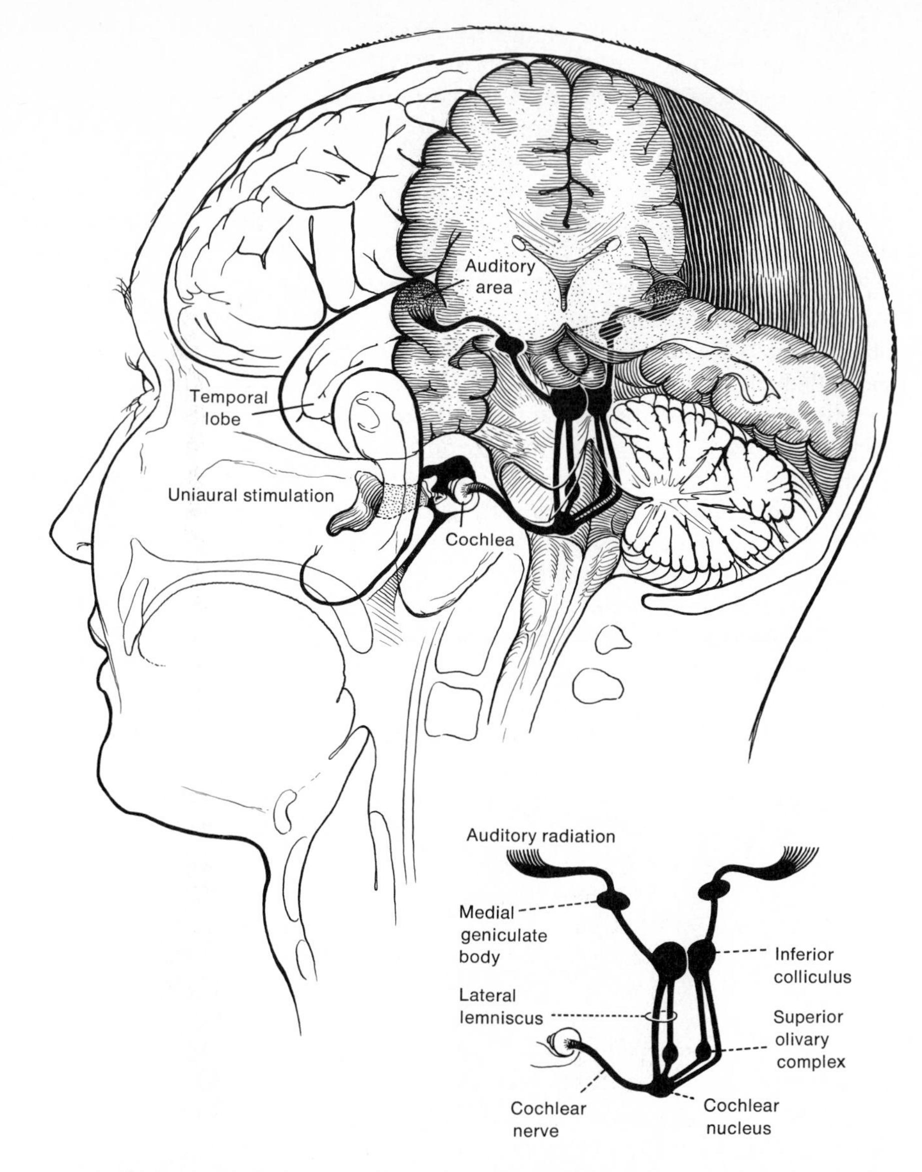

Figure 5-23. Auditory pathway. The impulses generated by displacement of the hair cells of the organ of Corti are carried along the corresponding nerve fibers to the auditory area of the cortex, where they are resynthesized to produce sounds.

inner ear. It is important to note that these potentials are not nerve impulses, and their exact relationship to the hearing process is not well understood, but it is considered to be the device whereby the physical energy of sound waves is converted into electrical energy, which is a transducer effect.

The second type of electrical potential is generated in the auditory nerve. Bekesy, by following the action potentials along the nerve pathway to the cortex of the brain, has demonstrated that stimulation of specific spots on the basilar membrane seems to be projected to corresponding spots in the auditory area of the cortex. This appears to be similar to the projection of images on the retina of the eye to the visual area of the brain.

Vestibular Apparatus

Just as the anterior part of the **vestibular apparatus,** the cochlea, serves for hearing, the posterior portion plays a role in the maintenance of equilibrium. The end organs of the posterior vestibular apparatus consists mainly of the three **semicircular canals,** the common sac into which these canals enter, the **utricle,** and another small sac, the **saccule** (Fig. 5-24).

The three membranous semicircular canals lie within the corresponding canals of the bony labyrinth but do not completely fill out the bony canal except at the ampullae.

The three semicircular canals lie in different planes: a horizontal or lateral, a frontal or superior, and a sagittal or posterior. Each dilated end of the membranous ducts is called the ampulla and terminates by opening into a membranous sac called the **utricle,** which is a division of the **vestibule.** The utricle connects by way of the endolymphatic duct with the second membranous division of the bony vestibule, the **saccule,** which in turn communicates directly with the scala media of the cochlea. The utricle and saccule, which are the two main membranous divisions of the vestibule, function in the perception of position in space in the vertical plane with relation to gravity (static function). The semicircular canals serve to distinguish active movements in other planes as opposed to the static system and are concerned with the perception of independent positions and

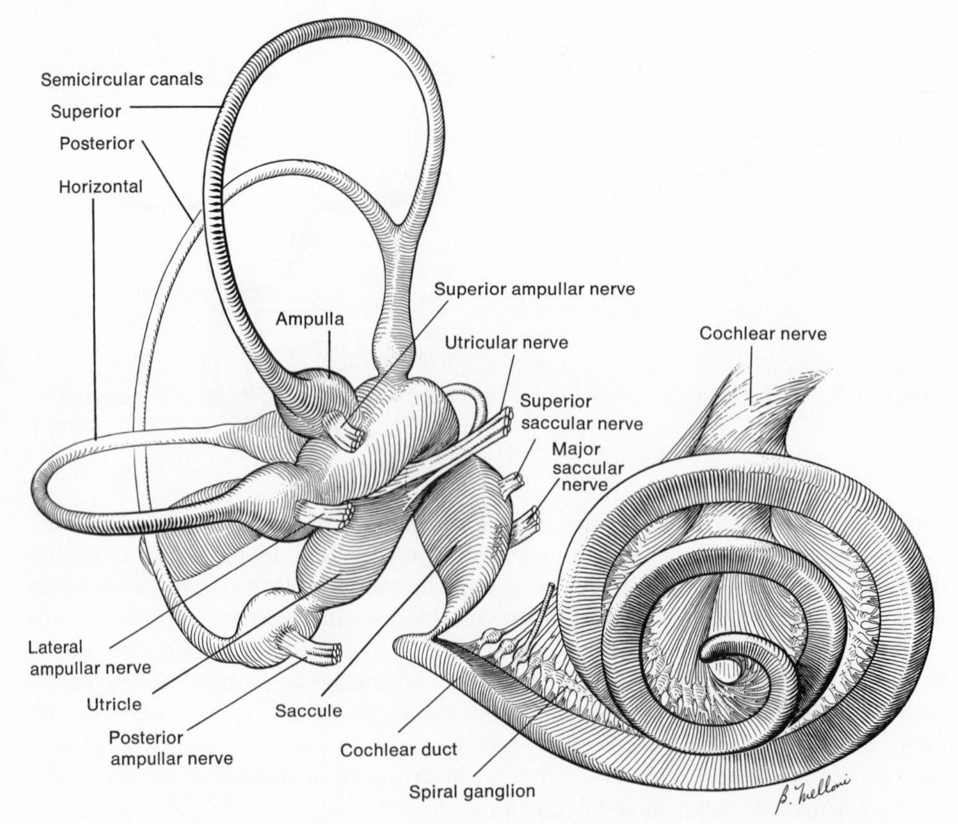

Figure 5-24. Membranous labyrinth from which all the nerves of the internal ear arise. The bony labyrinth has been removed so that both the anterior and posterior parts of the vestibular apparatus may be more clearly illustrated.

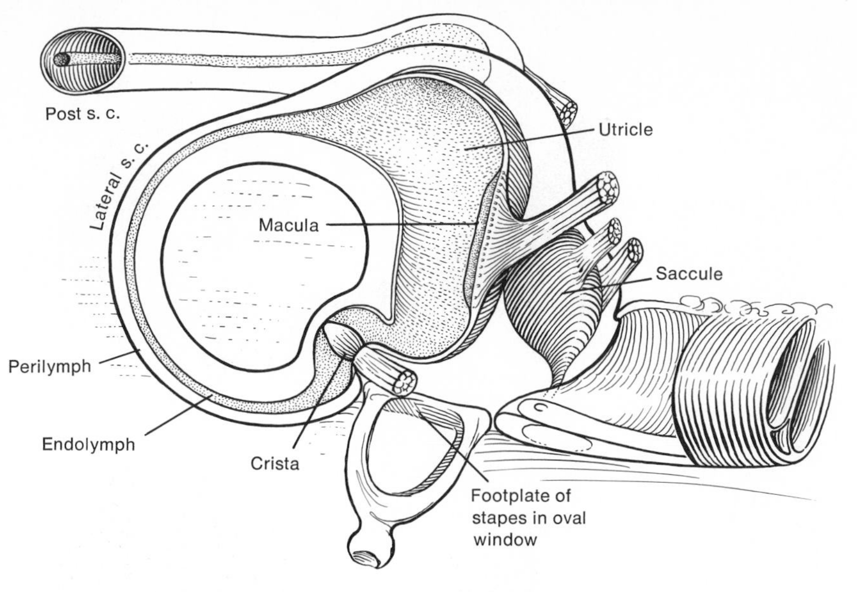

Figure 5-25. Cross section of the utricle and semicircular canal showing the location of the macula in the utricle and the crista ampullaris in the ampullae of the semicircular canals. The macula is the sensory area of the utricle.

movements, such as rotations, accelerations, head positions, and progressive motions in a straight line (kinetic function).

The membranous semicircular canal is covered on the inner surface by a layer of epithelium resting on a layer of connective tissue. The ampullae of the canals possess the real sense receptors, as each is lined by highly developed hair cells called the **crista ampullaris** (Figs. 5-25 and 5-26). The hairs dip into a gelatinous mass that sits on them like a hat, the **cupula.** The saccule and utricle also contain hair cells that dip into a gelatinous cupula overlying them, the entire mass being called the **macula** (Figs. 5-25 and 5-27). These hair cells contain the terminal nerve fiber of the saccular and utricular branches of the vestibular nerve (Fig. 5-28).

The Sense of Sight

The eye is a device by which the energy of a light pattern is converted into the energy of a nerve impulse that is conducted by the optic nerve to the visual cortex of the brain for interpretation as a visual image. The eye acts as a transducer for the specific purpose of converting the light energies to the electrical energies associated with the nerve impulse.

Eye

The **eye** appears to be a simple organ, especially since it is a thin-walled and hollow structure, but in reality it is a complex mechanism (Fig. 5-29). The essential anatomical features of the eye are a lens mechanism formed by the curved transparent front of the eye, the **cornea,** plus the **lens** inside, which together focus an image on the light-sensitive back wall of the eye, the **retina.** The lens separates the hollow interior of the eye into **anterior** and **posterior** cavities. The anterior cavity contains a fluid, the **aqueous humor,** and is divided into anterior and posterior chambers by the iris. The posterior cavity is filled with a jellylike substance, the **vitreous humor.** Figure 5-30 shows that the eyeball is composed of three layers. The outer, the **sclera,** is of connective tissue and quite tough and

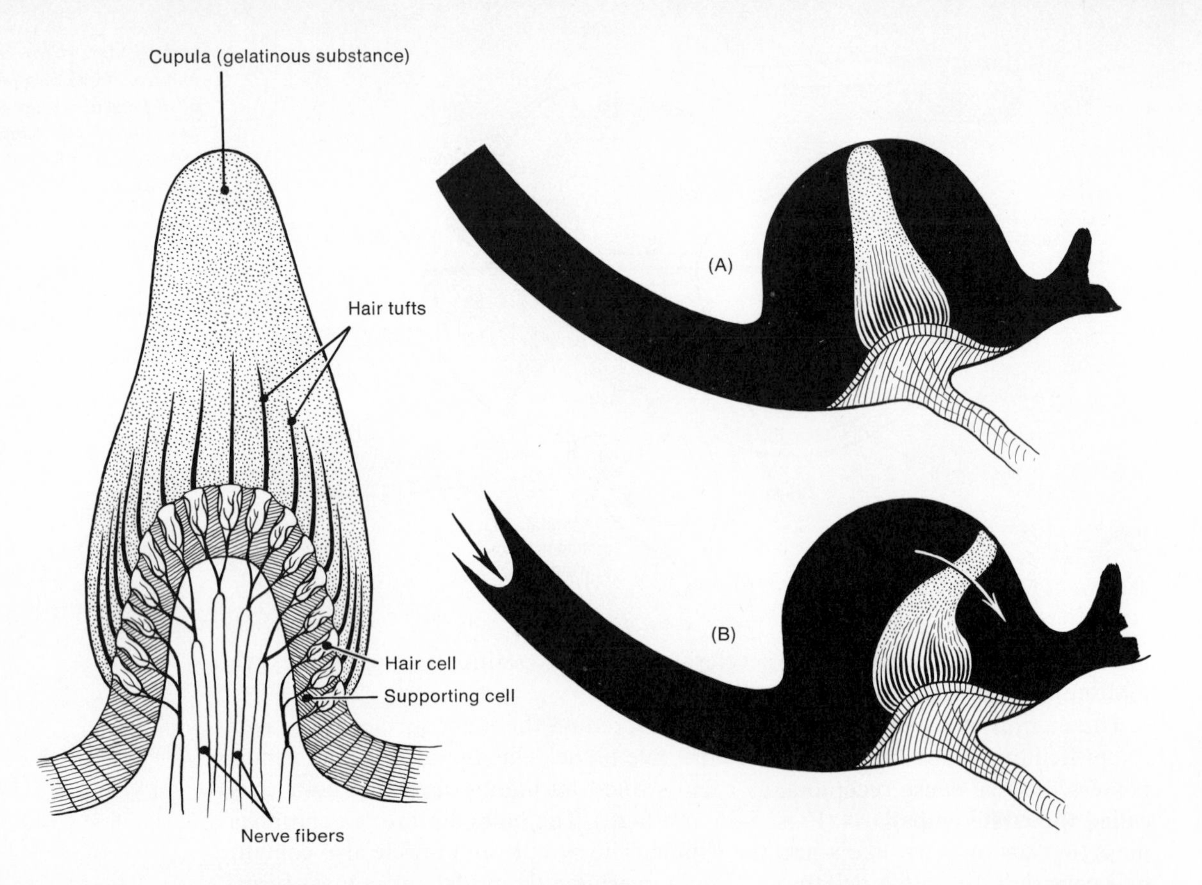

Figure 5-26. Organization of the crista ampullaris. A. *Cupula at rest.* B. *Movement of the endolymph in the semicircular duct displaces the cupula by the force of inertia, providing an adequate stimulus, which is picked up by the nerve fibers of the vestibular nerve.*

is commonly recognized as the "white of the eye." The anterior third of the sclera is transparent and forms the cornea. The middle layer of the eyeball, known as the **choroid,** contains the blood vessels, the ciliary muscle, and the ciliary nerve, which supplies the ciliary muscle. The ciliary muscle is attached to the choroid wall on one end and to the edges of the lens on the other; it can change the shape and focal length of the lens by pulling on it. In front, the choroid layer is separated from the cornea by a region of circularly arranged smooth muscle fibers, the **sphincter pupillae,** and radial fibers, the **dilator pupillae.** These muscles make up the **iris** (Fig. 5-31), which is the colored portion of the eye and acts as an adjustable light diaphragm. The iris has a perforation at its most anterior end called the **pupil.**

The innermost layer, the **optic retina,** is the light-sensitive area containing the receptors for sight. Microscopic sections of this remarkable mechanism show it to be composed chiefly of several layers consisting of a layer of pigment cells and three layers of neurons (Fig. 5-32). From the surface of the retina inward toward the optic nerve, the tissue is arranged as follows:

1. A *layer of pigment cells,* which serve, like the black paint on the inside of a camera, to absorb light, which otherwise would be reflected and diffused, causing blurring of the retinal image.

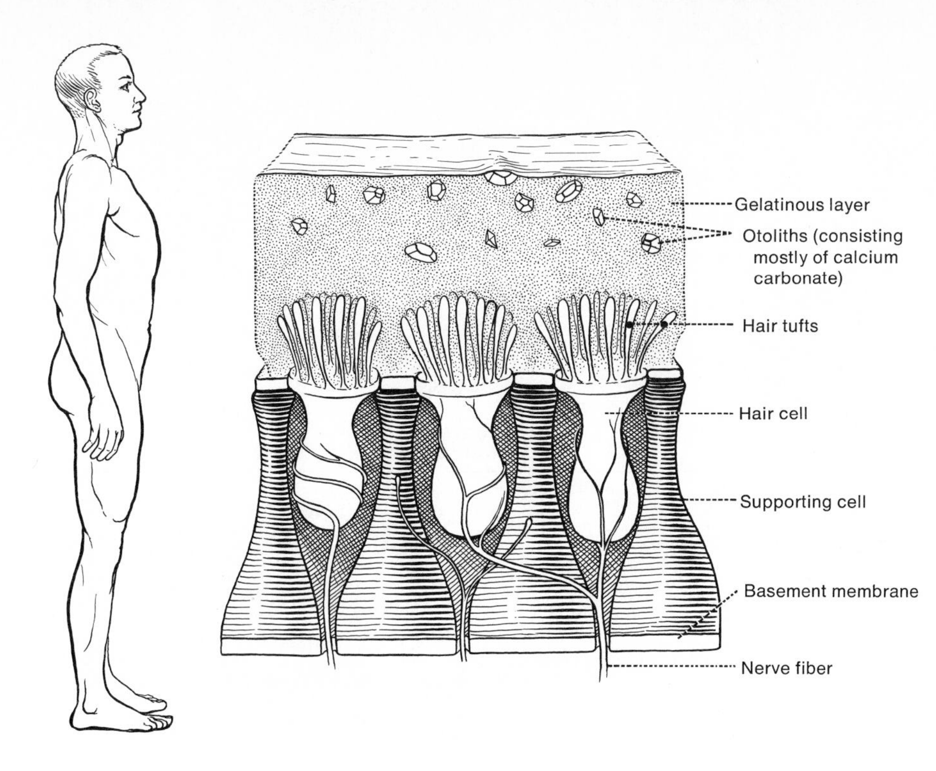

Figure 5-27. Organization of the macula and how it functions. The macula is covered by a gelatinous layer in which many small bony masses known as otoliths are embedded. The hair cells also project hair tufts into the gelatinous mass. The otoliths fall with gravity, distorting the gelatinous layer and hair tufts. The distortion stimulates the nerve fibers around the hair cells, which provides impulses that are transmitted by appropriate nerve tracts to the brain. Thus, the central nervous system is apprised of the relative position of the otoliths in the gelatinous mass and can control equilibrium.

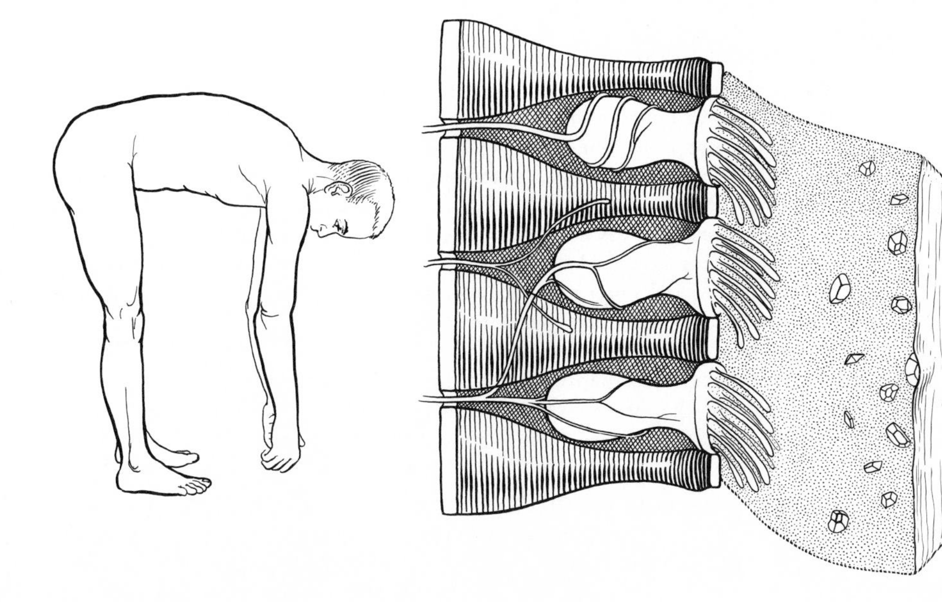

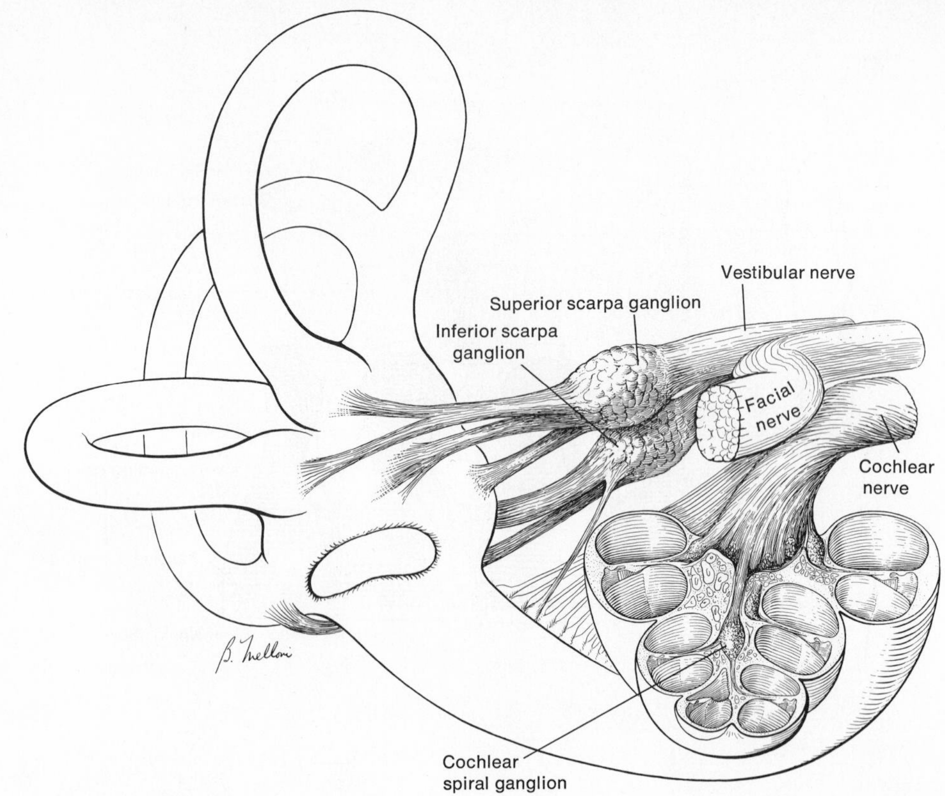

Figure 5-28. How the nerve fibers from the superior ampulla and lateral ampulla unite and, along with the fibers from the utricle and saccule, proceed to the superior ganglion of the vestibular nerve. The cochlear nerve originates at the organ of Corti. The vestibular and cochlear nerves unite to form the acoustic nerve.

2. A *receptor layer,* composed of cells that are called rods and cones because of their shape (Fig. 5-33). The rods are shaped like a cattail on its stalk; the cones are similar, but thicker at the base. The rod cells are more numerous than the cones, except at a spot in the middle of the retina where the central ray of light focuses. This spot, the **macula,** is composed entirely of cones (Fig. 5-34). The rods and cones are only part of their respective neurons. The nuclei of these cells are concentrated in an area referred to as the **outer nuclear layer.** The remainder of the rod-and-cone neuron extends beyond this region as fibers and terminates in an area referred to as the **outer plexiform layer,** where it synapses with bipolar nerve cells.
3. A *layer of bipolar neurons,* whose dendrites synapse with the rod-and-cone neurons in the outer plexiform layer. The nuclei of these bipolar cells are concentrated in a region referred to as the **inner nuclear layer** and their axons synapse with the third layer of neuron cells in the **inner plexiform layer.**
4. A *layer of ganglion cells,* whose cell bodies are concentrated in the **ganglionic layer.** The dendrites of these cells synapse with the bipolar cells in the inner plexiform layer; their axons run along the innermost surface of the retinal tissue, where they converge to form the optic nerve.

Optic Pathway

For a general view of the optic pathway, see Fig. 5-35. The region where the ganglion axons leave the eye as the optic nerve is referred to as the **optic disk** (Fig. 5-34). This area contains no visual elements and consequently is commonly

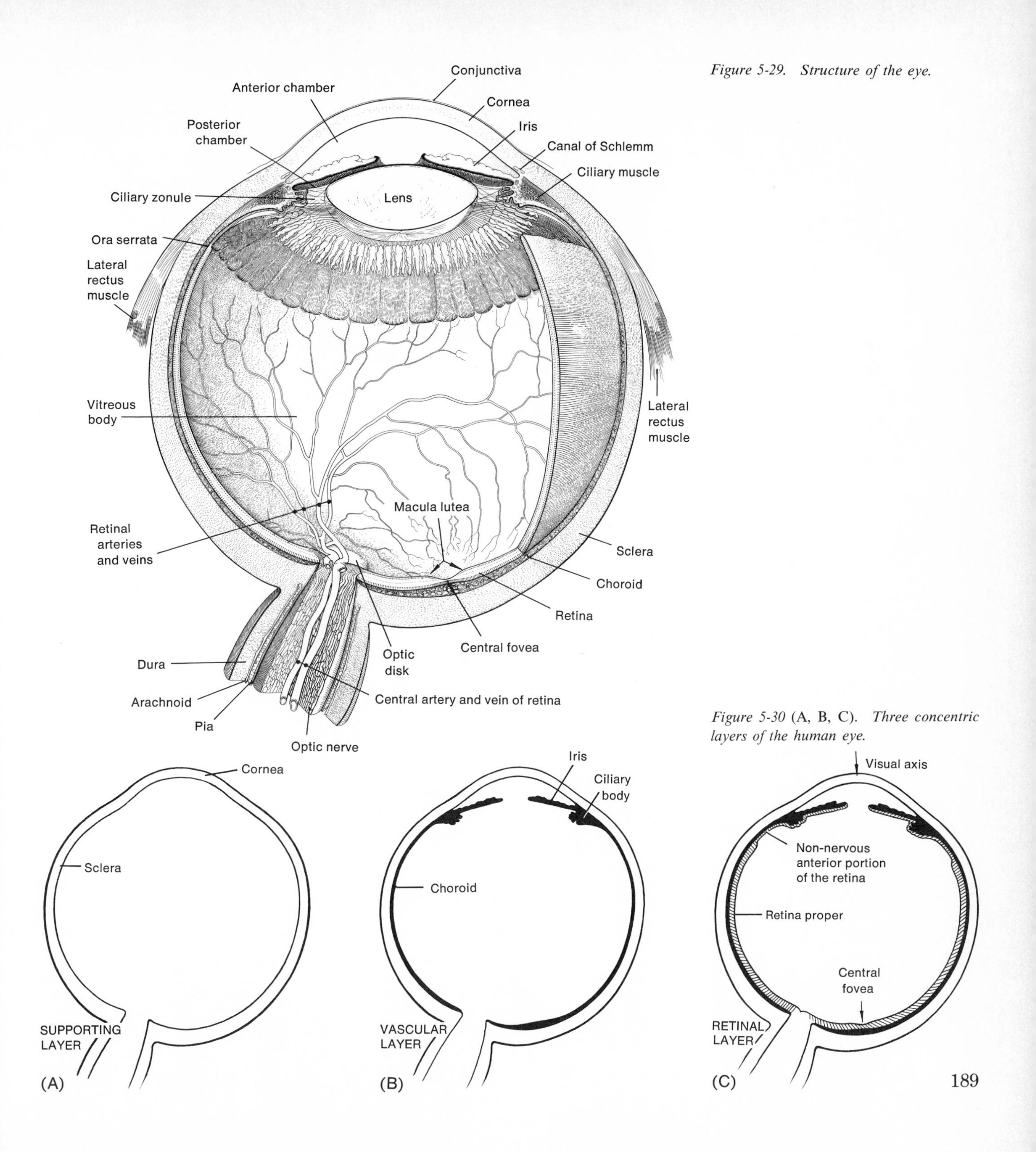

Figure 5-29. Structure of the eye.

Figure 5-30 (A, B, C). *Three concentric layers of the human eye.*

Figure 5-31. Magnified view of iris as seen from the front looking into the eye.

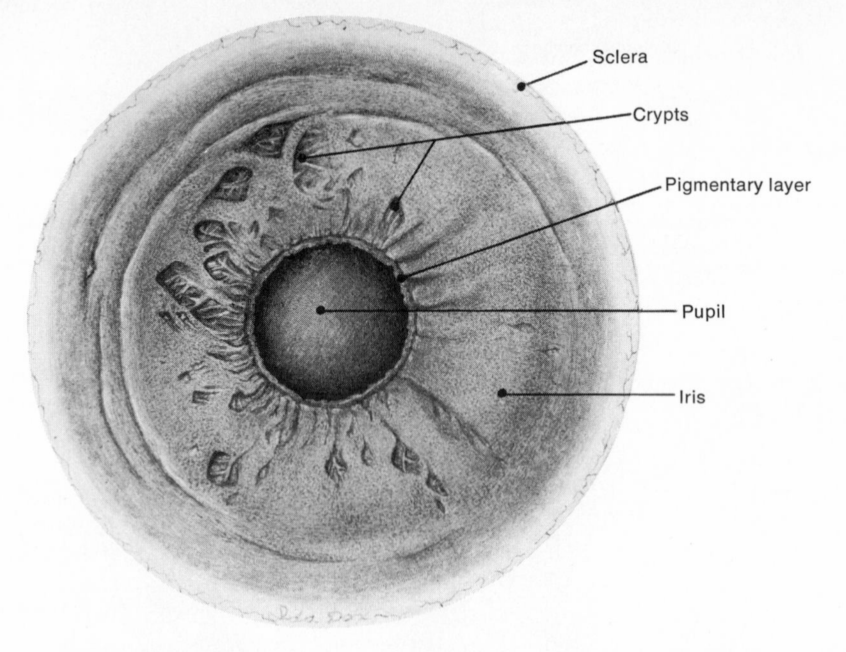

called the **blind spot** (Fig. 5-36). The axons, on leaving the back of the eyes to form the optic nerves, acquire a myelin sheath. The optic nerves run backward in the orbits, enter the brain cavity by the optic foramen, and converge to form a partial crossing at the **optic chiasma.** Only the fibers from the nasal half of the retina cross; those from the temporal half continue on the same side. The object of the crossing is to combine the fibers that are carrying the image from one side of the body, for the temporal side of one eye sees the same image that the nasal half of the other eye sees. The fibers of the optic nerves continue to run back horizontally as the **optic tracts** and terminate in the lateral geniculate bodies of the brain. Optic fibers spread out from the lateral geniculate body to form the **optic radiations** passing to the occipital visual receptive cortex (Fig. 5-37).

Figure 5-32. Microscopic section of the retina showing detail of the three layers of neurons: the rods and cones, the bipolar cells, and the ganglion cells.

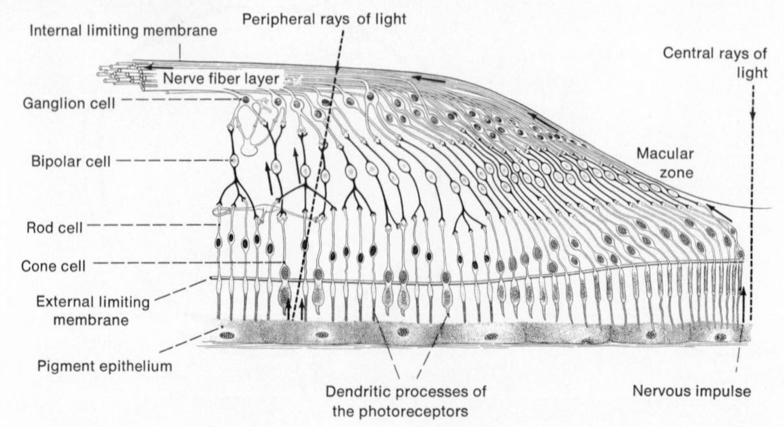

Some visual fibers bypass the lateral geniculate body and enter the superior colliculus. The superior colliculus is the optic reflex center of the brain stem because it has direct connections with other structures, such as the nucleus of Edinger-Westphal, which lies between the oculomotor nuclei, and sends fibers to the ciliary muscle, which thins the lens for distant vision. The superior colliculus also sends fibers to the sphincter pupillae, which reduces the amount of light entering the eye, thereby enabling the pupil to respond to the light. The spinal motor nuclei and the motor nuclei of the head and neck also receive fibers from the superior colliculus, which allows for immediate bodily response to sudden visual images.

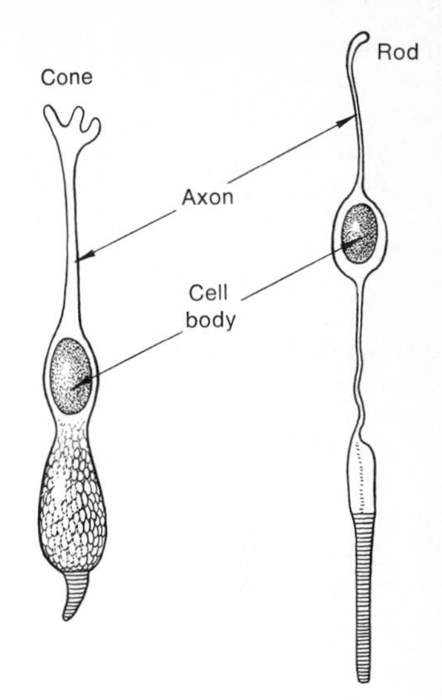

Figure 5-33. (Above) *Detail of rod-and-cone cell to show difference in shape.*

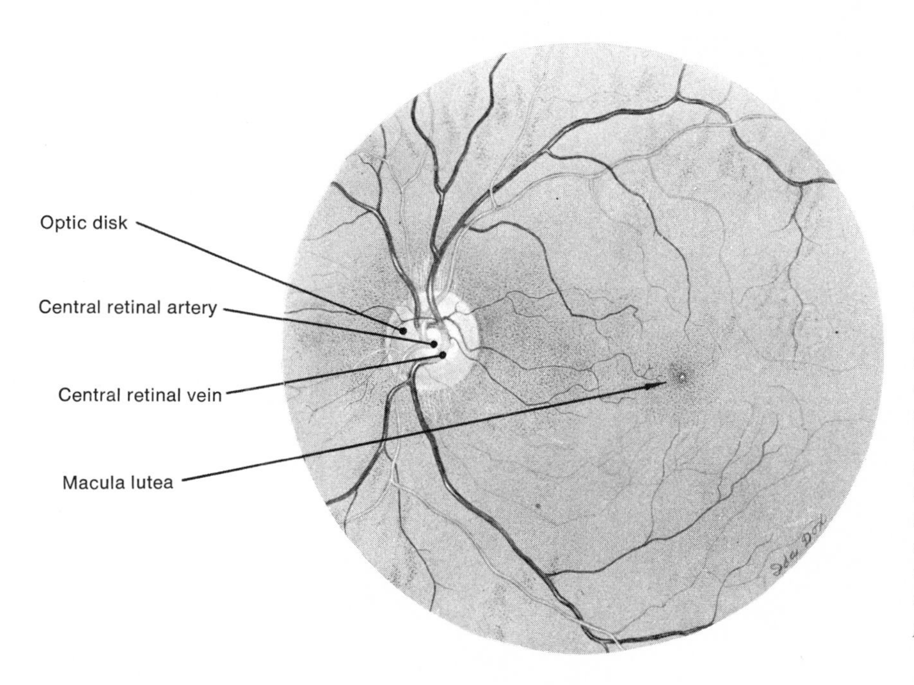

Figure 5-34. (Left) *View of the retina as seen through an ophthalmoscope. This view of the eyegrounds, or fundus, of the eye shows the macula lutea, a small yellowish area near the center of the retina. A slight depression in its center is named the fovea centralis. The site at which the optic nerve enters appears as a creamy white disk and is called the optic disk. It is also known as the blind spot because the light rays focused here cannot be seen; this is because there are only nerve fibers, and no rods or cones are present.* (*For proper orientation compare with Fig. 5-29.*)

Properties of Light Important to Vision

Light is an electromagnetic vibration that can act at times in a wave motion similar to sound and at other times like discrete particles of energy, with many of the properties commonly associated with mass particles. In the visual process light acts in both ways. In being refracted or diffracted from its normal pattern of traveling in a straight line, light is treated as a wave motion. On the other hand, whenever light is absorbed, as is common for all matter, light energy is transformed into chemical energy or heat, and in this case acts as a particle or quantum. It is this property that allows light to produce a chemical change in some substances, and this change is referred to as a photochemical change. This property of light is also very important to vision, as the retina is a specialized receptor responding to light through which the light stimulus is converted chemically into a nerve impulse.

Figure 5-35. Central visual pathways showing the course of impulses from the retinal quadrants to the visual cortex. Small inserts beside eyeball show projection of image within the visual field on the retina. Images on the right side of the subject are projected on the left side of the retina; those above, on the lower half of the retina, and so on; that is, the visual field is inverted on the retina. There is normally a region of maximum acuity in the central field, with a decline in the resolving power peripherally. The visual field of one eye overlaps that of the other, providing the basis for stereoscopic sight (notion of depth). The notion of depth can be given in monocular vision but is far more perfect in binocular vision.

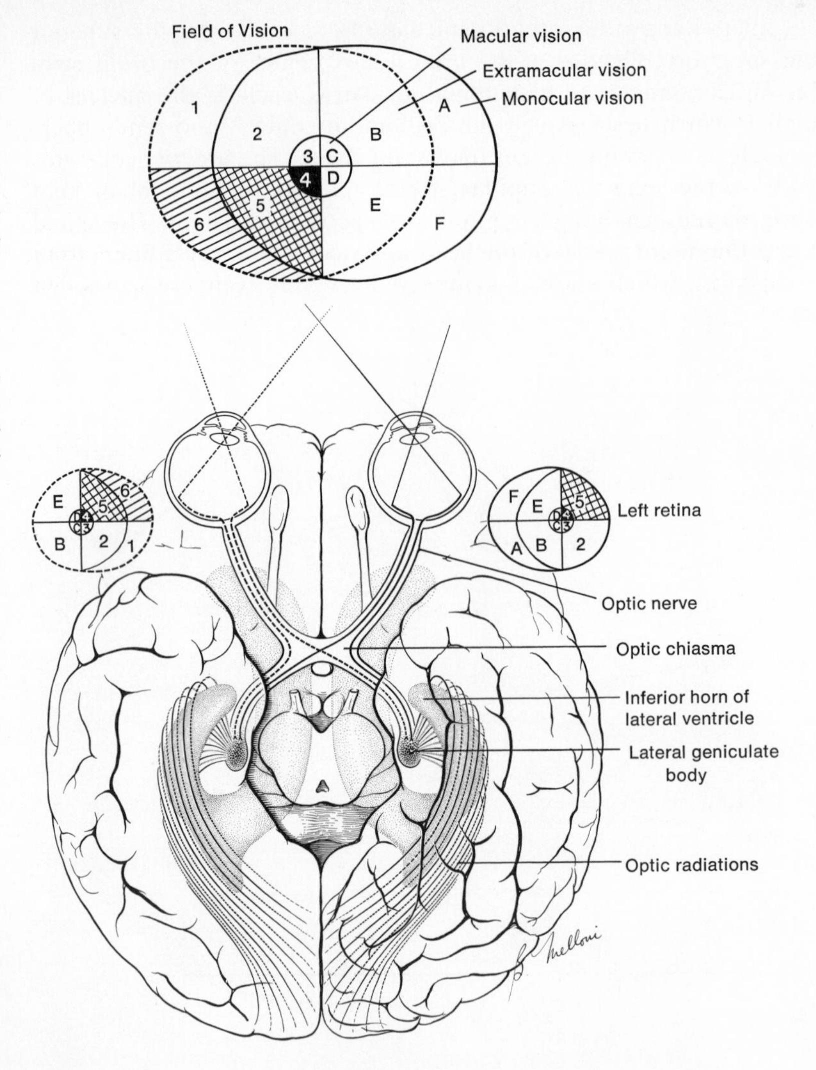

Figure 5-36. The blind spot. Close the left eye and focus the right eye on the cross. Move the figure slowly away from the eye and then toward the eye. At a distance between 15 and 20 cm, the flying ant is not seen.

+

Light waves exist in a number of different lengths the entire range of which is called the *electromagnetic spectrum.* Only a small fraction of the entire range of wavelengths is visible; these wavelengths vary in size between 7,500 and 4,000 Å or 0.000,028 to 0.000,014 of an inch. Light of wavelengths greater or less than this range is invisible to the eyes. Figure 5-38 shows the electromagnetic spectrum with approximate wavelengths of the various types of light waves.

Ordinary white light is a mixture of all the visible rays in the spectrum. A beam of light can be broken down into all its constituent wavelengths by passing it through a prism. A visible spectrum is produced as indicated in Fig. 5-38, which includes the longest visible wavelengths, which are red; shorter wavelengths than this are orange; the next wavelength is yellow, then green, blue, and indigo, and, finally, the shortest visible wavelength is violet.

Light waves commonly travel in straight lines, and the very fact that light casts sharp shadows demonstrates this. Unlike sound waves, light waves normally do not bend greatly as they pass the end of an obstacle, and therefore, areas of illumination coming from a source are more limited by a definite boundary than are sound waves. Light does bend around corners to a slight extent though, and this bending is called **diffraction.** Experiments show that in any kind of wave motion, the shorter the wave, the less diffraction, or bending, and the sharper

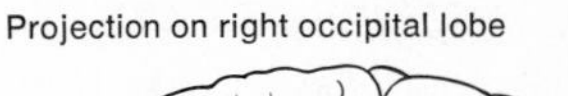

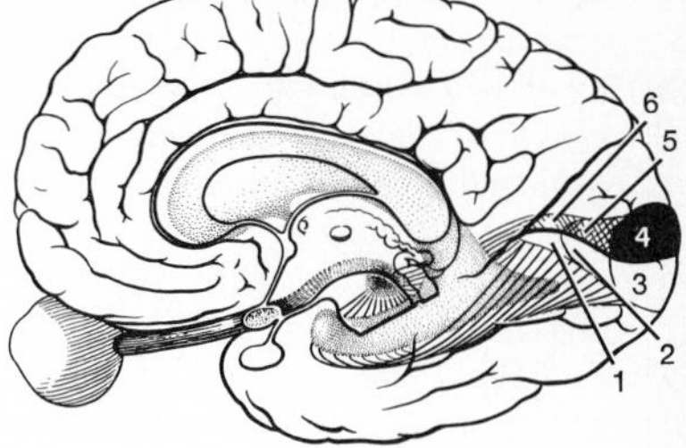

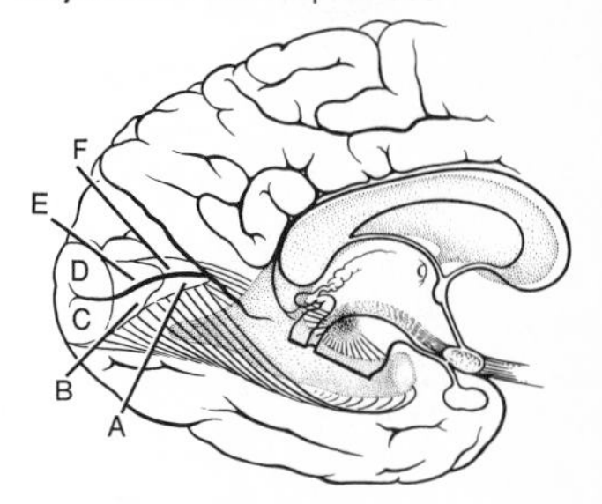

Figure 5-37. Projection of optic radiations to the occipital visual cortex. Corresponding visual quadrants 1, 2, 3, 4, 5, 6, and A, B, C, D, E, F of the visual field may be seen in Fig. 5-35.

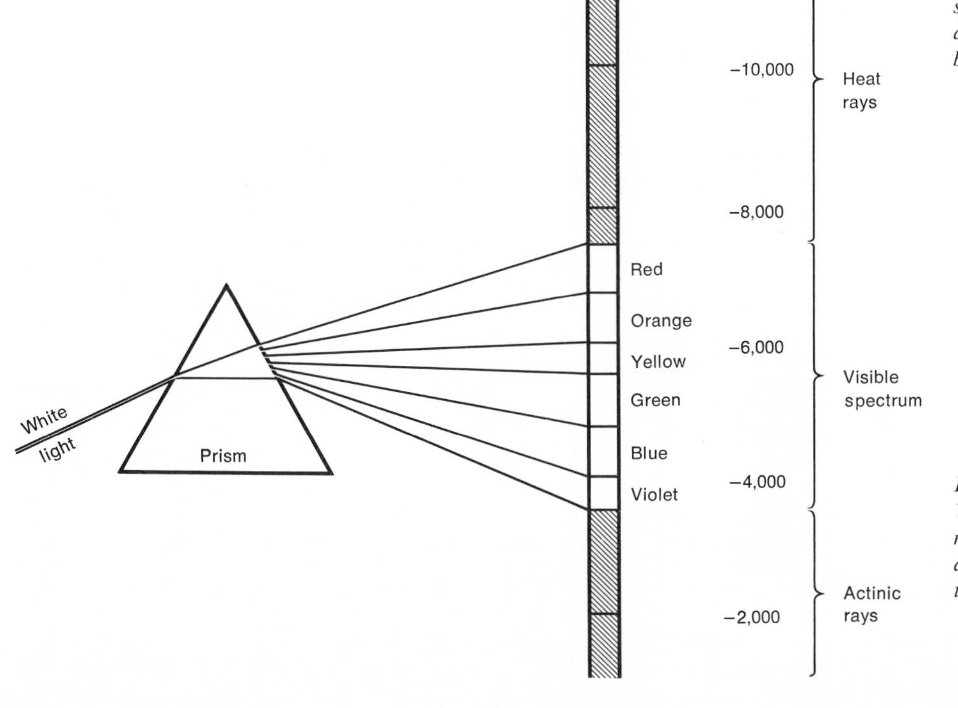

Figure 5-38. Electromagnetic spectrum. White light is a mixture of all the visible rays in the spectrum and can be broken into all its constituent wavelengths by passing through a prism.

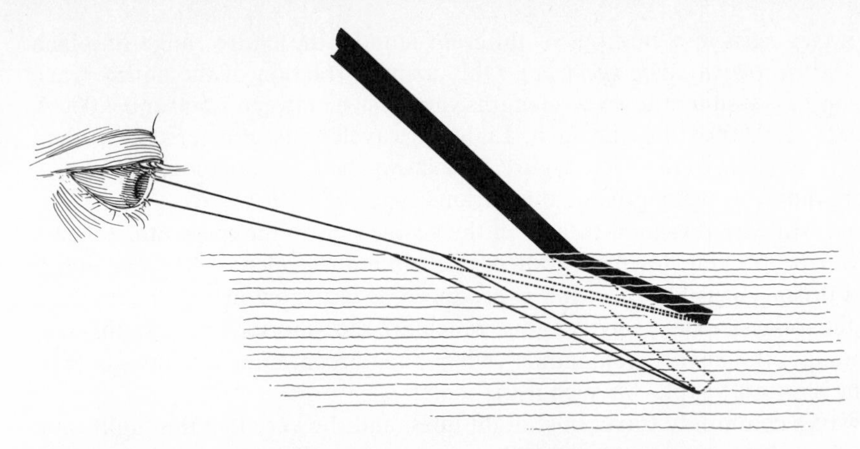

Figure 5-39. Demonstration of refraction. The apparent bend of a log when seen partly submerged in water is caused by the bending of light waves, which occurs when the velocity of the wave is changed as it passes obliquely from one medium, in which it is traveling, to another.

the shadow it casts. This is true in sound, where it can be demonstrated that tones of high pitch do not bend as much as tones of low pitch and therefore are cut off by obstructions more readily. In visible light, the wavelengths are extremely short, and bending is still less and therefore can easily be missed. This type of bending of waves as they pass the edge of an obstacle must not be confused with refraction. **Refraction** may be defined as the change in the direction of a light wave that occurs when the velocity of the wave is changed as it passes obliquely from one medium, in which the wave is traveling, to another. Light passes through either glass or water more slowly than it goes through air, so that when light passes obliquely from air into one of these media, it is bent. One of the most familiar examples of refraction is the appearance of a log that is partly in water (Fig. 5-39). Everyone has noticed the apparent bend or break of the log at the surface of the water. Figures 5-39 and 5-40 illustrate the properties of diffraction and refraction.

Optical Properties of Matter

When light falls on matter, it may be partially absorbed, transmitted, reflected, or scattered. The **absorption of light** involves molecular processes, which result in the production of heat. The energy possessed by the light may (1) cause individual atoms in the absorbing molecules of matter to vibrate along their axis

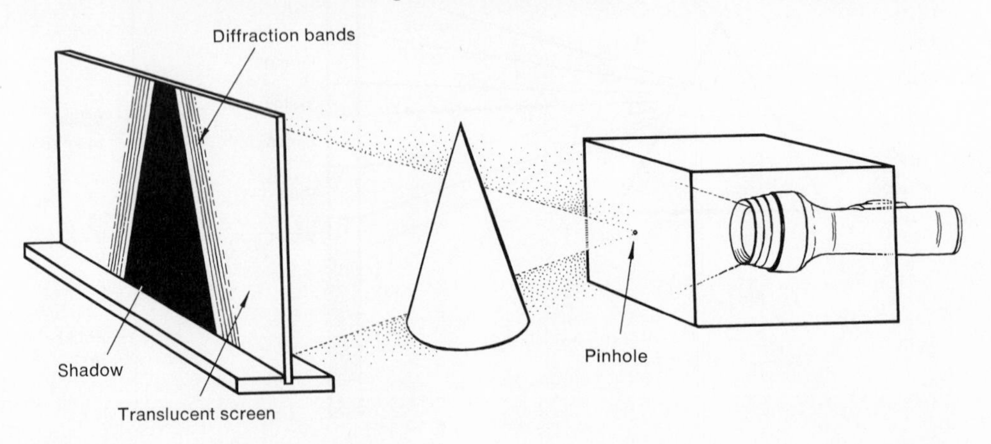

Figure 5-40. Diffraction of light.

of connection, (2) bring about changes in the energy levels of electrons within the atom, or (3) cause atoms to rotate about their axis of connection in the molecule. In this way energy is transferred from the light to the absorbing matter, and only those wavelengths of light which possess energies of a particular range of frequencies will be absorbed by any one substance. Absorbed light energy appears as heat, and with a few possible exceptions, some amount of selectivity of absorption is exhibited by all materials. The energy may be remitted in the form of light (transmission, Fig. 5-41), or again, the absorbing molecule may remain for some time as an activated molecule, capable of participating in chemical reactions that otherwise would be impossible. The primary action of light absorption may be represented by the formula

$$M + h \cdot c/\lambda = M^a,$$

where M is the molecule of the absorbing material; h is Planck's constant, equal to 4.13 electron volt sec; c is the velocity of light, and λ is the wavelength. M^a is the activated molecule, which is capable of entering into chemical reaction producing some biological change as an end process.

Figure 5-41. Transmission of light. When light falls on matter, it may be partly absorbed, transmitted, reflected, or scattered.

Reflection

The bouncing off of some light waves from the surface of a material on which light falls is referred to as **reflection.** Light is reflected from a plane or flat surface in such a manner that the angle of reflection Y is equal to the angle of incidence X (Fig. 5-42). This is known as the **law of reflection.** This phenomenon of reflection is what makes almost every object we see visible. Some materials reflect light waves better than others, and highly polished metal surfaces make the best reflectors. Silver is the metal considered to have the best reflecting properties. If the surface of the material reflecting the light rays is not flat, but curved, all the light rays reflected are brought together at some point, which is known as the point of **focus** (Fig. 5-42).

The optical system of the eye, which involves several lens systems, is normally considered for discussion as a simplified eye in that its refracting system is described as a single thick lens. The specific refracting system of the eye consists of the cornea and aqueous humor, the lens, which is biconvex, and the vitreous humor, acting as a single-surface concave lens. The iris, which is situated between the lens and the cornea, acts as a diaphragm stop in controlling the amount of light that enters the refracting system.

Since the lens system of the eye represents a curved surface, we will be dealing with the behavior of light rays as they pass through or are reflected at such a

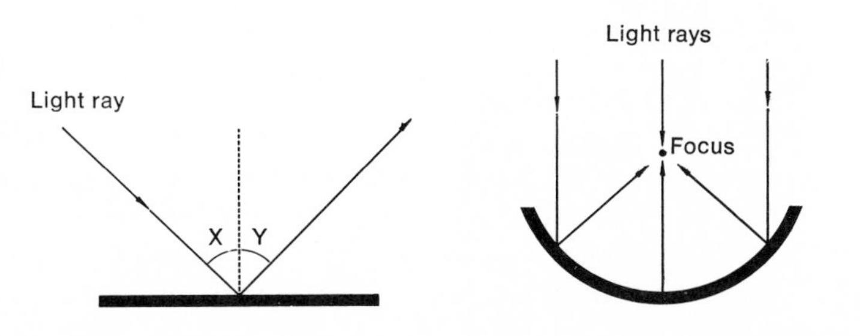

Figure 5-42. Law of reflection. If the surface of the material reflecting the light is curved, all the light rays reflected are brought together at some point, which is known as the focus.

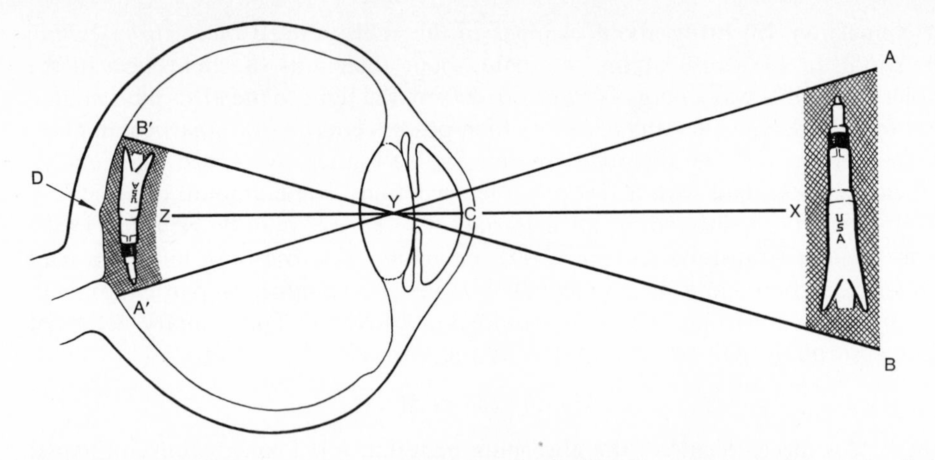

Figure 5-43. Formation of the retinal image in the reduced eye.

surface. This behavior is represented in Fig. 5-43. The cornea refracts the light rays in such a manner as to focus the rays slightly behind the cornea. The crystalline lens permits a sharp focus. The light rays from point *A* are refracted by both the cornea and the lens and focused at *A′*. Light from point *B* is focused by refraction at point *B′*. An image is formed on a plane parallel to the tangent of the surface at *C* and at a distance *CD* from the surface. Light from the object *AB*, which enters the surface perpendicularly, is not changed in direction. The path of such a ray *XYZ* is the optical axis of the system. As a result of the refraction of the light rays, the image appears inverted at the focal point on the retina. Human beings actually see everything upside down, but the sensory impulses reaching the brain are interpreted so that everything is perceived right side up.

Accommodation

Refraction of light rays depends not only on density differences of the medium through which the light is passing, but also on the curvature of the medium. The crystalline lens of the eye functions to assist the cornea in image formation and also permits a sharp retinal focus. It also serves the function of allowing changes in focusing power as the object distance changes or, in other words, as the eyes shift from viewing close objects to those that are farther away. Light rays coming from a distant object are brought to sharp focus exactly on the retina. This is due to the fact that in the normal eye viewing faraway objects, the ciliary muscles are relaxed. As a result, the elastic lens is forced to assume a thin, flattened shape as the ocular fluid causes internal pressure against the fibers of the suspensory ligament, which pulls the capsule surrounding the lens around its entire circumference. The flattened, thin lens with slight curvature has low focusing power. As one gets closer to the object, the light rays coming from it are more divergent rather than parallel, and the refraction of these divergent rays is brought to a focus behind the retina, producing a blurred image. This distorted image sets up a reflex action in which motor fibers send impulses to the smooth muscles of the ciliary body, causing them to contract. This contraction causes the ringlike ciliary body surrounding the lens to be pulled forward. The fibers of the suspensory ligament are loosened by this action, and the pull on the capsule is diminished. The elastic lens inside rounds up on both surfaces, increasing the curvature and the focusing power, bringing the image to focus on the retina since the greater focusing power of the thickened lens brings the light rays together at a shorter distance behind the lens. This change in the shape of the lens is called **accommodation,** and the reflex activity resulting in the change, the **accommodation reflex.**

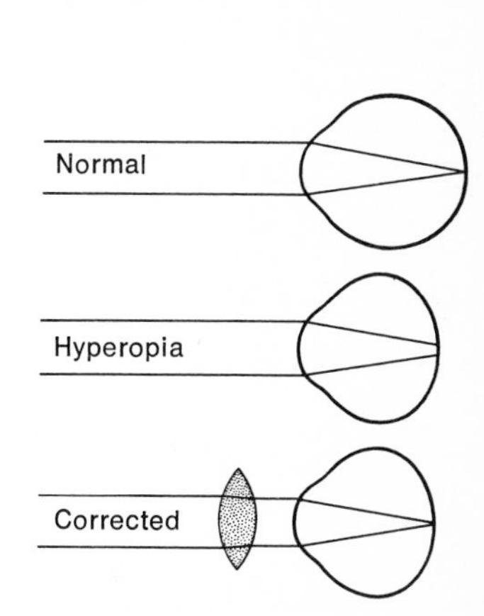

Figure 5-44. Course of light rays and formation of retinal image in the normal eye, in hyperopia (farsightedness), and in myopia (nearsightedness). The mechanism for correction by appropriate lenses is also shown.

The eye at rest is so constructed that when light from an object passes through the lens system of the eye, it is refracted, or bent, in such a way that it is focused on the retina. Light from objects of great distance or nearness is not focused on the retina and must be accommodated by changes in the thickness of the lens. The normal eye can see an object clearly at the **far point,** which is the most distant point at which an object can be seen clearly and is considered to be a distance of 20 ft. At this distance the rays of light reaching the eye from small objects are nearly parallel. The **near point** is the point nearest the eye at which an object can be seen clearly and is usually considered to be a distance equal to the width of the hand. The limit and amplitude of accommodation depend on the following factors:

1. The elastic quality of the lens; young people have lenses that can thicken or flatten out to a great extent, and therefore they can see over a wide range.
2. The condition or state of the ciliary muscles, which affects the lens thickness, and of the circular muscles of the iris, which control the amount of light entering the lens. In accommodating for near objects, the ciliary muscles are contracted, which allows the lens to round up to present a more curved surface; whereas the circular muscles of the iris contract to make the pupil smaller, thereby cutting down on the amount of light entering and also preventing light rays from entering near the periphery of the lens, where the light rays could be broken down into distinct colors, causing aberration. In focusing on distant objects, the lens is flattened and the pupil is dilated by the relaxation of the above-mentioned muscles.
3. The pressure of the ocular fluid on the lens, which can influence the lens in changing its shape.
4. The curvature of the cornea. This curvature may be so extreme as to bring the light rays to focus far in front of or so far behind the retina that it is physically impossible for the ciliary muscles to accommodate.
5. The length of the eyeball:
   a. The eyeball may be too short for the lens system. At short distances the image is focused so far behind the retina that the ciliary muscles cannot relax sufficiently to cause the lens to flatten out for proper focus (farsightedness).
   b. The eyeball may be too long for the lens system so that the light rays come to a focus so far in front of the retina that the ciliary muscle cannot contract sufficiently to cause the lens to present a curved surface that is capable of bringing the light rays to focus on the retina (nearsightedness) (Fig. 5-44).

Photochemistry of the Visual Process

Light is not able to cause response in any nerve cell except the highly specialized retinal receptors, the rods and cones. The most numerous are the rod cells, which are extremely sensitive to light and are responsible for the process of vision known as **scotopic vision.** Scotopic vision produces sensations of dark and light. The cone cells are less numerous and less sensitive to light and are found densely packed in a central retinal spot, known as the **fovea centralis.** The cones are responsible for the process of vision known as **photopic vision,** which is associated with the entry of large amounts of light into the eye and involves the discernment of image detail and the discrimination of colors. Therefore, the **fovea centralis,** which contains only cones, is an area responsible for the sharpness of vision and the appreciation of colors. The fovea centralis enables man to see the fine detail of objects observed, and therefore, many delicate and precise manipulations become possible because of it.

Rod Vision. Rhodopsin (visual purple) is a large protein molecule connected with ten small pigment molecules, called chromophore groups, which are chemically related to carotene, the yellow pigment of carrots, egg yolks, and butter. Rhodopsin is highly concentrated in the rod cells of the retina and in its unbleached form is a dark purple called **visual purple.** Pigmented substances are also found, but in very minute concentrations in the cones. When light strikes rhodopsin, it is bleached to a yellow pigment called **retinene** and a protein known as **opsin.** As a result of the chemical reaction

$$\text{Rhodopsin} \longrightarrow \text{Retinene} + \text{opsin},$$

energy is released and causes an impulse to be transmitted from the rods over the nerves leading to the visual cortex of the cerebrum, where it is interpreted as sensations of dark and light. The retinene produced in the above reaction is reduced under the influence of the enzyme retinene reductase to vitamin A. This passes by way of circulation to the liver, where it is stored.

In darkness rhodopsin is resynthesized by the retina. Vitamin A is taken up from the circulation by the retina, where it is converted back into retinene. This, in turn, combines with opsin to form rhodopsin. Since vitamin A is found in the rhodopsin molecule, a lack of this vitamin would prevent the formation of rhodopsin. Human beings whose diets are lacking in vitamin A suffer with "night blindness" because rhodopsin cannot be synthesized and there is nothing in the rods to react to extremely weak light rays that may enter the eye. In the presence of even moderate amounts of light, the bleaching of rhodopsin proceeds at a much faster rate than its resynthesis. In strong light the bleaching of rhodopsin is nearly complete, and as a result, scotopic vision (dark and light), which is dependent on rhodopsin, is virtually nonfunctional, and photopic vision (image detail and color) is used exclusively. In dim light this is practically the reverse of the above situation. Photopic vision plays little or no part in the visual processes in such light since resynthesis of rhodopsin occurs in the dark. Scotopic visual processes proceed and are used exclusively. It is for this reason that one experiences the difficulty in seeing on entering a dark theater from the lighted street. When a person is on the lighted street, most of the pigment within the rods has been broken down, and the dim light of the theater is not sufficient to oxidize the little rhodopsin that may be present, so that an impulse cannot be generated. However, in a few minutes we find that we are able to see without too much difficulty since sufficient rhodopsin has been resynthesized to react to the dim light present. This adjustment by the rods to the intensity of the illumination is referred to as **dark adaptation.** The reverse of this is true when we go from the dark theater to the light street. The rods contain a large quantity of visual purple. The bright light causes a sudden breakdown of rhodopsin, setting up an excessive number of impulses, which our cortex interprets as very bright light, making it difficult for us to see. In fact, it may be quite painful, and we squint to cut down the light entering the eye. Soon, most of the rhodopsin is bleached out, and fewer impulses are transmitted to the cortex. The eyes are said to be **light-adapted** with the final low rate of nerve impulses generated and normal vision returned.

A certain amount of light must reach the eye before there is any visual sensation. The amount of light necessary to produce visual sensation is called the **absolute visual threshold.** Hecht, Schlaer, and Pirenne have made accurate determinations of the absolute threshold of the eye and have shown that the rods are most sensitive to light of about 5,100 Å in length and that the range of absolute thresholds in energy units varies between 2.1 and 5.7×10^{-10} erg. From the equation for the energy value of individual quanta of light:

$$E = hv,$$

where h is Planck's constant equal to 6.62×10^{-27} erg sec and v is the frequency that is equal to velocity of light divided by its wavelength, we may calculate that the energy of a single quantum of light at wavelength 5,100 Å is 3.84×10^{-12} erg. Therefore, the number of quanta that must enter the eye in order for the sensation of vision to be produced varies from 54 to 148. Not all this light is absorbed by rhodopsin of the rod cells, since a large proportion of the light passing through to the retina is absorbed by the insensitive pigmented layer formed by the outer layer of cells of the retina. This pigment is **fuscin,** and if it were not for this absorption, much light would be reflected and scattered over the retinal area, so that a clear image would be almost impossible. As a result, Hecht and his co-workers have concluded that about 20 per cent of the light at 5,100 Å is absorbed by rhodopsin and that the number of quanta absorbed at absolute threshold is therefore 5 to 14.

Cone Vision. The cone cells are sensitive to bright illumination and also function in color vision (photopic vision). There are many different theories concerning the exact nature of the system responsible for the ability to see and distinguish different colors. All these theories assume that color vision is based on three primary colors, red, green, and violet, and involve the presence within cones of one or more photopigments for absorbing the three types of primary colors. At least one such substance is known; it is called **iodopsin.** Wald has demonstrated experimentally that a close relationship exists between rhodopsin, iodopsin, and a third pigment, found in a few animals, called porphyropsin, which involves slight variations in the protein and vitamin A portion of the molecule.

Of the various theories of color vision, the most acceptable is the Young–Helmholtz theory. This theory postulates the presence of three types of cones differentiated by the presence within the cones of three separate substances. In one group there would be a substance acted on more specifically by red light; in another group, a green-absorbing substance; and in still another group of cones, a substance reacting specifically to violet light. Each of these cones is affected by light of all wavelengths, but to a greater extent by some characteristic range of wavelengths. Long waves (red light) reaching the cones affect only those containing the substance specific for this wavelength, and impulses are initiated that are carried to the visual cortex and interpreted as red color. In the same way, green is seen if medium-length waves are absorbed, and violet, if short waves are absorbed. The color yellow is produced by the activity of the red- and green-cone pigments, and a combination involving green- and violet-absorbing substances produces the sensation of blue. In the same manner, the sensation of white results when all three types of cones are stimulated. A simple experiment that provides excellent evidence for the support of this theory was provided by Hecht. By placing a green filter over one eye and a red filter over the other, a yellow sensation is produced with both eyes open. With one or the other eye open, sensations of red or green are produced. If a yellow filter, which is a combination of red and green, and a blue filter, which is a combination of green and violet, are used, a sensation of white is produced.

Perception of different colors is not equally distributed over the retina. Experimentation has shown that distribution of cones is limited to the fovea centralis and an area a slight distance away from the fovea. As a result, at the periphery of the retina, when there is an absence of cones, no color sensation is evident and all objects present only a grayish appearance. Black is perceived when practically all the light rays are absorbed and is a sensation produced by the absence of stimulation.

Color Blindness

Color blindness is a condition in which there is a lack of one or more color-absorbing substances. According to the Young–Helmholtz theory, there are seven possible types. They are (1) red blindness, (2) green blindness, (3) violet blindness, (4) red-green blindness, (5) red-violet blindness, (6) green-violet blindness, and (7) red-green-violet blindness. The last one listed is called **achromatic,** or total, color blindness, and all objects appear to be black, gray, or white to those with this type. Red-green blindness is most common; violet blindness is very uncommon; and total color blindness is extremely rare.

Protective Structures of the Eye

The eye is protected by several anatomical structures. The strong **bony orbit** of the skull surrounds and protects the eyeball from mechanical injury except at its anterior end, which is protected by the eyelids, which are lined with conjunctival membrane. **Tarsal glands,** which are modified sebaceous glands, are located in the eyelids and open near their free edges. **Eyelashes** line the free edges of the lid and act to prevent dust and other foreign bodies from entering the eye. The eyelids are closed by the sphincterlike orbicularis oculi muscle, supplied by the fascial nerve. If the cornea, the conjunctiva, or the eyelashes are touched, reflex closure of the lid occurs (see p. 152). The eyelids are closed for a short time during sneezing, and in most instances they cover the eyes during sleep.

Lacrimal glands, which are situated in the upper and outer parts of the orbit, secrete **tear fluid** through ducts to the upper part of the conjunctiva. Tear fluid is a watery solution containing mostly sodium chloride and sodium bicarbonate. It has also been determined that tear fluid contains an enzyme, **lysozyme,** which

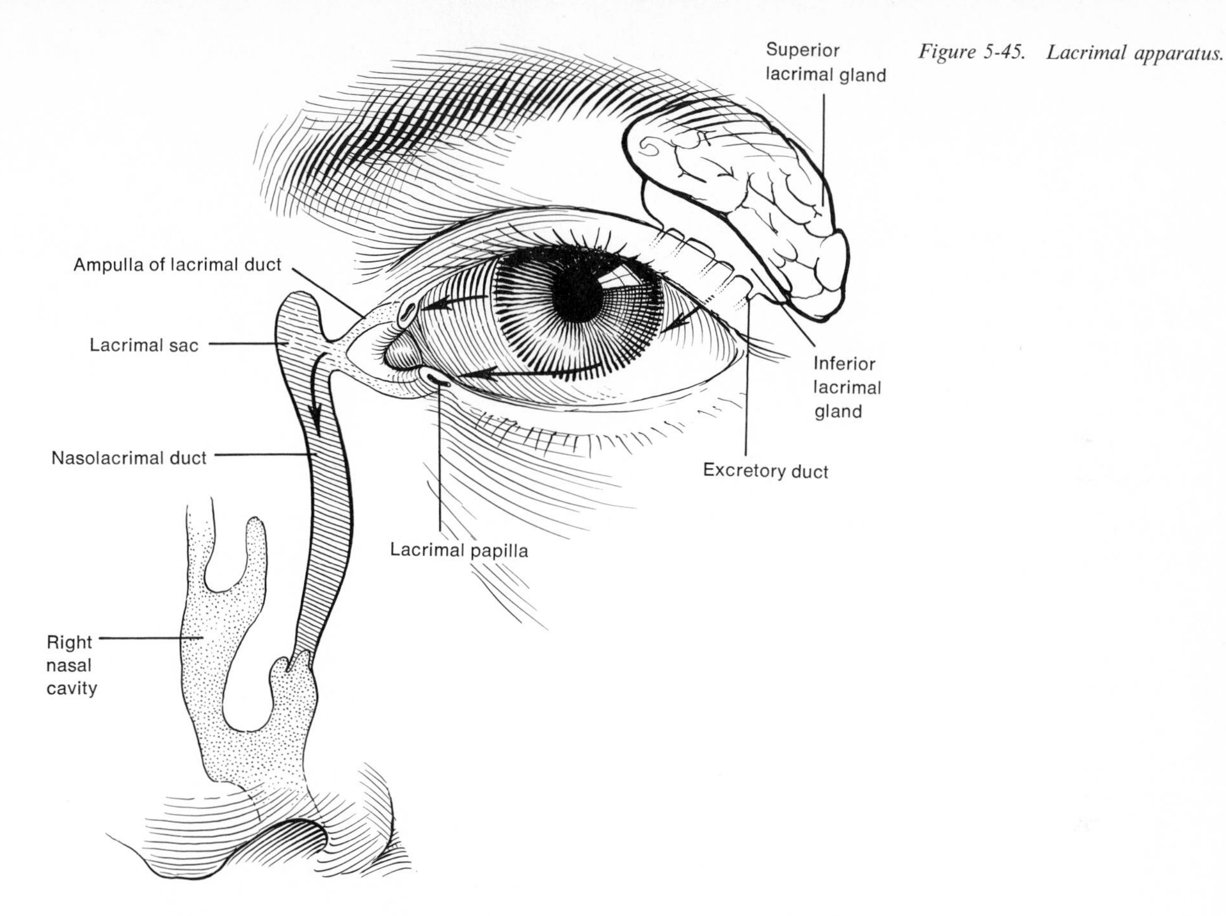

Figure 5-45. Lacrimal apparatus.

functions to kill bacteria, thus reducing the bacterial content of the conjunctival sac. Generally, the tear fluid serves to wash foreign bodies out of the eye and to dilute irritant materials. Tear fluid normally is secreted in amounts just sufficient to keep pace with its evaporation from the conjunctiva and is constantly renewed by frequent involuntary blinking movements. Blinking not only functions to renew the tear film over the cornea, but also protects the eye. If foreign particulate matter or other irritants get on or near the eye, excessive formation of tear fluid occurs. Excessive formation of tear fluid, which is known as crying, or lacrimation, occurs also in emotional circumstances or when a bright light is shone into the eyes. Excess fluid may drain by capillary action through minute openings, the **puncta lacrimalia,** located on the medial margin of each lid, into the **lacrimal canaliculi,** which lead into the lacrimal sac, and the nasolacrimal duct, which opens into the nose (Fig. 5-45).

References

Ammore, J. E. *Molecular Basis of Odor*. Springfield, Ill.: Thomas, 1970.

Best, C. H., and N. B. Taylor. *Physiological Basis of Medical Practice*. Baltimore: Williams & Wilkins, 1965.

Bosma, J. F. *Oral Sensation and Perception*. Springfield. Ill.: Thomas, 1970.

Brindley, G. S. "Central Pathways of Vision," in: *Annual Review of Physiology,* Vol. 32. Palo Alto, Calif.: Annual Reviews, 1970, p. 259.
Dittrich, F. L., and R. Entermann. *The Biophysics of the Ear.* Springfield, Ill.: Thomas, 1962.
Eldredge, D. H., and J. D. Miller. "Physiology of Hearing," in: *Annual Review of Physiology,* Vol. 33. Palo Alto, Calif.: Annual Reviews, 1971, p. 281.
Neff, W. D. *Contributions to Sensory Physiology.* New York: Academic, 1967.
Pirie, A., and R. Heyningen. *Biochemistry of the Eye.* Oxford: Basil Blackwell & Mott, 1956.
Prince, J. H., et al. *Anatomy and Histology of the Eye.* Springfield, Ill.: Thomas, 1960.
Rodieck, R. W. "Central Nervous System: Afferent Mechanisms," in: *Annual Review of Physiology.* Palo Alto, Calif.: Annual Reviews, 1971, pp. 203–240.
Squier, C. A., and J. Meyer. *Current Concepts of the Histology of Oral Mucosa.* Springfield, Ill.: Thomas, 1971.
Tamar, H. *Principles of Sensory Physiology.* Springfield, Ill.: Thomas, 1971.
Viers, E. R. (ed.). *The Lacrimal System.* St. Louis: Mosby, 1971.
Witkkovsky, P. "Peripheral Mechanisms of Vision," in: *Annual Review of Physiology.* Palo Alto, Calif.: Annual Reviews, 1971, pp. 257–280.

The Skin

The **skin,** or **integument,** is composed principally of epithelial and connective tissue, forming a tough, pliable covering for the body. It consists of two layers, a surface one, the **epidermis,** and an underlying thicker layer, the **dermis** (Fig. 6-1).

Epidermis

The term *epidermis,* when literally translated, means "the skin upon," which is exactly what it is—a tough outer covering about 1 mm thick. It is composed of a tough, horny mass of cells that no longer possess the essential properties of life. They are dead cells that are constantly worn away and replaced by cells coming from the deeper layers. The cells in the deeper layers are living and continually divide to produce new cells. As they push outward the new cells receive some nourishment from the capillaries below, but as they get farther away, this nourishment becomes inadequate, and the cells die. The protein material of the dying cells undergoes a change and in its new form is called **keratin,** a term derived from the Greek word *keras,* meaning "horny." Histologically the epidermis is composed of stratified squamous epithelium. The superficial portion is characterized by cornified cells and is known as the **stratum corneum.** This is the layer that is shed constantly. The deep portion of the epidermis is called the **stratum germinativum,** or germinating layer, in which mitosis of cells replaces those shed from the surface.

Dermis

The **dermis,** sometimes referred to as the true skin, constitutes the greater part of the total skin thickness. The skin ranges in thickness from 0.5 mm over the eyelids to 6 mm over the upper back, neck, palm of the hand, and sole of the foot. The dermis is divided into a superficial **papillary layer,** which contains many tufts of capillaries and which brings the blood into close proximity with the

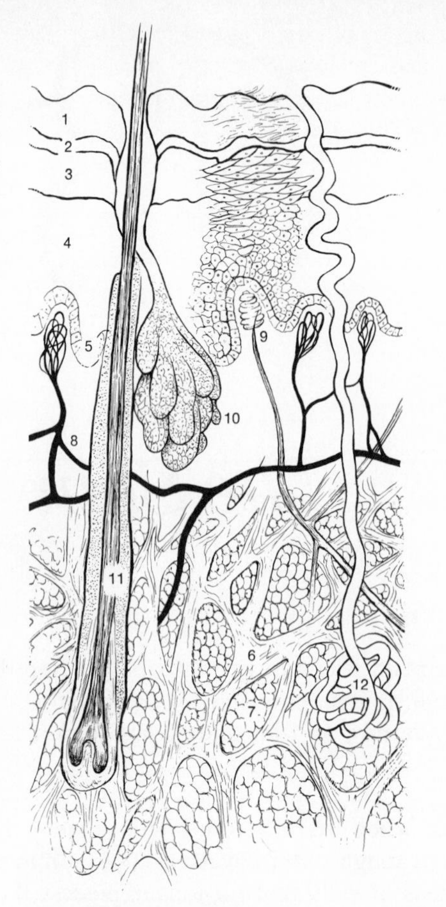

Figure 6-1. Section of human skin, perpendicular to the free surface (schematic). (1) Stratum corneum, (2) stratum lucidum, (3) stratum granulosum, (4) stratum germinativum, (5) papillary layer, (6) reticular layer, (7) fat cells, (8) blood vessels, (9) sensory receptors, (10) sebaceous gland, (11) hair follicle, (12) sweat gland.

germinating layers of the epidermis, and a deep layer, which is known as the **reticular layer.** This layer is a dense mass of interlacing white and yellow elastic connective tissue fibers and accounts for the toughness and strength of the skin. The dermis contains fat, numerous blood vessels, nerves, sensory receptors, hair follicles, and sweat and sebaceous glands.

Functions of the Skin

The skin is a versatile and unique organ; within it a multitude of intricate biological processes are continually occurring, designed to enable it to perform the numerous functions attributed to it. The primary function of skin is to protect the organism it covers. As a result of keratin formation in the stratum corneum, the skin becomes one of the most flexible, yet durable and resistant, structures known in nature. In this way it protects the deeper tissues from injury and drying and provides an excellent barrier against the penetration of bacteria. The skin serves a major role in maintaining a constant body temperature in warm-blooded animals, as the rate of heat loss from the body varies with changes in the size of skin vessels and the evaporation of sweat. The skin is an excellent excretory organ, confining itself mainly to the elimination of water together with some inorganic salts. The fact that some substances diffuse through skin, since it is permeable, contributes to its function as an absorptive structure. It is extremely limited in this capacity; with the exception of water, only a few chemical agents can pass this barrier. Common examples of such substances are mercury and methyl salicylate. In addition to performing the above-mentioned functions, skin is the organ for reception of touch, pain, temperature, and pressure, since it contains the receptors for these modalities of sensation and is rich in cutaneous sensory nerves. This particular function of skin is thoroughly discussed in Chapter 5.

Protective Function of Skin

The first line of defense of the skin against its environment is a thin, complex surface film, which consists of a mixture of materials secreted by sebaceous glands and sweat glands and the products of cornification. This film is well emulsified and spreads over the entire surface of the skin. It contains water, lactic acid, amino acids, urea, uric acid, and ammonia, which is derived from the sweat glands. The sebaceous glands contribute fats, fatty acids, and wax alcohols, and the cornification process gives rise to sterols, amino acids, phospholipids, and complex polypeptides. These products enable the film to act as an antiseptic, a neutralizer of acid and alkali, to interfere with absorption of toxic materials, to act as a lubricant of the horny layer, and to control hydration of this layer.

The next line of defense of the skin is represented by the horny layer, which is of prime importance in protection of the organism against the environment. The principal material found in this layer is keratin, a tough, fibrous protein, highly resistant to acid and alkali as well as common proteolytic enzymes. The horny layer is an almost complete physical barrier to electromagnetic waves, bacteria, fungi, parasites, and practically all noxious chemicals. The closely knit keratin mesh within the cells of the horny layer, and the tightly packed arrangement of these cells, prevents movement of most molecules through the intact epidermis. Chemicals of extremely small molecular structure probably pass through the surface of this layer but are held by the stratum corneum and are cast off since this layer is normally continuously shed. If the horny layer is broken, all physical and chemical agents can enter the body without restriction; even the smallest scratch increases the skin's permeability to such agents. The ability of the skin to repair injuries rapidly is an important part of its protective function.

A few bacteria normally reside on the skin surface at all times since the skin furnishes adequate nourishment and environmental conditions (moisture and temperature) for their growth. Although these resident microorganisms are not considered to be pathogenic, it is desirable that the rate of their reproduction be kept at a minimum because of the possibility that these bacteria might become invasive and the fact that body odor has been shown to be caused by bacterial decomposition of skin secretions. Most bacteria are not normally resident on the skin and rarely live to establish a foothold because of the low moisture content of the skin that usually is directly exposed to the exernal environment and the presence on the skin of naturally produced antibacterial and antifungal substances. Bacteria need adequate moisture for multiplication, and if they cannot multiply, they soon die. Not only does the secretion by the sebaceous glands and sweat glands of organic acids provide an acid environment that inhibits growth of bacteria, but also some of these acids, such as lactic and caprylic, have antibacterial properties, and others, such as capric and undecylenic, have antifungal properties. The lack of adequate moisture and the presence of specific antibacterial and antifungal organic acids are the two most important factors in making the skin unsatisfactory for the growth of most microorganisms, except those few relatively harmless species that constitute the normal resident types. Because of this, the likelihood that pathogenic organisms will invade the tissue is greatly reduced, if the other protective mechanisms, such as the structural barrier of the skin, should prove inadequate.

Thermoregulatory Function of Skin

Heat is continuously produced in the body by the oxidation of fats and carbohydrates. In warm-blooded animals the necessity to maintain constant body temperature makes it mandatory that the heat loss from the body be continuously adjusted to heat production. Since the skin acts as a large radiator surface, it is obvious that most of the heat must be lost through the surface, and since the blood is the primary medium of transfer of heat throughout the body, the complex vascular patterns found in the skin play a significant role in the regulation of heat loss. Skin abounds in **arteriovenous anastomoses,** especially the skin of the hands and feet, the lips, the nose, and the eyelids. The anastomoses are formed where arteries and veins make direct connection with each other without the benefit of their normal capillary connections (Fig. 6-2). The arterial portion of each anastomosis is often coiled into a ball-like structure, whereas the venous end is funnel-shaped. Interposed between these two vessels is a middle portion, which possesses a relatively thick muscular wall and is the most actively contractile part of the entire structure. It is these anastomoses that control the amount of blood flowing to the integument and therefore play a significant role in controlling heat loss from the surface of the body.

The center of control of this regulatory mechanism is located in the hypothalamic portion of the brain (Fig. 6-2). This control center consists of two parts. One part (antirise center) is stimulated by increases in the temperature of the blood circulating through it, and the other (antidrop center) responds to decreases in temperature of the blood. Nerve impulses generated by the center are transmitted to the anastomoses, causing the muscular portion to constrict or dilate and thereby controlling blood flow to the skin. If the **antirise center** is stimulated, the blood vessels of the skin are dilated, resulting in a greater flow of blood to the surface. In addition, impulses are sent to the sweat glands, increasing perspiration, and to the respiratory center, causing panting. All these physiological responses act to increase the rate of heat loss from the body, thus bringing the temperature to its constant value. Physiological responses to temperature decreases in blood, brought about by stimulation of the **antidrop center,** tend to reduce heat loss from the skin surface in several ways. Constriction of the blood vessels of the skin reduces the blood flow to the surface, while at the same time surface hairs are stimulated to become erect, which acts to increase the insulating layer of air surrounding the skin. Shivering is induced through the same mechanism to increase heat production.

Thermal sensory receptors located in the skin, which are stimulated by heat or cold, are connected with the central regulating mechanism discussed earlier.

These receptors aid in adjusting blood flow to the skin in response to changes in environmental temperature.

Heat loss through the skin surface takes place by four different physical mechanisms: radiation, convection, conduction, and evaporation. These, in turn, are influenced by the humidity, air flow, and temperature of the environment. **Radiation** represents transfer of energy from the surface of an object to the environment by that portion of the electromagnetic spectrum formed by the infrared rays. The skin, like any object that is warmer than its surroundings, radiates heat to the environment in the form of heat rays. Under comfortable conditions, radiation losses account for about 60 per cent of the total heat loss. When the environment is too warm [above 88° F (31° C)], the evaporation of sweat is sufficient to release heat by radiation and convection if the humidity is low, evaporation cools the body surface and balances heat production. For this reason, for environmental temperatures above 88° F, it is said that constant body temperature is not under **vasomotor control** but under **evaporative control.** The environmental temperature in which the blood flow to the skin is varied by vasomotor control to give skin temperatures such that heat loss equals heat production ranges between 66 and 88° F (between 19 and 31° C). Below 66° F, the environmental temperature is low enough to bring about constriction of the blood vessels. If it drops too low, maximal constriction of the blood vessels of the skin is brought about. Under these conditions the body has no control, and it cools as though it were an inanimate object.

Heat loss from **convection** results from the transfer of energy by means of moving air. When there is movement of environmental air over the skin surface, which is cooler than the skin temperature, the warmer air near the skin is continuously replaced by the cooler air of the environment, and heat is carried away from the body surface. Convection causes cooling not only by carrying warm air away, but also by increasing evaporation of perspiration. If the environmental air is warmer than the skin surface, heat cannot be lost in this way, but may be gained. This can be offset by heat loss due to increased evaporation.

Evaporation of water from the skin surface causes heat loss because transformation of water in vapor requires heat. At body temperature the heat required to evaporate 1 g of water from the surface of the body is 580 g cal. In the temperature range in which skin temperature is under vasomotor control, the heat loss by evaporation amounts approximately to 25 per cent of the total loss. Within this range of environmental temperature the skin is constantly covered by a thin layer of sweat, referred to as the *surface film.* This sweat is not visible, nor are we conscious of it, as it is almost immediately evaporated, and for this reason is called **insensible sweating.** It is the evaporation of this thin layer of fluid that accounts for the heat loss within the range 66 to 88° F (19 to 31° C). At an environmental temperature of about 88° F there is a general sweating reaction. The temperature at which this occurs is referred to as the *critical atmospheric temperature.* The skin temperature of resting subjects at which sweating is initiated ranges between 88 and 94° F. Sweating elicited by exercise starts at a lower skin temperature. This is because the deeper thermal receptors in the skin are concerned mainly with sweating and thus are more readily stimulated by internally

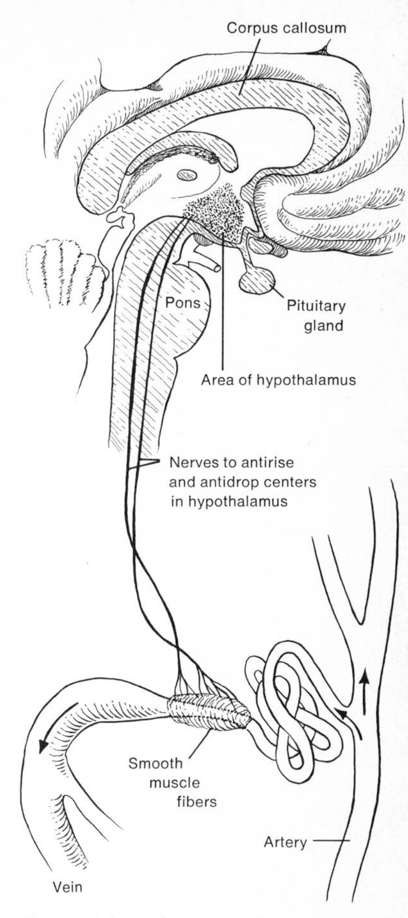

Figure 6-2. Thermoregulatory center of the brain. Nerve fibers from both antirise and antidrop centers are supplied to the arteriovenous anastomoses in the skin.

produced heat. This thermoregulatory effect of perspiring is useful only as long as the sweat can evaporate. In a dry, hot environment, evaporative cooling is a most efficient mechanism, but in an environment of high humidity, it becomes grossly inefficient because most of the sweat is not evaporated but is lost by running off the skin surface. Sweating then becomes a detriment because water and salt are lost in great quantities, and dehydration and salt deprivation set in.

Heat loss through the mechanism of **conduction** is rather difficult to measure because it depends on the area of contact and the nature of the substance with which the skin surface is in contact. Conduction may be defined as the transfer of heat from one medium to another when two media are in direct contact with each other. Experimentation has shown that heat loss in this manner is negligible because conventional clothing does not essentially change physical heat loss. Of the total heat lost from the body, 95 per cent is eliminated through radiation, conduction, convection, and evaporation. On the average, the daily heat loss from the body by radiation, conduction, and convection amounts to 1,900 cal, whereas evaporation accounts for approximately 400 cal—a total loss of 2,300 cal. The remaining 5 per cent of heat loss is due to heat carried by waste products, such as urine and feces, which accounts for about 50 cal in a 24-hour period. Approximately 75 cal of heat are spent in raising the temperature of inspired air to body temperature. The total quanity of heat lost daily from the body amounts to approximately 2,425 cal.

Excretory Functions of Skin—Sweat and Sebaceous Glands

If the skin is said to have an excretory function, it is because of the rich supply of sweat and sebaceous glands that are embedded in the dermis over various parts of the body.

Sweat Glands. **Sweat glands** were first described by Purkinje, in 1833, and later, in 1922, they were divided into two basic types by Schiefferdecker, called **eccrine** and **apocrine.** Eccrine glands, or the small sweat glands, are distributed all over the surface of the body with the lips, glans penis, and clitoris being the only areas free of these structures. Eccrine glands are simple, coiled tubular glands with the coiled portion situated deep in the dermis; the straight-tube portion of the gland passes up from the coil to the epidermis, where it coils again before terminating in an opening (Fig. 6-3). Each tubule consists of an excretory duct, the straight portion with its coiled epidermal ending, and a secretory portion, represented by the tightly coiled tube situated deep in the dermis. There are no morphological differences among the many eccrine sweat glands found in various regions of the body, although there is a difference in response to specific stimuli. The eccrine glands in the skin of the palms of the hands and soles of the feet respond very slowly and weakly, if at all, to heat stimuli, but very strongly to psychic and sensory stimuli. Note the sweating of the palms when you are nervous or frightened. On the other hand, the sweat glands on the forehead, the neck, the back and front of the body, and the back of the hands respond quickly and strongly to thermal stimuli and very weakly to sensory or psychic stimuli. These differences are brought about by the innervation of these glands from the central nervous system. Thermal sweating has its center in the hypothalamus, whereas emotional or psychic sweating is controlled by the **sensory** and **premotor cortex.**

The **apocrine** glands, or large sweat glands, do not occur over the whole body surface in man. They are found in the axillae (armpits), around the nipple of the breast, on the abdomen around the umbilicus, in the perineal and perianal regions, in the prepuce and scrotum of the male, in the perigenital region of the female, in the external ear canal, and in the nasal passages. The microscopic structure of apocrine and eccrine glands is similar, but apocrine glands are larger than eccrine glands, and their secretory coil is situated in the subdermis. In addition, the excretory duct of apocrine glands is wider and more tortuous and opens into the hair-follicle canal. Rarely do apocrine glands open directly onto the surface of the epidermis.

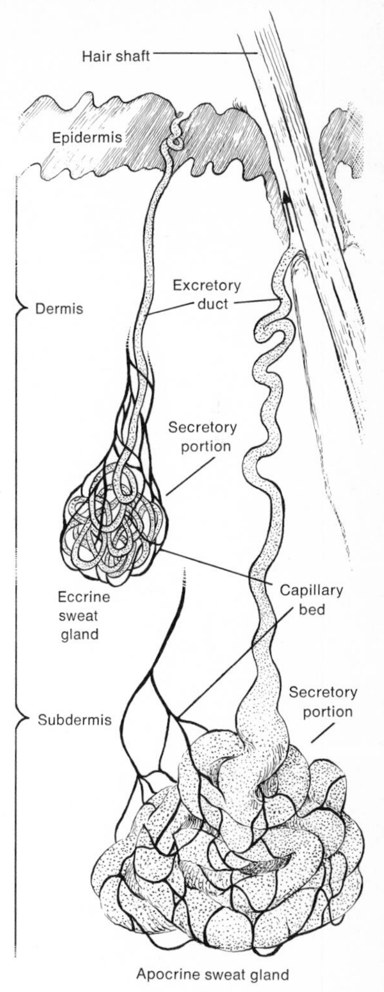

Figure 6-3. Eccrine and apocrine sweat glands showing the excretory duct, the tightly coiled secretory portion, and the vascular bed of each gland.

Eccrine Sweat. Eccrine glands are true secretory glands, producing the clear aqueous sweat utilized for heat regulation. If the environmental temperature rises above a critical level, between 88 and 90° F, there is a sudden outbreak of visible sweating over the whole body surface of man. This is called *reflex sweating*. At environmental temperatures below the critical temperature, the sweat glands continuously secrete slowly and intermittently. The droplets of sweat are so quickly evaporated that they are invisible. This sweating is referred to as *insensible sweat secretion*. Excitation of sweating by heat or exercise brings about an increase in the number of sweat glands functioning per unit of time as well as an increase in the output of single glands.

Volume. Under extreme conditions of heat, humidity, and work, as much as 3 gallons of sweat may be produced in a 24-hour period. Under basal conditions, a person in a perfectly air-conditioned environment will produce insensible perspiration amounting to at least 1 pint per day. Considerable quantities of fluid and dissolved substances thus can be lost through perspiration and, under various conditions, may increase so much as to cause severe dehydration and salt imbalance. Compensatory action to prevent such conditions from developing is brought into play immediately by the body. The output of urine by the kidneys, for example, is decreased. The great loss of water by perspiration evokes an insatiable thirst sensation, resulting in the consumption of large quantities of water so that the body water balance will be restored. The mechanism of thirst sensation is believed to be a reflex activity stimulated by the drying of the mucous membranes of the mouth and throat. When one becomes dehydrated, the quantity of saliva is drastically reduced and the mucous membranes are dry. Receptors in the membranes are stimulated by this condition, and impulses are sent to the thirst center of the brain. The individual as a result has the sensation of thirst and wants to take in more fluids.

When perspiration becomes marked, the quantity of salt lost from the body is also increased. In addition, as the thirst mechanism is brought into play, the salt concentration within the body fluids may be further diluted as large quantities of water are taken in. Distressing physiological reactions, such as weakness, nausea, vomiting, and often loss of consciousness, will result if the salt is not replaced. During hot, humid weather an increase in the intake of salt with the food is often advised, especially if considerable work or exercise is contemplated.

Composition. Eccrine sweat is a clear, watery secretion consisting of 99 per cent water and 1 per cent solid material. Half the solid material is inorganic salt, mainly sodium chloride, and half is organic material. Urea makes up the greatest percentage of organic materials. Creatinine, uric acid, ammonia, amino acids, glucose, lactic acid, and the water-soluble vitamins B and C have also been demonstrated in various minute amounts in eccrine sweat. The presence of lactic acid and the acid amino acids (dicarboxylic) and fatty acids in eccrine-gland secretion always makes freshly secreted sweat acid in its reaction, with a pH ranging between 4 and 6. Although the composition of sweat depends on material carried by the blood stream, it has been clearly demonstrated that sweat is not a simple filtrate of blood plasma but the product of an active secretory function of the cells of which the gland is composed.

Apocrine Sweat. Apocrine glands do not respond to thermal stimulation, and their secretions do not take part in thermoregulation. In general, apocrine gland function is periodic, responding promptly to mental stimuli, and experimentation has demonstrated that emotional situations stimulate these glands to excessive activity. The secretion of these glands is a thick white, gray, or yellowish solution that does not form a spherical droplet as eccrine sweat but forms a sticky cap on top of the external opening of the duct. Apocrine can be distinguished from eccrine sweat not only in its physical appearance, but also in its chemical composition. The milky droplets of apocrine secretion contain iron, proteins, reducing sugars, lipids, and ammonia. In addition, apocrine glands can secrete substances that add color to the sweat. The case of the hippopotamus, which is famed for sweating blood, has been cited many times[1]; when this animal was angered or excited, it actually excreted red apocrine sweat. Formation of blue sweat is attributed to the presence of the compound indoxyl, which on exposure to air is oxidized into the blue dye indigo.

Apocrine sweat contains many odorous substances that contribute to "body odor," whereas eccrine sweat does not. Indoxyl, for example, is an extremely malodorous compound. The odor of apocrine secretion may be described mainly as "caprylic," which is due to the presence of free volatile fatty acids. In addition, the presence of hydroxy acids and ammonia also contributes to the odor. Bacterial activity in the presence of these secretions contributes to the intensity and quality of the smell. Removal of apocrine and sebaceous secretion with warm water and soap effectively counteracts and suppresses the development of body odor in most persons. There are some persons who, despite scrupulous cleanliness, have intense and malodorous cutaneous excretions. The substances responsible for this smell are unknown. It is believed that an alkaline reaction of sweat can be responsible, since this type of environment can promote the activity of bacteria, which can result in the development of such odors. The treatment of such areas with mild acids usually influences the elimination of such unpleasant odors.

[1] L. Szodoray, "Heterotopic Apocrine Glands." *Orv. Hetilap.*, **4**:360 (1948). Quoted in *Excerpta Med.*, Sec. XIII, **3**:382 (1949).

Sebaceous Glands. **Sebaceous glands** are positionally closely associated with the hair follicle since they develop from the follicular epithelium of the hair. They are multilobed structures surrounding one half to two thirds of the circumference of the hair follicle and have direct connections with the hair root canal by a short duct (Fig. 6-4). In man sebaceous glands occur all over the skin surface where hair follicles are present and differ in size and shape in different regions. They are relatively small on most areas of the trunk and extremities, but quite large in the skin of the forehead, face, neck, and upper chest. The sebaceous glands produce an oily secretion called **sebum,** which forms the greater part of the lipid component of the surface film covering the skin surface. Sebum is formed within the larger inner cells of the gland, which become impregnated with fat. The cells finally degenerate because of this fat infiltration and break apart. The residue is conducted to the hair root canal by the duct of the gland, where it finally makes its way along the hair to the surface of the skin.

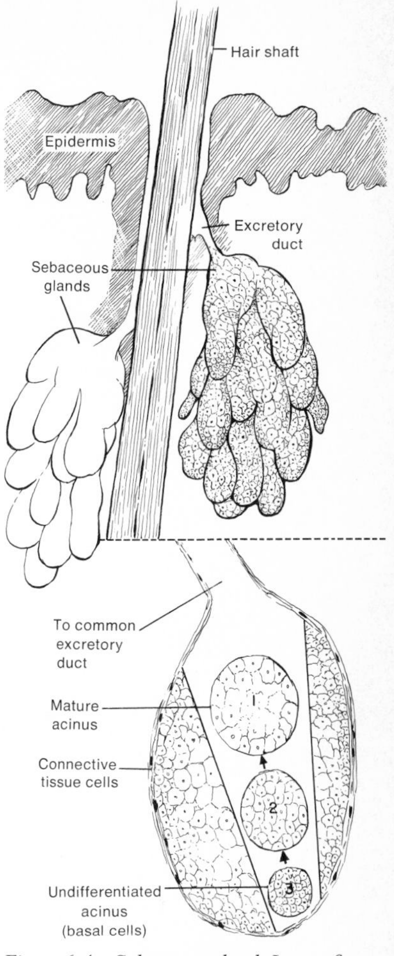

Figure 6-4. Sebaceous gland. Lower figure represents an enlargement of one lobe showing the development of a sebaceous acinus (1, 2, 3).

Sebum is different from tissue fats in being composed partly or entirely of waxes. The sebaceous gland must synthesize a number of unusual substances since waxes are not found anywhere else in the body. The chemical nature of sebum differs from species to species, cholesterol and large amounts of free fatty acids being the only substances that occur constantly. Human surface lipids, which are produced mainly by sebaceous glands, contain both free and esterified fatty acids. A large percentage of the surface lipid comprises a mixture of steroids and hydrocarbons. Cholesterol forms about 16 per cent of this type of matter. Squalene, an acyclic triterpene alcohol, is formed only in the sebum of man and the horse. Squalene is the principal hydrocarbon present, and it accounts for about 6 to 7 per cent of the total surface fat. Surface lipids also contain small amounts of vitamin E and appreciable amounts of acidic or phenolic compounds, which may function as antioxidants. Only a few glycerides and traces of phospholipids are present. The entire complex mixture of surface lipids is extremely hydrophilic. The amount of surface lipids on the skin is dependent on the size of the gland, the rate of secretion of sebaceous glands, and the wetness of the skin. Sweating appears to be the prime factor in the spread of sebum over the surface. The rate of excretion of sebum in man seems to be regulated in some manner by the relative thickness of the fat layer on the surface of the skin. If one starts out with a thoroughly defatted surface, sebum excretion sets in at a rapid rate, gradually declining, until it finally falls to a minimum when the fat layer has reached a certain thickness. The rapid excretion is resumed when the fat layer is removed again. It has been suggested that inhibition or cessation of secretion is mediated by the pressure of the sebaceous layer.

Hormonal regulation of sebaceous glands is well established. In human males, testosterone has been implicated as the major cause of pubertal acne. Experimental evidence also suggests the influence of progesterone administration on the pronounced enlargement of the sebaceous glands. This has given rise to the belief that in the female, acne is incited by progesterone, which is produced by the corpus luteum of the ovary. Adrenal androgens are also implicated as an inciting factor. Removal of the pituitary gland results in atrophy of the sebaceous glands, which may be counteracted only partially by progesterone and testosterone administration. Thus, it appears that the pituitary is necessary for the proper maintenance of sebaceous glands.

There is a great deal of morphological evidence that sebaceous glands in man are supplied by nerves. The fibers appear to be of sympathetic origin, and stimulation increases glandular activity. The fact that emotions cause the sebum to flow also points out the possibility of direct nervous action.

Sebum is liquid inside the duct and hair follicle. It diffuses up the follicular canal, and after reaching the surface, it spreads all over the skin and also penetrates between the horny lamellae of the stratum corneum. It solidifies at skin-surface temperature and normally appears on the skin and hair in a semisolid state. Sebum anoints the hair and keeps it from drying and becoming brittle. On the surface of the skin it forms a protective film that limits the absorption and evaporation of water from the surface.

Absorptive Function of Skin

Skin absorption refers to the penetration of substances from the outside into the skin and through the skin into the blood stream. The absorptive function of the skin is somewhat limited, since the waxy surface film, the tightly packed cornified cells of the stratum corneum, and the physiological barrier layer found between the horny layers and the granular layer underlying it provide efficient barriers against the penetration of most substances. Experiments have demonstrated that the skin is permeable to substances that are soluble in lipids and to substances in the gaseous state but practically impermeable to water and electrolytes. Chemical agents of small molecular dimensions probably pass directly through the epidermis, although the tightly packed arrangement of the cells of the horny layer would hold back large molecular aggregates. In addition, chemical substances may reach the dermis by way of the hair follicles into sebaceous glands and through the ducts of the sweat glands. Lipid-soluble substances probably enter mainly through the sebaceous glands, and the sweat duct appears to play a dominant role in the passage of water. The transfer of most chemical agents that can penetrate intact skin is probably by a path directly through the epidermis.

Experimental data on absorption of single compounds and groups of compounds lead to the following conclusions: Water is transmitted as water vapor, and there is diffusion inward and outward of equal amounts. Electrolytes applied to the skin in aqueous solutions do not penetrate the skin of mammals, or if they enter, they do so in small amounts through the sweat ducts and follicular canals. Lipid-soluble substances are rapidly and completely absorbed through the skin, and the absorption rate appears to be faster if the substance is somewhat water-soluble. An important group of these lipid-soluble compounds includes the phenolic compounds, such as phenol, salicylic acid, resorcinol, hydroquinone, and pyrogallol, which enter the skin with extreme ease from any vehicle. Estrogenic hormones, testosterone, progesterone, and deoxycorticosterone penetrate the skin rapidly and with great ease. The fat-soluble vitamins A, D, and K penetrate the skin with ease, and the carotenes are also easily absorbed. Of the lipid-soluble salts of heavy metals, mercuric bichloride is absorbed most easily through the skin. Salts of lead, tin, copper, arsenic, bismuth, antimony, and mercury, which are originally lipid-insoluble, may be transformed on or in the cornified layer into lipid-soluble substances by forming oleates on combining with fatty acids of the sebum.

Substances that are in the gaseous form at ordinary temperature easily penetrate the skin, with the exception of carbon monoxide. The mechanism for this transfer is assumed to be a simple diffusion of the gaseous molecules across the epidermis and through the sweat ducts and sebaceous glands. The amount and rate of transfer are dependent on the vapor tension of the gas inside and outside the skin, the temperature, and the water and fat solubility of the gas. Gases that are transmitted easily through the skin are oxygen, nitrogen, helium, carbon dioxide, ammonia vapor, hydrogen sulfide, hydrogen cyanide, and volatile aromatic oils. Physiologically, most important is the movement of oxygen and carbon dioxide. Goldschmidt et al.[2] have shown that in asphyxia, oxygen enters the skin in greater amounts and oxygenates the blood in the cutaneous vessels. It has also been demonstrated that the blood in the pulmonary capillaries will take up a greater oxygen supply from the lungs if the total skin surface is deprived of an oxygen supply, as in the experimental immersion of the body in nitrogen gas. It appears that if the body is deprived of a cutaneous oxygen supply, a compensatory mechanism is set up by the respiratory apparatus.

Skin Color

The color of the human skin is derived from a variety of chemical and physical properties associated with the skin structure. Five pigments are known to influence skin color. These are **melanin,** found mainly in the malpighian layer of the epidermis; **melanoid,** a pigment related to melanin but having a different absorption band of visible light; **carotene,** located mainly in the stratum corneum, as well as in the subcutaneous fat; and **oxyhemoglobin** and **reduced hemoglobin** in the blood vessels of the dermis. The distribution patterns and concentrations of these pigments influence skin color, as do the thickness and light-scattering properties of the overlying colorless tissue layers. In normal skin, melanin is usually present only in the malpighian layer of the epidermis. This pigment is the most important of the five in determining skin color. In heavily pigmented white skin and Negro skin, melanin is present in all the layers of the epidermis and is extremely abundant and evenly distributed in Negro skin.

Variations in the thickness of the skin can also modify skin color. Thin epidermis is more transparent than thick and allows the color of the blood pigments to express their color characteristics more readily. Persons with thin skin, as a result, have a ruddier complexion, and those with a thicker epidermis, which is less transparent, look yellower. The light-scattering qualities of the skin can also act as a factor modifying skin color. It is known that blue light is more readily scattered by tiny particles than red light. Thus, since the transparent stratum corneum scatters light only slightly, the deeper layers, where the dark-pigment granules are normally located, appear bluer because of the scattering effect of light in the overlying coarser tissue layer.

Melanin is a yellow to black pigment that functions not only in contributing

[2] S. Goldschmidt, B. McClone, and J. S. Donal, "Oxygen Absorption Through the Skin." *Am. J. Med. Sci.,* **187**:586 (1934).

color quality to the skin, but also, perhaps more importantly, in protecting the organism from ultraviolet light. In normal skin, exposure to ultraviolet light results in darkening of the melanin granules (tanning), as well as formation of new melanin. Specialized cells, **melanocytes,** present primarily in the skin and the eye, are the only cells in the body capable of forming melanin. They are able to do this as a result of the presence of the enzyme tyrosinase, which acts on the amino acid tyrosine, oxidizing it through red intermediate stages to a brown-black pigment, dihydroxyphenylalanine. Absence of the enzyme tyrosinase results in albinism.

Melanoid is thought, by some investigators, to be a degradation product of melanin. Others claim it to be the same as **hornfarbe,** a well-known dark oxidation product developing in keratin. **Carotene,** a yellow-orange pigment, is found in lipid-rich areas, such as the stratum corneum and the fat of the dermis and subcutaneous tissue. Carotene adds the strong yellow component to skin color but is definitely in greater concentration in the skin of women than in that of men.

Oxyhemoglobin imparts a red component to skin color and is especially evident in areas where there is a rich arterial supply, such as the skin of the face, neck, palms, soles, and nipples. **Reduced hemoglobin** contributes a bluish or purple character to skin color and is more evident in the lower parts of the trunk. Factors such as melanin concentration and skin thickness tend to suppress the hemoglobin-pigment color component effect.

Hair

Of all the animals in the vertebrate group only those in the class **Mammalia** possess hair. Hair is composed of keratinized cells compactly cemented together and is distributed over almost the entire body. It consists of a **root,** or portion below the surface of the skin, which penetrates deep into the dermis and is surrounded by a tube of epithelium, which is continuous with the epidermis, called the **hair follicle** (Fig. 6-5). The terminal ends of the hair follicle are enlarged into an onion-shaped region called the **bulb.** The bulb is somewhat invaginated at its base to form a cavity, which is filled with loose connective tissue called the **dermal papilla.** This papilla is highly vascular and provides nourishment for the growth of the hair. The grapelike clusters of sebaceous glands project laterally from the hair follicle into the surrounding dermis. Tiny smooth muscle fibers called the **arrector muscles** extend from the connective tissue layer of the skin into the side of the hair follicle just where the bulb portion or enlargement of the follicle begins. These muscles are innervated by sympathetic fibers only and contract under stresses of fright and cold, causing the hair that projects above the surface, the **shaft** portion, to stand up. They are also responsible for the appearance of "gooseflesh," as the contraction of these muscles causes the skin around the shaft to form slight elevations.

Hair Growth

The replacement of hair in animals usually occurs in cycles, for the hair follicles, from which the hair develops, do not produce hair continuously, but alternate with periods of rest (Fig. 6-6). The **hair cycle** may be defined as that period which encompasses the formation and growth of a new hair, followed by a resting stage, and terminating with the growth of another new hair from the same follicle. The first stage in the hair cycle is the active proliferation of cells by mitotic activity in the base of the bulb, called the **matrix.** These new cells from the matrix move to the upper part of the bulb, where they increase in size and become elongated. As they are pushed farther up the hair follicle, the cells become keratinized. This area of the follicle is called the **keratogenous zone,** and depending on the length of the follicle, it terminates at a level approximately one-third of the way between the tip of the papilla and the surface of the skin. The hair above this level loses its cellular composition and is then considered mature hair. The growth of hair is achieved chiefly by cell proliferation and by the increase in cell volume. As the cells move up from the matrix into the upper bulb, they increase in volume greatly as a result of an accumulation of hyaline granules and fibrils. At the completion of its growth cycle the actively growing hair is transformed into the dead **club hair.** This transitional phase is characterized by cessation of active proliferation of the matrix cells and a separation of these cells from the upper portion of the hair follicle. Following this stage is a resting stage, after which the hair cycle starts all over again. In man, the club hair is not immediately extruded from the follicle but remains there for about a week, at which time the new hair emerges, pushing the old club hair from its follicle.

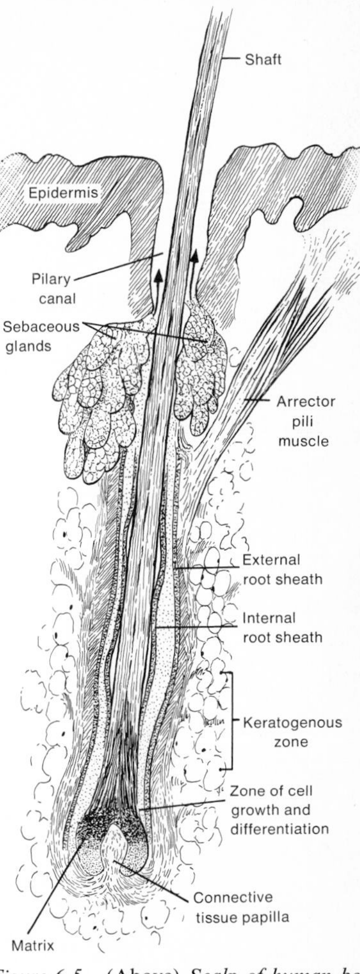

Figure 6-5. (Above) *Scalp of human being. Root of a hair in longitudinal section.*

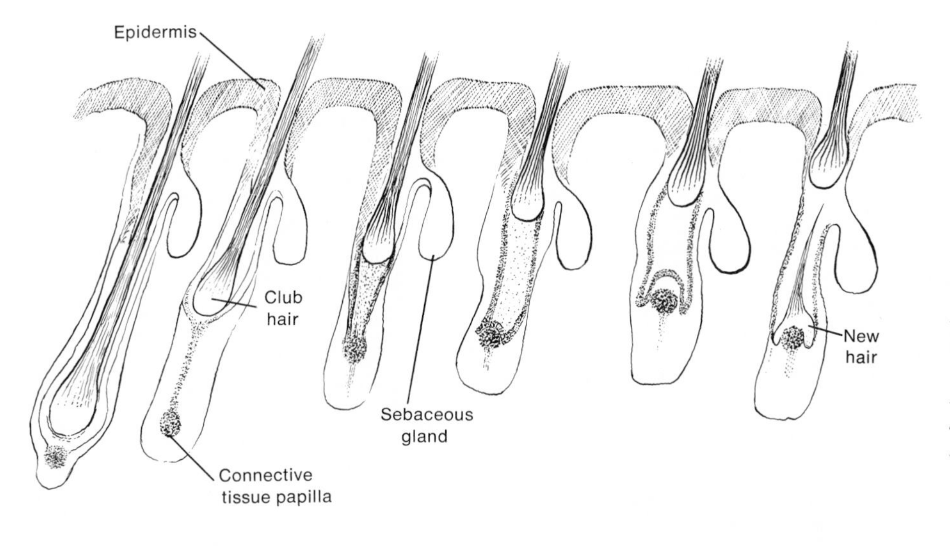

Figure 6-6. (Left) *Schematic representation of a hair cycle in which the formation and growth of a new hair occur.*

Functions of Hair

The hair of man and the hairy coats of other animals serve a number of functions. Hair serves a **protective function;** for example, the scalp hair forms a cushion around the head, protecting it from mechanical injury as well as from heat rays of the sun, and the eyebrows and eyelashes protect the eyes. The hairs in the nostrils and external ear canal protect these structures from the penetration of foreign bodies and dust particles. In furred animals the **thermoregulatory function** of hair is very important. In man, however, hair has lost its protective function against cold. The hair, with its large surface area, functions in the disposal of water from the surface by evaporation of sweat and also in the drainage of water from the skin after exposure to large concentrations, as in bathing or in rain. Hair serves a **sexual function,** in promoting the evaporation of apocrine sweat, and the accompanying characteristic odor that goes along with it provides a sexual attraction for the lower animals. Finally, acting like a lever, magnifying and transmitting touch stimuli to the touch receptors around the follicles, hair serves a **tactile function.**

References

Elden, H. R. (ed.). *Biophysical Properties of the Skin.* New York: Wiley, 1971.

Ferriman, D. *Human Hair Growth in Health and Disease.* Springfield, Ill.: Thomas, 1971.

Hardy, J. D., A. P. Gagge, and A. J. Stolwijk. *Physiological and Behavioral Temperature Regulation.* Springfield, Ill.: Thomas, 1970.

Marback, H. I., and H. S. Gavin. *Skin Bacteria and Their Role in Infection.* New York: McGraw-Hill, 1965.

Marples, M. S. *The Ecology of the Human Skin.* Springfield, Ill.: Thomas, 1965.

Montagna, W. *The Epidermis.* New York: Academic, 1964.

Nicoll, P. A., and T. A. Cortese, Jr. "The Physiology of Skin," in: *Annual Review of Physiology,* Vol. 34. Palo Alto, Calif.: Annual Reviews, 1972.

Rook, A. S., and G. S. Walton (eds.). *Comparative Physiology and Pathology of Skin.* Oxford: Blackwell, 1965.

Tregear, R. T. *Physical Functions of Skin.* New York: Oxford U.P., 1966.

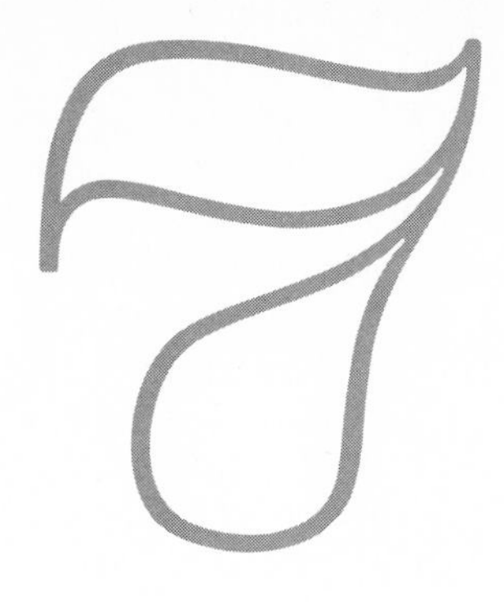

The Circulatory System: Mechanics of Circulation

The skeletal muscles and the nerves that supply them are referred to in general as the **neuromuscular system.** It is this particular system that enables the organism as a physiological unity to adjust itself to the external environment, secure food, and protect itself. The individual cells that make up this neuromuscular system must also be supplied with food and oxygen. They must also make adjustments to the internal environment and must be protected from the waste material they produce by its immediate removal. Special organs, such as the digestive, respiratory, and excretory organs, take care of these needs and in general are referred to as the **vegetative organs.**

The circulatory system (Fig. 7-1) acts as a connecting link between the neuromuscular system and the vegetative organs. The functions of the circulatory system are many and varied. It carries blood to the tissues to maintain an internal environment that is normal for the cells of the body. It must pick up oxygen from the lungs, carry it to the tissues, and remove their gaseous wastes, carbon dioxide. It must carry other waste products of cellular metabolism to the kidney and other excretory glands where they may be eliminated. The blood must absorb food material from the digestive tract and carry these energy- and tissue-building products to all the cells. The circulatory system helps to coordinate the activities of the various organs with one another by the transmission of those chemical regulators called **hormones.** It functions in the regulation of body temperature and in protecting the body against disease. Of the many functions carried out by the circulatory system, the most urgent is oxygen supply.

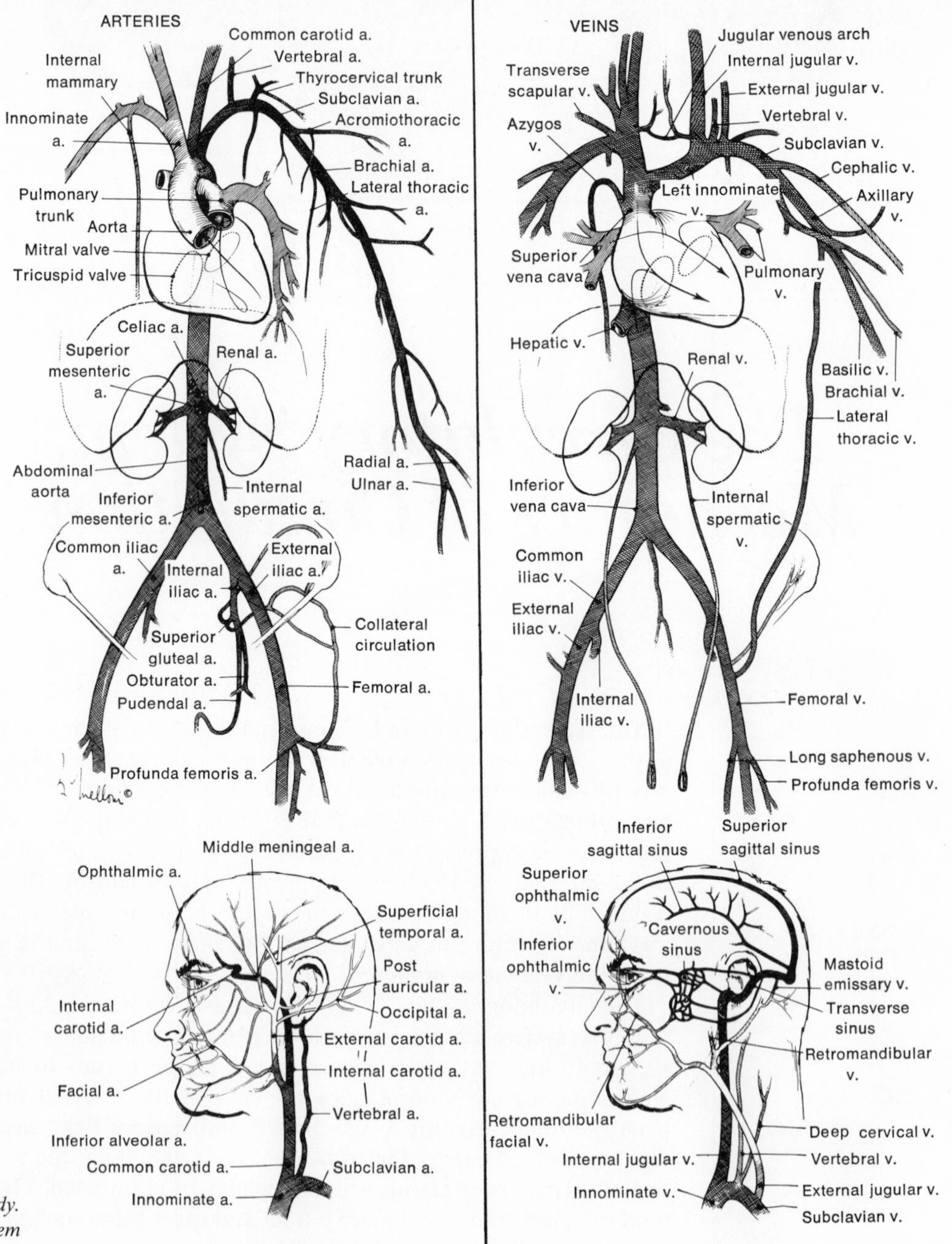

Figure 7-1. Arteries and veins of the body. Some of the main branches for each system are shown.

The Heart

In mammals the blood is kept circulating by two pumps contained within a single organ, the **heart.** One pump, which is represented by the chambers of the right side of the heart, receives blood from the systemic circulation and passes it on to the lungs, where it comes in contact with the alveolar air, picks up oxygen, and loses carbon dioxide. From here it is forced back to the second pump, which is represented by the chambers forming the left side of the heart. The circulation of blood from the right side of the heart to the lungs and back again to the left side of the heart is referred to as the **pulmonary circulation** and the pump as the **pulmonary pump.** The circulation of the blood from the left side of the heart to all the tissues of the body and back to the right side of the heart is referred to as the **systemic circulation.** The left chambers of the heart form the **systemic pump.**

Vessels called **arteries** carry blood away from the heart; the **veins** return the blood to it. The large arteries nearest the heart divide into smaller branches, called **small arteries,** which in turn divide into still smaller vessels, called **arterioles.** The arterioles lead into microscopic vessels of minute diameter, called **capillaries.** The capillaries branch profusely and form a network around the tissue supplied by the artery. It is through the walls of the capillaries that the exchange of materials between the blood and tissue takes place. Blood does not come in direct contact

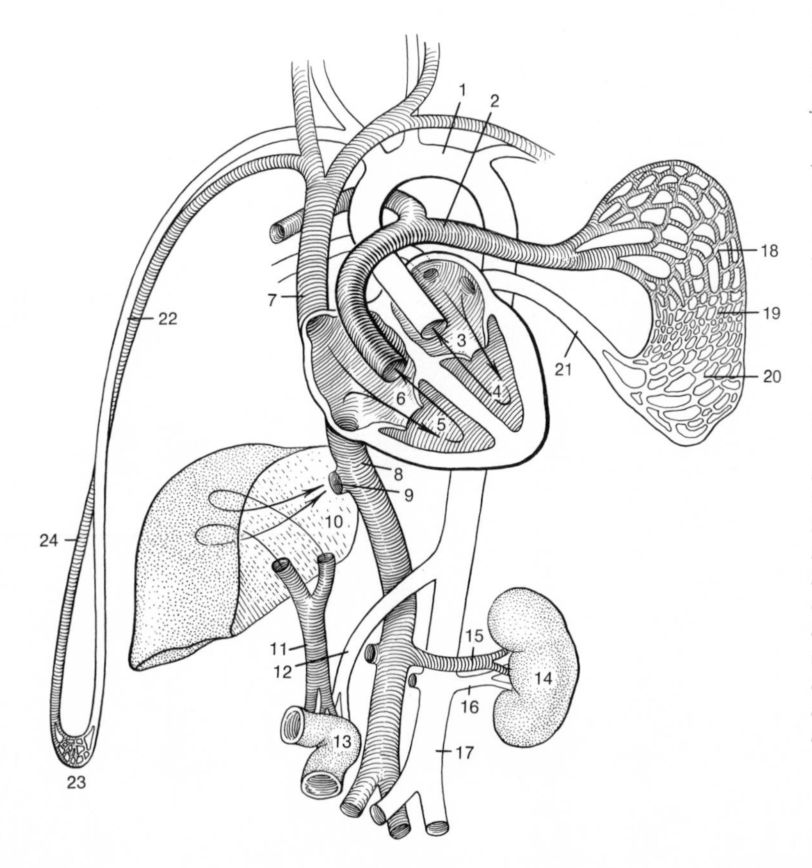

Figure 7-2. Schematic representation of the anatomical units of the cardiovascular system showing general circulation of the blood. Chambers of right side of heart (5, 6) *form the pulmonary pump, which forces blood to the lungs* (18, 19, 20) *by way of the pulmonary artery* (2). *The pulmonary vein* (21) *carries the oxygenated blood back to the left side of the heart* (3, 4), *which forces it into the systemic circulation by way of the aorta* (1) *and descending aorta* (17). *The kidneys* (14) *receive blood from the great artery by way of the renal artery* (16), *which in turn passes it back into the venous circulation through the renal vein* (15). *The alimentary tract* (13) *receives its arterial blood from the splanchnic artery* (12). *Venous blood from the alimentary tract must first pass through the liver* (10), *by way of the portal vein* (11), *before passing into the general circulation, by way of the hepatic vein* (9). *All the venous blood passes back into the right side of the heart by way of the inferior* (8) *and superior* (7) *venae cavae. Circulation of blood through the upper limbs is shown by the artery* (22), *capillaries* (23), and vein (24). *The arterial circulation is shown in white, the venous circulation, in crosshatching.*

with the individual cells, since it is always retained within the blood vessels. Each cell is bathed by its own liquid environment, the **tissue fluid.** The food and oxygen contained in the blood pass through the capillary wall into the tissue fluid and from there into the cells. Waste products produced by the cells (carbon dioxide, urea, creatinine, uric acid, sulfates) pass into the tissue fluid first and then into the blood, diffusing through the capillary wall. The blood and tissue fluid together are referred to as the **internal medium.** After the blood flows through the capillaries, it is collected into the smallest veins, called **venules.** The venules unite into larger vessels, called **small veins.** These combine to form larger veins until the largest veins in the body, which open into the heart, are reached. Veins are equipped at intervals with valves that prevent the blood from flowing in the opposite direction. Figure 7-2 schematically represents the anatomical units of the cardiovascular system.

The heart is an irregular, cone-shaped organ (Fig. 7-3) situated in the chest cavity mostly behind and to the left of the sternum in a space between the lungs called the **mediastinum.** The pointed end of the cone, the apex, is directed downward and to the left, and the broad flat base of the cone, which has several large vessels attached to it, is directed upward and to the right. The heart is divided into four chambers, called the **left and right atria** and **right and left ventricles.** Small triangular projections of the atrial chambers of the heart are referred to as the right and left **auricles.** The ventricles are the larger chambers and form the major portion of the heart as seen from the front, or ventral, surface. The atria are much smaller chambers and form the basilar portion of the heart as seen from the back, or dorsal, surface (Fig. 7-3).

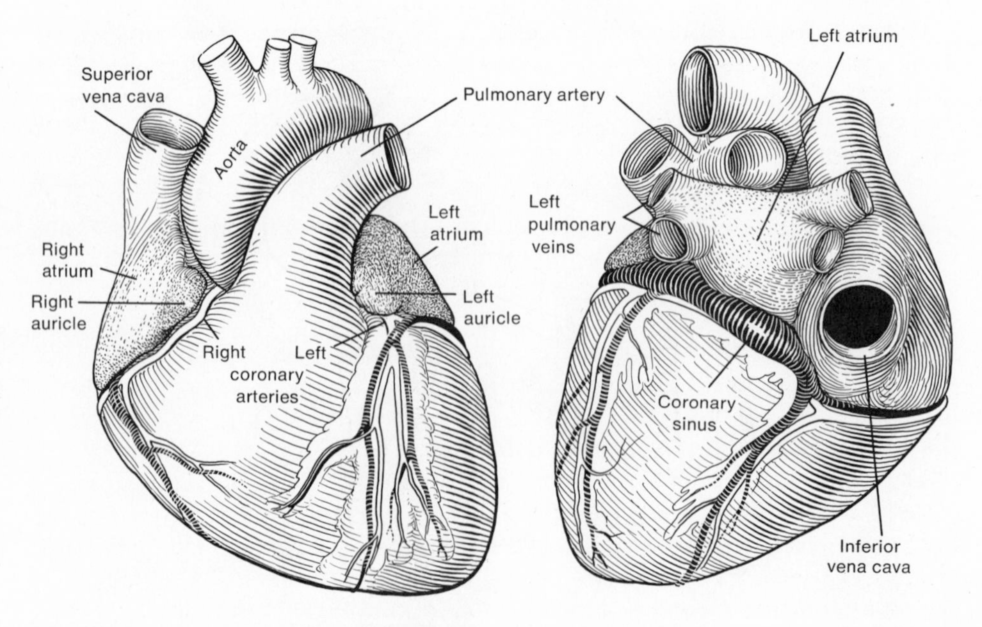

Figure 7-3. The heart—ventral, or anterior, surface to the left; dorsal, or posterior, surface to the right. Blood vessels supplying the heart musculature are shown also.

Blood Vessels Supplying the Heart

The wall of the heart is supplied with blood by the right and left coronary arteries. The **right coronary artery** leaves the aorta just as this structure exits from the left ventricle.

The artery runs on the surface of the heart along the right part of the coronary groove to the diaphragmatic surface, where it passes off several branches into the myocardium. The **left coronary artery** is much larger than the right and leaves the aorta in a similar position as the right coronary, but more to the left, to pass backward to the right of the pulmonary artery. It runs along the coronary groove to the diaphragmatic surface, where it breaks into numerous branches over the surface of the left ventricle.

Innervation of the Heart

The heart is innervated by both parasympathetic and sympathetic fibers. The final terminations of both types of fibers on heart muscle have not been described to the satisfaction of all observers. However, terminal unmedullated postganglionic fibers forming a rich, fine, intramuscular plexus around the muscle fibers have been described. The branches given off these plexi appear to terminate on the heart muscle cells by five knoblike endings which sometimes occur in groups of three knobs, designated as clover-leaf endings.

Preganglionic parasympathetic fibers of the vagus which supply the heart arise from an area of the medullary portion of the brain stem called the **dorsal efferent nucleus** (Fig. 7-4). The fibers pass to the heart wall to the region of the opening of the right precaval vein. After passing along the oblique vein of the right atrium they synapse with postganglionic fibers in the interatrial groove. They also synapse with scattered cell clusters on the ventral atrial walls and cells on the base of the left auricle. These postganglionic fibers end on the muscle of the atria and ventricles and also pass to the sinoatrial and atrioventricular nodes.

General visceral sensory fibers accompany the vagal motor cardiac innervation. These sensory fibers have their cells of origin in the **nodose ganglion** with their dendrites arising in the region of the heart where the great blood vessels enter. Their axons terminate in the gray matter of the medulla near the floor of the fourth ventricle.

The **sympathetic preganglionic** fibers have their cell bodies arising in the intermediolateral cell column of the first five or six thoracic levels of the spinal cord (Fig. 7-4). The fibers leaving the spinal cord (via white rami) ascend in the sympathetic chain so that the **postganglionic axons** arise from the superior, middle, and inferior cervical sympathetic ganglia as well as from the first five thoracic chain ganglia. The postganglionic fibers from the superior, the middle, and the inferior cervical ganglia which pass to the heart in the area of the cardiac plexus form the *superior, middle,* and *inferior cardiac nerves*. Those from the thoracic chain ganglia form the thoracic cardiac nerve (Fig. 7-4).

The sympathetic pathway, like the parasympathetic, is accompanied by general visceral sensory fibers. The cells of origin of this system are located in the first five spinal dorsal root ganglia. Their dendrites pass through the white rami into the chain ganglia and follow the various cardiac nerves to the heart. None passes through the superior cervical ganglia or accompany the superior cardiac nerves.

All the nerves on the right side innervate the right atrium, ventricle, and the

Figure 7-4. Autonomic innervation of the mammalian heart. The solid line represents the preganglionic fibers of the vagal (parasympathetic) nerve. The postganglionic fibers of the vagus are the terminal branchings of the nerve in the heart. The broken lines represent the sympathetic (accelerator) fibers. The preganglionic fibers are situated between the spinal cord and ganglia and end at cells in the inferior, middle, and superior cervical ganglia and also in the first to fifth ganglia of the thoracic chain.

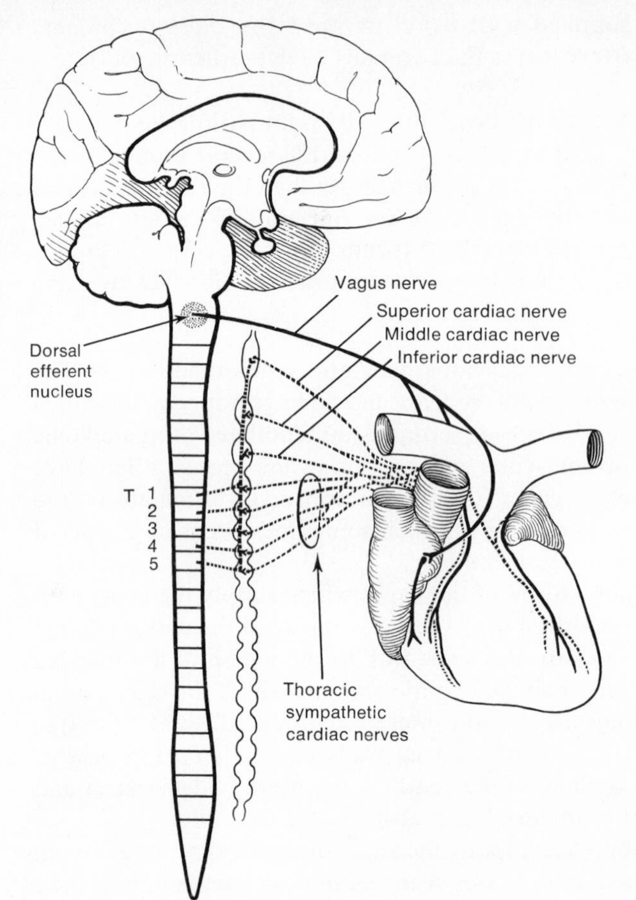

SA node, while those nerves on the left reach the left atrium, left ventricle, pulmonary veins, and AV node.

Cavities of the Heart

Blood is drained from the heart muscle by a large vein, the cardiac vein, which terminates in a large venous trunk, the **coronary sinus** (Fig. 7-3). This sinus, which is located in the coronary groove on a level with the pulmonary veins and the inferior vena cava, transfers the blood from the wall of the heart to the right atrium. There are numerous small veins that also transfer blood from the wall since they open directly into the atrium.

The right atrium is a thin-walled chamber containing three large openings (Fig. 7-5). It receives blood from the systemic circulation by way of two of these openings, which represent the entrance of the superior vena cava, bringing blood from the upper portions of the body, and the inferior vena cava, which drains blood from the lower portions. The right atrium transmits this blood into the right ventricle through the right **atrioventricular orifice.** This opening contains the **tricuspid valve,** which allows blood to pass from the atrium into the ventricle but prevents its passage in the opposite direction. The right atrium also contains

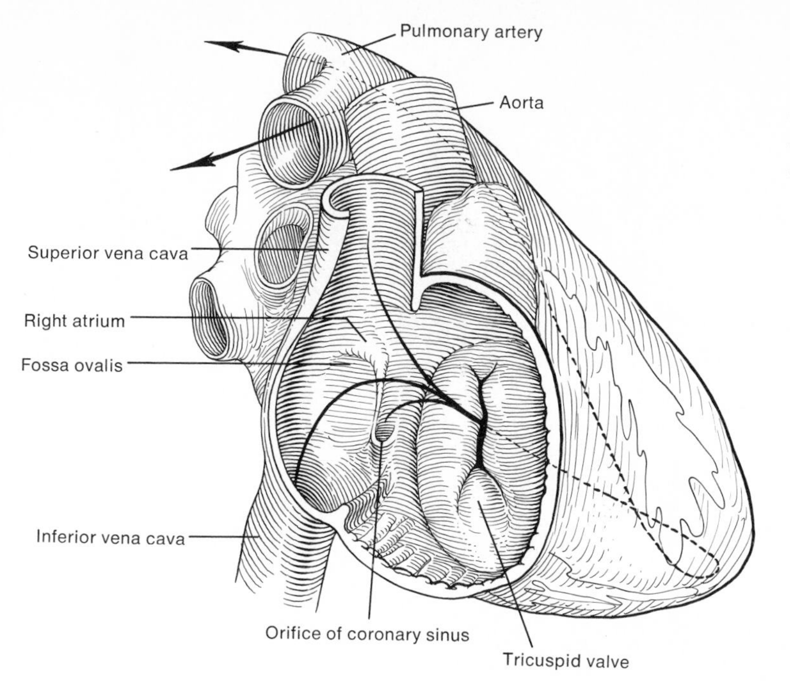

Figure 7-5. Heart with right atrium open to show inlets (superior vena cava, inferior vena cava, coronary orifice) and outlet (tricuspid valve). In addition, direction of blood after passing through the tricuspid into the right ventricle is shown.

the opening of the coronary sinus. The **right ventricle** (Fig. 7-6), which lies just below the right atrium, has a much thicker wall than either atrium. It has two openings, the right atrioventricular orifice in its roof and an opening that enters into the pulmonary artery, the **pulmonary orifice.** The pulmonary opening is guarded by the **pulmonary valve,** also called the **semilunar valve.** This valve is made up of three half-moon-shaped structures, which are arranged to prevent the backward flow of blood from the artery into the ventricle but offers no resistance to the passage of blood from the ventricle to the pulmonary artery. The interior of the ventricle wall contains many irregular muscular ridges, which are called the **trabeculae carneae** (small beams). The right ventricle forces blood through the pulmonary circulation (Fig. 7-2).

The **left atrium** lies, as did the right atrium, in the base of the heart but forming its left side. It is also a thin-walled chamber having four openings on its posterior wall for the **four pulmonary veins,** which bring oxygenated blood from the lungs into the left atrium. The blood passes from here to the left ventricle by way of the left **atrioventricular orifice,** which is guarded by the **bicuspid valve,** or **mitral valve** (Fig. 7-7).

The left ventricle (Fig. 7-8) has a much thicker wall than the right ventricle because it must drive the blood through the entire systemic circulation, which requires much more work. In addition to the atrioventricular opening in its roof, the left ventricle has another opening, the **aortic orifice,** which leads into the largest artery of the body, the **aorta.** This opening contains the **aortic valve,** also known as the semilunar valve (Figs. 7-9 and 7-10).

The tricuspid valve and bicuspid valve open when blood passes from the atria

Figure 7-6. Right ventricle laid open to show the pulmonary valve and the irregular muscular ridges (trabeculae carneae) in its wall. Circulation of blood from the right ventricle to the pulmonary artery is shown by the arrow.

Figure 7-7. Left view of the heart with the left atrium exposed.

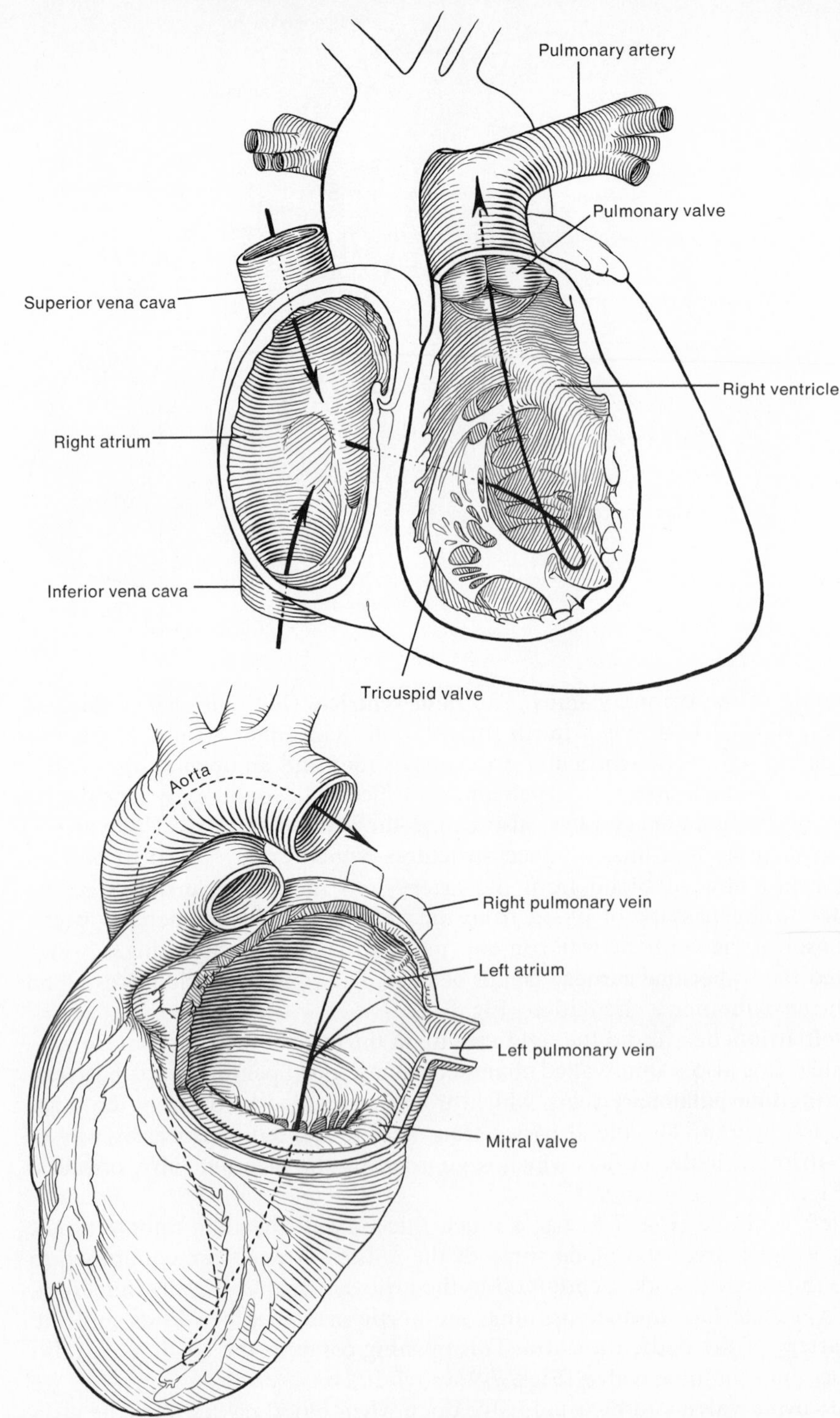

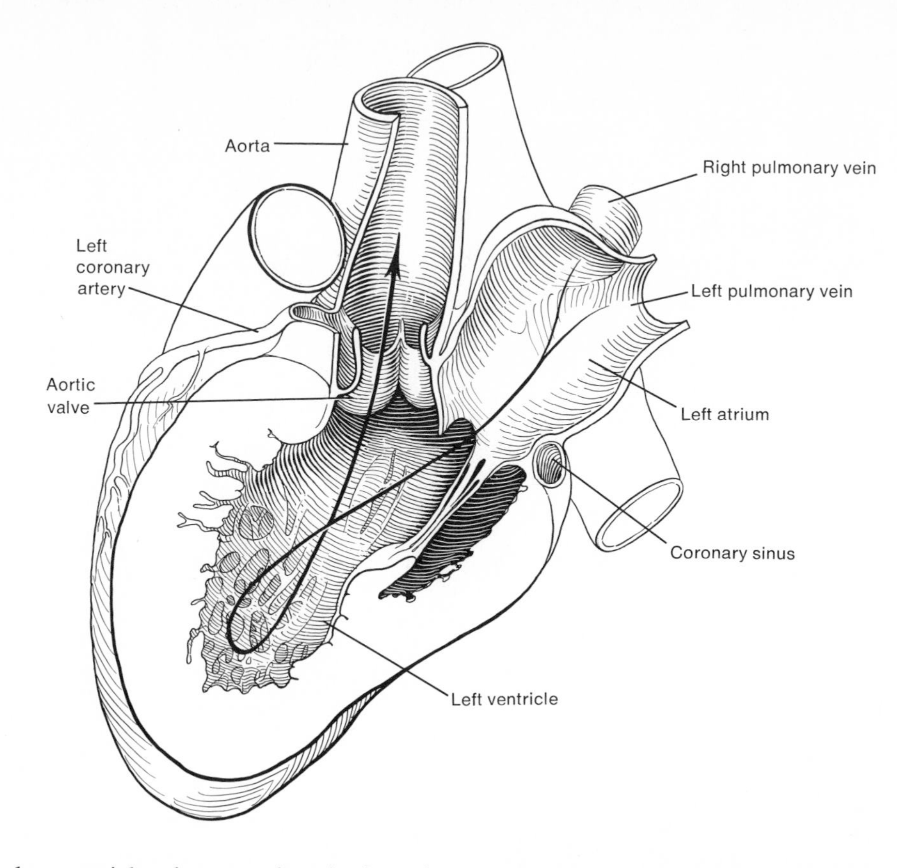

Figure 7-8. Left ventricle of the heart. Circulation of blood from left atrium to left ventricle into the aorta through the aortic valve is shown also.

into the ventricles, but are closed when the ventricles are contracted. To prevent the valves from bulging into the atria as a result of the great force exerted against the blood as the ventricles contract, strong cords, the **chordae tendineae,** which are attached to the valves at one end and to the papillary muscles (Fig. 7-11) in the ventricle wall at the other, shorten during contraction of the ventricle and pull down on the valves, keeping them closed together. This action prevents the valves from bulging into the atria or allowing blood to be ejected back into the atria.

The chambers of the heart are separated by **septa,** or walls. The septum between the right and left atria is characteristic since it is not equally thick throughout. The thinnest part is at the bottom of a poorly defined depression, the **fossa ovalis,** which marks the position of an embryonic connection, the **foramen ovale,** between the two cavities (Fig. 7-12).

Structure of the Heart Wall

The greater part of the thickness of the heart wall is composed of muscular tissue constituting the **myocardium,** which is much more abundant in the ventricles than in the atria. The muscular tissue of the myocardium is arranged in indefinite layers and runs in a complicated circular and oblique direction (Fig. 7-13). The wall of the atria is not continuous with that of the ventricles, as the musculature is interrupted between these two areas by fibrous tissue. The atrial and ventricular

Figure 7-9. (Far right) *Aortic valve opening into the aorta and some of the large arteries that branch off the aorta to supply the cranial region with oxygenated blood.*

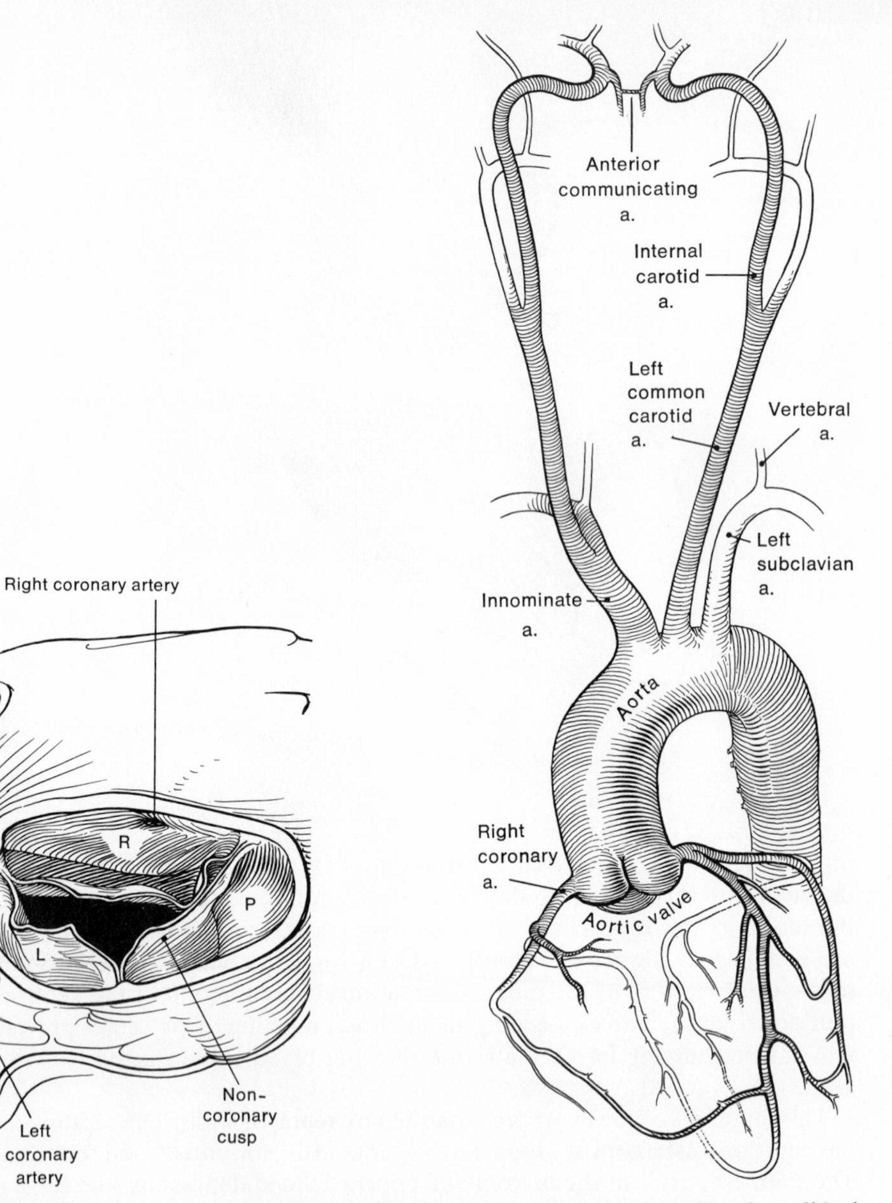

Figure 7-10. (Right) *Aortic valve as seen from inside the aorta showing detail of structure of its cusps.* L, *Left;* R, *right;* P, *posterior.*

walls are connected not by muscle, but by fibrous tissue and a bundle of modified muscle cells that arise in connection with the atrioventricular (AV) node. The AV node is discussed in more detail on page 238. The interior of the heart is lined by a thin, smooth, shiny membrane, the **endocardium.** This membrane consists of simple squamous epithelium and a layer of fibrous tissue and is continuous with the lining of the blood vessels and is also concerned in the formation of the valves. Inflammation of this lining as a result of bacterial infection is referred to as **endocarditis.** Externally, the heart is covered by a serous membrane, the **epicardium,** between which and the myocardium fat is usually stored. The epi-

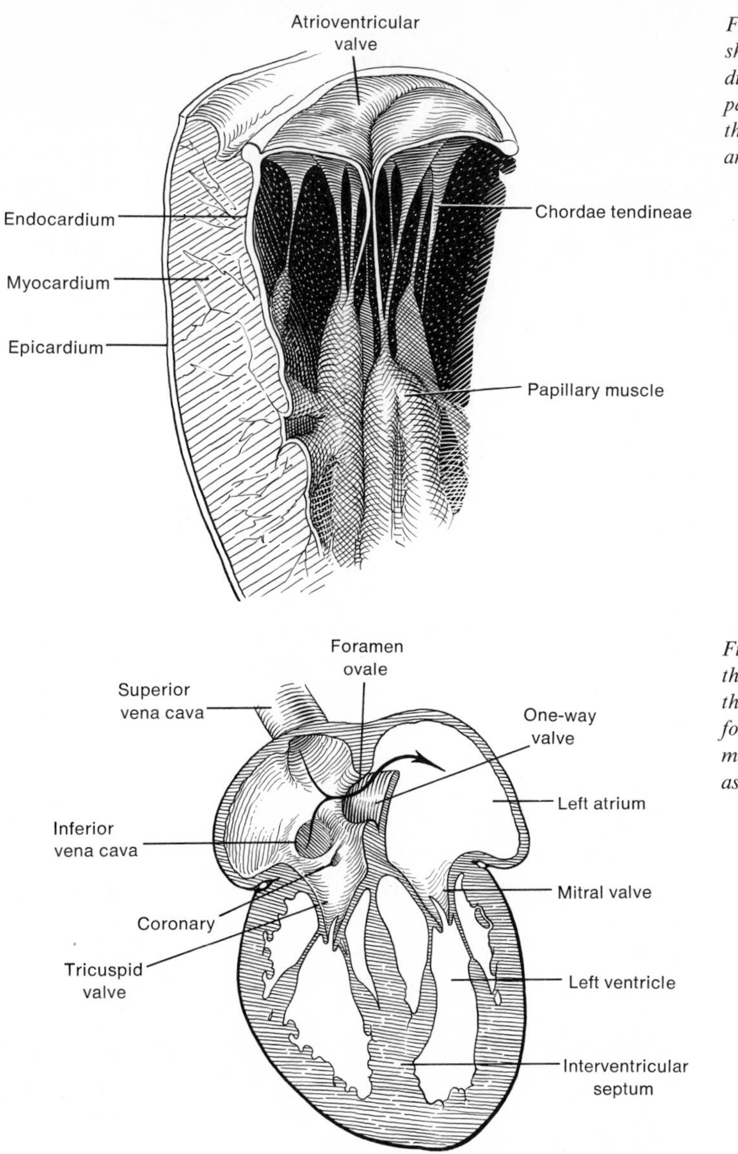

Figure 7-11. Enlargement of ventricle to show the attachment of the chordae tendineae to the valves on one end and to the papillary muscles in the ventricle wall on the other. The three layers of the heart wall are shown also.

Figure 7-12. Embryonic heart, showing the connection between the two atria, through the foramen ovale. In later life the foramen ovale becomes obliterated and remains as a poorly defined depression known as the fossa ovalis.

cardium is reflected back at the base of the heart to form a saclike covering for the heart, the **pericardium.** The potential space between the epicardium and the pericardium is filled with serous fluid, which keeps the opposed surfaces of the membrane smooth and slippery.

Characteristics of Heart Muscle

Heart muscle differs from other muscle types of the body not only in its histological make-up, but also in certain physiological characteristics that enable cardiac tissue to perform its function in an orderly manner. Heart muscle shows a uniqueness in its strength of contraction and its capacity for rhythmicality. The

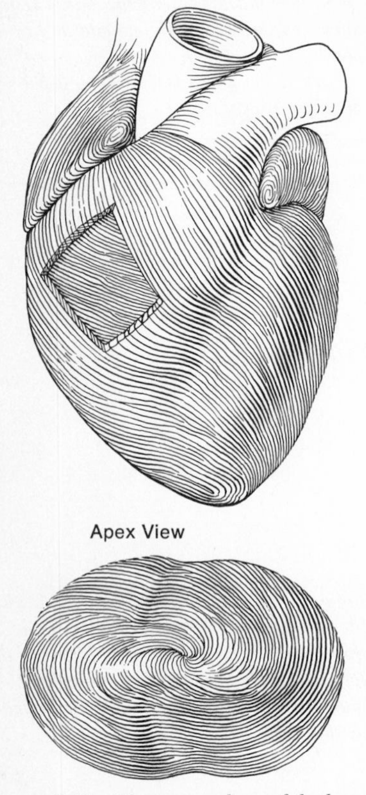

Figure 7-13. The myocardium of the heart is arranged in indefinite layers and runs in a complicated circular and oblique direction.

rhythmical activity is said to be inherent in the muscle itself and can be seen even in small pieces of cardiac tissue, cut away from the main mass. Heart muscle alternately contracts and relaxes in a regular clocklike manner in response to stimuli arising spontaneously from within. This spontaneous rhythmical contraction continues throughout the entire life of the muscle. Its unique strength of contraction comes from the peculiar arrangement of the muscle fibers in a pattern of spirals and whorls.

In common with all muscles, the heart has a refractory period, that is, a period in which the tissue will not respond to another stimulus. However, cardiac muscle differs even in this respect by possessing an unusually long one. The absolute refractory period of cardiac muscle lasts almost to the end of diastole. The heart cannot respond to another stimulus to contract until it has had an opportunity to fill again with blood. Sometimes an extra beat does occur; the heart will compensate for this extra beat by dropping the next normal beat. This is referred to as the **compensatory pause.**

In our discussion of muscular tissue we found that individual skeletal muscle cells follow the law of "all or none"; that is, if the cell is going to respond to the stimulus, it responds to its fullest extent. However, the skeletal muscle as a unit does not follow this law and may give graded responses depending on the strength of the stimulus. Cardiac muscle does not give the same result as a skeletal muscle; a minimal stimulus will invoke the same response as a maximal stimulus; that is, the heart responds to the "all or none" law, as it reacts with contracting, with constant strength, no matter how much the strength of stimulus is increased from threshold strength. This response in cardiac muscle is due to the fact that all its fibers constitute a syncytium, and therefore the excitatory process is always transmitted throughout the whole organ.

Like other muscles the heart will, within physiological limits, contract more strongly if it is stretched. If the heart is filled with a large volume of blood, which acts to stretch its walls, its contractions will be more powerful than when it is only partially filled. Starling expressed this reaction in his famous "law of the heart," which states that the amount of mechanical energy set free in passing from the resting to contracted state (strength of contraction) depends on the length of the muscle fiber. Increased length provides a greater area for more vigorous chemical action.

Histology of Cardiac Muscle

The histological unit of structure of cardiac muscle is a cell, rectangular in shape, with one or both ends having short, stubby processes (Fig. 7-14). The long, oval nucleus is centrally located with its long axis in the line of the long diameter of the cell. See page 80 for detailed description and Fig. 3-13A for an electron micrographic view.

The Blood Vessels

Arteries

The **arteries** are tubes, with relatively thick walls, branching from the main artery, the aorta, carrying blood to all parts of the body. The wall of an artery is considerably thicker and stronger than that of the corresponding vein, which returns blood to the heart. This arrangement is necessary because the pressure in an artery is always greater than that in a vein. The arterial wall may be

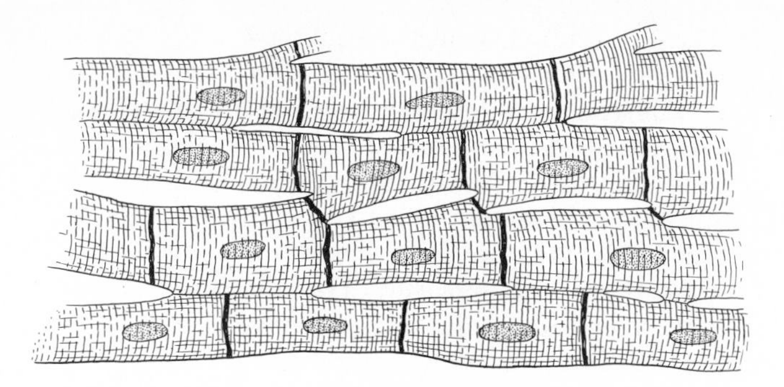

Figure 7-14. Longitudinal section of ventricular myocardium showing the histological unit of structure, the cardiac unit.

considered as composed of three coats of a more or less complex nature and differing slightly in structure in different arteries (Fig. 7-15).

The inner coat, or **tunica interna,** is composed of a lining of squamous epithelial cells (the endothelium), which rests on an elastic membrane of some thickness. This inner coat is rather delicate and easily torn, so that in injuries to arteries, it is likely to be broken.

The middle coat, or **tunica media,** is the thickest layer and is composed of elastic fibers and smooth muscle fibers, which are mostly circularly arranged. The various thicknesses of elastic fibers and muscle fibers are placed in more or less regular alternating layers.

The outer coat, or **tunica externa,** is composed chiefly of white fibrous connective tissue. In many arteries bands of smooth muscle fibers are present in this coat also and run for the most part in a longitudinal direction. The outer coat, being relatively inelastic and tough, limits the stretching of the artery and contributes greater strength to it.

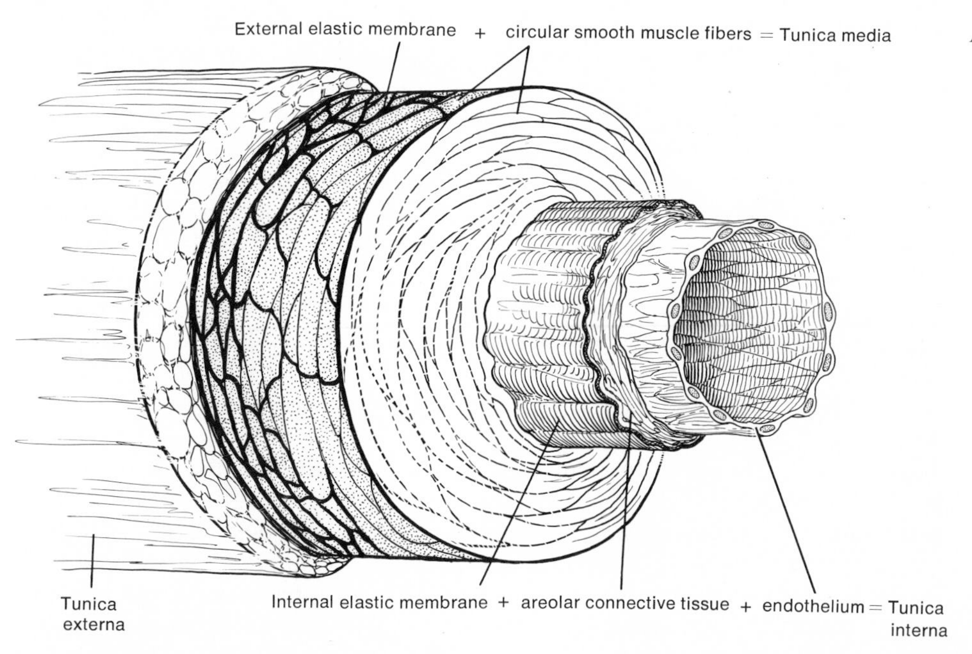

Figure 7-15. Structure of arterial wall.

The constitution of the arterial wall, especially of the middle coat, gives the arteries two important properties. First, they are very elastic, in the sense that they will stretch readily both lengthwise and crosswise when force is applied and return readily to their former size when the force is relieved. This elasticity is chiefily due to the elastic elements in the coats, but the smooth muscle fibers, being pliable, also contribute to the elasticity. Since the arteries possess such strong elastic walls, they do not collapse upon themselves but remain as open tubes. Second, the arteries by virtue of their muscular layers, are contractile. When their muscle fibers are stimulated by application of an electrical stimulus or under normal circumstances by nerve impulses passing to the artery by way of the vasomotor nerves, the circularly arranged muscles contract, narrowing the **lumen** of the vessel. This decrease in the size of the caliber of the vessel is called **constriction,** and the nerves that produce this effect, **vasoconstrictors.** An opening of an artery may be very narrow, very wide, or in an intermediate condition, depending on whether the muscle fibers are greatly contracted, not contracted at all, or only moderately contracted. When the opening is very wide, the artery is said to be **dilated** (Fig. 7-16) and the nerves that carry impulses into the arterial wall that inhibit stimulation of the muscle fibers are called **vasodilators.**

The smallest arteries of the body are called **arterioles.** The wall of these minute tubes, which break up into capillaries, is composed almost entirely of smooth muscle fibers and an endothelial layer (Fig. 7-17).

Capillaries

A **capillary** may be defined as a minute tubular passage hollowed out in connective tissue and serving as a connecting link between arterioles and venules. Capillaries are microscopic in size and are composed of a very thin layer of endothelium (Fig. 7-18). They branch and form a dense network throughout the area supplied by the artery (Fig. 7-19). It is through the walls of the capillaries that the exchange of materials between the blood and the tissue fluid takes place. The capillaries vary in size from region to region in the body; for example, in the lungs the capillaries are, on the whole, larger in caliber than in most other regions of the body. Some capillaries may be so narrow that a single red blood cell passes through them with difficulty, whereas others may be large enough to allow several blood cells to pass abreast. In addition, the same capillary will vary in size from time to time depending on existing conditions. Variations in diameter of the capillary lumen depend on the pressure of the blood within the capillary. The squamous epithelial cells forming the wall of the capillary are extensible, and the pressure of the blood within will distend the wall; likewise, since the wall is extremely thin, it will shrink and collapse when the pressure is removed. The capillary wall itself appears to be somewhat contractile, and under the influence of stimuli, the squamous cells show the property characteristic of all protoplasm in its ability to change its shape and form and thus influence the caliber of the tube.

The structure of the capillary appears to serve two purposes. First, the extensibility and elasticity of its walls permit it to adapt its caliber to the amount and force of the blood flowing into it. Second, and more importantly, its structure is adapted to carrying out the interchange of materials between blood and tissue,

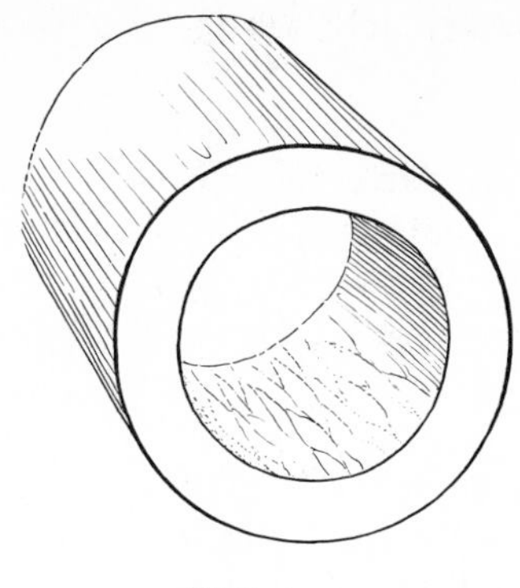

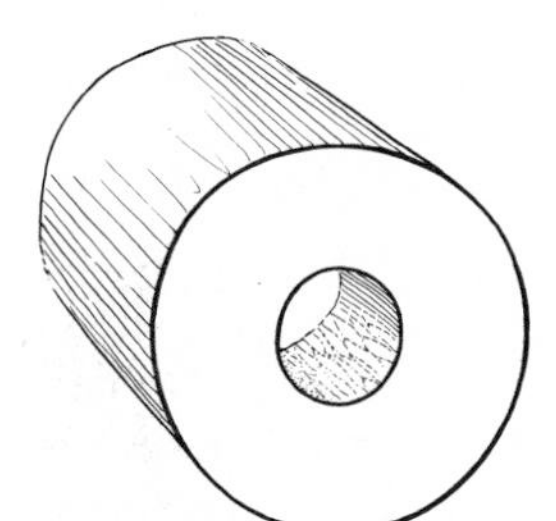

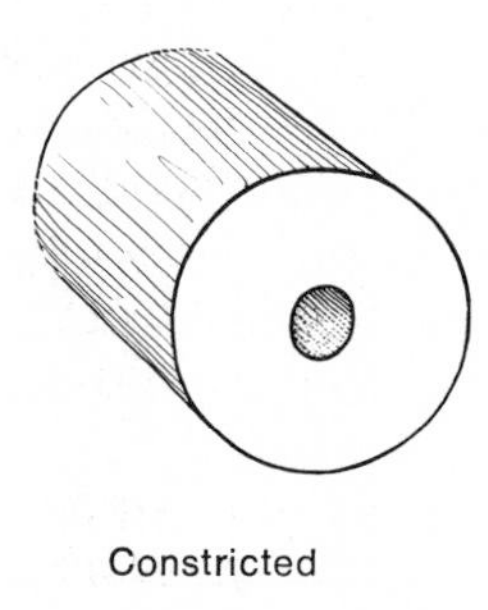

Figure 7-16. (Far left) *Change in size of lumen of blood vessel when dilated and constricted, in contrast to normal lumen.*

Figure 7-17. (Left) *Structure of arteriole wall. The fibroblasts and collagen fibers represent the external coat, the muscle fibers, the middle coat, and the endothelium, the internal coat.*

Endothelium
Elastic fibers
Muscle cell
Fibroblast
Collagen fiber

Figure 7-18. (Below) *Structure of a capillary.*

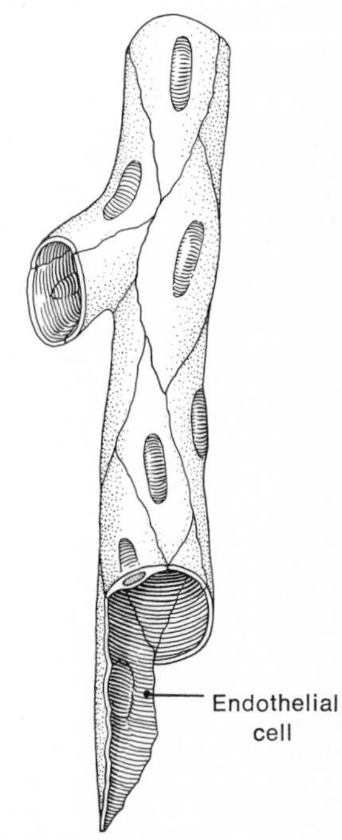

which can only take place through the walls of the capillaries. The features of the capillaries that facilitate this exchange are as follows:

1. The wall is extremely thin, consisting of a single layer of squamous epithelial cells.
2. The bore of the capillaries is quite minute, making it necessary for the blood to pass close to the capillary wall.
3. The capillaries are very numerous, which provides for a very large surface area through which diffusion can take place.
4. The blood moves through the capillaries at a very slow rate, which allows time for diffusion to take place.

The blood, after passing through the capillary network, is collected into the smallest veins, or venules.

Venules

The minute veins just emerging from the capillaries **(venules)** differ very little from the arteries of corresponding size. They usually have a wider base and have less muscular and elastic tissue in their walls. The venules unite to form small veins, which, in turn, combine to form larger vessels until eventually the great veins that open into the heart are reached.

Veins

Veins vary in structure in different parts of the body, but in general they differ from the corresponding arteries in having much thinner walls. In veins, the muscle and elastic tissue are less developed, and the inelastic white connective tissue layer is more developed (Fig. 7-20). For this reason veins are less extensible and elastic than arteries but are able to adapt themselves to considerable variations in the quantity of blood passing through them. However, not having a great deal of elastic power, veins cannot meet great variations in pressure. Also, in distinction to arteries, many veins, especially those of the limbs, are provided with valves (Fig. 7-21), which prevent the blood from flowing back into the capillaries. The veins of the viscera, the central nervous system and its membranes, and the bones do not possess valves. The walls of veins and arteries are supplied with blood by a system of small arteries, capillaries, and veins called **vasa vasorum.**

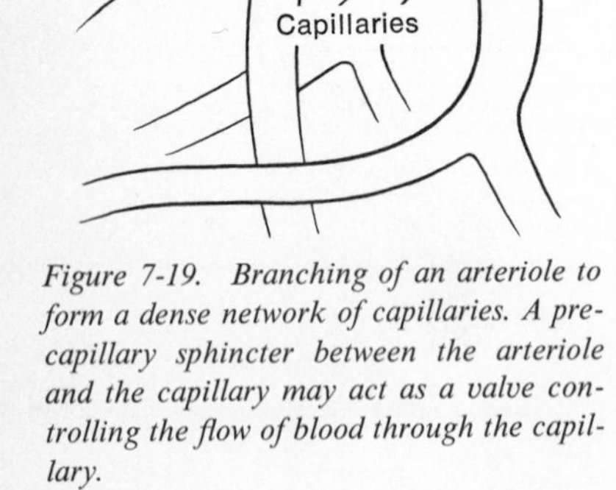

Figure 7-19. Branching of an arteriole to form a dense network of capillaries. A precapillary sphincter between the arteriole and the capillary may act as a valve controlling the flow of blood through the capillary.

Because of the direction of the flow of blood through the vessels, the arteries are said to **branch** and the veins are said to receive **tributaries.**

The arteries branching out from the aorta down to the great network of capillaries diminish in diameter as they divide. Wherever an artery divides into two branches, each branch is smaller in diameter than the artery from which it arises. Although the base of each division is less than that of the artery from which it arises, the two together are greater. Thus the summed cross-sectional area of the branches is greater than the cross-sectional area of the trunk. The cross-sectional area of the arterial system continues to increase from the aorta to the capillaries (Fig. 7-22). At the beginning of the capillaries the total cross-sectional area increases suddenly because the capillaries, although quite small, are very numerous. It has been estimated that the total cross-sectional area of the capillaries is several hundred times that of the cross-sectional area of the aorta.

The total cross-sectional area of the veins diminishes from the capillaries to the heart, but the great veins entering the heart have a total cross-sectional area nearly twice as great as that of the aorta.

Velocity of Blood Flow

The rate of flow of the blood through this system of closed tubes will be in inverse proportion to the total cross-sectional area at any level. As the total cross-sectional area of the arterial bed increases from the aorta to the capillaries, the rate of flow is decreased proportionately. At the capillaries, the rate of flow drops abruptly, as the total cross-sectional area is increased greatly by the multitudinous capillaries. In the veins a similar situation exists, but in the reverse direction, so that blood flows faster and faster as it approaches the heart.

Blood moves through the large arteries at a rate of approximately 250 mm per

second and drops down to a rate of less than 1 mm per second through the capillaries. In addition to the size of tubes affecting flow, the viscosity of the fluid within the tubular system as well as pressure will have an influence.

The relationship between flow in a long narrow tube, the radius of the tube, the viscosity of the fluid, and the pressure difference between two ends of the tube was studied by J. L. M. Poiseuille, who evolved a mathematical expression of this relationship which is now known as *Poiseuille's law*. This law states that

$$F = P \times \frac{\pi}{8} \times \frac{1}{u} \times \frac{r^4}{L} = \frac{P\pi r^4}{8uL},$$

where F = flow (volume passing per second at any point)
P = pressure difference between two ends of the tube
π = 3.14
8 = constant
u = viscosity
r = radius of tube
L = length of tube.

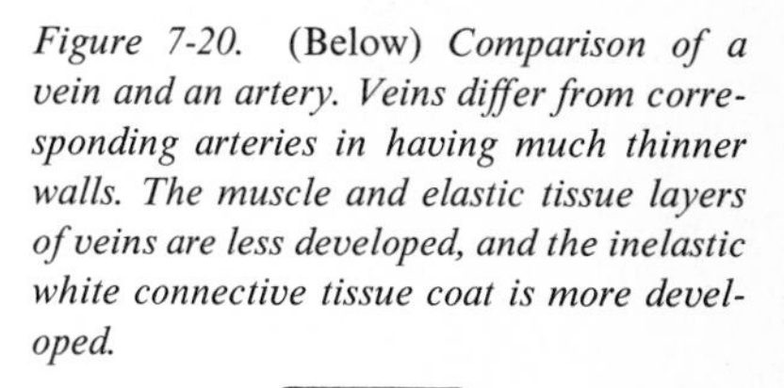

Figure 7-20. (Below) *Comparison of a vein and an artery. Veins differ from corresponding arteries in having much thinner walls. The muscle and elastic tissue layers of veins are less developed, and the inelastic white connective tissue coat is more developed.*

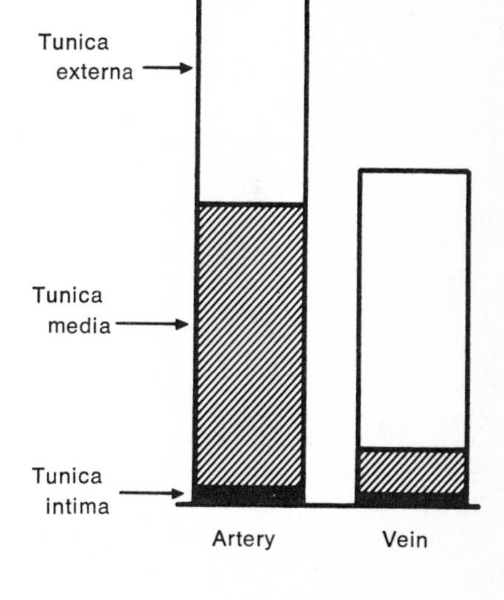

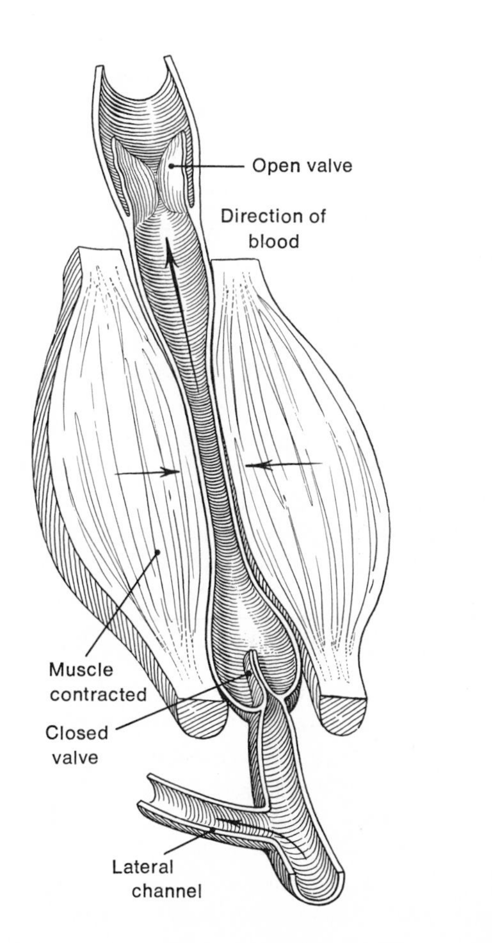

Figure 7-21. (Left) *Valves of a vein.*

Figure 7-22. *The total cross-sectional area of the arterial system increases from the aorta to the capillaries.*

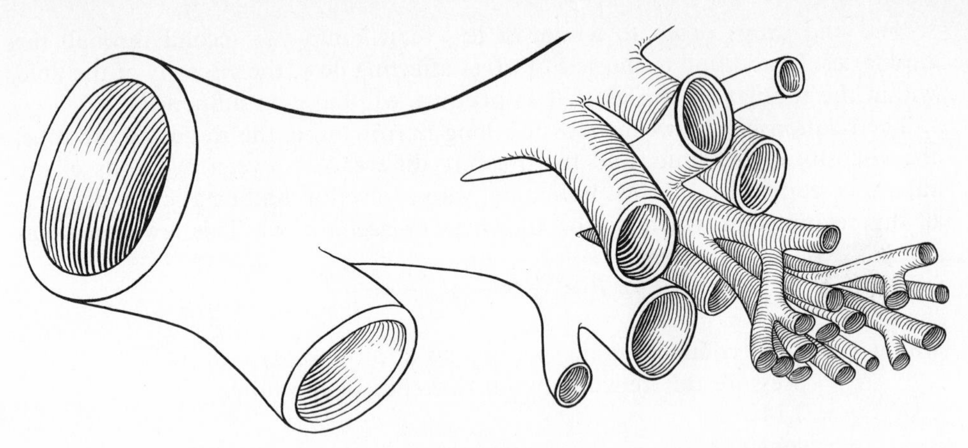

Flow is expressed in cubic centimeters if L and r are in centimeters, P in dynes per square centimeter, and u in poises per dyne second per square centimeter.

Resistance is another factor one must consider in an analysis of blood flow through the vascular system. The expression most frequently used for the determination of resistance in the cardiovascular system is

$$R = \frac{P}{F},$$

where R = resistance
P = pressure difference between two ends of a tube
F = flow, cubic centimeters per second

As the total cross-sectional area increases in a tubular system (the radius of the tube decreasing), the resistance increases and the flow decreases. Thus, we can say that flow varies inversely with the total cross-sectional area and resistance directly. Blood flow and resistance in the living organism are markedly affected by small changes in the caliber of the vessels.

Heart Action

The circulation of the blood from the heart through the pulmonary and systemic circulation and back to the heart again is kept going by the pumping action of the heart, the motive power of which is supplied by the contraction of its muscular fibers.

In the normal beat, the two atria contract at the same time, while the ventricles relax. When the ventricles are contracted, being synchronous in their action, the atria are relaxed. The phase of contraction is spoken of as **systole** and the phase of relaxation as **diastole.** Each atrium fills with blood from the great veins, the inferior and superior venae cavae, and the pulmonary veins during its diastolic period. When the atria are full of blood, a wave of contraction spreads over both structures, producing a sharp systole, forcing the blood from both atria into the ventricles. The ventricles receive blood from the atria during their diastolic phase and force it into the main arteries during ventricular systole. An alternate contraction and relaxation of the atria and ventricles followed by a short pause is called a **cardiac cycle.** Because of the arrangement of its muscle fibers in a spiral fashion, the heart undergoes a twisting, wringing motion as it contracts. As ventricular systole progresses, the apex is tilted slightly upward, and the heart twists somewhat on its long axis, moving from the left and behind toward the front and right, so that more of the left ventricle is displayed. As the systole is terminated and diastole sets in, the ventricles resume their previous form and position.

Sounds of the Heart

Two sounds have been ascribed to the cardiac cycle. These sounds can be heard quite distinctly if one places the bell of a stethoscope on the surface of the skin above the fifth intercostal space, about an inch below and a little to the median side of the left nipple. The first sound has been phonetically described as a **lubb** sound and, because of its booming quality, has been ascribed to the contraction of the ventricular muscle. In part, the sound is produced by the closure of the AV valves as the pressure rises in the ventricles.

The second sound, which is short and sharp and has been phonetically described as a **dupp** sound, occurs simultaneously with the closing of the semilunar valves, and thus it is probably generated by the closing of these valves. It is heard best when the bell of the stethoscope is held over the second right costal cartilage close to its junction with the sternum. It is at this point that the aortic arch comes closest to the surface and to which sounds generated at the aortic orifice would be best conducted. The cardiac cycle, as far as the sounds are concerned, could be represented by a lubb, dupp, pause.

Sometimes peculiar sounds that are referred to as murmurs can be perceived coming from the heart. These sounds occur in the heart when the blood flows through narrowed valvular orifices, as in atheromatous or sclerotic changes of the valve mechanism, or when blood is regurgitated back through an incompetent valve, as when the valvular orifice is greater in diameter than the spread of the valve leaflets. Many murmurs have no clinical significance and do not always mean a valvular lesion. They may occur when the flow of blood becomes turbulent within the heart with the formation of eddies, as may happen with an increase in the circulation rate or a decrease in the viscosity of blood. In such cases the flow easily becomes turbulent, and it is this turbulence that gives rise to the peculiar sounds we recognize as murmurs.

Duration of the Cardiac Cycle

As we have already seen, the cardiac cycle (Table 7-1) is made up of certain main events: (1) the systole of the atria, (2) the systole of the ventricles, (3) the diastole of the ventricles, which is represented by the time intervening between the cessation of contraction and the beginning of contraction again, and (4) the pause, or rest, of the whole heart, which comprises that period from the end of relaxation of the ventricles to the end of diastole of the atria and the beginning of their systole. During this period no part of the heart wall is undergoing any active change, neither contracting nor relaxing.

The average heart rate of man at rest ranges between 70 and 75 beats per minute. If we assume an average rate of 72 beats per minute, each beat with its short pause, which represents a cardiac cycle, would require about 0.8 second.

Table 7-1. *Summary of the Events of the Cardiac Cycle*

1. Cycle initiated by tonic discharge of the SA node.
2. Depolarization spreads over atrial muscle.
3. P wave of electrocardiogram (ECG), 0.08 sec.
4. Ejection of blood into ventricles. Contraction of atria—P-R interval of ECG, 0.1 sec.
5. Depolarization stimulus arrives at AV node and spreads rapidly through bundle of His and Purkinje fibers.
6. Atrial contraction ends—relaxation begins—lasts 0.7 sec.
7. QRS complex developed, 0.08 sec.
8. Contraction of ventricles begin.
9. T wave of ECG, repolarization begins.
10. Ventricular contraction ends—Q-T interval of ECG, 0.3 sec.
11. Ventricular relaxation begins, lasts 0.5 sec.
12. Overlap of ventricular relaxation and atrial relaxation—T-P interval of ECG—0.4 sec; called cardiac pause.
13. Cycle repeats; total cycle time 0.8 sec.

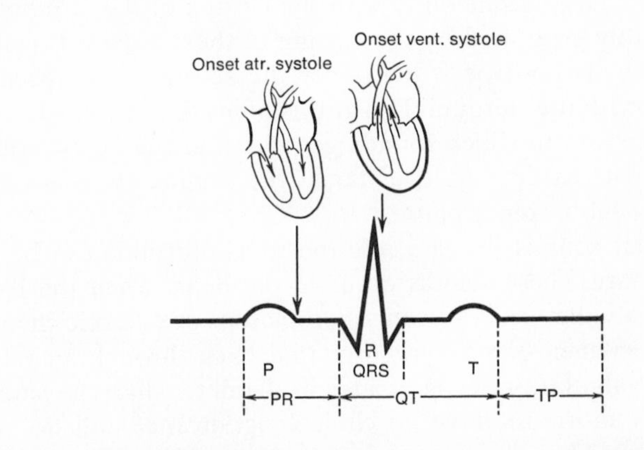

The duration of the individual events can be determined by graphically recording pressure changes in the atrium and ventricle during their diastolic and systolic phases and relating the times at which these changes occur to the waves of the electrocardiogram, which serves as a useful time scale. Atrial pressure changes as well as ventricular pressure changes may be measured by inserting long plastic catheters into their chambers and measuring the pressures electronically. Figure 7-23 schematically represents a composite of curves that would be obtained by the use of the method described above. The electrocardiogram is used as the time standard.

Atrial systole takes, on an average, 0.1 second. Ventricular systole requires approximately 0.3 second, and the pause, which represents the quiescent phase of the cardiac muscle, about 0.4 second, a total of 0.8 second. It is during the quiescent period that there is an overlapping of atrial and ventricular diastole. The total diastolic period of the atrium is 0.7 second, and that of the ventricle is about 0.5 second.

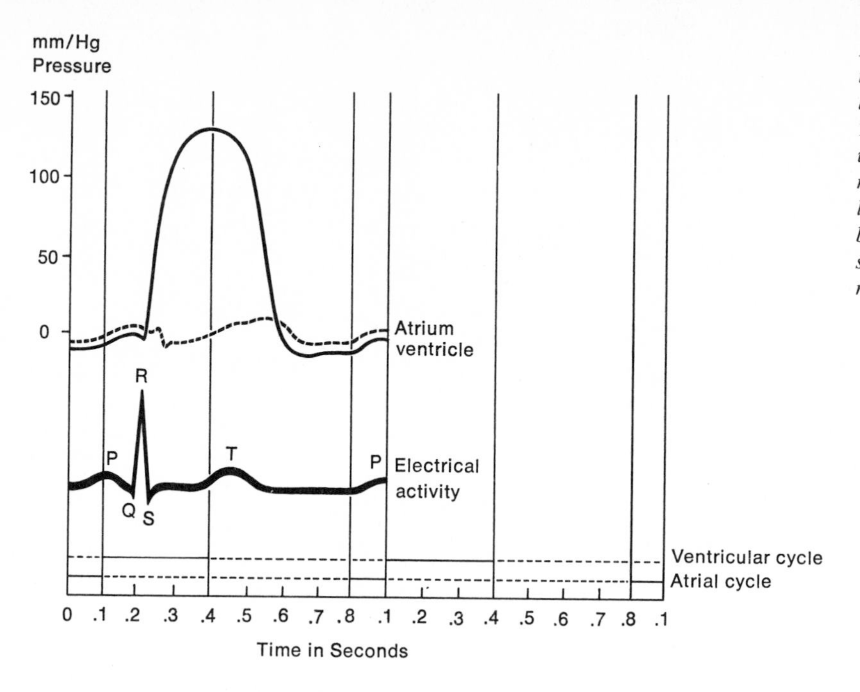

Figure 7-23. Cardiac cycle, showing electrocardiogram, atrial pressure (broken line), and ventricular pressure (solid line). The bottom two lines represent the ventricular and atrial cycles. The broken line represents the diastolic period, and the solid line, the systolic. It can be seen from the bottom two lines that between 0.4 and 0.8 second there is an overlapping of diastole, resulting in a 0.4-second quiescent period.

Cardiac Volume

Under resting conditions approximately 80 ml of blood is forced from each ventricle into the pulmonary and systemic circulation at each systole, giving a total **cardiac output** of 160 ml per beat. The output of the left ventricle into the aorta is referred to as the **stroke volume.** With an average stroke volume of 80 ml per beat, the amount of blood that is put out by the left ventricle per minute can be determined by multiplying this figure by the heart rate per minute. With an average rate of 72 per minute, 5760 ml (72 × 80 ml) of blood is pumped into the systemic circulation per minute. This volume of blood is referred to as the **minute volume.** A similar amount is forced into the pulmonary circulation, giving a total cardiac output of 11,520 ml. Variation in cardiac output will be determined by an increase or decrease in heart rate, in stroke volume, or in both. During exercise or excitement, the total cardiac output is greatly increased.

It is assumed that cardiac output is directly related to body size, so that it is usually expressed as output per minute per square meter of body surface. Cardiac output expressed in this way is referred to as the **cardiac index,** which is easily calculated if output, height, and total body weight are known.

The two most valid methods for determining cardiac output are the dye-dilution and direct Fick methods. The dye method requires the injection of a known amount of Evans blue dye into the venous circulation and a sampling of arterial blood at 1- or 2-second intervals for dye content. Construction of a time–concentration curve from the analyses will permit calculation of the area under the curve, which is related to the output of the heart per unit time.

The direct Fick method is based on the Fick principle, which states that the oxygen uptake per minute divided by the difference of oxygen content between

Figure 7-24. Nodal system in the mammalian heart.

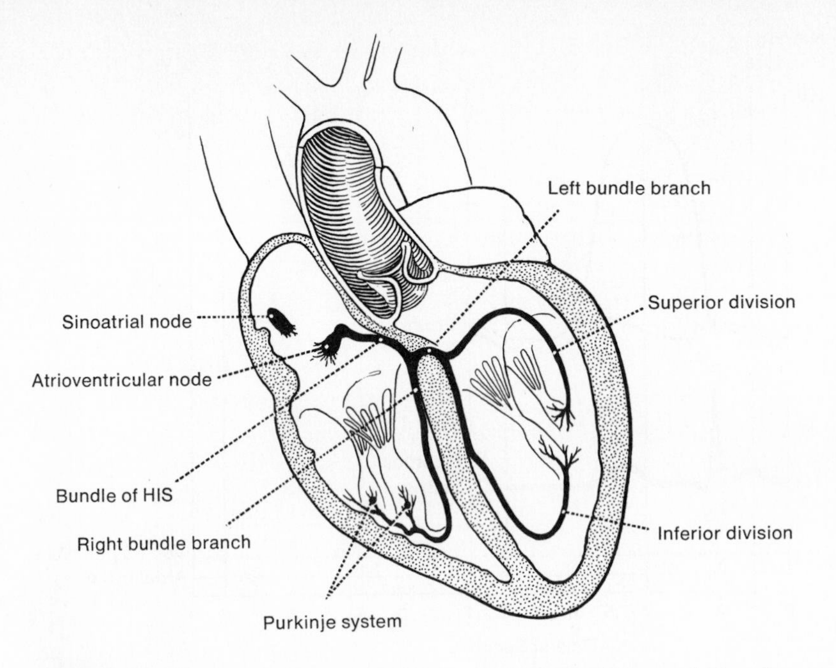

arterial and venous blood, per liter, yields the volume of blood moved by each ventricle in liters per minute.

Origin, Conduction, and Coordination of the Heartbeat

The mechanism whereby the heartbeat originates and the four chambers are coordinated is a fascinating one. It is clear that the origin of the heartbeat is not dependent on any connection with the rest of the body. This fact can easily be demonstrated in the laboratory by removal of a frog's heart from its body. It will continue to beat for hours if placed in a dish containing amphibian Ringer's solution. The heart of a mammal can also be removed and kept pulsating if it is maintained on an artificial circulation and at normal body temperature. The beats carried out by the isolated frog's heart or mammalian heart are in all respects identical with the beats executed by the heart in its normal condition within the living body. It may be concluded that the beat originates in the heart muscle itself, and therefore it is said to be **myogenic.**

It appears that the beat starts in the right atrium at a spot near the entrance of the superior vena cava called the **sinoatrial node** (SA node). This area of the mammalian heart corresponds with the **sinus venosus** of the frog heart.

The wave of electrical activity spreads very quickly over the entire atrial muscle, for the heart muscle cells are intimately fused together into a syncytium of tissue. This action wave is collected into another specialized area of tissue called the **atrioventricular node** (AV node), which is situated in the septal tissue between the right atrium and ventricle where it is passed, by way of the atrioventricular bundle, into the ventricles. The **atrioventricular bundle,** or **bundle of His,** is a band of specialized fibers, arising at the atrioventricular node, which pass down the interventricular septum and divide into two main branches, the **Purkinje fibers.** These fibers pass into the ventricles to be distributed to their walls. Figure 7-24 schematically represents the specialized conducting system of the heart. The conduction rate of the atrioventricular bundle is quite swift, approximately 500 cm per second. The rate of conduction through the ventricular muscle is relatively slow, about 50 cm per second. For this reason, although the size of the ventricle is quite large, all parts of the ventricles contract almost simultaneously.

The system of specialized tissue within the heart, the SA node, whereby the heartbeat originates and is conducted in such a way as to provide for the coordinated activity of alternate contraction and relaxation of the atria and ventricles is, in general, referred to as the **pacemaker** of the heart. It is this area wherein the beat originates and which experimental evidence indicates has a greater potentiality of beating than the other parts, in that, if isolated from the other parts, it beats more readily and with a quicker rhythm. In the intact heart the rate of beating is that of the fastest chamber. That the normal heartbeat actually starts in the sinoatrial node may be confirmed by the fact that if this structure is heated or cooled, the heart increases its frequency or slows, respectively. Other parts of the heart, if similarly treated, will not have this influence. Concerning the question of how the beat begins within the sinus, all that may be said is that it is the inherent physical and chemical constitution of this tissue that gives it that potentiality.

Electrical Changes in the Mammalian Heart—Electrocardiogram

A recording of the electrical changes accompanying the cardiac cycle is known as the **electrocardiogram (ECG).** With each contraction of the atria and ventricles, there arises an electrical-action potential followed by waves that are related to recovery. As we have seen, the heart contracts in an orderly sequence and derives its basic rate from the sinoatrial node. The events start here, and a wave of excitation travels over the atria, resulting in their contraction. The wave then activates the atrioventricular node, which in turn propagates the excitation down the bundle of His to the ventricles by way of the Purkinje fibers. The conduction of the excitation wave to the ventricular muscle results in ventricular contraction.

By placing electrodes on the atrium and ventricle, the characteristic action potentials of the electrical activity propagated by the pacemaker can be recorded. It is not necessary to place electrodes directly on the heart to record electrocardiograms. Since the body is made up of a conducting medium, the electrodes can be placed on the surface of the body to detect the heart's action potentials. The positioning of the surface electrodes is important, however, because the spread of excitation and recovery over the heart involves directivity. The heart's electrical forces, unlike other physiological phenomena, vary not only in direction but also in magnitude (Fig. 7-25). The chambers of the heart are hemispherical, and the particular region of the heart that is generating an electrical force at a given instant may be facing any direction. The characteristics of the electrical activity in a given electrode placement will be a function of the direction of the electrical force as well as of its magnitude. Therefore, there are places on the surface of the body that are more favorable for recording the electrocardiogram than others.

Particular electrode positions on the surface of the body have been adopted to standardize the recordings taken by some electrocardiographers (Fig. 7-26). The most commonly used are known as the three **standard limb leads** and are

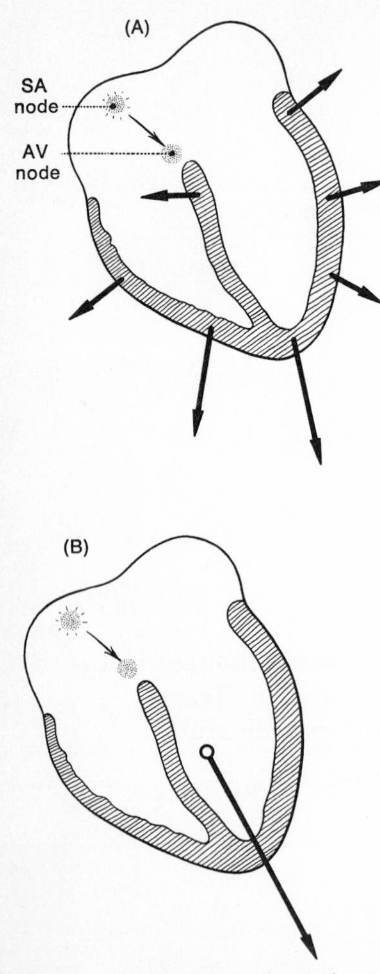

Figure 7-25. The heart's electrical activity varies in both direction and magnitude. A. *Instantaneous vectors during the QRS cycle.* B. *The average of instantaneous vectors. Some areas on the surface of the body are more favorable for recording the electrocardiogram than others.*

as follows: lead I—right arm and left arm; lead II—right arm and left leg; lead III—left arm and left leg.

The electrocardiograph is the instrument used to record the electrical activity of the heart. It consists of three basic components: the transducer, the amplifier, and the recorder. The transducer consists of the electrodes that pick up the very minute electrical signals generated by the active tissue and pass them to an amplifier, which amplifies these signals to an intensity adequate to energize the recorder being used. The recorder is merely a very large galvanometer with enough energy to drive an ink-writing stylus.

Development of Electrical Activity. The electrical potential that exists across the cell membrane of living cells is a consequence of its selective semipermeability. As we have discussed in Chapter 4, the cell membrane effectively separates a layer of positively charged ions from a layer of negatively charged ions. Under these circumstances the cell is said to be **polarized.** We have also seen that in nerve and muscle cells, the semipermeability of the membrane can be suddenly reduced by a stimulus, permitting a rapid flow of ions across the membrane. This results in the depolarization of the membrane in which a reorientation of ions takes place with the generation of an electrical current. It is this electrical potential, or voltage, that the galvanometer records. The propagation of depolarization along a nerve fiber serves for the transmission of the nerve impulse. In muscle, depolarization triggers the contractile mechanism within the cell, causing its contraction.

Depolarization of the atria produces electrical forces throughout the whole heart, which are recorded as waves.

Waves of the Electrocardiogram. In a typical record (Fig. 7-27), there are usually three clearly recognizable waves accompanying each cardiac cycle. The first wave is due to the depolarization of the atria, producing an electrical force that is recorded as a **P wave** in the electrocardiogram. This is a small upward wave and represents the action potential preceding atrial systole. Following this is usually a large upright triangular-shaped wave having a small downward wave at its base. This complex is known as the **QRS wave** and represents the depolarization of ventricular myocardium that precedes ventricular systole.

Immediately after depolarization, the cell restores the semipermeability of its membrane, in which another flow of ions takes place, resulting in the restoration of the polarized state. The flow of ions is much slower in restoring the polarized state, the process of which is called **repolarization.** The electrical potential generated by the atria during repolarization is too small to be recorded by body surface electrodes, and therefore, no deflection in the recording pen is seen after the P wave in the conventional tracing. When it is picked up, it is designated as the Pt or Ta wave. The third recognizable wave is a dome-shaped wave and can be directed upward or downward. It is called the **T wave;** produced just preceding ventricular diastole, it is the characteristic repolarization wave generated by the relatively large muscle mass of the ventricles.

Of paramount importance in the electrocardiogram are the time relationships between the various waves, since they are characteristic of heart action. The

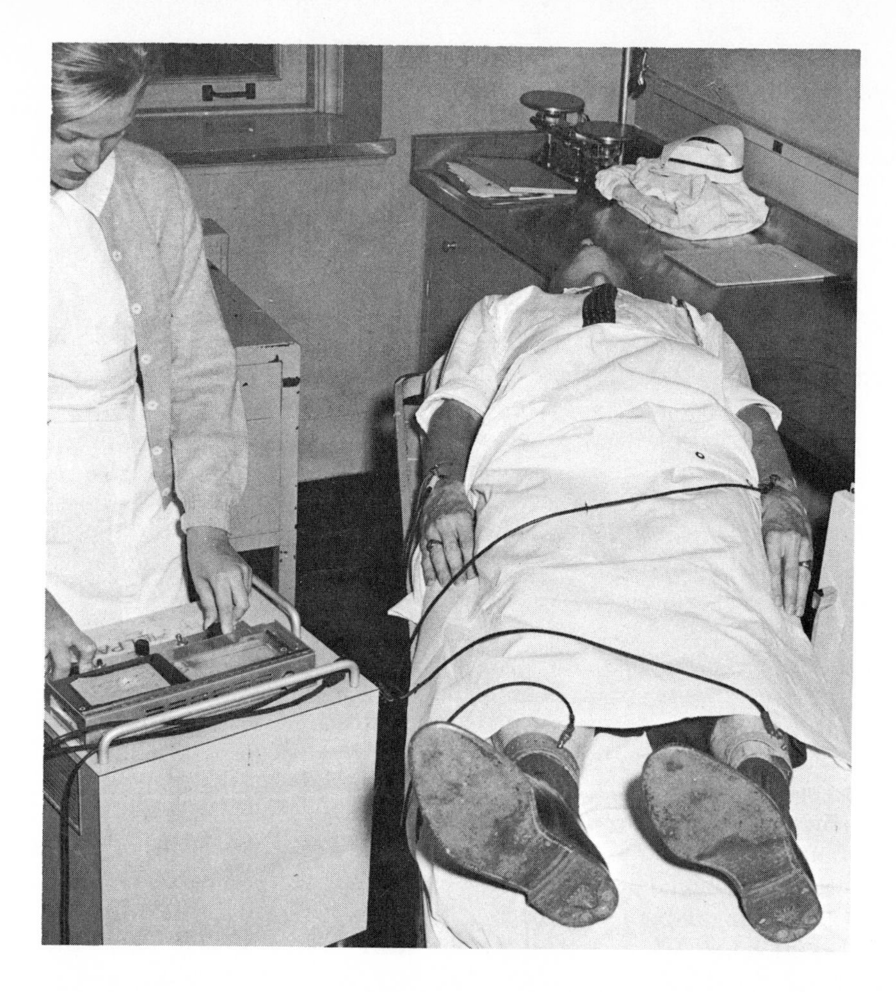

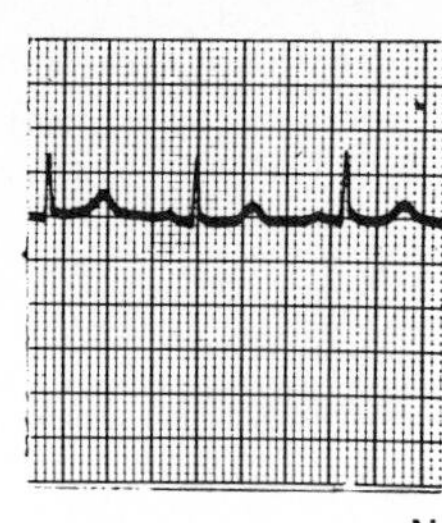

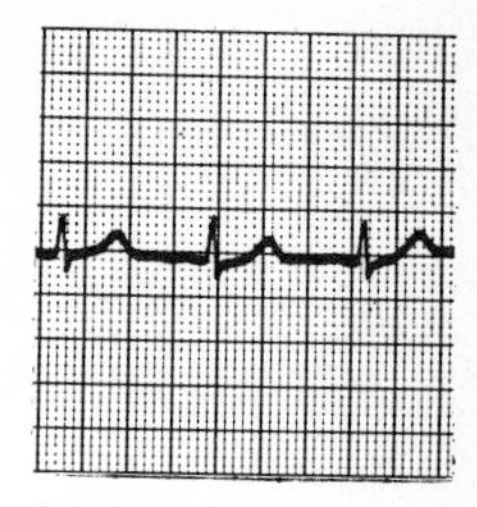

Normal

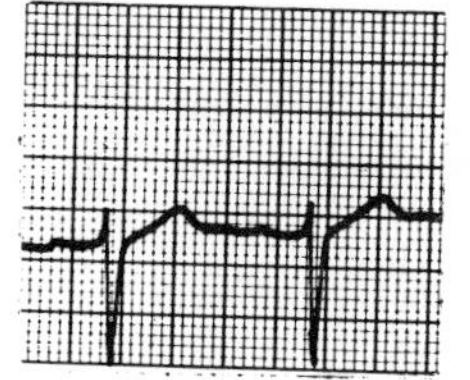

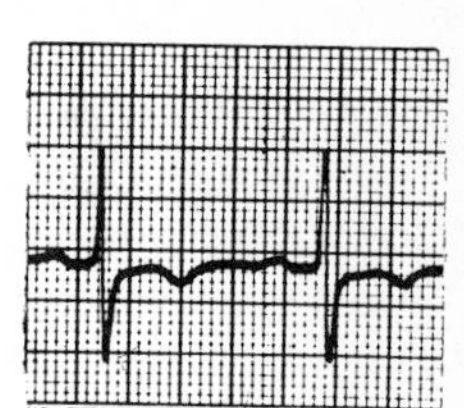

Abnormal

Figure 7-26. Photograph shows the placement of electrodes on the surface of the body for recording electrocardiograms. Upper graph shows a normal electrocardiogram; that below shows an abnormal type.

measurement scheme is as follows and reference should be made to Fig. 7-27. The **P-R interval** is measured from the beginning of P to the beginning of Q and represents the conduction time from the beginning of excitation of the atria to the beginning of ventricular excitation. The start of the P wave indicates the commencement of excitation in the atria. A few hundredths of a second later a rise in pressure lasting about 0.1 second develops in both atria and ventricles, which is due to atrial systole, and the delay is due to the latent period between the electrical and mechanical changes. The delay in conduction at the atrioventricular node is indicated in the ECG by the interval from the end of P to the start of Q.

The **QRS interval** is measured from the beginning of Q to the end of S and represents the time occupied by the excitation wave in passing over the ventricle. Mechanical consequences of ventricular contraction are not demonstrated until 0.05 second after the Q wave, which would indicate that the ventricular systole begins late in the downstroke of the R wave. Ventricular systole, which is represented by the **Q-T interval,** may be divided into the isometric period of contraction, in which there is rising tension but the lengths of the fibers remain unchanged (about 0.05 second), and the period of ejection, lasting through the rest of ventricular systole with a duration of 0.3 second. The QRS complex will reflect morphological changes of the heart, such as hypertrophy, atrophy, or dilation. It will also reveal abnormalities of interventricular conduction, but it will not reflect alterations in cardiac function.

The T wave is recorded simultaneously with the major mechanical events of the cardiac cycle. As it represents repolarization, which is the restoration of the semipermeability of the cell membrane, this would involve metabolic work by the cell. Any condition that produces alterations in cell metabolism, such as anoxia, acidosis, and shock, will result in alteration in the T wave.

Control of the Heart by the Nervous System

The origin of the heartbeat, as we have already discussed, is not dependent on the central nervous system, but the action of the heart can be modified by the state of mind or body so that there is some nerve control of heart action. The heart is connected with the central nervous system by means of two sets of nerve fibers from the autonomic system, parasympathetic fibers from the vagus, and sympathetic fibers from the cervical sympathetics (see p. 221 and Fig. 7-4).

The vagus is an inhibitory nerve, and stimulation of this nerve slows the heart by action on the pacemaker, resulting in a weak systole and prolonged diastole. The effect with an extremely strong stimulation can be so complete as to bring the heart, for a time, to a standstill. Stimulation also depresses the conduction from the atria to the ventricle by depressing the activity of the atrioventricular node. The action of the heart under normal circumstances is being continuously held down by a constant stream of impulses passing over the vagi to the heart. This condition is referred to as the **vagal tone** of the heart. When the vagi are cut in experimental animals, the heart rate rises, and after the administration of a parasympathetic inhibitor drug, such as atropine, the cardiac rate increases. In human beings, under atropine influence, the heart rate increases to 150 to 180 beats per minute.

Stimulation of the sympathetic fibers increases the heartbeat (chronotropic effect) by shortening the systolic and diastolic periods and increases the force (inotropic effect) of atrial and ventricular contraction. It appears that there is very little **sympathetic tone** or none at rest and that the increase in heartbeat **(tachycardia)** produced by excitement, exercise, and emotional states is due to sympathetic nerve discharge at those times.

Reflex Control of the Heart

The number of impulses passing along the vagi to the heart controls the normal heart rate. The center of this control is located in the medulla, which is under continual influence from vasomotor reflex activity, which allows the center to adjust cardiac activity to varying needs of the organism.

Sensory receptors called **baroreceptors** are located in the walls of the heart, in the aortic arch, in the carotid sinus, and in the superior and inferior venae cavae as they enter the right auricle and the left auricle at the entrance of the pulmonary veins. These receptors are stretch receptors which are stimulated by distention of the structure in which they are located. Their sensory fibers pass by way of the glossopharyngeal and vagus nerves to the cardioinhibitory center and the depressor portion of the vasomotor center. Impulses carried by these fibers excite the cardioinhibitory center and inhibit the automatic discharge of the vasoconstrictor nerves. This action produces a slowing of the heart with a decrease in cardiac output and a vasodilatation with a drop in blood pressure.

The reflexes which are of fundamental importance and constantly in play to ensure that the blood pressure and circulation are maintained at a suitable level are as follows:

1. *Cardiac Reflex*. Sensory fibers arising in the muscular wall of the heart and leading to the cardiac center of the brain are stimulated when the heart rate increases above normal. The impulses traveling to the cardioinhibitory center stimulate the vagus, and as a result the heart rate is slowed down.
2. *Carotid Sinus Reflex*. The carotid sinus, which is a small pocket found at the bifurcation of the common carotid into the internal and external carotid (Fig. 7-28), is supplied by sensory fibers from a branch of the glossopharyngeal nerve. These fibers are stimulated when the walls of the carotid sinuses are stretched, a condition that results with an increase in blood pressure. The sensory impulses entering the medulla reach the cardioinhibitory center, which, in turn, transmits impulses back to the heart through the vagus to slow the heart and decrease the force of contraction, resulting in a drop of pressure. When the pressure falls below normal within the carotid sinus, impulses are sent to the cardioaccelerator center, and the heart rate is increased by impulses transmitted back to the heart by way of the sympathetic fibers, resulting in an increase in pressure. There is no evidence that any special receptors of the carotid sinus are stimulated by low arterial pressure. The reflex changes that occur when the pressure drops below a normal level are due to cessation of receptor stimulation and thereby less inhibitory action. Normal heart action and pressure are due largely to impulses coming into the heart and blood vessels from the cardiac center through parasympathetic inhibitory and perhaps a little by sympathetic accelerator fibers. Any variation in blood pressure produces an inhibition or augmentation of these impulses. In augmentation, the volley of impulses generated in the carotid sinus varies in frequency with the degree of arterial pressure. The higher the pressure, the greater the number of nerve fibers stimulated and the greater the frequency of impulses in each fiber. Since these impulses impinge on the cardioinhibitory portion of the cardiac center, the vagus is stimulated. When blood pressure drops, fewer receptors in the carotid sinus are stimulated, resulting in a depression of vagal activity, which allows the normal automaticity of the heart to express itself, producing the opposite effect or an increase in heart rate with a concomitant increase in pressure.
3. *Aortic Reflex*. The arch of the aorta also has sensory nerve fibers running from its walls to the cardioinhibitory center. An increase in pressure will cause the heart to slow, as did the carotid sinus, and consequently there will be a fall in blood pressure. If the blood pressure falls below normal, cessation of impulses from this area will cause an acceleration of the heart and thereby an increase in blood pressure.
4. *Atrial Reflex*. It has been demonstrated that sensory vagal fibers are present in the walls of both caval veins and also in the wall of the right atrium. The receptors of these sensory fibers are stimulated by a distention of these cavities, as would result from a rise in pressure in these areas, causing impulses to pass up the vagus to the cardioinhibitory center in the medulla. The reflex response produced by the stretching of the walls of caval veins as well as the atrial receptors is similar to those produced by carotid sinus stimulation. There is bradycardia and vasodilatation with a decrease

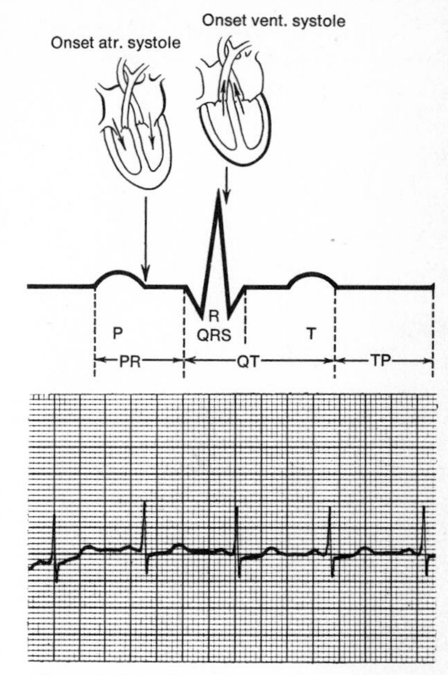

Figure 7-27. Typical record of an electrocardiogram (ECG). Lower graph represents actual tracing.

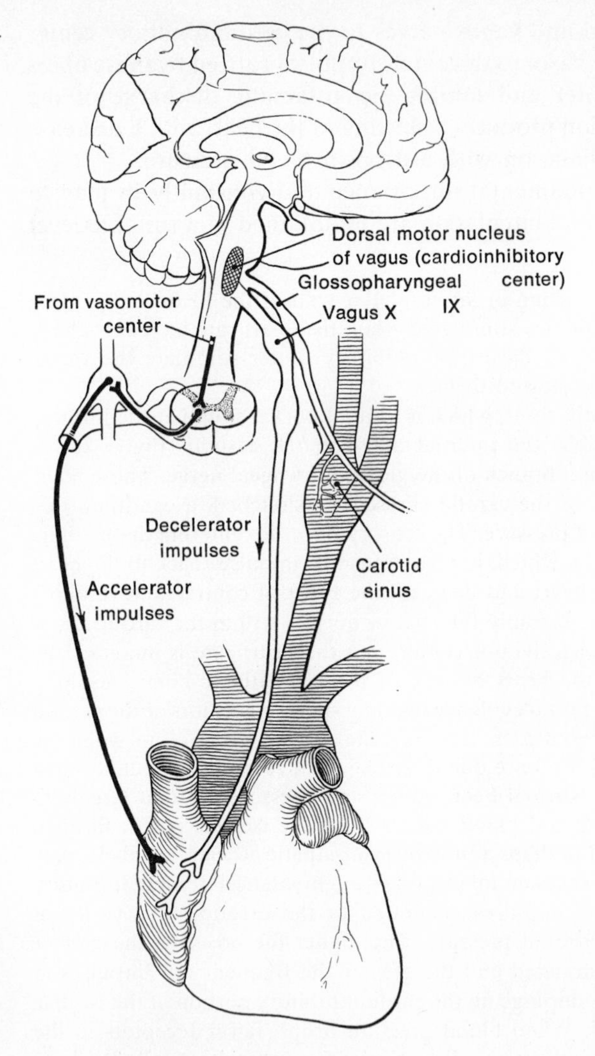

Figure 7-28. Carotid sinus and its connections to the central nervous system.

in cardiac output and a fall in blood pressure. In addition, there is venodilatation, which increases pooling in the veins and decreases venous return. This reflex mechanism probably prevents excessive rises in central venous pressure venous return.

5. *Bainbridge Reflex.* In 1915, Bainbridge demonstrated on anesthetized animals an increase in heart rate when the right atrium was stretched by the rapid infusion of saline or blood into this area. The rate was increased only if the initial rate was low. This response became known as the Bainbridge reflex and was believed to be a result of an inhibition of the cardioinhibitory center. As a result of the inhibition, the heart rate increases, which increases the blood output of the heart and thereby causes a drop in the pressure in the venae cavae and right atrium. Today, this effect is not believed to be a reflex but is most probably a direct effect of fluid injection on the cardiac muscle.

Other Factors Modifying Heart Rate

Although it appears that the main regulatory mechanism of the heart is nervous, it must be pointed out that other influences modifying heart action are at work. The beat of the heart, for example, is decidedly influenced by the metabolism of the heart muscle itself. The tissues of the heart, like all other tissues, need an adequate supply of blood with the proper chemical constituents. Any variation in quality or quantity of blood delivered to the heart will correspondingly affect the heart. Chemicals such as epinephrine affect the heart's beat by acting on its muscular fibers and the nerve fibers that innervate them. Carbon dioxide exerts a dilating effect. An excess of carbon dioxide and lack of oxygen produce a dilation of the heart with an increase in diastolic volume. Both carbon dioxide and oxygen also influence the action of the heart by way of the **chemoreceptors** in the carotid body. Any increase in the carbon dioxide or lactic acid concentration of the blood above normal levels will stimulate these sensory receptors which inhibit the cardioinhibitory center and the heart rate will increase. An increase in oxygen concentration will have the opposite effect. Variations in the amount of such ions as potassium, calcium, and sodium supplied to the heart musculature may modify the rhythm or the individual contractions, or both.

The physical or mechanical circumstances of the heart may also affect its beat. The contractions of cardiac muscle, like those of skeletal muscle, are enhanced up to a certain limit by stretch. A full ventricle will contract more vigorously than one less full. The limit at which such resistance is beneficial may be passed, and an overdistended ventricle will fail to contract at all.

Vasomotor Control of the Blood Vessels

Most blood vessels of the body are connected with the central nervous system by nerve fibers. These fibers are part of the autonomic system and influence the caliber of the vessels by their action, which varies the amount of contraction of their muscular coats. This direct nervous control of the diameter of the lumen of the vessels, especially those of the small arteries and arterioles, is one of the most important factors that determine the level of arterial pressure and the rate of flow in different parts of the systemic circulation (Figs. 7-29 and 7-30). These nerve fibers are referred to as **vasomotor nerves** and may be divided into two groups: **vasoconstrictors,** which are mainly of sympathetic origin, and **vasodilators,** which are mainly parasympathetic in origin. Stimulation of the vasoconstrictors produces narrowing of the caliber of the vessels, and it appears that these fibers are the means by which the central nervous system exerts a continued tonic influence and maintains a vessel "tone." These fibers have their central origin in the medulla in an area known as the **vasomotor center.** They leave the spinal cord by the anterior roots of the spinal nerves, from about the second thoracic to the second lumbar, to pass into the thoracic and abdominal chain of sympathetic ganglia. From here, the fibers proceed to their specific blood vessels by various nerves such as the splanchnic, cervical sympathetic, and hypogastric. General and local arterial tone is maintained and regulated by the transmission and distribution of impulses along these vasoconstrictor fibers (Fig. 7-29).

The vasodilator fibers appear to take origin in various parts of the central nervous system and are found in three major subdivisions of the peripheral nerves: the somatic sensory, the parasympathetic, and the sympathetic. The fibers proceed

Figure 7-29. *Stimulation of vasoconstrictors produces narrowing of the lumen of the vessels.*

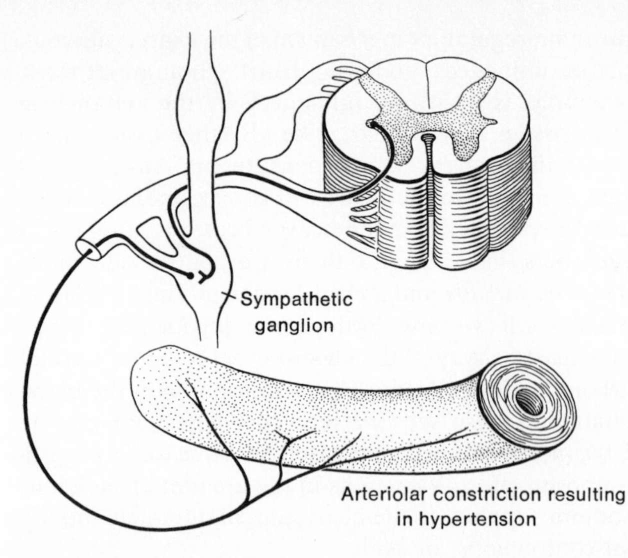

from their origins to their destinations in the muscular coats of the vessels by way of the dorsal roots and as part of the trunks and branches of various cranial and spinal nerves. The vasodilators do not continuously send impulses into these vessels, as do the vasoconstrictors, and, therefore, exert no tonic dilating effect. They are thrown into play generally as part of a reflex action, and their effects appear to be essentially local in nature, producing a dilation in the area the fibers control. The vascular areas under control of any set of fibers are relatively small so that activation of them produces little or no effect on the vascular system in general.

The vascular dilation produced by dorsal root fibers is called **antidromic** because it is brought about by impulses that travel in a direction the opposite of that

Figure 7-30. *Vasodilatation of vessels produced by ganglionic blockade.*

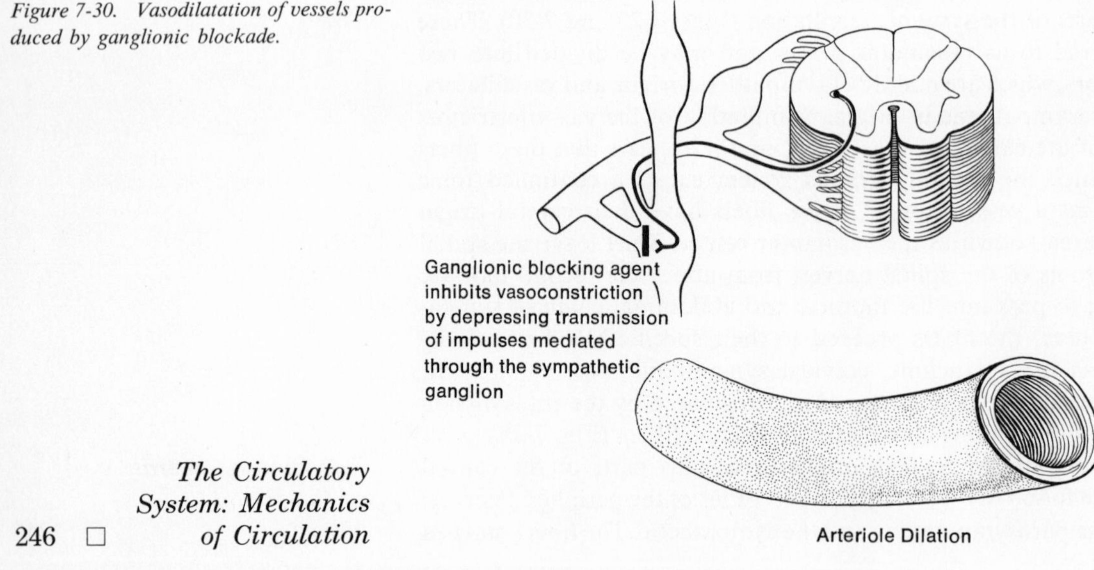

set up at a sensory ending. It appears that these fibers are distributed almost entirely to the skin and that they exert their action on the blood vessels by the release of a histaminelike substance. Parasympathetic vasodilator fibers are derived from the seventh and ninth cranial nerves, which are distributed to the blood vessels of salivary glands, pharynx, and brain. These fibers are concerned with the local regulation of blood flow in these parts. In addition, parasympathetic vasodilator fibers are also found in the pelvic nerves having their origin in the sacral segments of the spinal cord. Stimulation of these fibers results in the vascular dilation responsible for the erection of the penis and clitoris.

Sympathetic vasodilator fibers probably exist along with those nerves that contain vasoconstrictors. These vasodilator fibers originate in the cerebral cortex and pass through the medulla into the spinal cord without interruption. The vessels of the skeletal muscles are innervated by these fibers and although they travel with sympathetic vasoconstrictors which are adrenergic, these fibers are cholinergic and form the sympathetic vasodilator system. Stimulation of this system by exercise, or emotional stimuli such as fear and apprehension, produces vasodilatation in skeletal muscle. However, because there is concomitant vasoconstriction elsewhere, the blood pressure does not usually change, and may even rise.

Dilation may also be effected by an inhibition of vasoconstrictor impulses (Fig. 7-30). When the vascular area so affected is small, the effects are local, and more blood is distributed through the area. When the vascular area affected is large, the inhibition of tonic constrictor impulses may lead to a marked fall of general blood pressure. Inhibition of vasoconstrictor impulses is aptly demonstrated by the "blushing" reaction. Nerve impulses originating in the brain by certain emotional experiences produce a powerful inhibition of the vasomotor center of the brain, which controls the vascular areas of the head supplied by the cervical sympathetic. As a result, the muscular walls of the blood vessels of the head and face relax, producing a dilation, and the whole region becomes flushed with blood. Sometimes an emotion can produce the opposite effect, that is, "pallor" brought on by an undue arterial constriction, which is effected by the central nervous system and the cervical sympathetic.

The mechanism concerned with the opening and closing of capillaries is not well understood. Recent experimental work indicates that a muscle constituting a precapillary sphincter is present in the wall at the origin of each capillary. These muscular sphincters appear to be sensitive to nervous influences, to epinephrine, and to chemical conditions in the tissues in which they are located. It has been known for some time that the endothelial cells forming the wall of the capillary can contract passively, and this probably accounts for the changes in caliber of capillaries.

Little is known of the vasomotor changes in the large arteries and veins, but it is unlikely that the muscular coat of these vessels is without function. Observations on experimental animals indicate that the larger vessels can show marked contraction. In cases of diffuse peripheral dilation, constriction of large arteries always takes place. Rhythmical contractions of the great veins opening into the heart may be observed in many animals. Similar rhythmical variations have been observed in the portal veins. The size of blood vessels also can be altered by temperature. It appears from experimental evidence that these effects are to some extent direct and to some extent reflex activity. Cutting the nerve supply to a vascular area will not destroy the response of the blood vessels to changes in temperature. Lowering the temperature below normal causes blood vessels to constrict, and increasing the temperature above normal body temperature produces a dilation. It is uncertain whether the effects, in intact animals, are directly dependent on temperature changes, on secondary chemical and physical effects, such as changes in blood viscosity and tissue metabolites, or on reflex reactions of the dorsal root or central type.

Volume of Blood Within the Circulatory System

Although the blood participates in nearly every chemical and physiological process that takes place in the organism, its composition and volume are maintained practically constant despite continuous additions and losses of water and of metabolic and nonmetabolic substances. The term **blood volume** refers to the sum of the volume of cells and plasma in the circulatory system. In general the measurement of total bood volume involves a separate determination of plasma volume and/or cell volume. Plasma volume is usually determined by a dye method. In this method, first introduced by Rountree, Keither, and Geraghty in 1915, a known amount of a nontoxic, slowly absorbable dye—brilliant vital red—is directly injected into the circulation. The dye remains in the plasma long enough for the two to become thoroughly mixed. A sample is withdrawn, and the concentration of the dye in the plasma is determined colorimetrically by comparison with a standard mixture of dye and serum. From the extent of dilution of the dye in the sample, the plasma volume is determined. Today, the same method is used, but practically all measurements have been done with the blue dye T-1824 because hemolysis in the samples has less influence on the colorimetric determination with this dye than with brilliant vital red. The cell volume is usually obtained from the hematocrit value of venous blood by centrifugation. Total blood volume (BV) is calculated from the above determinations by the formula

$$BV = \frac{PV(\text{plasma volume}) \times 100}{100 - H(\text{hematocrit})}.$$

If, for example, the plasma volume was determined to be 2,750 ml and the hematocrit 45 per cent, the blood volume would be

$$BV = \frac{2{,}750 \times 100}{100 - 45} = 5{,}000 \text{ ml.}$$

Through many years of samplings on blood volume determinations of the normal population, investigators have found that the average normal value of 5,000 ml cannot be rigidly applied, for blood volume varies with such factors as age, sex, build, race, environment, and disease, and the range of individual determinations is considerable (Table 7-2). In the newborn infant the blood volume in terms of body weight is high, whereas in the second and third years of life it declines and then remains steady until the eleventh year. During adolescence blood volume keeps pace with growth, increasing until maturity. There appears to be no significant change in total volume with age from 20 to 90 years.

Blood volume also fluctuates with changes in posture and muscular activity. Standing, for example, can reduce the plasma volume appreciably as a result of increased venous and capillary pressure, which produces local **efflusion** of plasma from the capillaries in the extremities. Blood volume is always lower in the erect than in the horizontal position. Muscular activity reduces blood volume, the amount depending on the degree and duration of the exercise. Intensive physical training increases blood volume considerably, and blood volume decreases in normal subjects after prolonged bed rest. Bazett, Sunderman, and Scott showed that environmental temperature can influence blood volume. A warm environment can increase the plasma volume from 15 to 30 per cent; the blood volume is higher in summer than winter. Men have a higher blood volume than women. The lower blood volume in women is probably due to differences in body fat, as it has been known that the amount of excess body fat influences blood volume. In general, lean persons have a greater blood volume than obese persons. These samples point up the fact that average normal values of blood volumes cannot be strictly applied.

Blood Pressure

The volume of the blood, as we have already seen, remains relatively constant and, as a result of being contained within a closed system of tubes, exerts, according to its volume, a certain pressure against the walls of these tubes. At

Table 7-2. ***Cardiovascular Function*** **(*Normal Values*)**

Blood volume	7–9% body weight
Total	5–6 liters (5,000–6,000 ml)
Plasma volume (55% total)	2,750 ml
Cell volume (45% total)	2,250 ml
Cardiac output	5.0–6.0 liters/min
Heart rate	70–75 beats/min
Stroke volume	80 ml per beat
Blood pressure	
Right ventricle	Systolic: 22 mm Hg; diastolic: 0
Pulmonary artery	Systolic: 22 mm Hg; diastolic: 8
Left ventricle	Systolic: 120 mm Hg; diastolic: 5
Brachial artery	Systolic: 120 mm Hg; diastolic: 80
Brachial vein	Mean: 1.0 mm Hg
Capillary	Mean: 20–30 mm Hg
Rate of flow	
Aorta	300–500 mm/sec
Medium arteries	200–250 mm/sec
Arterioles	4–8 mm/sec
Capillaries	1 mm/sec
Venules	2–4 mm/sec
Great veins	200 mm/sec
Circulation time	20–25 sec (time for a volume to flow from left ventricle back to left ventricle)

each beat of the heart, which in man is about 72 times a minute, a certain quantity of blood, probably amounting to 80 ml, flows out of each of the ventricles with very great force into the aorta and with less force into the pulmonary artery. This discharge is relatively rapid, so that a brief period of time intervenes between successive discharges, causing the flow of blood from the heart into the arteries to be distinctly intermittent. At each beat of the heart, as much blood flows from the great veins into the right atrium as escapes from the left ventricle into the aorta.

When the finger is placed on an artery in the living body, such as the radial artery at the wrist, a sense of firmness is felt, and this firmness appears to be increased at intervals corresponding to the heartbeats. At each beat of the left ventricle the artery expands, or distends, and it is this expansion which we feel with our fingers that is called the **pulse.**

Experimentally, if we insert into an artery, for example the carotid, a long glass tube of small bore and hold this tube upright, the blood will rush into the tube and fill it with a column of blood of a certain height (Fig. 7-31). This demonstrates that as it is flowing through the carotid, the blood is exerting a certain pressure on the walls of the artery. Since there is nothing to counterbalance the pressure within the opened artery but the pressure of the atmosphere, the blood rises in the tube until its weight is equal to the pressure within the artery and remains at this height, oscillating slightly above and below this level with each heartbeat. In a dog, this column of blood will generally reach a height of 145 cm. Thus, the pressure that the blood exerts on the walls of the carotid of a dog is equal

Figure 7-31. Stephen Hales (1677–1761) performing the measurement of blood pressure. He tied an old mare to a gate and laid open the jugular vein. In it he inserted a glass tube 4 ft 2 inches long and observed the gradual rise of the bood column (December, 1731). (Pen and ink drawing. Courtesy Bettmann Archive. Copyright Otto L. Bettmann, 1954.)

to the pressure exerted by a column of a dog's blood 145 cm high. This is equal to the pressure of a column of mercury about 100 mm high. The pressure exerted by the blood on the walls of the blood vessels is referred to as the **blood pressure,** and in the experiment cited here, the blood pressure in the carotid of the dog is 100 mm Hg. Stephen Hales, the great English physiologist, was the first to demonstrate and determine blood pressure in this way, using the horse as the experimental animal.

Instead of using a long glass tube, if we insert a short T tube into the artery and connect this to a recording membrane manometer (Fig. 7-32), the blood

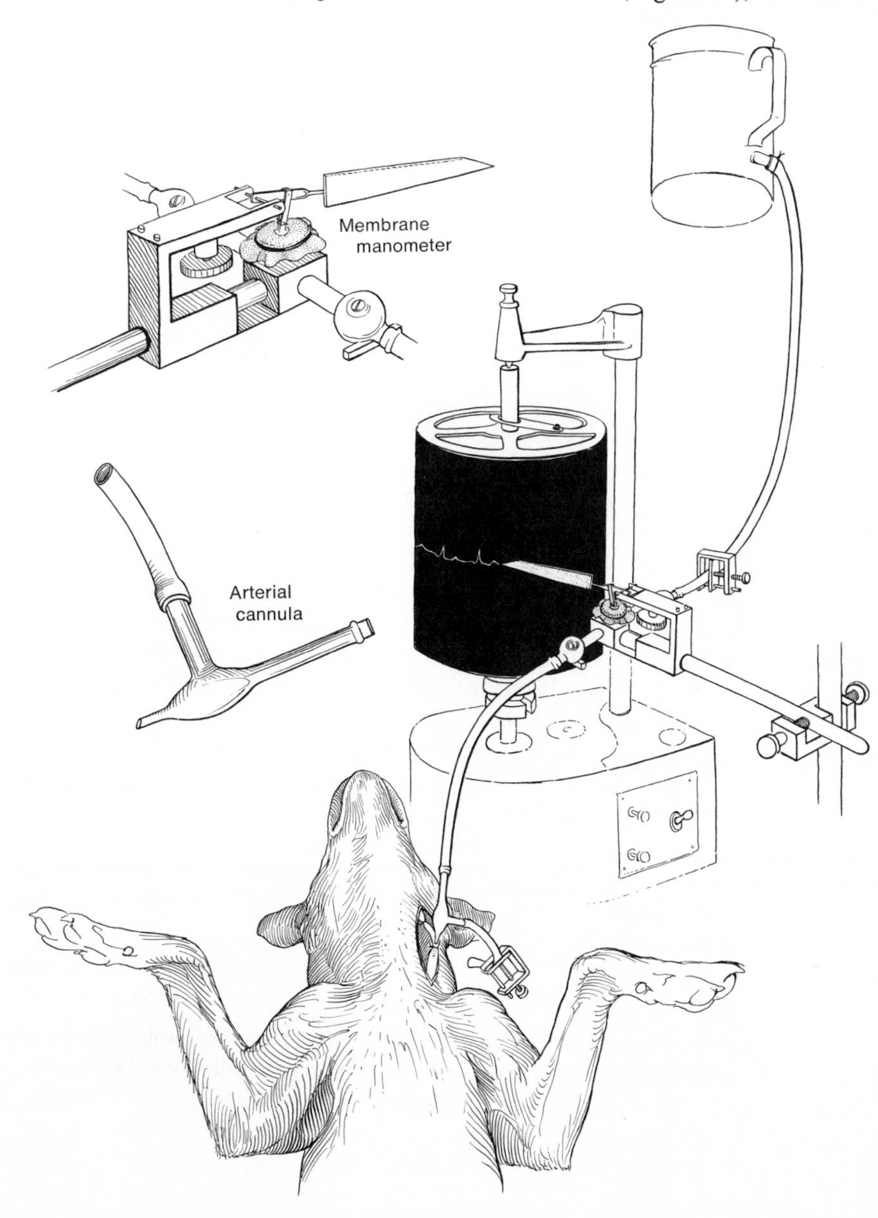

Figure 7-32. Arrangement for the direct recording of blood pressure from the carotid artery of the dog.

pressure can be more conveniently studied since it can be recorded on a moving surface, and a tracing such as that represented in Fig. 7-33 will be described. Each of the smaller curves (hh) corresponds to the heartbeat, the rise corresponding to the systole and the fall to the diastole of the ventricles. The larger undulations (rr) in the tracing are respiratory in origin and vary with the respiratory rate.

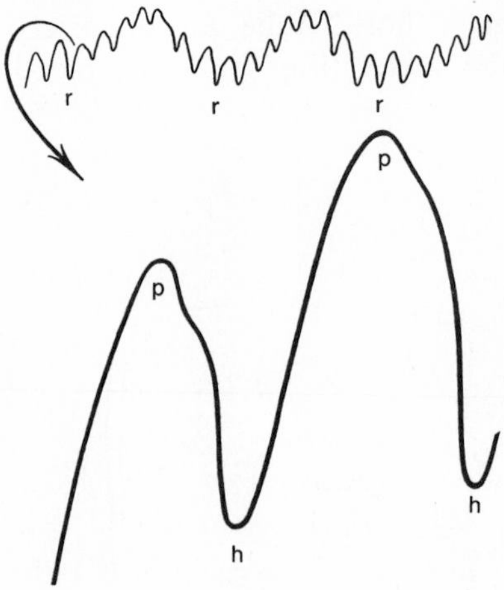

Figure 7-33. Tracing of blood pressure curve. Each of the smaller curves (hh) *corresponds to the heartbeat, the rise* (p) *corresponding to the systole and the fall to the diastole of the ventricles. The larger undulations* (rr) *in the top tracing are respiratory in origin and vary with respiratory rate.*

By the use of such a recording device, the mean pressure within the various arteries and veins can be studied. It appears that the pressure is high in all arteries but is greatest in the larger arteries near the heart and diminishes along the arterial tree as the arteries branch out toward the capillaries. In man, the mean pressure in the large arteries is approximately 100 mm Hg (Table 7-2). The manometer cannot be applied to capillaries, but the pressure in the capillaries can be determined in an indirect way. In the mammal the capillary blood pressure is approximately 20 to 30 mm Hg, which is greater than that in the veins but less than that in the arteries. There is a continued decrease of blood pressure from the aorta, through the arteries, arterioles, capillaries, and veins to the great veins entering the right atrium. The mean blood pressure is low in the veins—gradually decreasing until in the large veins near the heart, it may be lower than the pressure of the atmosphere and is said to be "negative."

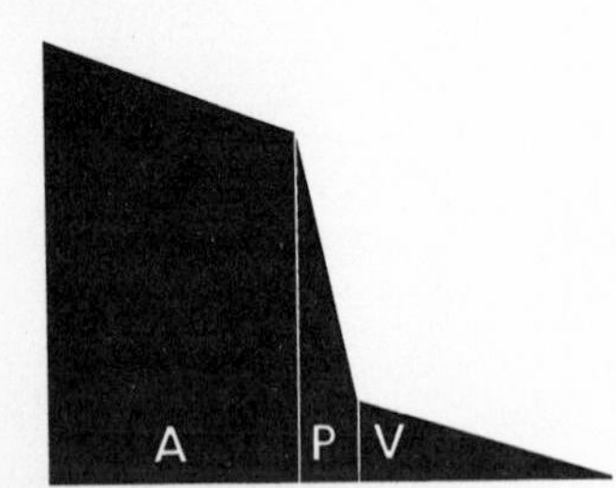

Figure 7-34. Graphic representation of the decline of blood pressure from arteries (A) *to veins* (V). *In man mean pressure in the arteries is approximately 100 mm Hg. In capillaries* (P) *the pressure falls to 20 to 30 mm Hg. In large veins near the heart it may be lower than atmospheric pressure, and in such cases it is said to be negative.*

In general, the decrease of blood pressure from the aorta through the systemic circulation to the right atrium is gradual. It will be found, however, that in the capillary bed the decline is quite swift; that is, the most marked fall of pressure takes place between the arterioles on one side of the capillaries and the venules on the other side. Figure 7-34 graphically demonstrates the decline of pressure from the arteries to the veins.

In the arteries the pressure is marked by oscillations corresponding to the heartbeats. Each oscillation consists of a rise, corresponding to the systole of the ventricle, and represents a rise in pressure. The blood pressure at this point is called the **systolic pressure,** and in man, it is approximately 120 mm Hg. This rise is followed by a fall corresponding to the diastole of the ventricle and represents a decrease of pressure below the mean. This pressure is referred to as the **diastolic pressure** and is approximately 80 mm Hg in human beings. Diastolic pressure represents the constant load that the arterial wall is called on to bear and the force or resistance that the ventricular contraction must overcome to throw open the aortic valve. The **mean pressure** is usually given as the average of the systolic and diastolic pressure and in man is approximately

$$100 \text{ mm Hg} \left(\frac{120 + 80}{2}\right).$$

The difference between the systolic and diastolic pressures is called the **pulse pressure.** In man, the pulse pressure amounts to about 40 mm Hg (120 − 80) and is clearly caused by the ejection of blood (80 ml) into the aorta during systole. Other things being equal, its magnitude will vary with the quantity of blood ejected by the left ventricle at each heartbeat.

Determination of Human Blood Pressure

Human blood pressure is not measured by direct arterial cannulization but by indirect methods. Three such methods have been developed; all employ an inflatable armband and a pressure-indicating device. The three methods are referred to as the **oscillatory,** the **palpatory,** and the **auscultatory.**

The oscillatory method involves placing a pressure cuff over an artery such as the brachial artery and observing the oscillations produced by the pulsations imparted to the cuff by the artery underneath. When the cuff pressure is higher than systolic pressure, the oscillations in pressure are abolished. As the pressure in the cuff is reduced, the oscillations become larger. The pressure at which these large oscillations are produced is regarded as the systolic pressure. As the pressure in the cuff is lowered still fruther, the oscillations become smaller, and the point at which this occurs rapidly is taken as the diastolic pressure.

The palpatory method can provide a measure of only the systolic pressure. The radial artery at the wrist is palpated with the fingers as the pressure in an arm cuff is increased to a point where the pulse disappears. The pressure in the cuff is then reduced until the pulse reappears; it is at this point that the pressure is recorded and taken as systolic pressure.

The auscultatory method is somewhat similar to the oscillatory but involves, instead of the visual oscillations produced by pressure changes, the appearance and disappearance of sounds, which are known as the **Korotkoff sounds,** with variations in cuff pressure. The technique involves the inflation of an arm cuff, placed around the arm at a level just above the elbow, to a pressure above systolic pressure. The bell of a stethoscope is placed over the brachial artery in an area just below the cuff. The pressure in the cuff is slowly reduced. As this reaches systolic pressure, the blood begins to spurt through the previously collapsed artery, producing a snapping sound as a result of the expansion and collapse of the wall of the artery with each beat. As the cuff pressure is further reduced, the sounds gradually become more faint as the arterial flow becomes less turbulent. The point at which the sounds almost completely fade is taken as the diastolic pressure.

The auscultatory method is the one most commonly used today and is con-

Figure 7-35. Determining blood pressure by the auscultatory method.

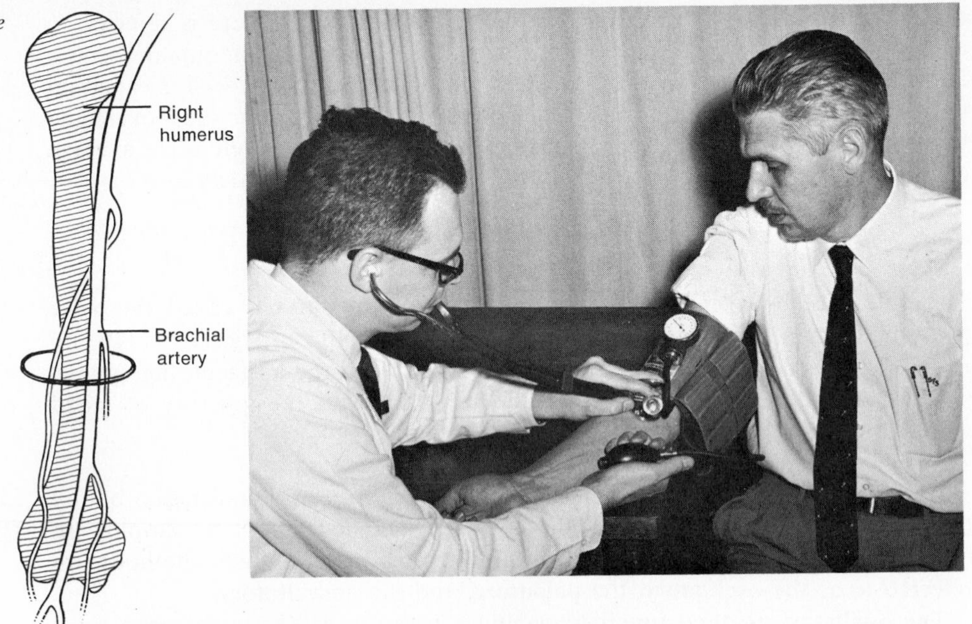

sidered to be more accurate than the other two. Figure 7-35 represents a typical **sphygmomanometer** being used in the determination of blood pressure by the auscultatory method.

Factors Responsible for the Maintenance of Normal Arterial Pressure and the Circulation of Blood

In the circulation of blood as well as the maintenance of blood pressure, the following factors are of fundamental importance:

1. Heart action
2. Blood volume
3. Peripheral resistance
4. Physical property of the walls of the large blood vessels
5. Viscosity

On the basis of these factors can be explained the main phenomena of circulation, such as the high pressure and pulsatile flow of the blood in the arteries, the steady stream through the capillaries, the low pressure and uniform flow in the veins, and, finally, the flow of blood back to the atrium.

Heart Action. The action of the heart will obviously determine cardiac output as well as the energy supplied in this output. These in turn will influence the circulation and the blood pressure. This action involves such factors as stroke volume, heart rate, and force of heart muscle contraction. The determination of the output of either ventricle per unit time, which is usually a minute, gives the volume of blood that is pumped into the arteries and the volume that determines blood pressure, as well as the volume that flows through the entire circulation in that time. The outflow of blood from the heart necessary to overcome the peripheral resistance must equal the inflow. An increase in rate increases diastolic pressure, whereas an increase in force of contraction increases systolic pressure. An increase in stroke volume also increases diastolic pressure.

Blood Volume. As we have already seen, the normal volume of blood amounts to about 5 liters, and this volume, contained within a closed tubular system, will exert a certain mean pressure on the walls of this tubular system. An increase in blood volume above normal causes an increase in pressure, since the pressure is directly proportional to the volume. A decrease in volume, such as occurs in shock or injury where there is a great loss of blood, causes the blood pressure to drop. In such instances, increasing the amount of circulating fluid by transfusion will help restore the volume and therefore the pressure.

Peripheral Resistance. We have already seen that the aorta branches into smaller arteries, and these branch still further into smaller arterioles until finally the very minute capillary bed is reached. These minute passages present a considerable resistance to the flow of blood through them. In fact, the friction and resistance are so great in the peripheral regions that passage of blood meets with the greatest difficulty, and in some areas no blood passes through these passages at all. The resistance to the flow of blood caused by the friction developed in so many minute passages is one of the most important factors influencing circulation. This friction is even greater in the arterioles than in the capillaries, since the flow is more rapid in the former. Friction decreases with a decrease in rate of flow. This friction generated in the small arteries and the capillaries is referred to as **peripheral friction** and the resistance it offers as the **peripheral resistance.** Not only does this resistance oppose the flow of blood through the minute vessels, but by working back along the whole arterial system, it represents the resistance, or force, that has to be overcome by the heart at each contraction of the ventricle.

The greater part of the peripheral resistance of the circulatory system is presented by the arterioles of the abdominal viscera. This area is referred to as the **splanchnic region.** Stimulation of the splanchnic nerve, a vasoconstrictor, causes an increase in the resistance to flow of blood through these vessels as a result of the narrowing of the lumen and consequently a reduction in the outflow of blood from the arterial system. The pressure will rise until outflow from the arterial system is balanced by the inflow. Dilation of these vessels, which represents a decrease in the peripheral resistance, causes the pressure in the arterial vessels to drop, since outflow in this instance becomes greater than inflow. It has been determined that if the vessels of the splanchnic area were fully dilated, they would have a volume capable of holding all the blood in the body. In such an event the blood pressure would fall to zero. In general this region controls normal blood volume, since the normal tone of these vessels, being under the influence of the vasomotor system, provides a vessel lumen size that will keep a proper volume of blood distributed through the circulatory system to maintain normal pressure.

Physical Properties of the Walls of the Large Blood Vessels. The arteries reaching from the ventricle to the region of peripheral resistance may be looked upon as a long stretch of elastic tubing made to withstand high pressures up to approximately 200 mm Hg or more. It is also this same elastic quality that the peripheral resistance can react through, on the intermittent force of the heart, which gives the circulation of blood its characteristic features. At the diastolic pressure existing under normal circumstances, the volume of blood contained within the arteries distends the walls of these arteries somewhat, and by virtue of their elasticity, the walls tend to recoil against this stretching force. At each beat of the heart the contents of the ventricle are thrown into the already overfilled system. Of this quantity only a fraction can pass through the resistance during the beat of the heart after completion of systole; the rest still remains on the arterial side of the resistance, the elastic wall of the artery having further distended to receive it. During the interval between this and the next beat, the elastic wall rebounds and, pressing on the blood within, tends to drive it onward through the peripheral resistance. The amount of blood remaining in the arterial area after the beat depends on the amount of resistance in the peripheral vessels and on the elasticity (distensibility) of the tube. The amount that passes through the peripheral resistance between beats depends on the degree of elastic recoil of the arterial wall. The flow on the venous side of the peripheral resistance is not the direct result of the beat of the heart, since all this force appears to be spent in developing and maintaining the distention of the arterial wall. It is the elastic recoil of the arterial wall acting as a subsidiary pump that forces a volume of blood through the peripheral resistance during a beat and in the succeeding interval equal to an amount it receives from the heart by the beat itself. In other words, the elastic reaction of the walls of the large arteries will have converted the intermittent flow observed in the arteries into a continuous flow on the venous side.

In summary, it can be seen that the arterioles and capillaries present an area of great resistance to the flow of blood from the arterial to the venous side. Because of this resistance the force of the contraction of the heart muscle is used in maintaining the arterial system in a state of great distention. This stretching of the arterial walls by the pressure of the blood thrown into them by the beating of the ventricles is the pressure we have referred to previously as the blood pressure. The elastic recoil of the overdistended arterial wall acts as a force by which the arterial system can empty itself by overflowing through the peripheral vessels into the venous system. The arterial system overflows at such a rate that as much blood passes from the arteries to the veins during each systole and diastole as enters the aorta at each systole, and it is for this reason that the flow is not intermittent but continuous in the veins.

Viscosity. Resistance to flow occurs in all the vessels of the vascular system, and, as has been stated, it is greatest in the arteriolar portion. Resistance to flow is dependent not only on the dimension of the vessels, which we have just discussed, but also on the character of the fluid. The character of the fluid is its **viscosity,** or thickness. The unit of viscosity is the **poise.** A centipoise (one hundredth of a poise) represents the viscosity of water at 20° C and is used as a convenient standard of comparison. Blood is some five times more viscous than water and owes its viscosity to the plasma proteins and the corpuscles suspended in it. The viscosity of blood varies greatly with the corpuscular content, since the corpuscles are responsible for most of the viscosity. Anything affecting the normal concentrations of these two factors will consequently alter the viscosity of the blood. The blood pressure varies directly proportionally to the viscosity, whereas the rate of flow is inversely proportional. An increase in viscosity increases blood pressure and decreases the rate of flow through the vascular system. An opposite effect is produced by a decrease in the viscosity.

Rate of Flow of the Blood

Previously we saw that the blood moves in the arteries from the heart to the capillaries with considerable speed. The velocity of the blood has been determined by various methods at different parts of the arterial system. They show that the velocity is greatest in the largest arteries, about 300 to 500 mm per second in man (Table 7-2), and diminishes as the arteries branch out to the capillaries. In the arterioles the rate is probably only a few millimeters per second. In the capillaries the rate is slowest of all, being about 1 mm per second. The flow is very slow in small veins but increases as these join into larger trunks, until in the great veins the rate is about 200 mm per second. Thus it can be seen that the velocity of the flow is inversely proportional to the total cross-sectional area of the vessels.

The total circulation time of a volume of blood going from the left ventricle and back again has been estimated approximately by determining the time it takes for an easily recognized chemical to appear after injection into the jugular vein of one side in the blood of the jugular vein of the other side. In man it takes approximately 20 to 25 seconds. Taking the rate of flow through the capillaries at about 1 mm per second, if any volume of blood had to pass through 10 mm of capillaries, half the whole time of its circulation would be spent in the capillaries. Since the exchange of materials between the blood and the tissue can only take place through the walls of the capillaries, it is obviously an advantage that the blood's stay in the capillaries be prolonged. The average length of a capillary is about 0.5 mm; therefore, about half a second is spent in the capillaries by any volume of blood and another half second in the capillaries of the lungs.

Factors Influencing the Flow of Blood Through the Veins

In the living body there are certain aids to the circulation other than the driving force of the heart, but these chiefly affect the flow along the veins back to the heart.

The veins are in many places provided with valves so constructed that they offer little resistance to flow into the veins from the capillaries but prevent its return toward the capillaries. The pressure of adjacent tissue surrounding the veins helps push the blood forward from one section of vein to another, toward the heart, so that this milking action serves as a pumping device driving blood from the lower extremities to the abdominal region. Various movements of the skeletal system help exert much pressure on the veins and therefore assist the circulation. The pressure exerted by the diaphragm as well as the movements of the alimentary tract promotes the flow along the veins in the abdominal region, therefore assisting the flow of blood from this region to the chest.

Respiration assists in circulating the blood back to the heart because it also acts as a pump. On each inspiration the pressure within the thoracic cavity is reduced to about -6 mm Hg below that of atmosphere; this tends to draw blood toward the chest. Each expiration tends to drive blood away from the chest. The arrangement of the valves of the heart causes this action of the respiratory pump to promote the flow of blood in the direction of the normal circulation, and this, therefore, makes room for more blood to flow from the great veins into the heart. All these factors help the flow of blood along the veins, but it must be remembered that they are helps only and not the real cause of the circulation. The ventricular contraction developing sufficient force to impart enough energy to drive the blood from the left ventricle to the right atrium is the main cause of circulation. Even when every muscle of the body is at rest and breathing is stopped for a period of time, the force imparted to the blood by ventricular systole is sufficient to drive the blood around its complete circuit.

References

Albert, S. N. *Blood Volume*. Springfield, Ill.: Thomas, 1962.

Barcroft, H., and H. J. L. Swan. *Sympathetic Control of Human Blood Vessels*. Baltimore: Williams & Wilkins, 1953.

Berne, R. M., and N. L. Matthew. *Cardiovascular Physiology*. St. Louis: Mosby, 1972.

Bloom, W., and D. W. Fawcett. *A Textbook of Histology*. Philadelphia: Saunders, 1968.

Bouthan, J. *ABC of the ECG: A Guide to Electrocardiography*. Springfield, Ill.: Thomas, 1970.

Burton, A. C. *Physiology and Biophysics of Circulation*. Chicago: Yearbook Publishers, 1965.

Dittmer, D. S., and R. M. Grebe. *Handbook of Circulation*. Philadelphia: Saunders, 1959.

Evans, J. R. *Structure and Function of Heart Muscle* (Monograph 9). New York: American Heart Association, 1964.

Ganong, W. F. *Review of Medical Physiology*. Los Altos, Calif.: Lange, 1965.

Goldman, M. J. *Principles of Clinical Electrocardiography*. Los Altos, Calif.: Lange, 1964.

Johnson, E. A., and M. Lieberman. "Heart: Excitation and Contraction," in *Annual Review of Physiology*, Vol. 31. Palo Alto, Calif.: Annual Reviews, 1971, p. 479.

Krayer, O. *Pharmacology of Cardiac Function*. New York: Macmillan, Inc., 1964.

Krough, A. *The Anatomy and Physiology of Capillaries*. New Haven, Conn.: Yale, 1922.

Malcolm, J. E. *Blood Pressure Sounds and Their Meaning*. Springfield, Ill.: Thomas, 1960.

Massey, F. C. *Clinical Cardiology*. Baltimore: Williams & Wilkins, 1953.

Mitchell, G. *Cardiovascular Innervation*. Baltimore: Williams & Wilkins, 1956.

Moulopoulos, S. D. *Cardiomechanics*. Springfield, Ill.: Thomas, 1962.

Randall, W. C. *Nervous Control of the Heart*. Baltimore: Williams & Wilkins, 1965.

Ruch, T. C., and H. D. Patton. *Physiology and Biophysics*. Philadelphia: Saunders, 1965.

Taccardi, B., and G. Marchetti. *Electrophysiology of the Heart*. Elmsford, N.Y.: Pergamon, 1965.

Zweifach, B. W. *Functional Behavior of the Microcirculation*. Springfield, Ill.: Thomas, 1961.

The Blood

Blood is a tissue that circulates, in what is virtually a closed system of blood vessels, to all other tissues of the body. The blood flowing through the vessels consists, under normal conditions, of a straw-colored fluid, the **plasma,** in which are suspended the red and white blood cells and the platelets. The **fluid fraction** makes up approximately 55 per cent of whole blood, and the formed elements, or **solid fraction,** make up 45 per cent (Fig. 8-1).

As the blood flows through the capillaries certain materials pass through the capillary wall into the tissue fluid, and certain materials of the tissue fluid pass through the capillary wall into the blood. There is thus an exchange of materials between the blood within the capillaries and the tissue fluid outside. A similar exchange of materials takes place at the same time between the tissue fluid and the tissue itself. The flow of materials from the blood to the tissue consists of nutritive substances and oxygen, which the tissue needs for repair, growth, and general metabolic activity. The flow of materials from the tissue to the blood consists of metabolic waste products, including the gas carbon dioxide, which must be eliminated from the body as soon as possible. From this it would appear that the composition and character of the blood would vary in different parts of the body and at different times, but the united action of all the tissues tends to establish and maintain an average uniform composition of the whole mass of blood. This chapter will be concerned with the main characteristics that are presented by blood, brought into a state of equilibrium by the common action of all the tissues.

Plasma

Plasma can be resolved by the clotting of the blood into **serum** and **fibrin** or by the centrifugation of whole blood in which the cellular portion is separated by gravity from the fluid portion. Plasma is about 91 to 92 per cent water. From this it is quite evident that water is the most abundant single compound of blood. The principal function of water is to act as a solvent and as a suspending medium for the various materials found in blood. Table 8-1 lists the major chemical components of the plasma.

Plasma Proteins

Plasma contains, in addition to water, a protein portion amounting to 7 to 7.5 g per 100 ml. The **plasma proteins** comprise the major part of the solids of the plasma. These proteins are actually a very complex mixture of simple and conjugated proteins and are usually separated into distinct fractions according to their solubilities by the use of various solvents and/or electrolytes. Three basic groups are found by these methods and are designated as albumin, globulin, and fibrinogen.

Albumin makes up 55 per cent of the total plasma protein, and there is evidence that more than one type of albumin is in plasma. One component, **mercaptalbumin,** which contains one free sulfhydryl group (SH), can be separated from the total serum albumins by crystallization. This fraction makes up approximately two thirds of the total.

The **globulin** fraction is a very complex mixture of mucoproteins, glycoproteins, lipoproteins, and metal-binding proteins. Mucoproteins are defined as those proteins which are in combination with more than 4 per cent of the carbohydrate hexosamine. Glycoproteins are those that contain less than 4 per cent hexosamine. These globulins are called the **alpha 1-** and **alpha 2-globulin** fractions. The lipoproteins consist of combinations of fat and protein, and 3 per cent of the plasma protein consists of this type, forming the **beta-globulin** fraction. Another fraction,

Figure 8-1. Centrifuged blood showing that the fluid portion makes up 55 per cent of whole blood, and the formed elements (solid portion) make up 45 per cent.

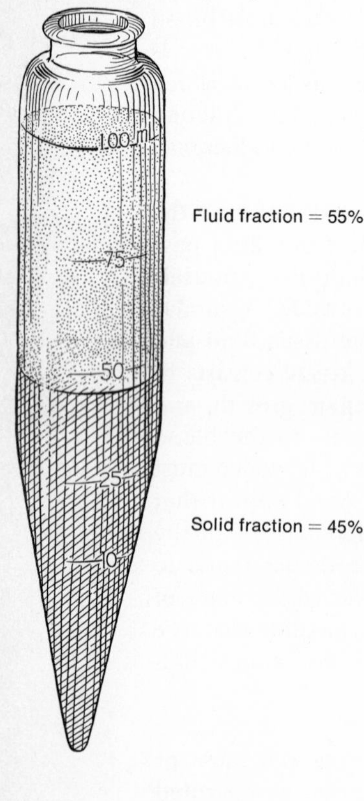

Table 8-1. ***Blood Chemistry (Normal Values/100 ml)***

Total protein	6.0–7.5 g
Albumin	3.5–5.5 g
Globulin	2.5–3.0 g
Fibrinogen	0.25–0.5 g
Nonprotein nitrogen	25–38 mg
Urea nitrogen	8–20 mg
Creatinine	1–2 mg
Uric acid	3–5 mg
Sodium	315–340 mg
Potassium	14–20 mg
Calcium	9–11 mg
Phosphorus	35–45 mg
Chloride	355–376 mg
Glucose	70–120 mg
Cholesterol	220–310 mg
Total lipid	400–1000 mg
Triglycerides	10–190 mg
Phospholipid	125–300 mg
Beta-lipoproteins	360–640 mg

known as **gamma globulin,** is the principal site of the circulating antibodies. The globulins in general make up about 44.8 per cent of the total plasma proteins.

The **fibrinogen** fraction of the plasma represents a small percentage of the total plasma–protein concentration, amounting to about 0.20 per cent.

Origin of Plasma Proteins. Conclusive information is still lacking as to the origin of plasma albumin and globulin. There is some evidence that most of the globulins and probably albumin are of hepatic origin, but certain fractions originate from other tissues. The reticuloendothelial system appears to be associated with the production of gamma globulin since it is definitely known to participate in the formation of antibodies. Lymphocytes and plasma cells are also believed to be sources of gamma globulin.

Whipple and his co-workers[1,2] have cited the liver as the sole source of fibrinogen. Their work demonstrated that those conditions which stimulate or depress liver function will also in equal degree stimulate or depress fibrinogen production.

Experimental evidence also indicates that dietary protein serves as a precursor of plasma protein. Whipple and his co-workers found that the regeneration of plasma proteins of dogs, after blood plasma depletion, occurred very slowly during starvation in contrast to the rate in those on a full diet. It was also demonstrated that there was a direct relationship between the regeneration of plasma protein and the quantity and quality of ingested proteins. A meat diet was found more beneficial for regeneration than a bread-and-milk diet. Egg and liver were less effective than meat and milk but more effective than casein and gelatin. In any case, it appears the the liver is essential in the normal regeneration of the plasma proteins since injury or disease of the liver is associated with a fall in plasma protein, and in experimentally plasma protein–depleted animals with such injuries the regeneration of these proteins is delayed. It appears that the maintenance of a minimal plasma protein concentration is essential to life, and with extreme depletion symptoms similar to shock develop.

Functions of the Proteins. The plasma proteins serve many important functions, one of the most important being the maintenance of the proper **osmotic pressure** between the circulating blood and the tissue spaces, which facilitates the exchange of materials between the blood and the tissue fluid. The total osmotic pressure of the plasma exceeds 6.5 atmospheres, or 4,940 mm Hg, and is due to a combination of inorganic salts, organic materials, and protein. The plasma proteins are responsible for about 25 mm of the total osmotic pressure, with serum albumin playing the most important role because it is the plasma protein with the highest osmotic pressure. The protein of the tissue fluid exerts an osmotic pressure of about 10 mm Hg, which leaves an effective pressure of the blood over that of the tissue fluid of 15 mm Hg (25 − 10). This pressure has the effect of drawing materials from the tissue fluid into the blood. Opposing this force is the pressure

[1] G. H. Whipple, C. W. Hooper, and F. S. Robscheit, "Blood Regeneration Following Simple Anemia." *Am. J. Physiol.,* **53:**151, 167 (1920).

[2] P. F. Hahn and G. H. Whipple, "Hemoglobin Production in Anemia Limited by Low Protein Intake." *J. Exp. Med.,* **69:**315 (1939).

Figure 8-2. *Forces that determine distribution of fluid between blood and tissue spaces.*

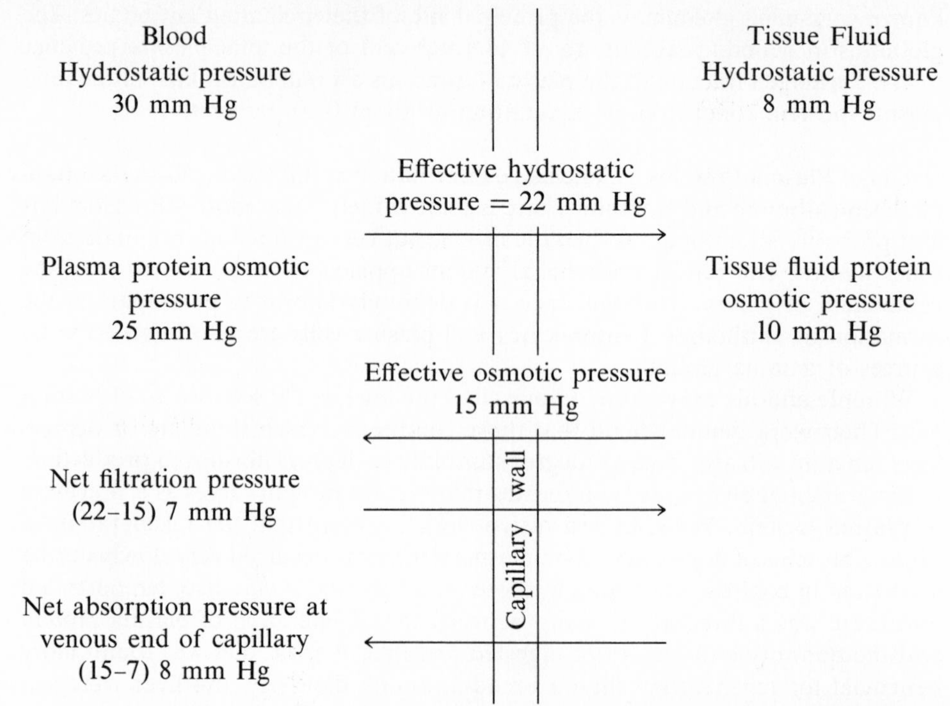

Figure 8-3. *Schematic representation of differences in hydrostatic and osmotic pressure.*

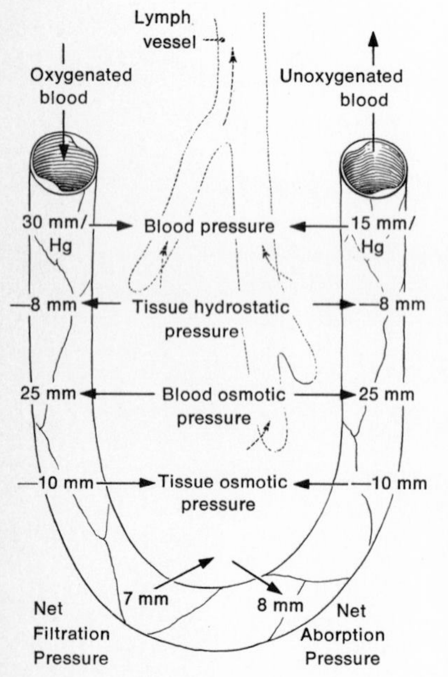

of the blood, which tends to force materials out of the circulation into the tissue fluid. This blood pressure, or hydrostatic pressure, amounts to about 30 mm Hg in the arteriole and drops to about 15 in the venule. The tissue fluid has a hydrostatic pressure of 8 mm Hg. The effective hydrostatic pressure at the beginning of the capillary is about 22 mm Hg (30 − 8), and at the venous end it is about 7 mm Hg (15 − 8). Therefore, on the arterial side the net result of the opposing hydrostatic and osmotic pressure is a 7 mm Hg (22 − 15) excess of **hydrostatic pressure,** which favors filtration of materials from the capillary to the tissue fluid. At the venous end of the capillary the **osmotic pressure** is in excess of about 8 mm Hg (15 − 7), which favors the absorption of material from the tissue fluid into the circulation. This mechanism, which accounts for the exchange of materials between the blood in the capillaries and the tissue fluid, is referred to as **Starling's hypothesis.**[3] Figures 8-2 and 8-3 schematically represent this exchange.

Any alteration of the balance between osmotic and hydrostatic pressure described above could result in the accumulation of fluid in the tissue space—a condition known as **edema.** This fluid can accumulate because of a decrease in plasma proteins or an increase in venous pressure.

The plasma proteins are also important in contributing to the viscosity of the blood, which is a factor, as we have already seen, in the maintenance of normal

[3] W. Zweifach and M. Intaglietta, "Fluid Exchange Across the Blood Capillary Interface." *Federation Proc.,* **25:**1784 (1966).

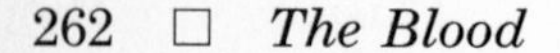

blood pressure. Another important function of the plasma proteins is their ability to assist in the **regulation of the pH** of the blood as a result of their amphoteric character, that is, their ability to act as both an acid and a base. At the normal pH of the blood the proteins act as an acid and combine with alkaline ions, mainly sodium.

Other functions of the plasma proteins include the action of fibrinogen in the **clotting mechanism;** the ability of the globulins to form specific **immunological structures** (antibodies) that are capable of reacting against organisms, their toxic products, and other chemical substances; the **transport of lipids,** including fat-soluble vitamins and steroid hormones and certain carbohydrates; and, finally, the function of the plasma proteins as a **food material** on which the body can draw in protein starvation.

Nonprotein-Nitrogen Substances in Plasma

Nonprotein nitrogen (NPN) is a term given to those substances in the plasma that contain nitrogen in their molecule but are not proteins. They include such compounds as urea, amino acids, creatinine, and uric acid. The remainder of the nitrogen present in blood after accounting for these substances is termed **rest nitrogen.** The total normal NPN amounts to 32 mg per 100 ml of plasma, which includes 19 mg of urea, 6 mg of amino acid, 1.2 mg of creatinine, 4 mg of uric acid, and approximately 2.1 mg of rest nitrogen.

Urea is the end product of protein catabolism and an excretory product in man, although in some lower forms of animal life, such as the elasmobranch fish, it has a definite function in maintaining normal heart action. Urea is a natural diuretic and has been used as a therapeutic agent in man in cases of edema.

The amino acids are the essential components of proteins and are formed when proteins are digested. These pass directly into the blood stream, where they are converted mainly by the liver into other products for immediate oxidation or storage.

Uric acid is the end product of nucleic acid metabolism and is filtered by the glomeruli of the kidney. It is partly reabsorbed by the renal tubules, so that in small quantities it is a normal constituent of blood. Blood uric acid is increased in gout, in certain cases of nephritis, and in leukemia.

Creatinine is an end product of creatine metabolism. A loss of water from the creatine molecule yields creatinine, a compound found in both blood and urine. The 24-hour excretion of creatinine in the urine of a given person is remarkably constant. An increase in the creatinine content of the plasma above normal is always indicative of kidney malfunction.

Inorganic Salts of Plasma. The salts of sodium, potassium, calcium, and magnesium are all found in plasma, with the salts of sodium being present in the highest concentration. Schmidt found from 400 to 500 mg of sodium in 100 ml of plasma. This included sodium chloride, sodium bicarbonate, sodium acid phosphate, and sodium alkaline phosphate. In addition to the basic ions listed above, the acid ions included carbonate, chloride, phosphate, and sulfate.

The functions of the above salts include their participation in the maintenance of the proper osmotic pressure, in the maintenance of the proper pH of the blood, and in the maintenance of the proper physiological balance between the tissues and the blood. Calcium ion is important as a catalytic agent in the clotting process.

Blood Sugar. Three types of sugar are found in the plasma. These are (1) the free, or crystalloid, (2) the loosely combined, or colloid, and (3) the firmly bound. The colloid sugar is believed to be in equilibrium with free sugar and to be easily converted into that form, and the firmly bound sugar is transformed into free sugar by enzyme activity.

The free sugar of the blood is generally considered to be glucose, $C_6H_{12}O_6$. The structural formula may be written

$$
\begin{array}{c}
\mathrm{C{=}O} \\
|\;\;\diagdown\;\mathrm{H} \\
\mathrm{H{-}C{-}OH} \\
| \\
\mathrm{HO{-}C{-}H} \\
| \\
\mathrm{H{-}C{-}OH} \\
| \\
\mathrm{H{-}C{-}OH} \\
| \\
\mathrm{CH_2OH}
\end{array}
$$

The normal values for blood sugar range from 70 to 120 mg per 100 ml of blood. The blood sugar is derived from carbohydrate and protein foods and from the glycerol fraction of fat. Most of the sugar does not remain in the blood but is temporarily stored as glycogen in the liver and muscles and as fat in the fat depots. The glycogen of the liver is converted into sugar and pushed into the circulation as it is needed by the body. This process, called **glycogenolysis,** is under the control of the hormone epinephrine and the sympathetic nervous system. The glycogen of the muscle probably does not contribute to the blood sugar, since it is used specifically for muscle contraction. As the cells of the body utilize the glucose, more sugar diffuses in from the tissue fluid, causing a lowering of the blood sugar. This is immediately brought up to the normal concentration by glycogenolysis. In healthy persons, a constant concentration of glucose is maintained by this mechanism. This complicated procedure will be discussed more thoroughly in Chapter 12.

Blood Lipids. The most important lipids of the blood are fats, phospholipids, and cholesterol. The **fats,** which are triglycerides of fatty acid, are of two basic structural types in the blood: (1) simple triglycerides, in which the fatty acid is the same in all three positions, and (2) mixed triglycerides, in which the fatty acids are different. The fatty acids contained in the fat of blood are largely the fatty acids of the food fats, varying from short-chain 4-carbon saturated acids, such as butyric acid of butter, to the 20-carbon arachidic acid of peanut oil. The fat appears in the blood in the form of finely divided particles about 1 μm in diameter. These fat particles are called **chylomicrons** and are present in the blood under ordinary conditions in small concentrations, their amounts increasing at the time of fat absorption through the intestinal tract. The fat in the blood is presumably food fat on its way to utilization or storage.

Phospholipids. The **phospholipids** are fats in which one fatty acid has been replaced by a phosphoric acid–base group. There are two types of phospholipids in human blood, and they appear to be present in equal amounts. They are **lecithin** and **cephalin.** Their structure may be represented as follows:

$CH_2{-}O{-}O{-}R$
$|$
$CH{-}O{-}O{-}R_1$
$|\qquad\qquad\quad O$
$CH_2{-}O{-}O{-}P{-}OH$
$\qquad\qquad\quad O{-}C_2H_4$
$\qquad\qquad OH{-}N{\equiv}(CH_3)_3$

Lecithin

$CH_2{-}O{-}O{-}R$
$|$
$CH{-}O{-}O{-}R$
$|\qquad\qquad\quad O$
$CH_2{-}O{-}O{-}P{-}OH$
$\qquad\qquad\quad O{-}C_2H_4 \cdot NH_2$

Cephalin

The basic group in lecithin is choline and in cephalin, amino ethyl alcohol:

$C_2H_4{-}OH$
$|$
$N{\equiv}(CH_3)_3$
$|$
OH

Choline

$CH_2{-}OH$
$|$
$CH_2 \cdot NH_2$

Amino ethyl alcohol

The phospholipids function primarily in the transport and utilization of fats in the body. During absorption of fat from the intestine, an increase in phospholipid content of the plasma has been found by all workers and appears to be definitely related to the incoming fat, since the composition of the blood lecithin changes with the fat absorbed. Studies of phospholipid formation and disappearance with radioactive phosphorus as a label show that the liver is virtually the only organ involved in synthesis and removal of blood phospholipids.

Cholesterol. **Cholesterol** belongs to the terpene group of compounds and is widely distributed in all cells of the body, particularly in nervous tissue. Cholesterol occurs only in animal fats; it is not found in plant fats. The formula for cholesterol has been satisfactorily worked out and is as follows:

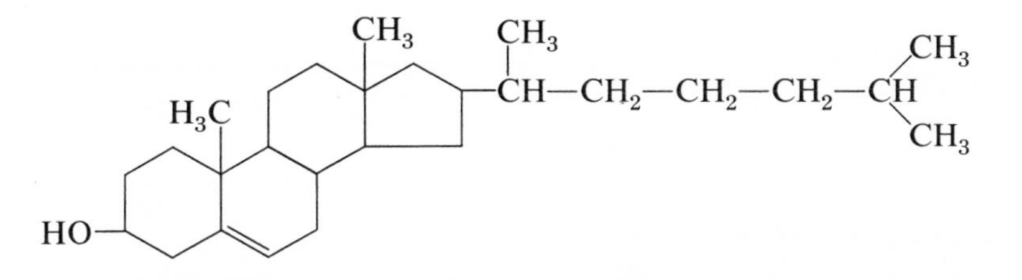

Egg yolk is the chief source of cholesterol taken in by way of the diet. The fats of meat, liver, and brain are also good sources. Cholesterol is easily absorbed from the intestine and appears in the blood both in the free form and combined as esters with fatty acids. Cholesterol can also be synthesized by the body. The liver is the principal organ for this synthesis, but other tissues, such as the adrenal cortex, skin, intestine, and testis, have also been shown to be involved in its production.

Little is known of the exact function of cholesterol in the blood or, for that matter, in the tissues. It appears that its main importance is its role as a physiological antagonist to the phospholipids. For example, cholesterol prevents hemolysis of the red cells, whereas lecithin promotes it. Cholesterol promotes water-and-oil emulsions, whereas the phospholipids promote the oil-in-water type. It also appears that cholesterol inhibits the settling out of the cells of the blood since it is electronegative and is highly absorbed to the surface of the blood cells.

The total concentration of cholesterol in the blood plasma ranges between 220 and 310 mg per 100 ml of blood. Cholesterol is apparently not burned by the body, so an excess must be excreted. The liver not only acts to synthesize cholesterol, but also excretes it in the bile as cholesterol or as cholic acid. Cholesterol may also be excreted directly through the intestines. Since this excretory mechanism is very efficient, it is difficult to produce an excess in the blood by diet, although if the cholesterol intake is great enough and persists long enough, high values may be produced. This excess cholesterol may be deposited in the connective tissue of the walls of arteries, producing the condition known as **atherosclerosis.** Experimental evidence supports the concept that atherosclerosis is a result of alteration in cholesterol metabolism. It is evident that this condition occurs more frequently in persons who subsist mainly on diets rich in animal fat and those who have diseases that are characterized by elevation of cholesterol levels, such as uncontrolled diabetes, hypothyroidism, and cirrhosis of the liver.

Formed Elements of the Blood

Forty-five per cent of the whole blood is composed of a solid fraction referred to as the formed elements. These elements include the **erythrocytes,** or red blood cells, the **leukocytes,** or white blood cells, and the **thrombocytes,** or blood platelets.

Erythrocytes

The red blood cells (Fig. 8-4) appear to be disk-shaped elements having a mean diameter of 7.2 μm and a thickness of 2.2 μm. They are also biconcave, giving the cell the advantage of equal and rapid diffusion of oxygen to its interior and a relatively large surface area for the absorption of gases. The adult cell contains no nucleus, having lost it during the maturation process in the red bone marrow. The cell membrane consists mostly of protein and lipid material. The protein is a combination of paraglobulins and nucleoproteins, and the lipid content is a mixture of lecithin and cholesterol.

The interior protoplasmic mass is referred to as the **stroma** of the cell and is made up of the same materials found in the membrane, or of materials similar to them. The red pigment hemoglobin is bound up in the stroma substance and makes up about 33 per cent of the cell volume. The remainder of the cell is made up of 2 per cent protein and fat and about 65 per cent water.

Erythropoiesis

Formation. In the embryo, the red blood cells are formed in the liver and spleen, and in the adult, they are formed, under normal conditions, in the bone marrow. The **hemopoietic function** of the bone marrow is limited to the vascular sinuses, which are so numerous in this structure. These sinuses have walls, like capillaries, and are similar to capillaries when collapsed, but when open, they have the diameter and lumen of large veins. The sinuses that are closed to the circulation at any one time are the actual site of formation of red cells. These sinuses run transversely through the marrow and are connected to capillaries that lead from arteries running longitudinally in the center of the marrow. The sinuses also lead to veins accompanying the arteries.

Thus the red blood cells are formed intravascularly in the marrow. The primitive red cell, or **erythroblast,** after being developed from the endothelium lining the sinuses, remains attached to the endothelium, giving rise to the older forms, normoblasts, reticulated red blood cells, and mature erythrocytes, which are found in the lumen of the sinus in groups or masses. The mature forms lose the property of adhesiveness and separate, at which time they are expelled into the circulation. Normoblasts and reticulocytes are thrown into the blood stream at regular intervals (Fig. 8-5).

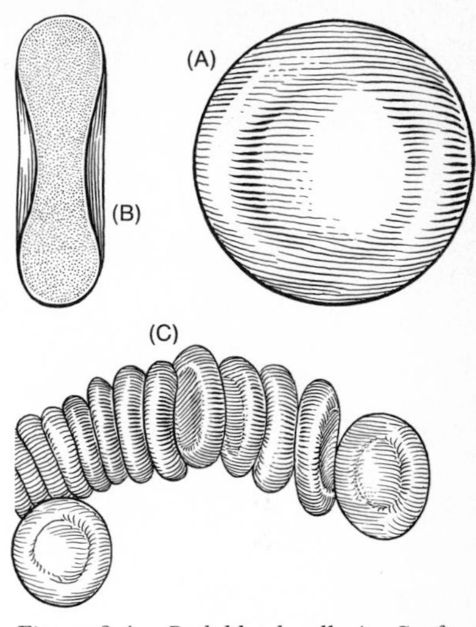

Figure 8-4. Red blood cell. A. *Surface view.* B. *End view—cell cut in half to expose stroma.* C. *Rouleau formation.*

Many factors can influence the formation of red blood cells. A great increase in the number of circulating red cells inhibits erythropoiesis. It is stimulated by anemia and also by low oxygen tension of the blood (hypoxia). It is belived that a hormone secreted by the kidneys called **erythropoietin** controls the production and release of red blood cells from the bone marrow. This hormone, which is a glycoprotein, also enhances hemoglobin synthesis. The formation of red blood cells appears to be subject to a feedback control.

Number of Cells. The average number of red cells in man is usually given as 5,000,000 per cubic millimeter of blood for males and 4,500,000 for females (Table 8-2). Normal values may be somewhat higher, and 6,000,000 is not unusual for a male. The cell count also varies about 1,000,000 per cubic millimeter per day. The high point of bone marrow delivery is in the morning and afternoon.

Figure 8-5. Stages in the formation of red blood cells.

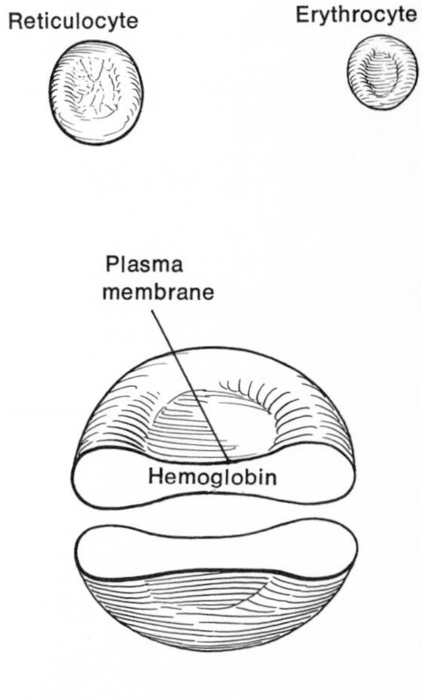

At birth the count is extremely high in the first 24-hour period, reaching 7,000,000 to 8,000,000 but dropping off to 5,000,000 within a few days. There are other factors that influence the count. In the terminal stages of pregnancy there is a drop of approximately 1,000,000; exercise is followed by a slight increase; emotional states also increase the red cell count. There is a slight increase after ingestion of a full meal. A rise in environmental temperature increases the count, and in cold weather or cold climates there is a moderate decrease. Altitude affects the count so that for each 1,000-ft rise of elevation, the lowering of the barometric pressure increases the red cell count approximately 50,000 cells. An increase of barometric pressure causes a decrease in the number of red cells. After a severe hemorrhage there is a decrease caused by loss of blood; after repeated small hemorrhages the count is elevated as a result of the stimulation of the bone marrow to greater activity.

Anemias. Pathological reduction in the number of red blood cells or the amount of hemoglobin is referred to as anemia. If the anemia is caused by disease processes that are limited to the organs and functions concerned with blood formation and destruction, it is called a **primary anemia.** If the anemia is a result

Table 8-2. ***Blood Components (Normal Values)***

Erythrocytes	
Males	5,000,000/mm^3; ±700,000
Females	4,500,000/mm^3; ±500,000
Reticulocytes	0.8%
Leukocytes	6000–10,000/mm^3
Neutrophils	60–65%
Eosinophils	2–4%
Basophils	0.5–1.0%
Lymphocytes	20–35%
Monocytes	3–8%
Platelets—thrombocytes	200,000–300,000/mm^3
Hemoglobin	14–16 g/100 ml; average, 15 g
Hematocrit (packed cell volume)	
Males	47%
Females	42%
Clotting time	5–15 min
Bleeding time	1–4 min
Sedimentation rate	Male: 0–12 mm/hr; female: 0–20 mm/hr
pH	7.35–7.45
Viscosity	5 times as viscous as water
O_2 carrying capacity	20 cc/100 ml

of disease processes of organs and functions that are not directly concerned with blood formation and destruction and is therefore purely a by-product or symptom, it is designated as a **secondary anemia.** The anemias are also classified into distinct types. Some of the more important and common ones are the following:

1. *Hemorrhagic anemia.* The hemorrhagic anemias are a result of a loss of blood and are divided into **acute** and **chronic** types. This type of anemia may be extremely mild or very severe, but certain characteristics are always found on blood examination. Red cells are reduced to any degree, and the hemoglobin concentration is also reduced in proportion to the reduction of red cells. Therefore, the color index is low, and this is one of the chief characteristics of this type of anemia.
2. *Hemolytic anemia.* In hemolytic anemia the red cells are destroyed, usually by some substance present in the blood stream or by action of hemolytic organisms. Malarial organisms, coal-tar products, and hemolytic substances the body produces are the main causative agents. Red cells are reduced to the same degree as hemoglobin. The number of reticulocytes is markedly increased, and fragility of the red cells is increased, this being one of the most important features of the blood picture. Color index may be normal or low.
3. *Aplastic anemia.* Aplastic anemia is a result of abnormalities in the blood-cell-forming mechanism. The activity of the red bone marrow is reduced by such factors as overexposure to X-rays or radioactive substances and poisoning with benzene or bacterial toxins. In most instances the bone marrow is replaced by fatty and fibrous tissue. The hemoglobin concentration is greatly reduced, as is the red cell count. Total count may not be more than 1,000,000 per cubic millimeter. The volume and color index may be normal or low.
4. *Pernicious anemia.* Pernicious anemia results from a defect in the formation of the

red cell. Castle and Minot[4] have demonstrated that defective formation of the red blood cell is due to the absence of a substance formed in the stomach known as the **intrinsic factor.** This factor acts on another factor, the **extrinsic factor,** which is derived from certain foods, producing the active liver substance known as the **antianemic factor,** or **erythrocyte-maturing factor (EMF).** This substance, formed in the stomach, is absorbed and modified in the intestine and stored in the liver to be delivered to the bone marrow when needed. The extrinsic factor is thought to be vitamin B_{12}. The absorption of vitamin B_{12} is dependent on the presence of the intrinsic factor, which is a combination of hydrochloric acid and a substance that appears to be a constituent of the gastric mucoprotein. This substance is found in the fundus and cardia of the stomach but not in the pyloric portion. Atrophy of the fundus and a lack of free hydrochloric acid are usually associated with the occurrence of pernicious anemia.

Typical cases of pernicious anemia may be recognized by the blood picture alone. The red cell count and hemoglobin count are low, but this is not considered diagnostic. The high volume index and color index, as well as the change in the size of the red cell from a normal 7.2 μm in diameter to a large 9.4 to 14 μm in diameter (macrocytosis), are essential for diagnosis. Minot and Murphy[5] first demonstrated the therapeutic effect of feeding liver to patients with pernicious anemia. Feeding such patients approximately $\frac{1}{2}$ lb of liver daily results in prompt and striking improvement within 3 to 5 days. Minot and his collaborators have also prepared a concentrated liver extract, which is highly effective and can be taken in one-half glass of water, orange juice, or tomato juice in place of the liver.

5. *Iron-deficiency anemia.* Iron deficiency causes a type of anemia in which the number of cells is not reduced, but their size and hemoglobin concentration are reduced, resulting in hypochromic microcytic cells. In this type of anemia, but not in others, iron treatment gives good results.

Destruction of Red Blood Cells. Red cells become old and must be destroyed. Red cells in circulation become damaged and also must be destroyed. Large cells called **hematophages,** or **macrophages,** in the spleen have the function of digesting these old and damaged red cells. It has been estimated that the duration of life of a red cell is approximately 120 days and that one fifteenth of the total number of cells is destroyed daily. It appears that damage to cells is due to stresses sustained by the cells in passing through the fine capillaries, particularly in contracting muscles.

The damaged cells or fragments and the old cells are taken up not only by phagocytic cells in the spleen, as already mentioned, but also by the **Kupffer cells** in the liver, the reticuloendothelial cells lining the walls of the capillaries, and other tissues of the body that ingest and destroy these cells with the formation of bile pigments. This is accomplished by the breaking of hemoglobin within the phagocytic cells into heme and globin. The heme is further split into iron and the pigment **hematoidin,** which becomes the bilirubin of the plasma. The iron is stored in the form of **hemosiderin** until it is needed for new hemoglobin synthesis. The bilirubin is excreted by the liver in the bile. The mechanism of

[4] W. B. Castle and G. R. Minot, *Pathological Physiology and Clinical Description of the Anemias.* (New York: Oxford U.P., 1936).

[5] G. R. Minot and W. P. Murphy, "Treatment of Pernicious Anemia by a Special Diet." *J. Am. Med. Assoc.,* **87**:470 (1926).

transfer of hemosiderin from the phagocytic cells to the newly developing cells is by the metal-combining beta globulin called **transferrin.**

The Spleen. The **spleen** presents the appearance of a dark-red spongy mass with a well-defined fibrous covering or capsule (Fig. 8-6). The organ is very soft, consisting of a spleen pulp, containing many phagocytic cells and whitish follicles of lymphoid tissue, the **malpighian bodies.** The large arteries supplying the spleen connect with the venous sinuses that deliver blood to the splenic vein and to the portal circulation and liver. The redness of the spleen is due to the abundance of red blood cells contained within it.

The functions of the spleen appear to be the removal from the circulation of damaged or old cells, formation of lymphocytes, and monocytes by the lymphoid follicles, and defensive immune reactions carried out by the lymphocytes and **plasma cells** of the spleen in the production of antibodies. In abnormal states it has been demonstrated that the spleen may remove excessive numbers of platelets from the circulation, causing **thrombocytopenia** (abnormal decrease in platelet count) which results in extensive hemorrhages **(purpura).**

In dogs, because of the large amount of smooth muscle in the capsule of the spleen, the reservoir or contractile function of the spleen is one of its most important. The spleen can be made to contract by many stimuli such as fear, pain, digestion, temperature variations, and exercise. Such contraction discharges extra blood into the circulation; thus blood volume and cellular concentration may be influenced by the action of the spleen. However, this function of the spleen is not important in man. In humans the skin and lungs are more important blood reservoirs and contain a large volume of blood at rest. During stress, when there is diffuse adrenergic discharge, constriction of the vessels in these organs and decreased blood storage in the splanchnic vessels and in the liver increases the volume of circulating blood in the muscles approximately 35 per cent. (The spleen

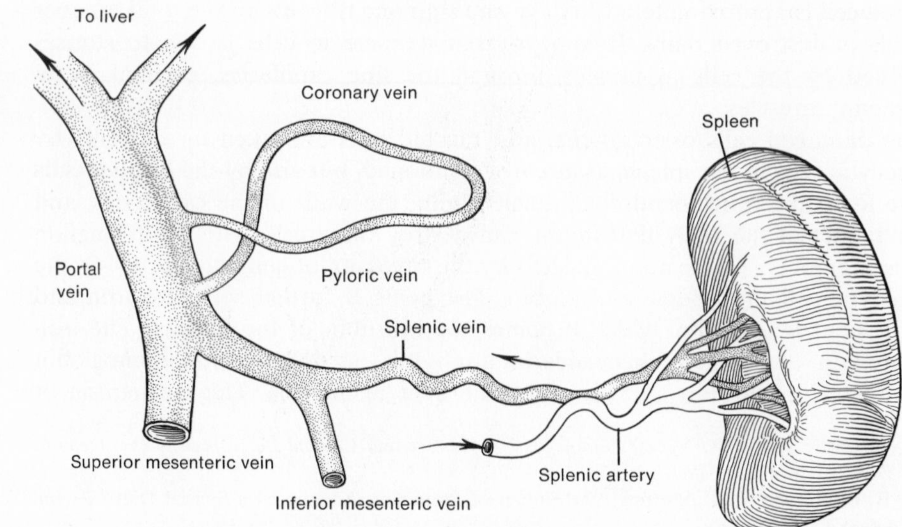

Figure 8-6. The spleen. The large arteries supplying the spleen connect with its venous sinuses, which deliver blood to the splenic vein and to the portal circulation and the liver.

is not essential to life, and in its absence its functions can be carried out by other tissues, such as the bone marrow, the lymph glands, and the liver.)

Polycythemia. **Polycythemia** is the term applied to the condition that exists when more than the normal number of red cells is found in the blood. Increases of more than 2,000,000 or 3,000,000 red blood cells per cubic millimeter above the normal count are considered clinically to indicate polycythemias; these are divided into the primary and secondary types. The primary type is that of unknown cause, and the secondary polycythemias are the result of known causes.

Polycythemias can result from a number of physiological causes. An increase in production of red blood cells caused by overactivity of the red bone marrow is one. A loss of fluid from the body, resulting in the concentration of cells and, therefore, an increase of the number per unit volume is another. Shock, profuse sweating, extreme diuresis, and loss of fluid through the lungs are causative agents of this type. The administration of certain drugs will also cause an increase in the red blood cells per cubic millimeter. Theoretically, polycythemia may be caused by a diminished destruction of red blood cells and also by an increased division of the cells themselves.

White Blood Cells

White blood cells or **leukocytes** (Fig. 8-7) are formed in both the lymph glands and the bone marrow. They are divided into two basic types; those that possess granules in their cytoplasm are called **granulocytes,** and those that are free of granules are referred to as the **agranulocytes.**

The granulocytes are formed in the bone marrow and are divided into three types on the basis of the staining reaction of their granules. These are the **neutrophils,** the granules of which stain with neutral stain, the **eosinophils,** staining with acid stain, and the **basophils,** staining with basic stain. The neutrophils are large cells with deeply staining, imperfectly divided or multilobed nuclei (Fig. 8-8). The cytoplasm is filled with fine lilac-colored granules. These cells constitute 60 to 65 per cent of the white blood cells. The eosinophils are approximately twice the size of normal red cells and have the same type of nucleus as the neutrophils. The cytoplasm is filled with coarse granules that stain a deep-red color with eosin (acid stain). The eosinophils make up about 4 per cent of the white cell count. The basophils (Fig. 8-9) are approximately the same size as the neutrophils. The cytoplasm shows coarse deep-blue or black granules after being stained with a basic stain such as methylene blue. The basophils are not present in any significant numbers in the adult circulation.

It has been demonstrated by radioactive labeling technique that the circulating granulocytic pool comprises only about half of the intravascular neutrophils. Many of the neutrophils are marginated along the walls of small blood vessels where they are in a state of rapid and continuous exchange with circulating cells. It is from these sites that they can be mobilized by exercise or epinephrine injection. From such studies it has also been found that the neutrophils leave the blood in a random manner rather than according to age. The average neutrophil spends 9.5 hours in the blood and probably has a life span of 6 to 9 days. The destruction of neutrophils at sites of infection are obvious; however, normal destruction of

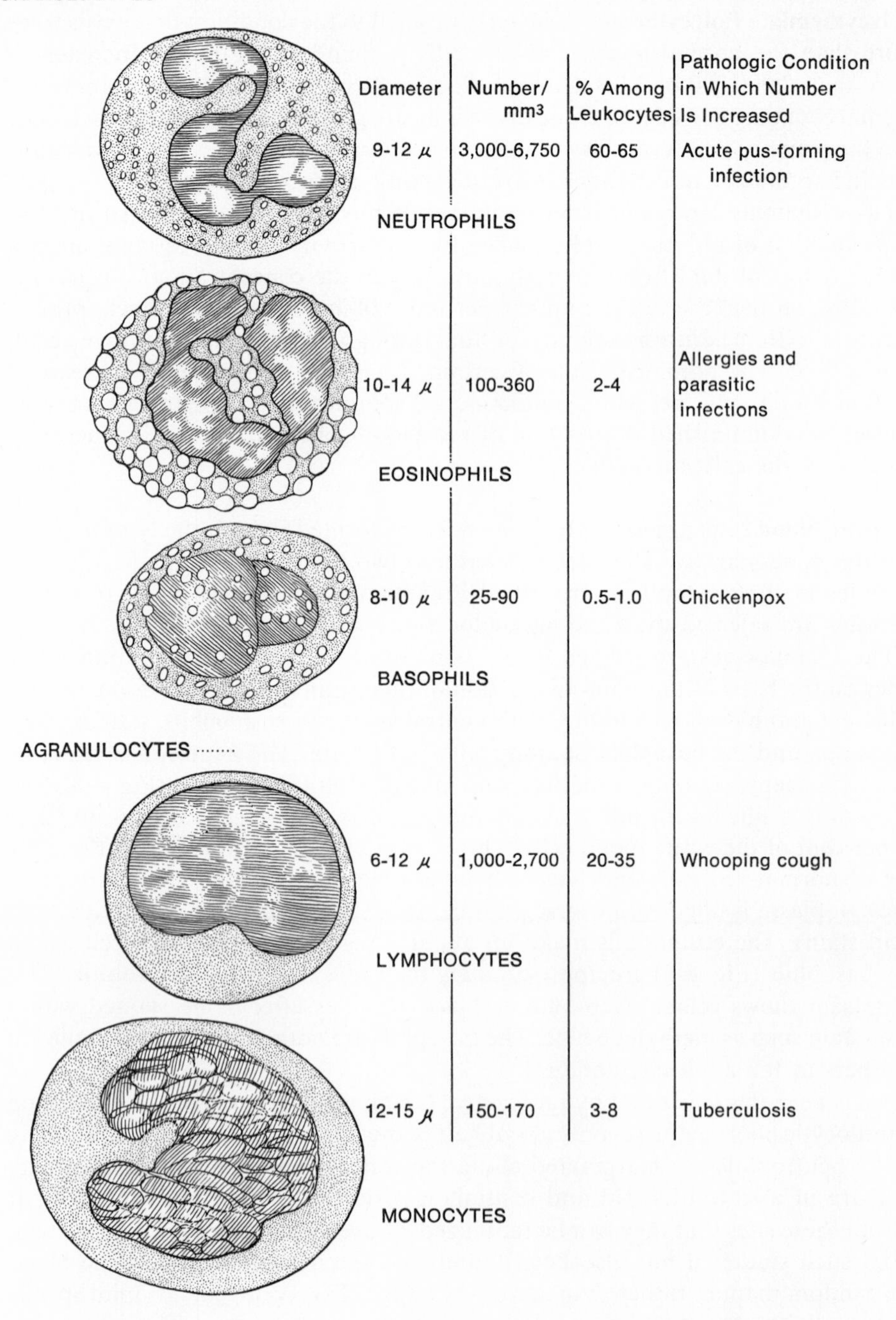

	Diameter	Number/ mm^3	% Among Leukocytes	Pathologic Condition in Which Number Is Increased
NEUTROPHILS	9-12 μ	3,000-6,750	60-65	Acute pus-forming infection
EOSINOPHILS	10-14 μ	100-360	2-4	Allergies and parasitic infections
BASOPHILS	8-10 μ	25-90	0.5-1.0	Chickenpox
LYMPHOCYTES	6-12 μ	1,000-2,700	20-35	Whooping cough
MONOCYTES	12-15 μ	150-170	3-8	Tuberculosis

Figure 8-7. Various types of white blood cells. Neutrophils contain lilac-colored granules. Eosinophils stain a deep-red color. The cytoplasm of basophils contains granules that stain deep blue or black with basic stains. Lymphocytes have a nucleus that stains a deep purple and a cytoplasm pale blue. The nucleus of a monocyte stains a pale blue, its cytoplasm a paler blue.

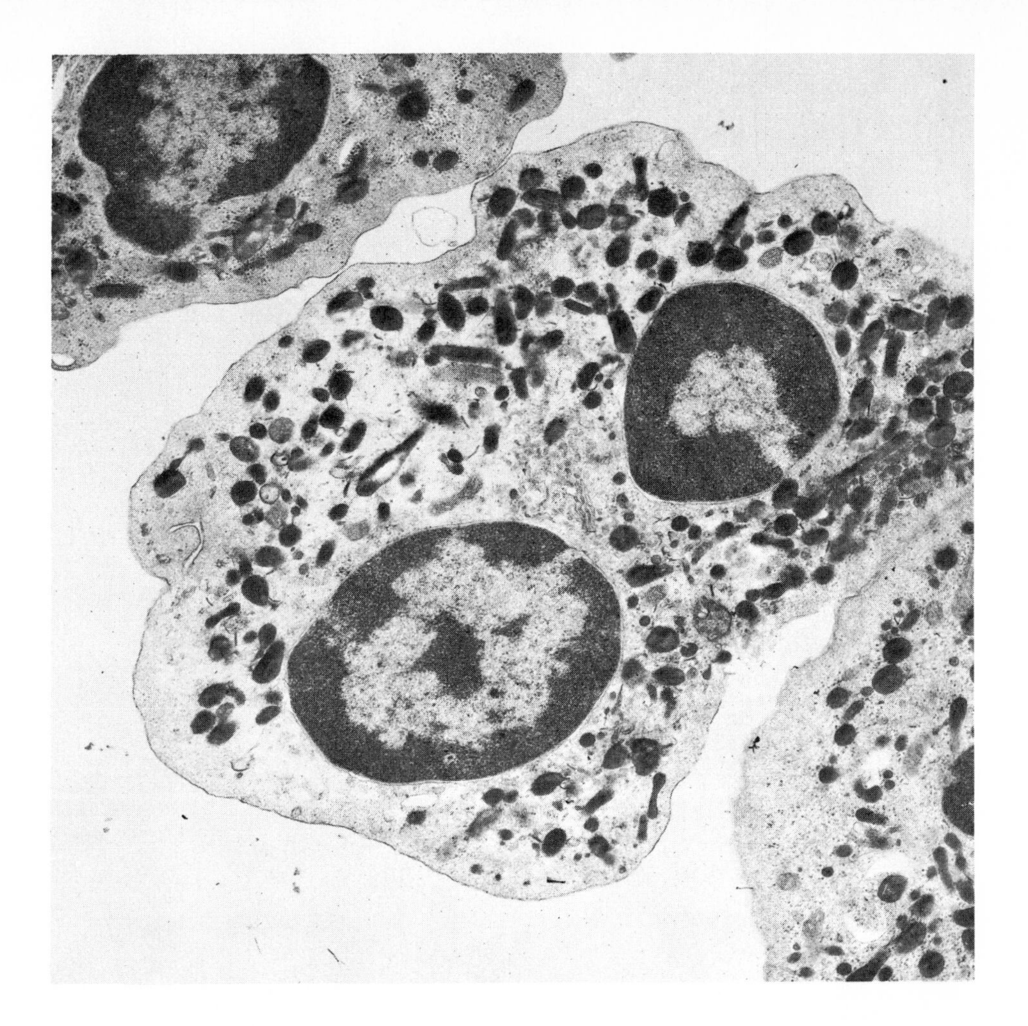

Figure 8-8. Human neutrophil magnified 23,352×. The multilobed nucleus appears as two distinct lobes due to sectioning. (Courtesy Timothy Mougel, Electron Microscopy Laboratory, Department of Zoology, University of Maryland, College Park, Md.)

these cells is not clear. The alveoli of the lungs, spleen, and liver appear to be areas of disposal in the normal organism.

There are three types of agranulocytes: **large lymphocytes, small lymphocytes,** and **monocytes.** The lymphocytes are formed in the lymph glands and in the spleen. Monocytes are formed in the marrow. The lymphocytes vary in size but are typically the size of a normal red cell or slightly larger. The nucleus stains a deep purple with Wright's stain and the cytoplasm a pale blue. Some authorities believe there is no point in differentiating large and small forms. The lymphocytes constitute from 20 to 35 per cent of the white blood cells. The monocytes (Fig. 8-10) are very large cells, being two to three times as large as normal red cells. The nucleus is large and oval and stains a pale blue. The cytoplasm stains a paler blue.

The neutrophils and the monocytes are actively motile and show phagocytic action. They are capable of leaving the blood vessels at will, a process known as **diapedesis,** to migrate to areas of the body being invaded by bacteria or foreign

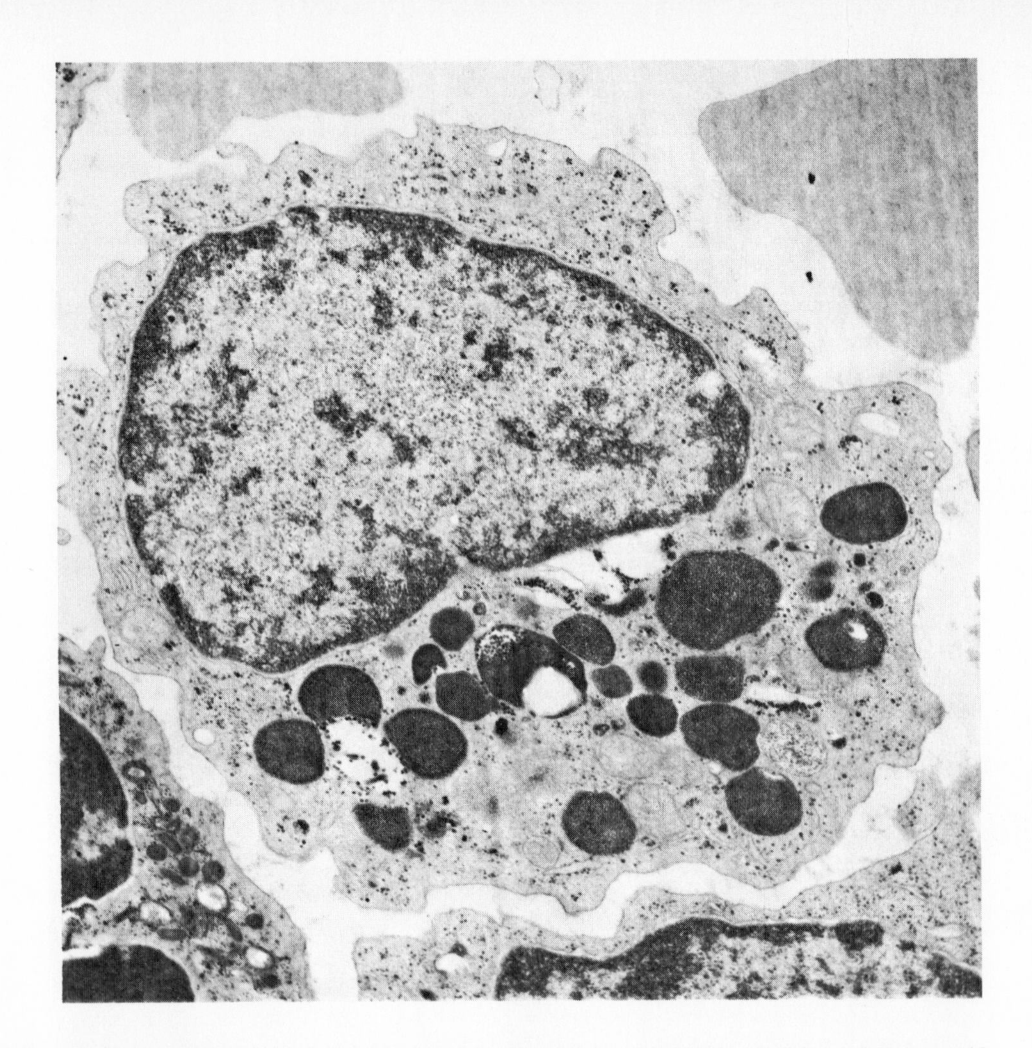

Figure 8-9. Electron micrograph of basophil from human blood (magnification 23,352×). (Courtesy Timothy Mougel Electron Microscopy Laboratory, Department of Zoology, University of Maryland, College Park, Md.)

particulate matter. Surrounding this material in amebalike fashion, they engulf and ultimately digest such particles. The neutrophils are microphagocytes; that is, they ingest bacteria and small particles. The monocytes are macrophagocytes in that they ingest larger particles such as worn-out red blood cells and other cells. Considerable study has been devoted to the mechanism involved in their migrations, and it is believed that some form of chemical attraction is involved.

The lymphocytes are not motile in ordinary preparations; under certain circumstances they have been shown to have ameboid movement, but are not phagocytic. There is strong evidence that the lymphocytes form antibodies and also possibly aid in a protective capacity by neutralizing toxic materials produced by bacteria. The exact function of eosinophils and basophils is not known. In certain diseases of the skin, in parasitic infections, such as trichinosis and hookworm, and in allergic states, such as asthma and hay fever, an increase in the eosinophil count (eosinophilia) is often found. This has led many workers to postulate that the eosinophil is intimately involved in the immune response of

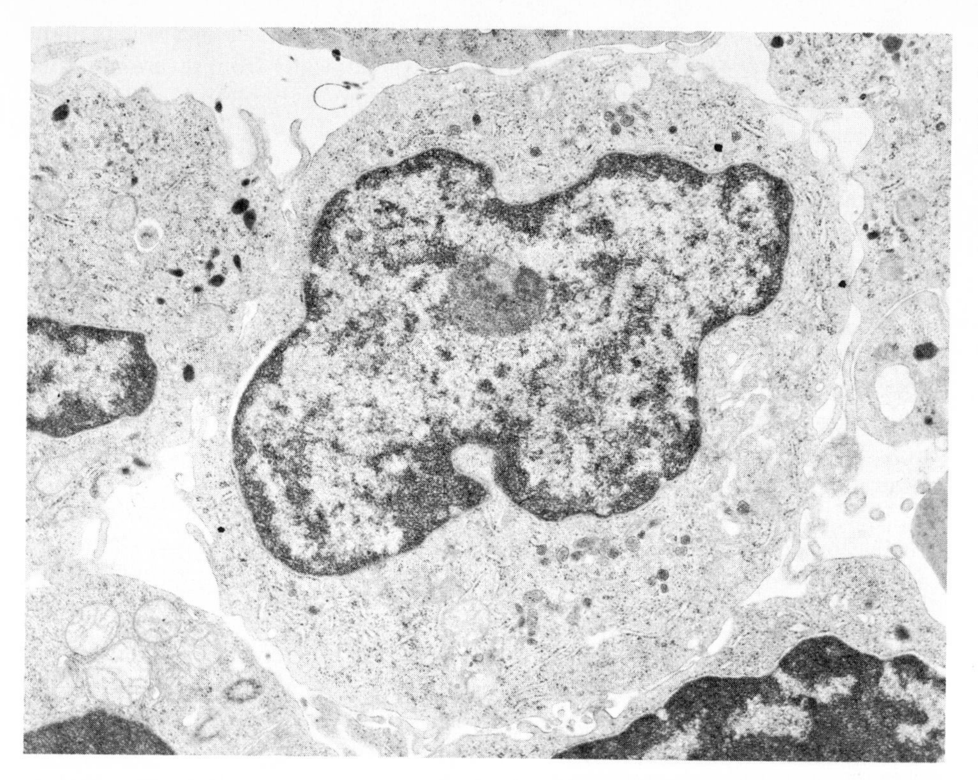

Figure 8-10. Human monocyte magnified 26,400×. (Courtesy Timothy Mougel, Electron Microscopy Laboratory, Department of Zoology, University of Maryland, College Park, Md.)

the body. Since eosinophils are attracted to sites of antigen–antibody reactions and also to sites of histamine and heparin injection (substances known to be present in basophils and mast cells) it has been hypothesized that these two cell types are closely associated with one another at sites of tissue injury.

The liberation of excessive adrenocorticotropic hormone by the pituitary, and cortisone by the adrenals, causes atrophy of the lymphoid tissues and destruction of some of the cells. There is a marked depression in the normal concentration of circulating lymphocytes and eosinophils (lymphopenia) under these conditions, which is also one of the signs of Selye's "alarm reaction."

There is recurring disagreement whether the blood basophil and tissue mast cell are the same cells in different sites or are cells with similar properties but of different origin. The studies of Hartman and his co-workers[6] show that both types of cells convert histidine to histamine and may be more than passive storehouses for this material. They both contain heparin and release histamine and heparin when the cell undergoes degranulation upon stimulation (antigen–antibody reactions). Some interesting studies now being carried on appear to link this degranulation with certain antigen–antibody interactions.

The normal white cell count varies from 6,000 to 10,000 per cubic millimeter. An increase in the number of white cells above 10,000 is referred to as a **leuko-**

[6]W. J. Hartman, W. G. Clark, and S. D. Cyr, "Origin of Blood Histamine." *Proc. Soc. Exptl. Biol. Med.,* **107:**123 (1961).

cytosis, and a decrease below 6,000 is called a **leukopenia.** Leukocytosis is found in most acute infections and also can result physiologically from severe muscular activity, digestion of a meal, emotion, massage, dehydration, and certain drugs. Leukopenia occurs in many diseases, such as typhoid fever, measles, influenza, dengue fever, many of the anemias, and malaria. It also will develop in chronic hemorrhage; malnutrition; lead, mercury, and arsenic poisoning; and also alcohol or morphine poisoning.

Leukemia may be defined as a disease characterized by widespread pathological proliferation of one or more of the white-blood-cell-forming tissues having an invariably fatal outcome. The leukemias are divided into two major types, the **myeloid** and the **lymphoid.** In the myeloid type there is an increase in the granulocyte-forming elements of the bone marrow, which gradually crowd out the red-cell-forming tissue, with the general pathology of anemia developing as an end result. In the lymphoid type, it is the lymphatic tissues that are increased in amount. The lymphatic enlargements include not only the superficial nodes of the neck, armpits, and groin, but also the lymphoid follicles of the nasopharynx, tonsils, tongue, and gastrointestinal tract. Anemia results from a crowding out of red-cell-forming tissue by the lymphocytes.

Blood Platelets

The **blood platelets,** or **thrombocytes** are formed in the bone marrow and appear to be derived from large cells called **megakaryocytes,** which make up about 1 per cent of the cells of the marrow. Fragments of they ytoplasm of the mega-
per cent of the cells of the marrow. Fragments of the cytoplasm of the megakaryocyte become detached and are pushed into the circulation as platelets. The normal count for adults ranges between 200,000 and 300,000 per cubic millimeter (Table 8-2). The blood platelet (Figs. 8-11 and 8-12) exists in the human blood as a definite morphological entity and is a nonnucleated disklike structure having an average diameter of 2 to 3 μm. The thrombocyte plays an important role in blood coagulation.

Blood Clotting

Blood remains fluid in the vessels during life, but when shed, it becomes viscid in a very short time; it continues to thicken until finally the whole mass becomes a complete gel of **fibrin** separated from a fluid portion, the **serum.** The clot itself is composed of a network of microscopic needlelike filaments known as fibrin, within which the cellular components of the blood are entrapped. The clear fluid portion, serum, that escapes from the main mass of material is essentially plasma with its fibrin content removed. Blood contains all the substances that are necessary for clotting, yet clotting does not occur under normal circumstances within the vascular system, but when blood is shed from the body, it is virtually impossible to prevent its coagulation unless chemical anticoagulants are added. This clotting is a protective mechanism whereby the excessive loss of blood from the body after injury can be controlled. Many theories have arisen on the mode of preservation of the fluidity of the blood in the vessels of the living organism and on the changes involved in the phenomenon of clotting of blood outside the body. There is still controversy as to the exact mechanism involved.

Mechanism of Clotting—Fibrin-Stabilizing System

The apparently simple change that occurs when blood clots is in reality a complex series of reactions involving a number of constituents of blood. These constituents are found in normal plasma and are believed to be the following:

1. **Fibrinogen,** a soluble protein of plasma and the precursor of fibrin. It is generally believed that fibrinogen is formed in the liver, as its normal concentration, which is about 0.2 to 0.4 g per 100 ml of plasma, is much reduced if the liver is isolated from the circulation or removed.
2. **Prothrombin,** another protein in the blood believed to be formed in the liver, acting as a precursor of **thrombin,** an enzyme necessary for the conversion of fibrinogen to fibrin. The liver requires vitamin K, a vitamin found in leafy green plants, to produce prothrombin. A vitamin K deficiency can lead to a reduction in the normal concentration of this protein, resulting in a bleeding condition.
3. **Calcium ion,** which is necessary to activate the conversion of prothrombin to thrombin. Substances that precipitate or bind calcium are anticoagulants, as blood does not clot in the absence of calcium. Calcium ion is present in the normal circulation and blood does not normally clot within the vessels; thus it is obvious that calcium is not the only agent necessary to convert prothrombin to thrombin.

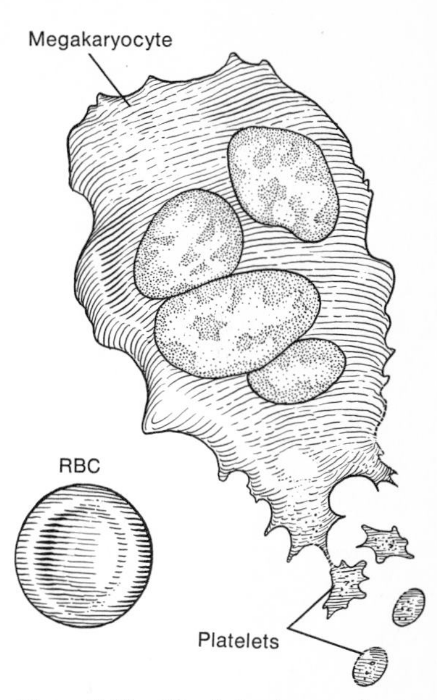

Figure 8-11. Blood platelets are fragments from large cells in the bone marrow called megakaryocytes. A red blood cell is included to show the size relationship of the three types of cells.

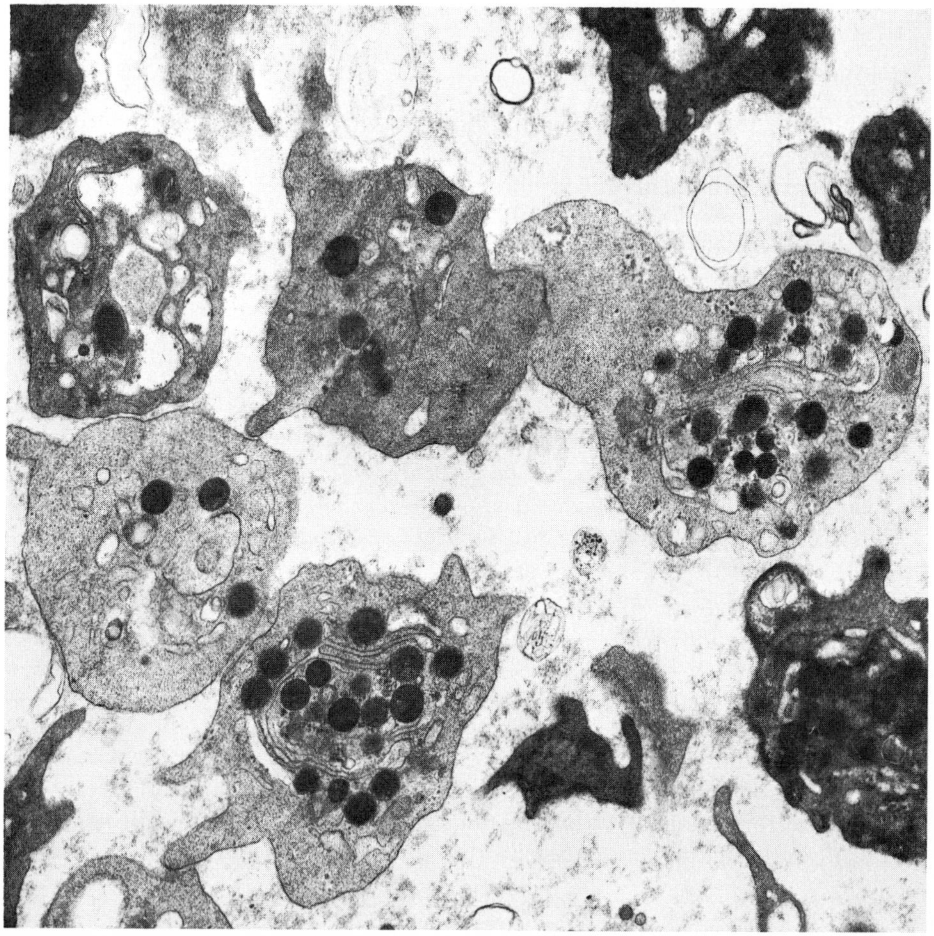

Figure 8-12. Human platelets magnified 29,124×. (Courtesy Timothy Mougel, Electron Microscopy Laboratory, Department of Zoology, University of Maryland, College Park, Md.)

4. **Thromboplastin,** a complex of protein and cephalin, which is not freely found in normal plasma but is contained in the blood platelets. It is also present in most tissue extracts—especially brain and lung tissue. According to widely accepted belief, thromboplastin is released from the platelets on injury and is the active agent that converts prothrombin into the active enzyme thrombin. Recent evidence indicates that thromboplastic activity can be detected in shed blood only when both platelets and an unknown substance called prothrombin A, labile factor, factor V, or Ac-globulin are present. It appears that this unknown plasma, **thromboplastic factor,** acts on the platelets in some way to form thromboplastin.

The coagulation process may be summarized as follows (Fig. 8-13):

(1) Platelets—tissue $\xrightarrow[\text{on injury}]{\text{disintegrate}}$ Inactive thromboplastin

(2) Inactive thromboplastin + thromboplastic factor ⟶ Active thromboplastin

(3) Prothrombin + active thromboplastin + calcium ⟶ Thrombin

(4) Fibrinogen + thrombin ⟶ Fibrinopeptides ⟶ Fibrin aggregate ⟶ Cross-linked fibrin

(5) Fibrin-stabilizing factor + thrombin ⟶ Active FSF (acts on step 4: Fibrin aggregate ⟶ Cross-linked fibrin)

The nature of this reaction is that fibrinogen, a soluble protein, is converted into an insoluble protein, fibrin, by the enzyme thrombin. It does this by hydrolyzing specific peptide fragments called fibrinopeptides from the parent protein. Fibrin, by virtue of its greatly reduced solubility in the physiological environment, forms extensive aggregates which give rise to a gel. The appearance of the first strands signals the onset of the clotting time of blood. These fibrin strands become strongly linked together by covalent bonds, forming a more stabilized clot showing great internal cohesion. This cross linking of fibrin threads appears to be an essential step in the formation of a normal clot which is capable of resisting the pressure of blood and thereby preventing secondary bleeding. It has been demonstrated that stabilization of the fibrin structure by cross links is the last stage of the clotting process and is due to the presence of a *fibrin-stabilizing factor* (*FSF*) in plasma. Activation of this factor is brought about by the enzyme thrombin, which thus serves a dual role in catalyzing two separate reactions which overlap in time.

Maintenance of the Normal Fluidity of Blood in the Body—Fibrinolysis

Some hematologists maintain that the fluidity of blood is maintained, even though materials that initiate clotting are liberated by the natural disintegration of blood platelets within the vessels or when tissue juices are absorbed slowly from wounds, by a specific anticoagulant secreted by the liver. Other researchers maintain that a substance normally present in plasma, which they call **normal antithrombin,** inhibits the action of thrombin that is produced under the circumstances referred to. Others hold that a normal antiprothrombin is present that inhibits the formation of thrombin. This material is identified with **heparin,** a powerful anticoagulant produced by the liver. Pickering and Hewitt[7] maintain

[7] J. W. Pickering and J. A. Hewitt, "Studies of the Coagulation of the Blood. Some Physicochemical Aspects of Coagulation." *Biochem. J.,* **15:**710 (1921).

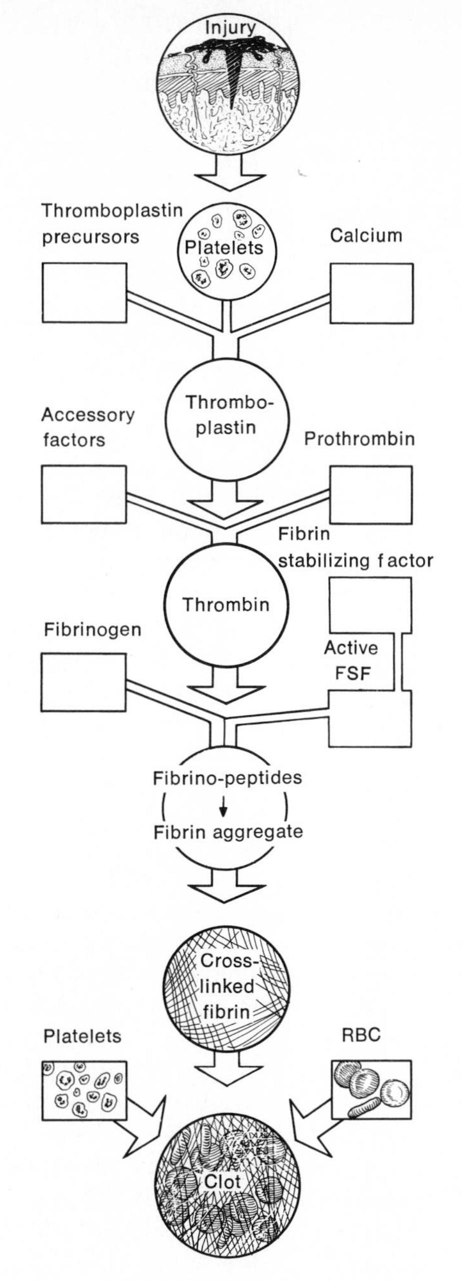

Figure 8-13. Formation of a clot.

that such a substance is probably not in a sufficiently free condition in the blood to control clottability.

A number of investigators suggest that clotting out of fibrin is a continuous process which is in dynamic balance with its dissolution (lysis). The mechanism that is concerned with this lysis of the clot has been referred to as the **fibrinolytic system** and the process as **fibrinolysis.** In plasma is found a substance called

plasminogen, which is secreted in part by the eosinophils. Plasminogen is converted into an active proteolytic enzyme called **plasmin,** which shows a certain selectivity for fibrin. Plasmin adsorbs selectively to fibrin, which is then digested by plasmin, resulting in fibrinolysis. Most tissues contain activators of plasminogen called **fibrinokinases.** The liver appears to be able to neutralize circulating plasminogen activators. It also produces a component normally present in the plasma known as **antiplasmin.** Antiplasmin and fibrin compete for plasmin, with fibrin having a somewhat higher affinity. It is believed that a thin layer of fibrin is continuously present on the endothelial lining of the blood vessels as a result of the balance that exists between fibrin formation and its lysis. An overbalancing fibrinolysin system may lead to hemorrhage, while an overbalancing coagulation system may result in a thickened fibrin meshwork that may trap large molecular-sized lipoproteins, particularly in the presence of lipemia and hypertension, thus initiating the formation of atheromatous plaques. The most important function of this system may be in the removal of intravascular thrombi and the dissolution of fibrin clots within the ducts of organs. The presence of plasminogen activator enzyme has been found in liquid excreta such as urine, tears, and milk, which probably aids in keeping channels of excretion open. Hemorrhage into these areas would result in the activation of the fibrinolytic system.

Fibrinolytic hemorrhage is becoming an increasingly recognized clinical syndrome. Extensive liberation of plasminogen activators from damaged tissue is most often involved. In liver disease there is a decreased production of antiplasmin and an impaired ability to neutralize plasminogen activators. Since inhibitors of the fibrinolytic system are now available, rapid diagnosis of fibrinolytic hemorrhage is of great importance.

Anticoagulants

The blood can ordinarily be maintained in a fluid state outside the body by addition of substances that interfere with the normal clotting process, **anticoagulants.** Compounds that precipitate or bind calcium will prevent clotting by interfering with the conversion of prothrombin into active thrombin. The citrates and oxalates of sodium and potassium are such agents and are commonly used in the laboratory to prevent coagulation. The conversion of fibrinogen to fibrin by thrombin can be prevented by a number of substances, such as heparin and Dicumarol. As we have seen, heparin is produced by the liver and by the mast cells in connective tissue about vessels. It is also present in other tissues of the body, particularly the lung, which is the main commercial source. Heparin requires the presence of the plasma thromboplatic factor to inhibit the conversion of prothrombin to thrombin and also to inhibit the action of thrombin, for in its absence heparin has no anticoagulant effect. Dicumarol is a substance derived from spoiled clover. It is not an anticoagulant in itself, but acts indirectly by inhibiting prothrombin formation in the body by acting as an antagonist to vitamin K. The addition of Dicumarol to blood outside the body will not prevent clotting.

Clotting Time

Clotting time is usually measured by collecting blood from a vein into Pyrex glass tubes 8 mm in diameter. One milliliter of blood is put into each of three tubes, and these are submerged in a bath at 37° C. Every 30 seconds the tubes are examined for the formation of a clot. The point at which the clot adheres to the walls of the tube is considered the end point. The clotting process is started when the platelets break up in coming in contact with the glass. The average time normally varies from 5 to 15 minutes. When platelets do not break up, as in hemophilia, the time is prolonged.

Bleeding Time

Bleeding time is the time required for cessation of bleeding from a small skin puncture. It is usually determined by puncturing the lobe of the ear with a lancet or needle; the blood droplets that form are dried by gently touching the wound with filter paper. When the paper is no longer stained, the bleeding has ceased. Normal bleeding time varies from 1 to 4 minutes. Whereas coagulation time involves only the fragility of platelets, bleeding time involves several factors: contraction of injured blood vessels, production of thrombokinase by injured tissue, and intravascular clotting and adhesion to the vessel wall. Discrepancies may result between the two tests on the same person; for example, in hemophilia the clotting time is extremely prolonged, but the bleeding time may be normal. The reason for this is that the response of a hemophiliac person to the pricking of an earlobe may be normal, owing to normal contraction of vessels, with clotting being produced by the thrombokinase of injured cells.

Hemophilia

Hemophilia is a condition in which abnormal bleeding occurs from wounds because of delayed coagulation of blood. It is an inherited trait that is limited to males and is transmitted as a sex-linked recessive by females. Blood from these persons clots very slowly, and the clot is quite fragile. The blood platelet count is normal or even slightly in excess of normal. If thromboplastin is added to such blood, it clots promptly, indicating that the defect may be associated with a lack of this substance in the blood platelets. It has also been found that normal platelet-free plasma corrects the clotting defect of hemophilic blood. Hemophilia also appears to be a disease associated with a deficiency of the plasma thromboplastic factor.

Blood Groups

Prior to the year 1901 attempts at transfusion of blood from one person to another often had disastrous results. No explanation for this phenomenon was available until 1901, when Landsteiner observed that the serum of normal persons contained **agglutinins** for the cells of other human beings. Agglutinins are those antibodies in a serum that cause clumping (agglutination) of the blood cells of individuals of the same species. Landsteiner also found that, in order to be susceptible to the agglutinins, the red cells must contain **agglutinogens** (antigens), with which the agglutinins react. The experimental facts indicated that two types of agglutinogens are found in human red cells; these were labeled A and B. On the basis of these two fundamental types, it was found that the reactions of serum and cells of human beings fall into four groups. In some blood the cells contain agglutinogen A, in others they contain B, in others A and B together, and in still others no agglutinogens are present. The four blood groups were designated A, B, AB, and O. It was found that in the blood of persons containing A agglutinogen, the sera contained agglutinin for B; in that of B type, the agglutinin

A was present. In those types containing A and B agglutinogen, no agglutinins were present; in the blood containing no agglutinogens, O, the sera contained the agglutinins for both A and B. In other words, an individual contains in his serum the agglutinin for the agglutinogens that are absent. Table 8-3 lists the four types of blood, the agglutinins in the serum, and the types of reactions that are observed when the sera and cells of numerous persons are mixed. The relative frequency of the various blood groups in the white population of America is also recorded.

Table 8-3. ***Blood Types***

Blood Group Agglutinogen on Red Cell	*Per Cent*	*Agglutinin in Serum*	*Blood Cells of Group*			
			O	*A*	*B*	*AB*
O (no agglutinogen)	40–45	ab	–	+	+	+
A	40	b	–	–	+	+
B	10–15	a	–	+	–	+
AB	5	O	–	–	–	–

In blood transfusions it is always desirable to use a donor of the same group as the recipient. In emergencies any donor can be used whose cells are not agglutinated by the serum of the recipient. Since the donor cells (usually 1 pint) are subjected to the entire volume of the recipient's serum, a reaction in which the donor cells would clump together in large masses, blocking the circulation in certain capillaries of the body, would be expected if the recipient's serum was of the type that could agglutinate the donor's cells. In addition, toxic products liberated by hemolysis of clumping cells result in severe damage. On the other hand, the serum of such a donor is diluted by the greater volume of the recipient's blood, and the cells of the recipient are not likely to be affected by the agglutinins in the donor's serum. Persons of blood group O are called universal donors because their cells, when injected into another person, will not agglutinate for lack of agglutinogen. Similarly, persons of blood group AB are called universal recipients since their serum, lacking any agglutinins, will not agglutinate the cells of any donor.

Determination of Human Blood Groups

The determination of blood groups can readily be carried out by the use of only two sera of blood groups A and B. Human grouping sera are obtained from persons of known groups A and B. Groups A and B are collected separately to avoid error and are allowed to clot. The serum is transferred to sterile containers and stored in small amounts in sealed ampules at 7° C.

The slide method is the one most frequently used in blood grouping and consists of dividing a glass slide in halves by a wax pencil (Fig. 8-14). The left side is marked *A* and the right side *B*. A drop of serum *A* is placed on the left and a drop of *B* on the right. A suspension of red cells of the person being grouped is prepared by adding one drop of blood to a test tube containing 2 ml of physiological saline, and a drop of this suspension is mixed with each of the sera on the slide. The reaction is hastened by mixing thoroughly, tilting the slide several times. The reactions that occur using human grouping sera are:

Group O: No agglutination of blood cells by either *A* or *B* serum
Group A: Agglutination of blood cells by *B* serum (contains a agglutinin)
Group B: Agglutination of blood cells by *A* serum (contains b agglutinin)
Group AB: Agglutination of blood cells by both *A* and *B* serum (indicates red cells are affected by both a and b agglutinins)

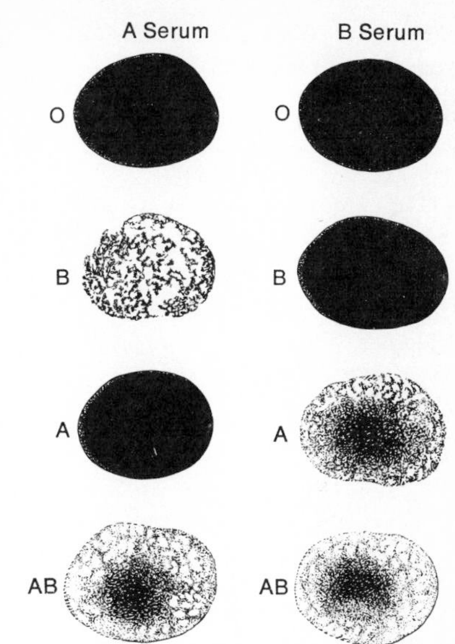

Figure 8-14. Slide method of blood grouping. A serum contains b agglutinins; B serum contains a agglutinins. Agglutination of blood cells in A serum indicates blood cells of B type; in B serum, of A type; in both A and B, AB type.

This method is not infallible, and it is better to match the donor's erythrocytes directly with the recipient's serum. Sometimes a **cross-matching** test is performed to ascertain the compatibility of the donor's and recipient's bloods. In a small test tube are placed 0.1 ml of a 5 per cent suspension of the donor's red blood cells and 0.1 ml of the recipient's serum. In another test tube is placed a 5 per cent suspension of the recipient's cells and the donor's serum. Both tubes are left for 20 minutes at 37° C. If incompatibility is present, both tubes will show agglutination.

In addition to the four main groups of human blood, O, A, B, and AB, there are agglutinogens A_1, A_2, M, and N. A_1 and A_2 are subgroups of group A and were discovered when absorption of group-B serum with the group-A cells of certain persons failed to remove agglutinins for the cells of certain other group-A persons. These subgroups are found also in certain persons of group AB and are the basis of some transfusion reactions. Landsteiner and Levine[8] established the presence of the M and N agglutinogens when certain sera from rabbits immunized against human blood contained agglutinins in addition to the a and b agglutinins. Three distinct types are recognized, depending on the presence of M, N, or a combination of M and N. There are no corresponding agglutinins in the plasma to the M and N agglutinogens, and therefore this factor need not be considered in blood transfusions. However, these factors can be used to determine paternity, as the following example demonstrates.

	Blood Group
Man 1	A, N
Mother	A, M
Man 2	A, M
Child	A, MN

Man 2 could not be the father, but man 1 could be.

Hereditary Aspects of the Blood Groups

Von Dungen and Hirschfeld definitely established the fact that blood groups A and B are inherited characteristics and follow mendelian laws. They postulated the existence of two pairs of genes, which they designated as Aa and Bb, that were responsible for the four blood groups. A and B genes are dominant over

[8]Karl Landsteiner and Phillip Levine, "A New Agglutinable Factor Differentiating Individual Human Blood." *Proc. Soc. Exptl. Biol. Med.,* **24:**600, 941 (1927).

Table 8-4. ***Genotypes of the Four Blood Groups***

Blood Group	*Genotype*
O	OO
A	AA or AO
B	BB or BO
AB	AB

the recessive genes a and b. In the O group, they further postulated, there is a lack of genes for blood type.

By calculation of the frequency of the genes in the general population, Bernstein has postulated that there are three genes and that the blood groups are inherited as a series of genes designated A, B, and O. A and B are dominant over O. From this he further postulated that certain genotypes are possible, as presented in Table 8-4.

A pair of genes is necessary for the expression of each genetic characteristic. This pair of genes, as expressed by the geneticist in written formula such as AA and AO, is called the genotype. Because each germ cell of the parent contains only one of three genes, A, B, or O, the fertilized egg can contain only those genotypes listed in Table 8-2. Table 8-5 demonstrates why only these genotypes could result on recombination of the genes in fertilization.

Table 8-5. ***Possible Genotypes of the Three Types of Genes, A, B, and O, with Fertilization of the Ovum***

Female	*Male Sperm Cell*		
Egg	*A*	*B*	*O*
A	AA	AB	AO
B	AB	BB	BO
O	AO	BO	OO

The presence of subgroups A_1 and A_2 probably implies the existence of five genes, A, A_1, A_2, B, and O. The blood groups M, N, and MN are probably transmitted by a separate pair of genes with the genotype possibilities of MM, NN, and MN.

Since blood cell agglutinogens are inherited in accordance with mendelian law, it is possible to apply these principles in cases of disputed paternity. For example, if both parents are blood type O, only children with blood type O can be conceived. Table 8-6 lists the possible genotypes of offsprings from parents of the various groups. In paternity cases only the absence of a gene type in the blood of the accused can be submitted as evidence.

Table 8-6. ***Possible Genotypes of Offspring***

Parent	*Possible Genotype of Offspring*
O × O	O
O × A	O, A
A × A	O, A
O × B	O, B
B × B	O, B
AB × AB	A, B, AB

RH Factor

As a result of experiments in which rhesus monkeys were immunized, that is, made to develop antibodies against the red cells of human beings, Landsteiner and Wiener[9] found another type of agglutinogen in human blood, which they called the **Rh factor** to indicate that rhesus monkeys were the source of their discovery. At present at least five different genes, designated as Rh_0, Rh_1, Rh_2, Rh^1, and Rh^{11}, are concerned in determining the Rh-positive type and the rh gene in determining Rh-negative types. It was found that the Rh factor is present in the blood of 85 per cent of persons of the white race, and in 15 per cent of the white population it is absent. If this Rh factor is present in any combination of the dominant Rh-positive genes, the blood type is designated Rh-positive; if the Rh factor is absent, the blood type is designated Rh-negative and the genic combination is rh rh. Corresponding agglutinins are not present even in Rh-negative blood. Agglutinins can be produced in an Rh-negative subject by the injection of Rh-positive blood, so that a later transfusion could cause the agglutination of Rh-positive injected cells in such persons who have developed the agglutinins from a previous transfusion. As with the other types, the absence or presence of the Rh factor is inherited. The developing embryo within an Rh-negative mother could be Rh-positive by inheritance from the father. Such an Rh-positive fetus could, by leakage of blood through the placenta, develop agglutinins within the mother's blood, which, in turn, by passing back into the fetal circulation, would lead to destruction of the red cells and death of the infant before birth or shortly after birth. This condition is known as erythroblastosis fetalis. Complete replacement of the infant's blood after birth with compatible blood can remove the agglutinins and save the lives of such infants.

Recently Freda et al.[10] have developed a way to neutralize the lethal effects of the Rh blood factor. They found that the injection of expectant Rh-negative mothers with gamma G-immunoglobulin within 72 hours of delivery neutralizes the sensitization these mothers develop to the Rh-positive blood of their babies.

The disorderly clumping of the red cells that is found in the agglutination reaction is to be distinguished from the normal tendency of red cells to come together with their broad surfaces in opposition, forming rolls of cells. These rolls, called **rouleaux,** consist of ten or twenty cells stacked together like a roll of coins (Fig. 8-4). The tendency to form rouleaux varies among different persons and largely determines the rate at which the red cells settle out when a volume of blood is allowed to stand (sedimentation rate). The rate at which a system of suspended particles settles out is directly proportional to the particle size; the larger the particle, the faster the sedimentation rate, other factors remaining constant. When the content of fibrinogen or globulin of the plasma is increased, as is the case in conditions such as pregnancy and most inflammatory diseases, the tendency to rouleau formation is also increased. This will result in a faster sedimentation rate as a consequence of the increase in suspended particle size (rouleaux); hence the measurement of sedimentation rate is of clinical interest.

[9] K. Landsteiner and A. S. Wiener, "On Presence of M Agglutinogens in Blood of Monkeys." *J. Immunol.,* **33:**19 (1937).

[10] V. J. Freda, J. G. Gorman, and W. Pollack, "Rh Factor: Prevention of Immunization and Clinical Trial on Mothers." *Science,* **151:**828 (1966).

Hemoglobin

Hemoglobin is the red respiratory pigment that is found in the blood and all of which is confined to the red blood cells. It is a conjugated protein, that is, a protein combined with other materials, consisting of **porphyrin, iron,** and **globin.** The structure has been demonstrated by analysis and proved by the synthesis of H. Fischer. Hemoglobin not only serves as the carrier of oxygen from lungs to tissues, but also functions as the most important of the blood buffers in the transport of carbon dioxide from tissues to lungs.

Porphyrins

Porphyrins are naturally occurring pigments, which are found throughout all plant and animal life. The porphyrin pigment consists of four pyrrole rings connected together in cyclic form (Fig. 8-15). Porphyrins have an affinity for various metals, and in combination with metals, they are known as metalloporphyrins. The metal complex of porphyrin occurs in many important biological compounds; for example, the green coloring matter of plants, chlorophyll, is porphyrin with magnesium. Hemoglobin, myoglobin, cytochromes, and iron-containing enzymes are porphyrins in combination with iron. The hydrogen atoms of the pyrrole rings may be replaced by various groups and radicals to form a large number of different compounds. **Heme** is an iron-containing porphyrin (Fig. 8-16) which has its hydrogen replaced by methyl, vinyl, and proprionic acid residues. It is the metalloporphyrin of the blood pigment. Other porphyrins have been identified in the feces, called **coproporphyrin,** and in the urine, in the disease porphyria, called **uroporphyrin.** The various porphyrins can be differentiated by the melting points of their methyl esters and by their absorption spectra.

Heme, the metalloporphyrin of the blood pigment, is capable of combining with various proteins. Such compounds are called hemochromogens. In combination with the protein **globin,** the resulting hemochromogen is known as hemoglobin.

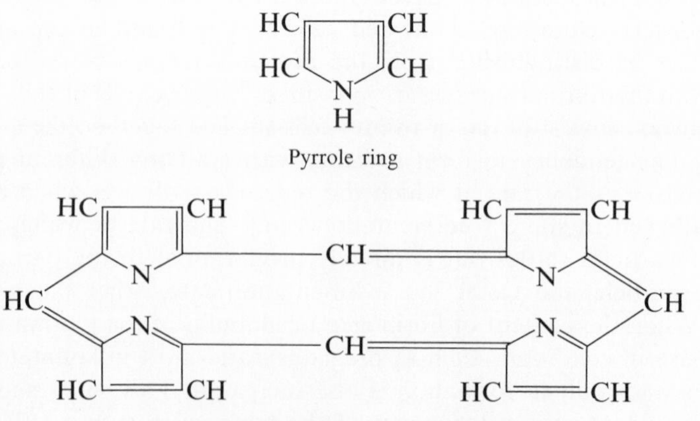

Figure 8-15. Structure of porphyrin.

Mammalian hemoglobin present in the erythrocyte is about 4 per cent heme and 96 per cent globin and has a molecular weight of about 68,000. There are four atoms of iron and thus four porphyrin groups in one molecule. Each metalloporphyrin is conjugated to a polypeptide and the four polypeptides are referred

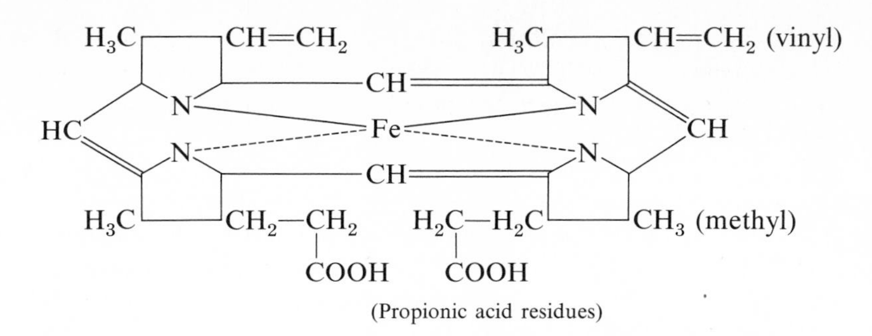

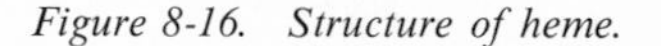
Figure 8-16. Structure of heme.

to as the globin portion of the molecule. Mammalian muscle hemoglobin (myoglobin) has a molecular weight of 17,000 and only one heme group per molecule. The amino acid composition of myoglobin also differs.

Hemoglobin occurs in several normal forms and is referred to as primitive (P), fetal (F), and adult (A). Primitive hemoglobin is present only in the earliest stages of fetal development and is soon replaced by fetal hemoglobin. The adult form begins to appear at the fifth or sixth month of embryonic development. At birth from 55 to 85 per cent of the hemoglobin is hemoglobin F, the remainder is hemoglobin A. At age four, practically all the hemoglobin is present in the adult form. Adult hemoglobin differs from the fetal form in that the latter has one half as many terminal valine residues in the polypeptide chains. Hemoglobin F has a greater oxygen-carrying capacity than adult hemoglobin. Human blood contains about 15 g of hemoglobin per 100 ml and each gram is capable of combining with 1.34 cc of oxygen.

Oxyhemoglobin

Hemoglobin can combine with oxygen because of the iron that is present in its molecule. Oxygen and hemoglobin react to form a chemical compound according to the equation

$$O_2 + Hb \rightleftharpoons HbO_2.$$

Hemoglobin not in combination with oxygen is called **reduced hemoglobin** and is symbolized by the letters Hb. Hemoglobin combined with oxygen is called **oxyhemoglobin** and is symbolized by the letters HbO_2. The combination of oxygen with hemoglobin is quite unstable; that is, no true oxide is formed because the iron is oxidized to the ferric state (Fe^{+++}), this property of combining with oxygen in such a loose fashion is lost. It appears that the protein portion of the hemoglobin molecule endows the iron with this unique property, for it is lost if the protein is denatured.

The capacity of the blood to absorb oxygen, that is, the ability of hemoglobin to change from the deoxygenated to the oxygenated state ($Hb + O_2 \longrightarrow HbO_2$), is called the **oxygen-carrying capacity** and is proportional to the hemoglobin concentration. Haldane and Hufner, in 1894, showed that 1 g of hemoglobin combined with 1.34 cc of oxygen or carbon monoxide measured at 0° C and 760 mm pressure. Thus, the oxygen-carrying capacity of 100 ml of normal human blood, which contains 15 g of hemoglobin, would be 20 cc (15×1.34).

The extent to which hemoglobin combines with oxygen depends on the partial pressure of oxygen (pO_2) in solution. When the pO_2 is high, most or all of the hemoglobin is combined with oxygen; when the pO_2 is low, little hemoglobin is combined with oxygen. The transport of oxygen by hemoglobin will be discussed more fully in Chapter 10.

Variations of Normal Hemoglobin

Various pigments may make their appearance in the blood under pathological conditions. When blood is treated with a variety of oxidizing and reducing agents, such as potassium ferricyanide, nitrite, chlorate, aniline, nitrobenzene, hydroxylamine, acetanilide, and pyrogallic acid, the iron that is in the ferrous (Fe^{++}) state in hemoglobin is oxidized to the ferric (Fe^{+++}) state and is known as **methemoglobin.** The hemoglobin cannot function in this state as a carrier of oxygen, and as a result, its presence in the blood denotes a decrease in the oxygen content of arterial blood. The blood becomes dark in color, giving rise to a type of cyanosis.

Reduced hemoglobin combines with hydrogen sulfide (H_2S) to form **sulfhemoglobin,** causing the blood to have a chocolate color. This condition is rare, and the factors responsible for its formation are not yet understood. Hydrogen sulfide is not normally absorbed from the intestinal tract in appreciable amounts, and it is believed that a strong reducing agent of unknown composition must be present in the blood to order for the minute amounts of hydrogen sulfide normally absorbed to convert hemoglobin into sulfhemoglobin.

Carbon monoxide, a poisonous gas formed as the result of incomplete combustion of carbon and produced by automobile engines and other gasoline motors, combines with hemoglobin in the same proportion as does oxygen. In combination with hemoglobin it forms a stable compound called **carbon monoxide hemoglobin,** or **carboxyhemoglobin,** imparting a very bright red color to the blood. As a result of its ability to combine with hemoglobin, carbon monoxide diminishes the amount of oxygen that can be carried by the arterial blood. Victims of carbon monoxide poisoning suffer from a lack of oxygen. The amount of oxygen deficiency and the degree of intoxication depend on the concentration of carbon monoxide in the air and the time of exposure. It has been estimated that a condition of collapse will be reached when sufficient air to contain 1 liter of the gas has been inspired.

Abnormal Hemoglobins

The existence of abnormal hemoglobins in the population was first noted by Pauling et al. in 1949.[11] Using electrophoretic techniques to determine the amino acid sequence in the polypeptide chains of hemoglobin, he showed that the hemoglobin of individuals suffering with sickle cell anemia was different. This difference was shown by the fact that of the 300 amino acids present in each polypeptide chain, one glutamic acid was replaced by a valine. Since amino acid sequence in the polypeptide chains of hemoglobin are genetically determined, mutant genes must exist to produce abnormal hemoglobin in the population. Sickle cell hemoglobin is designated as hemoglobin S. Other types of abnormal

[11] L. Pauling, H. A. Itano, S. J. Singer, and I. C. Wells, "Sickle Cell Anemia, a Molecular Disease." *Science,* **110:**543 (1949).

hemoglobin have been described using similar procedures and are designated as hemoglobin C, D, E, G, H, I, J, and K.

Destruction of Hemoglobin

When the old or damaged red cells are normally destroyed by the body, the hemoglobin is broken down by the cells of the reticuloendothelial system, for example, the Kupffer cells of the liver and the macrophages of the spleen, into heme and globin. The heme is converted into the bile pigment **bilirubin** by the opening of the porphyrin ring. The following structure has been proposed for bilirubin:

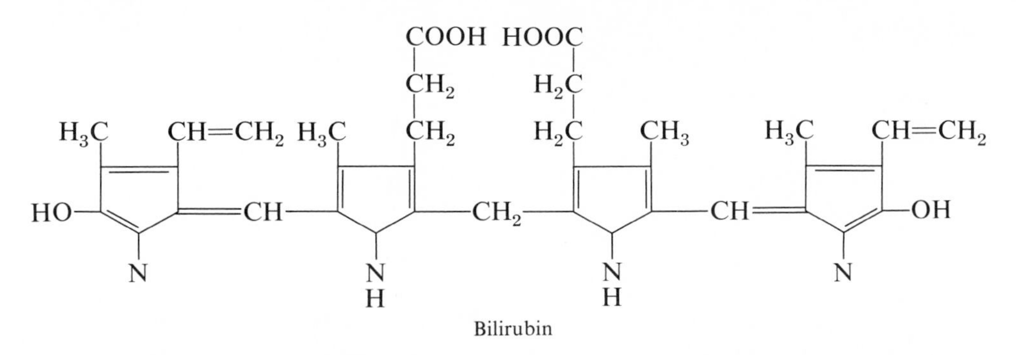

Bilirubin

The bilirubin is excreted by the liver into the bile. A small amount of bilirubin in loose chemical combination with protein is always found in the blood under normal circumstances. Its normal concentration is considered to be 0.1 to 1.0 mg per 100 ml of blood.

Acid–Base Balance of the Blood

The blood and the tissue fluid make up the internal environment of the body. In order for living cells to function normally, the reaction of the internal environment must be maintained relatively constant within comparatively narrow limits. When the term **acid–base balance** is applied to the blood, it refers more specifically to the chemical state resulting from the balance between positive ions (cations) and negative ions (anions). The positive ion content of the blood is made up almost entirely of Na^+, K^+, Ca^+, and Mg^+, and the negative ion content is made up of Cl^-, HCO_3^-, PO_4^-, and SO_4^-.

Acids are compounds of electronegative elements or groups plus ionizable hydrogen (H^+), and **bases** are compounds of electropositive elements or radicals plus ionizable hydroxyl group (OH^-). The acid–base balance of the blood is normally described in terms of **pH,** which is defined as the negative logarithm to the base 10 of the hydrogen-ion concentration:

$$pH = -\log (H^+).$$

The normal range of the blood pH is 7.3 to 7.45. Within this range there are daily variations, which are due to increased production of acids by exercise and normal metabolism of foodstuffs and increased bases due to the digestive processes in which hydrochloric acid secreted in the gastric juice leads to an increase in the bicarbonate content. Acid substances are produced in the following ways:

1. The oxidation of organic compounds within the cells gives rise to carbon dioxide, which combines with water to yield **carbonic acid** (H_2CO_3).
2. The breakdown of proteins gives rise to amino acids. Those containing sulfur in their molecule produce sulfuric acid (H_2SO_4) as a final product of disintegration.
3. The metabolism of fats sets free **fatty acids.**
4. Nucleoproteins and other phosphate-containing compounds give rise to **phosphoric acid** (H_3PO_4) in their disintegration. Nucleoproteins also give rise to the organic acid **uric acid.**
5. **Lactic acid** is formed in the course of carbohydrate metabolism, and oxalic acid is formed in the metabolism of some foods, such as cocoa and spinach.

Ammonia is the only basic substance found in any quantity. Other bases are set free in the metabolism of vegetable foodstuffs and in the secretion of hydrochloric acid by the digestive glands, as described above. With all these acids and some bases continually being produced, the blood still remains remarkably constant and even in disease may show very little variation. In the terminal stages of some diseases, the blood pH departs from the normal to a low and high of 6.8 and 7.8, respectively.

Concept of pH

All activities of the body are importantly affected by the hydrogen and hydroxyl-ion concentration of the fluids that surround the cells of the body. As we have seen, all acids are characterized by the formation of H^+ and all bases by the formation of OH^- ions. For example, an acid, when placed in an aqueous medium, will dissociate to a varying degree into its component ions, the strength of the acid depending on the degree of dissociation:

$$\text{Hydrochloric acid—HCl} \longrightarrow H^+ + Cl^-,$$
$$\text{Sulfuric acid—}H_2SO_4 \longrightarrow H^+ + HSO_4^-,$$
$$\text{Acetic acid—}CH_3COOH \longrightarrow H^+ + CH_3COO^-.$$

A base placed in an aqueous medium will dissociate into its corresponding ions, and as with the acid, its strength will depend on the degree of dissociation. Examples of such dissociation are as follows:

$$\text{Sodium hydroxide—NaOH} \longrightarrow Na^+ + OH^-,$$
$$\text{Potassium hydroxide—KOH} \longrightarrow K^+ + OH^-,$$
$$\text{Ammonium hydroxide—}NH_4OH \longrightarrow NH_4^+ + OH^-.$$

Water itself dissociates, although to a very slight degree, into its respective ions:

$$H_2O \longrightarrow H^+ + OH^-.$$

One of the most important properties of aqueous solutions is that the concentration of H^+ and OH^- ions in solution is such that their product is always equal to 0.00000000000001 g molecular weight per liter, which is a constant symbolized as K_w. This may be expressed in equation form:

$$(H^+) \times (OH^-) = K_w = 1 \times 10^{-14}.$$

1×10^{-14} is the expression of the above decimal figure in powers of ten.

In pure water the H^+ concentration is equal to the OH^- ion concentration, so that if

$$(H^+) \times (OH^-) = 1 \times 10^{-14},$$
$$(H^+) = 1 \times 10^{-7},$$
$$(OH^-) = 1 \times 10^{-7}.$$

Or the number of molecules dissociating is such that 1 liter, or 1,000 g, of pure water contains 0.0000001 g of hydrogen ions (1×10^{-7}) and an equal number of hydroxyl ions but has a weight of 0.0000017 g since the atomic weight of OH^- is 17 (O = 16, H = 1). Water is, therefore, chemically neutral in reaction.

From the definition of acids and bases given above, it is apparent that the addition of either acid or base to water will affect the ratio between the concentration of hydrogen and hydroxyl ions. The addition of one ion in excess will cause a lowering of the concentration of the other, since the product of the two is a constant. It becomes obvious, then, that the acidity or reaction of a solution depends on the concentration of H^+ present.

As we have seen, the H^+ concentrations in the physiological range are considerably less than 1. In order to express such concentrations in a convenient form, Sorenson introduced the concept of pH, that is, he expressed H^+ concentration as a negative logarithm instead of using decimals or expressions of decimals by powers of ten.

Example: The concentration of H^+ in pure water may be expressed as follows:

$$\text{Molar equivalent} = 0.0000001 \text{ mole}^{12} \text{ per liter,}$$
$$\text{By powers of } 10 = 1 \times 10^{-7} \text{ mole per liter,}$$
$$\text{(Logarithm) pH} = \log 1 \times 10^{-7} = 7.$$

In using pH, it must be remembered that a change of one unit in pH indicates a tenfold change in hydrogen-ion concentration and that an increase in pH means a decrease in hydrogen-ion concentration. Thus pH 4 is not one time as great as pH 5, but 10 times, and 100 times as great as pH 6.

Since pure water is neutral, having a pH of 7 in which the number of H ions is equal to that of OH ions and the product of both is 14, a pH scale has been devised that ranges from 1 to 14, with 7 being neutral. Table 8-7 shows relationships in this scale, with the concentration of H ions expressed as decimals and by powers of ten.

Blood Buffers

Although metabolic activity gives rise to many acids, the reaction of the blood remains remarkably constant and even in disease may show very little variation. The body has three mechanisms for maintaining pH. These are (1) the buffer systems of the blood; (2) the lungs, which excrete volatile acids, such as carbonic; and (3) the kidneys, which function to eliminate fixed acids.

The **blood buffers** are mixtures of weak acids with their salts of strong bases, which give them the ability to resist change in hydrogen-ion concentration. This ability is known as **buffer action** and is very important to the body because it

[12] A molar solution of an acid or base is one that contains in each liter an amount in grams of the acid or base equal to the molecular weight of that compound. For example: A molar solution of hydrochloric acid is one in which 36 g of hydrochloric acid is in solution in 1,000 ml of water (molecular weight of hydrogen is 1; of chlorine, 35).

Table 8-7. ***Scale of pH Showing H^+ Concentration Expressed by Decimals and by Powers of Ten***

	pH (− *log*)	*Decimal*	*Powers of Ten*
↑	1	0.1	1×10^{-1}
	2	0.01	1×10^{-2}
	3	0.001	1×10^{-3}
Acid	4	0.0001	1×10^{-4}
	5	0.00001	1×10^{-5}
↓	6	0.000001	1×10^{-6}
Neutral	7	0.0000001	1×10^{-7}
↑	8	0.00000001	1×10^{-8}
	9	0.000000001	1×10^{-9}
	10	0.0000000001	1×10^{-10}
Alkaline	11	0.00000000001	1×10^{-11}
	12	0.000000000001	1×10^{-12}
	13	0.0000000000001	1×10^{-13}
↓	14	0.00000000000001	1×10^{-14}

is part of the mechanism by which the normal pH of the body is regulated within the narrow limits of pH compatible with the life of most cells. The mixtures of the weak acids and their salts of strong bases are referred to as **buffer systems.** There is a total of six buffer systems in the blood. Three are associated with the plasma and include the following:

$$\frac{H_2CO_3}{NaHCO_3} = \frac{1}{20}.$$

This buffer system consists of **carbonic acid** and the salt that this acid forms with strong bases, **sodium bicarbonate.** At the pH of blood (7.4), a ratio of 1:20 must exist between the carbonic acid and bicarbonate fractions. Any tendency to an increase or decrease in H^+ activity will be met by an adjustment in the interaction of the two compounds above, as indicated below, to maintain this ratio: $H_2CO_3 \rightleftharpoons H^+ + HCO_3^- + Na^+ \rightleftharpoons NaHCO_3 + H^+$. Any alteration in the ratio will disturb the acid–base balance, which will be reflected in a change of the pH of the blood:

$$\frac{NaH_2PO_4}{Na_2HPO_4} = \frac{N1}{N4}.$$

This buffer system consists of a mixture of a weak acid, **acid sodium phosphate,** and its base, **alkaline sodium phosphate.** A ratio of 1:4 must exist between these two in order that the blood pH remain normal.

$$\frac{\text{H protein}}{\text{Na proteinate}} = \frac{\text{Acid protein}}{\text{Sodium proteinate}}.$$

As a consequence of their constitution of amino acids linked together, proteins contain a large number of acidic and basic groups. The acidic groups may be the terminal carboxyl group (—COOH) and the basic groups may be the amino group $(—NH)_2$. **Alanyl glycine,** which results from the linkage of the amino acids alanine and glycine, demonstrates how proteins can contain both groups.

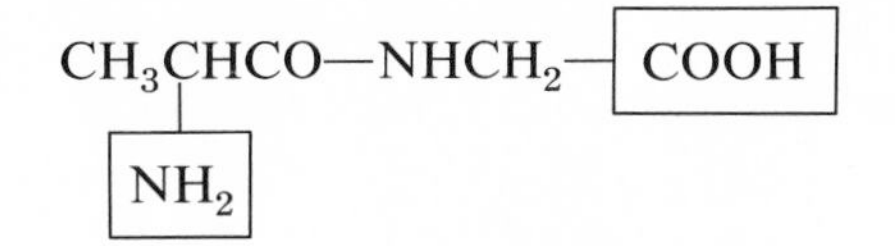

In acid solution, the protein acts as a buffer in that the basic amino group takes up excess hydrogen ions, forming $(—NH_3^+)$, whereas in basic solutions, the acidic COOH group gives up hydrogen ions, forming anions $(—COO^-)$.

At a given pH a protein will have the same number of cationic groups (NH_3^+) as anionic groups (COO^-). This is known as the **isoelectric point** of the protein.

Three of the buffer systems of the blood are associated with the red blood cells and consist of the following:

$$\frac{HHbO_2}{KHbO_2} = \frac{\text{Acid oxyhemoglobin}}{\text{Alkaline oxyhemoglobin}},$$

$$\frac{HHb}{KHb} = \frac{\text{Acid-reduced hemoglobin}}{\text{Alkaline-reduced hemoglobin}},$$

$$\frac{KH_2PO_4}{K_2HPO_4} = \frac{\text{Acid potassium phosphate}}{\text{Alkaline potassium phosphate}}.$$

Buffer Action. A buffer solution is a solution that protects against changes in pH that might result from the introduction of acid or alkaline into the solution. As we have discussed, buffers usually are composed of weak acids and their basic salts, which form more of the same weak acids or salts by combining with any strong acids or bases introduced into the solution. A reaction occurs, with the result that the strong acid combines with an equivalent of buffer salt and sets free an equivalent of buffer acid. For example, a solution contains the buffer system carbonic acid (H_2CO_3) and sodium bicarbonate $(NaHCO_3)$. If a strong acid, hydrochloric acid (HCl), is added to the solution it combines with the salt as follows:

$$HCl + NaHCO_3 \longrightarrow NaCl + H_2CO_3.$$

Since the carbonic acid ionizes to a much smaller degree than does the hydrochloric acid, the hydrogen ions are "tied up" and the resultant pH change in the solution is minimized.

If a strong base, sodium hydroxide (NaOH), is added to the buffer solution, it is neutralized by the buffer acid present to form more of the buffer salt:

$$NaOH + H_2CO_3 \longrightarrow NaHCO_3 + HOH.$$

Again the pH change is minimized, owing to the formation of the weakly ionized water.

It is evident from this that, since the salt of the buffer solution reacts when acid is added and the weak acid reacts when alkali is added, both the weak acid and its salt must be present if a buffer system is to offer resistance to pH change by the addition of either acid or alkali.

The buffer systems of the plasma take care of such fixed acids produced by metabolic processes as lactic, phosphoric, sulfuric, and hydrochloric and the keto acids, such as acetoacetic and beta-hydroxybutyric acid. Carbonic acid, which is a product of tissue activity, is produced in much greater quantities (800 to 900 g daily) than any other acid. This acid is buffered by the buffer substances of the red blood cell.

As blood passes from the arterial capillaries to the venous capillaries, carbon dioxide is absorbed from the tissue and diffuses into the red cell. Inside the red cell the carbon dioxide combines with water to produce carbonic acid.

At the same time the hemoglobin loaded with oxygen (oxyhemoglobin) gives up its oxygen to the tissue, becoming reduced hemoglobin. Reduced hemoglobin, being a weaker acid than oxyhemoglobin, gives up its store of alkali, which is then utilized to neutralize the carbonic acid. This function of hemoglobin supplies sufficient base for the carriage of more than half the total carbon dioxide of the blood. These changes are illustrated in the following diagram:

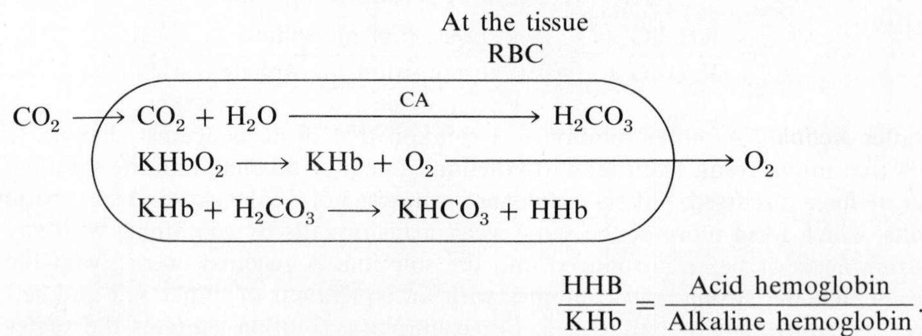

$$\frac{\text{HHB}}{\text{KHb}} = \frac{\text{Acid hemoglobin}}{\text{Alkaline hemoglobin}}.$$

When the blood returns to the lungs, the opposite situation prevails. As carbon dioxide diffuses out in the alveolar air, some alkali becomes excess. This alkali is then utilized to neutralize the acid oxyhemoglobin, which is being formed as oxygen diffuses into the cell and combines with reduced hemoglobin. These changes are illustrated on page 295. It must be remembered that there is never a complete change of all the oxyhemoglobin to reduced hemoglobin, so the reactions above actually represent a variation in the concentrations of the hemoglobin buffer pairs shown.

These buffers are responsible for approximately 60 per cent of the buffering capacity of the blood. The phosphate–buffer complex of the red blood cell accounts for about 25 per cent more.

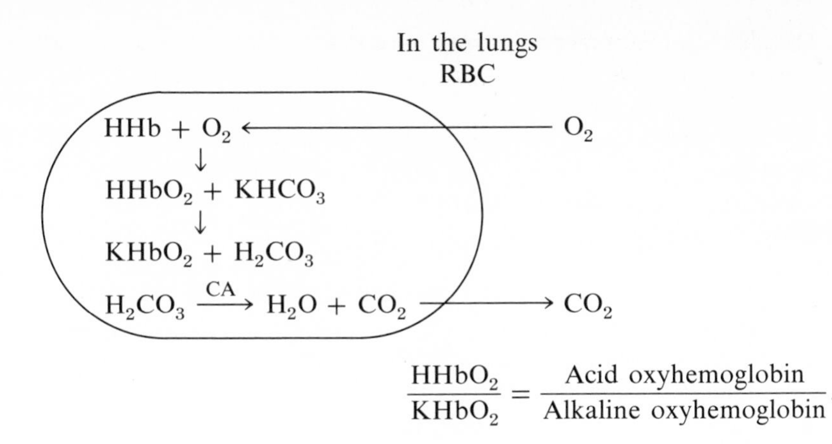

$$\frac{HHbO_2}{KHbO_2} = \frac{\text{Acid oxyhemoglobin}}{\text{Alkaline oxyhemoglobin}}.$$

Chloride Shift. Although most of the carbon dioxide is buffered by the red blood cell, it is carried as bicarbonate in the plasma. The mechanism for this occurs when, as a result of the buffering of carbonic acid by the alkali released from the reduced hemoglobin, potassium bicarbonate is produced. The concentration of HCO_3^- ion in the red cell is raised above that in the plasma, and as a result, the anions diffuse out of the cell into the plasma.

It is considered that under normal circumstances the red cell membrane is impermeable to sodium or potassium ion. As a result of the bicarbonate shift-out, the ionic equilibrium between plasma and the interior of the cell tends to become disturbed. Intracellular sources of potassium ion (alkali) are *indirectly* made available to the plasma by the shifting of chloride (anion) from the plasma into the red cell, where they combine. This allows more sodium ion, in the plasma, to combine with carbonate ion to form additional sodium bicarbonate. The **chloride shift,** as it occurs at the tissues between plasma and red blood cells, is as follows:

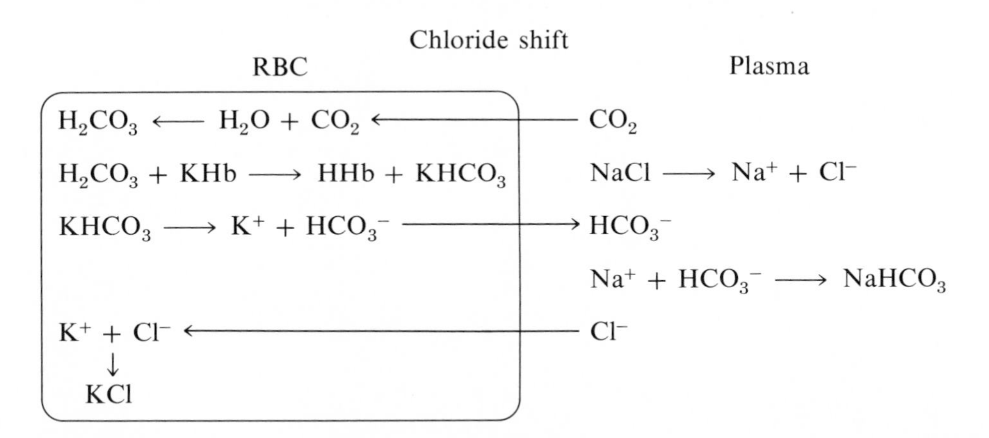

At the lung, when the reduced hemoglobin becomes oxygenated, the chloride shifts back into the plasma, liberating the potassium to buffer the acid oxyhemoglobin. The chloride again combines with the sodium in the plasma that has been liberated by the diffusion of carbon dioxide into the alveolar air.

Alkaline Reserve

The **alkaline reserve** is represented by the sodium bicarbonate concentration in the blood that is not combined with nonvolatile acid. We have already seen that all acids other than carbonic react with bicarbonate to set free carbon dioxide in the following manner:

$$\begin{pmatrix} \text{Fixed acid} \\ H_2SO_4 \\ \text{lactic} \\ H_3PO_4 \\ HCl \end{pmatrix} + NaHCO_3 \longrightarrow \text{Salt of the acid} + H_2CO_3$$

$$H_2CO_3 \overset{CA}{\rightleftharpoons} H_2O + CO_2.$$

The carbon dioxide is discharged from the lungs and thus minimizes the change in pH.

When carbon dioxide enters the blood through tissue metabolism, it is buffered by the base liberated from hemoglobin in the manner already described, and bicarbonate is formed in the plasma. The normal concentration of plasma bicarbonate (52 to 75 volume per cent), then, is a measure of the base left over after all acids other than carbonic acid have been neutralized. The bicarbonate is extremely important to the organism, since it constitutes a reserve of alkali readily available for the neutralization of any invading nonvolatile acid.

The pH of the plasma is dependent on the concentration of the mixture of carbonic acid and sodium bicarbonate in their normal ratio of 1:20. The relationship between this buffer system and the blood pH is best illustrated mathematically.

The equilibrium relation of **carbonic acid dissociation:**

$$H_2CO_3 \longrightarrow H^+ + HCO_3^-.$$

can be expressed as

$$\frac{[H^+][HCO_3^-]}{[H_2CO_3]} = K, \tag{8-1}$$

in which K is called the **equilibrium constant.** This equation states that the product of the hydrogen-ion concentration, $\mathbf{[H^+]}$, and the bicarbonate-ion concentration, $\mathbf{[HCO_3^-]}$, divided by the carbonic acid concentration, $\mathbf{[H_2CO_3]}$, is a constant (K) at equilibrium.

Equation (8-1) may be solved for H^+:

$$[H^+] = K\frac{[H_2CO_3]}{[HCO_3^-]}. \tag{8-2}$$

Taking the logarithm of both sides of the equation gives

$$\log [H^+] = \log K + \log \frac{[H_2CO_3]}{[HCO_3^-]}. \qquad (8\text{-}3)$$

Multiply both sides by -1:

$$-\log [H^+] = -\log K - \log \frac{[H_2CO_3]}{[HCO_3^-]}. \qquad (8\text{-}4)$$

By definition the negative of log [H] = pH, and the negative log of K = pK; therefore,

$$\text{pH} = \text{pK} + \log \frac{[HCO_3^-]}{[H_2CO_3]}. \qquad (8\text{-}5)$$

Equation (8-5) is known as the **Henderson-Hasselbalch equation;** it may be used to calculate the pH if one knows the separate concentrations of carbonic acid and bicarbonate. The pK for this particular buffer system is 6.1 (pK values for many buffer systems are listed in most biochemistry texts).

Normally the bicarbonate concentration in blood is approximately 26.0 mEq/liter with a normal range of 23–33 mEq/liter. (Although the bicarbonate in the above expression is shown to come from the dissociation of carbonic acid, the equation applies to all bicarbonate present in the blood regardless of its origin.) The concentration of carbonic acid is approximately 1.3 mM/liter. The blood pH may be calculated by inserting these figures in Equation (8-5):

$$\begin{aligned} \text{pH} &= 6.1 + \log \frac{26}{1.3} \\ &= 6.1 + \log 20 \\ &= 6.1 + 1.30 = 7.40 \text{ (normal blood pH).} \end{aligned}$$

The carbon dioxide content is a measure of the total quantity of carbon dioxide present in both forms (bicarbonate and carbonic acid), but does not indicate the ratio of the two. For this reason, a carbon dioxide analysis of the blood in itself does not reveal the nature of pH disturbances.

Any physiological condition increasing the numerator or decreasing the denominator of the fraction $H_2CO_3/NaHCO_3$ unaccompanied by an equivalent increase or decrease of the denominator or numerator, respectively, will decrease the pH and is termed an **acidemia.** In the same manner, any decrease in the numerator or increase in the denominator uncompensated by an equivalent change of the denominator or numerator will result in an increase of the pH and is termed an **alkalemia.**

The plasma bicarbonate may be greatly decreased, yet if the free carbonic acid is reduced to a corresponding extent and the normal ratio of 1:20 is maintained, the pH will show no appreciable change. This condition, however, is referred to as a **compensated acidemia,** or **acidosis.** Likewise, with an appreciable increase in the plasma bicarbonate above normal, with a corresponding proportional increase in the free carbonic acid so that the normal ratio of 1:20 is maintained, the hydrogen-ion concentration will show no appreciable change. This condition is referred to as a **compensated alkalemia,** or **alkalosis.**

The British National Research Council proposes that the terms *acidosis* and *alkalosis* be used to indicate decreases and increases of sodium bicarbonate from normal, and *acidemia* and *alkalemia* to indicate low and high pH. They state that it is probable that no general usage for the terms *acidosis* and *alkalosis* can be agreed on, and it is, therefore, better to avoid their usage.

Van Slyke has divided the described conditions into distinct types. These are as follows:

1. An increase in plasma bicarbonate level above normal without a parallel increase in carbonic acid. The result is an increase in the pH and therefore an alkalemia. This condition can result after the consumption of large quantities of sodium bicarbonate. It can also result from a loss of hydrochloric acid through persistent vomiting **(metabolic alkalemia).**
2. A decrease in the carbonic acid without a parallel decrease in the plasma bicarbonate, which results in an increase in the pH due to a loss of carbonic acid instead of an increase in sodium bicarbonate. This alkalemia can result from excessive ventilation (hyperpnea), resulting in a blowing off of excess carbon dioxide **(respiratory alkalemia).**
3. An increase in the carbonic acid concentration without a proportionate increase in the bicarbonate content. This results in a decrease in the pH of the blood, or an acidemia, and is usually produced when respiratory excretion of carbon dioxide is retarded. The condition is usually found in cardiac decompensation and impaired ventilation in the lungs. Breathing air rich in carbon dioxide can also produce this type of acidemia **(respiratory acidemia).**
4. A decrease in the plasma bicarbonate without a proportional decrease in the free carbonic acid, resulting in a decrease in the pH. This type of acidemia has been most frequently observed in cases of diabetic acidosis **(metabolic acidemia).**
5. An increase in the plasma bicarbonate with a proportionate increase in the free carbonic acid. In this condition the pH is normal, and this is therefore said to be **compensated alkalemia.** It can result after the administration of sodium bicarbonate in moderate amounts over a long period of time, or it can result from carbon dioxide retention, as in emphysema. In emphysema, which is an overexpansion of the alveoli of the lungs, the gas exchange in the alveoli is retarded, leading to a state of chronically increased carbon dioxide tension in the blood. The body raises the bicarbonate high enough to balance the increase in carbonic acid, and the pH remains normal.
6. A decrease in the available blood base with a proportional decrease in the free carbonic acid, resulting in a **compensated acidemia** condition with a normal pH. This condition can result from an alkali decrease as a result of acid increases in the blood, as in diabetes and nephrites or when one becomes conditioned to high altitudes.
7. Normal acid–base balance, in which the blood pH ranges between 7.3 and 7.45. The bicarbonate content at this level is between 52 and 75 volume per cent and the free carbonic acid is from 2.6 to 3.7 volume per cent, resulting in a ratio of $H_2CO_3/NaHCO_3$, or 1:20.

Role of the Kidney in Acid–Base Balance

The kidney affects acid–base equilibrium by aiding in the elimination of acids produced in metabolism, such as lactic, pyruvic, hydrochloric, phosphoric, and sulfuric and ketone bodies. These acids are, of course, partly neutralized by the plasma bicarbonate, but the kidney, in its ability to acidify the urine by exchanging hydrogen ion for sodium ion in the tubular filtrate, can conserve base for the alkali reserve of the body. The process of hydrogen-ion exchange is believed to be carried out in the distal tubule (see Chapter 14), where a sodium ion from disodium phosphate (Na_2HPO_4) is taken in by the tubule cell and replaced by a hydrogen ion to form acid sodium phosphate (NaH_2PO_4). This captured sodium ion is bound to HCO_3^- to form $NaHCO_3$, which is secreted into the blood.

The kidney is also able to manufacture the strong base ammonia from the amino acid glutamine by the action of the enzyme glutaminase as well as by removing amine groups (oxidative deamination) from many amino acids. This ammonia-producing mechanism is only brought into play when the production of acid by the body becomes excessive. The ammonia is used to neutralize the excess acid so that the reserves of fixed base will not be depleted. When alkali excess develops, the kidney aids in correcting the acid–base balance by excreting an alkaline urine.

References

Brewer, G. J. *Red Cell Metabolism and Function.* New York: Plenum, 1970.

Davenport, H. W. *The A.B.C. of Acid Base Chemistry.* Chicago: U. of Chicago, 1950.

Guyton, A. *Textbook of Medical Physiology.* Philadelphia: Saunders, 1968.

Kabat, E. A. *Blood Group Substances.* New York: Wiley-Interscience, 1956.

Krantz, S. B., and L. E. Jacobson. *Erythropoietin and the Regulation of Erythropoiesis.* Chicago: U. of Chicago, 1970.

Lemberg, R. *Hematin Compounds and Bile Pigments.* New York: Wiley-Interscience, 1949.

MacFarlane, R. G., and A. Robb-Smith. *Functions of the Blood.* New York: Academic, 1961.

Paulus, J. M. (ed.). *Platelet Kinetics.* New York: Am. Elsevier, 1971.

Putman, F. W. *The Plasma Proteins,* Vol. 1 and 2. New York: Wiley-Interscience, 1960.

Schiff, F., and W. C. Boyd. *Blood Grouping Technic.* New York: Wiley-Interscience, 1942.

Schumaker, V. N., and G. H. Adams. "Circulating Lipoproteins," in *Annual Review of Biochemistry,* Vol. 38. Palo Alto, Calif.: Annual Reviews, 1969, p. 113.

Seegers, M. H. *Blood Clotting Enzymology.* New York: Academic, 1967.

Weed, R. I. (ed.). *Hematology for Internists.* Boston: Little, Brown, 1973.

9

The Lymphatic System

In Chapter 8 it was noted that the internal environment consists of the blood and tissue fluid. The blood supplies the cells of the body with food and oxygen and removes their waste products. The tissue fluid, bathing all the cells of the body, acts as a connecting link between the blood and the cells, since the exchange of materials between the blood and tissue must occur across the tissue fluid. This **tissue fluid** is often referred to as **lymph** and is formed when certain parts of the plasma of the blood pass through the capillary wall into the tissue spaces, a process called **transudation.** The lymph is continually being drained from the tissue spaces through a system of tubules known collectively as the lymphatic system and returned to the blood.

Lymph Vascular System

In nearly every part of the body, connective tissue serves as the origin of the lymph vessels. These vessels may be said to begin as microscopic channels called **lymph capillaries** (Fig. 9-1), which take their origin from the multitudinous tissue spaces; they are closed vessels though and not in open communication with the fluid in the tissue spaces. These lymph capillaries, which are distributed in the same manner as the blood capillaries, are continuous with larger and larger vessels, until finally they all converge into the great **thoracic duct** (left lymphatic duct) and the **right lymphatic duct.** The great thoracic duct opens into the venous system at the junction of the left jugular and subclavian veins; the right lymphatic duct similarly opens into the junction of the right jugular and subclavian veins. Figure 9-2 illustrates the lymphatic system of the body.

The lymphatic vessels course through the connective tissues of the body like the veins, and their walls, like those of the blood vessels, are formed of three coats. The lymphatic vessels differ from veins in that they freely connect with one another, forming plexi, and they also enter into association with structures known as **lymph glands.**

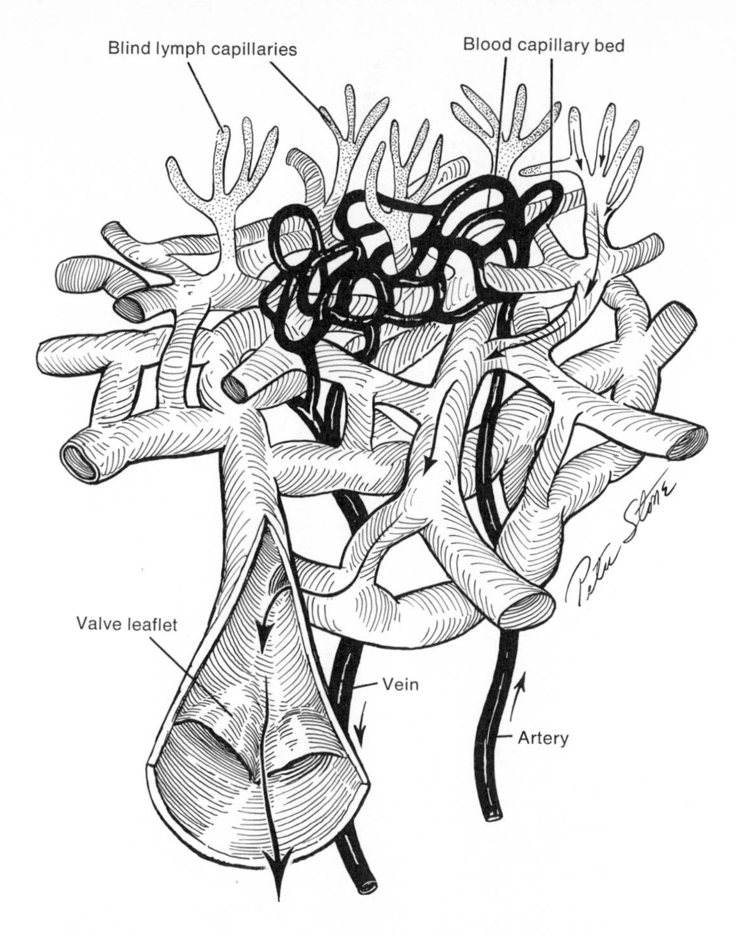

Figure 9-1. Lymph capillaries showing their termination in the tissue spaces as blind ends. Section through one indicates structure of valves.

Lymph Glands

Lymph glands, which are oval or bean-shaped aggregates of lymphatic tissue, are often called **lymph nodes** (Figs. 9-3 and 9-4), and they present a slight depression on one side, called the **hilus.** It is through the hilus that an artery and vein enter and leave; in addition, the several lymphatic vessels also exit, and since they carry lymph away from the glands, they are called **efferent lymphatics.** The node is surrounded by a capsule of fibrous connective tissue, which also radiates through the lymph node substance. These fibrous radiations, called **trabeculae,** divide the interior of the node into spaces called **lymph sinuses,** which are filled by masses of lymphoid tissue called **lymph follicles.** The follicular substance appears to be an area of interaction between the blood and the lymph, the latter entering the gland through the afferent lymphatic vessels, which penetrate in a scattered fashion on the convex side. White blood cells are present in the lymph in great numbers, most of which are lymphocytes. The follicular substance appears to be the site of their development since the central area of the follicle is a center of abundant mitotic activity. The excess cells that are a product of this activity pass into the lymph sinuses and leave the gland by the efferent lymphatic vessels.

Figure 9-2. *Superficial lymphatics of the body. The locations of the major lymph nodes are shown, as is the thoracic duct.*

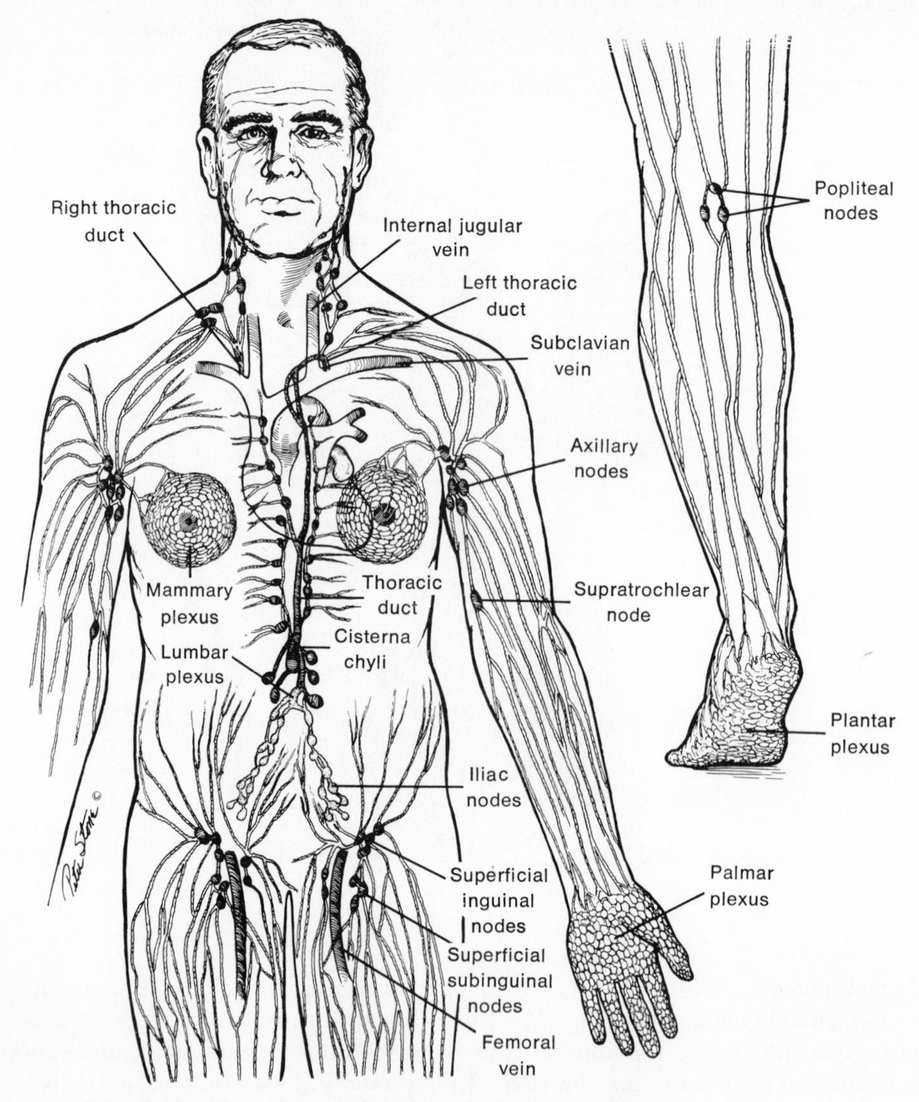

Experimentation has shown that lymph leaving a gland is richer in leukocytes than lymph coming into the gland.

Size and Location of Lymph Nodes

The lymph nodes vary considerably in size; some are as small as the head of a pin, and others are as large as a lima bean. The lymph nodes lie embedded in connective tissue, some being superficially located in the subcutaneous tissue and others being more deeply situated, usually in association with the large blood vessels. Lymph nodes usually occur in chains of two to twelve, although occasionally a node will exist alone. Lymph nodes are found on the back of the head and neck, under the rami of the mandible, around the sternomastoid muscles, and in the nasopharynx. A large group of nodes is located under the armpits and also under the chest muscles.

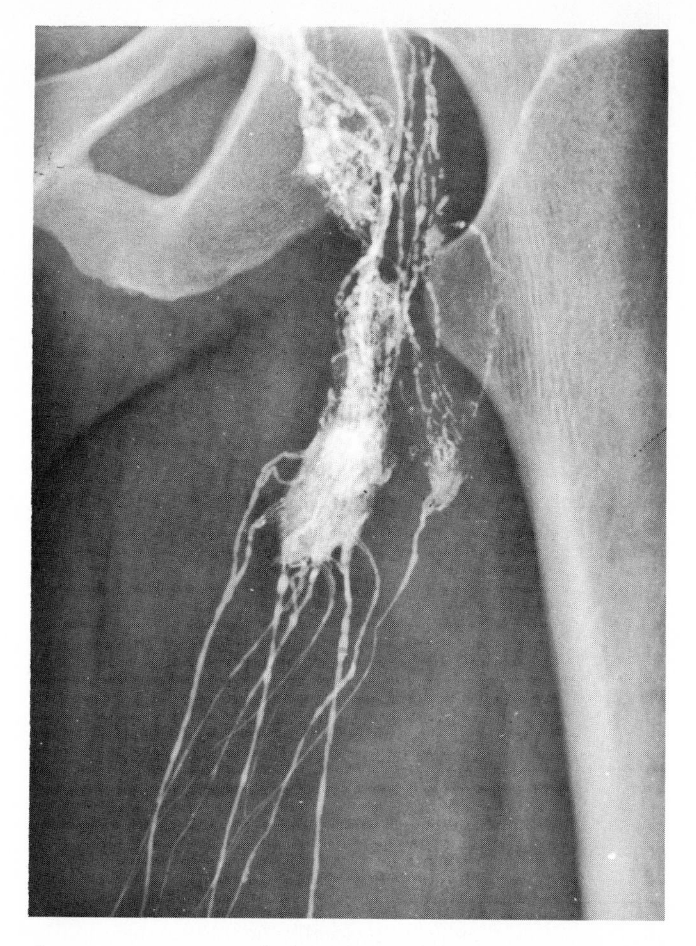

Figure 9-3. Magnification radiograph of normal left superficial subinguinal lymph node with afferent and efferent lymphatic channels in a 40-year-old woman. [Courtesy Harold J. Isard, Bernard J. Ostrum, and John R. Cullinan. From Medical Radiography and Photography, ***38**:108 (1962)*.]

The thorax contains a large group, some being situated in the thoracic wall and others being associated with the heart, lungs, thymus, esophagus, and trachea. A great number are massed in the groin region (Fig. 9-2).

Lymph Formation

Lymph is formed by the passage of material from the blood capillary into the tissue spaces, a process that is called **transudation.** It involves the processes of diffusion and filtration. **Diffusion** is the term applied to the passage of the molecules of one substance between the molecules of another substance to form a mixture of the two substances. For example, when two gases are brought into contact, the movement of the molecules of gas will soon produce a uniform mixture. If a dye is poured into a beaker of water, the water turns the color of the dye as the molecules of dye and water mingle with one another to form a uniform mixture of the two substances.

Theoretically, molecules of gases and liquids are constantly in motion and are able to move through membranes if they are permeable. **Osmosis** is a term applied to denote diffusion of such molecules through a semipermeable membrane. The direction of movement would be in both directions, but the molecules will travel

Figure 9-4. Structure of a lymph node.

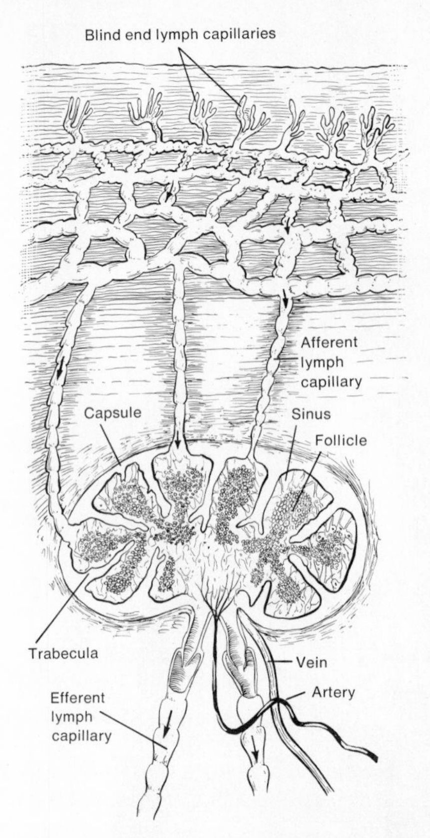

Figure 9-5. Diffusion, or osmosis, is not the direct result of fluid (hydrostatic) pressure. In the upper container the fluid pressure is equal on both sides of the membrane. However, the side containing the glucose solution becomes higher because of the osmotic flow of water from the weaker into the stronger solution.

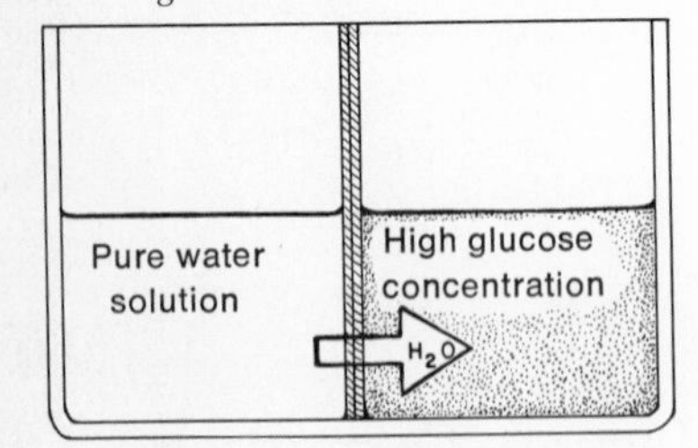

in greater numbers per unit of time from the place of higher concentration to the place of lower concentration until equilibrium is reached, at which time equal numbers of molecules will travel in each direction. Diffusion, or osmosis, although influenced by fluid (hydrostatic) pressure, is not the direct result of fluid pressure but may, on the contrary, be caused by differences of pressure on the two sides of the membranous partition and may work against fluid pressures (see discussion, p. 35). For example, when a solution of glucose is separated from a pure solvent such as distilled water by a membrane permeable to water but not to glucose, osmosis takes place whether the columns of fluid on both sides of the membrane are at the same level or not. If the columns are at the same level to start with (fluid or hydrostatic pressure the same on both sides of the membrane), the side containing the glucose solution becomes higher due to the osmotic flow of water from the pure solvent side into the glucose solution or from the side containing the greater concentration of water molecules (activity of the solvent molecules is greater) to the lower concentration (activity of the solvent molecules is less) (Fig. 9-5).

Filtration

Filtration may be defined as the passage of substances through a permeable partition as a direct result of pressure. Without differences of pressure, filtration does not take place, and the amount of material filtering through (filtrate) is dependent on the amount of pressure. Transudation is a result of both osmosis and filtration, since materials present in the tissue fluid, such as the proteins, could not be accounted for by osmosis alone because proteins of the blood plasma are indiffusible and yet the lymph contains a considerable quantity of these proteins. The blood flows through the capillaries with an approximate pressure of 30 mm Hg, but the pressure in the tissue spaces surrounding the capillary is much lower. As a result, this difference in pressure drives a certain quantity of the inorganic and crystalline constituents of plasma from the capillary into the tissue spaces along with a small quantity of proteins and a large quantity of water, while the red and white blood cells are excluded. At the venous end of the capillaries the pressure difference between the capillary and the tissue site is not as great, which acts to reduce the rate of filtration. By the same token, the plasma protein concentration is increased at the venous end by the loss of fluid to the tissue in the arterial capillaries, resulting in an increase of protein osmotic pressure, which acts as a "pulling" force to attract fluids into the capillary as well as to hold fluids within the vessel (Fig. 9-6). This action prevents excess loss of fluid from the capillaries and also aids in the return of water and crystalloids to the blood. Colloids cannot enter the capillary in this way but pass into the lymph capillaries along with water and crystalloids to be dumped back into the venous blood by way of the thoracic duct. Increases in hydrostatic pressure within the capillaries from any cause will increase transudation as well as interfere with return of substances to the capillaries. Under these conditions tissue fluid forms in greater quantities than can be carried off by the lymphatics, resulting in an excess accumulation of this fluid in the tissue spaces, a condition known as **edema.** In general, it can be said that two factors chiefly determine the amount of transudation: the pressure of the blood in the blood vessels and the condition of the walls of the capillaries in relation to the blood. Both factors will influence filtration, and the latter will influence osmosis.

Flow of Lymph

The lymph is projected along from the tissue spaces to the lymph capillaries by the processes of filtration, diffusion, and osmosis. The flow of lymph from the lymph capillaries to the venous system is maintained by differences in pressure that exist at the tissue site and the entrance of the great lymphatic ducts at the venous site. The tissue fluid is under greater pressure than the lymph in the lymph capillaries or the blood in the interior of the large veins of the neck. The lymph flows from a region of high pressure to a region of low pressure. The movements of respiration also aid in moving lymph through the lymphatics. Changes in pressure with respiration influence the lymphatics inside the thorax, causing them to act as pumps. The decrease of pressure in the thoracic cavity on inspiration expands the wall of the large lymphatic vessels, causing a sucking action on the lymph, which aids in its movement from the lower regions into the thoracic duct. There is abundant evidence that flow of lymph is increased by muscular activity and massage. Contraction of the skeletal muscle produces a milking action, forcing the lymph on toward the upper region of the body. The valves that are present in the lymphatic vessels (Fig. 9-7) prevent the lymph from flowing backward, in the same manner in which the blood is prevented from flowing backward in

Figure 9-6. How the difference in hydrostatic pressure between arterial capillaries and tissue fluid drives material from the capillaries into the tissue spaces. At the venous end of the capillaries the hydrostatic pressure is reduced, whereas the protein osmotic pressure is increased, causing materials to pass into the capillaries.

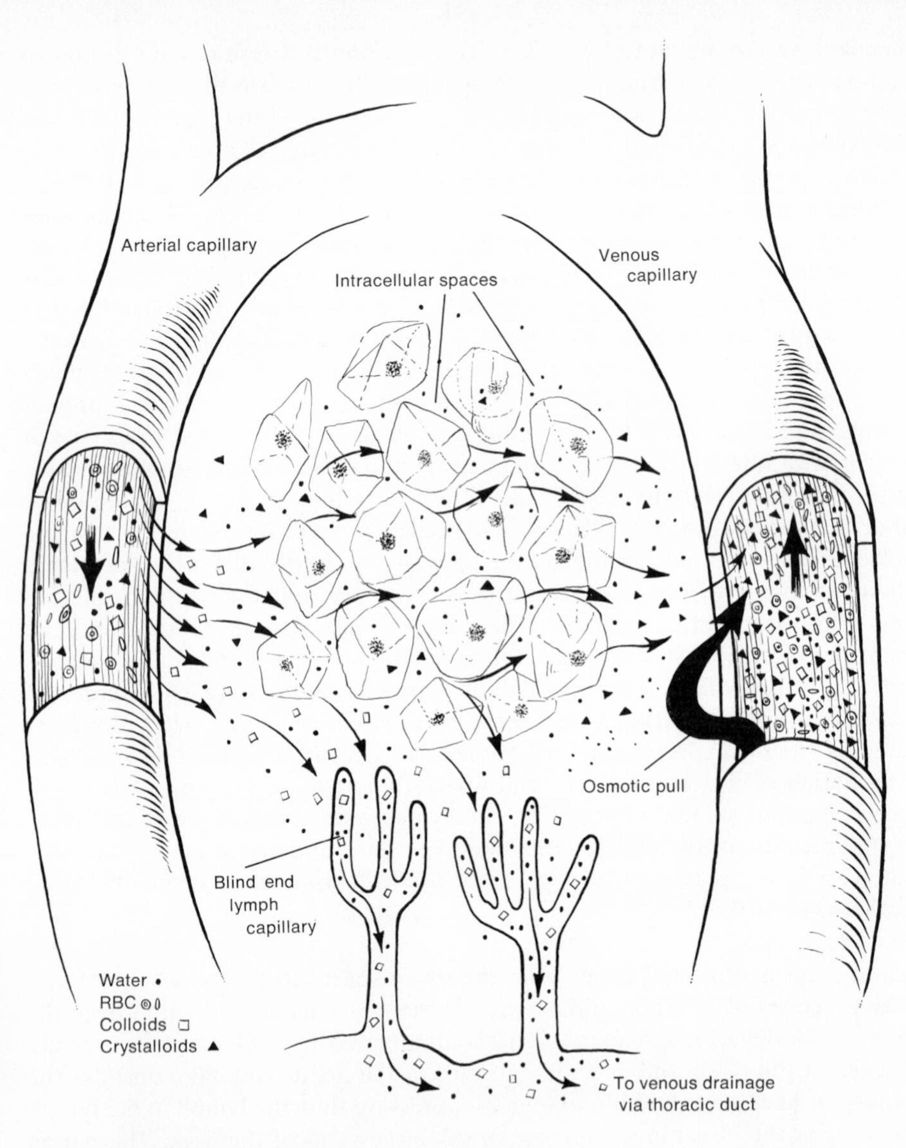

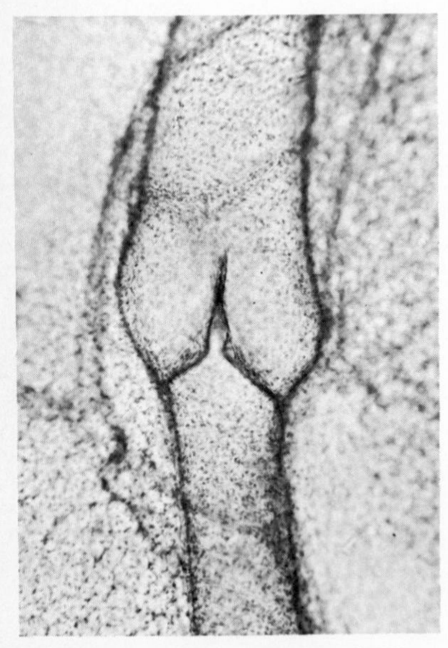

Figure 9-7. Valve in lymph vessel. (*Courtesy General Biological Supply House, Chicago.*)

the veins. In general, the amount of lymph flowing in a peripheral area is extremely variable and depends on the activity of that area, but under normal circumstances the flow is not very great even during activity.

Any condition that will produce an obstruction of the flow of lymph from the tissue spaces to the venous site, such as blockage of the lymph nodes, which occurs in infections of these areas, and changes in pressure differential between the tissue site and venous site, will cause an accumulation of lymph in the tissue spaces. This excessive accumulation of lymph in the tissue spaces we have already referred to as edema. Edema can also be caused by an excessive formation of lymph, the lymph accumulating faster in the tissue spaces than it can be carried away. This condition can be caused by changes in blood pressure associated with renal or cardiac disease, which in turn could affect capillary pressure. It also can be developed by any condition that increases the permeability of the capillary wall.

Composition of Lymph

The term *lymph* includes not only the fluid in the lymph vessels, but also the fluid that surrounds the tissue, the tissue fluid. The chemical composition of lymph varies a great deal depending on the source from which the sample being investigated is removed. For example, the lymph from the lower extremities is lower in protein concentration than that taken from the intestines or the liver. In addition, the lymph coming from the digestive tract during active digestion is much different from that coming from the same area when digestion is not going on. During digestion the lymphatic vessels of the intestinal tract, which are called **lacteals,** absorb the majority of fat from the intestine. The lymph in these vessels after a meal is milky white in appearance because of its high content of neutral fat and is referred to as **chyle.** The chyle, except for its high fat content, is similar in chemical composition to the lymph in other parts of the body.

The general characteristics of lymph are usually taken as those exhibited by a sample removed from the thoracic duct in an animal that has not been fed for a 24-hour period. Lymph obtained under these conditions is a clear, transparent or slightly opalescent fluid, which, like blood, clots when removed from the lymph vessels. Chemical examination shows that lymph contains prothrombin, calcium, and fibrinogen. No platelets are present, which perhaps is responsible for the slower clotting time of lymph in contrast to that of blood. The total solids are much less than those found in plasma, amounting to no more than 5 per cent. Lymph is 95 per cent water. The protein concentration of lymph varies considerably in different areas of the body, but that taken from the thoracic duct is about 3.2 per cent and is made up mainly of albumin, globulin, and fibrinogen. A certain amount of glucose appears to be always present, and several observers have found an appreciable quantity of urea. The concentration of salts and other substances that can permeate the blood capillary wall appears to be close to that observed in plasma. The calcium content is somewhat lower than in plasma, and the chloride content is somewhat higher.

Microscopic examination of lymph shows that it contains large numbers of leukocytes, which are mostly lymphocytes. Their numbers vary greatly with differerent animals and in the same animal under different circumstances. They range from 500 to 75,000 per cubic millimeter but in general, they are about as numerous in lymph as they are in the blood.

Serous Cavities

In mammals and other animals, large cavities, known as **serous cavities,** exist throughout the body and are considered as parts of the general lymphatic system, although they are not directly connected with the lymphatic system. These cavities are the pleural cavity of the chest, the pericardial cavity, the peritoneal cavity, and the cavities of the joints. They all contain serous fluid, which is actually lymph. The subarachnoid space surrounding the brain and spinal cord may also be regarded as a part of the lymphatic system, and it contains cerebrospinal fluid as a modified lymph.

Function of the Lymphatics

The functions of lymph appear to be similar in many respects to the functions of blood. The lymphatics act as a supplementary functional unit to the capillaries and veins in carrying tissue fluid from the tissue spaces to the veins. In areas of poor blood supply the lymph carries nourishment to cells and also carries their waste products away. The lacteals of the intestinal tract aid in the absorption of digested food, especially fats. The lymph system serves as a protective device in that inflammatory materials such as foreign proteins and bacteria which enter the body pass into the lymphatics, where they are ingested and partly destroyed by the leukocytes and macrophages that are normally present in this system. Since the lymph must pass through lymphatic glands, it must pick up gamma globulin formed there as a result of lymphocyte formation, and thus the lymph serves the body in a protective way by providing the blood with the means to form specific antibodies; that is, the gamma globulin formed on the formation or breakdown of lymphocytes has the ability to form specific antibodies under conditions of infection. Colloidal materials that are filtered out of the blood capillaries cannot enter them again because of the high pressure existing at the blood capillaries but they can enter the lymphatics. Thus another important function of this system is to return blood proteins from tissue fluids to the blood stream. Finally, since lymph bathes all cells of the body, it acts as an intermediary between the blood and the tissue, delivering to the cells the material they need to maintain normal function and carrying materials of their metabolic activity back to the blood.

The relationship of the lymphatic tissue to the **reticuloendothelial** system in general is of importance. This system is made up of those cells known as reticuloendothelial cells, which, although situated in many different parts of the body, bear certain similarities in morphology and in their function of phagocytosis. According to some, the term *reticuloendothelial system* should be restricted to include only those cells of the lymph nodes, spleen, liver, and bone marrow that have phagocytic properties and also the ability to differentiate into cells of the circulatory system.

Lymphoid Structures of the Nasopharynx

The lymphoid tissues of the pharnyx are so characteristically arranged that they form a ring consisting of the pharyngeal, tubal, faucial, and lingual tonsils as well as small nodes throughout the pharynx (Fig. 9-8). These lymph masses form a rich plexus throughout the pharynx and drain into the cervical lymph nodes and retropharyngeal nodes. The pharyngeal tonsil is present at birth but generally atrophies at puberty. The palatine tonsil is embedded in the lateral wall of the pharynx and is the lymphoid mass generally referred to as the **tonsil.**

The name *tonsil* is also often applied to the adenoid vegetation, or **adenoids,** in the forms of pharyngeal tonsil, third tonsil, or Luschka's tonsil. The word *adenoid* appears to have been proposed nearly two thousand years ago and will doubtless be permanently retained. The term *tonsil* is better restricted to the lymphoid tissue located between the pillars of the fauces (Fig. 9-9) and known more specifically as the palatine tonsils.

Lymphoid tissue is a normal constituent of mucous membranes, but in some instances it becomes pathological. Abnormalities in size (hyperplasias) can cause

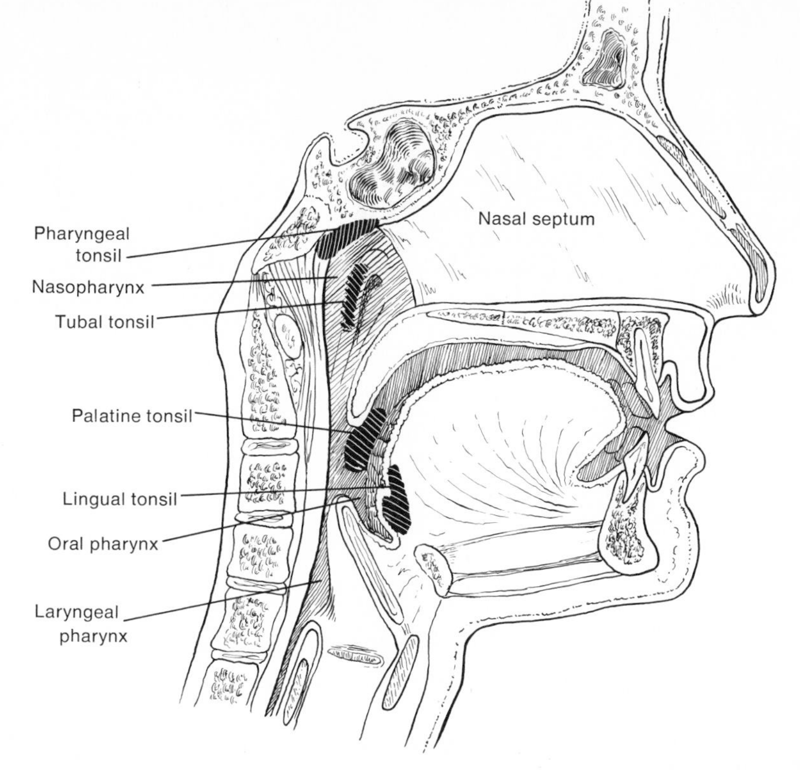

Figure 9-8. Sagittal section of head showing the location of the lymph masses in the pharynx and nasopharynx.

Figure 9-9. View of palatine tonsils from front showing their location between the pillars of the fauces.

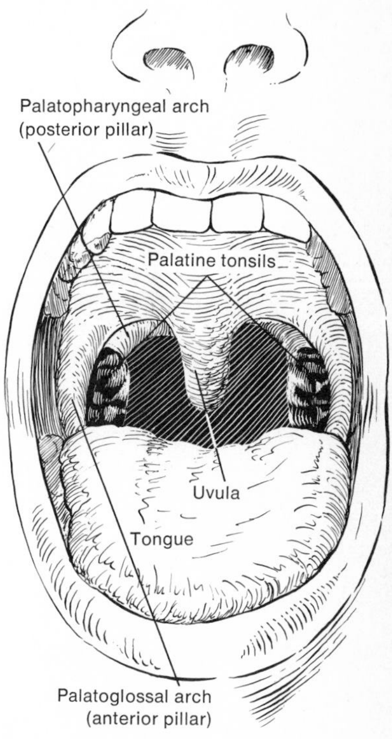

a good deal of disturbance, interfering with normal comfort and health. In addition, diseased lymphoid tissue—especially the tonsils and adenoids—provides a recognized avenue for invasion of the system by pathogenic germs. Since diseased tissue is no longer capable of performing its physiological function and is a detriment to health, removal by surgical operation must be resorted to; this is known as **tonsillectomy** for tonsils and **adenoidectomy** for adenoids.

The functions of the lymph structures of the nasopharynx differ in no way from those of the general lymphatic system. Most researchers are in agreement with Maximow and Bloom, who stressed that their participation in the formation of lymphocytes, which function in the attenuation of bacterial and viral agents, is the only established function that can be ascribed to them.

References

Abdou, I. A., et al. "Plasma Protein Equilibrium Between Blood and Lymph Protein." *J. Biol. Chem.*, **194:**15 (1952).

Bard, P. *Medical Physiology*. St. Louis: Mosby, 1965.

Bell, G. H., et al. *Textbook of Physiology and Biochemistry*. Baltimore: Williams & Wilkins, 1965.

Cope, O., and L. Rosenfeld. "The Lymphatic System," in: *Annual Review of Physiology*, Vol. 8. Palo Alto, Calif.: Annual Reviews, 1946, p. 297.

Dunker, C. K., and J. M. Yoffey. *Lymphatics, Lymph and Lymphoid Tissue*. Cambridge, Mass: Harvard U. P., 1941.

Fiore-Donati, L. (ed). *Lymphatic Tissues and Germinal Centers in Immune Response.* New York: Plenum, 1970.
Forder, L. L., et al. "Circulation of Plasma Proteins, Their Transport to Lymph." *J. Biol. Chem.,* **197**:625 (1952).
Kampmeier, O. F. *Evolution and Comparative Morphology of the Lymphatic System.* Springfield, Ill.: Thomas, 1969.
Kokhanina, M. I. "Reflex Effect from Chemoreceptors of the Excretory System on Lymph Flow." *Bull. Eksper. Biol. Med.,* **52**:18 (1961).
McMaster, P. D. "Conditions in the Skin Influencing Interstitial Fluid Movement, Lymph Formation and Lymph Flow." *Ann. N.Y. Acad. Sci.,* **46**:743 (1946).
Marshall, A. H. E. *An Outline of the Cytology and Pathology of the Reticular Tissue.* Springfield, Ill.: Thomas, 1956.

10

Respiration

The term **respiration** refers to the interchange of gases that takes place between an organism and its environment. For oxygen-consuming organisms the exchange involves the intake of oxygen and the elimination of carbon dioxide. Since both are gases, the exchange is mainly dependent on the physical process of diffusion rather than any active vital processes carried on by the tissues. Three types of respiration are recognized and are referred to as **external, internal,** and **cellular** respiration. External respiration involves the passage of oxygen from the air into the alveoli of the lungs and then into the blood. Internal respiration involves the passage of oxygen from the blood into the tissues, and **cellular respiration,** sometimes referred to as **biological oxidation,** refers to the actual utilization of oxygen by the cells of the body, resulting in the liberation of energy, some water, and carbon dioxide. Carbon dioxide passes mainly by diffusion from the tissues into the blood and, after transport to the lungs, from the blood into the air. The study of respiration, in the main, is concerned with three problems: (1) the mechanism by which a rapid interchange between the air and the blood is effected, (2) the means by which the blood transports oxygen and carbon dioxide to and from the tissues, and (3) the mechanism by which the cells take oxygen from the blood and give up carbon dioxide.

The Respiratory Passages

As we have already seen, external respiration is the exchange of gas between the air and the blood. This exchange takes place in the lungs, and, therefore, the term **pulmonary respiration** has also been applied to it. The lungs communicate with the external environment by a network of tubular passages, which, branching from the lungs to the outside, include the bronchioles, bronchi, trachea, larynx, pharynx, nasopharynx, and nasal cavity (Figs. 10-1 and 10-2).

The nasal cavities serve not only as structures that contain the special receptors

Figure 10-1. *Sagittal section of head, neck, and upper chest showing respiratory passages down to bifurcation of bronchus.*

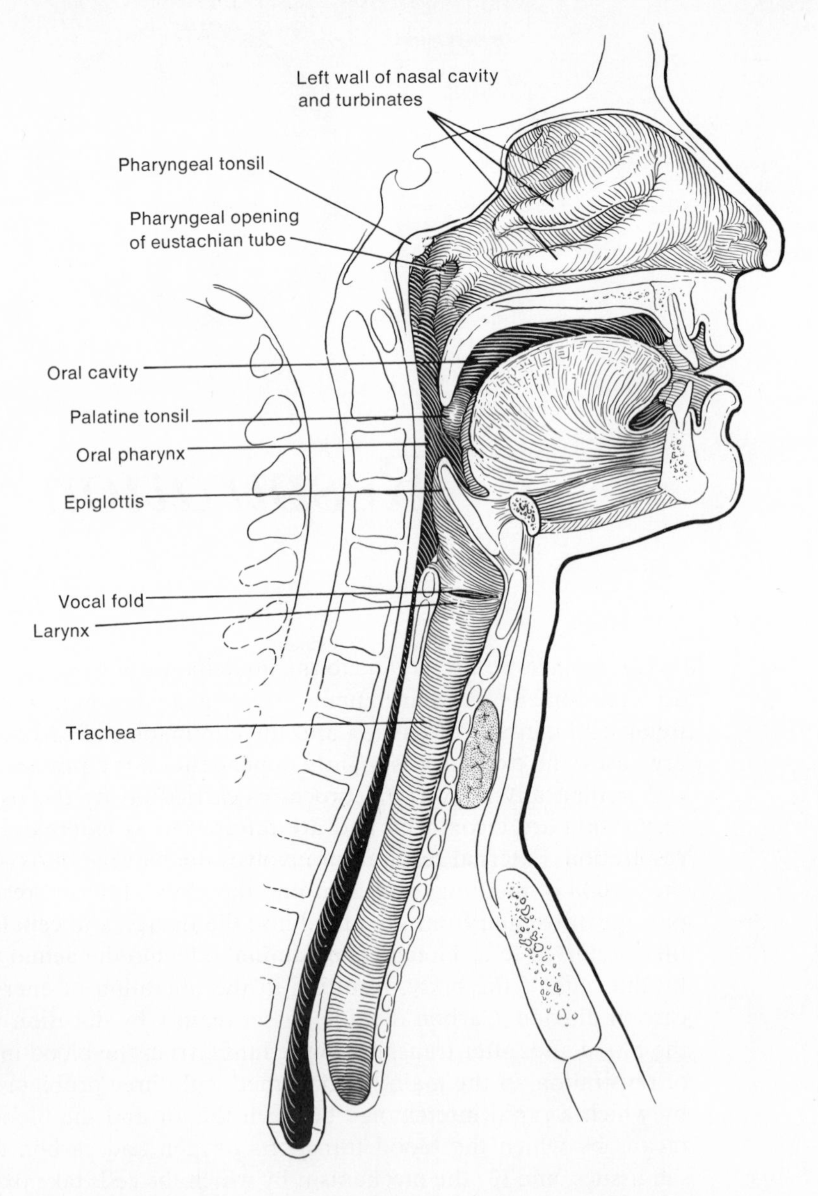

for the sense of smell, but also as functional passageways for air going to and from the lungs since they filter, warm, and moisten the air as it passes through. The nasal cavities open into the nasopharynx by two posterior nares, which in turn open into the pharynx, a common passageway for air from both the nose and mouth. The pharynx transmits air to the larynx, the organ of voice. The larynx, which will be described in more detail in the section dealing with voice, passes the air to the trachea.

The **trachea** (Fig. 10-2) is a cylindrical tube approximately $4\frac{1}{2}$ inches in length, having a mean diameter of 1 inch. Its wall consists mainly of connective tissue

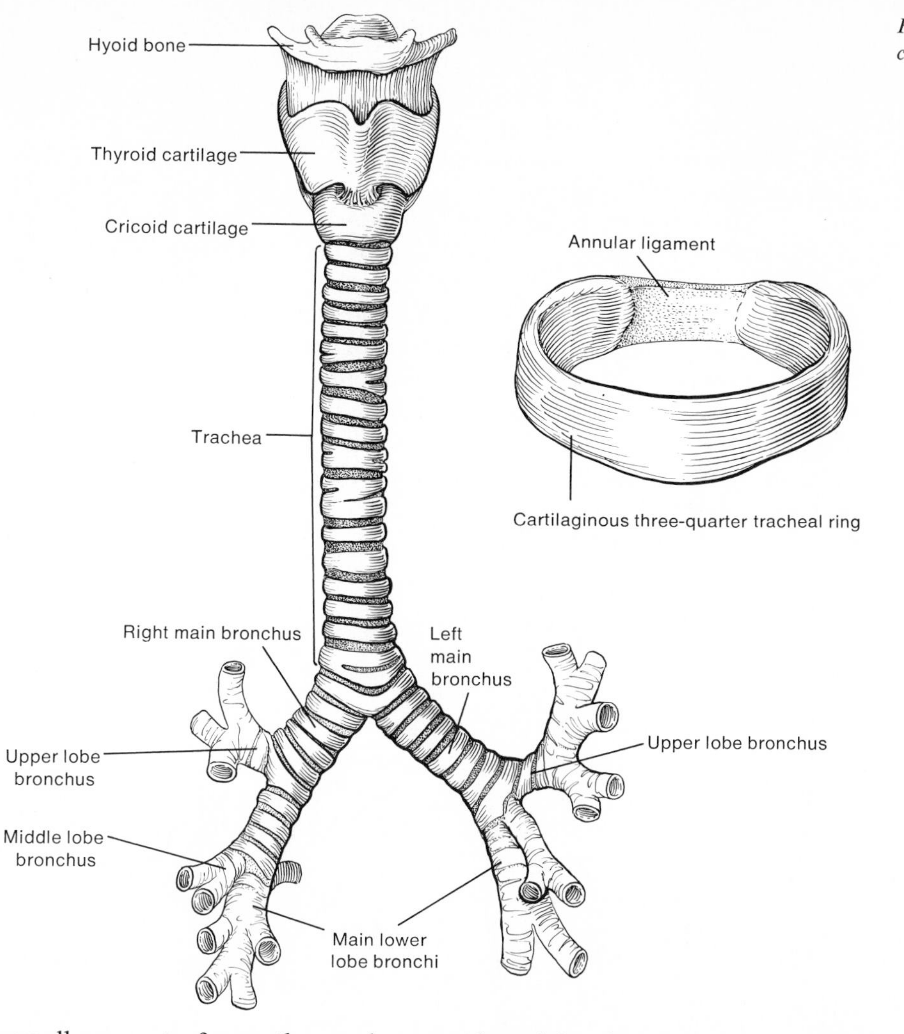

Figure 10-2. The human larynx and trachea.

and a small amount of smooth muscle strengthened by rings of cartilage. The cartilaginous rings do not encircle the trachea completely but are connected from behind by a tough membrane, the **annular ligament** (Fig. 10-2). The trachea, which is lined with a ciliated mucous membrane, serves as a large flexible tube, the lumen of which remains open at all times. The mucous fluid secreted by the goblet cells and small glands of the membranous lining help to enmesh foreign particles carried in by the inspired air, while the cilia are continually driving the mucus with its trapped particles upward toward the larynx and so into the pharynx. The role of the cilia in propelling mucus and waste material toward the mouth is of great importance. The cilia are influenced in their action by chemical changes in the blood and by substances applied locally; nerve impulses do not influence their action. The beating action of the cilia is depressed by cold and increased when the temperature of the cells is raised above normal. Efficiency of the cilia also depends in part on the viscosity and stickiness of the mucus secreted by the membrane. An increase in viscosity will decrease the rate and force of ciliary action, and a decrease in viscosity will have the opposite effect by increasing the rate and force.

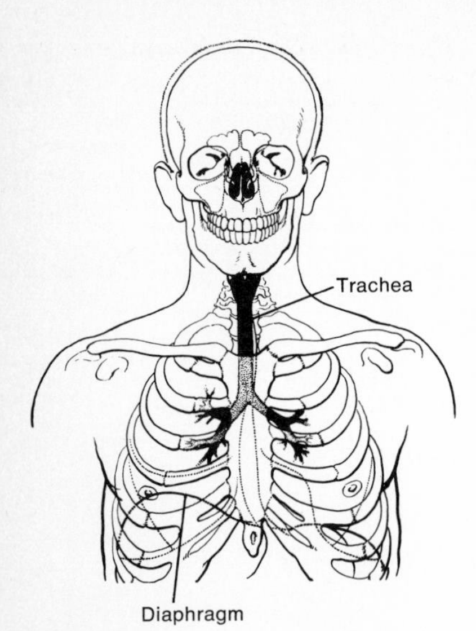

Figure 10-3. Position of trachea and diaphragm in relation to ribs.

The trachea begins at the lower end of the larynx, which is at the level of the seventh cervical vertebra and extends to the fourth or fifth thoracic vertebra where it terminates by dividing into the right and left **bronchi** (Fig. 10-3). As each bronchus enters the lung, it is accompanied by large blood vessels and surrounded by a number of lymph nodes. Each bronchus branches into lateral **bronchi,** going to the superior lobes of the lung, and ventral and dorsal bronchi, supplying the middle lobe. These bronchi continue to subdivide so that they become finer and finer, forming a complicated bronchial tree (Fig. 10-4). These fine branches are called **bronchioles;** the terminal ones, **respiratory bronchioles.** The smooth muscle is more highly developed in the walls of the latter than in any other part of the bronchial tree. The terminal bronchioles are 0.3 to 0.4 mm in diameter and do not vary up to the periphery of the lung, except where they give off branches called **alveolar ducts,** from which arise single **alveoli,** or **alveolar sacs,** which are considerably larger (Fig. 10-5).

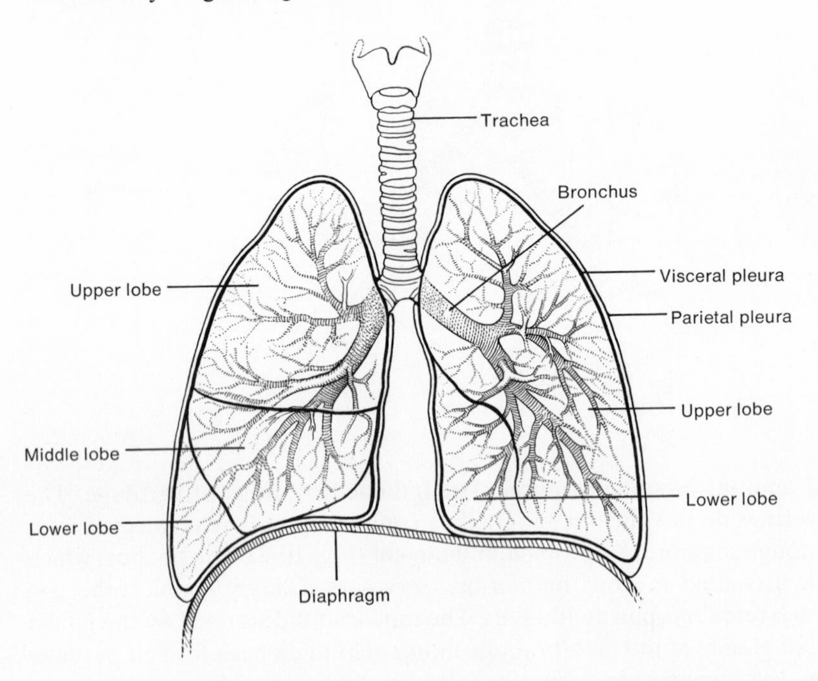

Figure 10-4. Bronchial tree in relation to lung showing lobes and membranes of lungs.

An alveolus is a small pocket about 0.25 mm in diameter; its thin wall is lined by a fine network of capillaries supported by connective tissue. It is through the thin wall of the alveoli that the actual exchange of gases between air and blood takes place. It has been estimated that the surface area presented by the lungs for respiration amounts to about 40 ft^2, but under ordinary circumstances only about 20 per cent of this surface is utilized at any one time.

The bronchial tree elongates and the bronchioles dilate on inspiration, followed

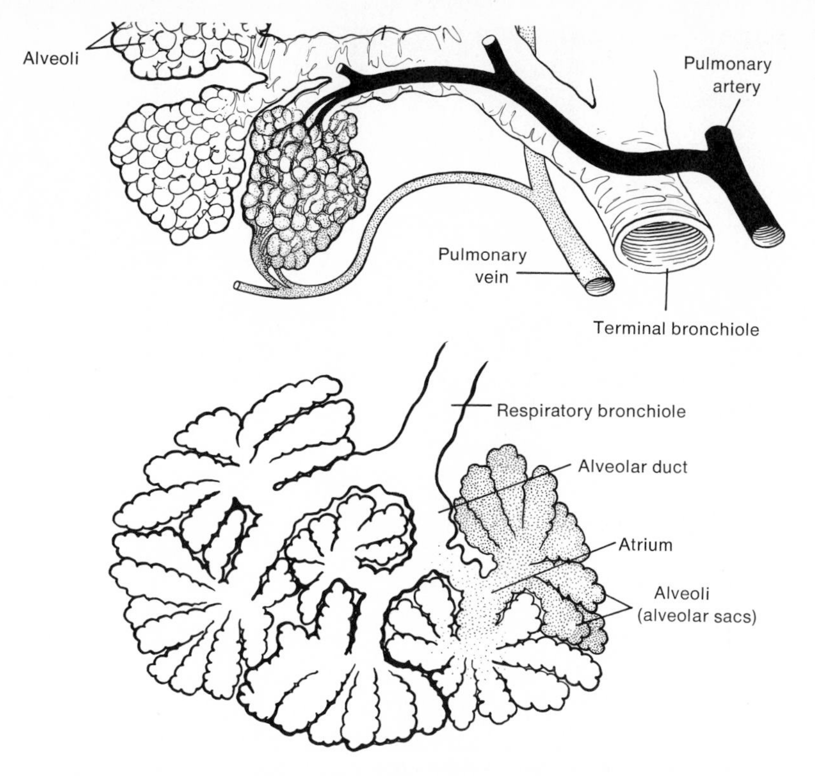

Figure 10-5. Structure of the alveoli showing the relationship of the terminal and respiratory bronchioles to the alveolar duct. The relationship of the pulmonary vein and artery to the alveoli is also represented. A lobule of a lung consists of a number of alveolar sacs with their bronchioles surrounded by connective tissue carrying blood vessels and lymphatics.

by shortening and constriction during expiration. The bronchioles also exhibit definite peristaltic movement, which assists in the propulsion of foreign materials toward the larger air tubes.

Three mechanisms serve in the expulsion of foreign material from the respiratory passages: (1) ciliary action, (2) peristaltic motion of the bronchioles, and (3) the cough reflex. The coughing act itself is elicited by irritation of the respiratory passages beyond the nose. A sneeze is elicited by irritation of the mucous membranes of the nose. The impulses responsible for a cough are set up by irritation of sensory receptors of the terminal fibers of the glossopharyngeal nerve and of the sensory endings of the vagus located in the larynx, trachea, and bronchi. Impulses that lead to a sneeze are set up by irritation of sensory receptors of the trigeminal nerves and probably also by excessive stimulation of the olfactory end organs. Both reflexes are characterized by a deep inspiration, followed immediately by a closure of the vocal cords, which do not open until the expiratory movement is under way, so that the outward movement of air is an explosive one, sweeping the offending material out of the respiratory passages.

The Lungs

The **lungs** take their origin as a diverticulum from the alimentary canal and may be considered a large, branched, specially modified gland lined with mucous membrane and consisting of a conducting portion and a secreting portion. The trachea, bronchi, and bronchioles represent ducts, and the secreting alveoli of an ordinary gland are represented by the air sacs, or pulmonary alveoli, which

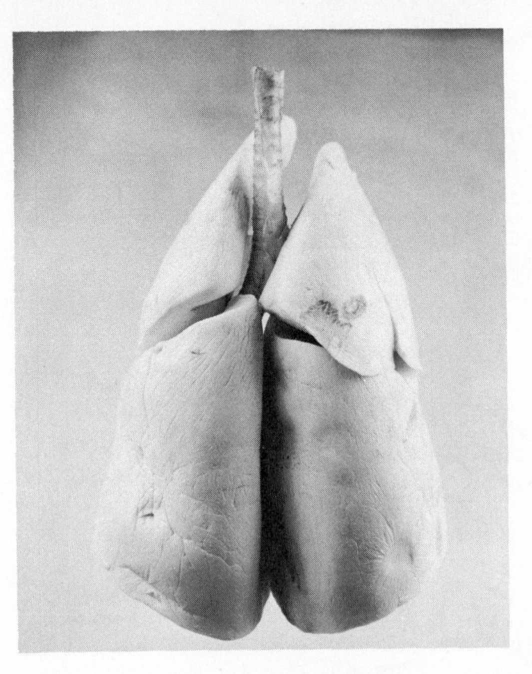

Figure 10-6. Dorsal view of dog's lungs. Right lung (right side of photo) is divided into three lobes. Left lung is divided into two lobes.

have lost most of their active secretory function. The lungs (Fig. 10-6) are cone-shaped organs which occupy the thoracic cavity and are separated into the right and left lung by the heart and other contents of the mediastinum. The right lung is larger and broader than the left and is divided into three lobes—superior, middle, and inferior. The left lung is smaller and narrower than the right and is divided into two lobes—superior and inferior.

The substance of the lungs is porous and spongy, containing a good deal of elastic tissue, and if permitted to do so, they will contract to a size much smaller than the thoracic cavity they occupy. Numerous blood vessels, lymphatics, nerves, bronchial tubes, and their alveoli make up the substance, which is divided into lobules (Fig. 10-5). A lobule consists of a number of alveolar sacs with their bronchioles surrounded by connective tissue carrying blood vessels and lymphatics. A number of lobules are bound together with their interlobular bronchi and still larger blood vessels to form a lobe, and several lobes join to form the lung.

The surface of the lung is covered by a serous membrane known as the **pleura,** which is reflected back, forming a covering also for the thoracic wall and the diaphragm (Fig. 10-4). The portion on the lung itself is known as the **pulmonary,** or **visceral, pleura;** that portion covering the chest cavity and the diaphragm is called the **parietal pleura.** Since the lungs fill the thoracic cavity under normal circumstances, the two pleura are in contact with each other so that the space between them, called the **pleural cavity,** is only a potential rather than an actual cavity. The membranes secrete a lubricating fluid, which prevents friction between the membranes and allows them to move easily on one another during respiration. The amount of serous fluid produced is normally small, and its absorption by the lymphatics keeps pace with its secretion. Inflammation of the pleural membranes, which is known as *pleurisy,* can cause friction, which increases the amount of serous fluid produced by the membrane. This results from irritation of the secretory cells and excessive transudation of fluid from the congested blood vessels.

Under normal circumstances the lungs are always in a state of distention, sometimes greater sometimes less because of their location in the airtight thorax, which is under a pressure less than that of atmospheric pressure. At the end of inspiration the intrathoracic pressure amounts to about 751 mm Hg; at the end of expiration the intrathoracic pressure has increased to 754 mm Hg; the range is 6 to 9 mm Hg below atmospheric pressure (−6 to −9 mm Hg). As a result of their elasticity and the negative pressures, the lungs fill the thoracic cavity. If air is admitted through the chest wall by puncture of the pleural membrane, the lungs immediately collapse and contract to a size much smaller than the thoracic cavity they occupy. Allowing air to enter the pleural cavity in this way is known as a **pneumothorax;** this is sometimes done to one lung when diseased, to assist its recovery.

Innervation of the Lungs

The nerves to the lungs are derived chiefly from the vagus. As the vagus passes around the root of the lungs, it gives off anterior branches, which form the anterior pulmonary plexus, and posterior branches, which form the posterior pulmonary plexus. Both of these plexi receive fibers from the sympathetic, especially from the first four thoracic ganglia, and also directly from the first, second, third, and fourth thoracic spinal nerves. The lower part of the trachea is supplied by fibers coming directly from the vagus. Some of the nerve fibers reaching the lung along the vagus nerve are motor fibers for the muscle cells of the bronchial passages and trachea. The majority of fibers are sensory, concerned with the regulation of respiration. The sympathetic fibers from the thoracic sympathetic chain function as vasoconstrictor fibers for the pulmonary vessels.

Mechanics of Respiration

Before birth the lungs contain no air; they are in a condition of collapse. This condition is called **atelectasis** and also involves the collapse of the bronchioles. The more rigid bronchi and the trachea are partly filled with fluid. When the chest expands with the first breath, the inspired air opens up some of these passages and succeeding breaths open the lungs more and more, until all the alveoli and bronchioles are included and the whole force of the expiratory act is directed toward driving out previously inspired air. In a newborn animal there is no negative pressure in the thoracic cavity. The lungs at rest are not distended, and injection of air into the pleura does not lead to contraction of the lung substance. This state of affairs is gradually established and is due to the fact that the thorax is growing more rapidly and thereby becoming relatively more roomy for the lungs contained within. Distention of the lungs in the adult occurs as a result of this increased capacity of the thoracic cavity and the ensuing drop in pressure.

The interior cavity of the lungs formed by the alveolar sacs and respiratory bronchioles is called the **intrapulmonic cavity** and is permanently open to the outside air by way of the bronchi and trachea. Intrapulmonic pressure is therefore the same as atmospheric pressure and is usually given as 760 mm Hg. A drop or increase in the intrapulmonic pressure will have the effect of sucking air into the lungs or expelling air until the intrapulmonic pressure is again in equilibrium with atmospheric pressure.

Breathing occurs in two phases, called **inspiration** and **expiration.** The act of breathing is directed toward changing the pressures within the lung by changing the size of the thoracic cavity. When all the muscles of respiration are at rest, a condition occurring at the end of a normal expiration, the size of the thorax may be considered normal and the intrapulmonic pressure equal to atmospheric pressure. An increase in the size of the thorax constitutes **active inspiration,** the result of which lowers the intrapulmonic pressure and causes air to rush into the lungs. Under normal conditions the thoracic cavity passively returns to its resting size after active inspiration, causing an increase in intrapulmonic pressure and an expulsion of air. This is known as **passive expiration.** The inspiratory and expiratory acts following one another form respiration.

Inspiration

On inspiration, the size of the thorax is increased by contraction of several muscles classed as **inspiratory muscles.** These include the **diaphragm** and the **external intercostals.** The scaleni, sternocleidomastoid, and pectoralis minor muscles are also inspiratory muscles, but they play a less important role in normal breathing. Contraction of the diaphragm, a dome-shaped muscle separating the thoracic cavity from the abdominal cavity, causes it to descend into the abdominal cavity, increasing the size of the thoracic cavity in a vertical direction (Fig. 10-7). Contraction of the external intercostal muscles elevates the sternum and the anterior extremities of the ribs, moving them upward and outward. This action causes an increase in the lateral and dorsoventral diameter of the thoracic cavity. This increase in volume of the thoracic cavity causes the elastic tissue of the lungs to expand and fill the extra space; as a result the pressure falls within the intrapulmonic cavity and air rushes in from the outside until intrapulmonic pressure is equal to the atmospheric pressure. The amount of air that passes into the intrapulmonic cavity in a normal inspiration amounts to about 500 cc, and this is termed **tidal air.** The same amount passes out in a normal expiration. In ordinary respiration the expansion of the chest never reaches its maximum. An additional thoracic expansion can be brought about by more forcible muscular contraction of the muscles mentioned above and by the action of additional muscles of the trunk, larynx, and pharynx. Such forced or labored inspirations lead to an inrush of an additional quantity of air before equilibrium is reached. This additional quantity of air is often called the **complemental air,** or **inspiratory reserve,** and amounts to an average of approximately 2,500 cc, or 3,000 cc if the tidal air is taken into consideration with the complemental. It has been determined that the intrapulmonic pressure may drop from 30 to 74 mm below atmospheric pressure with a strong inspiratory effort.

Expiration

On relaxation of the diaphragm and the accessory muscles of inspiration, the ribs and diaphragm return to their normal positions, and the thoracic volume decreases to its original size, causing an increase in the intrapulmonic pressure

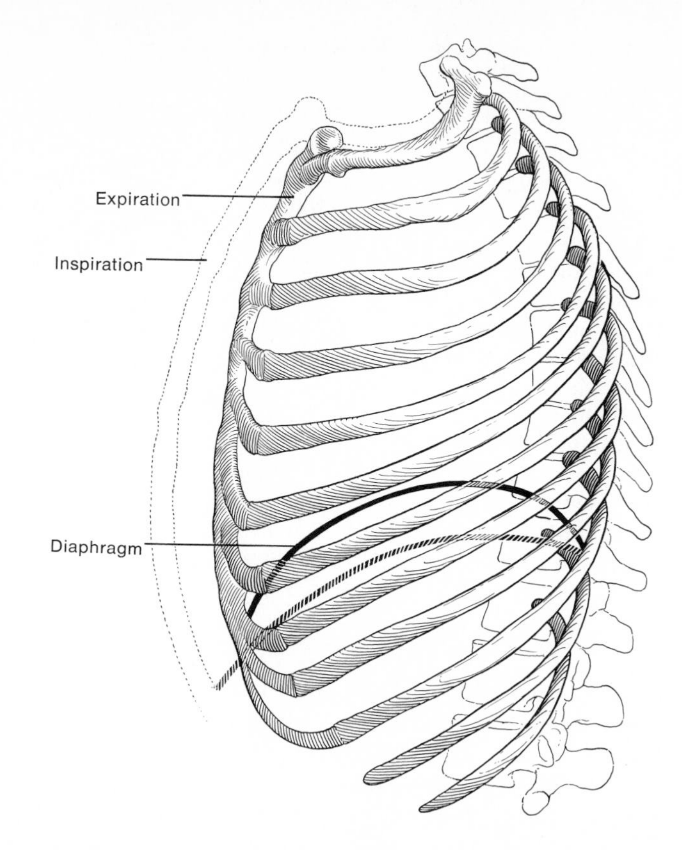

Figure 10-7. Position of the sternum, ribs, and diaphragm on inspiration and expiration.

above atmospheric so that air rushes out until equilibrium is reached. Normal expiration is considered passive because of the lack of active muscle contraction in this act and the fact that the elastic recoil of the lungs accounts for most of the action. However, in ordinary respiration the contraction of the thoracic cavity never reaches its maximum. Further decrease in thoracic volume or a **forced expiration** may be accomplished by the contraction of the **expiratory muscles,** which include the **abdominal muscles** and the **internal intercostals.** Contraction of the abdominal muscles forces the diaphragm farther up into the thoracic cavity, decreasing its ventral diameter. Contraction of the internal intercostals depresses the ribs and the sternum, decreasing the thoracic cavity in a dorsoventral and lateral direction. By use of these muscles, an active expiration can be accomplished with an additional quantity of air, the **expiratory reserve,** or **supplemental air,** being expelled. This amounts to about 1,000 cc in the average person if the tidal air is not included, or 1,500 cc if the tidal air is included. The positive pressure of a strong expiratory effort varies from 62 to 100 mm Hg. Even after the most forcible expiration, an additional quantity of air still remains within the lungs as a result of the normal distention of the lungs. This air amounts to about 1,500 cc and is termed the **residual air** (Table 10-1). The residual air is present in the lungs even after death but may be expelled when the pleural cavity is opened and the pressure on the two sides of the lungs becomes equalized.

Table 10-1. ***Pulmonary Function (Normal Values)***

Vital capacity	
Adult normal range	1,500–7,000 cc
Adult male	4,500 cc
Adult female	3,100 cc
Tidal air	200–700 cc; average, 500 cc
Inspiratory reserve	2,500 cc
Expiratory reserve	1,000 cc
Residual air	1,500 cc
Total lung capacity	Vital capacity and residual air
Respiratory rate (resting)	14–20/min; average, 17/min
Pulmonary ventilation (minute volume)	Tidal air × rate: average, 500 × 17 = 8,500; normal range, 3000–10,000
O_2 uptake	250 cc/min
CO_2 output	200 cc/min
Diffusion constant	23–43 cc/min
Respiratory quotient	$\frac{CO_2}{O_2}$ = 0.77–0.90; average, 0.85

When the lungs are collapsed as a result of opening of the chest, some air still remains entrapped within the alveoli as a result of the collapse of the bronchioles, which do not have rigid walls. This air cannot be expelled by ordinary means and is known as the **minimal air.** The minimal air is responsible for the characteristic buoyancy of pulmonary tissue; the lungs of a dead animal, once having taken air in, will float when placed in water and for this reason are popularly known as the "lights."

The total amount of air that can be expelled by the most forcible expiration possible after the most forcible inspiration amounts to about 3,500 to 4,500 cc in the average person. This is the sum of the complemental, tidal, and supplemental airs and has been called the **vital capacity.** The maximum volume of air that the lungs can expel may be divided into three portions:

Forced inspiration	1. Complemental air	2,500 cc (3,000 cc if tidal air is included)
Normal respiration	2. Tidal air	500 cc
Forced expiration	3. Supplemental air	1,000 cc (1,500 cc if tidal air is included)
	4. Vital capacity	4,000 cc

The complemental air is usually determined by measuring the amount of air that can be inspired by the most forcible effort, possible beginning at the end of a normal expiration. This volume would then include the tidal air. The supplemental air is usually determined following the most forcible expiratory effort possible after a normal expiratory effort has been made. This volume then would not include tidal air. Most of these volumes can be measured by using a modification of a gas meter called a *spirometer* (Fig. 10-8).

Vital capacity in normal adults varies from 1,500 to 7,000 cc. The average for the adult male between the ages of 18 and 30 may be accepted as 4,500 cc and

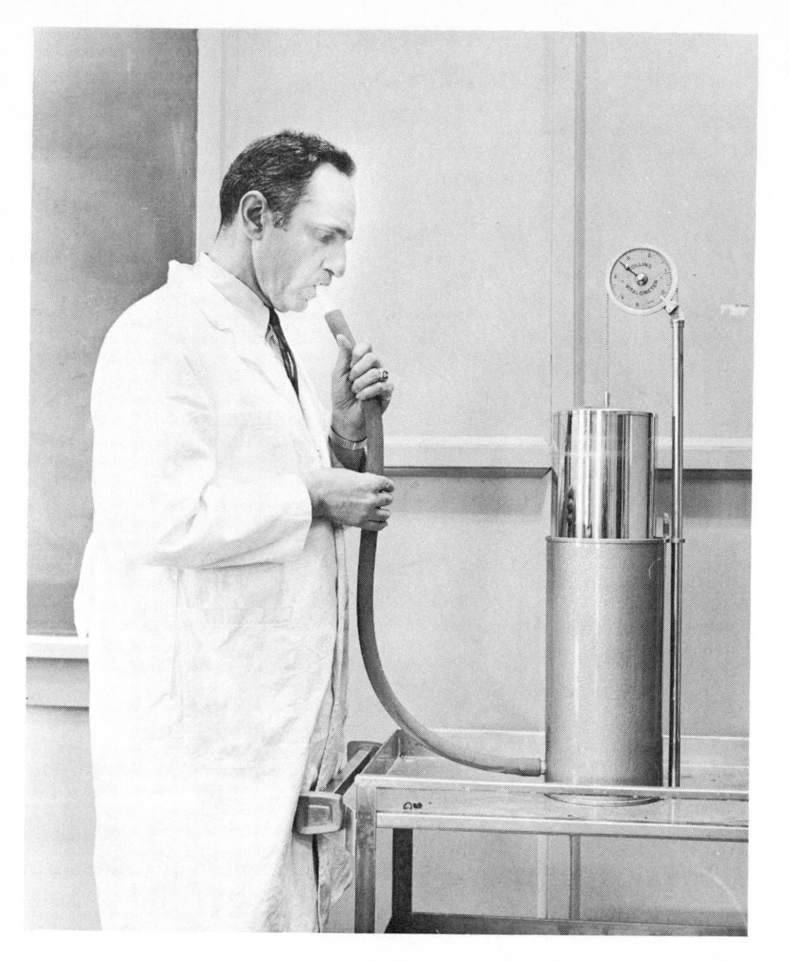

Figure 10-8. Determination of lung volumes using the vitalometer (spirometer).

for women in the same age range, 3,100 cc. The vital capacity together with the residual air comprises the **total lung capacity** (Table 10-1). The total lung capacity and its subdivisions vary with age, sex, and body size. The subdivisions of the measurable lung volumes will also show variations with body position. In the supine position the supplemental air decreases from the normal measurement for the standing position, and the complemental air increases and the vital capacity decreases. Variations in these measurable lung volumes may be a result of changes in the number of functioning lung lobules or of a limitation to the expansion of the chest cage because of body position or some other limiting factor.

Rate and Depth of Respiration

Graphic records of respiratory movements can be obtained in various ways. The simplest and perhaps the most generally useful is recording the movements of the air moving into and out of the lungs during quiet respiration with a recording spirometer. This apparatus is derived from the Benedict–Roth apparatus in general use for basal metabolism determinations (Figs. 10-9 and 10-10). Another method commonly used is recording the movements of the chest. In man

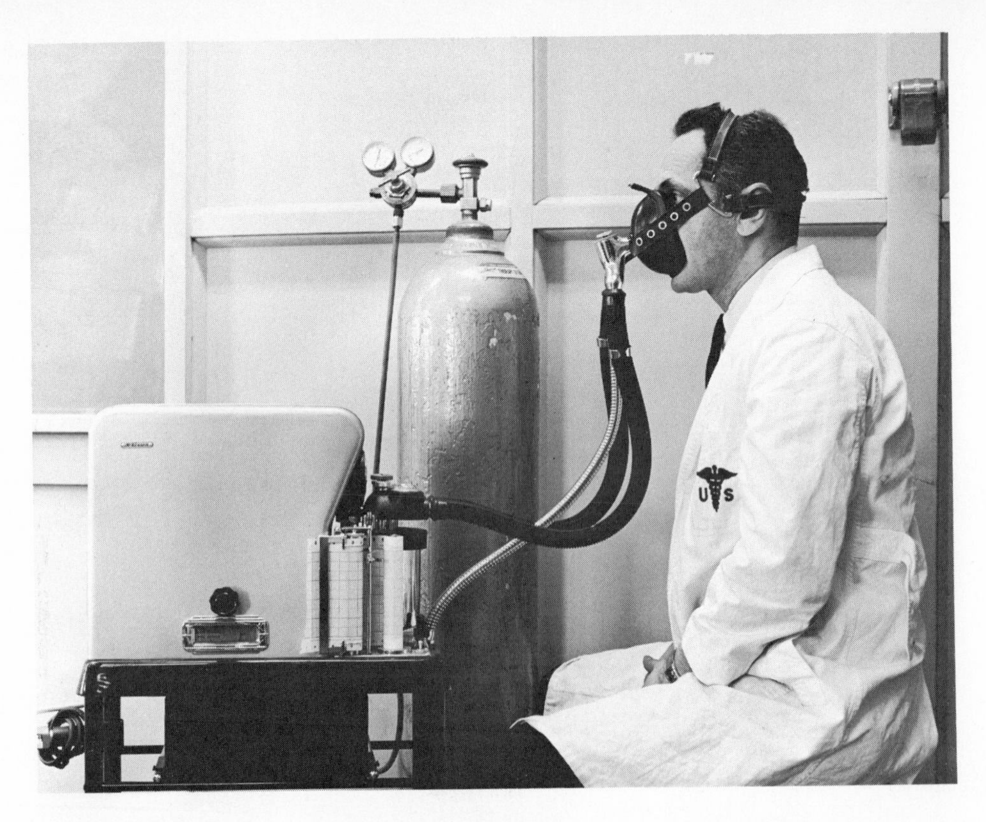

Figure 10-9. McKesson Metabolar in use to record respiratory movements graphically.

and larger animals the changes in the girth of the chest can be conveniently recorded by means of a pneumograph (Fig. 10-11). This consists of a hollow elastic cylinder, which can be strapped around the chest by a strap attached to each end of the cylinder. When the chest expands on inspiration, the ends of the cylinder are stretched and the bellows are pulled out, causing a decrease in the air pressure within the tube. When the chest contracts, as in expiration, the opposite occurs and the pressure is increased. If the interior of the cylinder is attached to a recording tambour, the lever of the tambour will be depressed on inspiration and elevated on expiration. Curves are obtained resembling the one shown in Fig. 10-12. As is shown, inspiration begins suddenly and advances rapidly, being followed immediately by expiration, which is rapid at first but tapers off slowly toward the end of the expiratory phase. In normal respiration the rate ranges between 14 and 20 per minute with the average taken as 17. The volume of each breath, that is, expiration or inspiration, is approximately 500 cc, with a range between 200 and 700 cc. The total amount of air taken in during 1 minute is called the **minute volume** of lung ventilation. The average minute volume can be determined by measuring the total volume of air inhaled during a certain time period and dividing this by the number of minutes or by multiplying the average rate per minute by the average depth of each inspiration. The volume of air inhaled during a certain time period can be determined by using a gas meter,

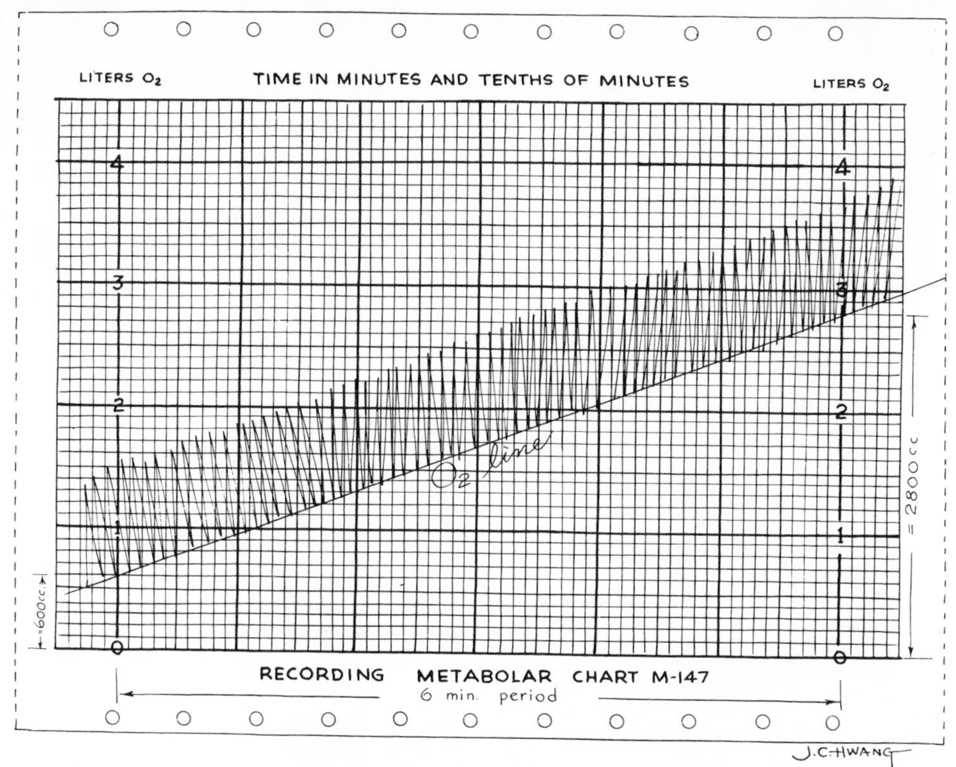

Figure 10-10. Typical record of the respiratory movements obtained from the apparatus represented in Fig. 10-9.

as shown in Fig. 10-13. During rest the minute volume varies from 3,000 to 10,000 cc, the average being taken as 8,500 (17 × 500) (Table 10-1).

The rate and depth of respiration are influenced by such factors as age, muscular activity, emotion, and heart action. The average rate during infancy is about 44 per minute, gradually decreasing to 26 per minute during childhood and to 17 per minute between the ages of 18 and 25 years. During muscular activity metabolism is increased—therefore more oxygen is required, which leads to increased rate and depth of respiration. The rate of respiration increases proportionately to the amount of muscular activity. When the activity is moderate and steady, the depth and frequency of respiration continue to increase for several minutes and then level off. This leveling off is referred to as the **steady state** and is normally reached within 2 to 5 minutes. When the muscular activity is severe, the rate and depth are increased throughout the entire period; at times the depth may decrease. Respiration rates ranging from 30 to 75 per minute have been observed during such activities as running and swimming. Excitement and other emotional states may increase the rate and depth of respiration by impulses reaching the respiratory center from the cerebral cortex or from the hypothalamus. Stimulation of the sensor receptors of pain, heat, or cold can also produce variations in rate and depth of respiration. Any change in the normal concentration of oxygen and carbon dioxide carried in the blood or a change in the hydrogen-ion concentration will be reflected in changes in the rate and depth of respiration since these conditions will stimulate the respiratory center either reflexly or directly.

Figure 10-11. Pneumograph in use for recording respiratory movements.

Figure 10-12. Typical graph showing the respiratory movements of man obtained by using the pneumograph. Upstroke is expiration; downstroke is inspiration. Each time mark represents 5 seconds.

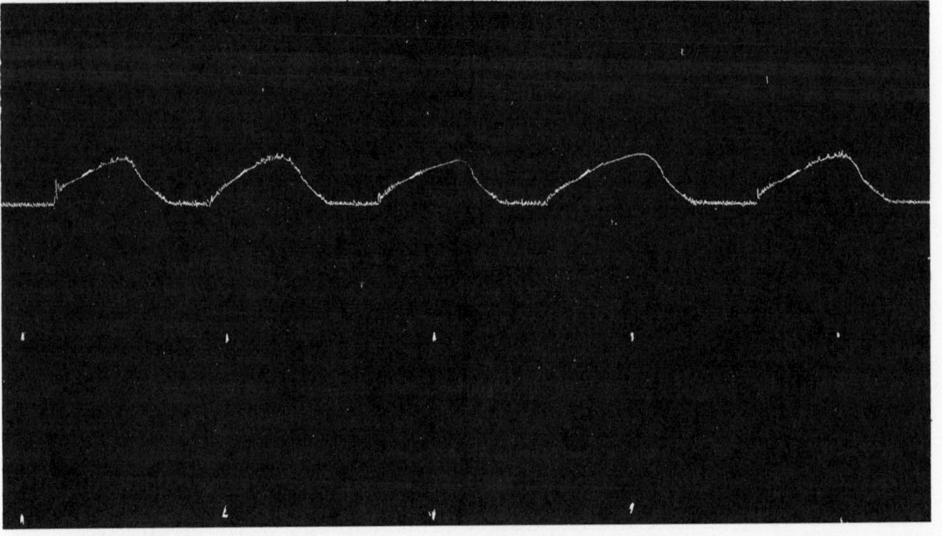

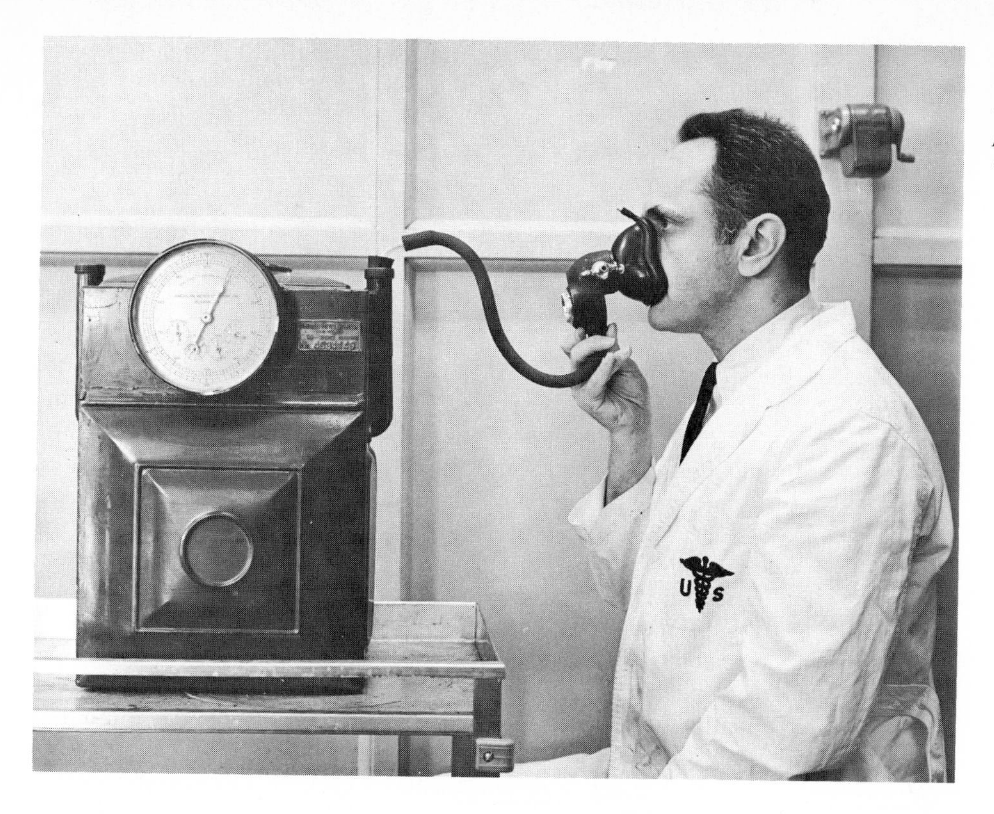

Figure 10-13. Apparatus for determining minute volume. The gas meter records the air inhaled by the subject. Exhaled air is passed off to the outside by way of the valve on the side of the mask.

Nervous Mechanism of Respiration

Although the diaphragm and all the other muscles used in respiration are voluntary muscles and can be controlled by direct effort of the will, breathing is an involuntary act and continues not only in the unconscious state, but even after the removal of all parts of the brain above the pons. Sections at various levels from the upper border of the pons down toward the medulla oblongata cause pronounced disturbances in respiration. Cutting through the medulla behind the tip of the calamus scriptorius causes cessation in breathing. The **center for respiration** in the brain thus appears to cover an extensive area, from the upper border of the pons to the lower third of the medulla. It is this center, sending coordinated impulses over efferent nerves, that brings about the coordinated muscular contractions of respiration. The respiratory center itself consists of two bilateral groups of nerve cells, one of which is mainly concerned with inspiration, the **inspiratory center,** and the other with expiration, the **expiratory center.** These two groups of neurons are found in the reticular formation of the medulla. Another group of neurons concerned with respiration is situated in the pons and is known as the **pneumotaxic center.** The action of the pneumotaxic center is to inhibit the inspiratory (apneustic) center and thus change the inspiratory movements into rhythmical movements characteristic of normal respiration. The inspiratory center is also under an inhibitory influence from the vagi. Pitts and his associates demonstrated that separation of the pneumotaxic center from the inspiratory center did not result in sustained inspiratory action unless the vagi were cut. When vagal influence was abolished by cutting of the vagi, sustained inspiratory movements (apneustic respiration) and complete cessation of rhythmical movements developed. The inspiratory center is therefore under a double inhibitory influence, either one of which is capable of converting the inspiratory movements into the rhythmical movements of normal respiration. The vagal impulses that inhibit the inspiratory center are initiated by the stretching of the lungs. Impulses originating in the inspiratory center pass out by way of the **phrenic nerve** to the diaphragm and of the **intercostal nerves** to the intercostal muscles to cause inspiration.

In summary, then, the activity of the respiratory muscles is coordinated and controlled by the nervous system, specifically consisting of a respiratory center contained within the brain stem covering an area from the upper border of the pons to the lower third of the medulla. The center itself consists of three definite areas: a pneumotaxic center, an inspiratory center, and an expiratory center. Normal respiratory movements are reflexly controlled, even though some voluntary control is possible. The sensory pathway for reflex control is by way of the vagus, which acts to inhibit the spontaneous activity of the inspiratory center. The phrenic and intercostal nerves are motor pathways from the inspiratory center to the muscles of respiration (Fig. 10-14).

Spontaneous respiratory activity is apparently dependent primarily on the inspiratory center. Impulses arising here spontaneously (tonic discharge) pass to the motor nerves supplying inspiratory muscles and also to the pneumotaxic center, which becomes excited and transmits impulses back to the inspiratory center, where they exert an inhibitory influence. The inspiratory center is also inhibited by impulses transmitted to it by way of the vagal nerves, which become stimulated on stretching of the lungs. This leads to the expiratory movement, which under normal circumstances is purely passive, that is, a relaxation of the muscles of inspiration.

The expiratory center does not discharge spontaneously (tonically) as does the inspiratory center but can be excited by way of sensory fibers that converge on it. Excitation of the expiratory center causes contraction of the muscles of expiration and at the same time inhibits the motor neurons of the inspiratory muscles. This results in a greater expiratory effort and an increase in lung deflation. Normally when breathing effort is increased, the extent of lung deflation is also increased by active stimulation of the expiratory center, which decreases the intrathoracic volume. It is most likely that compression receptors exist within lung tissue that transmit impulses inhibiting expiration in the same manner that stretch receptors inhibit inspiration. Thus, the degree of expiration is reduced by a reverse Hering–Breuer reflex (see p. 328). Figure 10-14 schematically represents the nervous control of breathing.

Chemical Control of Respiration

The respiratory center is also quite sensitive to changes in the carbon dioxide concentration of the blood flowing through it. Normal breathing in the resting subject, which is known as **eupnea,** depends primarily on the carbon dioxide tension of the arterial blood leaving the lungs. If the carbon dioxide tension increases above what is considered a normal tension of 38 mm Hg (50 ml per

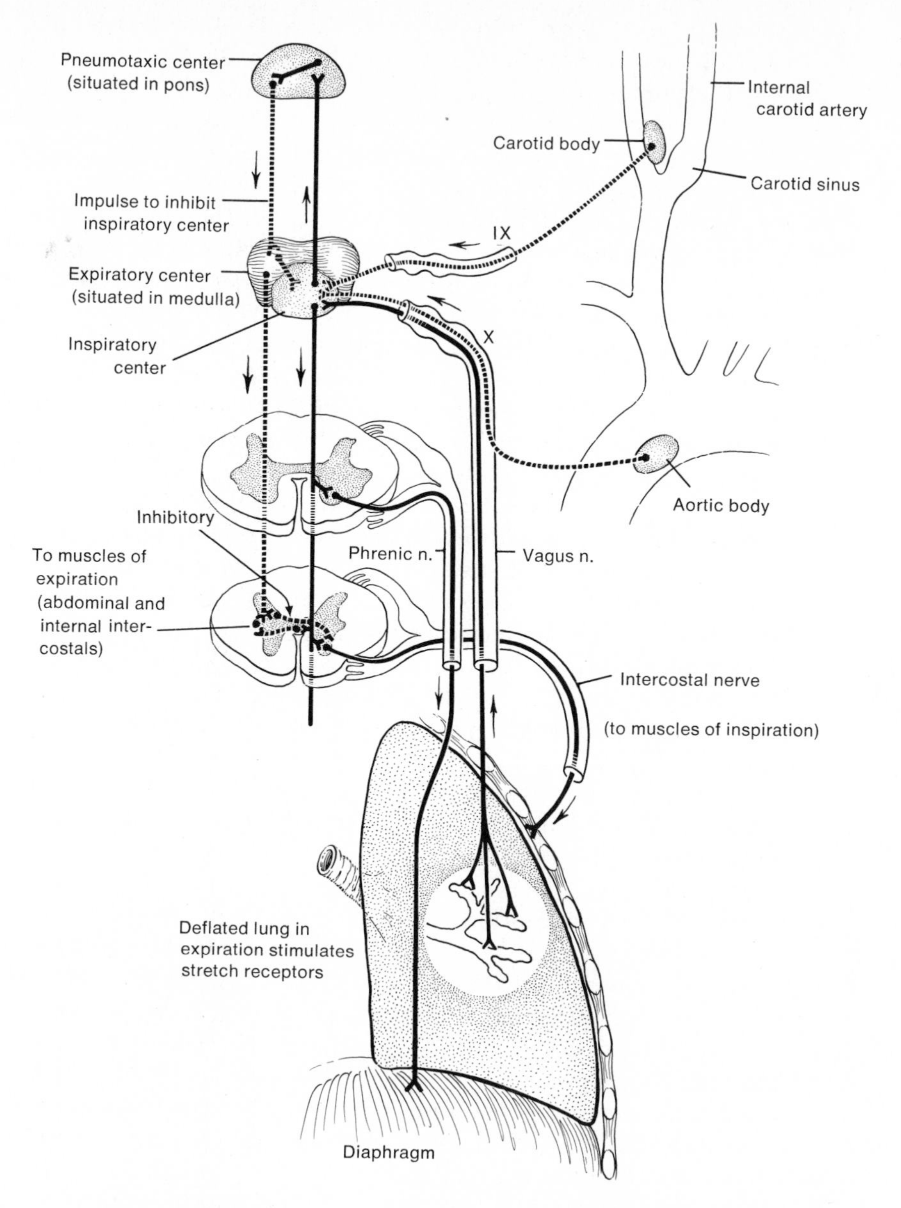

Figure 10-14. Schematic representation of the nerve control of breathing.

100 ml of blood), breathing is stimulated until the excess of carbon dioxide is eliminated and is restored to its normal level. On the other hand, if the carbon dioxide tension falls below normal, breathing is inhibited until sufficient carbon dioxide can accumulate to build it to its normal value.

Haldane and his co-workers showed that the respiratory center is so sensitive to changes in carbon dioxide tension that an increase of 0.3 per cent in the concentration of this gas in the alveolar air (which would mean an equivalent increase in the arterial blood as it is assumed that the carbon dioxide in alveolar air is in gaseous equilibrium with the carbon dioxide of the arterial blood) can double the volume of breathing. A decrease of 0.3 per cent below the level found during quiet breathing may cause apnea.

The oxygen tension of the blood plays a minor role as far as its influence on the respiratory center is concerned. This fact can be demonstrated very easily by rebreathing experiments. If a subject breathes deeply for a short period of time in order to overventilate the lungs (forced breathing) and his breathing is recorded, it can be seen from examination of the respiratory tracing that the period of forced breathing was followed by cessation of breathing. The cessation of breathing is known as **apnea.** It is a result of the washing out of carbon dioxide from the lungs during forced breathing so that the normal concentration of carbon dioxide falls well below the normal. A deficiency of carbon dioxide in the blood is known as **acapnea,** and this will exert an inhibitory effect on the respiratory center. If such a subject repeats the forced breathing but breathes into a paper bag, apnea will not develop because the gas mixture he is breathing contains excess carbon dioxide and therefore the alveolar carbon dioxide pressure does not drop and there is no drop in the arterial carbon dioxide concentration. In forced breathing, which is called **hyperpnea,** the first subjective reactions of the subject before apnea will be dizziness and faintness, resulting in some relaxation of his efforts. This probably arises from a disturbance of the circulation directly due to the too-vigorous movements of the chest. Also, a reduced carbon dioxide tension of the blood resulting from overventilation will cause constriction of the arterial blood vessels. For this reason it is best for the subject who is undergoing such tests to assume the supine position and to force depth rather than rate of breathing. It can be shown that a rise in oxygen pressure is not the stimulus to produce apnea, because gas mixtures containing a high percentage of oxygen have no inhibitory effect on breathing.

Reflex Control of Respiration

The activity of the respiratory center is influenced not only directly by carbon dioxide tension of the arterial blood, but also by reflex mechanisms (Fig. 10-15). Sensory impulses arising in different areas of the body can affect the center in various ways. For example, sensory impulses arising from stretch receptors located in the lungs and carried to the respiratory center by the vagus nerves inhibit inspiration, allowing expiration to take place. It appears the the function of these sensory impulses in normal breathing is to signal the depth of the inspiratory effort to the respiratory center and allow expiration to take place. It has also been demonstrated that an excessive expiratory effort sets up impulses in the vagus, which inhibits expiration and brings on inspiration. That these sensory impulses from the lungs play an important part in the control of respiration was first pointed out by Hering and Breuer in 1868, and therefore the action has been named the **Hering-Breuer reflex.** Thus, quiet breathing is a result of the following: The respiratory center inherently initiates inspiration. As the lungs expand, filling with air, the stretch receptors are stimulated, setting off sensory impulses in the vagus nerve, which become more intense as inspiration progresses. These impulses impinging on the inspiratory center inhibit its activity, and thus the muscles of inspiration relax and expiration takes place.

As expiration proceeds, impulses are set up by deflation of the lungs, which stimulate the inspiratory center. The center also comes under the influence of the stimulating effect of the build-up of carbon dioxide tension of the blood during the expiratory period. Thus inspiration is initiated again, and the cycle repeats itself, giving a rhythmical pattern to normal breathing. The rate of breathing is determined by the interval required for the neurons of the respiratory center to become sufficiently reactive to respond to the combined chemical and nervous stimuli operating at the time.

Another respiratory reflex, the **carotid and aortic reflex,** was discovered by Heymans and Heymans in 1927 when they demonstrated that impulses originating in the carotid sinus and aortic arch could influence the respiratory center. The carotid and aortic areas each contain two types of receptors; one type, **baroreceptors,** responds to mechanical stimulation, the other, **chemoreceptors,** to chemical stimulation. The pressoreceptors do not appear to serve any respiratory function under physiological conditions in the mammal. The chemoreceptors, which are contained in small glandlike structures, the carotid and aortic bodies (Fig. 10-14), when excited increase the rate and depth of respiration. The receptors are stimulated by a drop of oxygen tension of arterial blood, but not until it reaches a relatively low level. The average normal oxygen tension of arterial blood is about 100 mm Hg. An increase in oxygen tension has a depressant effect on the respiratory center. The chemoreceptors are also stimulated by an increase in carbon dioxide tension, but usually no reflex hyperpnea occurs until blood carbon dioxide is increased by more than 4 volume per cent. Chemoreceptors are also sensitive to increases in the hydrogen-ion concentration of the blood; however, respiratory responses to such changes are due to direct effects of increased hydrogen-ion concentration on the cells of the respiratory center. Under normal physiological conditions these receptors apparently play no role in the control of respiration.

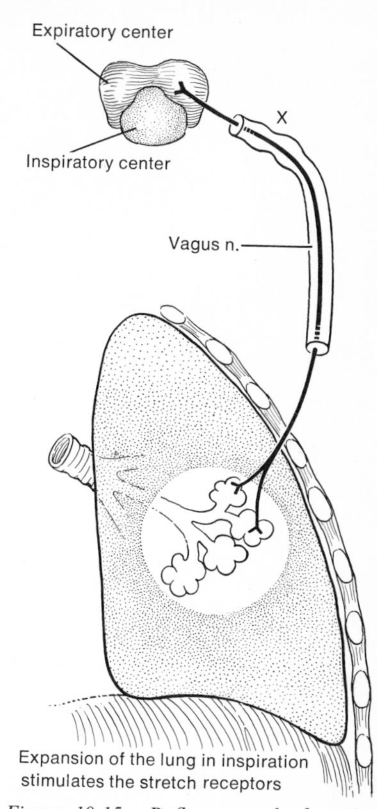

Figure 10-15. Reflex control of respiration. Expansion of lungs on inspiration excites stretch receptors, causing impulses to be carried to the inspiratory center, which produces an inhibition of the inspiratory center, allowing expiration to take place.

Reflex stimulation of the respiratory center may also be initiated by sensory impulses originating in the muscles and joints. Receptors located in the muscles and joints are stimulated by increased activity of these parts and probably play an important role in the increased respiration occurring during muscular activity. Sensory fibers from receptors of various kinds in the skin, especially for pain and temperature, lead to the respiratory center and, when stimulated, produce an increase in both rate and depth of respiration. Receptors in the mucous membranes of the respiratory tract initiate reflexes, which, when stimulated, modify respiration so as to protect the tract itself. During swallowing, respiration is reflexly inhibited by impulses running in the glossopharyngeal nerve, which are initiated in receptors located in the postpharyngeal wall. Sneezing and coughing are also modified respiratory acts and are reflexly initiated by stimulation of receptors in the respiratory tract. The sneeze and cough reflexes are discussed in detail on pages 315 and 333.

Voluntary Control of Breathing

Because the respiratory center has direct connection with the cerebral cortex, voluntary control of respiration for short periods of time is possible. The nerve cells giving rise to the voluntary impulses are in the motor areas of the cerebral cortex known as area 6a and 6b. Stimulation of area 6a increases the rate and depth of respiration, whereas area 6b appears to be inhibitory. The power to inhibit voluntarily the respiration is limited since the build-up of carbon dioxide concentration in the arterial blood during this period and its stimulatory effect on the inspiratory center will override the inhibitory influence. Thus, the muscles of inspiration contract despite the effort made to hold the breath. It appears that the breath can be held for about 40 seconds before involuntary control asserts itself.

Types of Breathing

There are two chief means by which the chest is enlarged in normal respiration: the descent of the diaphragm and the elevation of the ribs. The descent of the diaphragm causes a noticeable movement in the lower part of the chest and abdomen, and breathing in which this action is dominant is called **diaphragmatic breathing.** It appears to be characteristic of male breathing. The elevation of the ribs causes movement in the upper chest region, and breathing in which this action is the chief movement is called **costal breathing.** Costal breathing is characteristic in the female.

Respiratory Failure

Any agent that is capable of decreasing the activity of living cells will also effect a decrease in the activity of the nerve cells of the respiratory centers, and if the decrease is great enough, the nerve cells are rendered completely incapable of reacting to stimuli and respiration ceases. One common factor in all the conditions in which **respiratory failure** is likely to occur is oxygen lack in the cells of the center. In narcotic poisoning by such drugs as alcohol, ether, chloroform, morphine, barbiturates, and other depressants of the central nervous system, the oxygen lack is produced by the decreased pulmonary ventilation resulting from the direct depression of activity on the center. This depression of activity results from an interference with oxygen utilization within the cells of the center and from depression of the vasomotor center, which hinders circulation to the cells. All these effects are produced by the drugs mentioned in toxic dosages.

Other substances can also depress and, if in sufficient quantities, cause cessation of respiration by primarily interfering with tissue oxidations, because the respiratory nerve cells, like any other living cells, are depressed when their oxidative processes are deranged. Carbon monoxide, by binding up hemoglobin, prevents sufficient oxygen from being delivered to the cells. Cyanide poisoning produces respiratory failure by interfering with the utilization of oxygen by the cells.

Respiratory depression or failure can also result through direct injury or disease of the medulla or indirectly through increased cerebrospinal pressure. When cerebrospinal pressure exceeds that in the cerebral capillaries, the flow of blood is occluded and the entire brain suffers from acute anemia. The vasomotor center reacts to increase the blood pressure to overcome the obstruction, but if the cerebrospinal pressure continues to increase so that the vasomotor center cannot overcome it, both the respiratory center and vasomotor center suffer from anemia of increasing intensity, finally failing completely. Cerebrospinal pressure is raised in inflammatory disease and in the presence of tumors of the central nervous system. Toxemias, such as eclampsia and uremia, which do not involve the central nervous system directly, produce an increased cerebrospinal pressure and thus cause respiratory failure and vasomotor collapse in the manner described above.

Successful treatment of severe depression of the respiratory center or respiratory failure depends on the following two factors: (1) the effectiveness in overcoming the depression of the center by stimulation of the center, and (2) the prevention of irreversible damage to the cells of the body by oxygen lack until respiration and circulation become adequate. Adequate oxygenation of the blood in the lungs by artificial respiration, if breathing has stopped, and by administration of oxygen with or without 5 to 7 per cent carbon dioxide[1] will prevent irreparable damage to the body tissues and relieve the immediate danger of death. Proper circulation and blood pressure in the brain can be maintained by such drugs as epinephrine and ephedrine. These prevent anoxia in the cells of the center and enhance the removal of the toxic agent. When the oxygen supply is adequate, the administration of respiratory stimulants (analeptics), such as strychnine and caffeine, is more effective. Their effects are presumably due to some alteration of the neuronal protoplasm in such a way as to produce an increase in excitability.

Abnormal Types of Respiration

Changes in the normal depth and rate of respiration may be brought about by many factors. Such changes are considered abnormal, and the resulting respiratory patterns are classified into the following types:

Hyperpnea. **Hyperpnea** denotes an increase of either the rate or depth of respiration, or both, and thereby an increase in the quantity of air breathed. It can be produced by stimulation of the respiratory center by impulses coming from the cerebral cortex, as in excitement or other emotional states, by conditions that increase the demand for oxygen by the tissues, as in muscular activity, and by stimulation of sensory receptors in the skin, especially those mediating the sense of pain, heat, or cold. Hyperpnea usually results in a lowering of the carbon dioxide tension of the blood (acapnia), which results in a temporary cessation of breathing (apnea).

Dyspnea. **Dyspnea** is used to denote labored breathing or the sensation of respiratory distress. Dyspnea usually follows extreme hyperpnea that has not been able to satisfy the need for which it was instituted. Anything that interferes with the normal oxygen delivery to the cells and the elimination of carbon dioxide usually will give rise to the sensation of the necessity for increased respiratory effort. Dyspnea may be brought on by stimulation of the respiratory center directly by carbon dioxide excess or reflexly by a decrease in oxygen tension in arterial blood or an increase in the hydrogen-ion concentration due to fixed acids. Low

[1]In the past it was customary to add carbon dioxide to oxygen (air-breathing) tanks because of the stimulatory effect of carbon dioxide on the respiratory center. Near the end of World War II the U.S. Army Medical Corps issued a directive banning the use of such mixtures as a result of experiments that proved that carbon dioxide is positively harmful under abnormal conditions. The sensitivity of the neurons of the central nervous system to carbon dioxide is reduced under abnormal conditions, and cessation of breathing in respiratory failure is not due to a lack of adequate amounts of carbon dioxide but to inability of the center to respond to it.

oxygen tension may result from faulty oxygenation of blood in the lungs, as occurs in pulmonary disease such as pneumonia, asthma, emphysema, or obstruction (asphyxia), or from low oxygen tension in the inspired air, for example, at high altitudes or in nitrogen dilution. Interference with the transport of oxygen, as in circulatory failure, or a reduced oxygen-carrying capacity may also produce dyspnea, as will a reduced alkali reserve or retention of carbon dioxide.

Periodic Breathing. **Periodic breathing** is characterized by respiratory movements that gradually decrease both in extent and rapidity almost to the point of apnea and then become successively stronger until definite dyspnea is present. Breathing then gradually diminishes to the point of apnea, which lasts for several seconds, and then the cycle starts over again. Rhythmical waxing and waning of breathing, often with apnea occurring between periods, usually develops in patients with cardiac failure, uremia, pneumonia, and narcotic poisoning. This abnormal rhythm of respiration was first observed by Cheyne, but afterward was more fully studied by Stokes, so that it is called **Cheyne–Stokes respiration.** The abnormal rhythmical changes in the respiratory pattern are usually accompanied by corresponding changes in consciousness. During the decrease in depth and rate, consciousness gradually becomes cloudy and finally is lost during the period of apnea. As the breathing becomes more active, consciousness returns and during the dyspneic period is almost normal. The blood pressure also is likely to show characteristic changes, usually rising during the increase in breathing and falling during the decrease. It has been suggested that Cheyne–Stokes breathing is due to a depression of the sensitivity of the respiratory center to slight variations of carbon dioxide tension and is probably indicative of cerebral anoxia. The anoxia depresses the nerve cells of the center so that carbon dioxide tension rises far above normal before excitation occurs, resulting in excessive respiration, which does not cease until overventilation occurs. The resulting acapnia slows down respiration, finally ending in apnea, until carbon dioxide again rises above normal. The concomitant loss of consciousness during the apneic period is most probably a sign of cerebral asphyxia (anoxia).

Asthma. **Asthma** is a condition probably caused by an allergin acting directly on the smooth muscles of the bronchial tree, causing their constriction. This increases resistance to air flow, especially during expiration, which is further complicated by edema and secretion of mucus. A marked disturbance of respiration occurs, and the subject appears to be suffering from a sensation of suffocation. This sensation may be of circulatory origin in which the brain does not receive a sufficient amount of blood (cerebral ischemia) rather than a change in the carbon dioxide or oxygen tension of arterial blood. The marked increase in intrathoracic pressure slows venous return, and blocks flow in the plumonary vessels, and lowers the arterial pressure by decreasing cardiac output. Thus the cyanotic appearance of asthmatic subjects during an attack may be due to diminished blood flow rather than inadequate ventilation of the blood.

Emphysema. **Emphysema** is usually described as excessive expansion of the alveoli of the lungs caused by a chronic increased intrapulmonic pressure. The expansion of the alveoli often crushes and fragments other alveoli, thus diminishing the respiratory area. Destruction of the elastic tissue of the lungs hinders respiration also by diminishing the elastic recoil of the lungs, which normally forces air out.

Modified Respiratory Movements

Respiration, in addition to its function of gaseous exchange, also serves in other ways. Human beings, by means of modified and complicated respiratory movements, can express emotion through laughing, sobbing, yawning, and sighing. The air in respiration can also be made use of in a mechanical way to expel particulate matter from the upper air passages, as in a cough or sneeze. All these modified respiratory movements are essentially reflex in character, the origin of the stimulus determining each movement. Like the normal respiratory act, each of the above-listed acts may be carried out or influenced by a direct effort of the will.

Coughing consists of a long-drawn and deep inspiration followed by a complete closure of the glottis. A strong expiratory effort against the closing glottis follows, which suddenly opens, causing a blast of air to be drawn through the upper respiratory passages. The sensory impulses of this reflex act are initiated when a foreign body is lodged in the larynx or trachea or by the side of the epiglottis, which stimulates vagal endings. Coughing may also arise from stimuli applied to other sensory branches of the vagus, such as those supplying the bronchial passages and the auricular branch distributed to the external auditory meatus.

Sneezing involves movements that are basically the same as those in coughing except that it is usually produced by mild mechanical or chemical irritation of the sensory endings in the nasal mucosa and when the opening from the pharynx into the mouth is closed by contraction of the anterior pillars of the fauces and the descent of the soft palate, so that the force of the blast is driven entirely through the nose. The sensory impulses are carried by the nasal branch of the fifth cranial nerve (trigeminal).

Sighing is a deep and long-drawn inspiration chiefly through the nose, followed by a somewhat shorter, but large expiration.

Yawning is similarly a deep inspiration drawn through the widely open mouth and accompanied by an exaggerated depression of the lower jaw and frequently by an elevation of the shoulders. The exact cause is unknown, but it has been suggested that a lowered oxygen tension of the blood may be responsible. It is a response that is common to all vertebrates, and the act itself has a strong psychic effect in that it can be produced by the sight of others yawning.

Sobbing is an act in which a series of convulsive inspirations follow each other slowly with the glottis closing earlier than it should, so that little or no air enters into the chest, followed by a single prolonged expiration.

Crying consists of an inspiration followed by a whole series of short spasmodic expirations, the glottis being freely open during the whole time and the vocal folds being thrown into characteristic vibrations. Accompanying facial expressions are also part of this action.

Laughing consists essentially of the same movements as in crying, but the rhythm and the acccompanying facial expressions are different. Laughing and crying frequently become indistinguishable.

Hiccough consists of a sudden inspiration caused by spasm of the diaphragm during which the glottis suddenly closes so that further entrance of air is prevented. The striking of the air against the closed glottis produces the characteristic sound. Hiccoughing is usually a reflex act caused by irritation of the sensory endings in the gastrointestinal tract or other tissues of the abdominal cavity. The sensory pathways is by means of the gastric branch of the vagus. The spasm of the diaphragm is initiated by way of the phrenic nerve and the closure of the glottis by means of the inferior laryngeal nerve.

Speech will be covered in the section on voice.

Respiration in Exercise

The effects of muscular activity of the body demonstrate the intricacy of the ties that connect the respiratory system with almost all parts of the body. During muscular activity much of the oxygen in the blood is used up by the contracting muscles, while its carbon dioxide tension is increasing. This change in the blood causes an increase in the activity of the respiratory center to excrete the excess carbon dioxide, to obtain more oxygen, and to increase heat loss by way of the lungs. During severe exercise ventilation increases more than can be produced by inhalation of any concentration of carbon dioxide. Experiments have shown that after application of sphygmomanometer cuffs about the extremities to prevent return of venous blood, leg exercise will still increase ventilation. It has been suggested that receptors located in muscles and joints are stimulated by muscle action and reflexly influence the respiratory center, so that the respiratory movements are more than sufficient to compensate for the changes in the gases of the blood. The effectiveness of the increased respiration during exercise is enhanced by a concomitant increase in cardiac activity and an increase in the flow of blood through the pulmonary and systemic circulation. Without the increased circulatory activity, the increased respiration by itself during exercise would prove inadequate. Thus, the capacity for excessive muscular activity is determined by both the respiratory and circulatory systems working together in harmony rather than by either one alone. Increased respiration would be wasted effort unless it were accompanied by a quicker circulation, and an increased circulation would be of little use if not accompanied by increased ventilation. When the organism ceases to be able to meet the demands that the exercise is making on it, the subject becomes dyspneic, or is said to be "out of breath." Thus, it appears that the respiratory mechanism is not able to keep up, but in most instances the failure is more with the circulatory mechanism than with the respiratory system. When two men differ in their capacity for strenuous activity, such as running a race, the difference is usually in the ability of the heart to keep up with the increased respiratory movements and not in the ventilating capacity. Obviously, then, respiration involves two main factors: the respiratory mechanism and the circulation. The respiratory mechanism and its response in bringing oxygen to the blood can be influenced by anything that will affect its controlling center, the muscles of respiration, and the receptors that initiate respiratory reflex activity. The circulatory system that is involved in bringing the blood to the oxygen source as well as in delivering the oxygen to the tissues can be influenced in its activities by anything changing the action of the heart, the size of the vessels, the characteristics of the blood, and the utilization of oxygen by the tissues. For example, the amount of oxygen taken up from the lungs depends not only on the rate and depth of the respiratory and vascular pumps, but also on the capacity of the hemoglobin in the blood. An organism that is anemic from a loss of blood is thrown out of breath by a very slight exertion. This is not so much because of a weakness in the heart or respiratory system, but because, through lack of oxygen carriers, the tissues, including the cardiac and respiratory centers in the medulla, do not get sufficient oxygen.

Dead Air Space

The volume of air that fills the respiratory passages of the nose, pharynx, larynx, trachea, and bronchial tree is known as the **dead air space.** It amounts to about 150 cc. This passageway is only a conducting pathway for the air, since its wall is much too thick for any exchange between air and blood to take place. The walls of the alveoli, made up of a single layer of flattened cuboidal cells, form the respiratory surfaces of the lungs. This is the place where pulmonary gas exchange occurs, that is, exchange of gas between the air in the alveoli and the blood in the pulmonary capillaries. The air in the alveoli is termed **alveolar air,** and this simply means expired air that is in gaseous equilibrium with the arterial blood leaving the lungs. At the end of an expiration, the dead air space is filled with alveolar air. Outside air drawn in at inspiration mixes with the alveolar air in the dead space as it is drawn downward into the lungs. At the end of inspiration the dead air space is filled with fresh air. Since the tidal volume on the average is 500 cc, the air drawn into the alveoli at the end of inspiration is a mixture of 350 cc of fresh air and 150 cc of stale air. On expiration, the fresh air that remained within the dead air space is expired first (150 cc) followed by 350 cc of alveolar air, to make up 500 cc of tidal air expired. Thus at the end of an expiration the dead space is again filled with alveolar air (Fig. 10-16).

Figure 10-16. Dead air space. On inspiration, the first air that passes into the alveoli is 150 cc of stale air, which was left in the dead spaces from the previous expiration. This air is mixed with 350 cc of fresh air (if tidal air is 500 cc), so at the end of inspiration the alveolar air is a mixture of fresh and stale air, and the dead space contains 150 cc of fresh air.

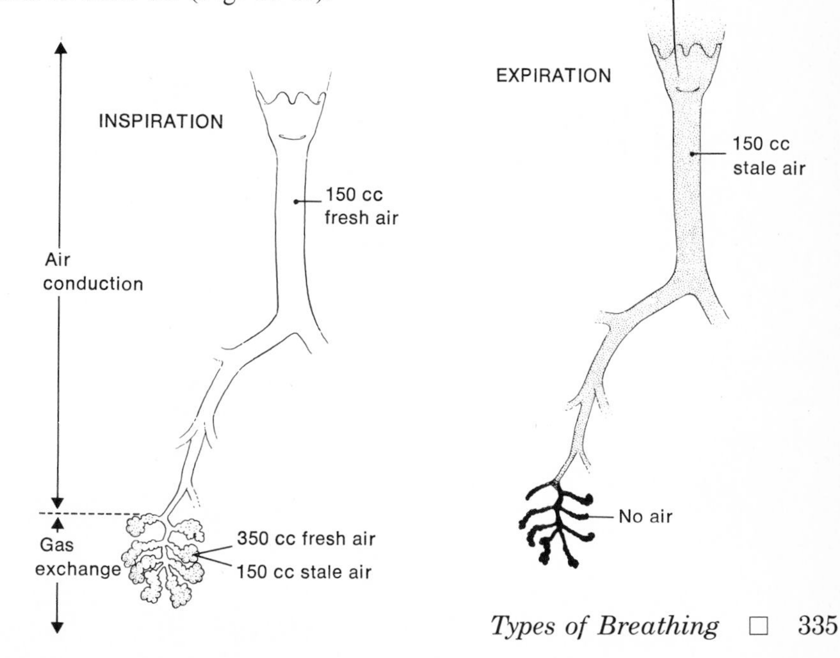

The volume of dead space varies, not only in different persons, but even in the same person with posture and other factors. During increased ventilation the dead space is increased, because the inspired air taken into the alveoli does not remain there long enough to come to equilibrium with the blood and the carbon dioxide of expired air is lower than normal. This produces an effect comparable to an increase in the dead space as by breathing through a long tube; that is, much of the room air inspired is physiologically useless since little fresh air gets into the alveoli.

Physical Exchange of Gases in the Lungs

The respiratory movements are devices by which the body can bring air into the lungs for a supply of oxygen and for the removal of carbon dioxide from the blood. Having considered the nature of the respiratory movements and their regulation, the next problem we must concern ourselves with is the factors concerned with the interchange of gases that takes place between the air in the alveoli of the lungs and the venous blood in the pulmonary capillaries. As it passes through the lungs, the blood takes up oxygen and gives off carbon dioxide and returns to the left atrium of the heart as arterial blood.

Composition of Atmospheric Air

The atmospheric air we inhale has the following composition:

Oxygen	20.96 per cent
Carbon dioxide	0.04 per cent
Nitrogen	79.00 per cent

The average normal adult consumes about 250 cc of oxygen and produces about 200 cc of carbon dioxide per minute when at rest. Any increase in activity increases the amount of oxygen consumed and carbon dioxide produced, and all these gases must pass through the alveolar membrane very rapidly because the blood passes through the pulmonary circulation in one second or less.

Composition of Expired Air

Expired air has the following composition:

Oxygen	15.8 per cent
Carbon dioxide	4.0 per cent
Nitrogen	80.2 per cent

In expired air, it can be seen that the difference in the percentage of oxygen is 5.16 and the difference in the percentage of carbon dioxide is 3.96. If we take the tidal air as 500 cc, the total quantity of oxygen expired in each breath is

$\frac{500 \times 15.8}{100} = 79$ cc, and the quantity of carbon dioxide is $\frac{500 \times 4.0}{100} = 20$ cc.

The composition of dry inspired air and expired air per breath (tidal air) in a normal adult at rest at room temperature is as follows:

	Inspired (Atmospheric), 500 cc			Expired, 500 cc		
	cc	*Per Cent*	*mm Hg*	*cc*	*Per Cent*	*mm Hg*
Oxygen	104.80	20.96	159.3	79	15.8	120.08
Carbon dioxide	0.20	0.04	0.3	20	4.0	30.40
Nitrogen	395.00	79.00	600.4	401	80.2	609.52
Total	500.00	100.00	760.0	500	100.0	760.00

Examination of this table indicates that there is a higher percentage of nitrogen in the dry expired air than there is in the dry inspired air. This increase is not due to an excretion of nitrogen by the body but represents the fact that the volume of carbon dioxide given out in any period is less than the oxygen taken in as a result of some of the oxygen's being used for the oxidation of hydrogen contained in food material, producing water instead of carbon dioxide.

Composition of Alveolar Air

Alveolar air samples can be collected for analysis by having a subject expire forcibly through a short length of rubber hose having a bore about 1 inch in diameter. At the end of this forced expiration a sample of the air is extracted from the tubing, near the mouth, with a hypodermic syringe. The air sample is then placed in a Haldane–Henderson gas-analysis apparatus and analyzed for its content of oxygen and carbon dioxide. At the end of expiration just prior to inspiration, alveolar air has the following composition:

	cc	*Per Cent*	*mm Hg Tension*
Oxygen	70	14.0	100
Carbon dioxide	28	5.6	40
Nitrogen	402	80.4	573
Water vapor			47
(tidal air)	500	100	760

Alveolar air is in equilibrium with arterial blood at this point. Immediately after inspiration these values are diluted with incoming air; that is, 500 cc of alveolar air on inspiration has the same composition as 350 cc of inspired air mixed with 150 cc of alveolar air. The latter figure represents the amount present in the anatomical dead air space. The composition of alveolar air at the end of an inspiration would be approximately as follows:

	Inspired Air	*Dead Air Space*	*Alveolar Air*	
	cc	*cc*	*cc*	*Per Cent*
	350.00	150.00	500.00	
Oxygen	73.26	21.00	94.26	18.85
Carbon dioxide	0.14	8.40	8.54	1.70

A comparison of the gaseous content of alveolar air at the end of inspiration and at the end of an expiration reveals how much oxygen and carbon dioxide in cubic centimeters per 100 is exchanged between the alveolar air and blood in the pulmonary capillaries:

Gaseous Content Alveolar Air (cc per 100 cc) at End of			
	Inspiration	*Expiration*	*Exchange with Blood*
Oxygen	18.85	14.00	4.85
Carbon dioxide	1.70	5.60	3.90

When the gases of the inspired air come in contact with the alveolar membrane of the lung, it is assumed that the exchange of gases takes pace in accordance with the usual physical laws of diffusion. Thus, the gas passes through the membrane and into the blood or from the blood into the alveoli in accordance with the difference in the pressure of that particular gas on either side of the membrane. The gas pressure in the blood as well as the alveoli is usually expressed in terms of partial pressures[2] of that gas, that is, pO_2 or pCO_2. In blood the oxygen tension (pO_2) is the pressure of the dry gas with which the dissolved oxygen in the blood is in equilibrium. Similarly, the carbon dioxide tension (pCO_2) is the pressure of the dry gas with which the carbon dioxide in the blood is in equilibrium. The significance of the composition of alveolar air will be better understood by reference to the following table:

Tension mm Hg	*Venous Blood mm Hg*	*Alveolar Air mm Hg*	*Arterial Blood mm Hg*
Oxygen	40	100	100
Carbon dioxide	46	40	40

As the blood circulates through the lungs, the carbon dioxide in the blood, having a higher pressure than the alveolar air, diffuses out into the alveoli until no further pressure difference exists between blood and alveoli. If the pressure of carbon dioxide in the alveolar air is 40 mm Hg, the pressure of carbon dioxide in the arterial blood in the pulmonary circulation will also reach this value. By the same token, the alveolar air, having a greater pressure of oxygen than the

[2]In a mixture of gases, each gas exerts its own partial pressure. For example, the partial pressure of oxygen at sea level would be 159 mm Hg. This is determined in the following way: Since the total pressure of air at sea level is 760 mm Hg and the oxygen content of air is approximately 21 per cent, the partial pressure of oxygen (pO_2) is $760 \times 0.21 = 159$ mm Hg. Similarly, the partial pressure of carbon dioxide (pCO_2) is $760 \times 0.04 = 0.3$ mm Hg. In the alveoli of the lung the pressure must be corrected for the vapor pressure of water in the alveolar air, which is equal to 47 mm Hg at 37° C. The corrected total pressure would be $760 - 47 = 713$ mm Hg. The partial pressure of oxygen in the lung would be approximately 100 mm Hg. The partial pressure of carbon dioxide in the lung is therefore about 40 mm Hg ($713 \times 0.056 = 39.9$).

incoming venous blood, will give up to the blood a quantity of oxygen sufficient to establish an equilibrium of gas between the two sides of the membrane. This means that the oxygen pressure in the alveolar air is the same as that of arterial blood leaving the lungs. The tension of nitrogen is the same in both venous blood and lung alveoli (573 mm Hg), indicating that the gas is physiologically inert.

Mechanism of Gaseous Exchange Between the Lungs and the Blood

The average normal adult consumes about 250 cc of oxygen and produces about 200 cc of carbon dioxide per minute when at rest. The exchange of these gases occurs in the lungs between the blood in the pulmonary capillaries and the air within the alveoli; in the course of this exchange the gases have to pass through two membranes, the walls of the alveoli and the walls of the capillaries. The transfer of gases between the blood and alveolar air is accomplished by diffusion, that is, a passage of gas molecules from a place at which the pressure of the gas is higher to one at which it is lower. The fact that this exchange is a matter of diffusion and not a matter of active secretion is shown in the gaseous equilibrium that is always reached between the alveolar air and the arterial blood. If in any situation the partial pressure gradients of arterial blood were greater than alveolar air, it could be supposed that active secretion was taking place.

The alveolar surfaces of the lung are well suited to permit free diffusion of gases between the blood and air since the membranes forming the barrier to diffusion offer very little resistance. In addition, the surface area provided for respiration is extremely large. It has been estimated that the total area of the capillaries in the alveolar walls of the human lungs available for respiratory exchange is 140 m. Under normal circumstances only a fraction of this area is utilized in the respiratory act.

It is obvious that the amount of gas that diffuses through the pulmonary membrane is dependent on the difference in pressure on the two sides of the membrane (alveolar air and blood). The amount (cubic centimeters) passing through the membrane *per minute* for each millimeter difference in tension is referred to as the **diffusion constant** (Table 10-1). In normal resting adults the values for oxygen range from 23 to 43 in men and are somewhat lower in women. During exercise the rate of diffusion ranges from 37 to 56, the increase probably being due to the increased rate of blood flow through the lungs. Carbon dioxide diffuses at a much swifter rate than oxygen, probably because it is more soluble. Its diffusion constant is about 20 to 30 times that of oxygen. The rate at which a gas will pass through a membrane by diffusion depends not only on the difference in pressure of the two sides, but also on the solubility of the gas in the fluid substance of the membrane.

The rate at which oxygen diffuses into the blood varies considerably among different persons. Differences in the available respiratory area at any one time may account for these variations, as well as differences in the thickness of the respiratory membranes and differences in the effectiveness of alveolar ventilation.

Transport of Oxygen by the Blood

After the exchange of gases has occurred, the blood is considered to be in an oxygenated form known as **arterial blood,** having a partial pressure of oxygen of about 100 mm Hg and a carbon dioxide tension of 40 mm Hg. At these tensions, because there is no chemical reaction between oxygen and plasma, its uptake by plasma is governed by Henry's law; that is, the amount *physically dissolved* is related directly to the partial pressure of O_2. One milliliter of plasma at 38° C takes up 0.00003 cc of O_2 for each increase of 1 mm Hg pressure. The corresponding volumes dissolved by the blood at the pressures existing in the alveolar air and at 38° C are approximately as follows:

Oxygen	0.3 cc per 100 ml of blood
Carbon dioxide	3.0 cc per 100 ml of blood
Nitrogen	0.8 cc per 100 ml of blood

The actual quantities of oxygen and carbon dioxide carried in the blood are much greater. The excess is carried in chemical combination rather than simple solution and serves as a reservoir from which additional gas molecules are liberated to pass into the plasma when the tension of the gas is reduced, or stored when the tension is increased. Gases in chemical combination do not exert a pressure and, therefore, cannot diffuse.

The transportation of oxygen by the blood from the lungs to the tissues is due mainly to the ability of **reduced hemoglobin** to combine in a loose fashion with oxygen, resulting in the production of **oxyhemoglobin:**

$$Hb + O_2 \longrightarrow HbO_2.$$

Hemoglobin is a protein that contains iron in the ferrous state, four molecules to each molecule of hemoglobin. Oxygen combines with the iron of the hemoglobin, two atoms of oxygen reacting with one atom of ferrous iron (Fe++). Thus, the formula above may be more accurately represented by the equation

$$4Hb + 4O_2 \rightleftharpoons Hb_4O_8.$$

The globin portion of the hemoglobin molecule endows it with the property of forming a reversible combination with oxygen. When the protein is denatured, hemoglobin loses this ability, and the iron is oxidized to the ferric form (Fe+++). Hemoglobin reversibly combined with oxygen is spoken of as **oxygenated hemoglobin,** or oxyhemoglobin.

The quantity of hemoglobin in the blood of an adult ranges between 15 and 16 g per 100 ml. Each gram of hemoglobin can combine with 1.34 cc of oxygen. Consequently, 100 ml of blood has the capacity of carrying 20 cc (15 × 1.34) or 21.4 cc (16 × 1.34) of oxygen in chemical combination. This figure is referred to as the **oxygen-carrying capacity** of the blood. Since the blood leaving the lungs is normally not fully saturated with oxygen but only 97.5 per cent saturated, the oxygen content of arterial blood is somewhat less than the figures noted. The reason for this is that not all the alveoli of the lungs are opened at the same time, so that some blood circulates through the lungs without picking up oxygen. The ability of hemoglobin to function as an efficient carrier of oxygen depends on the fact that the amount of oxygen held by it in chemical combination varies directly with the tension of oxygen and inversely with the tension of carbon dioxide. In the lungs the oxygen tension is high and the carbon dioxide tension is low, whereas in the venous blood the situation is just the opposite, enabling oxygen to enter the blood and carbon dioxide to diffuse out. This loss of carbon dioxide (acid) makes for a greater alkalinity of blood in the pulmonary capillaries, which enables hemoglobin to combine with more oxygen at the same oxygen tension. It has also been determined that the reduced temperatures at the lung site enhance the formation of oxyhemoglobin from reduced hemoglobin. Three factors influence the ability of hemoglobin to pick up oxygen in the pulmonary capillaries. They are (1) the tension of oxygen and carbon dioxide, (2) the pH of the blood, and (3) the temperature. It must also be apparent that the oxygen-carrying capacity of the blood is largely a function of the hemoglobin concentration.

Association and Dissociation of Oxyhemoglobin

The unique property of hemoglobin to combine with oxygen reversibly may be illustrated by the graph represented in Fig. 10-17, which is referred to as the oxyhemoglobin **association** or **dissociation curve.** In plotting such a graph, per cent saturation of hemoglobin is used instead of milliliters of oxygen to take into account the varying hemoglobin concentrations found in individuals under varying conditions, that is, normal anemia, polycythemia, and so forth.

$$\text{Per cent saturation of Hgb} = \frac{O_2 \text{ combined with Hgb}}{O_2 \text{ carrying capacity}} \times 100$$

From an examination of the graph we can determine the following properties of hemoglobin:

1. At 100 mm Hg, the partial pressures of oxygen normally existing in arterial blood, the hemoglobin is 97.5 per cent saturated with oxygen. If the oxygen-carrying capacity is 20 ml of O_2 per 100 ml of blood, hemoglobin holds in chemical association 19.5 ml

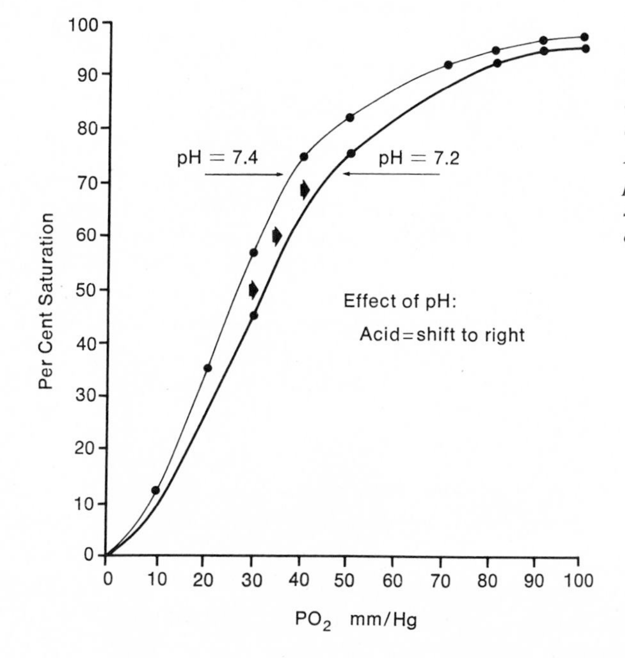

Figure 10-17. Oxygen hemoglobin dissociation curve at pH 7.4, temperature 38° C. The effect of a change of pH toward the acid pH 7.2 on the hemoglobin dissociation curve shows the curve shifting to the right. A change in pH toward the more alkaline pH 7.6 would have an opposite effect of shifting the curve to the left of the pH 7.4 curve (not shown).

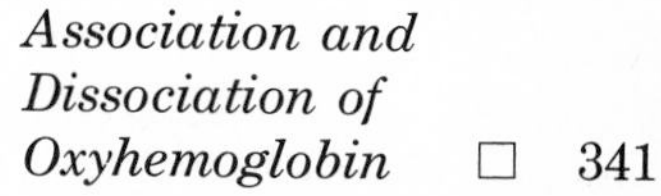

of O_2. Experimentation has shown that hemoglobin becomes 100 per cent saturated when the partial pressure of oxygen is about 250 mm Hg. At 100 per cent saturation the oxygen associated with hemoglobin increases only 0.5 ml per 100 ml of blood. Since maximal hyperventilation with air can only raise the partial pressure of oxygen within the alveoli to about 130 mm Hg, such action cannot add much more oxygen to the blood.

2. The curve runs almost horizontal between oxygen tensions of 100 and 70 mm Hg. It can be seen that a decrease in pO_2 from 100 to 90 mm Hg decreases the saturation only 1 per cent and the amount of oxygen associated with hemoglobin decreases 0.2 ml/100 ml of blood. From 100 to 70 mm Hg pO_2 the per cent saturation drops only to 94.5 per cent and from 100 to 90 to 92.7 per cent. This results in a decrease of oxygen associated with hemoglobin of 0.58 ml and 0.96 ml per 100 ml of blood, respectively. This not only enables man to live in a variety of different altitudes higher than sea level without much reduction in the amount of oxygen carried by the blood, but it is also a protective device enabling blood circulating through the lungs to load oxygen despite a relatively large decrease in pO_2.
3. Further scrutiny of the graph shows that between oxygen pressures of 40 and 10 mm Hg the curve drops off very sharply, giving the oxygen dissociation curve its characteristic S shape, referred to as a *sigmoid curve*. At the tissue site the pO_2 is usually in this range and because the amount of oxygen hemoglobin can hold depends upon the partial pressure of oxygen, the blood supplying active tissue dissociates and gives up oxygen. At a partial pressure of 40 mm Hg hemoglobin can hold only 75 per cent of its oxygen, and at a pO_2 of 20 mm Hg, only 35 per cent. Thus, the tissues are protected within this portion of the curve because a relatively small decrease in oxygen partial pressure allows them to withdraw large amounts of oxygen. As we have already seen, venous blood leaving the tissue site has a partial pressure of oxygen of 40 mm Hg and a saturation of 75 per cent (normal at rest).
4. The curve is capable of shifting from the normal either to the right or left. Changes in pH and temperature can cause such shifts. A decrease in the pH as a result of either an increase in the lactic acid concentration or an increase in carbon dioxide levels of the blood causes the curve to shift to the right. The graph shows such a shift with a decrease of the pH to 7.20. It can be seen that the flat part of the curve does not shift as much as the steep portion. This means that even though the blood in the pulmonary capillaries may be more acid, very little change in oxygen loading is affected, but the acidification of blood in the tissue unloads more oxygen than normally. One can see from the curve that at a blood pO_2 of 30 mm Hg and a pH of 7.4, hemoglobin is 57 per cent saturated with oxygen; at the same pO_2 but a pH of 7.2, hemoglobin holds only 45 per cent, which represents the extra oxygen available for use by the tissues. The effect of a change in the acidity on the oxyhemoglobin association curve is known as the **Bohr effect.** An increase in temperature has a similar effect on the dissociation curve. A decrease in temperature from normal has the opposite effect. In extreme cold oxyhemoglobin may not dissociate at all—which may explain the bright redness associated with exposed parts of the body such as the ears and nose in winter.

Hemoglobin, acting as a reservoir of oxygen, delivers to the tissues as much of the gas as they require. Tissue requirements are expressed in the tensions of oxygen and carbon dioxide in the tissues. The more active the tissue, the lower the tension of oxygen and the higher the tension of carbon dioxide to which the arterial blood in the tissue capillaries is exposed. Carbon dioxide diffuses very readily into the blood and helps to expel the oxygen by combining with hemoglobin to form a **carbamino compound,** which reduces the oxygen-binding power of hemoglobin, and also by reacting with water to produce carbonic acid; the increased acidity diminishes the amount of oxygen with which hemoglobin can combine at a given oxygen tension or reduces the ability of hemoglobin to hold oxygen. Oxygen leaves the blood in the tissue capillaries to diffuse into the tissue fluid, not only because the oxygen tension outside the capillaries is lower than that inside, but also because of the reduction in the ability of hemoglobin to hold oxygen as a result of the combination of carbon dioxide and hemoglobin and the increased acidity at the tissue site. It has also been demonstrated experimentally that the release of oxygen from oxyhemoglobin is accelerated by increased temperature. The temperature at the tissue site is higher than that at the lung site. Under physiological circumstances this action of increased hydrogen-ion concentration and of carbon dioxide on the liberation of oxygen from oxyhemoglobin is an important factor in the delivery of oxygen to the tissues.

Supply of Oxygen to the Tissues

The arterial blood reaches the capillaries with an oxygen tension of 100 mm Hg and a carbon dioxide tension of 40 mm Hg. The blood contains 19.3 cc of oxygen per 100 ml at this tension with the reaction $Hb + O_2 \rightleftharpoons HbO_2$ in equilibrium; of this, about 0.3 cc is in physical solution. This equilibrium is upset when the blood reaches the tissue capillaries as a result of (1) the low oxygen tension in the tissue fluid, which is lower than that of the plasma because oxygen is constantly diffusing from the former into the cells where it is utilized, and (2) separation of the blood from the tissue fluid only by the very thin, permeable capillary wall. The oxygen tension of the tissue fluid in a resting subject is about 35 mm Hg. A difference in oxygen tension of 65 mm Hg drives the oxygen in solution in the plasma into the tissue fluid. This drop in oxygen tension in the plasma immediately leads to a dissociation of oxygen from combination with hemoglobin, which diffuses out of the red cell into the plasma and then into the tissue fluid. At the same time carbon dioxide diffuses from the tissue fluid into the capillaries because its tension in the former is about 46 mm Hg and in the arterial blood only 40 mm. Some of this carbon dioxide combines with hemoglobin, which decreases its ability to hold oxygen, and some forms carbonic acid. The increased acidity also decreases the ability of hemoglobin to hold oxygen, so that more oxygen is free to diffuse out into the plasma. By the time the blood flows through the capillaries and reaches the venules, it contains about 15 cc of oxygen per 100 ml of blood and has an oxygen tension of 40 mm Hg. The arterial blood contains approximately 19 cc of oxygen per 100 ml and the venous blood 15 cc of oxygen per 100 ml, and thus only 4 cc of it is given up to the resting tissue. The percentage of available oxygen utilized by the tissue is called the **coefficient of oxygen utilization,** and from the above it can be seen that resting tissue utilizes approximately 21 per cent of the available oxygen. In general, the tissue takes out of the blood only as much oxygen as it needs to maintain its metabolic activity. As tissue activity increases, the need for oxygen is greater, and consequently the coefficient of oxygen utilization is increased. This increase in oxygen uptake by the tissues during activity is mitigated primarily by the greater reduction in oxygen tension of the tissue fluid, possibly even to zero, if the cells are using oxygen to a great extent. At the same time the carbon dioxide tension and hydrogen-ion concentration of the tissue fluids are increased. Consequently, much more oxygen diffuses out of the red cell into the tissue fluid. The increase in temperature resulting from the excess activity of the tissue also helps in the dissociation of more oxygen from hemoglobin. Active tissue may take up as much as 80 per cent of the available oxygen from the arterial blood, so venous blood emerging from tissue undergoing exercise may contain less than 10 cc of oxygen and as little as 5 cc per 100 ml.

Other factors also play an important role in supplying oxygen to the tissues when there is a greater demand for oxygen by the tissue. These are cardiac action, rate of blood flow through the capillaries, and the number of capillaries open at any one time. For example, during exercise the tissue requires great quantities of oxygen. This increase in oxygen need may be met by increasing the rate or volume of total blood flow through the tissue. This can be brought about by increased heart action, resulting in a greater cardiac output and increased arterial pressure. In addition, blood vessels to the active tissue dilate, and capillaries previously closed become open. The total amount of oxygen furnished and yielded to the tissue by these circulatory changes may be enormous.

Transport of Carbon Dioxide

The total amount of carbon dioxide that can be extracted from the arterial blood of normal subjects ranges from 50 to 53 volume per cent. At rest the amount in the venous blood is from 54 to 60 volume per cent (cc of CO_2 per 100 ml of blood = volume per cent). During muscular exercise the venous blood may contain as much as 70 to 75 cc of carbon dioxide per 100 ml. The amount of carbon dioxide that can be dissolved in the plasma as calculated from the data on the solubility of this gas would be about 3 volume per cent at the carbon dioxide tension of 40 mm Hg which exists in the alveolar air or arterial blood. It is obvious that the major fraction of carbon dioxide must be carried, as oxygen is, in chemical combination. It has been demonstrated that approximately 10 cc of carbon dioxide is chemically combined with the amino group of hemoglobin as carbamino compound. Although this constitutes only 20 per cent of the total carbon dioxide, it is an important factor in the exchange of this gas because of the relatively high rate of the reaction:

$$\underset{\text{Free amino group of hemoglobin}}{HbNH_2} + CO_2 \rightleftharpoons \underset{\text{Carbamino compound}}{HbNHCOOH}.$$

A small amount of carbon dioxide is carried as **carbonic acid.** As carbon dioxide diffuses from the tissue fluid into the blood plasma and then into the red cells because of the differences in pressures, it combines with water in the presence of the enzyme **carbonic anhydrase** (CA) to form carbonic acid. The reaction for this is

$$CO_2 + H_2O \overset{CA}{\rightleftharpoons} H_2CO_3. \qquad (10\text{-}1)$$

Carbonic acid dissociates inside the red cell into hydrogen ion and bicarbonate ion:

$$H_2CO_3 \rightleftharpoons H^+ + HCO_3^-. \qquad (10\text{-}2)$$

The bulk of the carbon dioxide in the blood is carried as **bicarbonate** in combination with sodium or potassium. The mechanism for this is as follows: After bicarbonate ions are formed inside the erythrocyte, most of them diffuse out of the red cell into the plasma and an equal number of chloride ions (Cl^-) diffuse into the cells in exchange. This shifting of chloride ions into the cell to satisfy the ionic equilibrium is referred to as the **chloride shift.** As the bicarbonate content of the plasma increases, the chloride content decreases.

The amount of carbon dioxide physically dissolved in the blood is not large, but it is important because any change in the concentration of the carbon dioxide in physical solution, which is directly proportional to the carbon dioxide tension, will cause a shift in the equilibrium:

$$CO_2 + H_2O \overset{CA}{\rightleftharpoons} H_2CO_3 \rightleftharpoons H^+ + HCO_3^-.$$

At the tissue site, because of the higher carbon dioxide tension in the tissue fluid (46 mm Hg), carbon dioxide diffuses rapidly into the plasma. As a result, the tension of the plasma immediately rises and carbon dioxide diffuses into the red cells, causing the formation of bicarbonate ions because of a shifting to the right of the above reaction.

The prompt formation of bicarbonate from carbon dioxide reduces the carbon dioxide tension and permits more carbon dioxide to diffuse into the red cell to produce more bicarbonate. At the same time oxygen is being relased from oxyhemoglobin, resulting in an accumulation of reduced hemoglobin. Reduced hemoglobin is a weaker acid than oxyhemoglobin. Therefore, it can combine with more hydrogen ions than oxyhemoglobin. Potassium ion is now available to combine with chloride ion, which is shifting in from the plasma, and sodium ion is available in the plasma to combine with the bicarbonate ion, which is shifting out of the cell. The amount of carbon dioxide that can be bound with hemoglobin is increased when oxygen becomes dissociated from hemoglobin. All these reactions reach an equilibrium at the carbon dioxide tension existing in venous blood, which is 46 mm Hg at rest and 60 mm Hg or more during exercise, and occur rapidly enough to be completed during the 1 second or so that the blood spends in the tissue capillaries. These reactions occurring at the tissue site may be graphically represented as in Fig. 10-18.

When the venous blood reaches the lungs, all the above processes are reversed. The carbon dioxide tension of the venous blood (46 mm Hg) is higher than that of alveolar air (40 mm Hg), and carbon dioxide diffuses out of the plasma into the alveoli. The reduction of carbon dioxide tension in the plasma causes the reaction $H_2CO_3 \overset{CA}{\rightleftharpoons} H_2O + CO_2$ to occur since its equilibrium point was 46 mm Hg. Carbon dioxide is liberated from the bicarbonate within the red cells, and the gas diffuses out into the plasma and then into the alveoli until the equilibrium point between venous blood and alveolar air is reached, that is,

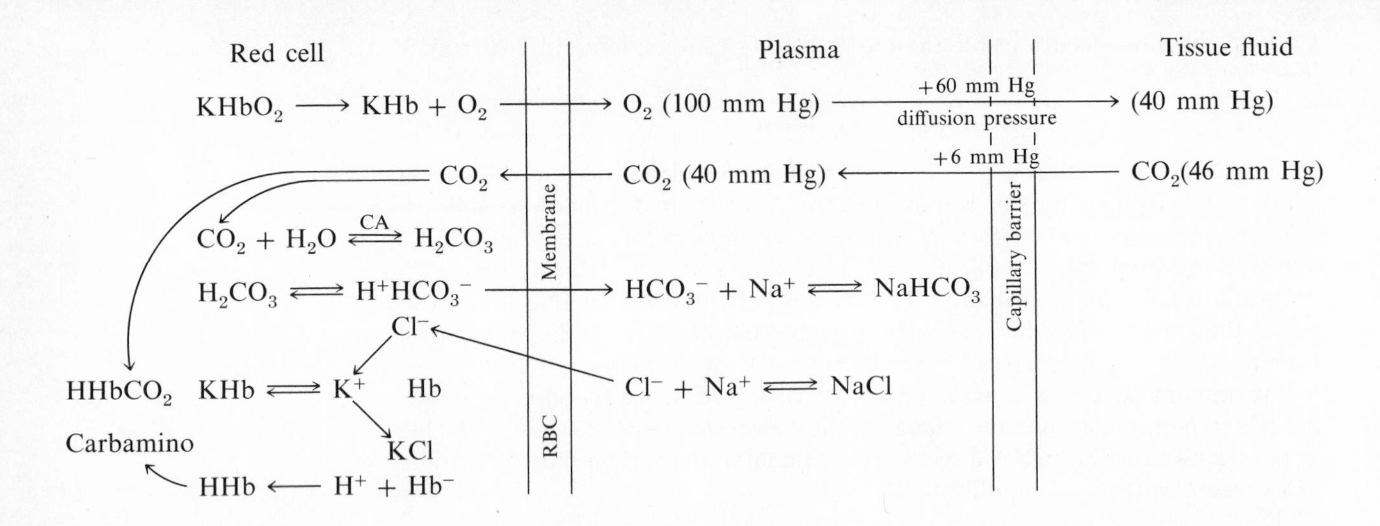

Figure 10-18. Exchange of carbon dioxide and oxygen at tissue site.

Figure 10-19. Exchange of oxygen and carbon dioxide at the lung site.

40 mm Hg (carbon dioxide tension in the alveoli). At the same time, oxygen is diffusing into the plasma and combines with hemoglobin in the red cells, making hemoglobin a strong acid. The oxyhemoglobin associates with the potassium ion, and a corresponding number of free chloride ions diffuse out into the plasma, which enables an equal number of bicarbonate ions to diffuse into the red cell. The bicarbonate ions are immediately dissociated to yield more carbon dioxide, which diffuses out into the alveoli until the reaction comes to equilibrium. The carbamino compound also gives up some of its carbon dioxide when hemoglobin becomes oxygenated and therefore more acid because the NH_2 group loses some of its ability to bind carbon dioxide under these conditions. The reactions that occur in the lungs resulting in the removal of carbon dioxide from the blood are represented in Fig. 10-19.

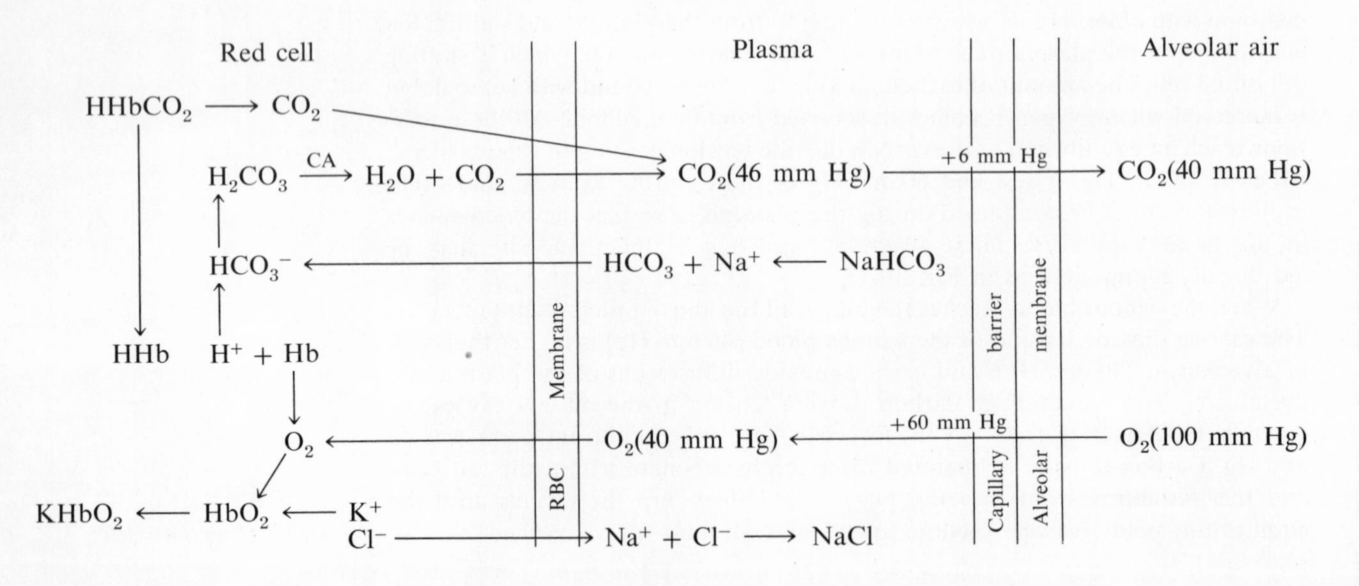

From what has already been discussed, it is obvious that carbon dioxide is carried in the blood in the following ways: (1) 10 per cent physically dissolved in the plasma, (2) a small amount as carbonic acid, (3) about 20 per cent as carbamino compound, and (4) about 70 per cent as bicarbonate. The amount of carbon dioxide expelled from the blood in the lungs has been estimated at 4 cc per 100 ml of blood, or 200 cc per minute in a subject at rest (cardiac output, 5,000 ml per minute). Since it takes 1 second for the blood to circulate through the pulmonary capillaries, 3.3 cc of carbon dioxide is expelled per second. Because 70 per cent of the carbon dioxide is derived from the fraction carried as bicarbonate, this obviously swift exchange would be impossible if it were not for the presence of the enzyme carbonic anhydrase within the red cells, for the rate at which equilibrium is attained in the reaction $H_2CO_3 \rightleftharpoons H_2O + CO_2$ is too slow to account for the amount of carbon dioxide exchanged in the period of 1 second. Carbonic anhydrase is a zinc protein complex that specifically catalyzes the dissociation of carbon dioxide from carbonic acid. This enzyme is also found in muscle tissue, in the pancreas, and in spermatozoa. The parietal cells of the stomach as well as the cells of the kidney tubules contain large quantities of carbonic anhydrase. It is believed that in the stomach it is involved with the secretion of hydrochloric acid.

Utilization of Oxygen by the Tissues

External respiration, which is the exchange of gases occurring between the air in the alveoli and the blood in the pulmonary capillaries, and internal respiration, which is that exchange occurring at the tissue site, are only mechanisms whereby oxygen is supplied to the tissues. The utilization of oxygen and the production of carbon dioxide by the tissues have been termed **cellular respiration,** or **biological oxidation,** and represent only the final stages of biological oxidation. Many other oxidative changes,[3] which involve energy transfer without the aid of oxygen, precede it. Within the cell, oxygen combines with the end products of digestion, such as glucose, fatty acids, and amino acids. Certain other products, such as alcohol, drugs, and poisons, are also oxidized, but the reactions from which the energy for organ activity is derived are the oxidations of the simple food materials. The utilization of oxygen by living cells depends entirely on a series of enzymes that are present within the cells. Those enzymes that act to add oxygen to a substance whereby the original substance is changed in some characteristic way

[3]Oxidation originally meant an addition of oxygen to a substance in such a way as to increase the proportion of oxygen in its molecule, which resulted in a change in the substance itself. Oxidation can also be carried out without the aid of oxygen, for example, reactions that increase the proportion of oxygen in the molecule by the removal of hydrogen. Dehydrogenation (removal of hydrogen) of a substance, which results in an increase in the proportion of oxygen in the molecule, is in a sense the same as adding oxygen to the molecule. The opposite of these reactions, that is, the removal of oxygen or the addition of hydrogen, which results in a decrease in the proportion of oxygen in the substance, is known as **reduction.** Every oxidation reaction in the living cell is accompanied by a simultaneous reduction. Since transfer of electrons is involved in all oxidation–reduction reactions, oxidation may also be defined as removal of electrons with an increase in valence, and reduction as a gain in electrons with a decrease in valence. Since electron transfer is difficult to visualize in complex biological reactions, the concept of oxygen addition or hydrogen removal is more commonly used.

are called **oxidases.** Most oxidases are conjugated proteins whose prosthetic groups contain metals such as copper, ion, or zinc. Most oxidase reactions involve the transfer of hydrogen to oxygen to form water. An example of such a reaction occurring in the tissues is the oxidation of reduced cytochrome with the formation of water and oxidized cytochrome:

$$O_2 + \text{reduced cytochrome} \xrightarrow{\text{cytochrome oxidase}} H_2O + \text{oxidized cytochrome.} \quad (10\text{-}3)$$

In general, the above type of reaction occurring in the tissues may be written

$$\underset{\text{Reduced substance}}{O_2 + \text{M-H}_2} \longrightarrow \underset{\text{Oxidized substance}}{H_2O + \text{M-O}}.$$

The liberation of electrons or hydrogen from a substance is brought about through the activity of **dehydrogenases.** The action of the dehydrogenases is to accomplish oxidation without the addition of oxygen, by removal of hydrogen from an oxidized substance. These types of reactions are extremely common in biological systems and were extensively studied by Thunberg and his associates. Thunberg found that the common dye methylene blue would easily accept hydrogen and be reduced to a colorless liquid. Methylene blue can act as a visible indicator of hydrogen transfer. By injection of such dyes in body tissues, it was found that hydrogen transfer occurs as long as certain specific enzymes are present. By the use of such methods, a considerable number of dehydrogenases have been discovered. Many dehydrogenases require activators, or coenzymes, which are present in the intact cell. Two such coenzymes in animal tissues are coenzyme I (NAD) and coenzyme II (NADP). An example of dehydrogenase activity is the oxidation of lactic acid to pyruvic acid:

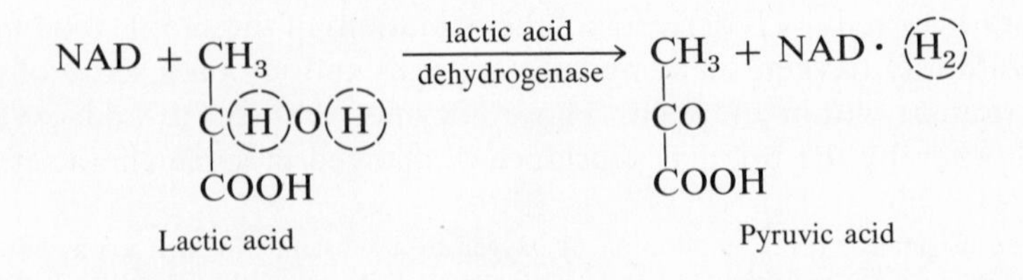

This reaction, then, represents an oxidation without the benefit of oxygen; that is, the proportion of oxygen in the lactic acid molecule is increased by the removal of hydrogen. The changed substance is known as *pyruvic acid.* Coenzyme I acted as the hydrogen acceptor.

In summary, the biological oxidations occurring in the tissue, which are often referred to as true respiration, involve hydrogen and electron transfer through a series of enzyme systems, as exemplified in Equation (10-4), before the actual utilization of molecular oxygen takes place. Examples of electron-transfer reactions occur in the **citric acid cycle:**

$+2H$

$-2H$ Lactic acid

Pyruvic acid $\xrightarrow[+ H_2O - CO_2 - 2H]{}$ (11) Acetic acid

(1) $\downarrow + CO_2$

Oxalacetic acid

(2) $\downarrow$ active acetate — (12) CoA acetate (active acetate)

Citric acid $\leftarrow$

(3) $\downarrow - H_2O$

cis-Aconitic acid

(4) $\downarrow + H_2O$

Isocitric acid

(5) $\downarrow$ isocitric dehydrogenase
$-2H$
NADP $\longrightarrow$ NADP $\cdot H_2$

Oxalosuccinic acid

(6) $\downarrow$ oxalosuccinic decarboxylase
$-CO_2$

Ketoglutaric acid

(7) $\downarrow +H_2O$
$-2H$
$-CO_2$

Succinic acid

(8) $\downarrow$ succinic acid dehydrogenase
$+H_2O$

Fumaric

(9) $\downarrow$ fumarase
$+H_2O$

Malic

(10) $\downarrow$ malic acid dehydrogenase
$-2H$
NAD $\longrightarrow$ NAD $\cdot H_2$

Oxalacetic acid

From the above series of chemical reactions it can be seen that, in reactions (5), (7), (8), (10), and (11), hydrogen is given off. These five pairs of hydrogen electrons are transferred to molecular oxygen to form water through the oxidase system, as exemplified by Equation (10-3). Of the five molecules of water formed, four can be accounted for as used within the cycle, as indicated by reactions (4), (7), (9), and (11). The remaining molecule of water plus one given off by the reaction at 3 result in the formation of two molecules of water in excess.

Carbon dioxide is given off at reactions (6), (7), and (11). This provides three molecules of carbon dioxide in excess to be liberated by the tissues. If we represent the oxidation of pyruvic acid by an empirical formula, we find that for every $2\frac{1}{2}$ molecules of oxygen utilized by the tissues, two molecules of water are formed and three molecules of carbon dioxide are liberated:

$$\underset{\text{Pyruvic acid}}{C_3H_4O_3} + 2\tfrac{1}{2}O_2 \longrightarrow 3CO_2 + 2H_2O.$$

The ratio of the volume of carbon dioxide produced by a tissue to the volume of oxygen absorbed is called the **respiratory quotient,** or **RQ,** of that particular tissue:

$$RQ = \frac{\text{Volume of } CO_2 \text{ produced}}{\text{Volume of } O_2 \text{ absorbed}}.$$

The ratio of these volumes as determined for the body as a whole is derived from the amount of oxygen consumed and carbon dioxide given off by a person per unit time. The subject of the respiratory quotient will be dealt with in more detail in Chapter 12.

Anoxia

As we have seen, the tissue receives its oxygen supply through the blood and is able to utilize the oxygen by virtue of the enzyme systems contained within the cells. In the respiratory function, the blood acts simply as a transporting system for the oxygen, picking it up as it flows through the lungs and discharging it at the tissue site. The lungs function to bring oxygen from the environment into contact with the blood. The oxygen supply to the tissue, therefore, is dependent on the following: (1) the efficiency with which blood is oxygenated in the lungs, (2) the efficiency of the blood in delivering oxygen to the tissues, and (3) the efficiency of the respiratory enzymes within the cells to transfer hydrogen to molecular oxygen. Any faulty action of these three factors may result in the tissues' not receiving an adequate supply of oxygen. A lack of sufficient oxygen for normal tissue function is referred to as **anoxia.**[4] This condition, which is also called **oxygen want** or **oxygen lack,** is classified into specific types depending on the factor or factors that produced the oxygen deficit. If the anoxia results from a defective oxygenation of blood in the lungs, it is called *anoxic anoxia.* If the blood does not transport sufficient oxygen because of a decrease in its ability to combine

[4]The terms *anoxia* (no oxygen) and *hypoxia* (low oxygen) are frequently used interchangeably. Anoxia is more commonly used to denote a general decreased oxygen concentration in the body.

with oxygen, that is, a lowered oxygen-carrying capacity, the condition is called *anemic anoxia*. If the delivery of oxygen is hindered by a slowing of the circulation, the condition is called *stagnant anoxia*. Last, if the respiratory enzymes are involved, that is, if something has modified them so that they have lost their ability to function, the condition resulting is called *histotoxic anoxia*.

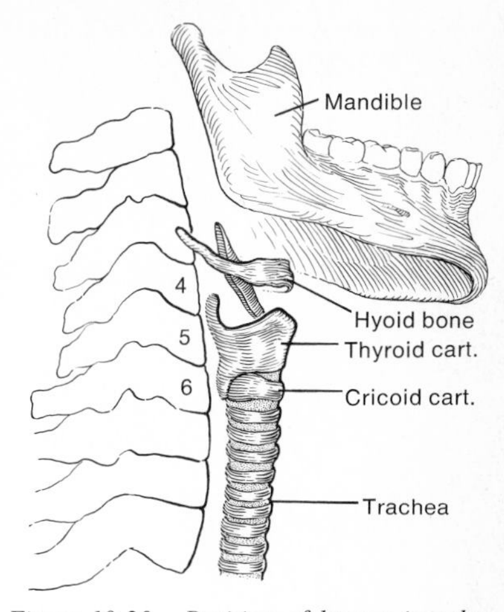

Figure 10-20. Position of larynx in relation to cervical vertebrae.

Anoxic anoxia is produced when the oxygen tension in the inspired air is lowered, as for example at high altitudes or when the atmosphere is diluted with inert gases. Diseases or abnormalities of the respiratory organs may limit their function as oxygenating structures and therefore causes anoxia. In pneumonia, asthma, emphysema, and pulmonary edema the blood cannot pick up its normal oxygen store. Obstruction of the air passages as in choking or drowning, where there is water in the lungs, can also produce a lack of oxygen at the tissues. Anoxic anoxia usually manifests itself in the form of labored breathing (dyspnea), an increase in the amount of reduced hemoglobin in the blood resulting in a bluish tinge to the skin called **cyanosis,** and, if the anoxia is prolonged, mental disturbances in the form of mania (excessive excitement and hysterical activity).

Anemic anoxia can result, as the name implies, from anemia of any type. Any drug or chemical that binds to hemoglobin, such as nitrates, sulfanilamides, carbon monoxide, and nitric oxide, to name a few, so that the oxygen-carrying capacity is lowered, will also produce this type of anoxia.

Stagnant anoxia results from a slowing of the circulation. The blood is oxygenated normally as it flows through the alveoli; it picks up its normal oxygen load and has its normal oxygen tension, but the tissue does not get sufficient oxygen because the flow of blood through the capillaries is extremely slow. Stagnant anoxia can occur in any condition that slows the circulation, such as heart failure, shock, and venous block.

In **histotoxic anoxia,** the respiratory enzymes of the tissues are affected so that the ability of the tissues to utilize oxygen is reduced or completely lost, even though the oxygen supply is normal. Posions such as cyanide or sulfide result in this type of anoxia because they inhibit the action of the cytochrome oxidase system by combining with the iron in the oxidase molecule. Cyanosis is not manifested by a subject poisoned in this way because the venous oxygen content approaches the arterial content and the concentration of reduced hemoglobin is lower.

The Larynx: Physiology of Speech

Previously we saw that the larynx and the trachea were essential parts of the respiratory passages, receiving inspired air from the nose and mouth and passing it on to the bronchi and lungs. The larynx is, however, more than a passageway for air; it acts as a valve to prevent the passage of food materials into the airways below, and more importantly, it is modified to control the expulsion of air from the lungs in the production of sound, the basis of speech.

The larynx (Fig. 10-20) lies in the neck anterior to the esophagus in an area covered by the fourth, fifth, and sixth cervical vertebrae. In its medial portion a prominence formed by the fusion of the thyroid cartilage plates can be seen externally. This laryngeal prominence is commonly referred to as the "Adam's apple."

The structure of the larynx (Fig. 10-21) admirably adapts it for the various functions it carries out. The cartilages in its wall give it a fairly rigid form; its muscles act on the cartilages and ligaments to modify its opening, and its mucosal membrane lining with its special folds contributes to the function of the larynx to produce sound.

There are six types of cartilage in the wall of the larynx. Three are single and three are in pairs. The former are the **thyroid, cricoid,** and **epiglottic** cartilages, and the latter are made up of two **arytenoid,** two **corniculate,** and two **cuneiform** cartilages. These various cartilages are united by true joints, the joints being strengthened by various membranes and ligaments. The **thyrohyoid membrane,** which suspends the larynx from the hyoid bone and attaches below to the upper border of the thyroid cartilage, is a fibroelastic membrane. It forms the median **thyrohyoid ligament** at its midline. The **conus elasticus** is an elastic membrane that unites the thyroid and arytenoid cartilages from above to the entire arch of the cricoid cartilage below. Another membrane, the **vestibular membrane,** is a submucosal sheet of connective tissue having an attachment in the angle of the thyroid cartilaginous plates and at the sides of the epiglottis. Figure 10-21 schematically represents some of the membranes and ligaments of the larynx.

The superior opening of the larynx, called the **glottis,** is guarded by the epiglottic cartilage, or **epiglottis.** This cartilage is a very thin, flexible, leaf-shaped structure, the apex of which is referred to as the **petiolus.** The inferior opening of the larynx corresponds to the inferior border of the cricoid cartilage. The cavity of the larynx is found between these two openings, and stretching across its sides are the vocal cords. The vocal cords consist of two pairs: the **superior,** or **false, vocal cords** and the **inferior,** or **true, vocal cords (vocal folds).** The false vocal cords are ligamentous bands extending from the thyroid cartilage to the arytenoid cartilages, forming the superior and medial wall of the **laryngeal ventricle.** The laryngeal ventricle is a blind sac found between the true and false vocal cords (Fig. 10-22). The vocal folds (true vocal cords), also called the **chordae vocales,** are membranoligamentous bands that stretch across the cavity of the larynx from

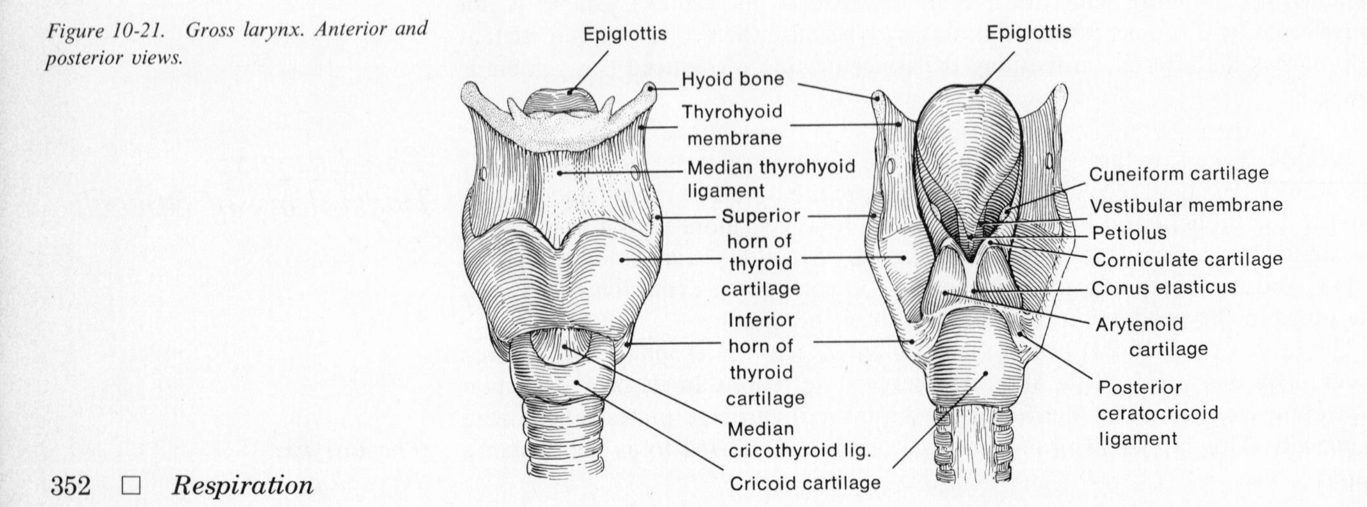

Figure 10-21. Gross larynx. Anterior and posterior views.

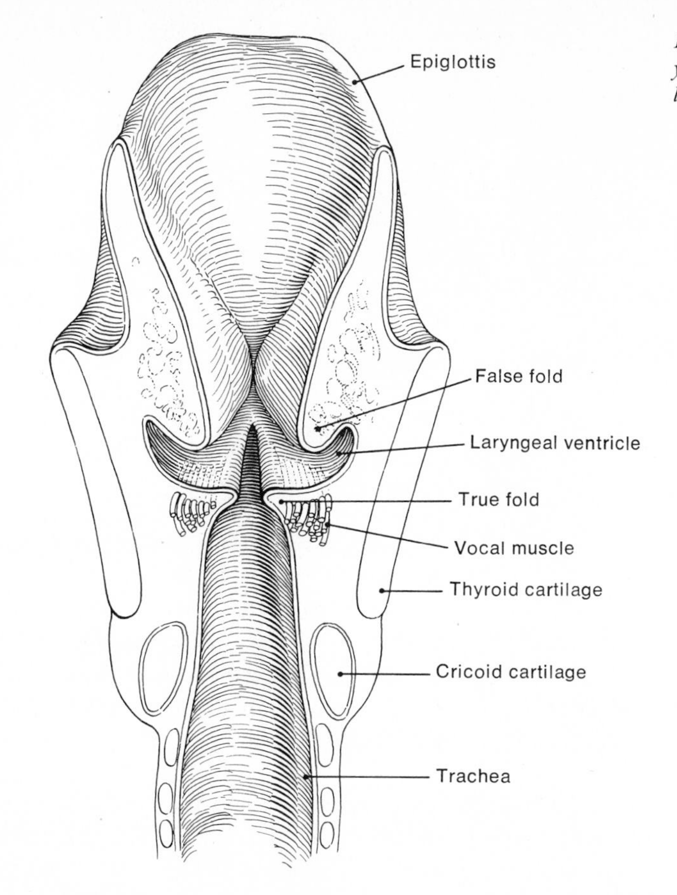

Figure 10-22. Longitudinal section of larynx showing location of vocal folds and laryngeal ventricles.

the thyroid cartilage to the anterior borders of the arytenoid cartilages (processus vocalis). The vocal folds are lubricated by a mucous secretion from the mucous lining of the laryngeal ventricle.

The muscles of the larynx are divided into two groups: respiratory and phonatory. The **respiratory muscles** consist of a pair of **posterior cricoarytenoid muscles,** which lie on the posterior surface of the cricoid cartilage (Fig. 10-23). On contraction, they draw the vocal folds away from the midline (abduction) of the laryngeal cavity, thereby opening the glottis and allowing air to pass freely through, during respiration. The **phonatory muscles** consist of the **anterior cricothyroid muscle** situated on the external surface of the larynx, the oblique and transverse **interarytenoid muscles,** and the **cricothyroarytenoid,** or **vocal, muscle.**

The action of the anterior cricothyroid muscle is to cause tensing of the vocal folds. It accomplishes this by pulling the arch of the cricoid cartilage upward, which forces the arytenoids backward, thereby stretching the vocal folds, which are attached to the vocal processes of each arytenoid. The interarytenoid muscles, because of their insertion into the anterolateral surfaces of each arytenoid, exert their pull toward the midline, causing a lateral-to-medial sliding movement and resulting in adduction of the posterior third segment of the vocal fold. The

Figure 10-23. *Posterior surface of larynx showing location of cricoarytenoid muscle. Contraction of muscle draws the vocal folds away from the midline.*

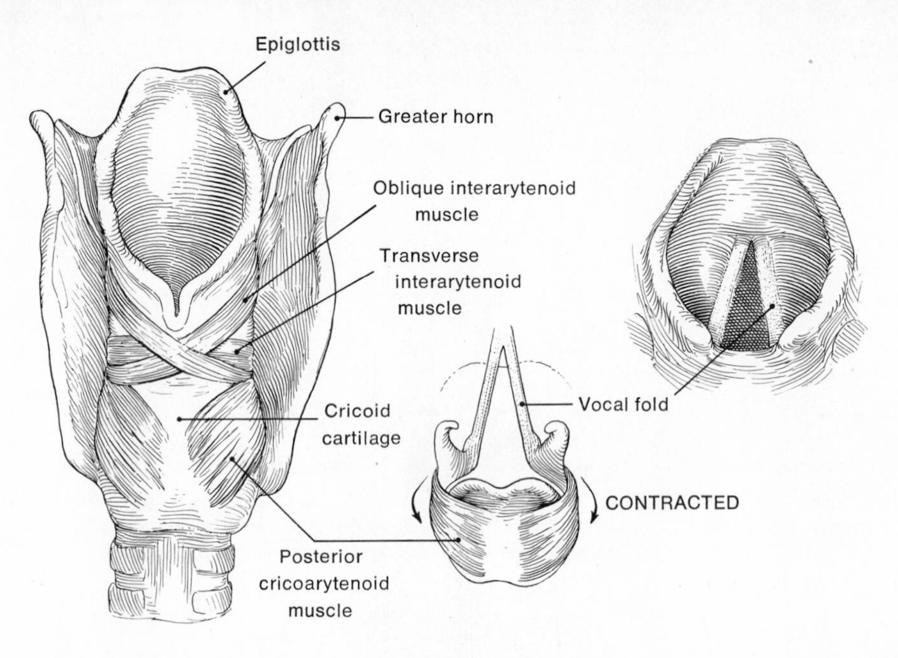

cricothyroarytenoid muscles exert their pull mostly on the anterolateral surface of the arytenoid inward, causing an adduction or closing of the anterior two thirds of the folds. Figure 10-24 schematically represents the actions of these muscles on the vocal folds.

A certain group of muscles, called the **sphincter muscles of the larynx,** when acting together can cause the glottis to close. This group of sphincteric muscles plays an accessory role in phonation under normal circumstances, but in states of motor dysfunction, it may play a major role. The muscles making up this group are the aryepiglotticus, the external and internal thyroarytenoid, and the arytaenoideus or posticus.

Most of the laryngeal muscles for phonation and respiration are supplied by nerve fibers from the inferior laryngeal branch of the vagus. If this nerve is

Figure 10-24. *Action of the anterior cricothyroid muscle and interarytenoid muscle on the vocal folds* A *and* B. *By contraction of the anterior cricothyroid, the vocal folds are tensed.* C *and* D. *Adduction of the vocal folds is brought about by contraction of the oblique and transverse interarytenoid muscle.*

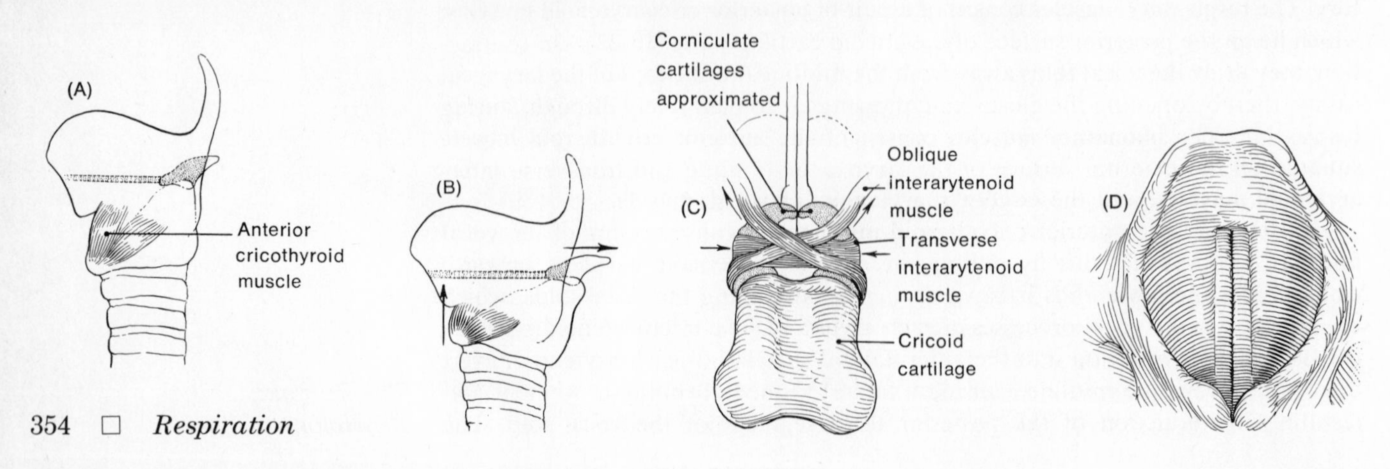

severed, the voice is lost, since the adduction of the vocal folds can no longer be effected. The cricothyroid muscle, which is the chief tensor muscle of the vocal folds and, therefore, the most important muscle of the larynx, is innervated by the superior laryngeal branch. When this nerve is severed on one side, the corresponding vocal fold is relaxed, and it becomes impossible to produce high notes.

Phonation, or the production of audible sounds by the action of the vocal organs, can be voluntary or involuntary. **Involuntary phonation** is exemplified by grunting, groaning, sighing, coughing, and sneezing. **Voluntary phonation** is usually associated with communicative activity, which includes three types of vocal expression: talking, whispering, and singing. These sounds, which we call the voice, are caused by a composite action by which blasts of air driven by variations in the expiratory movements throw into vibration the vocal folds, which cause the column of air above them to be set also into vibration. The vibrating airwaves are resonated and articulated by contact with the pharynx, nasal cavity, palate, tongue, teeth, and lips. Since the sound is generated in the vocal folds, they and other structures, such as the laryngeal muscles and cartilages, which affect their condition are referred to as the **vocal apparatus.** The chambers above the vocal folds including the laryngeal ventricle and false vocal cords constitute a subsidiary vocal apparatus to modify the sound originating in the vocal folds.

Voice, like the other sounds, has three common attributes: intensity, pitch, and quality. Intensity, or loudness, of the sound depends on the force of the expiratory blast. Pitch is related primarily to the frequency of vibration, which depends on the length and tension of the vocal folds. The length of the vocal folds varies only with age and sex; however, the tension is more variable. The lower limit of tone is produced by the longest fold during its relaxed state. Since the adult male has longer vocal folds than the female, he possesses a deeper voice. The voices of a soprano, tenor, and baritone depend on the respective lengths of their vocal folds. Problems of pitch in speech are in the main due to variations in the tension of the vocal folds. The quality, or timbre, of a sound depends on the frequency and character of the overtones that accompany any fundamental note sounded. Timbre of the voice is determined by such factors as sex, racial development, environment, and mental and physical growth. The physical quality of the vocal folds is perhaps the most important factor in determining quality.

Speech is made up of vocal sounds, which are produced in two ways: (1) by vibrations of the vocal folds and (2) by expiratory blasts, which are interrupted or modified in various ways as they pass through the throat and mouth without vibrating the vocal folds. The former are the only real vocal sounds and are referred to as **vowel sounds;** they differ in character only by the prominence of certain overtones brought about by the shape assumed by the mouth and pharyngeal passages and orifies as the vibration of air passes through them. The following vowels depend on the shape and form of the resonating cavities; *a* as in *father* is produced when the mouth is widely opened so that the cavity is funnel-shaped; *e* (*a* in *cat*) is sounded by retracting the lips and raising the larynx, the mouth cavity taking on the form of a broad flask; *i* as *ee* in *feet* is produced in a similar way, but the mouth cavity is somewhat shorter; *o* is sounded when the mouth cavity is somewhat flask-shaped with the lips being protruded so that the sounding is elongated; *u* (*oo*) is produced when the resonating cavity reaches a form of greatest length by depression of the larynx and protrusion of the lips as far as possible. When the vowels are properly said or sung, the posterior nares are closed by the soft palate so that no air passes out of the nose. This can be demonstrated by holding a flame before the nostrils. When the posterior nares are not properly closed when the vowels are sounded, the flame will be blown out and the tones acquire a nasal character.

Sounds produced without the vibrations of the vocal folds are consonants and are classified according to the place at which the expiratory air is modified or interrupted. Thus, for **labial** consonants the expiratory blasts are modified at the lips. If the air is modified at the teeth by the movement of position of the front part of the tongue with reference to the teeth or hard palate, the consonant produced is referred to as **dental.** If the air is modified in the throat by the movement or position of the root of the tongue in reference to the soft palate or pharynx, the sounds produced are referred to as **guttural** consonants. Some consonants, such as *m*, *n*, and *ng*, must have vibrations of vocal folds as a basis with the usual vibrating air passing through the nose instead of the mouth. These are called **resonants,** or **nasals.**

Whispering is a special form of speech in which laryngeal action is suppressed and audibility is reduced to a minimum. It is effected chiefly by the lips and tongue.

Singing is a highly specialized form of phonation, the notes of which conform to certain rhythmical inflections and modulations.

References

Aviado, D. M. *The Lung Circulation*. Elmsford, N.Y.: Pergamon, 1965.

Bell, G. H., et al. *The Textbook of Physiology and Biochemistry*. Baltimore: Williams & Wilkins, 1965.

Caro, C. G., (ed.). *Advances in Respiration*. Baltimore: Williams & Wilkins, 1966.

Comroe, J. H., *Physiology of Respiration*. Chicago: Yearbook Publishers, 1965.

Dejours, P. (ed.). *Respiration Physiology*, Vol. 1, No. 1. Amsterdam: North-Holland, 1966.

Dittmer, D. S., and R. M. Grebe. *Handbook of Respiration*. Philadelphia: Saunders 1958.

Drinker, C. K. *The Clinical Physiology of the Lungs*. Springfield, Ill.: Thomas, 1955.

Dubrul, E. L. *Evolution of the Speech Apparatus*. Springfield, Ill.: Thomas, 1958.

Haldane, J. S., and J. G. Priestley, *Respiration*. New Haven, Conn.: Yale, 1935.

Hall, I. S. *Diseases of the Nose, Throat and Ear*. Baltimore: Williams & Wilkins, 1956.

Hoops, R. A. *Speech Science*. Springfield, Ill.: Thomas, 1960.

Miller, W. S. *The Lung*. Springfield, Ill.: Thomas, 1950.

Morley, M. E. *The Development and Disorders of Speech in Childhood*. Baltimore: Williams & Wilkins, 1957.

Peters, R. N. *The Mechanical Basis of Respiration*. Boston: Little, Brown, 1969.

Piper, J., and P. Schild. "Respiration: Alveolar Gas Exchange," in: *Annual Review Physiology*, Vol. 33. Palo Alto, Calif.: Annual Reviews, 1971, p. 134.

Simpson, J. F., et al. *A Synopsis of Otorhinolaryngology*. Baltimore: Williams & Wilkins, 1957.

Slonin, N. B., and L. H. Hamilton. *Respiratory Physiology*. St. Louis: Mosby, 1971.

West, J. B. *Ventilation/Blood Flow and Gas Exchange.* Oxford: Basil Backwell & Mott, 1966.
———. "Respiration," in: *Annual Review of Physiology,* Vol. 34. Palo Alto, Calif.: Annual Reviews, 1972, p. 193.

The Digestive System

The neuromuscular system, as we have seen, enables an animal to secure its food, protect itself, and, in general, maintain itself with respect to its environment. This chapter is concerned with the manner in which a complex organism such as man, after securing its food, is able to prepare the bulk carbohydrates, fats, and proteins into a form that can be used by the individual cells that form the body. Most of the cells require that their nutrients, which represent an energy source, be presented in a form already dissolved in the tissue fluid. The problem presented to the body, then, is to break down these complex foodstuffs into molecules small enough to pass through tissue membranes so that they can enter the blood stream and be delivered in a soluble form to the various tissues. In fact, it is this type of activity that is known as **digestion,** which may be simply defined as the breaking down of insoluble complex food materials into simple absorbable forms. This is the special function of the organ system in the body known as the **digestive tract** (sometimes referred to as the **alimentary canal,** or **gastrointestinal tract**). Those aspects of digestion which will be considered are (1) the anatomical components associated with digestion; (2) the highly specific character of the chemical changes within the tract, activated by particular enzymes; (3) the delicate timing and coordination of the various steps in digestion; and (4) the transportation of digested foodstuffs through the intestinal wall into the blood.

Design of the Digestive Tract

The general design of the human digestive tract is shown in diagrammatic form in Fig. 11-1. It is essentially a tube beginning at the posterior region of the mouth and extending throughout the length of the body. In the embryo it is relatively straight and uncomplicated, but as development proceeds, the tube becomes coiled in the abdominal region as a result of its rapid increase in length in contrast to

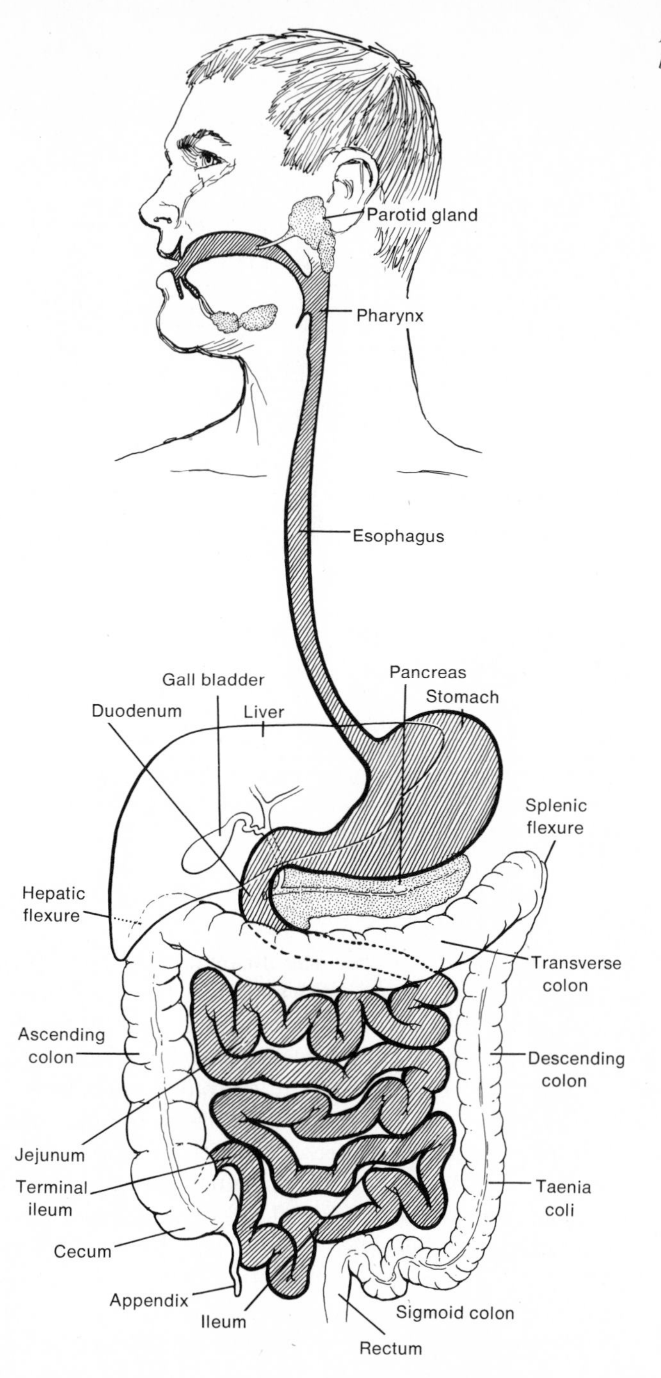

Figure 11-1. Digestive tract of the human being.

the development of other parts of the body. In the adult the alimentary tract attains a length of some 25 ft. The uppermost portion of the tube, known as the **pharynx,** opens into the nose and mouth, or oral cavity. The portion associated

Figure 11-2. *Sagittal section of head and neck showing pharynx and esophagus in relation to other parts.*

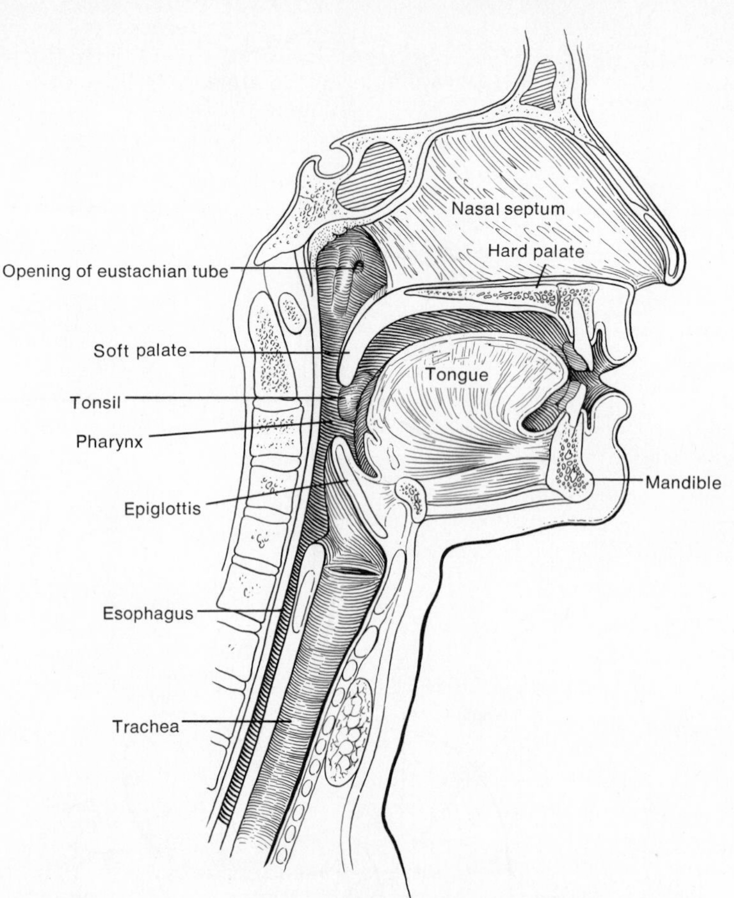

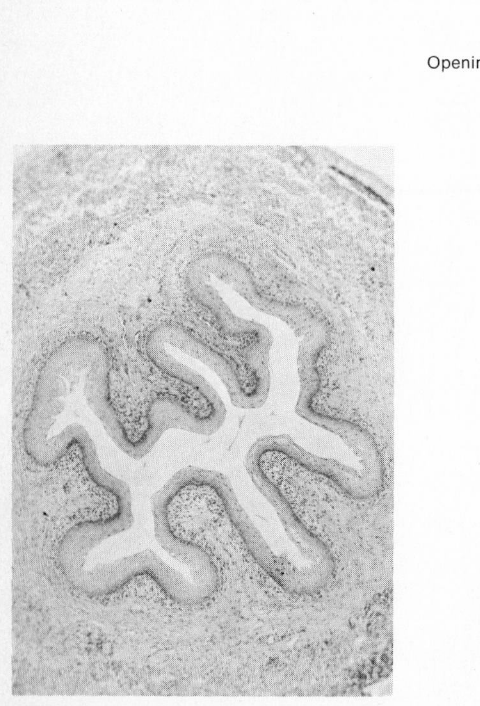

Figure 11-3. Cross section of mammalian esophagus showing stratified squamous epithelium, glands of submucosa, and two layers of muscularis. (Courtesy Ward's Natural Science Establishment, Inc., Rochester, N.Y.)

with the nose is known as the **nasopharynx,** and that portion associated with the mouth is known as the **oral pharynx.** The nose and mouth are separated from one another by the **hard palate,** which is continuous behind with the **soft palate.** The soft palate projects backward and reaches almost to the posterior wall of the pharynx; when elevated, it is capable of shutting off the nasopharynx during the passage of food from the mouth into the oral pharynx. At its lower end, the pharynx communicates freely with the first part of the air passage, the larynx, forming the **laryngeal pharynx,** and in back of the larynx the pharyngeal tube becomes the **esophagus** (Fig. 11-2). The pharynx is a tube common to both the respiratory and digestive systems. The **esophagus** begins at the lower end of the pharynx at the level of the sixth cervical vertebra. It is a collapsed, smooth muscular tube (Fig. 11-3) having a total length of approximately 10 inches. It occupies a position in the midline with its upper 2 inches in the neck and the lower 8 inches in the thorax. The esophagus leads into the abdominal parts of the digestive tract, consisting of the stomach, small intestine, and large intestine. The **stomach** is the most dilated portion of the digestive tract and has an average capacity of about 1 quart (Fig. 11-4). It is a J-shaped structure; Figure 11-5 illustrates the various divisions of this structure. The cardiac orifice has no true sphincter valve, but the last part of the esophagus, which enters into the stomach, offers some sphincter action. The pyloric orifice is guarded by a very powerful sphincter formed by an increase in the normal circular musculature of the tube. This valve permits fluids to be squirted out into the small intestine but prevents the exit of solids. The wall of the stomach is very muscular and serves as a churn to help break up and mix the food with the gastric juices. The mucous membrane lining of the stomach is relatively thick and is divided into areas of 1 to 6 mm in diameter by a system of furrows (Fig. 11-6). Each of these areas appears to be further divided by tiny grooves into irregular convoluted ridges. The invaginations produced by the furrows are called the **gastric pits** or rugae. The mucous membrane is occupied by a multitude of glands, which open into the bottom of the gastric pits. Besides the **mucous glands,** the gastric glands consist of two types of cells, the **chief cells** (zymogenic cells), which secrete pepsin and other enzymes, and the **parietal** cells, which secrete hydrochloric acid (Figs. 11-7 and 11-8). These gastric glands are abundant in the body and fundus of the stomach but are absent from the pyloric regions, where only mucous glands are found.

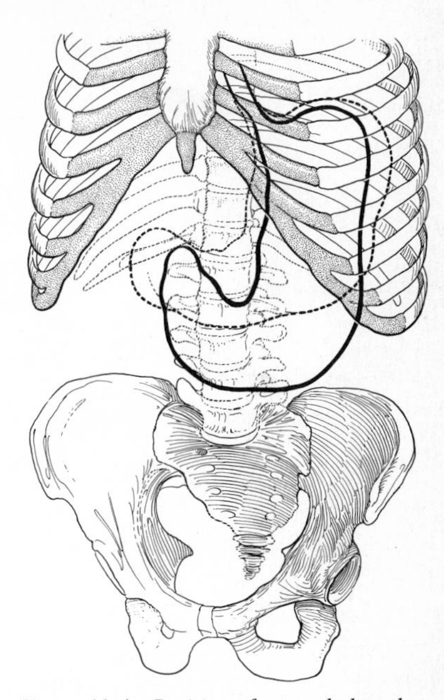

Figure 11-4. Position of stomach, based on X-ray studies. Dotted line represents position in reclining position, solid line on standing.

The pyloric canal leads into the **small intestine,** which is a tube about 23 ft long and divisible into three portions, the **duodenum,** the **jejunum,** and the **ileum,** each gradually passing into the other. Their structures, although showing some differences, are generally the same. Figure 11-9 illustrates a cross-sectional view through the small intestine.

The duodenum is the shortest part of the small intestine, being only 10 inches long. It is relatively thick-walled, and its mucous membrane, thrown into deep folds, possesses large **duodenal digestive glands** (Brunner glands). These large glands extend into the submucosa and are most numerous near the pylorus of the stomach. The bile duct and pancreatic duct open into the duodenum, so that digestive juices secreted by the liver and pancreas are added to the duodenal digestive juices.

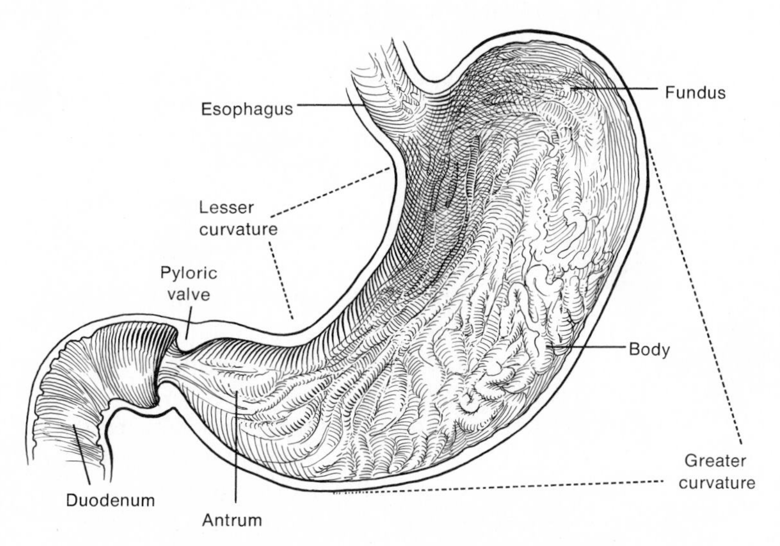

Figure 11-5. Interior of stomach showing its various divisions.

Figure 11-6. Schematic representation of gastric mucosa showing its division into areas of 1 to 6 mm by a system of furrows.

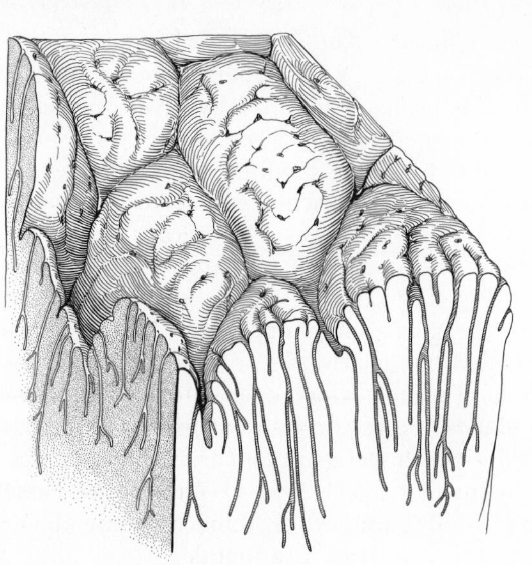

Figure 11-7. Gastric glands from the fundus of the stomach.

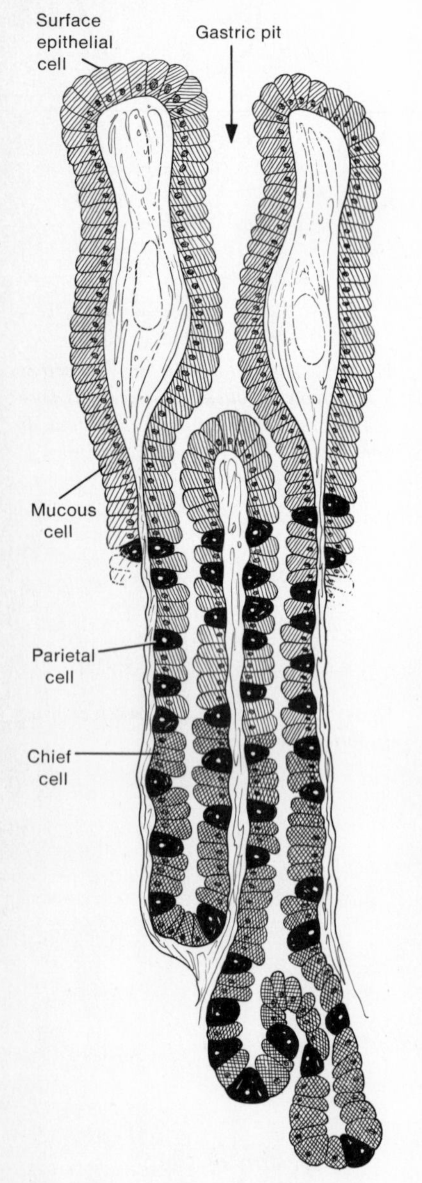

The next 22 ft of small intestine is made up of the **jejunum** and the **ileum.** This section (Fig. 11-10) shows a high degree of motility, and the lumen of the tube gradually decreases from a diameter of nearly 2 inches at the duodenum to 1 inch at the end of the ileum. The mucous membrane surface of this portion, like that of the duodenum, is greatly increased through the presence of deep circular folds, which form the **valves of Kerckring** and **villi** (Fig. 11-11). Villi are microscopic fingerlike processes that project into the lumen of the small intestine. Their enormous numbers give a nubby appearance to the interior surface and a velvety feeling to the touch. Each villus (Fig. 11-12) contains an arteriole and venule and a lymph vessel called a **lacteal.** Villi serve as structures for the absorption of digested foods. At the bases of the villi are the openings of the intestinal glands—the **crypts of Lieberkühn** (Fig. 11-13), which secrete the intestinal digestive enzymes.

The small intestine terminates by opening into the first part of the **large intestine,** called the **cecum** (Fig. 11-14). The opening is guarded by a valve, the **ileocecal valve,** which prevents the regurgitation of the cecal contents into the ileum (Fig. 11-15). The cecum possesses a large lumen, exceeded in size only by that of the stomach, and is further characterized by the presence of a diverticulum, the **vermiform appendix** (Fig. 11-14). The appendix is a blind sac, opening into the cecum a little below the ileocecal orifice and having a length ranging from 2 to 8 inches. The richness of lymphoid tissue in this structure gives it an appearance similar to the tonsils, and it has often been called the "tonsil of the abdomen." It is frequently infected and then, like the tonsils, must be removed surgically before it becomes a menace to the organism.

The portion of the large intestine that extends between the ileocecal valve and the point where it makes a left-angled turn is known as the **ascending colon** (Fig. 11-16). It is a narrower segment than the cecum and lies along the right side of the posterior abdominal wall. It becomes the **transverse colon** at a point near the liver where the ascending colon makes a left turn, the bend being known as the **right hepatic flexure.** The transverse colon is the largest and most mobile part of the large intestine. It stretches across the abdominal wall, where it terminates in the **left flexure** opposite the spleen. At this point the large intestine becomes the **descending colon.** The descending colon, which is the narrowest portion of the large intestine, passes inferiorly from the left splenic flexure to the brim of the pelvis, where it becomes the **sigmoid colon.** The sigmoid colon, sometimes referred to as the **pelvic colon,** is characterized by its S-shaped appearance; the sigmoid colon begins at the brim of the pelvis on the left side and passes across to the right and upward and then back and down to end at the median line opposite the third segment of the sacrum, where it is continued as the **rectum.** The rectum continues downward and forward in the concavity of the sacrum and coccyx, terminating about 2 inches below and in front of the tip of the coccyx. The tube is about 6 inches long, and it is generally smaller in caliber than the sigmoid colon. The rectum continues downward, passing through the medial borders of the levator ani muscles to become the anal canal, which, after a course of about an inch and a half, opens to the exterior at the anal orifice, which is referred to as the **anus.** Two sphincters guard the anal orifice. An **internal sphincter,** which is a thickening of the circular layer of smooth muscle of the anal canal, is involuntary in its action, being under the influence of the autonomic nervous system. It is located in the uppermost portion of the anal canal. The **external sphincter,** a voluntary skeletal muscular ring surrounding the anal orifice, is the lower and more important sphincter. The retention of the rectal contents is controlled by the external sphincter; it is a muscle that is under voluntary control and is continuously active except during defecation.

The mucous membrane of the large intestine does not form folds as it does in the small intestine, except in the rectum. It is also devoid of villi, and therefore its internal surface is relatively smooth. The glands of Lieberkühn are somewhat different in the large intestine, being straighter and longer. There is also a greater concentration of goblet cells. The muscular coats of the large intestine are similar to those of the small intestine except in the arrangement of the outer longitudinal

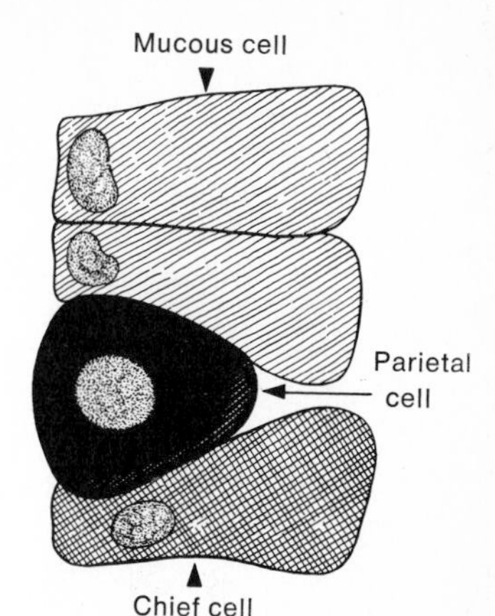

Figure 11-8. Detail of parietal cell between a mucous cell and a chief cell.

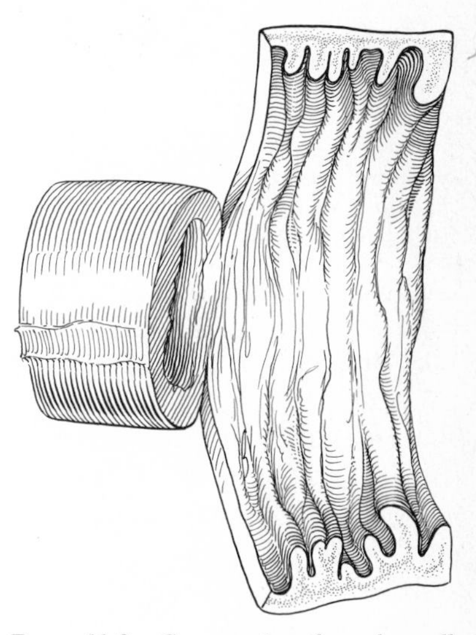

Figure 11-9. Cross section through small intestine.

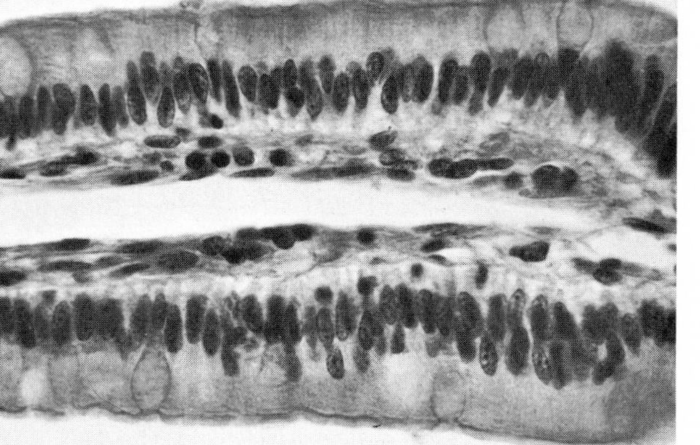

Figure 11-10. Human jejunum. (Courtesy General Biological Supply House, Chicago.)

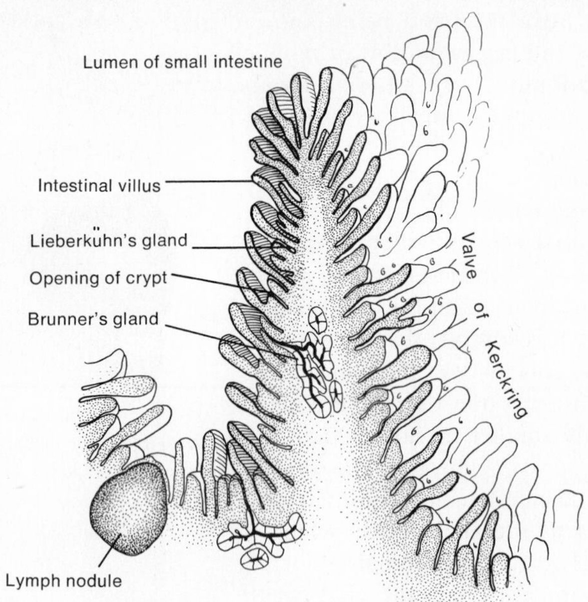

Figure 11-11. Magnified view of a portion of the wall of the small intestine showing the valves of Kerckring, villi, and Brunner's gland.

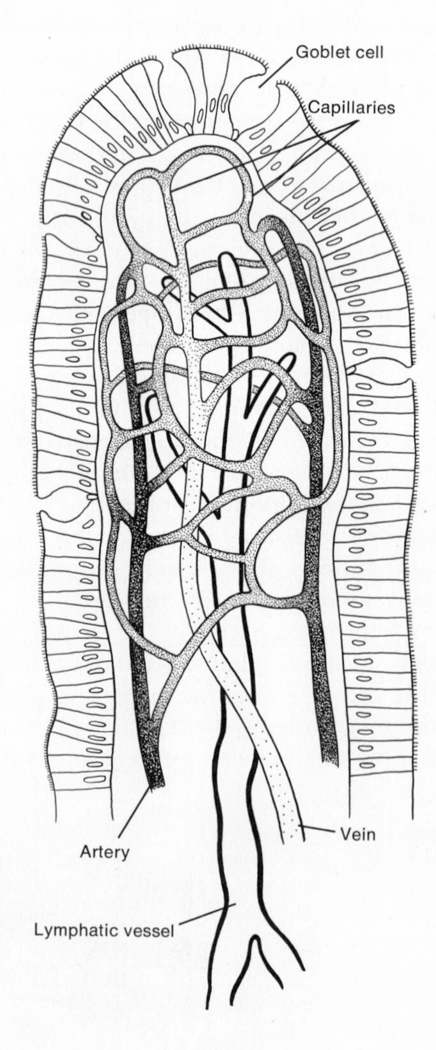

Figure 11-12. Top part of an intestinal villus showing the arrangement of its vascular elements.

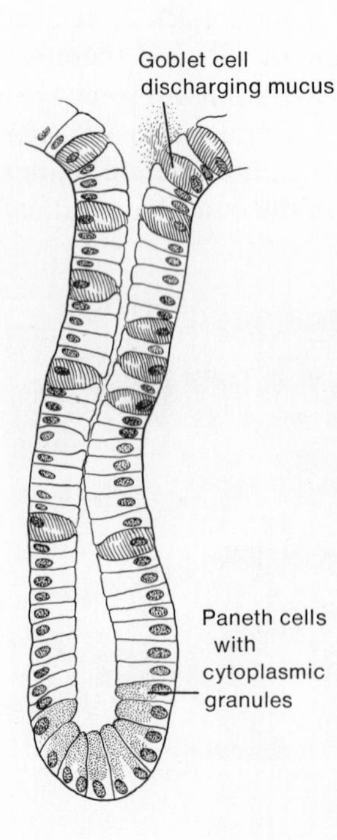

Figure 11-13. Crypt of Lieberkühn. At the bases of the villi are the openings of the intestinal glands, which secrete the intestinal digestive enzymes.

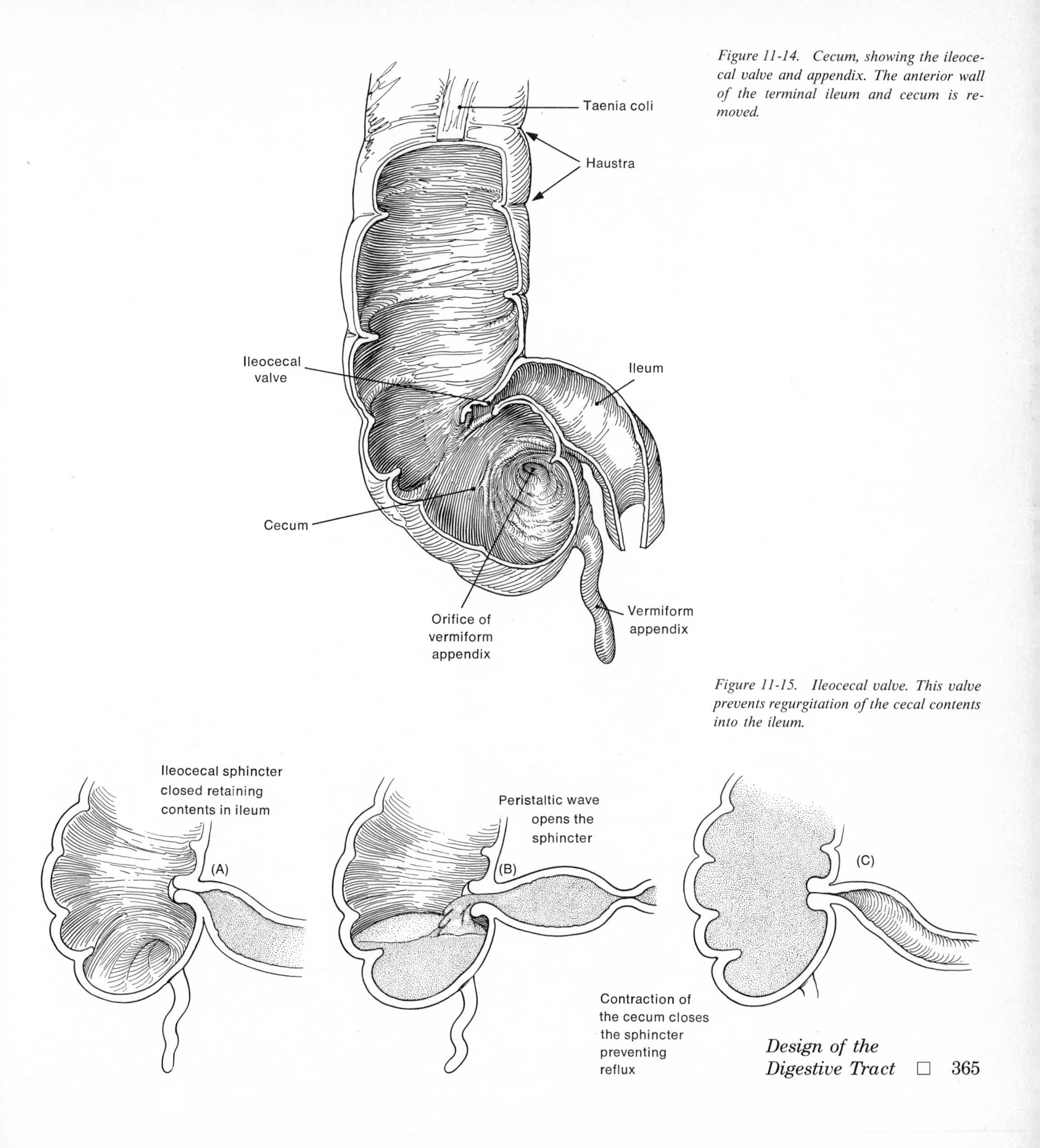

Figure 11-14. Cecum, showing the ileocecal valve and appendix. The anterior wall of the terminal ileum and cecum is removed.

Figure 11-15. Ileocecal valve. This valve prevents regurgitation of the cecal contents into the ileum.

Figure 11-16. Colon (large intestine).

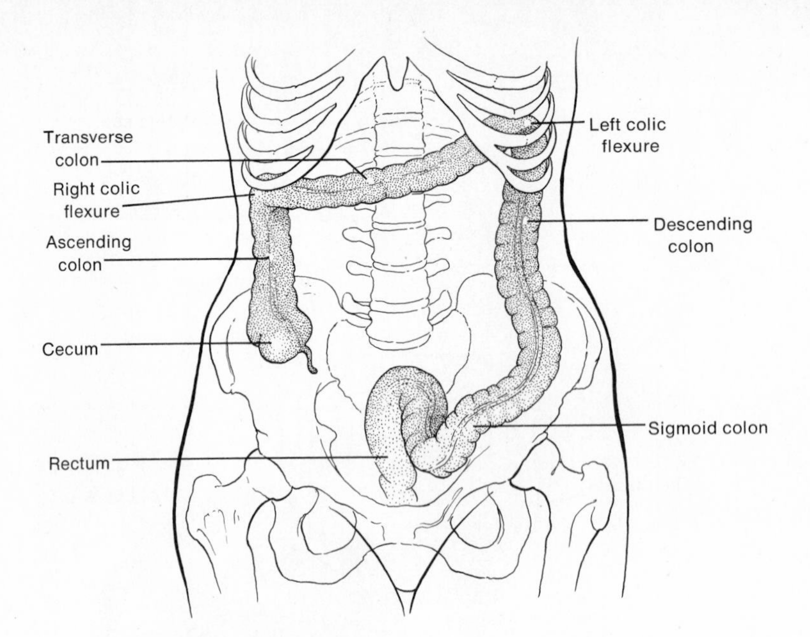

Figure 11-17. Section of colon showing the structure of its wall diagrammatically.

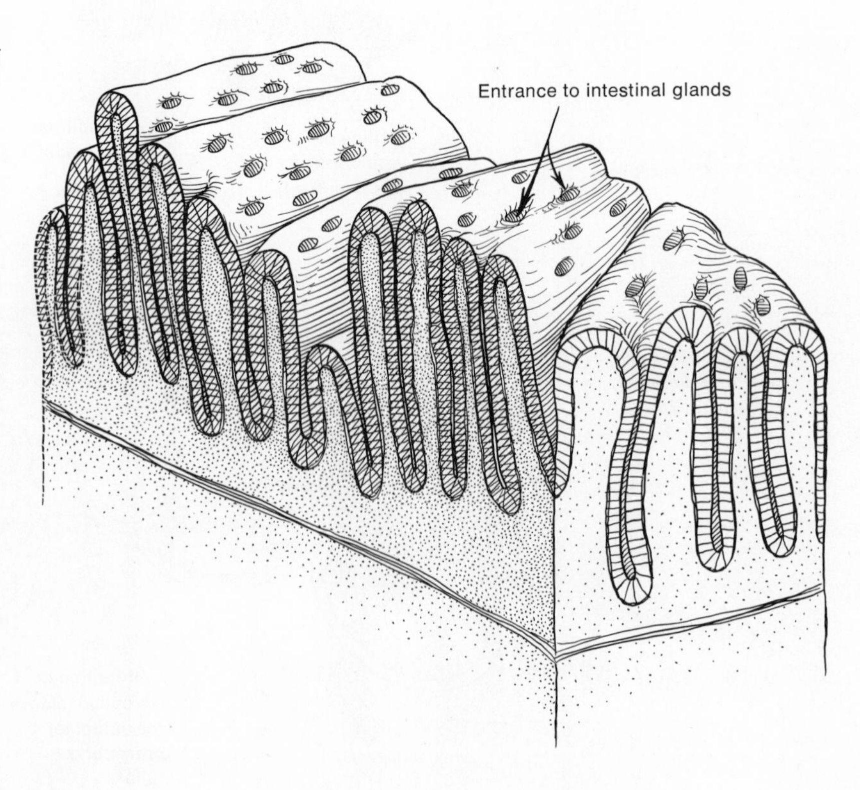

layer. In the colon the longitudinal coat does not surround the wall but consists of three muscular bands known as the **taenia coli.** Beginning at the base of the appendix and being shorter than the colon itself, they pucker the wall of the colon so that it consists of a series of pouches, of **haustra** (Fig. 11-14). Figure 11-17 diagrammatically shows the structure of the wall of the large intestine.

Nerve Supply of the Gastrointestinal Tract

The nerve supply seems to be the same for all parts of the alimentary tract and has been studied extensively in the small intestine. It has been demonstrated that the nerve supply consists of two parts: an **intrinsic** portion, which is represented by nerve cells and fibers originating and located in the wall of the intestine itself, and an **extrinsic** portion, represented by fibers of the vagus nerve and the postganglionic fibers of the sympathetic (Fig. 11-18). The intrinsic nerve mechanism consists of a series of plexuses composed of large or small groups of nerve cells and bundles of nerve fibers. A series of such plexuses is located in the narrow space between the circular and longitudinal muscular layers of the intestinal wall and is known as **Auerbach's plexus.** Similar plexuses are located in the submucosal layer and form the **plexus of Meissner.** This network of nerves within the walls of the digestive tube is responsible for the spontaneous movements of the intestinal

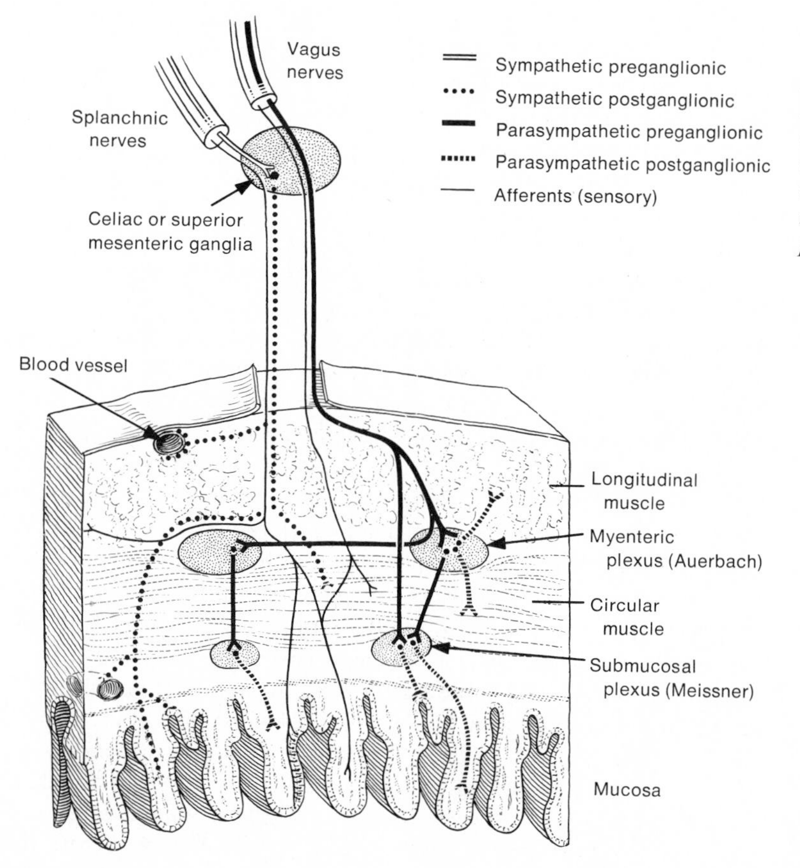

Figure 11-18. Cross section of the intestinal tract showing its innervation (enteric nervous system). The extrinsic portion consists of nerve fibers of both parasympathetic and sympathetic origin. The intrinsic portion is represented by nerve cells and fibers originating and located in the wall itself made up of Meissner's and Auerbach's plexuses.

tract even after the extrinsic network has been cut; that is, they are somewhat similar in function to the nodal system found within the heart. Some investigators claim that the plexuses represent complete reflex arcs even though all their neurons are supposedly motor in nature and no local sensory fibers exist. The axons of these intestinal neurons appear to divide into two branches. One branch receives sensory stimuli and transmits the impulse generated to the other branch without having to pass through the cell body. Just what role these plexuses play in controlling the gastrointestinal movement is not clear at present.

The extrinsic nerve supply consists mainly of vagal fibers, which, after entering the intestinal wall, terminate in tiny branchings on the cells of both Meissner's and Auerbach's plexuses. The sympathetic fibers, which are branches of the splanchnic nerve originating in the celiac ganglion, do not enter into synaptic junction with the cells of these plexuses but appear to branch profusely in the muscular layers to form intramuscular plexuses and finally to terminate on the smooth muscle cells. The sympathetic fibers also appear to supply the blood vessels within the intestinal wall. The extrinsic nerve supply functions to regulate the intrinsic neuromuscular mechanism that determines the movements of the digestive tube.

Accessory Structures of the Gastrointestinal Tract and Their Secretions

Several glands functioning importantly in the digestive process, which are derivatives of the digestive tract in their embryonic development and which are still connected by their ducts to its wall, are the **liver** and the **pancreas.** The **salivary glands,** which open into the oral cavity, also produce secretions that function in the digestive process and therefore make the glands accessory structures of the digestive system. These glands may be appropriately considered at this time.

Salivary Glands

There are three pairs of **salivary glands,** which are named according to their location. They are the **parotid,** the **submaxillary** or **mandibular,** and the **sublingual** (Fig. 11-19). They secrete a liquid, the **saliva,** into the oral cavity when sensory nerve endings in the oral mucous membrane are stimulated mechanically, thermally, or chemically and also as a result of certain psychic or olfactory stimuli. The saliva secreted by these glands helps prepare the food for digestion in the stomach and intestine. Many small salivary glands, **buccal glands,** are located in the mucous membrane of the oral cavity and continually secrete saliva, which acts to moisten and lubricate the membranes of the mouth and pharynx. Figure 11-20 shows a section through a typical large gland. The saliva collected from the oral cavity is a mixture of the secretions of the various salivary glands.

Saliva is a viscous, colorless, opalescent liquid containing water, salts, some protein, chiefly mucin with small amounts of albumin and globulin, and an enzyme known as **ptyalin,** or **amylase.** The reaction of saliva varies from weakly alkaline to weakly acid, the saliva ranging in pH from 6.0 to 7.9. Its quantity ranges between 1,000 and 1,500 ml in a 24-hour period for an adult. The quality and quantity of saliva collected from the oral cavity depend on many factors. The type of stimulation, diet, age, time of day, disease, and so on, can change the secretion; even the secretion of one gland may change considerably with variations in the stimulus. All salivary glands do not secrete the same type of saliva, and this salivary secretion can also vary with the gland that is participating predominantly at any one time. Thus from the parotid gland, which contains large concentrations of serous cells, the secretion is a serous or albuminous watery liquid, lacking mucin but containing high concentrations of salts, proteins, and ptyalin and therefore having a high digestive power. The other glands, containing mucous cells as well as serous cells, secrete saliva that is more viscid since it contains more mucin. The small glands, containing only mucous cells, elaborate a viscid secretion that consists almost exclusively of mucin.

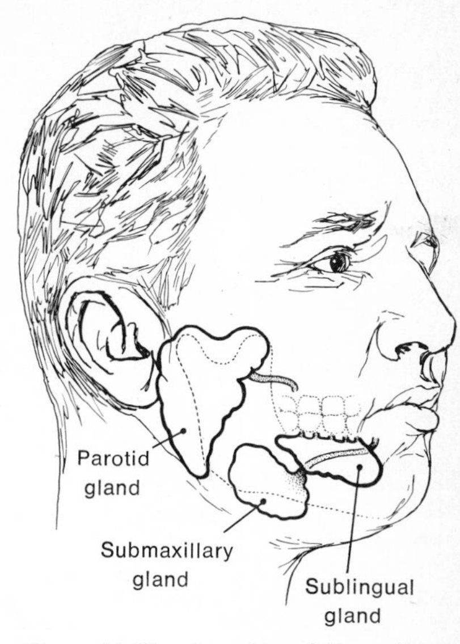

Figure 11-19. Location of the salivary glands, showing those on the right lateral side.

The secretion of saliva is governed by the autonomic nervous system, the glands having innervations from both the sympathetic and parasympathetic systems (Fig. 11-21). Ordinarily the secretion of saliva is a result of a reflex stimulation of the secretory nerves controlled through the salivary center in the medulla. The **parotid gland** receives **sympathetic fibers** from the **first** and **second thoracic spinal nerves** by way of the **superior cervical ganglion. Parasympathetic fibers** from the **ninth cranial (glossopharyngeal)** pass to the parotid by way of the otic ganglion. The **submaxillary** and **sublingual glands** are supplied by **sympathetic fibers** that reach them after synapsing in the superior cervical ganglion. **Parasympathetic fibers** of the **seventh cranial nerve (facial),** the chorda tympani, synapse with postganglionic fibers in the submandibular ganglion and pass to the secreting cells of the submandibular and submaxillary glands.

Stimulation of the parasympathetic or an injection of acetylcholine leads to copious secretion of a serous saliva, which is maintained for a long period. Stimulation of the sympathetic or an injection of epinephrine produces a scanty flow of viscous saliva. Injection of such drugs as acetylcholine, pilocarpine, and physostigmine causes secretion of saliva because synaptic transmission in the parasympathetic system is by way of acetylcholine. Therefore, these drugs enhance parasympathetic activity by acting as a stimulus (acetylcholine) or preventing the

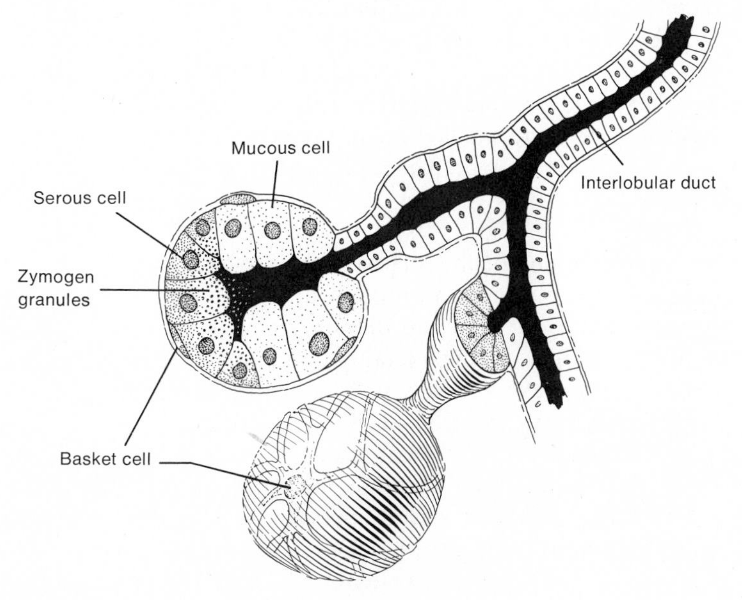

Figure 11-20. Section through a typical large salivary gland.

Figure 11-21. Nerve supply to the salivary glands.

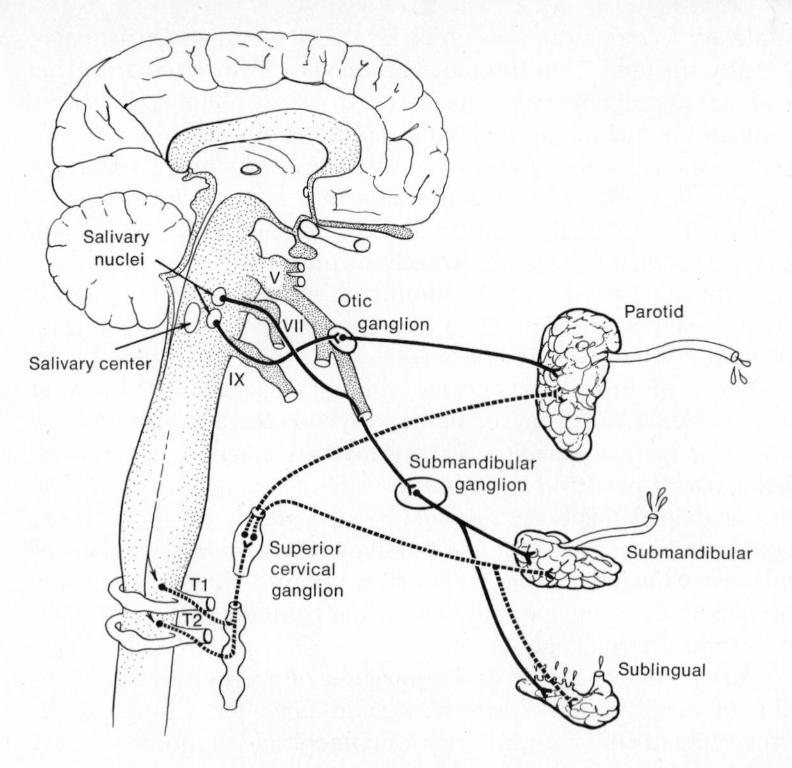

destruction of acetylcholine normally produced in the body (pilocarpine, physostigmine). Epinephrine and ephedrine cause a scanty viscid flow as a result of sympathomimetic activity. Atropine blocks the action of acetylcholine on the secretory gland and thus inhibits secretion.

The quantity and type of secretion are also modified by the type of food that acts as the stimulus. Acid-producing foods as well as dry foods reflexly produce a rather large volume of serous saliva. Milk and other cold liquids produce a small flow of viscid saliva. Meat produces a copious flow of saliva that is extremely high in enzyme content. The mechanisms that produce these adjustments are not known.

The Liver

The **liver** is one of the largest organs in the body. It is situated mainly on the right side, lying under and protected by the lower ribs in contact with the undersurface of the dome of the diaphragm. It is a complex organ and is involved very importantly in the metabolic activity of the organism. The liver will be considered more completely in Chapter 12. In this chapter we will be concerned mostly with its function in the secretion of bile.

The secretion of bile is continuous, and it is passed on into the intestine by way of the gallbladder through the common bile duct, which opens near the pylorus. The **gallbladder** is a reservoir for bile with a capacity of about 50 ml. It lies in a depression at the edge of the visceral surface of the liver, and when it is full, its size and shape resemble those of an elongated pear (Fig. 11-22). The

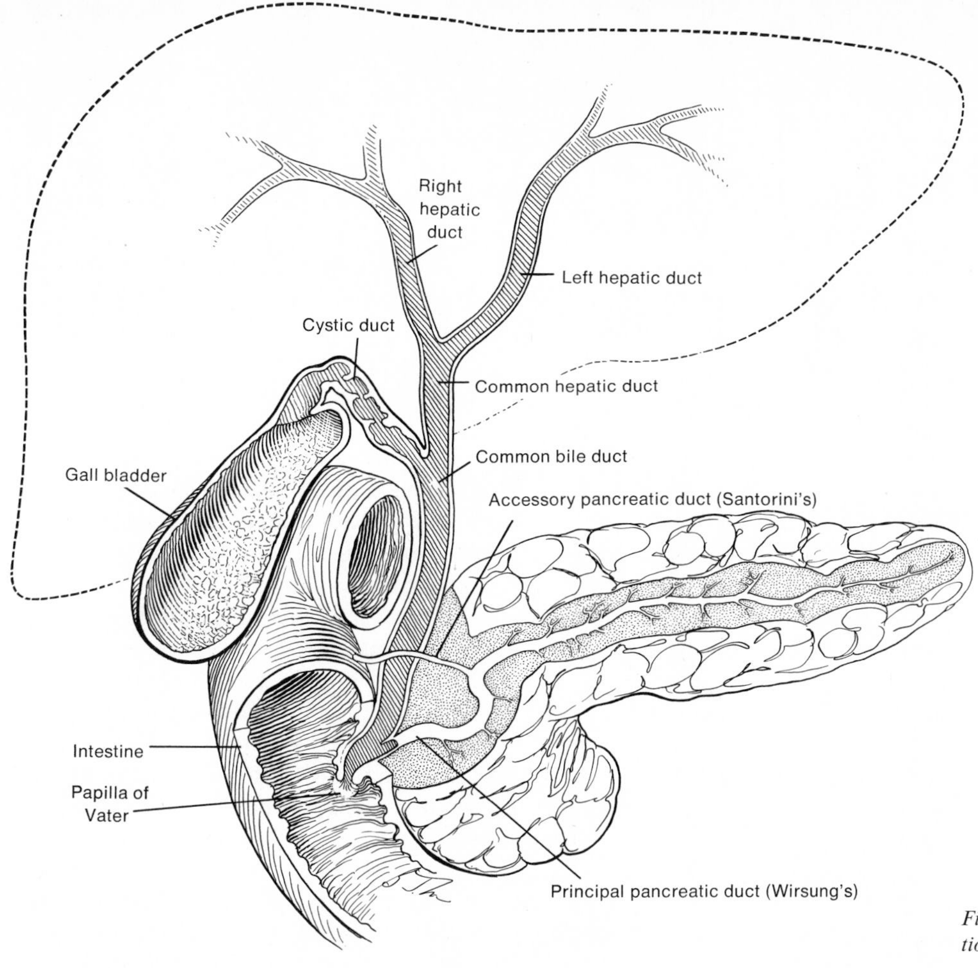

Figure 11-22. Gallbladder and its connections to the liver and intestine.

duct of the gallbladder, known as the **cystic duct,** joins the **hepatic duct,** which descends from the liver to form the **common bile duct.** The common bile duct is joined by the duct of the pancreas just as it enters the duodenum. The **sphincter of Oddi** guards the common entrance and is normally closed except during a meal, at which time the sphincter relaxes and bile is squirted into the intestine when the gallbladder contracts. The mucous membrane of the gallbladder is honeycombed in structure and functions mainly in the concentration of bile salts. Precipitation of these bile salts in the gallbladder leads to the formation of gallstones. Since the gallbladder is not a vital structure, its removal is often resorted to when gallstones form.

Bile is a viscous, yellow-brown or greenish fluid, alkaline in its reactions and bitter in its taste. It consists principally of water with varying amounts of bile salts, bile pigments, cholesterol, and inorganic salts. It is secreted continuously by the parenchymal cells of the liver at a rate of 500 ml in a 24-hour period.

The bile salts, which are the sodium salts of glycocholic and taurocholic acids, are formed in the liver by the conjugation of glycine and taurine with cholic or deoxycholic acid. The bile pigments, bilirubin and biliverdin, are responsible for the color of the bile. The green color is due to biliverdin and the yellow color to bilirubin. The bile pigments result mainly from the breakdown of the hemoglobin of the red cells, which occurs mainly in the liver but to some extent in the spleen and other tissues. The first step in the formation of bile pigments appears to involve an oxidative opening of the porphyrin ring of hemoglobin to produce carbon dioxide and an open-ring compound. The opening of the rings makes it possible to split off the iron portion very easily, and apparently the globin and the iron become detached to produce biliverdin, which may then be reduced to form bilirubin.

Cholesterol is an excretory product in the bile since its amount varies with the amount in the blood. Any reduction in the bile salt concentration causes cholesterol to be deposited in solid form in the gallbladder; it is normally kept in solution by the bile salts as a choleic acid complex. Insoluble cholesterol deposited in the gallbladder constitutes one variety of gallstone. The inorganic salts are chiefly sodium chloride and bicarbonate. The salts of potassium, calcium, and magnesium are also present with some phosphate and sulfate.

Secretion of bile from the liver can be stimulated by bile salts and the hormone **secretin,** which is secreted by the intestinal glands. The emptying of the gallbladder, which is dependent on the contraction of its wall, is not under the control of the nervous system; that is, it is not a reflex phenomenon but is dependent on the hormone **cholecystokinin.** When acids or partly digested proteins or fatty materials enter the duodenum, they stimulate the mucosa to liberate cholecystokinin into the blood stream, which causes a contraction of the gallbladder and a relaxation of the sphincter of Oddi.

The Pancreas

The **pancreas** is the most important source of digestive juices and, next to the liver, is the largest gland connected with the digestive tract. It is a soft, pink-white organ that lies behind the posterior parietal peritoneum, adhering at one end to the middle portion of the duodenum and extending transversely across the abdomen to the spleen. In the adult it is approximately 6 inches in length and about 1 inch in width and varies in weight from 65 to 160 mg. The glandular substance consists of an exocrine portion (Fig. 11-23), which secretes certain digestive juices that are carried to the duodenum by its duct, the pancreatic duct. The duct of the gland communicates with the bile duct to enter the duodenum in common with it (Fig. 11-22) and discharges its secretion into the duodenum through the sphincter of Oddi. In 29 per cent of the population the pancreatic duct enters the duodenum separately. Besides producing the pancreatic digestive juice, certain cells arranged within the pancreas in definite groups and called the **islets of Langerhans** manufacture an endocrine secretion, insulin, which plays an important role in the control of carbohydrate metabolism of the body. This function of the gland will be considered in Chapter 12.

Pancreatic juice is a clear, colorless liquid having a pH of about 8. In man

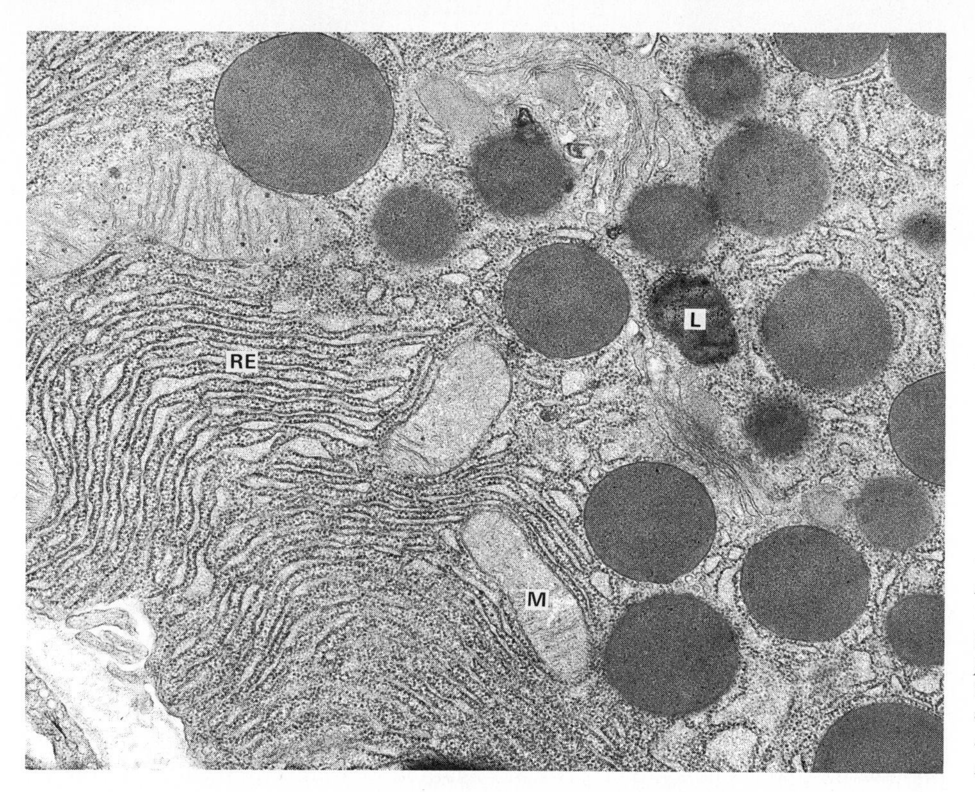

Figure 11-23. Exocrine portion of rat pancreatic cell (magnification 38,250×). The large spherical structures are Zymogen granules. L, *Lysosome;* RE, *rough endoplasmic reticulum;* M, *mitochondria. (Courtesy Timothy Mougal, Electron Microscopy Laboratory, Department of Zoology, University of Maryland, College Park, Md.)*

the pancreas produces about 1,200 to 1,500 ml of digestive juice per day. It is mostly water, containing about 1 per cent inorganic salts, the most important being sodium chloride and sodium bicarbonate. It also contains the salts of calcium, potassium, and magnesium. The large concentration of sodium bicarbonate in pancreatic juice gives it the degree of alkalinity capable of neutralizing equal volumes of acid gastric contents that pass through the pylorus into the duodenum.

The organic constituents of pancreatic juice are mainly enzymes, which are secreted in the main as inactive precursors. One example of these enzymes is **trypsinogen,** which is converted into active **trypsin** by the enzyme enterokinase, secreted by the duodenal mucosa. **Chymotrypsinogen** is another enzyme precursor, which is activated by trypsin to form active **chymotrypsin.** The enzyme **carboxypeptidase** is also present, as well as **lipase** and **amylase.** The pancreatic juice also contains important nucleases. The action of these enzymes will be considered in the discussion of the chemistry of digestion (p. 379). If the pancreas is removed, or if certain diseases affect its function, the digestion of fats and nitrogenous substances is seriously impaired.

The pancreas is supplied with nerve fibers from both the parasympathetic (vagal fibers) and sympathetic (celiac fibers). Stimulation of parasympathetic fibers causes the pancreas to produce a smaller amount of juice, which is thick and rich in enzymes. Stimulation of sympathetic fibers supplying the pancreas produces a scanty flow of watery juice. The chief control of the gland appears to be through a chemical mechanism. Bayliss and Starling in 1902 gave the first clear proof of this when they found that an extract from the mucosal lining of the duodenum could produce a copious flow of pancreatic juice when injected into the blood stream of animals. The effect was attributed to the action of the hormone **secretin,** which is produced by the intestinal glands. Another chemical mechanism was described by Harper and Raper in 1943. They isolated from the duodenal mucosa a factor called **pancreozymin,** which, when injected into the blood stream, produces a secretion of pancreatic juice rich in enzymes. It appears that the secretion of pancreatic juice is brought about through reflex activity initially but is heightened and continued when the acid contents of the stomach reach the duodenum and cause a liberation of secretin and pancreozymin from the duodenal mucosa.

Motility of the Gastrointestinal Tract

Several types of motion are associated with the digestive tract. These help to prepare the food for digestion as well as to move it along the digestive tube. When food enters the mouth, impulses that are set up in the tactile nerve endings in the membrane of the oral cavity, as well as in the taste buds and olfactory areas, indicate whether the food is acceptable or not. If acceptable, the motions of **mastication** (chewing) and **deglutition** (swallowing) follow. The bulk food is broken up in the mouth by the grinding action of the teech in the chewing process. At the same time the saliva, which is stimulated to flow more copiously, is mixed up with the food to make a soft mass, or bolus, suitable for swallowing. Chewing is both a reflex and a voluntary act. When a piece of food is taken into the mouth, the teeth are voluntarily closed on it, but the pressure on the teeth and gum causes the jaws to open reflexly. The jaws are again closed, and the process continues as long as there is anything left in the mouth to chew.

Swallowing

Swallowing, or deglutition, although a complex process, requires only a few seconds and is both a voluntary and involuntary act. In the first stage of swallowing, the mouth is closed, and the bolus of food is forced back into the oral-pharyngeal region by an upward and backward movement of the tongue caused mainly by the contraction of the **mylohyoid** and **styloglossus muscles.** The soft palate is elevated during this time and approaches the posterior pharyngeal wall to close off the nasopharynx. The larynx rises as the bolus passes over the back of the tongue, which forces the bolus against the epiglottis. At this point the continuation of the process of swallowing is an involuntary act. The opening of the larynx into the pharynx is closed off by the action of the sphincter muscles surrounding it and by the cappinglike action of the epiglottis as it folds over the laryngeal orifice. This action prevents the bolus of food from entering the respiratory passages and guides it into the esophagus. As the food enters the esophagus, the ring of muscle fibers at the mouth of the esophagus, which form a sphincter, relaxes, allowing the bolus to pass through. At this point the larynx drops to its original position, and the epiglottis unfolds. The esophageal sphincter contracts, closing off the entrance and preventing air from entering the esophagus during inspiration. If a number of swallowing movements are made in rapid succession, the soft palate and epiglottis do not return to their initial position until the series is completed.

Peristalsis

The bolus of food in the esophagus reflexly initiates the movement of **peristalsis,** which moves the mass of food down to the stomach. In general, there is one esophageal peristaltic wave for each swallowing movement. The waves of contraction that carry the food along the esophagus consist of a coordinated contraction and relaxation of the ring of muscle fibers in the esophageal wall; that is, the muscle fibers above the food mass contract while the fibers below the bolus of food relax (Fig. 11-24). Each wave travels at the rate of about 2 inches per second and is reflexly initiated by the food's contracting the upper pharyngeal wall. The sensory impulses are transmitted by the trigeminal (V), glossopharyngeal (IX), and vagus (X) cranial nerves to the medulla, which controls the onward passage of the wave of contraction. Water and semiliquid food pass rapidly down the esophagus to the stomach mostly by the action of gravity; the peristaltic wave coming down later aids in moving the fluid through the sphincter guarding the entrance into the stomach.

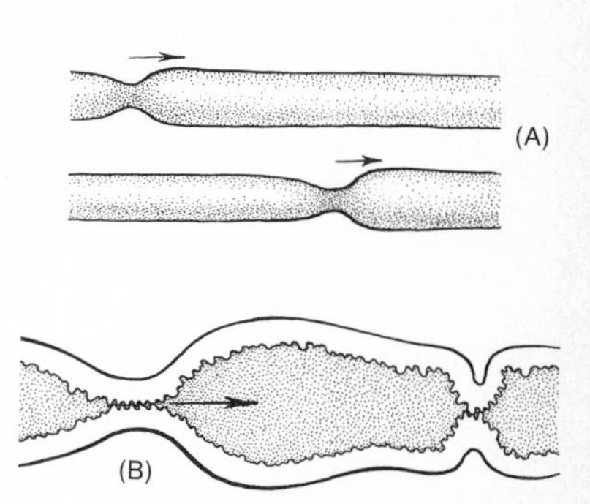

Figure 11-24. Peristaltic wave. A. *External appearance of intestinal loop undergoing peristalsis.* B. *Longitudinal section.*

With each bolus of food swallowed, a small amount of air is also taken in, which forms the "air bell" in the fundus of the stomach. **Eructation,** or belching, is the expulsion of this swallowed air from the esophagus or stomach. In certain highly nervous persons, air swallowing, which is known as **aerophagy,** leads to actual distention of the stomach and its accompanying distress. Persons who eat hurriedly and gulp their food usually suffer with such stomach distress because of the excess air that is swallowed and not because of gases produced by any excessive fermentation in the stomach. Any reflux of the acid gastric contents into the lower end of the esophagus irritates the mucosal membrane of that structure, and the subject has a burning sensation commonly referred to as **heartburn.** In some persons it appears that the esophageal mucosal fold in this area, which acts as a valve, is lax and loses its ability to prevent reflux of gastric contents.

Gastric Filling and Motility

When food reaches the lower esophagus the cardia of the stomach relaxes and the food enters the stomach. The difference between intraabdominal and intrathoracic pressure during inspiration intermittently holds up solid food just above the cardia so that filling is a gradual process. As succeeding amounts of food enter the stomach, the preceding boluses are pushed toward the greater curvature. Upon filling, the smooth muscle fibers of the stomach wall are stretched without increase in tension, allowing the stomach wall to relax, an act that facilitates the reception of more food. The relaxation of the stomach wall at this time is referred to as **receptive relaxation.** The peristaltic activity that arises when the stomach is full is independent of that which occurs in the esophagus. At the beginning of digestion the waves of motion do not involve the entire stomach but begin in the body of the stomach and travel toward the pylorus. Approximately three peristaltic waves occur per minute, being produced by contractions of the circular muscle fibers of the stomach. As digestion proceeds, the waves of motion begin closer to the cardia, and finally after several hours the waves originate at the esophageal orifice and pass over the entire structure. This movement serves to mix the food with gastric juice, which begins the digestion of protein and also serves to churn the food into finer particles.

The activity of peristalsis as it occurs in the stomach is controlled by both chemical and nervous mechanisms. Innervation of the gastric wall is by way of autonomic pathways. Stimulation of the vagus (parasympathetic) stimulates peristaltic motion, and sympathetic stimulation mainly inhibits activity. If the nerve supply to the stomach is cut, after a period of time peristalsis returns, indicating that contraction is not dependent on the impulses carried by the vagus and sympathetic nerves. Like other portions of the digestive tube, the gastric wall contains both the Meissner and Auerbach plexuses, which probably are the sites of origin of gastric contraction. They probably play an important role in coordinating the rhythmical movements of the stomach.

Gastric motility is also influenced by certain secretions produced by the intestinal mucosa. It has been demonstrated that certain food materials, such as carbohydrates and fats, on entering the duodenum cause a release of the hormone **enterogastrone** from mucosal membrane. When released into the circulation, this chemical inhibits the vigor of gastric peristalsis.

It also appears that there are osmoreceptors in the duodenum which sample the material leaving the stomach. When this material is not isotonic, enterogastrone is released and gastric motility is inhibited.

Gastric Emptying

At one time it was believed that the emptying of the stomach's content into the duodenum was controlled by the pyloric sphincter. Since excision of the pylorus does not alter the rate at which the stomach empties, this position is no longer tenable. More recent work has shown that the antrum, pylorus, and upper duodenum apparently function as a unit. Contraction of the antrum is followed by a sequential contraction of the upper pyloric region and the duodenum. The contraction of the pyloric segment tends to persist slightly longer than the duodenal contraction, which prevents regurgitation from the duodenum back into the pylorus. The gastric contents are squirted a bit at a time into the duodenum.

The rate at which the stomach empties into the duodenum is regulated neurally as well as hormonally. Gastric pressure increases when there is an increase in contraction of the stomach wall, which usually occurs on ingestion of food. Any agent that diminishes motility will decrease gastric pressure and thereby increase the length of time of gastric evacuation. Factors that increase duodenal contraction, such as distention or mechanical irritation of the duodenum, inhibit gastric motility through a neurally mediated mechanism, the **enterogastric reflex.** Products of protein digestion and hydrogen ions bathing the duodenal mucosa also initiate this reflex. The release of enterogastrone from the duodenal mucosa by the presence of fats and by carbohydrates to a much lesser degree inhibits gastric motility, causing a decrease in gastric pressure and thereby a retardation of gastric evacuation. Food rich in carbohydrate leaves the stomach in a few hours. Protein-rich food leaves more slowly and emptying is slowest after a meal containing a large amount of fat. Within 2 to 6 hours after the ingestion of food, the stomach has emptied almost all its contents into the small intestine. Excitement is said to hasten gastric emptying and fear to slow it. Cutting the vagus nerve also slows gastric emptying. Most of the mechanisms that inhibit gastric emptying also inhibit secretion of pepsin and hydrochloric acid.

Regurgitation. Under certain conditions, instead of gastric contents being evacuated into the intestinal tract, they are forcibly ejected through the mouth. This action is referred to as **regurgitation,** or **vomiting.** This is a reflex movement, which can be initiated in many different parts of the body. Irritation of the stomach by food or drugs is the most common cause of regurgitation. Excessive distention and compression or irritation of the intestine, appendix, bile ducts, and other abdominal viscera can also initiate this movement. Impulses initiated in this manner are transmitted by sensory nerves to the vomiting center located in the medulla, where coordinated impulses are sent out over various motor fibers to the stomach, diaphragm, and abdominal muscles. The vagus nerve is the pathway to the stomach, the phrenic nerve to the diaphragm, and the spinal nerves to the abdominal muscles. Inhibitory impulses are also sent from the center to the cardia and esophagus. The action of vomiting is not due to peristalsis occurring in the opposite direction but is caused by the relaxation of the esophagus and cardia while there is a simultaneous contraction of the abdominal muscles and diaphragm. The end effect of these coordinated actions is to increase the intragastric pressure by the squeezing action of the diaphragm and abdominal muscles, which causes ejection of the contents of the stomach through the mouth. If the cardia and esophagus are not relaxed, ineffectual vomiting movements occur.

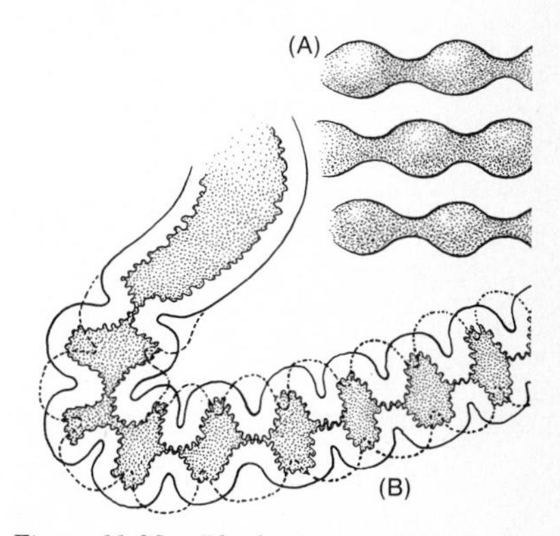

Figure 11-25. Rhythmic segmentation. A. *The annular constrictions in the same loop of intestine at different intervals.* B. *The segmentation's mashing action in a loop of intestine.*

Movement in the Small Intestine. In the small intestine three distinct types of movement are recognized. These have been described as segmentation, normal peristaltic waves, and pendular movements.

Segmentation is the first movement observed in intestinal action. A stretch of resting intestine suddenly shows a regular series of annular constrictions, so the entire loop of intestine appears as a multilobed structure (Fig. 11-25). Each lobe then constricts while the formerly constricted portion relaxes. This alternate constriction and relaxation of stretches of the intestinal tract is repeated six to ten times per minute. This movement does not push the intestinal contents along the tract but serves to mix the contents with the intestinal juices as well as bring the contents in closer contact with the absorbing surface.

The movement of pushing the intestinal contents onward is produced by **peristaltic waves,** which are the same as those already described in the esophagus and stomach. These are contractions of the circular muscle of the intestinal wall behind a mass of intestinal contents while the circular muscles in front of the mass relax (Fig. 11-24). In the intestine two types of peristaltic waves are recognized, slow and fast. Bayliss and Starling in 1900 described the slow waves that progress along the intestine at the rate of 2 cm per minute. Alvarez described waves that are distinct from the slow waves in that they travel at rates of 2 to 25 cm per second without any preceding relaxation and may pass along the whole length of the small intestine. Alvarez called these fast waves **peristaltic rushes.**

All the above movements lead to the movement that may be described as **pendular;** that is, the intestinal loop begins to swing as a result of its anatomical position and the imparted energy of the movements of segmentation and peristalsis. Pendular movements aid in mixing the intestinal contents with the digestive juices.

Peristalsis is not very active in the most posterior portion of the ileum but is increased by the ingestion of food. The ileocolic sphincter relaxes before a peristaltic wave, thus allowing the intestinal contents to pass on to the colon, and then contracts, preventing the regurgitation of the contents of the colon into the ileum.

Movement in the Large Intestine. In the large intestine of man two types of peristaltic movement have been described: **propulsive** and **nonpropulsive.** In general the movements of the large intestine are few. They serve to churn the contents and help in the absorption of water. Several times in a 24-hour period, usually after a meal or in the course of defecation, a strong wave of contraction beginning in the transverse colon travels toward the descending colon, pushing all the contents of the colon into the pelvic colon and sigmoid. Here the contents remain until the movement of **defecation** occurs. Defecation is a reflex movement that is initiated when the rectum is distended or stimulated by movements of the feces within it. The sensory impulses travel along the pudendal and pelvic nerves to the spinal center, situated in the lumbosacral segments of the spinal cord (second, third, and fourth sacral segments). The motor pathway from the spinal cord is by way of autonomic fibers. Sympathetic impulses have an inhibitory effect on the walls of the colon (antiperistaltic) and cause contraction of the internal and external sphincter. Parasympathetic fibers arise in the sacral segments of the spinal cord and carry impulses, which increases the motility of the colon and relaxes the internal sphincter but causes contraction of the external sphincter.

Stimulation of the rectum brings into play the parasympathetic pathway, which causes a strong wave of peristaltic contraction of the rectum and relaxation of the internal anal sphincter, causing emptying of the large intestine. Simultaneous contraction of the diaphragm and abdominal muscles aids in the emptying process. Because the external sphincter is a striated muscle under voluntary control, this emptying of the rectum may be voluntarily inhibited even when the reflex action is set into motion. Intense emotions, such as fear, fright, or extreme excitement, inhibit the act of defecation by excessive firing of the sympathetic.

The rectum appears to be able to distinguish between flatus and feces purely on the basis of the degree of distention of the rectum. Feces produce more distention than flatus, and when these sensations rise to consciousness, a voluntary decision is made, depending on the degree of sensation, whether to maintain the contraction of the external sphincter or relax it.

Normal motility of the colon may give rise to one or more bowel movements per day. If the materials in the colon are moved through too slowly, there is excessive water absorption and the matter becomes hard and dry, giving rise to less frequent and more difficult evacuations, a condition referred to as **constipation.** On the other hand, if the contents of the colon are moved too rapidly, there is less water absorption and the material reaches the rectum in a more fluid condition. This gives rise to the condition of more frequent and loose evacuations, referred to as **diarrhea.** It appears that the consistency of the evacuated matter is a more adequate criterion of normal colon motility than the number of movements.

Laxatives, also known as **cathartics,** are used quite frequently to stimulate colon movements. The laxative action in some instances is produced by the chemical-stimulating effect some materials, for example, certain foods and drugs, have on the lower intestinal tract. Other laxatives owe their action to their ability to prevent water absorption from the large intestine, resulting in the filling of the colon with a watery bulk, which is very effective in promoting peristalsis. The various cathartic salts, such as magnesium sulfate, are able to do this since they are impermeable to the intestinal membrane and thereby hold water in the colon and also draw it from the cells by osmosis. Oils have a similar action except that they act by coating the intestinal wall, physically preventing water absorption, and are also effective in softening the relatively dry fecal material.

Chemistry of Digestion

The act of digestion involves the splitting of the large complex molecules of food substances, such as carbohydrates, fats, and proteins, into smaller, simpler, absorbable molecules. The process as it is carried on in the body involves the action of many enzymes and several specific areas of the digestive tract. According to the area in which digestion is carried on, the digestive process may be classified as salivary digestion, which occurs in the mouth, gastric digestion, or intestinal digestion.

Salivary Digestion

The secretion in the mouth known as saliva is mixed thoroughly with the food materials by the action of mastication. The only digestive enzyme present in saliva, ptyalin (amylase), being brought into intimate contact with the food material, acts on starches to break them into less complex forms known as **erythrodextrins.** The food remains in the mouth too short a time for the digestion of starch to proceed very far, but the action of the enzyme is still maintained even after the food is swallowed, until the acid gastric juice penetrates the food mass. The mucin contained in saliva lubricates the food, which facilitates its passage down the esophagus. In addition, saliva functions as a solvent for certain constituents of food and therefore subserves the sense of taste, since the taste buds can be stimulated only when these constituents are in solution. Saliva also aids in keeping the mouth and teeth clean.

The starches are polysaccharides (carbohydrate), and those that are normally found in the mammalian diet are named amylose, amylopectin, and glycogen. They are all molecules consisting of many glucose units linked to each other in a repeating fashion and are hydrolyzed in the presence of the enzyme amylase (ptyalin) to residual polysaccharides; that is, the links between glucose units are broken in several areas to produce simpler components. The first product of the action of amylase on starch is **soluble starch,** which gives a blue color with iodine. As the action continues and smaller molecules are formed, the color reaction with iodine changes from blue to red. These smaller fragments of the original complex polysaccharide, called **dextrins,** differ in molecular size and complexity, and since they give a red color with iodine, they are called red sugars or **erythrodextrins.** The erythrodextrins are further split into smaller fragments until free **maltose** is produced. The changes described may be graphically represented as follows:

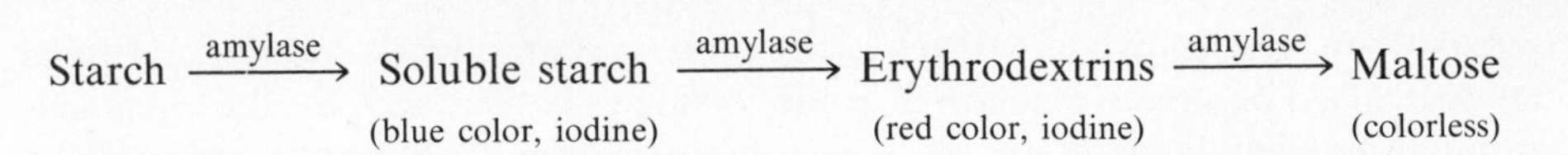

The enzyme amylase functions in alkaline, neutral, or faintly acid solutions to hydrolyze starches to dextrins and then to maltose. As the acid gastric juice is mixed with the foodstuffs in the stomach, the action of amylase is stopped as a result of the increased hydrogen-ion concentration produced in the stomach by the free hydrochloric acid.

Gastric Digestion

After leaving the mouth, the food is carried by peristaltic movements of the esophagus to the stomach. Here it is further physically broken up by the churning action of the stomach and chemically acted on by the gastric secretion called **gastric juice.**

Normal gastric juice is a thin, light-colored fluid, which is acid in its reaction. The acid reaction is due to free hydrochloric acid, which is equivalent to a solution of 0.16 *N* hydrochloric acid. In addition to hydrochloric acid, gastric juice contains various salts and the enzymes **pepsin, rennin,** and **lipase.**

The function of hydrochloric acid is to provide a medium in which the proteolytic enzymes can act most satisfactorily. It is produced by the parietal cells of the gastric mucosa, and its mode of production has best been explained by a theory proposed by Davenport and elaborated on by Davies and Ogston. The theory is based on the fact that large quantities of the enzyme carbonic anhydrase are present in the parietal cells and relatively low quantities in other cells of the gastric mucosa. As in blood cells, this enzyme catalyzes the reaction of the hydration of carbon dioxide to carbonic acid:

$$H_2O + CO_2 \xrightarrow{CA} H_2CO_3. \quad (11\text{-}1)$$

which dissociates into hydrogen ion and bicarbonate ion:

$$H_2CO_3 \longrightarrow H^+ + HCO_3^-. \quad (11\text{-}2)$$

The bicarbonate ion shifts out into the blood stream, and chloride ion shifts into the parietal cell to take its place. The chloride thus becomes available for secretion along with hydrogen ions into the gastric juice. The energy required for the transport of these ions through the parietal cell membrane, which is against a concentration gradient, is supposedly derived from the aerobic oxidation of such compounds as glucose.

Pepsin is the enzyme in gastric juice that specifically acts on proteins to break their long, complex chain formation into shorter fragments called **proteoses** and **peptones.** These fragments are later broken down to free amino acids by the proteolytic enzymes of the pancreatic and intestinal juices. Pepsin is secreted by the chief cells of the gastric mucosa in the form of an inactive precursor called **pepsinogen,** which is converted into the active pepsin on coming into contact with the hydrochloric acid of the stomach. Pepsin works best in the acid medium provided by the gastric contents. Its activity is stopped in an alkaline environment.

The fact that the stomach wall, itself a protein, is not digested by the enzyme pepsin or irritated by the high concentration of hydrochloric acid is attributed to the buffering action of the mucus secreted by the mucous glands and to the presence of antienzymes in the cells of the intestinal mucosa, which protect them from digestion by proteolytic enzymes. At death the surface epithelium of the stomach and intestine is rapidly digested.

Rennin is an enzyme that coagulates protein, and its main function in gastric digestion is the curdling of milk. It is not certain that this enzyme is present in the gastric juice of adults, but since the action of this enzyme requires a less acid environment than is found in the adult stomach, if it is present, it must be almost inactive. It is definitely found in the stomach of young children and healthy infants. The optimum pH for rennin is from 5 to 6, a pH value normally found in the stomach of the very young. The action of rennin is to convert the insoluble casein of milk into its soluble form, paracasein, which is then precipitated in the presence of calcium ion to form the curdle.

Gastric lipase is a fat-splitting enzyme which has limited action at the normal acidity of the stomach. Its optimum activity is manifested in solutions, the concentrations of which show a low acidity. It may function more importantly as a fat-splitting enzyme in the stomachs of young animals, where gastric acidity is lower.

Gastric juice also contains an **intrinsic factor,** which acts on certain components of the food, probably vitamin B_{12}, to form the **erythrocyte-maturing factor,** or **antianemic factor,** which is necessary for the normal development of red blood cells.

The components of gastric juice and their actions may be summarized as follows:

Gastric Juice	*Action*
Hydrochloric acid	Provides proper medium for action of pepsin; aids in coagulation of milk in adults
Pepsin	Splits proteins into proteoses and peptones
Lipase	Acts in slight digestion of fats, especially butter and egg yolk
Rennin	Coagulates or curdles the protein of milk
Intrinsic factor	Acts on certain components of food to form the antianemic factor

In human beings, gastric secretion is carried on continuously, and a volume of 2 to 3 liters in 24 hours is considered average. The rate of secretion is under the control of both nervous and chemical mechanisms. Vagal stimulation causes the glands to secrete a highly acid juice containing much pepsin. Stimulation of the sympathetic supply to the stomach causes vasoconstriction and a decrease in glandular secretion, although this response may vary.

The ingestion of food increases the flow of gastric juice, and in general, the gastric secretion under these conditions may be divided into three stages, or phases, called cephalic, gastric, and intestinal.

The **cephalic phase** is reflexly initiated by stimuli arising in the cortex from the sight of food or from thoughts of an appetizing meal or in the olfactory area from the smell of food and from the taste of food. This phase is readily inhibited by emotions such as fear, anger, anxiety, and sensual thoughts.

The **gastric phase** is initiated by the entrance of food in the stomach. Certain foods are capable of provoking the continued secretion of gastric juice by their ability to stimulate the gastric mucosa, especially the pyloric portion, to secrete a hormone, **gastrin.** Gastrin is carried by the blood stream to the gastric glands, stimulating these structures into further action. Such foods as meats, soups made from meats, alcohol, sodium bicarbonate, and the products of gastric digestion are able to invole this reaction more readily than carbohydrates or fats. The gastric phase of gastric secretion is due to the liberation of gastrin by the stimulating effect on the gastric mucosa of certain food materials, especially of protein origin.

It has been demonstrated that local mechanical stimulation of food against the gastric mucosa has very little effect in producing an increased gastric secretion.

The **intestinal phase** of gastric secretion is initiated when the partly digested protein products of gastric digestion reach the small intestine. These products stimulate the intestinal mucosa to release an intestinal hormone, **intestinal gastrin,** which stimulates the gastric glands to continue their secretion.

As gastric digestion is carried on, the food is changed to a semifluid mass of uniform consistency called **chyme.** This material is passed on in small quantities through the pyloric sphincter into the duodenum, where most of the digestion of foodstuffs begins. It normally requires about 6 hours for the stomach to digest its food completely and empty itself of all constituents, but this time may vary considerably with the type of food ingested and the emotional state of the subject. Carbohydrates pass through the stomach quickly. Proteins pass through more slowly, and fatty foods pass through the slowest. Emotional states especially fear and anger, can cause digestion and emptying to be delayed, so that food may still remain in the stomach after 24 hours.

Intestinal Digestion

As it is passed into the small intestine, the chyme is mixed with the pancreatic juice secreted by the pancreas, the bile from the liver, and the secretion from the glands of the intestinal mucosa, called **intestinal juice.**

The **intestinal juice,** or **succus entericus,** which is produced by the intestinal glands, has a definite alkaline reaction with a pH of 7.6 or more and contains a large number of enzymes, which complete the digestive process. The most important of these enzymes are

1. *Aminopeptidase* (*erepsin*), which splits fragments of protein more complex than proteoses and peptones into free amino acids.
2. *Amylase,* which digests starch to maltose.
3. *Maltase,* which acts on maltose to form the monosaccharide glucose.
4. *Lactase,* which splits the disaccharide lactose into galactose and glucose.
5. *Sucrase,* which splits cane sugar (sucrose) into fructose and glucose.
6. *Lipase,* which is responsible for the breakdown of fat.
7. *Nucleases,* which splits nucleic acids, into nucleotides. Nucleotides are further split into phosphoric acid and nucleosides by enzymes called *nucleotidases* also present in intestinal juice.
8. *Enterokinase,* which activates the inactive trypsinogen in pancreatic juice to trypsin.

The main stimulus for intestinal secretion is the local mechanical effect produced by the chyme and chemical agents found in it. It appears that a hormone is released from the intestinal mucosa that stimulates the intestinal glands to secrete. This hormone, called **enterocrinin,** has been extracted from the intestinal mucosa.

Pancreatic juice, which is also found in the small intestine, is a colorless fluid having a strong alkaline reaction with a pH range of 8.0 to 8.9. The secretion of pancreatic juice is reflexly produced by the liberation from the intestinal mucosa of secretin and pancreozymin under the influence of the acid chyme, which soon becomes neutralized by the alkaline pancreatic juice and by the intestinal juice and bile.

The major protein-splitting enzymes found in pancreatic juice are trypsin, chymotrypsin, and carboxypeptidase. **Trypsin** acts to split proteins into proteoses, peptones, polypeptides, and amino acids. Its action is very similar to that of pepsin, except it is able to go as far as to liberate some free amino acids. **Chymotrypsin** appears to have the same action as trypsin but the specificity of acting on proteins whose linkages involve the carboxyl group of tyrosine and phenylalanine. **Carboxypeptidase** splits off amino acids from the terminal portion of polypeptides.

Pancreatic juice contains an enzyme like ptyalin, called **pancreatic amylase,** which possesses a greater digestive power than that found in the saliva and hydrolyzes starch to maltose. The enzyme maltase, also a constituent of pancreatic juice, splits maltose into glucose.

A lipase is also found in pancreatic juice and is called **pancreatic lipase.** It is the most important fat-splitting enzyme in the digestive tract because it is capable of splitting the neutral fats of the food into **fatty acids** and **glycerol.** An example of this type of activity would be as follows:

$$\underset{\text{Tripalmatin}}{C_3H_5(OOCC_{15}H_{31})_3} + 3H_2O \xrightarrow{\text{lipase}} \underset{\text{Palmitic acid}}{3C_{15}H_{31}COOH} + \underset{\text{Glycerol}}{C_3H_5(OH)_3}.$$

The facility with which pancreatic lipase is able to attack the fat molecule depends on the surface exposed by the fat. The more highly emulsified the fat is, the more surface is exposed to enzyme action and the more rapid is the enzyme reaction. The bile, which is produced by the liver and which contains many salts, subserves the digestion of fat by lipase since it aids in the emulsification of fats by lowering surface tension of the fatty film surrounding fat particles. Closer contact between the lipase and the fat particle is possible.

Important nucleases are also present in pancreatic juice. These function to split nucleic acids into nucleotides.

Through the action of the enzymes of the pancreatic and intestinal juice and by the emulsifying action of the bile salts, the bulk of digestion is carried out in the small intestine. Thus, the chyme enters the intestine from the stomach and contains nonabsorbable forms of foodstuffs such as fat, protein and protein derivatives, and carbohydrates including starch; these are then converted to their simple absorbable forms. Proteins are completely digested to **amino acids.** Fats are broken down into their absorbable forms of **fatty acids** and **glycerol,** and carbohydrates are finally split into simple monosaccharides such as **glucose** and **fructose.** These products of digestion usually reach the ileocolic sphincter in 3 or 4 hours, and the last of the meal may leave the ileum 8 or 9 hours after ingestion. The digestive process in the small intestine is summarized in Table 11-1.

By the time the intestinal contents arrive at the beginning of the large intestine, digestion and the absorption of the products of digestion are almost complete. The large intestine secretes a juice that is rich in mucus but contains no enzymes. It is a thick, opalescent alkaline fluid, which functions as a lubricant to facilitate the movement of the colonic materials and as a protective covering for the mucosa.

The chyme, which passes from the small intestine into the large intestine, is greatly reduced in bulk by the absorption of most of its water and some soluble substances through the wall of the colon. This absorption occurs mainly in the cecum and ascending colon but can occur all along the tract down to the rectum. This function of the large intestine is most important in maintaining the water balance of the body, even though most of the reabsorption of water occurs in the small intestine. This property makes it possible to introduce fluids into the

Table 11-1. ***Summary of Digestion in the Small Intestine***

Gland	*Secretion*	*Action*
Pancreas	Pancreatic juice	
	Trypsin	Proteins ⟶ Proteoses and peptones
	Chymotrypsin	Proteins ⟶ Proteoses and peptones
	Carboxypeptidase	Polypeptides ⟶ Amino acids
	Amylase	Starch ⟶ Maltose
	Maltase	Maltose ⟶ Glucose
	Lipase	Fats ⟶ Fatty acids and glycerol
	Nuclease	Nucleic acids ⟶ Nucleotides
Liver	Bile	
	Bile salts	Emulsification of fat
Intestinal gland	Succus entericus	
	Erepsin (aminopeptidase)	Protein fragments ⟶ Amino acids
	Amylase	Starch ⟶ Maltose
	Maltase	Maltose ⟶ Glucose
	Lactase	Lactose ⟶ Galactose and glucose
	Sucrase	Sugar cane ⟶ Fructose and glucose
	Lipase	Fats ⟶ Fatty acids and glycerol
	Nucleosidase	Nucleosides ⟶ Pentose sugar and organic base
	Nucleotidases	Nucleotides ⟶ Phosphoric acid and nucleosides
	Enterokinase	Trypsinogen ⟶ Trypsin

body when necessary by way of the rectum. Various minerals, such as salts, calcium, and iron, can also be introduced into the body in this way, as well as simple carbohydrates (glucose) and drugs such as atropine.

The material remaining in the large intestine, usually for a period of about 36 hours, is prepared for elimination by the action of bacteria that are normal inhabitants of this area. They will ferment the remaining carbohydrates with the release of hydrogen, carbon dioxide, and methane gas. The major part of the gas found in the large intestine, however, is derived from the air swallowed with the ingestion of food. Any cellulose taken in with fruits and vegetables is digested in the large intestine by these bacteria. The colonic bacteria also act on any residual protein and amino acids to form **indole** and **skatole.** These substances contribute to the characteristic odor of feces and result specifically from the action of bacteria on the amino acid tryptophan. Compounds like tyramine and histamine may be formed from bacterial action on the amino acids tyrosine and histidine, respectively. Tyramine is a vasoconstrictor and will cause elevation of the blood pressure. Histamine is a vasodilator and can cause a lowering of the blood pressure. The intestinal bacteria also play a significant role in nutrition since certain of the vitamins are synthesized by these microorganisms. A substantial proportion of the daily requirements of riboflavin, nicotinic acid, biotin, folic acid, and vitamin K is synthesized in the large intestine.

At the end of about 36 hours, the chyme in the large intestine has become solid or semisolid and is known as **feces.** The color of feces is due to the presence of **stercobilin,** derived from the bile pigments biliverdin and bilirubin. Both these pigments are reduced to urobilinogen in the large intestine. Some of the urobilinogen is absorbed through the intestinal wall into the blood, where it passes by way of the portal vein to the liver. The liver excretes some of this in the bile and passes some into the general circulation to be carried to the kidney and excreted in the urine. The portion of urobilinogen not absorbed in the large intestine is converted to stercobilin, the brown pigment of the feces.

If the bile pigment cannot enter the intestinal tract because of obstruction of the common bile duct by a stone or tumor, the feces are pale in color. Under these circumstances, the bile backs up into the liver and passes back into the blood stream. As a result the bilirubin in the blood rises above its normal concentration, and this imparts a yellow color to the skin. This condition is known as jaundice and, as produced in the manner described, is said to be **obstructive jaundice.** An excessive rise of bile pigment in the blood will also be reflected in the color of the urine as a greater quantity of urobilinogen will be excreted by the kidney. The urine becomes dark green-yellow in color.

An excessive production of bile pigment by the liver will cause a change in the color of the feces as a result of an increase in the amount of stercobilin. The feces may become dark brown to black in color. In addition, more urobilinogen is absorbed through the intestinal tract into the blood stream and will finally show up in the urine. If the concentration in the blood is greatly increased, the subject will also show jaundice. Disease conditions that result in excessive destruction of red blood cells, that is, hemolysis of red cells, with a consequent increase of hemoglobin to be processed, produce the changes described above. The jaundice resulting is called **hemolytic jaundice.** Jaundice conditions resulting from disease of the liver itself (hepatitis) are referred to as **hepatic,** or **toxic jaundice.**

Intestinal Absorption

Most of the absorption of the digested foodstuff takes place in the small intestine, as this structure is especially adapted to carry on this process. The absorption of certain drugs occurs through the oral mucous membrane, and small amounts of water, simple salts, glucose, and alcohol can be transferred through the gastric mucosa into the blood stream, but the transfer of most materials across the epithelial boundary of the gut is primarily a function of the small intestine.

We have seen that the process of digestion breaks proteins into amino acids, carbohydrates into simple sugars, and fats into fatty acids and glycerol. Approximately 90 per cent of the ingested foodstuffs is absorbed in the course of its passage through the approximately 25 ft of small intestine. Two pathways are open for these products of digestion to pass from the interior of the alimentary canal into the body itself. One of these entrances is the capillaries of the portal system, which lead by way of the portal vein directly to the liver. Here the absorbed food material is subjected to certain powerful influences before it finds its way to the right side of the heart. The second area of entry is by way of the lacteals, which are the lymphatic vessels of the small intestine. From here the absorbed food material passes into the general circulation by way of the thoracic duct after having undergone only such changes as it encounters in the lymphatic system.

The structure within the small intestine responsible for the absorption of digested food material is the villus. The number of villi is enormous. Each villus is a fingerlike process containing an arteriole, a capillary plexus, a venule, and a lymphatic vessel (lacteal). The venules lead into the veins of the hepatic portal system, and the lacteals, into those of the thoracic duct. Absorption is facilitated by this arrangement, since the villi provide a large surface area through which absorption can take place. The blood passing through the capillaries of the villi is brought into close association with the food material within the intestine. Segmental movements occurring in the small intestine not only act to increase blood and lymph flow through this area, but also aid in the absorption process by bringing the food material into close contact with the intestinal mucosa. Finally, the great length of the small intestine provides sufficient time for the absorption of food materials to take place.

The passage of substances through the intestinal mucosa into the blood and lymph is not entirely explicable on the basis of the physical laws of diffusion, osmosis, and filtration, but appears to depend on an active transport mechanism involving considerable energy exchange. In addition, the intestinal epithelium appears to possess the ability to select certain substances to be absorbed (selective absorption) while excluding others. For example, sodium chloride is readily absorbed, but sodium sulfate is not; many simple sugars are more readily absorbed than others whose molecular weights are smaller. Energy is required for the active transport of materials through the intestinal barrier, where osmotic equilibria have been satisfied. This energy is probably provided by a phosphorylating mechanism dependent on an energy-rich phosphate compound such as adenosine triphosphate

(ATP). Selective absorption is probably dependent on specific enzyme systems within the intestinal epithelial cells.

Carbohydrate Absorption

It has been shown that the main products of carbohydrate digestion are the simple monosaccharides, chiefly the 6-carbon sugars, glucose, fructose, mannose, and galactose. Sugars are absorbed from the intestine into the blood of the hepatic portal veins. Some fourteen monosaccharides have now been found that are actively absorbed against a concentration gradient. Glucose and galactose are absorbed more rapidly than fructose or mannose. The rapid and continuous absorption of such sugars as glucose and galactose, even against an osmotic gradient and independent of their concentration in the intestinal lumen, indicates that a mechanism other than diffusion or osmosis is operating. This mechanism, which is probably a phosphorylation reaction, converts these sugars to glucose or galactose phosphate involving an enzyme **hexokinase** and a high-energy phosphate source, which is presumably ATP. Sugars, such as mannose and fructose, that are not phosphorylated in the body are absorbed by the slower process of simple diffusion. The opposite process of dephosphorylation is supposed to occur as the sugars leave the intestinal epithelium and enter the portal blood. Analysis of portal blood shows only sugar in the free state and not sugar phosphates. If the enzyme hexokinase is poisoned by the injection of phlorhizin, glucose and galactose cannot be phosphorylated and therefore can be absorbed only by diffusion, which reduces their rate of absorption through the intestinal barrier considerably. It also appears that in order to be phosphorylated a sugar must have a hydroxyl group at the carbon-2 position.

Experimental evidence indicates that the sodium ion and the "sodium pump"[1] are apparently essential for the active transport of sugars. Replacing the sodium ion with lithium or magnesium abolishes active sugar transport.[2] Digitalis and similar drugs inhibit the active transport of sugars, probably by preventing sodium-ion transport with the resulting effect on sugar transport.

Protein Absorption

During the process of digestion most proteins are broken down into their constituent amino acids, and it is believed that under normal circumstances only the amino acids can be absorbed directly from the intestine. The amino acids pass into the portal circulation and are carried to the liver before entering the general circulation. This is shown by the rise in amino acid content of the portal blood during the absorption of a protein meal. Like carbohydrates, amino acids are transported through the membrane both actively and passively. L-Amino acids

[1]Sodium pump: Living cells usually contain more potassium than sodium, whereas the fluid environment surrounding the cells contains more sodium. This difference in ion concentration usually was explained on the basis that the cell membrane was impermeable to sodium and potassium ions. The origin of this unequal distribution is not known. Today, however, the high potassium concentration and low sodium concentration found in cells, especially nerve, muscle, and red blood cells, is thought to depend on the activity of a pump in the cell membrane that uses energy to accumulate potassium and expel sodium.

[2]T. Z. Csaky and L. Zollicoffer, "Ionic Effect on Intestinal Transport of Glucose in the Rat." *Am. J. Physiol.*, **198:**1056 (1960).

are absorbed more quickly from the gut than D-amino acids, suggesting the possibility of an enzymatic transmission through the intestinal barrier for the former, whereas the latter are absorbed by passive diffusion. Recent investigations[3,4] show that neither of these concepts is entirely correct. The active transport of an amino acid may be inhibited by the presence of one or more other amino acids. For example, it has been demonstrated that the amino acid D-methionine can be absorbed against a concentration gradient and that such absorption is susceptible to metabolic inhibition and the presence of L-methionine. Thus, it appears that the D and L forms of the amino acids can share the same transfer mechanism, but the D form has a lower affinity for it than the L form. It also appears that the mucosal cell has the ability to store amino acids since the mucosal concentration can be as high as nine times that in the surrounding medium. Storage within the intestinal epithelial cells has been demonstrated for L-tyrosine, L-tryptophan, and L-phenylalanine. Samey and Spencer[5] have suggested that the passage of amino acids into the mucosal cell from the intestinal gut takes place through the active process, while the outward diffusion from the intestinal membrane into the circulation is passive.

There is no doubt that intact protein molecules can pass through the intestinal mucosa. However, this occurs only in certain persons in rather special circumstances, and the mechanism of transport is obscure. **Pinocytosis,** a mechanism whereby the cell takes up material by a drinking process (that is, materials are taken up discontinuously in droplets that are engulfed or sucked in by the cell) has been suggested.

Large, intact protein molecules are known to be antigenic, that is, to stimulate immunological responses (formation of antibodies). When the concentration of these antibodies has been increased sufficiently by repeated absorption of small amounts of intact protein, the reaction between the antibodies and foreign protein (antigen) may produce asthma or skin rashes (hives), and the subject is said to be sensitive or allergic to the protein. Those persons in whom an immunological response to ingested protein occurs must therefore be able to absorb some undigested protein.

Absorption of Fats

Ingested fats are hydrolyzed by the intestinal lipases in the presence of bile salts into glycerol and fatty acids. This action is not always complete, and some fat remains and can be absorbed as neutral fat if it is dispersed in very fine particles not more than 0.5 μm in diameter. Bile salts aid in the dispersing of unsplit fat and thus function importantly in their absorption. Glycerol, because it is water soluble, is easily absorbed from the intestine, but the fatty acids form a complex with the bile salts before they can be absorbed directly through the lipoid membrane of the intestinal cells. It has been demonstrated that the intesti-

[3] E. C. C. Lin and T. H. Wilson, "Transport of L-Tyrosine by the Small Intestine in Vitro." *Am. J. Physiol.*, **199:**127 (1960).

[4] R. P. Spencer and A. H. Samey, "Intestinal Transport of L-Tryptophan in Vitro." *Am. J. Physiol.*, **199:**1033 (1960).

[5] A. H. Samey and R. P. Spencer, "Accumulation of L-Phenylalanine by Segments of Small Intestine." *Am. J. Physiol.*, **200:**505 (1961).

nal mucosa contains a lipase that recombines the long-chain fatty acids with glycerol to form a **long-chain triglyceride.** These long-chain triglycerides are made up of those fatty acids with more than twelve carbon atoms, such as palmitic (C_{16}), myristic (C_{14}), and stearic (C_{18}) acids. Within the mucosal cell the triglycerides are incorporated into chylomicrons and pass directly into the lacteals as neutral fat and thence into the blood by way of the thoracic duct to be deposited in the fat depots of the body. The short-chain fatty acids, such as lauric (C_{12}), and decanoic (C_{10}), pass unchanged through the intestinal barrier directly into the portal blood.

About 10 to 15 per cent of the fat in lymph is in the form of phospholipid, suggesting that phosphorylation may be involved in fat absorption, but the exact role it plays in fat transport is not understood. It has been suggested that lecithin (phospholipid) is formed, which increases the solubility of fat and thus promotes its absorption. It has also been suggested that phosphorylation acts to stabilize fat droplets in the presence of the protein of the blood plasma. Table 11-2 summarizes the absorption of fat through the intestinal epithelium.

Table 11-2. ***Absorption of Fats***

Form Absorbed	*Intestinal Epithelium*	*Route of Absorption*
Glycerol }	Recombined with long-chain fatty acid ⟶ Triglycerides	As chyle in lymph vessel
Long-chain fatty acids }	Some phosphorylated ⟶ Phospholipid	Portal blood
Short-chain fatty acids	Short-chain fatty acid	Portal blood
Particulate fat, less than 0.5 μm	Neutral fat	As chyle in lymph vessel

Absorption of Cholesterol and Other Substances

The absorption of cholesterol takes place mainly in the upper small intestine of man. It is incompletely absorbed; only 40 to 60 per cent is absorbed in contrast to dietary fat, which is normally 95 per cent absorbed. It has been demonstrated,[6] by means of cannulations of the thoracic duct and the administration of labeled cholesterol, that dietary cholesterol is absorbed mainly by the lacteals and that the major portion is esterified while passing through the intestinal barrier. Related plant sterols are not absorbed from the intestine, with the exception of ergosterol, which is absorbed after it has been converted by irradiation to vitamin D. Vitamins A and D are absorbed along with fats. Vitamin K is absorbed only when bile salts are present. In cases where bile is not secreted into the intestinal tract, vitamin K is not absorbed. Plasma prothrombin levels, then fall, since its production depends on this vitamin. Lack of vitamin K in the body produces an increase in clotting time and a tendency in such persons for hemorrhage to occur.

[6] L. Hellman, E. L. Frazell, and R. S. Rosenfeld, "Direct Measurement of Cholesterol Absorption via the Thoracic Duct in Man." *J. Clin. Invest.*, **39:**1288 (1960).

Water and Ion Absorption

Water absorption through the intestinal barrier appears to be mainly a passive process, following the osmotic forces developed by the over-all exchange of materials between the intestinal epithelial cells and the fluid in the lumen. Experimental evidence in dogs and cats[7] indicates that water movement through the intestinal membrane also occurs in the absence of or against high osmotic gradients. This would suggest that some active transport mechanism, requiring an energy source, also participates in water absorption.

The transport of various ions across the intestinal barrier seems to be mainly a process depending on an active mechanism, although passive diffusion does occur. Different areas of the small intestine vary in their ability to transport ions. For example, it has been demonstrated that glycolysis provides a source of energy for ion transport by the jejunum since, if glycolysis is prevented in the mucosal tissue, transport of ions practically ceases. In the ileum this has very little effect. Work carried out by Gilman and Loelle[8] shows that sodium is actively transported across the intestinal mucosa against an electrochemical gradient and is mainly dependent on the presence of glucose as an energy source. It has also been demonstrated that calcium is actively transported against a concentration and potential gradient and that both active and passive transport is enhanced by vitamin D. Phosphate is absorbed at all levels of the small intestine but most effectively from the ileum. The absorption of iron through the intestinal mucosa appears to be regulated by the amount of iron stored in the body. When iron stores are depleted, iron absorption from the gut increases, and it decreases when the iron content of the body is above normal. The regulatory mechanism for iron absorption is still obscure, although work is being carried out in search of specific hormones that may play a vital role.

References

Armstrong, W., and A. S. Nunn. *Intestinal Transport of Electrolytes, Amino Acids, and Sugars*. Springfield, Ill.: Thomas, 1971.

Babkin, B. P. *Secretory Mechanism of the Digestive Glands*. New York: Harper (Hoeber), 1950.

Blond, K., and D. Haler. *The Liver*. Baltimore: Williams & Wilkins, 1950.

Bortoff, A. "Digestion; Motility," in: *Annual Review of Physiology,* Vol. 30. Palo Alto, Calif.: Annual Reviews, 1972, p. 219.

Brooks, F. P. *Control of Gastrointestinal Function*. New York: Macmillan, Inc., 1970.

Code, C. F., *et al. An Atlas of Esophageal Motility in Health and Disease*. Springfield, Ill.: Thomas, 1958.

Conway, E. J. *The Biochemistry of Gastric Acid Secretions*. Springfield, Ill.: Thomas, 1952.

Glass, G. B. J. *Introduction to Gastrointestinal Physiology*. Englewood Cliffs, N.J.: Prentice-Hall, 1968.

Gray, C. H. *Bile Pigments in Health and Disease*. Springfield, Ill.: Thomas, 1961.

James, A. H. *The Physiology of Gastric Digestion*. Baltimore: Williams & Wilkins, 1957.

Lerche, W. *The Esophagus and Pharynx in Action*. Springfield, Ill.: Thomas, 1950.

[7] B. E. Vaughan, "Intestinal Electrolyte Absorption by Parallel Determination of Unidirectional Sodium and Water Transfer." *Am. J. Physiol.*, **198:**1235 (1960).

[8] A. Gilman and E. S. Loelle, "Substrate Requirements for Ion Transport by Rat Intestine in Vitro." *Am. J. Physiol.*, **199:**1025 (1960).

McHardy, Gordon. *Gastroenterology*. New York: Harper (Hoeber), 1962.
Palmer, W. L., and J. B. Kersner. "The Esophagus," in: Sodeman, W. A. (ed). *Pathologic Physiology*. Philadelphia: Saunders, 1956.
Shepard, R. S. *Human Physiology*. Philadelphia: Lippincott, 1972.

Metabolism

The energy required by an animal for growth, maintenance, and function is derived mainly from the oxidation of organic foodstuffs within the cells of the body. These organic nutrients, as we have already seen, are prepared for such utilization by the digestive system by which they are put into absorbable forms to pass through the intestinal mucosa into the blood and lymph. The field of metabolism concerns itself with what happens to these food materials after they are absorbed and includes their circulation to and the chemical changes within the body cells. In general, "metabolism" is usually the term applied to the over-all series of chemical reactions taking place within the body cells in which oxygen is consumed (aerobic organisms) and carbon dioxide and other wastes are universal end products.

In metabolic activity two main types of chemical reactions occur:

1. Reactions occur in which the absorbed foods, such as glucose, amino acids, and fats, are further broken into simpler forms by the process of oxidation. Carbon dioxide, water, nitrogenous wastes, and other substances are usually the end products of such reactions with a release of energy. Because this energy release is in the form of heat, these reactions are said to be **exothermic,** and the heat may be determined and measured in units of calories. This phase of metabolism in which absorbed foods are burned or oxidized is referred to as **catabolism.**
2. Reactions occur in which the absorbed nutrients are used by the cell to construct or synthesize more complex compounds, such as tissue proteins, tissue carbohydrate (glycogen), and cellular secretions such as hormones and enzymes. These reactions usually use up energy instead of releasing it and therefore are said to be **endothermic** reactions, dependent for their energy on the catabolic phase of metabolism. This building-up phase of metabolic reactions is referred to as **anabolism.**

The Liver

Our foods consist of six kinds of materials—carbohydrates, fats, proteins, vitamins, inorganic salts, and water—and the study of metabolism is usually separated into these types, as will be done here. However, before considering the metabolism of the various types of organic nutrients, it may be well to consider one organ of the body that plays a very important and prominent role in metabolic processes. In Chapter 11 we discussed the important activity of the liver in the secretion of bile, but important as the secretion of bile may be, the liver's other metabolic functions are of still greater importance.

The liver is the largest and most complex organ in the body, and in more highly developed animal species it is a vital organ; that is, some of its functions are necessary for life. The whole organ (Fig. 12-1) is covered by a capsule of connective tissue, which also extends throughout the organ, dividing the several lobes of the liver into its basic structure, the lobule. The lobule (Fig. 12-2) consists of a series of hepatic cells that radiate out from the central venule to form a polygonal-shaped structure of sufficient regularity to make a histological section appear to be a mosaic. Each lobule is supplied with a dense network of capillaries originating from the portal vein. Arterial blood is supplied by the hepatic arteries, which anastomose with the portal venous capillaries. The liver is unique in this

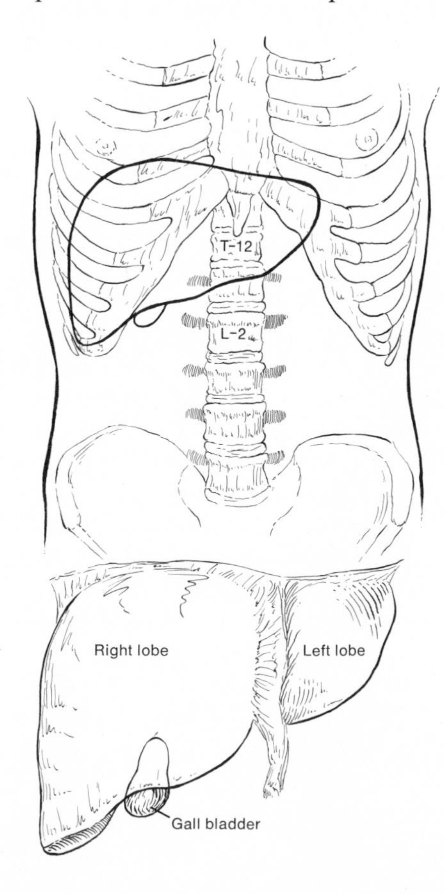

Figure 12-1. Liver, showing that it occupies chiefly the upper right hypochondriac and epigastric regions directly beneath the diaphragm. Lower diagram shows anterior view of liver removed from the body.

way because it receives a double blood supply. In general, about 70 per cent of its blood supply is from the portal veins, and about 30 per cent is from the arterial system. The arterial blood provides oxygen to the hepatic cells and supporting tissue. The portal blood, draining the gastrointestinal tract and containing the absorbed products of digestion, as well as possible toxic material of intestinal origin, must also pass through the liver before reaching the general circulation. A system of capillaries and open spaces spreads from the lobules to surround a series of hepatic cells (cords) to conduct blood to the central hepatic vein, by which blood leaves the liver. The liver also has a generous supply of lymphatic vessels, which drain into the numerous lymph nodes associated with this structure.

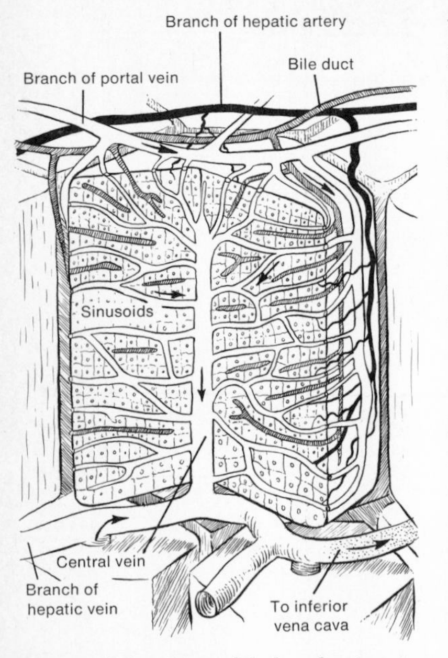

Figure 12-2. Liver lobule, showing its venous and arterial blood supply. The relation of liver cells to sinusoids and bile canaliculi is shown also.

Three types of cells are associated with liver tissue. These are (1) the cells of the biliary duct system, (2) the Kupffer cells, and (3) the hepatic cells. It is the last-named cell that is responsible for the majority of functions of this organ and is involved importantly in carbohydrate, protein, fat, mineral, and vitamin metabolism, as well as heat production and detoxication of certain poisons and drugs circulating in the blood. The Kupffer cells, as we have already seen, act to remove foreign bodies and dead and injured red blood cells from the circulation. They are also important in antibody production and bile formation.

In man the hepatic cells are relatively large, measuring 20 to 30 μm in diameter. They are polyhedral or roughly cubical in shape, each cell containing a large rounded nucleus, although some may have two or more. Every cell in a lobule is in contact with a neighboring cell and with one or more blood vessels. Where the surface of the cell makes contact with a blood vessel, it is generally separated from the wall of the blood vessel by a lymph space. The appearance of hepatic cells changes greatly depending on the physiological activity. Sometimes the cell substance appears dense, compact, and of fairly uniform texture, the cell being relatively small. At other times, especially after a meal rich in carbohydrates, the cell substance appears large and bulky because of the accumulation of glycogen. Very frequently, especially after a few days' fast or after a meal rich in fats, the hepatic cell is filled with fat globules.

The bile ducts consist of the large **hepatic duct,** leading from the liver, the **cystic duct,** leading from the gallbladder, and the **common bile duct,** formed by the junctions of the first two and leading into the duodenum. The walls of the ducts are made up of an inner layer of mucous membrane consisting of an epithelium of columnar cells resting on a connective tissue base. This layer is surrounded by a well-developed muscular coat, formed by an inner layer of circularly arranged muscle fibers and an external layer of muscle fibers longitudinally arranged. The wall of the gallbladder has essentially the same structure. Both the gallbladder and the ducts are capable of undergoing peristaltic contraction by which the rapid flow of bile into the intestine is brought about.

In man the bulk of the absorbed food is transported first to the liver before being distributed by the general circulation to the various tissues of the body. The arrangement of the intestinal circulation makes this possible. After arterial blood passes through the dense network of capillaries supplying the intestinal wall, especially those of the intestinal villi, it is finally collected into veins, which eventually join into the hepatic portal vein that leads directly into the liver (Fig.

12-3). Blood leaving the intestinal tract, containing most of the absorbed organic nutrients, can travel only to the liver. This arrangement makes the liver the major distributing organ of food stuffs to the body and the regulator of not only what kinds, but also what quantities, of nutrients.

Carbohydrate Metabolism

During the process of digestion, carbohydrates such as starch and sugar are hydrolyzed to simple sugars called *monosaccharides*. It is in this form that they are absorbed through the wall of the small intestine mainly into the blood stream. Not all monosaccharides can be utilized for an energy source of the cell. Of the three principal monosaccharides absorbed, that is, glucose, fructose, and galactose, glucose is of major importance. Both galactose and fructose are converted to glucose by the liver.

Blood Sugar. After absorption of carbohydrate food is completed, the blood sugar level may reach 120 to 130 mg per 100 ml of blood. After several hours, the glucose level declines from these values and is found to be between 70 and 100 mg per cent. This level of blood sugar is maintained between meals even though glucose is constantly used by the tissue. Even during prolonged fasting the percentage

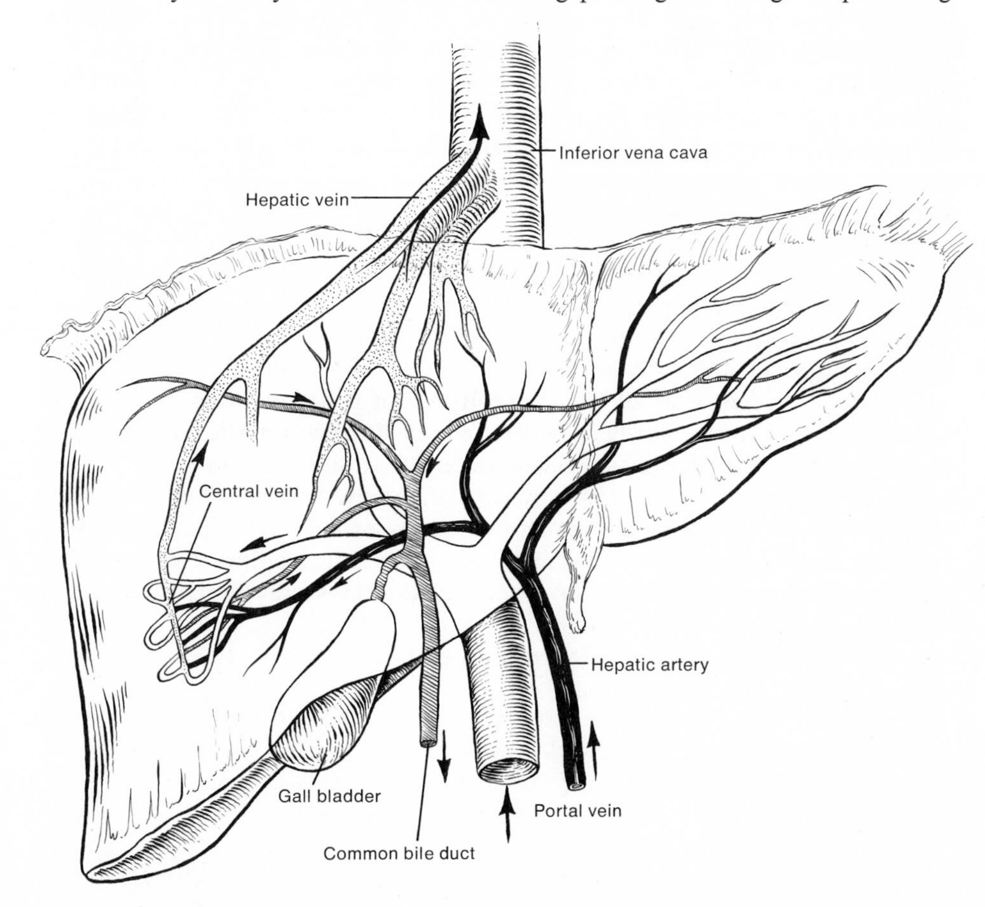

Figure 12-3. Vascular arrangement of the liver.

does not fall much below 70 to 100 mg. This constancy of the blood sugar is of great significance, for it indicates the necessity for a certain percentage of sugar in the blood for the normal functioning and even the life of the organism. The liver plays a vital role in maintaining the relative constancy of the blood glucose concentration. The amount of glucose taken out of the blood by the tissues is counterbalanced by the amount of glucose liberated into the circulation by the liver. This balance may be graphically represented in the following way:

$$\text{Food carbohydrate (glucose, galactose, fructose)} \xrightarrow[\text{vein}]{\text{hepatic portal}} \text{Liver (glycogen)} \xrightarrow[\text{glucose}]{\text{blood}} \text{Cells.}$$

It is the liver's ability to take the incoming carbohydrate from the digestive tract, that is, glucose, galactose, and fructose, and synthesize it into glycogen (animal starch) that enables the liver to carry out this regulatory role for maintaining blood sugar. The glycogen formed in the liver is then broken down under the influence of enzymes present in the liver to glucose and liberated into the circulation at a rate sufficient to maintain a blood sugar level of 70 to 100 mg per cent. It is clear, therefore, that the glycogen represents a storehouse for carbohydrate coming from the food, to be liberated during a period when no carbohydrate is available. The process whereby glycogen is formed from carbohydrate is referred to as **glycogenesis.** The breakdown of glycogen to glucose is known as **glycogenolysis.** As the blood glucose level rises in the general circulation above that which is considered normal, from whatever cause, glycogenolysis decreases and glycogenesis increases.

The skeletal muscle tissue also absorbs glucose from the blood and stores it in the form of glycogen, but the glycogen that accumulates in these tissues is not reconverted to blood glucose, as the muscle tissue does not possess the same enzyme systems (phosphatases) found in the liver.

The glycogen of muscle is broken down by a series of chemical reactions to lactic acid, a process referred to as **glycolysis,** but it never forms free glucose. Glycolysis occurs anaerobically in skeletal muscle during exercise. Under normal circumstances part of the lactic acid produced is oxidized by molecular oxygen to carbon dioxide and water with a release of energy. Part may be reconverted to glycogen in the muscle, as the reactions of glycolysis are reversible. If muscular activity is severe, excess lactic acid may be excreted into the blood stream to be converted to glycogen by the liver. These relationships may be represented as follows:

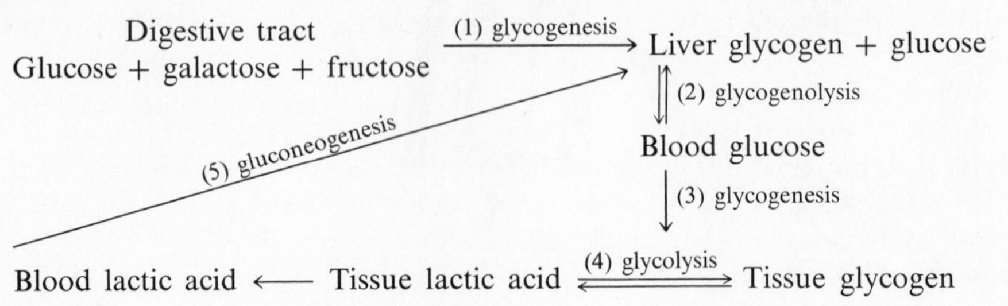

In general the blood glucose is obtained from three sources: (1) absorption of carbohydrate from the diet; (2) hydrolysis of glycogen stored in the liver; and (3) conversion in the liver of noncarbohydrate precursors, such as certain amino acids, glycerol obtained by hydrolysis of neutral fat, and compounds produced in the metabolic breakdown of glucose. The last of these processes is referred to as **gluconeogenesis.** The liver is also able to convert some of the glucose to fat, which is stored in the fat depots throughout the body for later use and is known as **lipogenesis.**

Regulation of the Blood Sugar Level. The mechanisms that regulate the concentration of sugar in the blood involve not only the liver, but also other organs, such as the kidney, adrenals, pancreas, and nervous system.

When the blood sugar rises to relatively high levels, a condition known as **hyperglycemia,** the kidneys actively participate in restoring the sugar level to normal by excreting the excess in the urine. Glucose is continually being filtered by the glomeruli of the kidney but is reabsorbed back into the blood stream by the renal tubules in an amount consistent with its normal concentration. Any excess will be eliminated by the kidney tubule, causing an increase in the sugar concentration of the urine, a condition known as **glycosuria.** This usually occurs when the blood sugar levels exceed 170 to 180 mg per 100 ml of blood. Glucose reabsorption by the kidney tubule is discussed in more detail in Chapter 14.

Specialized cells within the pancreas known as the *islets of Langerhans* secrete the hormone *insulin.* Insulin will reduce blood sugar by stimulating those processes that cause glucose to leave the blood stream. It does this in the following four ways:

1. Stimulates conversion of glucose to glycogen in the liver (glycogenesis)
2. Stimulates conversion of glucose to fat in the liver (lipogenesis)
3. Inhibits the production of carbohydrate from fats and proteins (glyconeogenesis)
4. Stimulates carbohydrate oxidation in the tissue (glycolysis)

An excess of insulin thus could lower the content of the sugar in the blood well below normal levels. This condition of a lower sugar content of the blood is known as **hypoglycemia;** it can result in the development of symptoms of nervousness, of hunger, and of muscular weakness. The symptoms become more marked if the percentage falls lower than 50 mg per cent, resulting in convulsions and even in death unless glucose is administered. Experimental evidence indicates that the secretion of insulin by the pancreas is stimulated by a rise in the concentration of the blood sugar, the increased concentration having a direct effect on the cells of the islets of Langerhans. It is also thought that a nervous reflex action is involved, but its mechanism is unknown.

A lack of insulin in the body resulting from the removal or disease of the pancreas could produce the opposite effect of an excess of insulin, that is, a relatively high concentration of glucose in the blood (hyperglycemia), a condition known as **diabetes.** Sugar diabetes is characterized by a decreased ability of the tissues to utilize carbohydrates and a breakdown of the glycogenic and lipogenic function of the liver. As a result, the blood sugar level is elevated, and the glycogen stores of the liver are depleted. The liver is severely strained to produce glycogen by gluconeogenesis, which leads not only to excessive breakdown of protein of body cells and emaciation, but also to the accumulation of toxic waste products, such as ketone bodies, from excessive fat metabolism.

That the nervous system plays a part in carbohydrate metabolism was first demonstrated by Claude Bernard when he stimulated mechanically the floor of the fourth ventricle in rabbits and produced hyperglycemia and glycosuria. This area was referred to as the **glycogenic center.** Stimulation of the great splanchnic nerve or of the hepatic nerves has similar effects. Thus glycogenolysis appears to be reflexly controlled from a nerve center in the brain, which probably is under the influence of sensory impulses coming from nerve fibers originating in the muscle, so that increased muscular activity will call forth a greater discharge of sugar from the liver to provide more fuel energy.

The adrenal glands also play a part in carbohydrate metabolism. This can be shown by the fact that removal of these glands results in a lowering of the blood sugar concentration and development of the symptoms associated with hypoglycemia. It is the hormone epinephrine, secreted by the medulla of the adrenals, that stimulates the breakdown of liver glycogen to glucose (glycogenolysis). Epinephrine also stimulates the breakdown of muscle glycogen to lactic acid (glycolysis). When the blood sugar level falls below 70 mg per cent, the excitation of the glycogenic center of the brain causes a firing of sympathetic nerve fibers, which causes the adrenal glands to liberate epinephrine. The epinephrine stimulates the breakdown of liver glycogen to glucose, which counterbalances the fall of the blood sugar concentration. Its action on the muscle causes a greater production of lactic acid, which may be carried to the liver for synthesis to glycogen. Thus, it may be said that the sugar reserves of the body are mobilized by the action of epinephrine in two ways: (1) by stimulating gluconeogenesis, it provides more glycogen for glucose production; and (2) by stimulating glycogenolysis, it provides more glucose for cellular utilization.

Thyroxine is another hormone that has an influence on carbohydrate metabolism, but its mode of action is not too well understood. It is believed to render liver glycogen a bit more unstable so that it takes less epinephrine than normal to cause glycogenolysis, thereby causing a hyperglycemic state to be developed much more readily. Anterior pituitary hormones also influence carbohydrate metabolism. It appears that two separate actions are involved as a consequence of the release of two separate hormone fractions. The first, an action produced by the adrenocorticotropic hormone, has the effect of potentiating the action of the growth hormone. The second action appears to be associated with the growth hormone of the anterior pituitary, which, by its action of depressing carbohydrate oxidation in tissues with a concomitant accumulation of glycogen stores in the skeletal muscles, depresses liver carbohydrate metabolism.

Carbohydrate metabolism may be summarized as follows:

1. Food carbohydrate in the form of glucose, fructose, and galactose is delivered to the liver by way of the hepatic portal circulation after absorption from the small intestine.
2. In the liver excessive glucose is converted into glycogen, a process known as *glyco-*

genesis, to be stored until the normal blood sugar level drops below 70 to 120 mg per 100 ml. The conversion of glucose to glycogen involves its phosphorylation, with ATP acting as a phosphate donor in the presence of the enzyme **glucose hexokinase** to produce an ester of glucose, which cannot diffuse across the cell membrane.

$$\text{Glucose} + \text{ATP} \xrightarrow{\text{glucose hexokinase}} \text{Glucose-6-phosphate} + \text{ADP}.$$

This reaction is reversible in the liver because another enzyme, a **phosphatase,** present only in the liver, enables this organ to liberate free glucose from glucose-6-phosphate which can diffuse out again into the blood. Fructose and galactose are similarly phosphorylated as the first step of their conversion to glycogen:

$$\text{Galactose} + \text{ATP} \xrightarrow{\text{galactose hexokinase}} \text{Galactose-1-phosphate} + \text{ADP}.$$

$$\text{Fructose} + \text{ATP} \xrightarrow{\text{fructose hexokinase}} \text{Fructose-6-phosphate} + \text{ADP}.$$

All these esters are then converted into glucose-1-phosphate as follows:

$$\text{Galactose-1-phosphate} \xrightarrow{\text{galactokinase}} \text{Glucose-1-phosphate}$$

$$\text{Glucose-6-phosphate} \underset{}{\overset{\text{phosphoglucomutase}}{\rightleftarrows}} \text{Glucose-1-phosphate}$$

$$\text{Glucose-6-phosphate} \rightleftarrows \text{Fructose-6-phosphate}.$$

Glucose-1-phosphate, in the presence of the enzyme phosphorylase and phosphate ion, is linked to other glucose molecules until a chain of eighteen glucose units is produced. Glycogen is a polysaccharide composed of some eighteen glucose units. The formation of glycogen from glucose is facilitated by insulin.

3. Conversion of glucose to fat or lipogenesis by the liver is another way that extra blood glucose is utilized. The exact steps by which sugars are converted into fatty acids are not yet known. Vitamin B_1 (thiamine) appears to be essential, as well as insulin.
4. Liver glycogen is formed from certain amino acids, such as alanine, glycine, serine, cysteine, and glutamic acid, which form keto acids when they have their amine group (NH_2) removed. Pyruvic acid, which is one of these keto acids, may be converted into glycogen. Glycogen may also be formed from fat, although the evidence that glycogen is formed in the liver from fat is not conclusive. The production of glycogen from substances other than glucose constitutes the process of gluconeogenesis.
5. The liver glycogen, synthesized in the various ways described above, is continuously being broken down into glucose by the process of glycogenolysis in the following way:

$$\begin{array}{l}\text{Glycogen}\\ \quad\Big\updownarrow \text{ phosphorylase} + H_3PO_4\\ \text{Glucose-1-phosphate}\\ \quad\Big\updownarrow \text{ phosphoglucomutase}\\ \text{Glucose-6-phosphate}\\ \quad\Big\downarrow \text{ phosphatase (in liver only)}\\ \text{Glucose.}\end{array}$$

6. Glucose diffuses out into the blood and is delivered to the tissue, where it may be further oxidized into carbon dioxide and water with a release of energy. In muscle tissue, glucose may also be synthesized into glycogen, but this glycogen is not directly available as a source of blood glucose since no enzyme is present to re-form glucose from glucose-6-phosphate. In muscle, the glycogen breaks down to pyruvic and lactic acid, a process usually referred to as glycolysis. Some of the lactic acid is oxidized to carbon dioxide and water with a release of energy by way of the citric acid cycle. (See Chapter 10 for details of glycolysis and the citric acid cycle.) Oxygen does not enter into any of the metabolic reactions until hydrogen is released from the dehydrogenation reactions, which occur in the citric acid cycle. Energy is released along the whole series of enzyme systems involved in the transfer of hydrogen to oxygen, and water is formed as a final end product of carbohydrate metabolism. Carbon dioxide is produced during the citric acid cycle as a result of its removal from the carboxyl group of a carboxylic acid (decarboxylation). Little energy is released in this reaction.

A portion of the lactic acid produced by muscle glycolysis may be converted back to glycogen (gluconeogenesis) in both muscle and liver.

Fat Metabolism

Transport of Fat from the Intestine. The digestion and absorption of fats have already been discussed. As we have seen, fat enters the system in the form of neutral fat, fatty acids, phospholipids, and cholesterol. The lymphatic vessels of the intestinal region transport about 60 per cent of the absorbed fat, principally neutral fat in the form of finely emulsified particles called **chylomicrons.** This fat is carried to the general blood circulation by way of the thoracic duct. Glycerol and fatty acids, along with the phospholipids, enter the blood vessels of the portal system and pass directly into the liver. After ingestion of fat, the fat content of the blood increases considerably, and the plasma takes on a milky appearance because of the numerous very fine fatty globules suspended in it. This **lipemia** may last up to 6 or 7 hours, after which time the fat content of the blood returns to the fasting level. In healthy subjects the total blood fat concentration varies greatly, ranging from 600 to 1,200 mg per 100 ml. Physiological increases in blood fat are found under the following circumstances: (1) after ingestion of a meal containing high concentrations of fat; (2) in prolonged fasting; (3) after exercise; (4) during pregnancy and lactation; (5) in subjects under the influence of alcohol; (6) in some diseases, such as diabetes, nephrosis, and anemia.

Fats in the blood come not only from the fat absorbed from food fat, but also from the conversion of carbohydrate and protein into fats.

Functions of Fat. After absorption fat is oxidized by the tissues, metabolized by the liver, incorporated into the tissues as part of its structure, deposited as storage fat in the fat depots of the body, and excreted in the milk and sebaceous secretions of the skin.

Fat that is incorporated into the tissues is referred to as **constituent fat.** This fat is constant and is not changed by fasting. Terroine[1] called this fat the "element constant"; it has a high iodine number and a large proportion of phospholipid.

Storage fat differs from constituent fat in that it varies considerably with the

[1] E. F. Terroine, "Fat Metabolism." *Ann. Rev. Biochem.*, **5**:227 (1936).

diet and is diminished by fasting. Terroine referred to the storage fat as the "element variable"; it is mainly made up of glycerides.

Oxidation of Fats. Fats are completely oxidized to carbon dioxide and water with a release of energy. This can be represented by the reaction

$$2C_{51}H_{98}O_6 + 145O_2 \longrightarrow 102CO_2 + 98H_2O + 9.4 \text{ cal per gram of fat.}$$

The chemical reaction of fat metabolism, whether the fat is neutral fat, phospholipid, or fatty acid, is believed to occur mainly in the liver by a process that consists essentially in the oxidation of long-chain fatty acids. The neutral fats must first be broken into their fatty acid constituents and glycerol.

The glycerol portion of the fat molecule follows a different pathway than the fatty acids. It is converted into glycogen by the liver after being phosphorylated.

The fatty acids found most commonly in the organism are **palmitic acid,** which has 16 carbon atoms in its chain, $CH_3(CH_2)_{14}COOH$; **stearic acid,** with an 18-carbon-atom chain, $CH_3(CH_2)_{16}COOH$; and **oleic acid,** with an 18-carbon-atom chain,

$$CH_3(CH_2)_7CH = CH(CH_2)_7COOH.$$

The oxidation of these acids was first studied by Knoop, who developed the classical theory of beta oxidation. Knoop postulated that the fatty acids are oxidized at the beta carbon atom (the third carbon atom counting from the carboxyl group), and a substance with two carbon atoms is separated from the fatty acid chain, leaving a shorter fatty acid component. By a series of successive oxidations, the long fatty acid chain loses its carbon atoms two at a time. For example, stearic acid, which has 18 carbon atoms, would be oxidized in the following way:

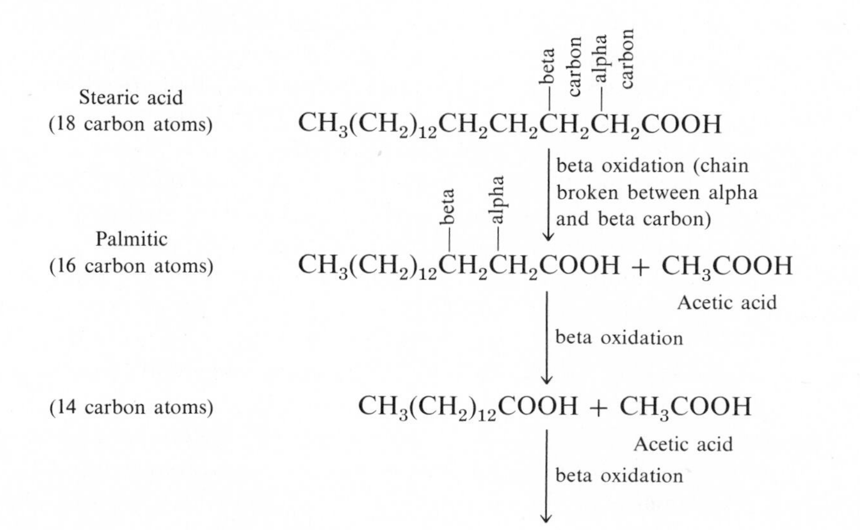

. . . until nine 2-carbon acetic acid fragments are formed.

McKay[2] showed that the 2-carbon atom fragments formed in beta oxidation combine with one another to form **acetoacetic acid:**

$$CH_3COOH + CH_3COOH \xrightarrow{\text{condensation}} \underset{\text{Acetoacetic acid}}{CH_3CO \cdot CH_2COOH} + H_2O.$$

Acetoacetic acid is readily converted by the addition of hydrogen to the molecule (reduction) into **beta-hydroxybutyric acid.**

The reaction can be written in the following way:

$$\underset{\text{Acetoacetic acid}}{CH_3CO \cdot CH_2COOH} + 2H^+ \longrightarrow \underset{\beta\text{-Hydroxybutyric acid}}{CH_3CHOHCH_2COOH.}$$

It is activated by a specific enzyme, beta-hydroxybutyric dehydrogenase, and a coenzyme, nicotinamide-adenine dinucleotide.

Acetoacetic acid can also be decarboxylated to produce acetone:

$$\underset{\text{Acetoacetic acid}}{CH_3CO \cdot CH_2COOH} \xrightarrow{-CO_2} \underset{\text{Acetone}}{CH_3COCH_3.}$$

Another theory of fatty acid oxidation is called the **multiple alternate oxidation theory.** In this process palmitic acid, for example, which is a 16-carbon-atom fatty acid, would split into four acetoacetic acid molecules directly, and not give only one molecule as in the process of beta oxidation. Acetoacetic acid would then be converted by reduction into beta-hydroxybutyric acid and by decarboxylation into acetone.

Ketone Bodies. The three end products resulting from the oxidation of fatty acids in the liver are acetoacetic acid, beta-hydroxybutyric acid, and acetone. These three substances are called **ketone,** or **acetone, bodies.** Under normal conditions of metabolism the ketones are produced in very small quantities. Normal values for ketones in the blood range from 1.5 to 2.0 mg per 100 ml, and less than 1 mg of ketones is excreted in the urine in 24 hours.

The condensation reaction by which ketones are formed is reversible, so that oxidation of ketones could proceed by way of the Krebs cycle on the re-formation of acetic acid. Both acetic acid and acetoacetic acid have been shown to condense with oxalacetic acid to enter the Krebs cycle for complete oxidation.

When an excessive amount of fat is catabolized to meet the energy requirements of the body, more acetic acid is produced to be oxidized. When the production of acetic acid molecules exceeds the capacity of the tissue to oxidize them, they condense to form ketone bodies. The ketone bodies then accumulate in the blood and are excreted in the urine. This accumulation of ketone bodies in the blood above normal concentrations is referred to in general as **ketosis.** The increase in the blood ketone is more specifically referred to as a **ketonemia** and that in the urine as a **ketonuria.**

[2]E. M. McKay, R. H. Barnes, H. O. Carve, and A. N. Wick, "Ketogenic Activity of Acetic Acid." *J. Biol. Chem.,* **135:**157 (1940).

An increase of ketone bodies is observed when the organism is undergoing complete fasting, when there are excess fats and insufficient carbohydrates in the diet, in severe diabetes, and after an injection of anterior pituitary extract. It appears that the liver is the principal organ for ketone body production, as is demonstrated by the following observations:

1. When fatty acids are added in vitro to surviving liver slices, ketone bodies are formed; but when fatty acids are added to other tissues, ketone bodies are not formed.
2. Harrison and Long,[3] on analyzing various tissues, found that liver has a higher concentration of ketone bodies than any other tissue. It has also been demonstrated that blood leaving the liver has a higher concentration of ketone bodies than blood entering it.
3. In animals that have had the liver removed, injection of anterior pituitary is not followed by an increase in ketone bodies, as would be expected in the normal animal.

Ketone bodies formed in the liver are not normally oxidized there but are carried to the tissues, especially muscle and kidney, to be completely oxidized. This fact was demonstrated by Mirsky[4] when he added ketone bodies to minced muscle tissue and found that they were completely oxidized to carbon dioxide and water. The administration of glucose and insulin diminishes the formation of ketone bodies (ketogenesis) but does not modify their oxidation by the tissue (ketolysis).

Generally it is believed that the oxidation of fatty acids and the production of ketone bodies can occur only in the liver. However, recent experimental work[5] with rat tissue indicates that the peripheral tissues not only oxidize ketone bodies, but also are able to oxidize directly long-chain fatty acids completely to carbon dioxide and water and energy.

Effect of Ketosis. From what has already been said, it should be clear that an accumulation of ketone bodies can be due either to an abnormally high rate of formation or to an abnormally low rate of disposal. Since the two ketones, acetoacetic acid and beta-hydroxybutyric acid, are moderately strong organic acids, they must be neutralized with alkali in order to preserve the pH of the blood and tissue fluid. If the organism persists in generating more ketone bodies than can be utilized, as is the case in uncontrolled diabetes, complications result as a direct effect of a depletion of the alkali reserve and its concomitant acidosis. Coincident with the depletion of alkali that results from the excretion of the sodium salts of acetoacetic acid and beta-hydroxybutyric acid in the urine, large quantities of fluid are also lost in the urine. A tendency to nausea and vomiting causes additional fluid loss, and depression of the central nervous system leading to loss of reflexes and ultimately to coma interferes with normal water intake. All these factors complicate the acidosis by producing a severe state of dehydration. Dehydration and acidosis due to excess ketone body concentration in the

[3] H. C. Harrison and C. H. Long, "The Distribution of Ketone Bodies in Tissue." *J.Biol. Chem.,* **133:**209 (1940).

[4] I. A. Mirsky, "The Nitrogen Sparing Action of Glucose in Normal, Phlorhizinized and Depancreatized Dogs." *Am. J. Physiol.,* **120:**386 (1937).

[5] M. E. Volks, R. H. Millington, and S. Weinhouse, "Oxidation of Endogenous Fatty Acids of Rat Tissue in Vitro." *J. Biol. Chem.,* **195:**493 (1952).

Figure 12-4. *Development of a fat cell. At the right is a section of fat tissue showing typical signet-ring appearance of fat cells.*

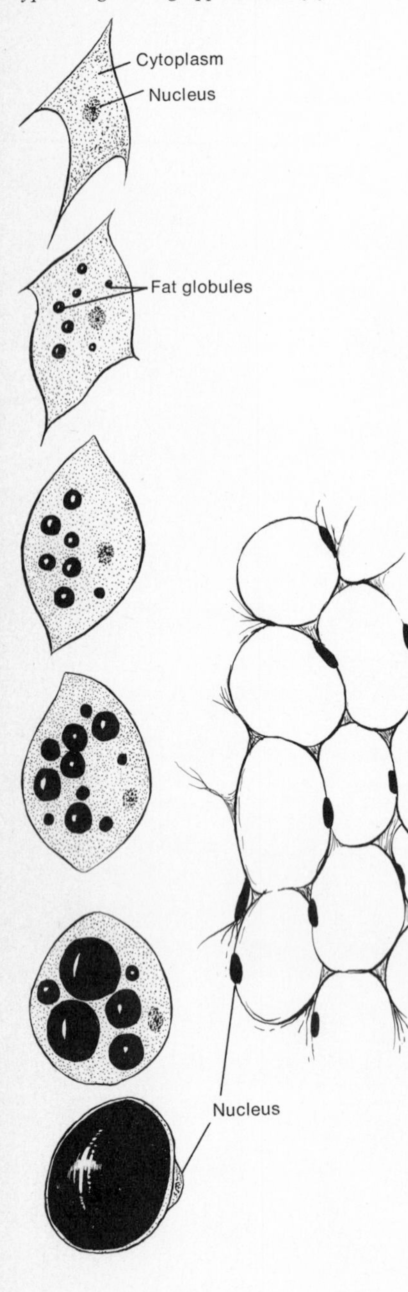

blood are distinguishable from those due to other causes by the presence of ketonemia and ketonuria. In addition, severe ketosis may be detected by the odor of acetone on the breath and in the urine.

Reduction of Ketosis. On the basis of what has been said, the aim in the treatment of ketosis is to restore the body chemistry to normal. In order to do this, two defects must be corrected:

1. Normal carbohydrate metabolism must be restored in order to offset the accumulation of ketone bodies due to excess fat metabolism. When this defect in carbohydrate metabolism is a result of insufficiency of insulin, as it is in cases of diabetes, this may be corrected by injections of insulin. When ketosis is due to lack of sufficient carbohydrate in the diet, as in starvation ketosis, this may be corrected by carbohydrate alone.
2. The sodium and water loss from the body must be replaced. These effects are commonly obtained by intravenous injection of isotonic solutions of a suitable sodium salt, such as sodium chloride, sodium bicarbonate, or sodium lactate.

Storage of Fat. In addition to being oxidized, fats are stored in the body as **adipose tissue** in various areas called **fat depots.** A small piece of adipose tissue, when examined under low power with the microscope, appears to be made up almost entirely of rounded masses of highly refractive material closely packed together. These rounded masses, which stain an intense black with osmic acid, a reaction of fat, are arranged in irregular lobules between which small amounts of connective tissue carrying blood vessels can be seen. Under high power it becomes obvious that fat tissue is made up of a collection of fat cells held together by a minute quantity of vascular connective tissue. The fat cell (Fig. 12-4) is a cell in the cytoplasm of which fat has collected to such an extent that the cell is almost wholly transformed into a large vacuole filled with fat. The cell substance is reduced to a thin envelope of the vacuole, thickened at one part where the nucleus has been pushed on one side by the gathering fat. If fat tissue is fixed and stained and the fat removed by solvents, what was previously visible as a rounded mass of fat now appears to be a sheet of hollow cells, that is, a mass of cells in which the cell substance has been transformed into a single large vacuole. The cell cytoplasm is reduced to a thin envelope or cell membrane, except at one part, where a thicker disklike portion is present, in which a rounded oval nucleus is located. The fat cells under these circumstances take on the appearance of signet rings and for this reason have often been referred to as *signet-ring cells.* About 50 per cent of the stored fat is found in the subcutaneous tissues, 12 per cent around the kidney, 10 to 15 per cent in the mesenteric tissue, 20 per cent in genital tissue, and 5 to 8 per cent between the muscles. The stored fats in various depots vary from one part of the body to the other in gross appearance, microscopic structure, and chemical composition. The melting point of human fat, for example, varies from 0.5 to 41° C in the different stores. In general it is usually figured at 37° C, owing to its high content of oleic acid. The depot fat is composed mainly of the glycerides of stearic, palmitic, and oleic acids, which make stored fat of the "hard type," that is, fat having a high melting point and low iodine number.

Stored fats come mainly from the food fats but may also consist of fat converted from carbohydrate, which also causes the deposition of the hard-type fat. A high-protein diet increases the cholesterol ester content of the liver but decreases fat storage in general.

Experimental work with animals fed with several fatty acids containing deuterium in their molecule, which would make it possible to identify them in the organism, showed that the greater part of those fatty acids with more than ten carbon atoms are first stored and later utilized. Fatty acids with a shorter chain are directly oxidized or are converted into higher acids. It has been demonstrated that one fatty acid is readily converted into another by the liver. For example, stearic acid, an 18-carbon fatty acid can lose two carbon atoms, giving palmitic acid, with sixteen carbons in the chain. Palmitic acid can be lengthened again into stearic acid or broken into acids with shorter chains.

The fact that carbohydrate food could be converted into fat to be stored was first demonstrated scientifically by Liebig in 1852 and later by Lawes and Gilbert in 1859. A young suckling pig was slaughtered and its body fat and protein analyzed; another animal from the same litter was fed on a high-carbohydrate diet and later was slaughtered and analyzed. The increase in body fat of the animal was more than could be accounted for by the protein and fat content of the feed. Thus, the fat must have been produced by the conversion of carbohydrate. According to some workers in the field, certain vitamins in the B complex are necessary for lipogenesis from carbohydrate. Thiamine is especially important, and riboflavin, pyridoxine, and nicotinic acid play a part but are less important. Experiments in which glucose labeled with deuterium has been fed to animals show that only 3 per cent of the glucose fed is stored as glycogen, whereas 30 per cent is converted into fatty acids.

Proteins can also be a source of storage fat, as experiments with animals on an exclusively protein diet have demonstrated. Hard fat is produced from protein conversion, as it is from carbohydrate, and the process is enhanced by pyridoxine. However, not much body fat has its origin in protein, because deamination of amino acids gives rise to fatty acids with short carbon chains that are easily and directly oxidized.

The liver is not as important a storage organ for fat as it is for carbohydrate and protein. Usually the liver contains from 3 to 5 per cent fat. One-half to one-third of this fat is made up of phospholipids; the rest is neutral fat and cholesterol. However, depending on circumstances, the liver stores variable amounts of fat, and under certain pathological conditions, up to 50 per cent of the liver weight can be fat.

The fat content of the liver depends on the balance between fat formed in the liver and fat secreted from the liver. Fat in the liver has its origin from food fats, other fat depots, and synthesis of fat in the liver. Fat that disappears from the liver is either broken down and oxidized in the hepatic cells and cells of the peripheral tissues or secreted into the blood and transported to other fat stores. In certain cases an excessive amount of fat accumulates in the liver, a condition called **fatty infiltration,** which eventually disturbs the function of the hepatic cells.

Among the principal causes of excess fat in the liver, the following can be mentioned:

1. Whenever the ability of the liver to store, produce, or metabolize carbohydrate is impaired, there may be excess fat in the liver. In patients with diabetes, for example, in which hepatic storage of glycogen is reduced, fatty livers are characteristic.
2. Dietary factors may also induce the development of a fatty liver. The feeding of high-cholesterol diets or high-fat, low-protein diets induces a fatty liver. Excessive fat in the liver often follows a deficiency of **lipotropic substances** in the diet, that is, a substance that prevents abnormal accumulation of fat in the liver.
3. Toxic factors, such as poisoning with carbon tetrachloride, chloroform, phosphorus, lead, arsenic, alcohol, and phlorhizin, all lead to fatty livers. Because of the impairment to carbohydrate metabolism in the liver from these poisons, the demand for an alternative source of energy stimulates the transport of large quantities of fat to the liver.
4. Infections or degenerative processes that affect the liver, such as infectious hepatitis, necrotic hepatitis, and yellow fever, also produce fatty infiltration of the liver.

Lipotropic and Antilipotropic Substances. **Lipotropic substances** prevent the accumulation of excess fat in the liver and diminish the fat content of the liver when it is excessive. The principal lipotropic substances are choline, homocholine, inositol, betaine, methionine, proteins containing methionine, and certain substances extracted from the pancreas. Most of these substances have lipotropic activity because they possess methyl groups (CH_3) that are readily given up to other substances; that is, they are methyl donors. Dragstedt et al.[6] have isolated an active principle from pancreatic juice known as **lipocaic,** which literally means "fat burning." According to Chaikoff and Enteman[7] there is an antifatty liver factor (AFL) in pancreatic tissue and pancreatic juice, which has proteolytic activity and permits utilization of methionine in proteins.

There are differences in the chemical mechanisms by which the lipotropic factors produce their effect on the fatty liver. Choline and inositol accelerate the formation of phospholipids from liver fat, which are easily secreted from the liver. That choline is incorporated into the phospholipid molecule has been demonstrated by administering choline labeled with radioactive phosphorus, which is later found in lecithin (a phospholipid) in the liver. When there is not sufficient choline in the body, fat accumulates in the liver because it cannot be converted to phospholipid and thus transferred to the blood (mobilized). The addition of as little as 1 mg of choline per day in the food of rats is sufficient to prevent excessive deposition of fat in the liver. Betaine and methionine and proteins containing methionine donate methyl groups that are used in the synthesis of choline in the body. It is this ability to donate methyl groups that makes these substances lipotropic. Preformed choline need not be provided in the diet, since it can be synthesized in accordance with the need. The synthesis of choline in

[6] L. R. Dragstedt, J. van Prohaska, and H. P. Harnu, "Observations of a Substance in Pancreas (a Fat Mobilizing Hormone) Which Permits Survival and Prevents Liver Changes in Depancreatized Dogs." *Am. J. Physiol.,* **117**:175 (1936).

[7] I. L. Chaikoff and C. Enteman, "Antifatty-Liver of the Pancreas." *Advan. Enzymol.,* **8**:171 (1948).

the body requires the amino acid glycine or serine and occurs in the following way:

Serine is decarboxylated to produce ethanolamine:

$$\underset{\text{Serine}}{HOCH_2{-}CHNH_2{-}COOH} \xrightarrow{-CO_2} \underset{\text{Ethanolamine}}{HOCH_2{-}CH_2NH_2}$$

Methyl groups from the donors, such as betaine or methionine:

$$\underset{\text{Betaine}}{\begin{matrix} COO{-} \\ | \\ CH_2{-}N^+(CH_3)_3 \end{matrix}} \qquad \underset{\text{Methionine}}{CH_3{-}S{-}CH_2{-}CH_2CHNH_2COOH}$$

are transferred to ethanolamine by the process of transmethylation:

$$\underset{\text{Methionine}}{3CH_3{-}S{-}CH_2{-}CH_2CHNH_2COOH} + \underset{\text{Ethanolamine}}{HOCH_2{-}CH_2NH_2} \longrightarrow$$

$$\underset{\text{Homocysteine}}{3H{-}S{-}CH_2{-}CH_2{-}CHNH_2COOH} + \underset{\text{Choline}}{HOCH_2{-}CH_2\overset{+}{N}(CH_3)_3\,OH^-}$$

The fatty liver produced by anterior pituitary extract is a result of the accumulation in the liver of fat coming from other fat stores.

Antilipotropic substances are those that oppose the lipotropic effect of choline and other lipotropic substances. When these substances are present in high concentrations, they create conditions favorable to fatty infiltration of the liver. Cystine, thiamine, cholesterol, and biotin are considered to be antilipotropic substances. The fatty liver provoked by cystine or thiamine is probably due to excessive synthesis of fatty acids by the liver.

Mobilization of Fat. Stored fat does not remain inert in the tissues, as might be supposed from its name, but is continuously being removed for utilization and replaced by new fat coming into the organism. Studies in which fatty acids labeled with deuterium (heavy hydrogen) have been used indicate that a large part of the ingested fat, even when the diet is calorically inadequate, is first deposited in the storage depots before being catabolized. When the reserves of fat are called on to provide energy, large amounts of fat appear in the liver, which represents the first step in its metabolic breakdown. Thus, the fat depots of the body, which originally received most of the ingested fat, contribute this fat for energy needs of the body by way of the liver. It appears that the mobilization of depot fat to the liver is dependent on a hormonal mechanism that involves both the anterior pituitary and the adrenal cortex. Levin and Farber[8] believe that although adreno-

[8] L. Levin and R. K. Farber, "Hormonal Factors Which Regulate the Mobilization of Depot Fat to the Liver." *Recent Progr. Hormone Res.*, 7:399 (1952).

corticotropic hormone (ACTH) is part of the system that controls the release of fat from the fat depots, a separate pituitary factor called **adipokinin** is also active. It is assumed that these pituitary hormones, together with the corticosteroids from the adrenal cortex, act within the fat depots to activate enzyme systems involved with fat mobilization. The transport of depot fat to the liver occurs in the form of neutral fat, fatty acids, phospholipids, and cholesterol esters, with the lipoproteins of the plasma globulins being the principal vehicles for their transport. About 8 per cent of the plasma protein consists of combinations of lipid and protein. Three per cent of this lipoprotein is formed by the beta globulins. The beta globulins carry much more fat than the alpha globulins, and most of this is in the form of free and esterified cholesterol and phospholipids. Such combinations of fat with protein provide a means by which the relatively insoluble fat can be transported in a predominantly aqueous medium such as blood plasma.

Cholesterol. **Cholesterol** is a sterol found only in fats of animal origin. Egg yolk is the chief source of cholesterol, although the fats of meat, liver, and brain are good sources. The absorbed sterols, both free and esterified, are transported chiefly by the lymph to the blood. The total concentration of cholesterol in the blood plasma is highly variable, averaging in the adult about 200 mg per 100 ml. The cholesterol concentration of the blood is not only a result of ingested cholesterol but is due also to synthesized cholesterol within the body. The liver is the principal organ for this synthesis, but other tissues, such as the adrenal cortex, skin, and testis, have also been shown to synthesize cholesterol. Acetic acid serves as a direct precursor of cholesterol.

The liver both adds cholesterol to the blood and removes it from the blood. It excretes it in the bile as cholesterol or as cholic acid. It is probable that the blood cholesterol levels are the result of the balance that exists between cholesterol excretion and synthesis. The condition in which the cholesterol level of the blood is increased is known as **hypercholesteremia** and is probably caused by an inadequate rate of excretion.

The sterols have many important functions in the body, for they enter into the formation of hormones, vitamins, and the cellular structure of some organs. They can also prove to be detrimental, since the deposition of cholesterol and other lipids in the connective tissue of the walls of arteries produces degenerative changes in the vessels known as **atherosclerosis.** Atherosclerosis is known to be more prevalent among groups subsisting on diets rich in animal fat and also occurs in diseases that result in elevated cholesterol levels, such as uncontrolled diabetes and hypothyroidism.

Protein Metabolism

Dietary protein is broken down by the digestive tract into amino acids, which are absorbed from the intestine into the portal vein by which they are transported to the liver. Since proteins are constituted of numerous units known as amino acids and the body can utilize protein only in this particular form, protein metabolism is mainly the metabolism of amino acids. Amino acids are fatty acids in which one of the alpha hydrogen atoms has been replaced by an amine group (NH_2). Therefore an amino acid has a basic (NH_2) group and an acid carboxyl (COOH) group. The simplest amino acid is glycine, which is the amine of the fatty acid acetic acid, or alanine, which is the amine of the keto acid pyruvic acid.

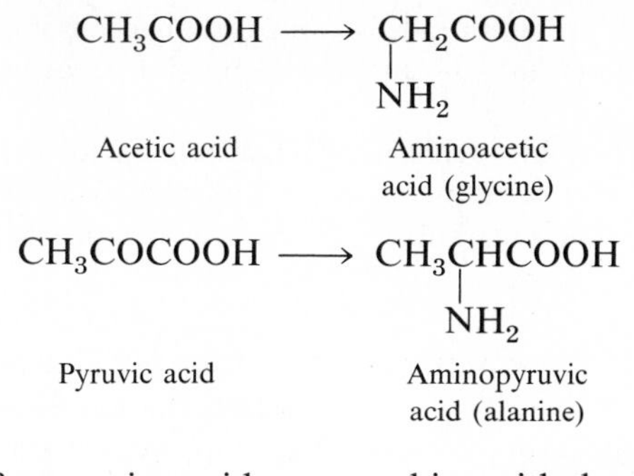

The basic group of one amino acid may combine with the acid group of another with a loss of water to form a union of two or more amino acids. For example,

$$\underset{\text{Alanine}}{\underset{\large NH_2}{CH_3\overset{|}{C}HCO}}\;\boxed{OH + H}\;\underset{\text{Glycine}}{NHCH_2COOH} \longrightarrow \underset{\text{Alanylglycine}}{\underset{\large NH_2}{CH_3\overset{|}{C}HCO \cdot NHCH_2COOH}} + H_2O.$$

The union CO · NH is known as a **peptide linkage.** The reverse process may take place in which water is added at the peptide linkage, causing the separation of amino acids, a process known as **hydrolysis.**

$$\underset{\text{Alanylglycine}}{\underset{\large NH_2}{CH_3\overset{|}{C}HCO \cdot NHCH_2COOH}} + H_2O \longrightarrow \underset{\text{Alanine}}{\underset{\large NH_2}{CH_3\overset{|}{C}HCOOH}} + \underset{\text{Glycine}}{\underset{\large NH_2}{\overset{|}{C}H_2COOH}}.$$

In the above ways amino acids can be synthesized into tissue proteins and tissue proteins hydrolyzed into amino acids. The normal level of amino acid nitrogen in the systemic circulation ranges from 4 to 6 mg per 100 ml. The injection of a solution containing a mixture of amino acids directly into the blood stream is followed by their rapid disappearance, as they are quickly taken up by the tissues, particularly the liver, intestine, and kidney. The most common amino acids found in human plasma are the following: glutamine, alanine, lysine, valine, cysteine, glycine, proline, leucine, arginine, histidine, isoleucine, threonine, serine, tryptophan, phenylalanine, methionine, tyrosine, glutamic acid, and aspartic acid.

Fate of the Absorbed Amino Acids. After absorption, the amino acids undergo one of the following processes:

1. They may be temporarily stored in the tissues, especially the liver and muscle.
2. They may be utilized in the formation of proteins specific for each animal and for each tissue (protein anabolism).
3. They may have their amine group (NH_2) removed, a process called **deamination,** and

it may be transferred to other substances, a process called **transamination.** Such reactions are of considerable biological importance, since deamination is the first step in the metabolic breakdown of these compounds, and transamination is a method whereby the synthesis of amino acids from carbohydrate intermediates can occur.

4. The deaminized residues may be converted into glucose or into acetoacetic acid and then into fat.
5. Tissue proteins and amino acids may be disintegrated and converted into other substances with the formation of waste products such as urea and creatinine (protein catabolism). The liver and the kidney are the principal organs involved in the regulation of protein metabolism.

Protein Storage. Protein or amino acids, unlike fat and carbohydrates, are not stored as such in the tissue to any great extent. Reserves of protein accumulate in the liver and possibly in the muscle and the blood plasma, but there is no satisfactory demonstration of protein storage. Increased body proteins observed in growth, or when there is an increase in muscle size as a result of exercise, cannot be considered as storage proteins but as an increase in body mass.

Tracer studies using amino acids or proteins labeled with isotopic nitrogen, ^{15}N, show that ingested amino acids or proteins constantly replace the tissue proteins and that continuous synthesis and hydrolysis of proteins is occurring in the cells, particularly in organs such as the liver and muscle. Thus, the organism depends on a continual supply of dietary protein for normal functioning, and when the losses of nitrogen from protein catabolism are equal to the nitrogen intake from dietary protein, the subject is said to be in **nitrogen balance.** If the intake of nitrogen is greater than the output, nitrogen is gained, and the subject is said to be in a **positive nitrogen balance.** This occurs during growth, in pregnancy, in hypertrophy of muscle tissue due to muscular exercise, and in convalescence after a wasting disease. If the loss of protein nitrogen from the body is greater than the intake, there is a loss of nitrogen, and the subject is said to be in **negative nitrogen balance.** A negative balance occurs in cases of undernourishment and of excess protein catabolism, which occurs in wasting diseases and fever. In these cases large quantities of protein are necessary to compensate for the loss and to maintain nitrogen balance. The adult organism in nitrogen balance will excrete the same amount of nitrogen in a 24-hour period that has been taken in. For example, if a subject receives 15 g of nitrogen, he will excrete this amount in the urine. Protein nitrogen is retained in the organism only when there is a positive nitrogen balance, as occurs in the circumstances previously mentioned.

Control of Protein Metabolism. Some endocrine glands, such as the anterior pituitary, thyroid, and gonads, have some influence and appear necessary during certain periods of life to maintain protein anabolism and catabolism. Removal of the anterior pituitary or the thyroid stops growth, and the formation of protein in the body ceases completely. The administration of anterior pituitary extract and of the male hormone testosterone stimulates the retention of nitrogen and increases the formation of protein. Excess thyroid secretion and some of the corticoadrenal hormones increase protein catabolism and therefore nitrogen excretion.

Amino Acids. *Deamination of Amino Acids.* Amino acids from both the ingested protein and the tissue protein are rapidly taken up from the blood by tissues, especially the liver and kidney, where they are deaminated. In deamination the alpha NH_2 group of the amino acids is split off, resulting in a deaminated residue (that is, a keto acid) and ammonia. Deamination is catalyzed by a specific enzyme, amino acid oxidase, which aids in the oxidative removal of the amino group.

$$\underset{\text{Amino acid (alanine)}}{CH_3\underset{\displaystyle NH_2}{\underset{|}{C}}HCOOH} + \tfrac{1}{2}O_2 \xrightarrow{\text{amino acid oxidase}} \underset{\text{Keto acid (pyruvic acid) + ammonia}}{CH_3\underset{\displaystyle O}{\underset{\|}{C}}COOH + NH_3}.$$

The resulting keto acids are readily oxidized by the cells to carbon dioxide and water. The process probably involves the liberation of carbon dioxide from the keto acid, resulting in the formation of a fatty acid. The fatty acid is oxidized further according to the usual pathways for the breakdown of these substances. The reactions may be written as follows:

$$\underset{\text{Pyruvic acid}}{CH_3COCOOH} + \tfrac{1}{2}O_2 \longrightarrow \underset{\text{Acetic acid}}{CH_3COOH} + CO_2,$$

$$\text{Acetic acid} + O_2 \longrightarrow CO_2 + H_2O.$$

Keto acids, such as pyruvic acid, can also be converted into lactic acid and into glucose. Because the process of partial oxidation after deamination gives a fatty acid, this suggests the possibility that amino acids play a role in fat storage.

Oxidative deamination as described above is apparently not the only mechanism in the body for the removal of amino groups from amino acids, for it appears that the removal of amino groups can also be accomplished by a transaminating mechanism.

Transamination and Synthesis of Amino Acids. It has long been recognized that amino acids could be formed in the body from corresponding keto acids. Studies with rats showed the ability of the cells to use keto acids to replace certain of the essential amino acids in the diet for growth purposes. It appeared that those keto acids could acquire amino groups with the production of the corresponding amino acids.

More concrete evidence that amino acids can be formed from corresponding keto acids was reported in 1937 by Braunstein and Kritzmann, whose observations revealed what appears to be the prime mechanism by which the removal of amino groups from amino acids is effected. These investigators discovered that in the presence of extracts of a variety of tissues, for example, liver, muscle, heart, and kidney, a reaction occurred that they described as an intermolecular transfer of amino groups, a process they called transamination. In this process the amino group is transferred from an animo acid to a keto acid with the formation of a keto acid corresponding to the original amino acid and an amino acid corresponding to the original keto acid. This mechanism therefore plays a part both in the breakdown of amino acids to yield keto acids and in the formation of new amino acids from keto acids. These reactions are catalyzed by the **transaminase enzymes** of the liver with the aid of the coenzyme pyridoxal phosphate. The equation below may be used to represent this general reaction.

It will be seen that the reaction involves the transfer of the amino group of one amino acid to a keto acid with the formation from the latter of an amino acid and the generation of a new keto acid.

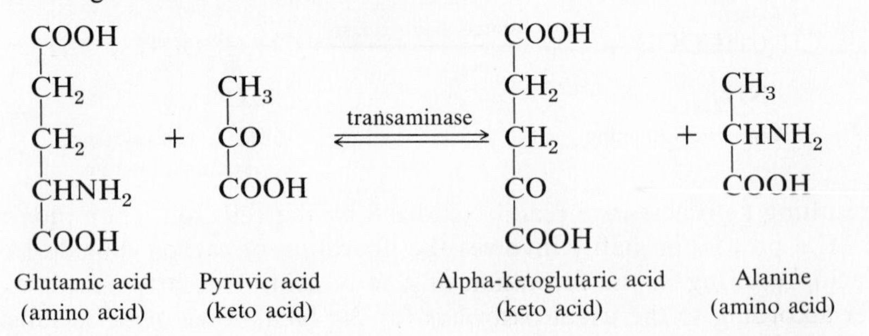

Braunstein and Kritzmann found that a number of keto acids could act as receptors of the amino groups of glutamic acid and that amino acids other than glutamic readily yielded their amino groups to alpha-ketoglutaric acid.

After deamination, the fatty acid residue, as we have already seen, is readily oxidized into carbon dioxide and water, or it may be converted into glucose or into acetoacetic acid. The amino acids that can be converted into glucose after deamination are called **glycogenic** and consist of glycine, alanine, serine, threonine, cystine, aspartic, glutamic, ornithine, and proline. Those amino acids that can be converted into acetoacetic acid are called **ketogenic** and include leucine, isoleucine, tyrosine, and phenylalanine.

Synthesis of Proteins. Amino acids are utilized by the cells in the building of tissue proteins. During growth, pregnancy, and hypertrophy of muscle tissue, additional proteins are formed from the amino acids taken from the blood. The exact method of protein synthesis is discussed in Chapter 1. Generally the amino acids are taken up by the tissues and converted into a common unit that provides the basis for the synthesis of protein. This process involves the rapid breaking and re-forming of peptide linkages (CO · NH) during which time amino acids are taken out of the tissue protein molecule and new amino acids of the same kind are inserted. The tissue proteins, therefore, are not in a static state but are constantly changing as the continual exchange of amino acids takes place. Sprinson and Rittenberg, using glycine labeled with ^{15}N, have determined the time it takes for half the tissue protein to be renewed. In man, the half-life of the total body protein is 80 days; that of muscle, 158 days; and that of the proteins of liver and plasma, about 10 days.

End Products of Protein Metabolism. The major end product of protein metabolism in man is urea, although ammonia, creatinine, and uric acid are also formed. **Urea** ($CO[NH_2]_2$) is a diamide of carbonic acid and is derived principally from the ammonia released from the deamination of amino acids. In 1931 Krebs demonstrated that the addition of amino acids to rat liver slices resulted in a measurable increase in urea formation. The following year, in an extension of this study, Krebs and Henseleit[9] showed that the addition of ornithine or citrulline to liver slices increased the formation of urea, especially in the presence of ammonia. On the basis of their observations Krebs and Henseleit proposed the following series of reactions as explanation of urea formation:

1. The combination of ornithine, carbon dioxide, and ammonia to form citrulline and water:

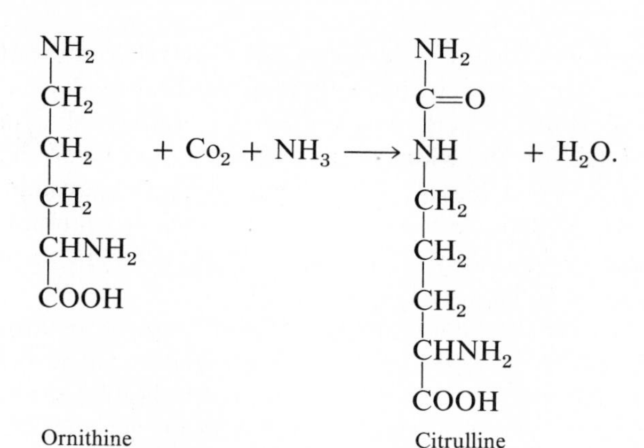

2. The combination of citrulline with ammonia to form arginine and water:

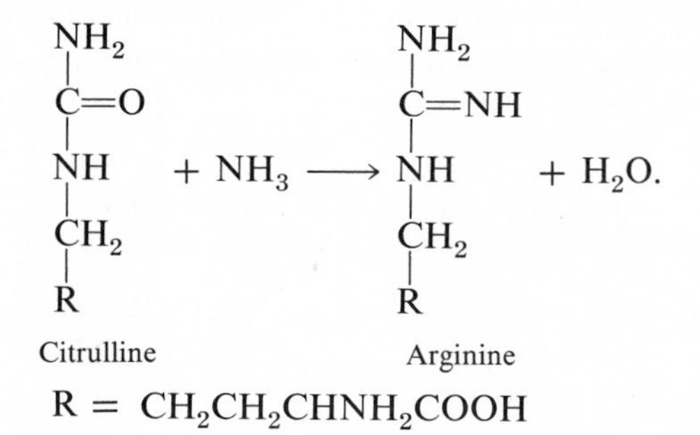

3. The hydrolysis of arginine into ornithine and urea, the reaction being catalyzed by the enzyme arginase:

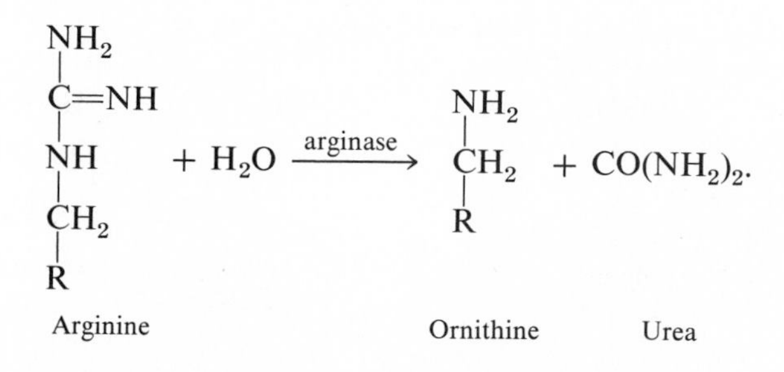

[9]H. A. Krebs and K. Henseleit, "Untersuchungen über den Harnstoffbildung in Tierkörper." *Z. Physiol. Chem.*, **210**:33 (1932).

The ornithine then becomes available for another cycle.

Ammonia is formed by the deamination of amino acids and also from urea in the following manner:

$$\underset{\text{Urea}}{(NH_2)_2CO} + \underset{\text{Salt}}{2NaCl} + \underset{\text{Carbonic acid}}{H_2CO_3} + 2H_2O \longrightarrow \underset{\text{Ammonium chloride}}{2NH_4Cl} + \underset{\text{Sodium bicarbonate}}{2NaHCO_3}.$$

Kidney tissue appears to be the main site for the occurrence of the above reactions and plays an important role in the regulation of the acid–base balance of the body. After the injection of acids or the excess production of acids in the body, ammonia formation by the kidney is increased. The ammonium chloride is eliminated by the kidneys, leaving the sodium bicarbonate available for neutralization of acids other than carbonic acid. The content of ammonia in the blood is very low, amounting to only 0.1 mg per 100 ml, although the amount found in the urine may be as high as 1.0 mg per day.

Creatine and **creatinine** are also end products of protein metabolism. Creatine is found chiefly in the muscles, although small traces can be demonstrated in the brain, testes, blood, and other organs. It is present both in the phosphorylated state as phosphocreatine, an important compound in muscular contraction, and in the free state. Creatinine is a by-product of creatine metabolism and is formed when water is removed from creatine. The output of creatinine appears to be constant in any one person regardless of the amount of protein in the food. The origin of creatine has been established by metabolic studies using isotope techniques and is believed to be as follows:

1. A transamination reaction takes place between glycine and arginine to form guanidoacetic acid, which occurs in the kidney but not in liver or heart muscle:

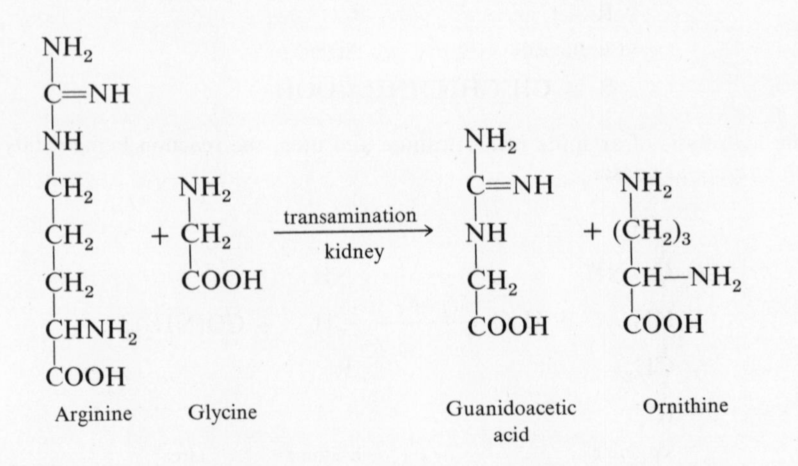

2. Guanidoacetic acid is methylated in the liver, during which methionine acts as the methyl donor to form creatine:

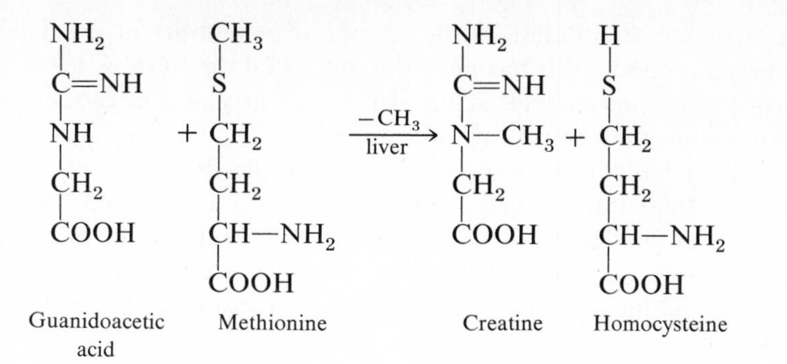

3. Creatinine is the anhydride of creatine, the reaction being irreversible in the body and apparently preliminary to the excretion of most of the creatine:

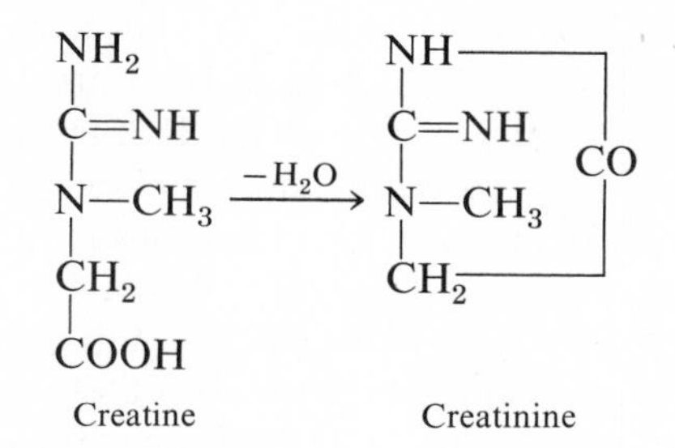

The urine of normal healthy adult males contains creatinine but no creatine. Creatine occurs normally in the urine of children along with creatinine and also in the urine of adults in whom breakdown of muscle tissue is occurring, as during fevers and starvation and in hyperthyroidism. It is also excreted in the urine of women during menstruation and during and after pregnancy. Creatine excretion is also a characteristic of certain muscular diseases, the degree of creatine elimination being directly proportional to the degree of muscular disability.

Uric acid as an end product of protein metabolism in man is negligible; most of the uric acid eliminated in the urine, which varies from 2.5 to 5 mg per 100 ml, comes from the metabolism of nucleic acids. In birds and reptiles uric acid takes the place of urea as the chief end product of protein metabolism.

Specific Dynamic Action of Proteins. During the absorption of dietary protein, there is an increase in the metabolic rate. This is referred to as the **specific dynamic action (SDA)** of proteins. For every 100 calories produced in basal conditions, 130 cal are produced if the subject ingested the equivalent of 100 cal of meat protein. It appears that the SDA of proteins is dependent on the process of deamination that takes place in the liver, since it is known that experimental removal of the liver suppresses deamination processes and the SDA. Six amino acids are known to be responsible for the specific dynamic action of proteins. These are glycine, alanine, leucine, glutamic acid, tyrosine, and phenylalanine. The intensity of the SDA produced by proteins depends on the percentage of these amino acids present in the molecule. The extra energy expended during protein metabolism does not seem to depend on the utilization of the deaminated residue or on the stimulation of the secretions or motility of the digestive tract during protein digestion. It has been proposed that the SDA is due to the amino groups of the six amino acids referred to previously and to certain products of the intermediary metabolism of those acids.

Fats and carbohydrates also produce a specific dynamic action. For every 100 cal of carbohydrate ingested, 104 cal in heat are produced and 106 cal for every 100 cal of fat. The SDA of fats has been attributed to the increase of blood in the veins during fat absorption (fat plethora). That of carbohydrate is also attributed to plethora, which increases glucose consumption and its conversion into glycogen.

Water Metabolism

Water is an important constituent of every cell in the body and composes well over 50 per cent of the cell's substance. The total body water is equal to about 40 to 65 per cent of the total body weight, with an average of 55 per cent for adult males and 47 per cent for adult females. The water of the body is classified into extracellular water and intracellular water. **Extracellular water** is that water found in the plasma and tissue fluid and makes up about 25 per cent of the total body water. The **intracellular water** is that water inside the cells of the body and makes up about 75 per cent of the total body water. It has been estimated that the **total body water** of a male subject weighing 154 lb is approximately 42 liters, intracellular water making up approximately 31.5 liters of that amount, and the extracellular water amounting to 10.5 liters. Of the extracellular water, the interstitial fluid accounts for 7.3 liters of that amount, and that in the circulatory system, 3.2 liters. Although the body water is distributed among three spaces, the intracellular, the interstitial, and the vascular space, there is a rapid and continuous exchange of fluid between them. The membrane separating the various spaces is freely permeable to water but may be selective to substances dissolved in it. The movement of water between the cells and interstitial space and between the interstitial space and the vascular space is controlled mainly by osmotic forces; that is, a dynamic equilibrium exists between the various compartments, with water passing freely from the fluid of lower osmotic pressure to that of the higher. The osmotic forces are maintained by the substances dissolved in the body water. These substances (solutes) are important not only in directing fluid distribution, but also in maintaining acid–base balance. They are normally divided into three classes of substances: (1) inorganic ions, (2) organic substances of large molecular size, and (3) organic substances of small molecular size.

The **inorganic ions** make up the largest quantity of dissolved materials in the body fluid and are the most important in the maintenance of the normal distribution of water and in its retention in the body. The ionic composition, expressed in units of milliequivalents,[10] of each of the three chief subdivisions of the body fluid is given in Table 12-1.

[10] It is customary to express reacting ions in concentration units of equivalents, since one chemical equivalent of any substance is exactly equal in chemical reactivity to one equivalent of any other substance. Since the quantities of dissolved ions are so small in body fluids, the **milliequivalent** (1/1000) is used. To convert the ion concentration from milligrams per 100 ml to milliequivalents per 1,000 ml (1 liter), multiply the number of milligrams per 100 ml by 10 to determine the milligrams per liter.

Table 12-1. ***Ionic Composition of Plasma, Interstitial Fluid, and Intracellular Fluid***

Extracellular Fluid				*Intracellular Fluid*	
Plasma		*Interstitial Fluid*		*Cell Fluid*	
Positive Ions (cations) (+)	mEq/liter	Cations (+)	mEq/liter	Cations (+)	mEq/liter
Na^+	145	Na^+	138	Na^+	14
K^+	5	K^+	5	K^+	155
Ca^{2+}	5	Ca^{2+}	5	Mg^{2+}	25
Mg^{2+}	3	Mg^{2+}	3		
Total (+)	158	Total (+)	151	Total (+)	194
Negative Ions (anions) (−)		Anions (−)		Anions (−)	
HCO_3^-	27	HCO_3^-	27	HCO_3^-	10
Cl^-	105	Cl^-	108	Organic phosphate	
HPO_4^{2-}	2	HPO_4^{2-}	2	PO_4^{2-}	110
SO_4^{2-}	1	SO_4^{2-}	1	SO_4^{2-}	20
Organic acids	6	Organic acids	6	Protein	55
Protein	16	Protein	2		
Total (−)	157	Total (−)	146	Total (−)	195

It is clear from Table 12-1 that the composition of interstitial fluid is similar to that of plasma, except that chloride ion practically replaces protein in the negative-ion column. It may also be seen that the fluid inside the cell differs in ionic composition from the extracellular fluid in that potassium is the principal cation (+) in contrast to sodium in extracellular fluid. In addition, phosphate ion rather than chloride ion is the principal anion (−).

The ionic composition of extracellular fluid may be regarded as being built around sodium and chloride ion and on the fact that a normal concentration of Na^+ in the plasma of 138 to 145 mEq per liter implies normal osmotic concentration of the extracellular fluid. The principal ions of cell fluid are potassium, magnesium, phosphate, and protein. The phosphate exists chiefly in the form of organic phosphate as adenosine triphosphate and phosphocreatine. The concentration of inorganic phosphate inside the cell is similar to that outside the cell, that is, 2 to 3 mEq per liter. Although cell membranes are freely permeable to

---

Divide the milligrams per liter by the milliequivalent weight of the element, which is the millimolecular weight divided by the valence. For example, if the normal concentration of calcium is 10 mg per 100 ml, there are 100 mg (10 × 10) in 1 liter. The molecular weight of calcium is 40, which is also its millimolecular weight if an equivalent gram weight were dissolved in 1 liter of solution. Dividing the millimolecular weight by calcium's valence, which is 2 (from the periodic table), we convert this to its milliequivalent weight of 20—($\frac{40}{2}$). The milligrams per liter is then divided by the milliequivalent weight to get milliequivalents per liter, that is, $\frac{100}{20} = 5$ mEq per liter.

most ions, it appears that they are not functionally permeable to sodium and potassium since there are such large differences in the concentration of Na^+ and K^+ ions on opposite sides of the cell membrane. The reasons for this are not clearly understood, but from the standpoint of osmotic forces in directing the movement of water and in controlling the water content of the body, sodium and potassium are the most important, since it is the shifting of water from one compartment to the other that serves to adjust differences in intracellular and extracellular osmotic pressures produced by these ions in solution. If the solute is able to pass through the membrane, it will naturally migrate through it and distribute itself equally in the water of the various compartments. Under these conditions the dissolved ions have no influence on the distribution of water. In the case of Na^+ and K^+, the membranes separating the compartments are functionally impermeable to them. The distribution of water on both sides of the membrane is affected by the water-retaining tendency exerted by the solute, which is restricted through selective permeability to one side of the membrane. The higher the concentration of ions on one side of the membrane, the greater the osmotic pressure and therefore the greater the tendency to retain water. The effect of osmotic pressure on the distribution of fluid in the tissue spaces is also discussed on pages 261–262 and 303.

Since water is freely diffusible across the membrane separating the extracellular and intracellular compartments, its movement, as we have already seen, is determined by the changes in concentration of the osmotically effective ions on either side. The principal ions involved in water distribution are sodium, which is almost entirely confined to the extracellular space, and potassium, which is confined to the intracellular space. Changes in sodium concentration are the most common cause for shifting of water. This is the reason that sodium intake is restricted in order to control excessive accumulation of water in the tissue spaces in various pathological states. Under certain conditions potassium ion leaves the cells, especially in cases of excessive vomiting and diarrhea. If replacement of lost ions is made by sodium salt alone, sodium migrates into the cells to replace the lost potassium. The deficiency of potassium leads to serious alterations in cellular metabolism and normal functioning of the organism.

The organic compounds of small molecular size, such as glucose, urea, and amino acids, diffuse across membranes quite freely and therefore are not important in the distribution of water. However, if such compounds are present in large quantities, they aid in retaining water and thus can influence total body water.

The organic compounds of large molecular size, such as the plasma proteins, profoundly influence the transfer of water from one compartment to another but not usually total body water.

In contrast to the selective permeability of the cell membrane separating the intra- and extracellular ions, there is a relatively free exchange of all ions across the capillary wall, that is, the membrane separating the vascular fluid compartment from the interstitial compartment. However, the plasma proteins, which normally cannot diffuse across the capillary membrane, exert an osmotic influence, which effects the transfer of fluid through this barrier. When the plasma protein concentration decreases from its normal level, fluid moves from the blood to the interstitial spaces, resulting in edema, a condition we have already discussed. The osmotic effect exerted by the plasma proteins helps maintain the proper water content of the vascular compartment and also prevents any excessive elimination of water from the body in its passage through the kidney.

Water Intake. Water must be supplied to the individual regularly to provide for its use in formation of tissue and tissue products and to compensate for the losses through excretions from the body. Two main sources of water are liquids **(preformed water)** and most solid foods. The average amounts of these waters are

Liquids imbibed	1,000 ml
Water from solid food	1,000 ml

Some water is also formed within the body when the solid foods, such as the carbohydrates, fats, and proteins, are oxidized in the process of metabolism. This water is called the **water of oxidation,** or **metabolic water,** and amounts to approximately 500 ml. It has been demonstrated that the oxidation of 100 g of carbohydrates yields 55 g of water; of 100 g of fat, 107 g of water; and of 100 g of protein, 41 g of water. Thus the total availability of water to the individual, or water intake, would be approximately 2,500 ml.

Water Loss. Water is lost to the body by excretions through various routes: from the kidney in the form of urine, from the skin in the form of sweat, from the lungs as water vapor in the expired air, and from the intestines in the feces. The daily output of water from these various sources of an average person in a temperate climate would be as follows:

Urine	1,300 ml
Sweat	650 ml
Lungs	450 ml
Feces	100 ml
Total	2,500 ml

A person maintaining a normal content and distribution of water through the various fluid compartments by proper intake and excretion of water is said to be *in water balance.* The tissue cell is in positive balance with regard to water when it accumulates it and in negative balance when it loses it. A negative water balance may result when water intake is restricted for any reason or when losses are excessive. Under these circumstances the extracellular fluid becomes concentrated and hypertonic to the cells. As a result water shifts from the cells to the interstitial spaces to compensate.

The earliest symptom of intracellular dehydration is severe thirst, which is reflexly initiated by the drying out of the mucous membranes of the mouth and throat. The kidney attempts to compensate for water deprivation by concentrating the urine and excreting a smaller volume. Under more severe conditions of water depletion, nausea, vomiting, and loss of coordination develop. Dehydration and its symptoms may be alleviated by the administration of water by mouth or by glucose solution intravenously.

The condition of positive water balance, or **overhydration,** occurs when large amounts of electrolyte-free solutions are administered to patients or when large amounts of water are ingested, resulting in a deficit of ions in the extracellular fluid. Great losses of electrolytes through excessive vomiting or from prolonged diarrhea can also cause overhydration. The resulting hypotonicity of the extracellular fluid causes water to pass into the cells, which are now hypertonic to the fluid surrounding them, resulting in an intracellular edema. The loss of water from the extracellular space by its shifting into the interior of the cells is followed by a compensatory shift of water from the plasma compartment to the tissue spaces. The resulting decrease in blood volume causes a fall in blood pressure, slowing of the circulation, and its concomitant impairment of kidney function. This overhydration of the tissue cells, which is in reality a form of dehydration, is usually corrected by the administration of proper isotonic saline solution.

Mineral Metabolism

The animal body requires seven principal minerals to maintain the normal physiological function and health of the protoplasmic structure. These are calcium, potassium, sodium, chlorine, phosphorus, sulfur, and magnesium. Seven other minerals are necessary in trace quantities: iron, copper, iodine, manganese, zinc, cobalt, and fluorine. The body's needs and the functions of most of these minerals are discussed in Chapter 13. Here we will be concerned mainly with the absorption and metabolism of the principal minerals of the animal body.

Calcium. **Calcium** is the mineral most likely to be deficient in the human being, because the ability of different persons to utilize calcium in food varies considerably. It is also absorbed with difficulty from the intestinal tract. Calcium being soluble in acid media, it is more easily absorbed in those regions of the intestine where the contents are slightly acid. Alkaline solutions tend to precipitate calcium, decreasing its absorption through the intestinal barrier. Vitamin D promotes the absorption of calcium from the intestine. Thus, when the intake of vitamin D is low, calcium absorption is defective. Large amounts of free fatty acids also impair calcium absorption, since they react with free calcium to form insoluble calcium soaps. In addition, excessive phosphate impairs calcium absorption, since insoluble calcium phosphate is formed, especially in the more alkaline areas of the intestinal tract.

Calcium is found in the plasma fraction of the blood in normal concentrations ranging from 9.5 to 11.5 mg per 100 ml. None is present in the red blood cells. More than 50 per cent of the calcium in the blood is in the form of diffusible calcium ions, and the remainder is in the form of a nondiffusable protein complex, probably with serum albumin. It appears that the blood calcium level is controlled by the vitamin D content of the diet in that low vitamin D content decreases calcium absorption and the level of blood calcium drops; conversely, when the vitamin D content is large, the level of blood calcium also increases.

The blood calcium level is also controlled by the parathyroid glands. The hormone secreted by this gland profoundly influences the metabolism of calcium. The hormone functions in the mobilization of calcium from the bones. If the parathyroids are removed surgically, as is sometimes done in surgical operations for the removal of the thyroid, the calcium level of the blood falls and may drop below 7 mg per 100 ml. If it reaches such a level, the condition of tetany results. If the gland becomes hyperactive, producing excessive amounts of hormones, as it may as a result of a tumor, the blood calcium level is raised and may rise to 15 to 22 mg per 100 ml. The increase is not due to increased absorption but results from the mobilization of calcium from the bones and soft tissue as a result of increased osteoclastic (bone-destroying) activity. As a result, the bones become weak, easily fractured, and deformed, a condition known as **osteitis fibrosa.** In addition, large amounts of calcium are excreted in the urine, and calcium salts may be deposited in the walls of the larger blood vessels. A deficiency of calcium in the diet or a defective absorption of calcium due to low vitamin D content or a combination of both in growing children is characterized by a faulty calcification of bones. The condition is known as **rickets** and is common among children maintained on poor diets, especially diets lacking in sufficient quantities of milk and food products high in calcium and phosphorus content. **Renal rickets** is caused by a tubular defect in the kidney in which the normal reabsorption of calcium from the tubular filtrate is inhibited and large quantities of calcium are excreted in the urine. Vitamin D administration does not correct this decrease in blood calcium level.

Normally only small quantities of calcium are excreted in the urine because it is highly reabsorbed by the kidney tubule. Most of the calcium is excreted in the feces and represents unabsorbed dietary calcium and that which is excreted by the intestinal mucosa, especially that of the distal ileum and the colon.

Phosphorus. **Phosphorus** in the form of inorganic phosphate is absorbed readily from the intestinal tract if the medium is slightly acid. In alkaline solutions it tends to be precipitated in the form of an insoluble salt, chiefly calcium phosphate.

The normal concentration of phosphorus in the blood is approximately 40 mg per 100 ml. A large part of this phosphorus is in the form of organic phosphate and is more abundant in the red cells than in the plasma. Phosphorus is also present as inorganic phosphate and lipid phosphorus. The former is present mainly in the plasma and is that fraction which is of the most clinical significance. The lipid phosphorus, like the organic phosphate, is more abundant in the red cells.

The metabolism of phosphorus is intimately associated with that of calcium, as the ratio of calcium to phosphate in the diet is of some importance in regulating the absorption and excretion of each of these minerals. Any change in the ratio of calcium to phosphorus from 1:1.5 for adults and 1:1.0 for children will affect the retention or excretion of each; that is, if either element is given in excess, the excretion of the other is increased. Thus, when the blood phosphorus level is increased, a compensatory decrease in the calcium level is effected. In hypoparathyroidism, where the blood calcium level is decreased, blood phosphorus levels are increased. In hyperactivity of the parathyroid glands, the blood phosphorus levels are low as a result of the increased plasma calcium concentrations.

Several different types of enzymes called phosphatases occur in many tissues of the body, especially bone, cartilage, and kidney. Their function is to break down simple organic phosphates, such as phosphoglucose and phosphoglycerol, so that free inorganic phosphate is liberated.

Phosphorus is excreted mainly by the kidneys; a small portion is excreted in the feces. In nephritis the elimination of inorganic phosphate is impaired, resulting in its retention in the blood, which causes an elevation in plasma phosphorus concentration. This may produce acidosis, which is common in nephritis and also contributes to the lowered plasma calcium.

Magnesium. Approximately 70 per cent of the **magnesium** in the body is combined with calcium and phosphorus in the complex of bone salt. The remainder is in the body fluids and soft tissues. Magnesium is found mainly inside cells. The magnesium content of normal blood ranges from 2 to 4 mg per 100 ml. The magnesium content of muscle is about 21 mg per 100 g; it is essential for the functioning of certain enzyme systems such as phosphoglucomutase. Magnesium is excreted from the body in the urine and feces.

Potassium. Like magnesium, **potassium** is found mainly in the intracellular fluid and in normal concentrations of about 155 mEq per liter. The concentration of potassium ion in the blood ranges from 15 to 20 mg per 100 ml. Within the cell it functions by influencing acid–base balance and osmotic pressure and maintains the normal distribution of water between the various fluid compartments of the body. It is also a very important constituent of the extracellular fluid because it influences muscular activity, especially heart muscle. Changes in its normal concentrations in extracellular fluid can cause paralysis of skeletal muscle and abnormalities in conduction and activity of cardiac tissue.

Most of the potassium ingested is absorbed through the intestinal barrier and passed into the general circulation to be distributed throughout the fluids and tissues of the body. The excess of potassium is mainly excreted by the kidney. High serum potassium accompanied by high intracellular potassium, a condition known as **hyperkalemia,** occurs in diseases that affect kidney function, excessive dehydration, and shock. A symptom of elevated extracellular potassium is decreased heart rate, which is usually followed by vasomotor collapse and cardiac arrest. Other symptoms may be numbness, tingling of the hands and feet, and weakness of the muscles of the extremities and those of respiration. Elevated plasma potassium may be corrected by the administration of deoxycorticosterone (DOCA), which is an adrenocortical hormone that increases the excretion of potassium by the kidneys. Overactivity of the adrenal cortex can thus induce a potassium deficit, known as **hypokalemia.** The reduction in the extracellular potassium causes potassium to be transferred from the intracellular fluid to the extracellular space to compensate for the loss. This potassium is also quickly removed by the kidneys. Potassium deficiency is also likely to develop in conditions involving losses from the gastrointestinal tract, as in diarrheas and gastrointestinal fistulas and also after prolonged administration of solutions into the circulatory system that do not contain potassium. Symptoms associated with hypokalemia are increased heart rate, dilatation of the heart, muscle weakness, irritability, and paralysis. The administration of a solution of potassium chloride containing 1.8 g per liter intravenously may alleviate the low serum potassium, or a 1 to 2 per cent solution given by mouth to adults, divided into doses so that 4 to 12 g of potassium chloride is administered in a 24-hour period, is effective.

Sodium. **Sodium** is the chief positive ion of the extracellular fluid, and most of the sodium in the body is confined to this space. It occurs as chloride, phosphate, and carbonate and in combination with simple organic acids, such as lactic and hydroxybutyric. Sodium comprises approximately 93 per cent of the basic ions in the blood.

The body's main source of sodium is the sodium chloride used in seasoning, and it is easily absorbed from the intestinal tract. It has been estimated that about 10 to 30 g of salt is ingested each day; very little salt is found in the feces, except in diarrhea, when much of the sodium escapes reabsorption. After absorption, the sodium is distributed throughout the body, mainly in the extracellular fluid. The concentration of sodium ion in normal plasma is 330 mg per 100 ml. The metabolism of sodium, like that of potassium, is influenced by the adrenocortical hormones. In an insufficiency of the hormone, increases in the excretion of sodium in the urine are noted with resultant decreases in plasma sodium. The adrenocortical steroids facilitate sodium retention. Under normal conditions 90 per cent of ingested sodium is excreted in the urine in the form of chloride and phosphate. A low plasma sodium occurs in Addison's disease because of the lowered activity of the adrenal cortex in secreting its hormones in this condition. Sodium depletion may also develop in nephritis because of the poor reabsorption of sodium under these circumstances. Since sodium functions importantly in the maintenance of the osmotic pressure of the body fluid, thus protecting the body against excessive fluid loss, and also in the preservation of normal irritability of nerve and muscle, changes in its normal concentration in the fluid compartments of the body are reflected in these areas. Sodium depletion, with its concomitant dehydration, is associated with such symptoms as muscular cramps, nausea, headache, weakness, the vascular collapse. Excessive sweating can also cause the loss of considerable sodium in the sweat. A special need for sodium chloride replacement thus exists in any condition causing sodium depletion, for example, in occupations where excessive sweating is a factor and in nephritis, adrenal insufficiency, and protracted diarrhea.

Overactivity of the adrenal cortex can lead to increased plasma sodium because the adrenocortical hormones increase the retention of this ion. Injections of cortisone, ACTH, or deoxycorticosterone also can induce an increased sodium plasma level. In most instances the rise in the concentration of sodium is masked because of the concomitant retention of water.

Chlorine. **Chlorine** occurs in the diet as chloride and almost entirely as sodium chloride. In general the metabolism of chloride ion is essentially the same as that of sodium, so that anything that affects the metabolism of one will also affect the metabolism of the other. Unlike sodium and potassium, chloride ion freely diffuses through the cell membrane and thereby plays a special role in the blood by the action of the chloride shift (see p. 295). The chloride ion is also essential in water balance and osmotic-pressure regulation. In the parietal cells of the gastric mucosa the chloride ion plays an important role in the formation of gastric hydrochloric acid, since the latter is believed to originate from the carbonic acid in the presence of the enzyme carbonic anhydrase. The reaction may be written as follows:

$$H^+ + HCO_3^- + Na^+ + Cl^- \xrightarrow{CA} HCl + NaHCO_3.$$

After a meal the urine is usually alkaline; this is believed to be due to the formation of extra sodium bicarbonate in the process of hydrochloric acid formation, as cited above. This transient alkalinity of the urine is referred to as the **alkaline tide.**

Sulfur. **Sulfur** is widely distributed throughout the animal body, mostly in combination with cell protein. The main sources of sulfur for the body are the two sulfur-containing amino acids, cystine and methionine. Inorganic sulfur, as either elemental sulfur or its salt form, sulfate sulfur, is not known to be utilized.

The importance of sulfur-containing compounds in detoxication mechanisms within the body is well known. The conjugation of toxic compounds, which involves in some instances the combination of these substances with sulfur compounds, followed by their elimination in the urine as sulfur conjugates, is one of the ways the body is able to detoxicate metabolites such as indole, skatole, and various phenolic compounds. These sulfate conjugates excreted in the urine make up most of the organic sulfate fraction of the total urine sulfur. Sulfur compounds used for conjugation reactions are cystine, methionine, and sulfate. Sulfur is also important in tissue respiration in the form of sulfhydryl groups (SH). Many enzymes depend on a free SH group for maintenance of activity; that is, converting the SH group to the S—S linkage will inactivate them, and also reduction will cause a reactivation. In fact, the conversion of two SH radicals to S—S is probably an important oxidation-reduction system in the body, as exemplified by the oxidation of cysteine to cystine.

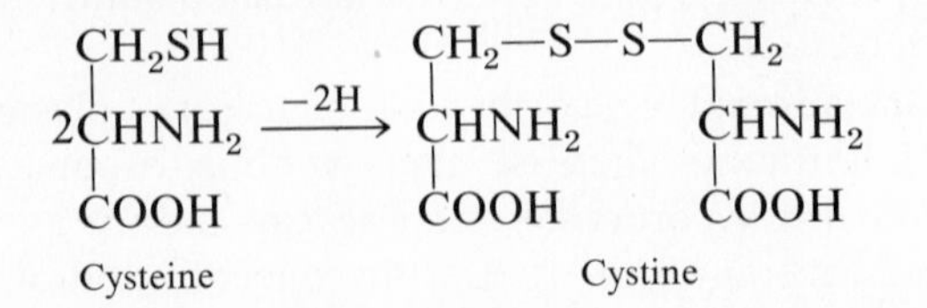

Other compounds of sulfur found in the body in addition to cystine and methionine are glutathione, insulin, thiamine, coenzyme A, biotin, and the sulfuric acid in cartilage, tendon, and bone. Small amounts of inorganic sulfate in the form of sodium and potassium are present in blood and other tissues. Organic sulfur is mainly oxidized to sulfate and excreted in that form or in the ethereal form of conjugated sulfates.

Energy Balance

The first law of thermodynamics, which is the **law of conservation of energy**, holds that any system in a given condition contains a definite quantity of energy, and when the system undergoes change, any gain or loss of its internal energy is equal to the loss or gain in the energy surrounding the system. The study of the body's input and output of energy is the application of this law to biological systems and is called the **energy balance of metabolism.** The **input of energy** is derived from the oxidation of foodstuffs and varies considerably according to the amount and type of material being utilized. The output of energy is expended for the functional activities of the various organs (heartbeat, respiration) and tissues (synthesis of proteins, hormones, and various secretions), for the performance of mechanical work, and for the heat necessary for the maintenance of body temperature.

The potential energy of a foodstuff can be determined by measuring the amount of heat produced when the foodstuff is burned in an apparatus called the **bomb calorimeter** (Fig. 12-5). It consists of three vessels, one inside the other. The outermost vessel is a heavy metal case that can resist internal pressure. Another vessel, which sits in the first, is insulated with cork to resist heat loss. A third vessel placed in this one and filled with water is the one in which the bomb is placed. The bomb contains a weighed amount of foodstuff and is filled with oxygen under pressure. The foodstuff is ignited electrically, and the amount of heat produced is measured by determining the rise in temperature of the water surrounding the bomb. The unit of heat used is the kilocalorie, or large calorie, indicated by capital C, which is the amount of heat required to raise the temperature of 1 kg, or 1,000 g, of water from 15 to 16° C. This method of determining the energy content of foodstuffs is a direct method, and the values derived from the complete combustion of the three basic foods, in this manner, are as follows:

1 g of carbohydrate yields	4.1 C
1 g of fat yields	9.3 C
1 g of protein yields	5.4 C or 4.1 C

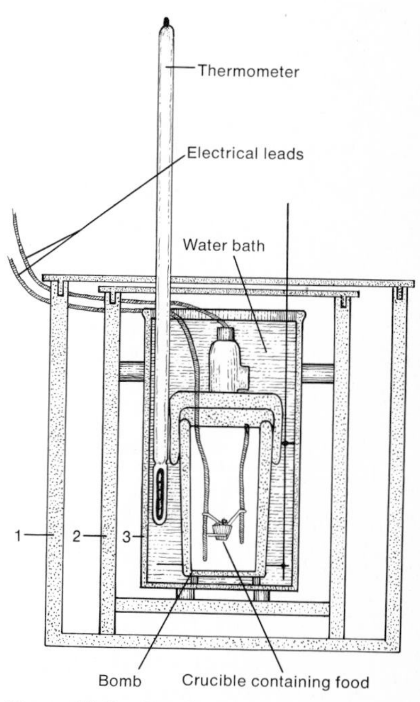

Figure 12-5. Bomb calorimeter. The innermost vessel (3) *contains water in which is immersed the bomb containing the crucible with the food to be burned. The food is ignited by the electrical leads. The rise in the temperature of the water after the combustion of the foodstuff in the bomb is measured by the thermometer.*

In the bomb calorimeter, protein is oxidized to carbon dioxide, water, and oxides of nitrogen, but in the body it is not completely oxidized since part of it is converted to urea, which is excreted. Urea has an appreciable energy potential, and when this energy is subtracted from the figure of 5.4 for the complete oxidation of protein in the bomb calorimeter, a physiological value of 4.1 C is obtained for protein.

Another method of determining the energy content of foodstuffs has been devised by Benedict and Fox.[11] This is an **indirect method,** and it consists in measuring the oxygen used in burning the food material and then converting this value into calories.

The same methods can also be used to measure the energy output by the animal body. It can be determined by the direct method by placing the animal in a calorimeter (Fig. 12-6) or indirectly by measuring the amount of oxygen consumed

[11] F. G. Benedict and F. L. Fox, "A Method for the Determination of Energy Values of Foods and Excreta." *J. Biol. Chem.,* **66:**783 (1925).

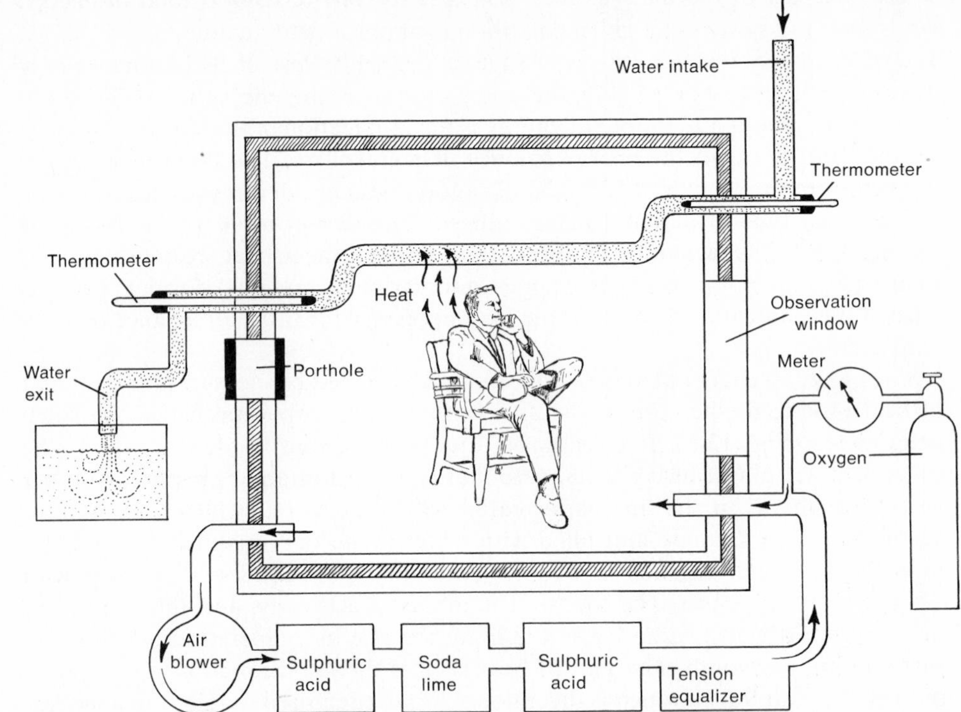

Figure 12-6. Atwater–Benedict respiration calorimeter. This apparatus is used for measuring the energy output of the animal body by the direct method. The heat produced by the subject is absorbed by the water, its temperature change being noted from the thermometer located at the intake and the one at the outlet. Food may be introduced and excreta removed by way of the porthole. Air is recirculated through the chamber after passing over sulfuric acid and soda lime to absorb moisture and carbon dioxide. Oxygen is added to the system as the air passes into the chamber.

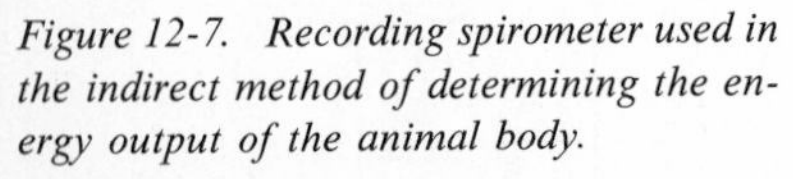

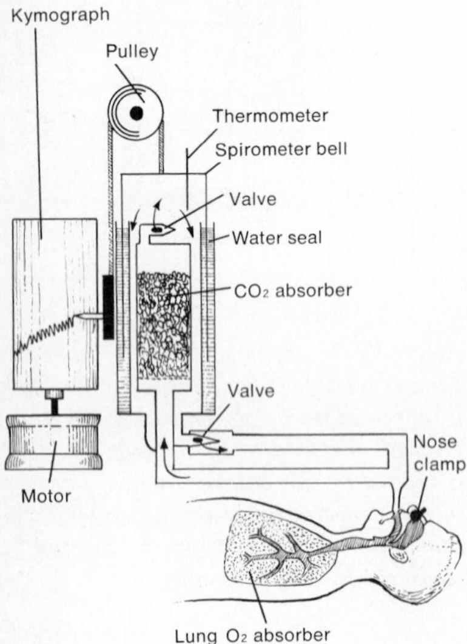

Figure 12-7. Recording spirometer used in the indirect method of determining the energy output of the animal body.

or the amount of carbon dioxide given off and then converting these values into calories. In general the indirect method is much simpler and more commonly used clinically. The apparatus normally used is the recording spirometer of the Roth–Benedict type (Fig. 12-7), which consists of a gas container (spirometer bell) that is filled with oxygen. The top of the bell is connected by a cord running from a pulley to a writing lever, which records the level of the gas container on a rotating kymograph. The spirometer bell is connected by a system of rubber tubes to a mouthpiece through which the subject must breathe when his nose is closed by a clip. As the subject breathes in oxygen from the spirometer bell, his expired air passes through a valve into a container of soda lime, which absorbs the carbon dioxide, and then back into the spirometer bell. As oxygen is consumed, the level of the spirometer bell falls, and from the record on the kymograph the amount of oxygen consumed in a given period can be measured (Fig. 12-8). Before this can be converted into the actual energy evolved, it is necessary to know how many large calories are produced when 1 liter of oxygen is consumed.

Caloric Value of Oxygen

The caloric value of 1 liter of oxygen depends on the type of food being oxidized. It has been determined that when carbohydrate is oxidized in the body to carbon dioxide and water, for each liter of oxygen used, 5.047 C are produced; thus, the caloric value of oxygen used in the oxidation of carbohydrate is said to be approximately 5. For the combustion of human fat the caloric value of

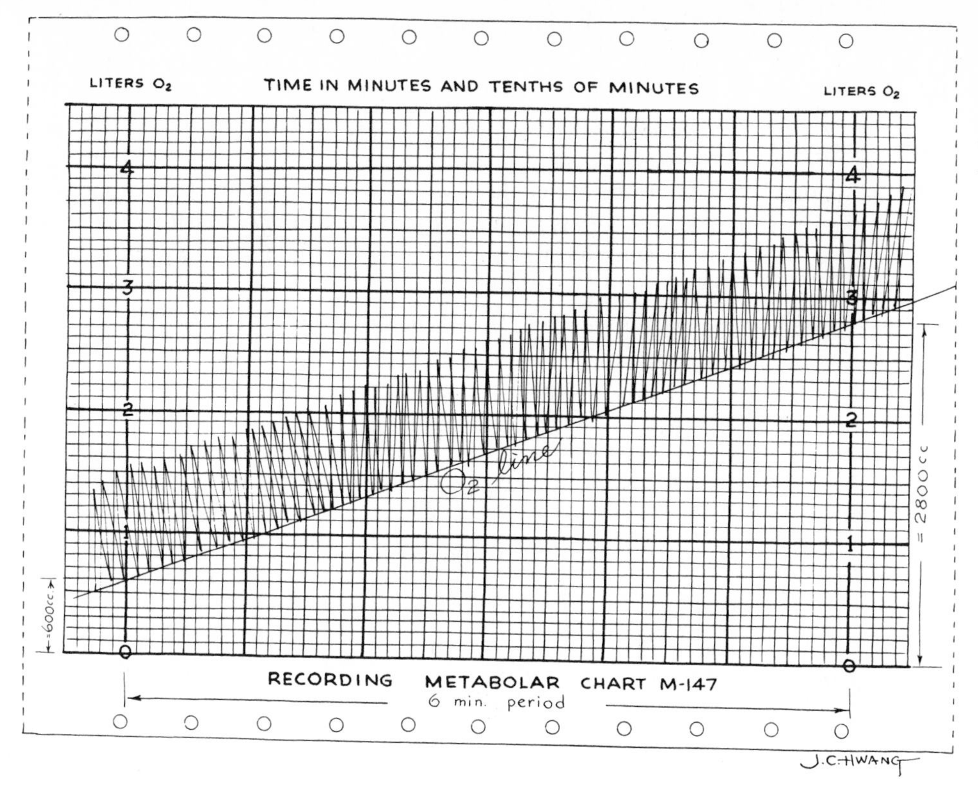

Figure 12-8. Typical recording from a spirometer of the Roth–Benedict type. Rise in oxygen line in 6 minutes is 2,800 cc − 600 cc = 2,200 cc, or 2.2 liters, of oxygen consumed in 6 minutes; in 1 hour, 22 liters.

oxygen is 4.725 per liter, and for protein the value obtained is 4.5 C. Depending on the types and proportions of the foodstuffs being metabolized, it is obvious that when a liter of oxygen is used by the body, between 4.5 and 5 C are produced. It is assumed that the body oxidizes a mixture of carbohydrate, fat, and protein, and therefore an average caloric value of 1 liter of oxygen would depend on the proportion of each of these foodstuffs being oxidized. To understand this point we must consider the respiratory quotient and its implication in the derivation of the **caloric value of 1 liter of oxygen,** assumed under normal metabolic activity to be equal to 4.825 C.

Respiratory Quotient

The respiratory quotient (RQ) is defined as the ratio of the volume of carbon dioxide produced to the volume of oxygen used during the same period of time. The volume of oxygen required and the amount of carbon dioxide produced in the burning of foodstuffs outside or within the body depend on the elemental composition of the substance. The respiratory quotient for carbohydrate, fat, and protein can be determined from equations representing the oxidative splitting of these compounds into their final products.

When carbohydrate is oxidized, since its molecule always contains twice as many hydrogen as oxygen atoms, the proportions necessary to form water, extra oxygen is required only for the formation of carbon dioxide. That the respiratory quotient will be 1 is demonstrated by

$$\underset{\text{Glucose}}{C_6H_{12}O_6} + 6O_2 \longrightarrow 6CO_2 + 6H_2O,$$

$$\text{RQ carbohydrate} = \frac{CO_2}{O_2} = \frac{6}{6} = 1.$$

In the oxidation of fat, in the molecule of which there are many more hydrogen atoms than oxygen, extra oxygen is required not only for the oxidation of carbon, but also for the oxidation of hydrogen. For this reason more oxygen will be consumed than carbon dioxide liberated, and the RQ for fats will be less than 1. For example, when the fat triolein is oxidized, the RQ is 0.71, as is evident from

$$\underset{\text{Triolein}}{C_{57}H_{104}O_6} + 80O_2 \longrightarrow 57CO_2 + 52H_2O,$$

$$\text{RQ fat} = \frac{CO_2}{O_2} = \frac{57}{80} = 0.71.$$

For tripalmitan, another fat, the RQ would be

$$\underset{\text{Tripalmitan}}{2C_{51}H_{98}O_6} + 145O_2 \longrightarrow 102CO_2 + 98H_2O,$$

$$\text{RQ} = \frac{102}{145} = 0.703.$$

In general the RQ for food fat, which is composed chiefly of the glycerides of oleic, palmitic, and stearic acids, is taken as 0.70.

The RQ of protein is more difficult to calculate because some of the oxygen consumed in its oxidation remains combined with nitrogen, along with some of the carbon of the constituent amino acids, and is excreted as nitrogenous waste in the urine and feces. Determinations of the RQ of protein must be based on accurate analyses of the ingested protein and the excreted nitrogenous wastes. Experimentation has shown that the RQ of protein, when calculated indirectly from the quantities of carbon and hydrogen involved in the oxidative process, would be 0.80 since 77.5 liters of carbon dioxide would be produced and 96.7 liters of oxygen utilized:

$$\text{RQ protein} = \frac{CO_2}{O_2} = \frac{77.52}{96.70} = 0.802.$$

For a subject on an ordinary mixed diet the respiratory quotient represents the oxidation of a mixture of carbohydrate, fat, and protein and is taken to be 0.85. An RQ of 0.85 represents an average RQ value under normal metabolic conditions. Table 12-2 lists the various metabolic constants previously discussed.

Significance of RQ. When the values of oxygen consumed and carbon dioxide exhaled by a subject at any one time are determined, the RQ obtained indicates the kind of foodstuff being metabolized. When carbohydrates alone are burned, an RQ of 1.0 is obtained, and when fat alone is burned, the RQ falls to 0.70.

Table 12-2. *Metabolic Constants*

Food Oxidized	*Large Calories Liberated per Gram*	*RQ*	*Caloric Value of 1 liter of Oxygen*
Carbohydrate	4.1	1.00	5.047
Fat	9.3	0.707	4.735
Protein	4.1	0.802	4.463

It would appear that the highest obtainable RQ under physiological conditions would be 1 and the lowest 0.70. However, under abnormal conditions these limits may be exceeded. The conversion of quantities of carbohydrate to fat within the body would tend to raise the RQ above 1.0. In this case a substance rich in oxygen (carbohydrate) is converted into one containing less (fat) with a net release of oxygen, which is available for the metabolic needs of the body, thereby reducing the amount of oxygen required from the outside through inhalation. As a result there will be more carbon dioxide produced in proportion to the amount of oxygen consumed. Respiratory quotients greater than 1.0 are rarely encountered and are confined chiefly to gluttons and fattening hogs. In fasting states and in uncontrolled diabetes the RQ falls from normal and comes close to 0.70, which is indicative of increased fat catabolism. In severe uncontrolled diabetes, respiratory quotients lower than 0.70 have been reported. This type of RQ reflects the conversion of protein to carbohydrate, wherein greater quantities of oxygen are consumed without increases in carbon dioxide output, because the carbohydrate molecule being synthesized contains more oxygen than the protein can supply.

In addition to oxidative processes, other factors also affect the value of the respiratory quotient. Increases in the respiratory quotient above normal are seen in hyperventilation; in excessive acid formation, such as lactic acid formation; in exercise and convulsions; in ketosis, as in diabetes; and in acid retention, as in nephritis. All these conditions cause an excessive release of carbon dioxide in the alveolar air and blood without an increased formation of carbon dioxide. The respiratory quotient is usually decreased by hypoventilation; by excessive loss of acid, as in protracted vomiting (hydrochloric acid); and in alkalosis produced by alkali therapy. These conditions lead to a retention of carbon dioxide without any change in the oxidation process of the body.

As we have already mentioned, the RQ indicates which food substances are being burned in the organism, and since the body burns the three principal kinds simultaneously, the resultant RQ reflects this fact. Because the amount of heat produced by 1 liter of oxygen varies according to the substance oxidized, the RQ therefore must be known in order to determine accurately the caloric value of the oxygen absorbed. An average caloric value of 4.825 C per liter of oxygen is assumed, corresponding to an RQ of 0.82, which is commonly observed in subjects in the basal state.

In clinical practice a **nonprotein RQ** is used in which the volumes of oxygen used and of carbon dioxide produced in protein metabolism are obtained by calculating from the excreted nitrogen the amount of protein metabolized in the body during a definite time period. These values are deducted from the volumes of the total respiratory exchange during the same time period. For example, a subject eliminates 14.16 liters of carbon dioxide per hour and absorbs 16.32 liters of oxygen per hour. The same subject excretes 0.6 g of nitrogen per hour. From protein-combustion experiments it is known that for each gram of nitrogen produced in combustion of protein, there is an output of 4.76 liters of carbon dioxide, and 5.94 liters of oxygen is absorbed. Therefore, in the example given, the amount of carbon dioxide given off and oxygen absorbed for the oxidation of 1 g of protein would be

$$0.6 \times 4.76 = 2.86 \text{ liters of } CO_2 \text{ per hour,}$$
$$0.6 \times 5.94 = 3.56 \text{ liters of } O_2 \text{ per hour,}$$
$$\text{Protein RQ} = \frac{2.86}{3.56} = 0.80.$$

The volumes of gas involved in carbohydrate and fat metabolism can be calculated by subtracting the above gas volumes from the total amount of oxygen used and carbon dioxide given off by the subject during a time period of 1 hour:

$$14.14 - 2.86 = 11.30 \text{ liters of } CO_2 \text{ per hour,}$$
$$16.32 - 3.56 = 12.76 \text{ liters of } O_2 \text{ per hour,}$$
$$\text{Nonprotein RQ} = \frac{11}{13} = 0.85.$$

A table analyzing the oxidation of mixtures of carbohydrate and fat on the basis of the nonprotein RQ was prepared by Zuntz and Schumburg. If only mixtures of carbohydrate and fat are oxidized, the respiratory quotients vary between 0.707 (when only fats are oxidized, with 1 liter of oxygen having a heat value of 4.686 C) and 1.00 (when only carbohydrates are oxidized by the body, with 1 liter of oxygen having a caloric value of 5.047 C). Table 12-3 lists some of these values and shows the proportion of each of the foodstuffs burned at different RQ's.

The caloric value of a liter of oxygen at a nonprotein RQ can be calculated from the following equation developed by Lusk[12]:

$$C = 4.686 + \frac{(\text{RQ} - 0.707) \times 0.361}{0.293}.$$

If carbohydrate alone is being oxidized, the RQ = 1, so that

$$C = 4.686 + \frac{(1 - 0.707) \times 0.31}{0.293}$$
$$= 4.686 + 0.361 = 5.047$$

which is the caloric value of 1 liter of oxygen used in the oxidation of carbohydrate (Table 12-2).

The RQ must be known in order to obtain accurately the caloric value of the

[12] G. Lusk, *The Elements of the Science of Nutrition,* 4th ed. (Philadelphia: Saunders, 1928).

Table 12-3. ***Caloric Value of 1 liter of Oxygen at Various Respiratory Quotients***

Nonprotein RQ	*Percentage of Total Oxygen Consumed*		*Calories per Liter of CO_2*
	Carbohydrate	*Fat*	
0.707	0	100.00	4.686
0.75	14.70	85.30	4.739
0.80	31.70	68.30	4.801
0.85	48.80	51.20	4.862
0.90	65.90	34.10	4.924
1.00	100.00	0	5.047

oxygen absorbed. If the RQ is not known, an average caloric value of 4.825 C per liter of oxygen is used, which corresponds to an RQ of 0.82, the RQ most commonly observed in subjects in a basal state. The volume of oxygen consumed during a certain time period multiplied by this value gives the heat output during the same period.

The percentages of oxygen used for the burning of fat and carbohydrate can be calculated from the nonprotein RQ by mathematical analysis. For example, if the nonprotein RQ is equal to 0.90, approximately 33 per cent of the oxygen is used for the oxidation of fat and 67 per cent for the oxidation of carbohydrate. Or, in other terms, 33 per cent of the total heat production is obtained from fat and 67 per cent from carbohydrate. The following mathematical relationship can be used for such determinations:

$$\begin{aligned}\text{Nonprotein RQ} &= x(\text{RQF}) + 1.00 - x(1.00)\\ &= x(0.70) + 1.00 - x(1.00)\end{aligned}$$

where $x(0.70)$ equals the percentage of fat oxidized and $1.00 - x(1.00)$ is equal to the carbohydrate oxidized. In our example of an RQ of 0.90, the following would apply:

$$0.90 = x(0.70) + 1.00 - x(1.00)$$

$$0.90 - 1.00 = 0.70x - 1x$$

$$0.10 = 0.3x$$

$$x = \frac{0.1}{0.3} = 0.333 = 33\% \text{ fat}$$

$$100 - 33.3 = 66.7\% \text{ carbohydrate.}$$

Determination of respiratory quotients, although not giving any information concerning the intermediate steps in the conversion of consumed oxygen to carbon dioxide has greatly advanced our understanding of such conditions of abnormal carbohydrate and fat metabolism as diabetes and of the physiology of food utilization.

Basal Metabolism

The total energy output by the human organism can be differentiated into two parts: (1) a portion necessary to maintain normal physiological functions, for example, heartbeat, circulation, respiration, muscle tone, and glandular secretion; and (2) a variable portion needed in the maintenance of body temperature, work, and the absorption of food materials. This second type of energy and the rate of oxygen consumption depend on the muscular activity, the nature and amount of food eaten, changes in environmental temperature, and emotional states. The first portion is relatively constant for persons of the same age and sex and is usually expressed as the amount of energy output per unit of surface area per hour. This value is called the **basal metabolic rate (BMR)** and is expressed as a plus or minus percentage of the standard normal value (taken as 100 per cent) of a subject of the same age and sex (Table 12-4). Variations of plus or minus 15 per cent are accepted within normal limits.

Certain conditions are necessary for the measurement of the BMR. These are as follows: (1) the subject should have had nothing by mouth for at least 12 hours before the test, (2) the subject should not be in a condition of emotional excitement but must be mentally and physically relaxed (usually a half hour of bed rest before testing is used to ensure these conditions), (3) the subject must be awake but in the recumbent position during the test, and (4) the environmental temperature should be between 20 and 25° C.

The following example represents a typical BMR determination: The subject is a man aged 35 years, weighing 70 kg (154 lb), and 170 cm in height. From Figure 12-9 it will be seen that his surface area is 2.0 m^2. Figure 12-8 shows his

Table 12-4. ***Standard Basal Metabolic Rate****

Age (yr)	*Average Calories per Hour per Square Meter of Body Surface*		Age (yr)	*Average Calories per Hour per Square Meter of Body Surface*	
	Males	*Females*		*Males*	*Females*
5	53.0	51.6	20–24	41.0	36.9
6	52.7	50.1	25–29	40.3	36.6
7	52.0	49.3	30–34	39.8	36.2
8	51.2	48.1	35–39	39.2	35.8
9	50.4	46.9	40–44	38.3	35.3
10	49.5	45.8	45–49	37.8	35.0
11	48.6	44.6	50–54	37.2	34.5
12	47.8	43.4	55–59	36.6	34.1
13	47.1	42.0	60–64	36.0	33.8
14	46.2	41.0	65–69	35.5	33.4
15	45.3	39.6	70–74	34.8	32.8
16	44.7	38.5	75–79	34.2	32.3
17	43.7	37.4			
18	42.9	37.3			
19	42.1	37.2			

*W. M. Boothby and J. Berkson, "Studies of the Energy of Metabolism of Normal Individuals." *Am. J. Physiol.*, **116**:468, 485 (1936); **121**:669 (1938).

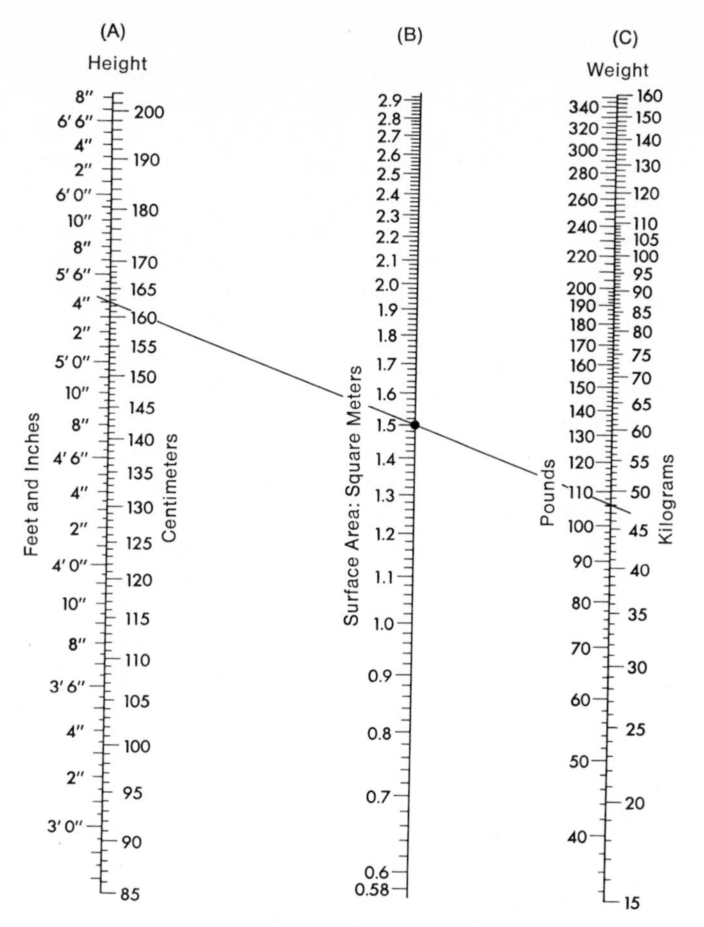

Figure 12-9. Nomogram for calculating surface area. A line joining the height on the left-hand scale (A) *to the weight scale* (C) *cuts scale* (B) *at the surface area predicted by DuBois's formula, given on page 434.*

record on the McKesson Metabolar. The straight line drawn along the lower edge of the tracing shows that during a measured period of 6 minutes the spirometer bell fell through a distance equivalent to 2,800 cc of oxygen. Since the oxygen-consumption line did not begin at zero but at a level representing 600 cc, the rise in the oxygen line in 6 minutes is 2,800 − 600 = 2,200 cc, or 2.2 liters in 6 minutes, or 22 liters (2.2 × 10) in 1 hour. This volume must be corrected to standard temperature and pressure (760 mm Hg, +273° C) by use of the following formula. The observed barometric pressure was 745 mm Hg, and the environmental temperature 21° C.

$$
\begin{aligned}
V_{\text{stp}} &= V \times \frac{P}{760} \times \frac{273}{273 + t} \\
&= 22 \times \frac{745}{760} \times \frac{273}{294} \\
&= 22 \times 0.980 \times 0.928 \\
&= 22 \times 0.909 \\
&= 19.9 \text{ liters of oxygen per hour.}
\end{aligned}
$$

By multiplying the corrected gas volume by 4.825, calories per hour is derived; $20 \times 4.825 = 96.50$ calories per hour. The surface area of the subject was 2 m^2; therefore, the calories per square meter per hour would be

$$\frac{96.50}{2} = 48.25 \text{ calories per square meter per hour.}$$

In determining the basal metabolic rate, the Roth–Benedict respiratory apparatus (Fig. 12-7) or McKesson Metabolar is usually employed. This apparatus has been previously described (p. 426). In clinical practice, the BMR can be estimated with accuracy by measuring the oxygen consumption for two 6-minute periods. These volumes are corrected to standard temperature and pressure. The average oxygen consumption, which is determined from the oxygen-consumption chart (Fig. 12-8), is multiplied by 10 to convert it to an hourly basis. Since the carbon dioxide is not measured, the respiratory quotient cannot be ascertained, but an RQ of 0.82 is usually assumed. For this RQ the caloric value of oxygen is 4.825. The hourly oxygen consumption is multiplied by 4.825, which represents the heat production in calories per hour. This is converted to calories per square meter of body surface per hour by dividing calories per hour by the subject's surface area. This is usually done because it is considered a more accurate expression of heat production, since basal metabolic rate is proportional to the area of the body surface. The surface area in square meters can be determined by using a formula proposed by DuBois,

$$S = W^{0.425} \times H^{0.725} \times 0.007184,$$

where S is the surface area in square meters, W the nude weight in kilograms, and H the height in centimeters. By application of this formula, charts have been made by which the body surface of a subject can be easily determined. A nomogram (Fig. 12-9) is an example of such a chart; it relates height and weight to surface area.

The normal BMR for the sex and age group of this subject, as shown in the standard table (Table 12-4), is 39.2 C per square meter per hour. The subject's BMR, which is above normal, is then reported as a plus percentage of this normal standard.

$$\text{BMR} = \frac{39.2 - 48.25}{39.2} \times 100 = 23\% \text{ or } +23.$$

A BMR between a plus or minus 15 is considered normal; our sample subject, then, would not be considered normal. In hyperthyroidism the BMR may reach plus 50 to plus 70, whereas in hypothyroidism BMR figures of minus 30 to minus 60 are not unusual.

After food is absorbed, the metabolic activities of the body are stimulated, with the result that the heat production increases. Thus, a person with a basal metabolism of 48.25 C per square meter per hour and a surface area of 2.0 m^2 would have a caloric output of 48.25×2; that is, 90.50 C per hour, or 90.5×24, which is 2,172 C per day. The metabolism of a quantity of food having a caloric content of 2,172 C would raise the subject's energy output approximately 10 per cent, or to 2389 C per day. This increase in metabolism due to absorption of foods is called the **specific dynamic action of food** and varies according to the type of food ingested. For proteins, it is about 30 per cent of the basal metabolism, and for fat and carbohydrate it is 4 or 5 per cent. On a mixed diet the specific dynamic action of foods increases the metabolism about 10 per cent and must be taken into account when determining the caloric requirement. Additional food must also be given to cover the expenditure of energy in the activities of everyday life, which will vary with the individual and his occupation. In Chapter 13 we will consider the caloric requirements of man in more detail.

References

Comar, C. L., and F. Browner (eds.). *Mineral Metabolism,* Vol. 2. New York: Academic, 1964.

Davidson, C. S. *Liver Pathophysiology.* Boston: Little, Brown, 1970.

Devel, H. S. *The Lipids.* New York: Wiley-Interscience, Vol. 1, 1951; Vol. 2, 1955; Vol. 3, 1957.

Fallis, B. D. *A Textbook of Human Histology.* Boston: Little, Brown, 1970.

Florkin, M., and E. H. Statz (eds.). *Lipid Metabolism.* New York: Am. Elsevier, 1970.

Greenberg, D. M. (ed.). *Metabolic Pathways: Lipids, Steroids, and Carotenoids.* New York: Academic, 1969.

———. *Metabolic Pathways: Nucleic Acids, Protein Synthesis and Coenzymes.* New York: Academic, 1970.

Hess, G. P., and J. A. Rupley. "Structure and Function of Proteins," in: *Annual Review of Biochemistry,* Vol. 40. Palo Alto, Calif.: Annual Reviews, 1971, p. 1013.

Jeanrenaud, B., and D. Hepp (eds.). *Adipose Tissue,* New York: Academic, 1970.

Mikal, S., *Homeostasis in Man: Fluids, Electrolytes, Proteins, Vitamins, and Minerals.* Boston: Little, Brown, 1969.

Nair, P. P., and D. Kritchevsky. *The Bile Acids: Chemistry, Physiology, and Metabolism.* New York: Plenum, 1971.

Plummer, D., *An Introduction to Practical Biochemistry,* New York: McGraw-Hill, 1972.

Stumpf, R. K., "Metabolism of Fatty Acids," in: *Annual Review of Biochemistry,* Vol. 38. Palo Alto, Calif.: Annual Reviews, 1969, p. 159.

Tipson, R. S., and D. Horton (eds.). *Carbohydrate Chemistry and Biochemistry,* New York: Academic, 1970.

13

Nutrition

Very closely allied to metabolism is the science of nutrition. To many, nutrition is the same as food consumption, but eating is really only the first step, since digestion, absorption, and distribution of organic nutrients to individual cells must occur before these food elements become useful to the organism. The science of nutrition more specifically involves the various chemical and physiological activities that transform organic nutrients into body elements. It is also concerned with the relation of these organic nutrients, which come from our food, to the physical well-being of the organism. Studies have shown that the chemical composition of the animal body is vitally influenced by the composition of its food. Therefore, it would be useful to consider first the composition of the animal body, with special reference to man.

Composition of the Body

The gross composition of the mature body, not including the fat content, is approximately 75 per cent water, 20 per cent protein, and 5 per cent mineral. There is such variation in the fat content among individuals and in the same person at different times that the body constituents are usually expressed on a fat-free basis. If fat is taken into consideration, a mature man in a good state of nutrition would have the following approximate composition: water, 59 per cent; protein, 18 per cent; fat, 18 per cent; and mineral, 4.3 per cent. Carbohydrate makes up less than 1 per cent of the body composition at any given moment.

These substances that make up the gross composition of the body are not equally distributed throughout the various organs and tissues but are found more concentrated in certain tissues than in others, depending on function. For example, water is found in every part of the body but is present in the amount of 90 to 92 per cent in blood plasma, whereas muscle contains approximately 72 to 78 per cent, and bone approximately 45 per cent. The enamel of the teeth contains only 5 per cent water. Proteins are present in every cell of the body, but they are present in greater amounts in such structures as muscles, tendons, and connective tissues. Fat is present in practically all tissues of the body but is most abundant as adipose tissue, occurring under the skin and around the intestines, lungs, and heart. The small amount of carbohydrate in the body is found principally in the liver, muscles, and blood. The inorganic salts in the body are also found in varying amounts in different parts. Calcium, which is present in the largest amount, is concentrated mainly in the bones and teeth. Phosphorus, in combination with calcium, is mainly concentrated in the skeletal structures; small amounts are present in fats, proteins, carbohydrates, and nucleic acids. Sodium and chloride predominate in extracellular fluids, and sulfur occurs throughout all tissues of the body as part of the protein molecule. Potassium is the major mineral in muscle. Magnesium is widely distributed throughout the body, and iron, an essential constituent of the hemoglobin of red cells, occurs in lesser amounts throughout the other tissues. Many other elements, such as iodine, copper, zinc, manganese, and cobalt, are present in the tissues of the body and are known to be necessary for life. Some minerals, such as boron, silicon, bromine, aluminum, and nickel, are known to be present normally as part of the chemical composition of the body, although their specific function, if any, is not known.

Foods

Man is an omnivorous animal subsisting on a much more varied diet than most other animals. Man requires not only (1) a certain quantity of food, but also (2) a considerable variety of special food ingredients for the maintenance of his cellular constituents.

Man's food consists of the following classes of materials: carbohydrates, fats, proteins, inorganic salts, vitamins, and water. These materials subserve three main functions: (1) production of energy, (2) replacement and addition of new protoplasm, and (3) maintenance of the normal environment of protoplasmic material and of normal metabolic activity.

Carbohydrates

The food **carbohydrates** consist of sugars and starches, of which there are many different kinds. Sugars and starches are delivered to the liver and tissues in the form of simple monosaccharides, mainly glucose, fructose, and glactose. Glucose is the most common of all and is the only one used as a source of energy by the tissues. Fructose and galactose are converted by the liver to glucose. In addition, sugars are used to some degree in the structural make-up of cells and also may be stored as a reserve energy supply in the form of animal starch referred to as **glycogen.** Glycogen is chiefly stored in the liver and in muscle tissue.

The carbohydrate requirement varies with the individual, but in general it is usually recommended that no more than 66 per cent of the total caloric intake be obtained from this source. The average adult energy requirement, according to the Food and Nutrition Board of the National Research Council,[1] for a mature man between the ages of 18 and 35, weighing 154 lb, 5 ft 9 inches tall, and engaged

[1] *Recommended Dietary Allowances,* A Report of the Food and Nutrition Board (Publication 1146). Washington, D.C.: National Academy of Sciences, 1964.

Table 13-1. ***Recommended Daily Dietary Allowances,* Revised 1963***

FOOD AND NUTRITION BOARD, NATIONAL ACADEMY OF SCIENCES–NATIONAL RESEARCH COUNCIL DESIGNED FOR THE MAINTENANCE OF GOOD NUTRITION OF PRACTICALLY ALL HEALTHY PERSONS IN THE UNITED STATES (ALLOWANCES ARE INTENDED FOR PERSONS NORMALLY ACTIVE IN A TEMPERATE CLIMATE)

	Age† years from to	*Weight kg (lb)*	*Height cm (in.)*	*Calories‡*	*Protein g*	*Calcium g*	*Iron mg*	*Vitamin A Value IU*	*Thiamine mg*	*Riboflavin mg*	*Niacin Equiv.§ mg*	*Ascorbic Acid mg*	*Vitamin D IU*
Men	18–35	70 (154)	175 (69)	2,900	70	0.8	10	5,000¶	1.2	1.7	19	70	
	35–55	70 (157)	175 (69)	2,600	70	0.8	10	5,000	1.0	1.6	17	70	
	55–75	70 (154)	175 (69)	2,200	70	0.8	10	5,000	0.9	1.3	15	70	
Women	18–35	58 (128)	163 (64)	2,100	58	0.8	15	5,000	0.8	1.3	14	70	
	35–55	58 (128)	163 (64)	1,900	58	0.8	15	5,000	0.8	1.2	13	70	
	55–75	58 (128)	163 (64)	1,600	58	0.8	10	5,000	0.8	1.2	13	70	
	Pregnant (2nd and 3rd trimester)			+ 200	+20	+0.5	+5	+1,000	+0.2	+0.3	+3	+30	400
	Lactating			+1,000	+40	+0.5	+5	+3,000	+0.4	+0.6	+7	+30	400
Infants**	0– 1	8 (18)		kg × 115 ±15	kg × 2.5 ±0.5	0.7	kg × 1.0	1,500	0.4	0.6	6	30	400
Children	1– 3	13 (29)	87 (34)	1,300	32	0.8	8	2,000	0.5	0.8	9	40	400
	3– 6	18 (40)	107 (42)	1,600	40	0.8	10	2,500	0.6	1.0	11	50	400
	6– 9	24 (53)	124 (49)	2,100	52	0.8	12	3,500	0.8	1.3	14	60	400
Boys	9–12	33 (72)	140 (55)	2,400	60	1.1	15	4,500	1.0	1.4	16	70	400
	12–15	45 (98)	156 (61)	3,000	75	1.4	15	5,000	1.2	1.8	20	80	400
	15–18	61 (134)	172 (68)	3,400	85	1.4	15	5,000	1.4	2.0	22	80	400
Girls	9–12	33 (72)	140 (55)	2,200	55	1.1	15	4,500	0.9	1.3	15	80	400
	12–15	47 (103)	158 (62)	2,500	62	1.3	15	5,000	1.0	1.5	17	80	400
	15–18	52 (117)	163 (64)	2,300	58	1.3	15	5,000	0.9	1.3	15	70	400

*The allowance levels are intended to cover individual variations among most normal persons as they live in the United States under usual environmental stresses. The recommended allowances can be attained with a variety of common foods, providing other nutrients for which human requirements have been less well defined. See text for more detailed discussion of allowances and of nutrients not tabulated.

†Entries on lines for age range 18–35 years represent the 25-year age. All other entries represent allowances for the midpoint of the specified age periods, i.e., line for children 1–3 is for age 2 years (24 months); 3–6 is for age $4\frac{1}{2}$ years (54 months); etc.

‡Tables 13-2 and 13-3 show calorie adjustments for weight and age.

§Niacin equivalents include dietary sources of the preformed vitamin and the precursor, tryptophan. 60 mg of tryptophan represents 1 mg of niacin.

**The calorie and protein allowances per kilogram for infants are considered to decrease progressively from birth. Allowances for calcium, thiamine, riboflavin, and niacin increase proportionately with calories to the maximum values shown.

¶1,000 IU from preformed vitamin A and 4,000 IU from beta-carotene.

in moderate activity, is approximately 2,900 calories, and (2) for the average woman between the ages of 18 and 35, weighing 128 lb and 5 ft 4 inches tall, is 2,100 calories.

Energy released by the various foodstuffs is measured in units of calories. A **calorie** may be defined as the amount of heat required to raise 1,000 g of water (1 liter) 1° C, from 15 to 16° C. This is known as a *large calorie,* or *kilocalorie.* A large calorie, usually capitalized (Calorie), is 1,000 small calories or 1 kilocalorie. In nutrition the C of large calorie is often not capitalized. Since all three foodstuffs (protein, fat, and carbohydrate) furnish calories, it is obvious that the amounts of each of these nutrients in a given food determine the caloric value of that food. The body is capable of forming carbohydrate from fats and protein; thus, it would appear unnecessary to have carbohydrate in the diet. However, experimental evidence indicates that the organism must have some carbohydrate in its diet if it is to remain in a healthy state.

From our studies of metabolism we know that glucose is oxidized into carbon dioxide and water, with energy being released in the process. This reaction may be empirically written as

$$\underset{\text{Glucose}}{C_6H_{12}O_6} + 6O_2 \longrightarrow 6CO_2 + 6H_2O + \text{energy}.$$

It must be remembered, however, that glucose is oxidized by way of the Krebs cycle (see p. 47) and involves a series of biochemical reactions.

One gram of carbohydrate when oxidized is known to yield 4.1 large calories. Thus, if he could subsist on pure carbohydrate, the average man would have to consume 707 g (1.55 lb) of carbohydrate per day in order to supply the necessary energy (2,900 cal) to maintain his normal physiological processes and daily activity; the average woman, 512 g (1.13 lb). However, many nutrients are now recognized as essential for the maintenance of general health and vigor in human nutrition. Table 13-1 lists the daily dietary requirements and allowances recommended by the Food and Nutrition Board, National Research Council.

The total number of calories or the amount of energy needed to support the energy expenditure for the internal work of the body, such as respiration, growth, circulation, and muscle tone, and to carry on daily activities is dependent on the amount of activity in which a given person is engaged. If more calories are consumed from our food than are needed to balance the expenditure, the extra calories will be converted into fat to be stored in the fat depots of the body. As a result the body weight will be correspondingly increased. If fewer calories are consumed than are needed to meet the energy requirements, the extra calories will be provided from the fat stores, and the weight of the individual will be decreased. When there is a balance between the intake of food energy and the energy expenditure, there is no increase or decrease in body weight, and the individual is said to be in caloric balance. Thus, a desirable weight can be maintained by making proper adjustments in food intake. Tables 13-2 and 13-3 list desirable weights for men and women as prepared by the Metropolitan Life Insurance Company.[2]

Individuals vary considerably from the standard average individual on whom the tables of the National Research Council are based, and thus a method for estimating personal requirements for calories would be practical.

Determining Caloric Requirements for Individuals

The energy requirements of an individual may be looked upon as consisting of two fractions: (1) that fraction which is required just to maintain the normal physiological functions for the living processes, that is, respiration, circulation, and muscle tone; and (2) the extra energy required to carry out daily activities.

The first of these fractions of energy can be measured by determining the energy exchange of a fasting, resting subject. This energy requirement is sometimes referred to as the **basal metabolic rate (BMR).** The determination is usually made in the morning after a good night's sleep, no food having been eaten after the dinner meal the evening before, and after the subject has been physically and mentally at rest for at least $\frac{1}{2}$ hour. The oxygen consumption is measured for a

[2] *How to Control Your Weight.* (1958: Metropolitan Life Insurance Company).

Table 13-2. ***Desirable Weights for Men of Ages 25 and Over***

Height (with Shoes on, 1-inch Heels)		*Weight (in Pounds According to Frame, in Indoor Clothing)*		
Feet	*Inches*	*Small Frame*	*Medium Frame*	*Large Frame*
5	2	112–120	118–129	126–141
5	3	115–123	121–133	129–144
5	4	118–126	124–136	132–148
5	5	121–129	127–139	135–152
5	6	124–133	130–143	138–156
5	7	128–137	134–147	142–161
5	8	132–141	138–152	147–166
5	9	136–145	142–156	151–170
5	10	140–150	146–160	144–174
5	11	144–154	150–165	159–179
6	0	148–158	154–170	180–184
6	1	152–162	158–175	168–189
6	2	156–167	162–180	173–194
6	3	160–171	167–185	178–199
6	4	164–175	172–190	182–204

Courtesy the Metropolitan Life Insurance Company.

Table 13-3. ***Desirable Weights for Women of Ages 25 and Over***

(FOR GIRLS BETWEEN 18 AND 25, SUBTRACT 1 LB FOR EACH YEAR UNDER 25)

Height (with Shoes on, 1-inch Heels)		*Weight (in Pounds According to Frame, in Indoor Clothing)*		
Feet	*Inches*	*Small Frame*	*Medium Frame*	*Large Frame*
4	10	92–98	96–107	104–119
4	11	94–101	98–110	106–122
5	0	96–104	101–113	109–125
5	1	99–107	104–116	112–128
5	2	102–110	107–119	115–131
5	3	105–113	110–122	118–134
5	4	108–116	113–126	121–138
5	5	111–119	116–130	125–142
5	6	114–123	120–135	129–146
5	7	118–127	124–139	133–150
5	8	122–131	128–143	137–154
5	9	126–135	132–147	141–158
5	10	130–140	136–151	145–163
5	11	134–144	140–155	149–168
6	0	138–148	144–159	153–173

Courtesy the Metropolitan Life Insurance Company.

6-minute period and corrected to standard conditions of temperature and pressure. The oxygen consumption is multiplied by 10 to convert it to an hourly basis and then multiplied by 4.825 cal, the heat production represented by each liter of oxygen consumed. This figure, 4.825, was determined by DuBois[3,4] after a large series of measurements and is generally used in determining basal metabolism from the oxygen consumption alone; when this figure is multiplied by the oxygen consumed per hour, the energy production in calories per hour is derived. This is usually converted to calories per square meter of body surface per hour by dividing calories per hour by the subject's surface area. A more thorough discussion of basal metabolism can be found in Chapter 12. An example of this type of calculation is as follows. A man, 35 years of age, 170 cm in height and 70 kg in weight, consumed 1.2 liters of oxygen in 6 minutes (corrected to STP 0° C and 7,600 mm Hg):

$$1.2 \times 10 = 12 \text{ liters of } O_2 \text{ per hour}$$

$$12 \times 4.825 = 58 \text{ cal per hour}$$

Surface area $= 1.81 \text{ m}^2$ (from DuBois's formula)[5]

$$\text{BMR} = 58/1.81 = 32 \text{ cal per square meter per hour.}$$

The normal BMR for a person of the age and sex of the subject is obtained from standard tables, which were established by Harris and Benedict[6] in a large series of determinations made on normal persons. The actual rate is then compared with the average normal, and the rate is expressed as a plus or minus percentage of the normal. Plus or minus 15 per cent from the standard average normal is usually taken as the normal limit.

However, if we assume that the sample subject is normal, the energy requirement for this person just to maintain life processes would be

$$58 \times 24 = 1396 \text{ cal per day}$$

or

$$32 \times 24 = 768 \text{ cal per square meter per day.}$$

The second fraction of our daily requirement of energy is for the activities an individual pursues during his daily existence (work, play, etc.). Since individuals vary in their activity and each activity varies more or less in energy expenditure, determinations have been made of the caloric requirements for various types of activities. Table 13-4 lists some common activities and their actual energy requirement per pound of body weight.[7] By keeping a detailed record for a 24-hour period of the various activities a subject pursues, we may determine the total

[3] E. F. DuBois, *Basal Metabolism in Health and Disease.* (Philadelphia: Lea, 1936), p. 494.

[4] D. DuBois and E. F. DuBois, "Clinical Calorimetry." *Arch. Internal. Med.,* **17**:863 (1916).

[5] Surface area of human beings is calculated from a formula developed by DuBois (*op cit.*) $A = W^{0.425} \times H^{0.725} \times 0.007184$, where A = surface area in square meters, W = weight in kilograms, and H = height in centimeters.

[6] J. A. Harris and F. G. Benedict, *A Barometric Study of Basal Metabolism in Man* (Publication 279). (Washington, D.C.: Carnegie Institution of Washington, 1919.)

[7] W. S. Spector, *Handbook of Biological Data.* (Philadelphia: Saunders, 1956).

Table 13-4. ***Energy Cost of Various Activities****

Activity	Cal/min
Men	
1. Supine, basal	1.17
2. Supine, basal	1.19
3. Lying, at ease	1.50
4. Sitting, at ease	1.80
5. Sitting, calculating	1.78
6. Sitting, reading	1.98
7. Sitting, writing	1.91
8. Standing, at ease	1.98
9. Standing, relaxed	1.25
10. Dressing	4.00
11. Washing hands, face, neck; brushing hair	2.74
12. Brushing clothes	2.57
13. Cleaning shoes	3.49
14. Dressing, washing, shaving	3.56
15. Walking (indoors), 2.4 mph	4.3
16. Walking (indoors), 3.0 mph	5.1
17. Walking (outdoors), 4.0 mph	8.2
18. Walking (outdoors), 4.2 mph	9.1
19. Walking (outdoors), 4.4 mph	9.5
20. Walking (outdoors), 4.6 mph	9.9
21. Walking (outdoors), 4.8 mph	10.7
22. Walking up and down stairs, 97/min	8.4
23. Walking up and down stairs, 116/min	9.3
24. Climbing 15-cm stairs, 14.8 m/min	9.8
25. Climbing 15-cm stairs, 17.6 m/min	10.3
26. Climbing ladder, 17-cm step, 50° angle, 9.1 m/min	7.7
27. with 50 lb	14.3
28. Climbing ladder, 17-cm step, 90° angle, 11.9 m/min	11.5
29. with 50 lb	25.4
30. Walking, hard snow, 6 km/hr	11.9
31. Walking, loose snow, 20-kg load, 4 km/hr	20.2
32. Walking, snowshoes, soft snow, 4 km/hr	13.8
33. Skiing, level hard snow, 6 km/hr	9.9
34. Driving car	2.8
35. Driving motorcycle	3.4
36. Bicycling, 5.5 mph	4.5
37. Bicycling, 9.4 mph	7.0
38. Bicycling, 13.1 mph	11.1
39. Rowing, 33 strokes/min	19.0
40. Rowing, 22 strokes/min	12.3
41. Peeling potatoes	2.7
42. Laboratory work	3.2

Table 13-4. ***Energy Cost of Various Activities****

Activity	Cal/min
43. Washing dishes	3.3
44. Making beds	7.0
45. Cleaning windows	3.7
46. Miscellaneous office work, sitting	1.6
47. Miscellaneous office work, standing	1.8
48. Shoemaking, shoe repair	2.7
49. Shoemaking, shoe manufacturing	3.0
50. Locksmith, working	2.5
51. Tailor, cutting	2.6
52. Tailor, pressing	3.9
53. Armature-winding	2.2
54. Radio mechanics	2.7
55. Printing, hand compositor	2.2
56. Printing, printer	2.2
57. Watch and clock repairer trainee	1.6
58. Light assembly line	1.8
59. Draftsman	1.8
60. Light machine work (engineering)	2.4
61. Typewriter mechanic trainee	2.1
62. Medium assembly work	2.7
63. Sheet-metal worker	3.0
64. Machinists (engineering)	3.1
65. Plastic-molding	3.3
66. Joiners	3.6
67. Turners	3.7
68. Toolroom workers	3.9
69. Machine-fitting	4.2
70. Casting lead balls in mold	4.8
71. Loading chemicals into mixer	6.0
72. Unloading battery boxes from oven	6.8
73. Shoveling, 8-kg load, 1-m lift, 12/min	7.5
74. Hewing with pick	7.0
75. Pushing wheelbarrow, 57-kg load, 4.5 km/hr	5.0
76. Bricklaying	4.0
77. Mixing cement	4.7
78. Stonemasonry, shaping stones	3.8
79. Plaster-lathing	3.1
80. Plastering walls	4.1
81. Carpentry, measuring wood	2.4

Table 13-4. *Energy Cost of Various Activities**

Activity	Cal/min
Women	
82. Supine, basal	.98
83. Sitting	1.09
84. Standing	1.11
85. Dressing and undressing	2.30
86. Washing, dressing, undressing	3.30
87. Walking, 2.8 mph	2.00
Walking 2.8 mph with 20-lb load:	
88. Load carried with shoulder yoke	1.94
89. Load carried on one shoulder	2.07
90. Load carried in two bundles in either hand	2.19
91. Load carried on tray in front of body	2.25
92. Load carried on tray in front of body with strap around shoulder	2.28
93. Load carried on head	2.49
94. Load carried on hip	2.72
95. Horizontal walking, 1.1 mph	1.99
96. Horizontal walking, 2.2 mph	2.84
97. Horizontal walking, 3.4 mph	2.90
98. Skiing, level hard snow, moderate speed	10.80
99. Skiing, uphill hard snow, maximum speed	18.60
100. Washing dishes, top of pan 42 inches from floor (height of subject, 66 inches)	1.37
101. Washing dishes, top of pan 32 inches from floor (height of subject, 66 inches)	1.53
102. Paring potatoes (sitting)	1.23
103. Paring potatoes (standing)	1.29
104. Beating batter (standing)	1.43
105. Kneading dough (standing)	2.04
106. Step up 7 inches	2.27
107. Arm reach and trunk bend to 3 inches above floor (average height of subject, 62 inches)	2.88
108. Arm reach and knee bend to 3 inches above floor (average height of subject, 62 inches)	4.12
109. Arm reach and body pivot 36 inches above floor (average height of subject, 62 inches)	1.75
110. Arm reach 72 inches above floor (average height of subject, 62 inches)	1.82
111. Hand-sewing sheets, 18 stitches/min	1.18
112. Hand-sewing sheets, 30 stitches/min	1.25
113. Darning	1.26
114. Crocheting, 32 stitches/min	1.27
115. Knitting, 23 stitches/min	1.29
116. Machine-sewing (foot-operated)	1.43
117. Washing clothes by hand	2.69
118. Rinsing clothes	2.42

Table 13-4. ***Energy Cost of Various Activities****

Activity	Cal/min
119. Drying clothes (in extractor)	2.08
120. Wringing clothes, by hand	2.21
121. Putting up and removing clothesline	2.14
122. Hanging up clothes from basket on floor	2.63
123. Ironing, standing (high board)	1.64
124. Ironing, standing (normal board)	1.69
125. Washing floor on knees	1.63
126. Sweeping floor	1.85
127. Vacuum-cleaning rug (moving 1 ft/sec)	1.63
128. Vacuum-cleaning rug (moving 3 ft/sec)	2.58
129. Bed-making, stripping	5.40
130. Typing, electric, 40 wpm	1.31
131. Typing, mechanical, 40 wpm	1.48
132. Typing, 59 wpm	1.34
133. Typing, 115 wpm	1.72
134. Leather-tooling (reclining)	1.13
135. Leather-tooling (sitting)	1.28
136. Leather-stamping (sitting)	1.33

*Modified from table in *Handbook of Biological Data,* edited by Spector (Philadelphia: Saunders, 1956). Courtesy National Academy of Sciences.

energy expenditure for those activities from the table. Adding this energy expenditure to the energy required for basal metabolism gives the total calories required per day. Thus, if our standard subject, who is moderately active, requires 3,200 cal per day and has a basal metabolic expenditure of, say, 1,396 cal, his daily activities require 1,804 cal:

1,396 cal—basal expenditure
1,804 cal—daily activities
3,200 total caloric expenditure.

Taylor[8] suggests the following easy way of quickly estimating personal requirements for calories each day. Based on the number of calories recommended by the National Research Council for men of different ages engaged in moderate activity and weighing 154 lb (3,200 cal) and for women weighing 128 lb (2,800 cal), the caloric requirement on a per pound basis would be as follows:

$$\text{Men: Moderate activity } \frac{3{,}200}{154} = 21 \text{ cal per pound}$$

$$\text{Heavy work} = 25 \text{ cal per pound}$$

$$\text{Sedentary activity} = 16 \text{ cal per pound.}$$

[8]Clara M. Taylor, *Food Values in Shares and Weights.* (New York: Macmillan, Inc., 1959).

Women: Moderate activity $\frac{2,300}{128} = 18$ cal per pound

Heavy work = 22 cal per pound

Sedentary activity = 14 cal per pound.

By multiplying the desirable weight (see Table 13-2 or 13-3) by the caloric requirement per pound listed above, the total caloric requirement can be established. For example, a man who is 6 ft tall with a medium frame should weigh 161 to 173 lb. By multiplying this desirable weight by the number of calories required per pound, which varies with the type of activity the individual is pursuing, the total caloric requirement would be

$$161 \times 21 = 3,381 \text{ cal per day}$$

or

$$173 \times 21 = 3,533 \text{ cal per day.}$$

It has been suggested that if much weight is to be lost, the caloric intake should be set arbitrarily at 1,200 to 1,400 cal per day.

Nutritionists maintain that it is almost impossible to obtain all the minerals and vitamins needed each day for good nutrition if the food intake is less than 1,200 cal. A good reducing diet is one that contains all the essential nutrients, such as protein, inorganic salts, and vitamins, but is low in calories. An increase in weight may be effected by increasing the caloric requirement above that which is just necessary to maintain body weight.

Calories are essential for the maintenance of general health and vigor in all animals. We have seen that carbohydrates are our main source of calories, but fats are also a good source. In addition to the essential need of calories by the organism, nine other nutrients are considered essential and are specifically recommended by the Food and Nutrition Board of the National Research Council. These include protein, calcium, iron, vitamin A, thiamine, ascorbic acid, riboflavin, niacin, and vitamin D. Foods that usually furnish these materials will also provide the other nutrients known to be essential.

Fats

Fats and fatlike materials are important constituents of most foods, such as meat, milk, cheese, nuts, and fish, and evidently supply some of the energy during normal activity. Such food fats as butter, margarine, and egg yolk are concentrated energy foods and a good source of vitamin A. In general, fat and carbohydrate appear to be interchangeable as a source of energy or heat for the body, but some cells do discriminate between the two. Cells of the central nervous system require carbohydrate for their energy source, and in diets lacking carbohydrate, the liver begins to carry on a greater gluconeogenesis, using protein material as a basic source of raw materials. This can be detected by the increase of nitrogen being excreted in the urine as a result of nitrogen release when proteins are utilized. If amino acids, which are capable of being converted to glucose, are supplied to a person whose diet is lacking in carbohydrate, excess protein loss from the body as a result of gluconeogenesis is eliminated. For every gram of fat oxidized in the body, 9.3 large calories of energy are released. It is obvious that fat releases more than twice as much energy per unit weight as carbohydrate does. In addition to being utilized as an energy source, fat is also considered to be mainly an important reserve store of energy for the body. Cholesterol and phospholipids are important structural components of the cell and therefore of great importance in the normal functioning of the cell. The cell membrane contains large concentrations of these lipids, which aid in the transport of certain material through the cell membrane. Small quantities of fat appear to be essential in the diet to permit the absorption of the fat-soluble vitamins from the small intestine. Factors that interfere with the digestion or absorption of fat from the small intestine will also interfere with the absorption of these vitamins, for example, vitamins A, D, E, and K.

Little information is available concerning the human requirement of food fat. We have already seen that fats can be produced from proteins and carbohydrates. However, certain specific requirements of fatty acids have been demonstrated in laboratory animals such as the rat. Burr[9] showed that rats maintained on fat-free diets developed dry, scaly skin, irregularities in reproduction, and failure of growth. The addition of fat-soluble vitamins to their diets could not cure this deficiency; so it was due to the lack of certain fatty acids. Further experimentation indicated that if one of three unsaturated fatty acids—linoleic, linolenic, or arachidonic—was supplied to such rats, the deficiency could be either prevented or cured. It was demonstrated that the animal had a specific need for these unsaturated fatty acids. For this reason they are referred to as the **essential fatty acids.** The specific use to which these unsaturated acids are put is not known, although they are important constituents of the phospholipid molecules, which are especially abundant in nervous tissue. The human need for these same fatty acids is not known. However, they are so widely distributed in plant as well as animal fats that a deficiency would not be likely to appear even if the diet were extremely low in fat.

Proteins

Food **proteins** must be broken down into their constituent amino acids before they can be utilized by the cells. There are 20 to 25 amino acids found in the body, which can be combined in different ways to produce different proteins. This synthesis of protein by the body from amino acids is one of the most important functions food protein serves, since growth and replacement of worn and injured tissue are not possible unless protein or amino acid nitrogen is available to the animal. Carbohydrate and fat are unable to replace protein in this growth function, although protein can be converted by the body to carbohydrate or to fat. Since protein can also supply all the energy needed by the body, by virtue of its having the same energy content as carbohydrate, one would conclude that protein by itself would be sufficient to maintain the normal functioning of the body. However, the body also has requirements for such substances as vitamins and inorganic salts, in order to maintain its cells in a state of good health and function. In addition, experimentation has shown that a high-protein diet in laboratory animals leads to disturbed kidney function and early death.

[9] G. O. Burr, "Significance of the Essential Fatty Acids." *Federation Proc.,* **1**:224 (1942).

In addition to the calories protein furnishes, its necessity for growth and replacement of worn tissue, and its ability to be converted to fat and carbohydrate, protein functions in other ways in the maintenance of good health. It is essential for normal reproduction and lactation in the adult and in the construction of hormones, enzymes, antibodies, and other secretions produced by the organism.

Some of the amino acids needed by the cell for its normal metabolic activity can be synthesized by the cell itself. Others, which are absolutely necessary for growth and replacement of worn tissue and which are necessary for the synthesis of other amino acids, cannot be synthesized by the body. The amino acids that are absolutely necessary for the proper maintenance of protoplasm and cell function and that cannot be synthesized by the body are called the **essential amino acids.** Rose[10] was one of the first research workers to show that there are ten amino acids that must be present in the diet if normal growth is to occur in the rat. He found these amino acids to be tryptophan, isoleucine, methionine, threonine, histidine, arginine, lysine, leucine, valine, and phenylalanine. Human experiments have shown that man's requirements are very similar, with the exception that histidine and arginine are not essential. The essential amino acids for human beings with their formulas are listed in the order of their complexity:

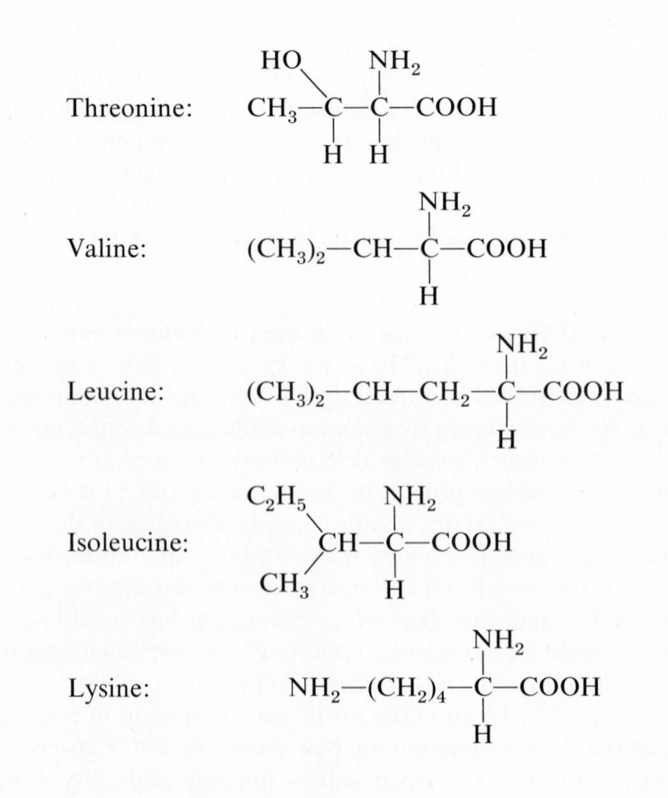

[10]W. C. Rose, "The Nutritive Significance of the Amino Acids." *Physiol. Rev.*, **18:**109 (1938).

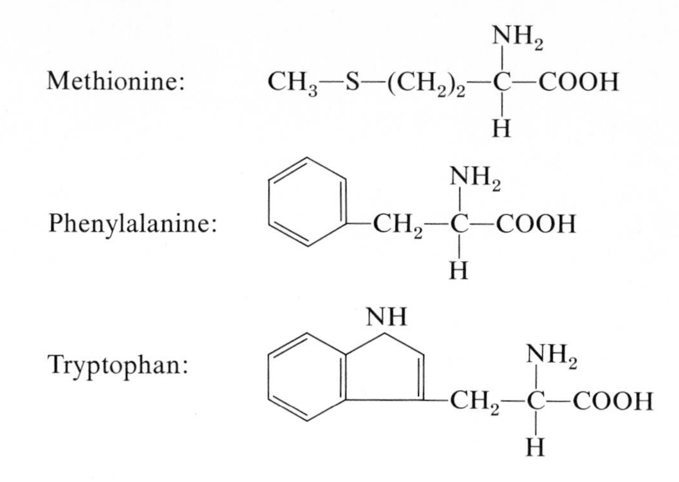

As we have already seen, if growth is to occur, the diet must contain an appropriate amount of protein. Even if growth is not taking place, as in the adult, small amounts of protein must be supplied to maintain normal cell function and to replace the protein nitrogen lost from the normal wear and tear of the tissue which is excreted by the body mainly in the urine. The Food and Nutrition Board of the National Research Council recommends 1 g of protein per kilogram (2.3 lb) of body weight per day for men. This amount provides sufficient dietary protein to replace that which is lost through normal wear and tear and also some extra to provide for an unusual variation in protein utilization and for emergency situations.

Some food proteins are better sources for the essential amino acids than others. On the basis of their essential amino acid content and their ability to make possible optimum growth, proteins are qualitatively classified as to their **nutritive,** or **biological, value.** If all the essential amino acids are present and in good proportions, a protein is concerned to be of high biological value. Food proteins considered to be of excellent quality are found in milk, cheese, meat, fish, eggs, and poultry. Vegetable proteins, which contain smaller amounts of the essential amino acids, are not considered to be as high a quality as are proteins of animal origin. In addition, animal proteins are believed to occur associated with important vitamins, such as B_{12}, which contributes to their being of better quality than those of vegetable origin.

Essential Minerals

We have already seen that the body contains a large number of mineral elements, which occur both in combination with each other and in combination with organic constituents. Many of these mineral elements, such as calcium, phosphorus, sodium, potassium, chlorine, magnesium, iron, sulfur, iodine, manganese, cobalt, copper, and zinc, have been proved to perform essential functions in the body and thus must be present in food. The fact that the total quantity of these elements increases as growth occurs indicates that they must become a part of the organic structure of the human body and therefore must be essential for normal function and growth. Other elements, such as aluminum, arsenic, silicon, and nickel, also occur consistently in the human body, but there is no laboratory proof of its need for them. Of all the minerals, those most likely to be deficient in the human diet are calcium, phosphorus, iron, and iodine.

Calcium. **Calcium** and phosphorus usually occur in the body combined with each other and are closely associated with each other in metabolism. An inadequate supply of either in the diet limits the nutritive value of both. Ninety-nine per cent of body calcium is localized in the bones and teeth, and only 1 per cent is found in the muscle and other soft tissues and in the blood. Calcium is present in bone as one of the apatite group of minerals, composed of calcium phosphate and calcium carbonate arranged in a definite proportion and crystal lattice. It has the formula $CaCO_3 \cdot nCa_3(PO_4)_2$, in which n is not less than 2 or more than 3. The ratio of calcium to phosphorus in bone remains remarkably constant at about 2:1.5. Not only is bone important as a structural framework for the body, but it also serves as a calcium reservoir. The calcium in bone is in dynamic equilibrium with the constituents of the body fluids and other tissues, and a constant exchange takes place. When the calcium requirement of an individual increases, as during pregnancy and lactation, the calcium is readily mobilized, usually from the trabecular portion of bone.

The concentration of calcium in the blood ranges from 9 to 11 mg per 100 ml. The blood calcium is entirely in the plasma with very little, if any, in the erythrocytes, and it occurs mainly in the soluble ionized form with some bound with protein. Calcium functions in the body not only as a structural component of the skeleton and teeth, but also in other major roles. It plays a vital role in the preservation of the normal neuromuscular irritability since its presence is necessary for normal functioning of nerve tissue. A decrease in the calcium level of the blood causes an irritability of nerve tissue, resulting in muscle twitching, cramps, and finally convulsions, a series of symptoms known as **tetany.** Tetany is usually a result of abnormal calcium metabolism and occurs in hypofunction of the parathyroid, vitamin D deficiency, and alkalosis and as a result of the ingestion of large concentrations of alkaline salts. The calcium blood level, since it is regulated by the parathyroid hormone, will be lowered if this gland is not secreting enough hormone. Vitamin D is necessary for the normal absorption of calcium from the intestinal tract, so a decrease or lack of vitamin D will interfere with its absorption. Alkaline materials decrease the solubility of calcium salts and presumably interfere with their absorption, leading to decreased blood calcium levels. Conversely, an acid reaction in the gut increases the solubility of calcium salts and aids in their absorption. Concentrations of calcium above the normal range depress nerve irritability, resulting in drowsiness and extreme lethargy. **Hypercalcemia** may be found after overdosage with irradiated ergosterol and in hyperparathyroidism. In addition to the above functions, calcium is also important in the clotting of blood, in the coagulation of milk in the stomach, and in the maintenance of the normal heart rhythm.

The Food and Nutrition Board recommends a daily intake of 800 mg of calcium to maintain normal levels in the adult and from 1.2 to 1.4 g for growing children.

Milk, cheese, shellfish, egg yolk, and green vegetables such as broccoli and kale are good food sources of calcium.

Phosphorus. The **phosphorus** content of the body varies from about 14 g at birth to approximately 670 g in the adult. About 80 per cent of this phosphorus is found in the bones. The remaining 20 per cent is widely distributed in the body and is an essential component of every cell. Phosphorus has more functions than any other mineral element in the body. We have already noted that phosphocreatine and the phosphorylated sugars are important in muscle metabolism. The inorganic phosphates make up a buffer system of the blood and also play a possible role in the formation of hydrochloric acid by the gastric glands. Phospholipids, such as lecithin, sphingomyelin, and cephalin, are important constituents of nervous tissue. In addition, phosphorus is an important component of many enzyme systems and is involved with the storage and transfer of energy within the cells of the body in such high-energy phosphate compounds as adenosine triphosphate (ATP).

Normal blood phosphorus levels range from 35 to 45 mg per 100 ml. Ten per cent of this amount is in the form of inorganic phosphate, which represents the phosphorus available for chemical reactions. Normally the blood phosphate varies inversely with the blood glucose.

The Food and Nutrition Board recommends a daily allowance of phosphorus for adults of $1\frac{1}{2}$ times that of calcium, whereas for growing children, an intake equal to that of calcium is recommended. Meat, fish, poultry, and eggs are excellent sources of phosphorus. Milk and other dairy products and nuts and legumes are considered good sources.

Iron. **Iron** is found principally in the hemoglobin of the blood, this substance accounting for 66 per cent of the total iron in the adult. Iron is also a component of myoglobin (muscle hemoglobin) and of the cytochromes and other enzyme systems found in the tissues. Iron can be demonstrated histologically in the chromatin of all cells. The iron compounds of the body are intimately associated with oxygen transport and cellular respiration. Thus, the hemoglobin of the blood serves to carry oxygen to the tissues, and myoglobin in the tissues serves as a temporary oxygen acceptor and reservoir.

The amount of iron in the body at birth is about 0.4 g and in the adult, about 4.5 g. About 1 g of iron in the adult is stored iron, found principally in the liver and spleen in the form of a protein complex. This iron is readily mobilized when the need arises. The Food and Nutrition Board recommends a daily intake of 7 to 12 mg for children and 10 to 15 mg for adults. Normal losses of iron from the body occur by way of shed hair and shed epithelial tissue and mucosal cells and also by way of the sweat, urine, fecal, and biliary excretions. Loss of blood through injury or disease or blood donation contributes considerably to iron loss from the body. Iron deficiency in man results in a type of anemia in which the number of red cells is either normal or reduced, but the total quantity of circulating hemoglobin is always subnormal. Thus, each red cell has a reduced hemoglobin content and, in comparison with a normal erythrocyte, appears quite pale.

For this reason the name *hypochromic anemia* has been applied to this type, which may result from an inadequate amount of iron in the diet or faulty absorption of iron from the intestinal tract. Diseases or injuries that are accompanied by loss of blood, such as bleeding ulcers, hookworm infestation, and bleeding hemorrhoids, usually result in iron depletion and therefore hypochromic anemia. Since the oxygen-carrying capacity of the blood is reduced as a result of the low hemoglobin concentration, weakness, headache, shortness of breath, and a tendency to fatique easily are common symptoms of iron deficiency in man.

The following foods are good sources of iron: meat, liver, heart, shellfish, egg yolk, beans, and legumes. Dried fruits, nuts, cereal products, and leafy green vegetables are fair sources.

Iodine. **Iodine** is an essential mineral for man since it is the basic component of the thyroid hormone. Although the mature animal body is estimated to contain less than 0.00004 per cent iodine, if this minute amount is not maintained in the diet, difficulties in metabolism result. This occurs because the thyroid hormone regulates the metabolic rate of the individual, and output of the hormone by the thyroid gland is dependent on the availability of iodine. In the absence of sufficient iodine, the thyroid gland, which normally oxidizes the iodide ion it receives from the circulation to iodine and incorporates it into the amino acid tyrosine, which is then converted to thyroxine, increases its activity in an attempt to compensate for the deficiency. The overactivity results in enlargement of the gland, a condition known as simple or endemic goiter and caused primarily by a lack of iodine. In such instances the hair and skin of the adult show premature aging, and mental and physical sluggishness may develop. In all cases there is a decrease in the basal metabolic rate. The removal of the thyroid or underactivity of the gland in early life results in a stunting of growth and retardation of mental and sexual development. A deficiency of thyroxine is known as **hypothyroidism,** and an excess is called **hyperthyroidism.** Although it is known that iodine deficiency will cause goiter, this is not the only cause. However, where the occurrence of goiter shows that attention to iodine nutrition is needed, the most practical method is to use a special source, such as ionized salt or sodium or potassium iodide.

It is known that iodine occurs in traces in all rocks, soils, water, and dust of the earth. However, certain regions are deficient in iodine and are usually high in incidence of simple deficiency goiter. Among natural foods that contain good quantities of iodine are seafoods, cod-liver oil, vegetables grown on iodine-rich soils, and the sea plant kelp. The major source of iodine in the United States is ionized salt. The minimum requirement of iodine for normal maintenance is not known, but the levels that are high enough to prevent goiter range from 0.15 to 0.30 mg per day. During puberty and pregnancy the need for iodine is increased because the demand for thyroid hormone increases during these periods.

Sodium, Potassium, and Chlorine. **Sodium, potassium,** and **chlorine** are essential for the normal functioning of the body. These minerals are found almost entirely in the fluids and soft tissues of the body, where they maintain osmotic pressure and acid–base balance and play important roles in water metabolism. Dietary deficiencies of these minerals are rare under ordinary circumstances. The Food and Nutrition Board states that 5 g of **sodium** chloride is more than sufficient to maintain normal requirements, whereas normal intake in the diet may range from 2 to 20 g daily. Practically all the sodium, except that found in bone, is found in the extracellular fluids, where it plays an important role in the maintenance of the normal hydrogen-ion concentration of the body fluids and the normal distribution of water between the various compartments (see Chapter 12).

In contrast to sodium, **potassium** is confined mainly to the intracellular fluid and is the chief cation (+) in that area. Potassium influences the contractility of all types of muscle and is necessary for cardiac action and transmission of nerve impulses. Deficiencies of potassium result in muscular weakness, increased nervous irritability, and cardiac irregularities. Excessive amounts of potassium are toxic, having an inhibitory action on muscle activity and nerve transmission and, in excessive concentrations, leading to cardiac arrest. It has also been determined that potassium is necessary for growth and activates some enzyme reactions, such as the phosphorylation of creatine. The dietary need for potassium is approximately the same as that for sodium. However, dietary potassium deficiency does not exist under ordinary circumstances because all nutrients, especially plant products, are so rich in this element. Normal food intake will always supply the element in excess of need; like sodium, potassium is readily absorbed, and the excess over body needs is immediately excreted.

Chlorine is found in large concentrations both within and outside the cells of the body tissues and is the chief anion (−) of the extracellular fluid. Approximately 15 to 20 per cent of the chlorine of the body appears to be in organic combinations. Chlorine in the form of sodium chloride makes up approximately 66 per cent of the acidic ions of the blood. This indicates its important role in sharing the function of maintaining the acid–base balance of the blood. Chlorine is also secreted by the gastric mucosa as free hydrochloric acid and aids in the regulation of osmotic pressure in blood and tissues. In addition, chlorine plays an important role in the maintenance of normal cardiac action.

Since sodium chloride (salt) serves as a condiment as well as a nutrient, the intake tends to be highly variable and frequently in excess of need. When the intake of sodium and chlorine is at a minimum, the body makes adjustments by decreasing the excretion of these minerals in the urine. This is also true for potassium. When the intake is great, the excess over body needs is eliminated by a corresponding increase in excretion in the urine, which also increases the water requirement. The kidney is the main regulatory organ controlling the concentration of minerals in the blood.

Copper. Experimental work has demonstrated that a small amount of **copper** is necessary, along with iron, for hemoglobin formation. The role of copper in hemoglobin formation is incompletely understood, but it is believed that copper stimulates the absorption of iron and plays a role in the synthesis of iron into hemoglobin and cytochrome compounds.

Copper is also necessary for melanin pigment formation in the body since it is a component of the enzyme tyrosinase and is probably associated with other oxidation–reduction enzymes of the cells. A store of copper is normally present in the liver and spleen.

The daily requirement of copper is about 2 mg. However, copper is so widely distributed throughout various foods that even poor diets contain the amount considered necessary for normal function. It has been determined that an adult can ingest up to 20 mg of copper per day without developing toxic symptoms. Excellent sources of copper are eggs, whole-wheat flour, beans, beets, liver, oysters, peas, fish, kale, asparagus, and spinach.

Sulfur. **Sulfur** occurs as a constituent of many proteins, being contained in the amino acid methionine, cysteine, and cystine. Insulin is an example of an important protein that contains sulfur. Vitamins such as thiamine and biotin also contain sulfur. Although sulfur is an essential constituent of body tissues and secretions, it must be supplied in the diet in the form of organic complexes, that is, methionine and/or cystine, because in its inorganic form it is not useful for these purposes. Deficiencies of sulfur are usually caused by inadequate intake of the sulfur-containing amino acids methionine and cystine. Such deficiencies may cause anemia, negative nitrogen balance, inhibition of hair growth, and necrosis of the liver. Some good dietary sources of methionine and cystine are beef, liver, lamb, heart, fish, poultry, eggs, cheese, beans, and peas.

Other Elements. Fluorine, magnesium, zinc, and manganese are all found in the body tissues and fluids, although not usually in amounts sufficient to affect the acid–base balance of the body. It has been estimated that the adult human body contains about 2 g of **zinc** and approximately the same amount of **fluorine.** Animal experiments have indicated that these elements are not indispensable. Although a deficiency of fluorine substantially increases the incidence of dental caries, it has also been established that higher concentrations of fluorine in domestic water supplies of 1 to 1.5 mg per liter may produce mottled tooth enamel. Zinc is an important component of the enzyme carbonic anhydrase. It is also necessary for normal growth. Because of the wide distribution of zinc in foods, it is improbable that zinc deficiency ever occurs in man. It has been established that excess dietary zinc may cause fibrosis of the pancreas and hypochromic anemia. **Manganese** is almost as widespread as copper throughout the tissues of the body and is important as an activator of several enzymes active in the Krebs cycle. It has also been demonstrated to be an important adjunct to hemoglobin synthesis and an essential mineral for growth, reproduction, and lactation. No evidence of manganese deficiency in man has yet been reported since the element is widely distributed in food of plant and animal origin. The human requirement of this mineral is unknown. **Magnesium,** which is a component of soft tissues as well as bone, is present in the adult body in an amount of approximately 21 g. The proper balance between magnesium and calcium in such tissue as heart, skeletal muscle, and nervous tissue is necessary to maintain normal function. The muscle tissue contains three times as much magnesium as calcium. In addition to being necessary for the functional integrity of the neuromuscular system, magnesium is also necessary to maintain normal structure of growing tissue.

Magnesium participates in bone formation and is present in bones as $Mg_3(PO_4)_2$ and $MgCO_3$, as part of the complex salt of the bony tissues. It is also an important activator of phosphorylytic enzymes. Like the manganese requirement, the magnesium requirement for man has not been determined, although diets that supply 250 to 350 mg keep healthy adults in balance. Deficiencies of this mineral may result in lesions of the kidney, hyperirritability, tetany, increased heart rate and sinoatrial block, and changes in teeth and bones. A dietary deficiency of magnesium has never been cited in man and would be extremely rare, for magnesium, like many of the other minerals, is widespread in the various food materials used by man.

Water

Among the various foods, **water** may be considered one of the more vital nutrient requirements of the body. It has been observed that the body can lose practically all its fat and more than half its protein and still survive, but a loss of one tenth of its water results in death. It is obvious that mammals will live longer without food than without water. It has already been mentioned that water makes up more than 50 per cent of the composition of the animal body and that many tissues contain 70 to 90 per cent of this substance. Its properties as a solvent make it a medium in which almost all metabolic reactions of the body take place. Its high specific and evaporative heats are important in temperature regulation. Its functions in connection with the transportation of nutrients, with secretions and excretions, and in maintaining the blood pressure have already been discussed in Chapter 7. The fluid compartments of the body and the dynamics of water exchange have been discussed under water metabolism in Chapter 12.

The body needs water not only for tissue formation and gland secretions, but also to replace that lost by excretion through the intestinal tract, kidneys, lungs, and skin. The Food and Nutrition Board recommends a daily intake of 2,500 ml under conditions of moderate temperature and exercise. However, it is obvious that the requirement will vary according to the magnitude of the factors that govern the losses. There are no deleterious effects from an excessive consumption of water, except under pathological conditions, but water intake is ordinarily kept within normal bounds by the thirst mechanism, which does not require conscious regulation.

Vitamins

Vitamins belong to a group of potent organic compounds, which occur in minute proportions in natural foods and are essential in small quantities in order for specific metabolic activity to occur within the cells of the body. Vitamins cannot be synthesized directly by the body and must be obtained from the diet or other sources, such as bacterial synthesis within the intestinal tract or synthesis within the body from precursors, or **provitamins.** These compounds are not important as energy sources but are necessary for the maintenance of physiological well-being, a fact that was demonstrated as far back as 1881 when Lunin, working with artificial diets, recognized that a diet consisting of only purified carbohydrate, protein, fat, and minerals would not sustain life. He expressed the view that there must be other substances essential for life in addition to the basic nutrients. Vitamins, in contrast to the other nutrients, are required in relatively small quantities. However, since they are found in varying quantities in different foods and no one food contains all of them in sufficient quantity to satisfy human requirement under normal conditions of food intake, vitamin deficiencies are not rare. The condition induced by the absence of sufficient quantities of any of the vitamins from the diet is called **avitaminosis.** In fact, the discovery of vitamins came about from many isolated observations that certain diseases could be cured by the administration of specific foods. For example, scurvy, a disease in which the capillaries become fragile, resulting in a bleeding condition, while wounds heal with difficulty and the patient grows weak and eventually dies, has been known since the time of Hippocrates (400 B.C.). Sixteenth-century Canadian Indians cured scurvy by drinking water in which pine needles had been soaked. Several centuries later the Scottish physician James Lind tried fresh fruits and vegetables as a cure. He found that oranges and lemons brought about improvement most quickly. In 1795 the standard treatment for scurvy was a daily ration of lemon or orange juice. Other diseases, such as rickets and beriberi, which were common among people who existed for long periods on limited diets, were also found to be curable when the diets of such patients were improved. Much of the more precise knowledge of these factors found in natural foods that were able to alleviate symptoms of certain diseases has been gained during the past 50 years from animal experiments with purified diets to which specific nutrients were added. Research workers can produce vitamin deficiencies in animals experimentally, and thus they have been able to show how several of these factors play an important role in metabolizing foods and maintaining normal functions of various organs and tissue structures. The use made of small animals, and more recently of bacteria, as aids in isolating, identifying, and assaying the respective vitamins is remarkable. Since the discovery of the first vitamin approximately 50 years ago and the coinage of the term *vitamine* by Funk in 1912, knowledge in this field has greatly advanced. The most recent significant attainment has been in the chemical identification and recognition of the specific biochemical roles of various vitamins in human metabolism. New ones are being discovered each year, and the subject of vitamin nutrition is becoming more complicated. There are some thirteen vitamins for which specific physiological functions have been worked out, and their chemical nature has been established so definitely that they are generally accepted as distinct dietary essentials. The probability that there are still undiscovered vitamins must be recognized. However, a discussion of nutrition can deal only with those that are well established. For more recent knowledge of vitamins, the student must follow the new contributions as they appear in the various research journals.

The vitamins have been divided into two groups on the basis of solubility. McCollum first proposed the name **fat-soluble** for the factor found in butter and **water-soluble** for the factor concerned with beriberi, since the first was extractable from foods with fat solvents and the second from foods with water. Today, the known fat-soluble vitamins are **A, D, E,** and **K,** and they are found in foods in association with fats. The fat-soluble vitamins are absorbed through the intestinal barrier along with the dietary fats, and conditions interfering with fat absorption will also interfere with their absorption. The water-soluble vitamins include thiamine (B_1), riboflavin (B_2, G), nicotinic acid (niacin), pyridoxine (B_6), cyanocobalamine (B_{12}), pantothenic acid, inositol, biotin (H), folic acid (M), ascorbic acid (C), and choline.

Fat-Soluble Vitamins. *Vitamin A.* **Vitamin A,** which is a high-molecular-weight alcohol, exists in two forms, A_1 and A_2. A_1 is the vitamin commonly referred to as A and is found in mammals and saltwater fish. Its structure is as follows:

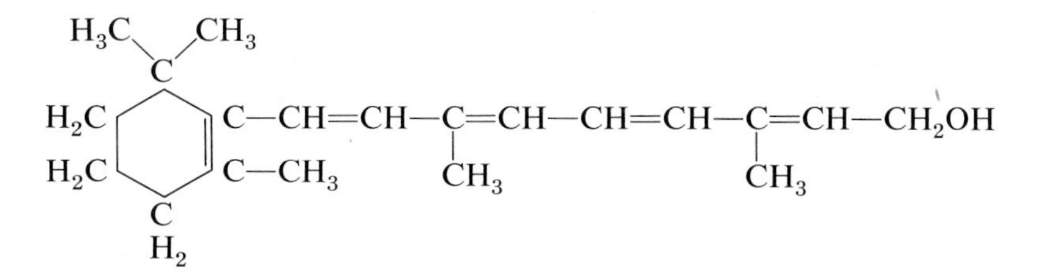

Vitamin A_2 (dehydroretinol) is closely related to A_1 but is found only in freshwater fish. It has not been isolated in pure form, nor can it be completely separated from vitamin A by any known means; it differs from A in that it has an additional double bond in its molecule and a modified terminal group. There is no indication at present that A_2 is important in animal nutrition. Large quantities of vitamin A are found as precursors or as provitamin A in compounds known as carotenoids, which are the orange and yellow pigments of fruits and vegetables. Vitamin A is formed in the body from these precursors, or provitamins, upon hydrolysis. Good sources for vitamin A in the form of precursors would be dark-green and yellow vegetables and yellow fruits. For vitamin A as such, which is obtained only from animal sources, eggs, liver, milk, butter, and cheese are good sources. Fish-liver oils are the richest natural sources of this vitamin. Animals usually ingest more provitamin A than is required, but only about 70 to 80 per cent of this quantity is converted into vitamin A. It appears that the body has a special regulatory mechanism that converts the precursors to vitamin A according to body needs.

Absorption of vitamin A and the carotenoids from the intestinal tract requires bile salts and fats. A lack of adequate fat in the diet, as well as biliary and pancreatic disease, will interfere with proper absorption of these compounds.

Since vitamin A is essential for the maintenance of normal health and growth of epithelial cells, it is to be expected that a deficiency would result in abnormalities of this tissue. Prominant in vitamin A deficiency is the atrophy of the normal epithelial cell layer followed by a thickening and keratinization of the cells. Such changes are first observed in the mucosal lining of the mouth and respiratory tract, which are later followed, if the deficiency is prolonged, by similar structural changes in the epithelium of the eyes, ductless glands, urethra, kidney, and intestinal tract. In man, there is also a generalized dryness of the skin and hair. **Xerophthalmia,** a condition characterized by a drying of the cornea and conjunctiva with ulceration, is particularly prevalent in children who are A-deficient over a prolonged period.

Vitamin A is necessary for the activity of visual purple (rhodopsin), which is the compound that breaks down in the chemical processes that underlie vision. The pigmented layers of the eye contain large amounts of vitamin A. A lack of vitamin A results in faulty synthesis of rhodopsin and therefore a decreased ability for dark adaptation (night blindness). It appears that vitamin A is also involved in the growth of developing bone and teeth, since it is apparently essential for the normal activity of the osteoblasts and osteoclasts. Laboratory experiments have demonstrated that deficiencies of vitamin A in growing animals result in faulty bone and teeth formation.

Since man and animals do not possess the ability to synthesize the vitamin, it must be supplied in the foods ingested as vitamin A per se or as the carotenes (provitamin A). The Food and Nutrition Board has recommended a daily allowance of 5,000 international units (IU) for adults, 1,500 IU for infants, and 4,500 IU for adolescents. An international unit is based on the activity of an international standard preparation, and 1 IU is equal to 1 USP (*United States Pharmacopeia*) unit. For vitamin A, 1 unit represents that amount of A_1 activity contained in 0.6 μg in international standard beta carotene.

Excessive vitamin A (hypervitaminosis) can cause drying and peeling of skin, loss of hair, enlargement of the liver and spleen, and excessive bone fragility.

Vitamin D. The **D vitamins** are related to the sterols and, because they are able to prevent and cure rickets, are also known as antirachitic factors. Two of these compounds are important, and these are known as D_2 and D_3. D_2 is also known as **ergocalciferol** and is formed when the plant sterol, **ergosterol,** is activated by irradiation with ultraviolet rays. It is a white, crystalline material, soluble in fat and stabile to heat, acids, alkalies, and oxidation. It has the formula

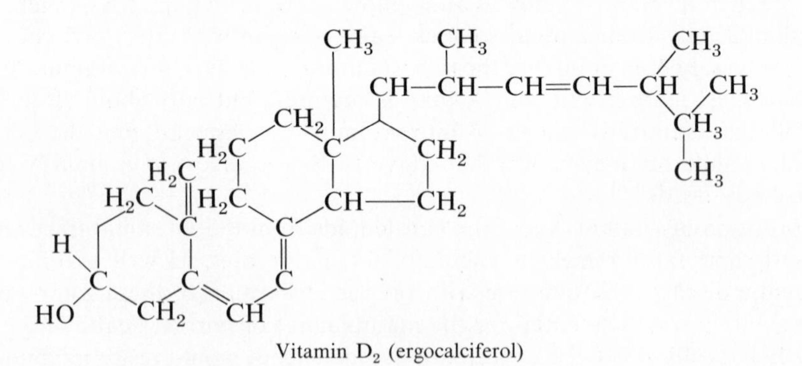

Vitamin D_2 (ergocalciferol)

It has been shown that the skin of man and animals contains large amounts of provitamin D (7-dehydrocholesterol), which can be converted to vitamin D_3 by exposure of the skin to sunlight. The amount of vitamin D formed in this manner is greatest during the summer months, which also accounts for the increased vitamin D content of milk and eggs during the summer and the lowered content in the winter. D_3 is also a white, odorless, crystalline compound, soluble in fats and in the common fat solvents, such as ether, chloroform, and acetone. Its formula is

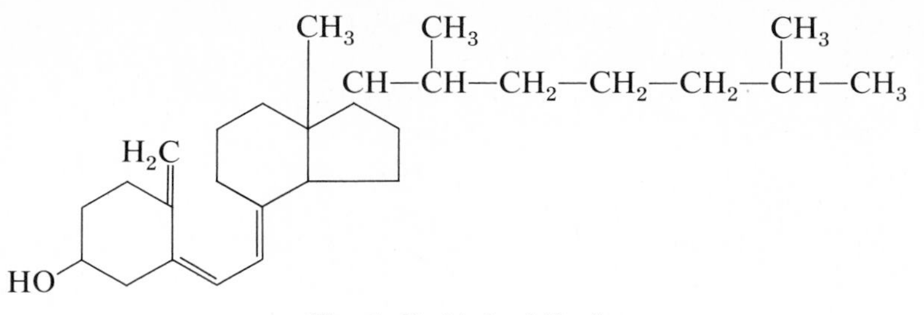

Vitamin D_3 (cholecalciferol)

D_2 and D_3 are of equal biological value in man.

Most food contains little vitamin D, and the requirements for this vitamin are mostly met by the artificial enrichment of suitable foods, such as milk, butter, margarine, and bread. Most of the milk consumed today is fortified so that 1 qt contains 400 units of vitamin D. Good dietary sources of vitamin D are mainly of animal origin and include the following: egg yolk, liver, butter, and saltwater fish that have a high content of body fat. Examples of such fish are salmon, herring, and sardines.

Vitamin D is absorbed through the wall of the small intestine along with the dietary fats. Bile and moderate amounts of fat facilitate its absorption. The vitamin passes into the circulating blood and is distributed to all parts of the body. Normal human blood serum contains 50 to 135 IU, or USP units, per 100 ml. Unlike vitamin A, which is stored in the liver, vitamin D has no special storage area, although certain tissues, especially those rich in fats, such as the brain, liver, and spleen, contain substantial amounts. In fish, vitamin D is stored in the liver and to a lesser degree in the viscera. It has been demonstrated that vitamin D is excreted by way of the bile into the intestinal tract and that some of it is destroyed. No vitamin D is excreted in the urine.

The principal function of vitamin D is to increase the absorption of calcium and phosphorus from the intestinal tract. The vitamin also appears to aid in regulation of the blood calcium level and also may promote the conversion of inorganic phosphorus to organic phosphorus in bone. The exact mechanism whereby vitamin D carries out its function is unknown, but it has been suggested that it activates the alkaline phosphatase enzyme of bone and the intestinal mucosa. These phosphatases catalyze the removal of phosphate from hexosophosphate and glycerophosphate, thus liberating inorganic phosphate, which influences the deposition of calcium phosphate in the bone and may also operate in a similar manner in intestinal absorption.

A deficiency of vitamin D is characterized by a lack of activation of alkaline phosphatases with resultant effects on phosphate metabolism in the intestine and bone. In fact, increases in alkaline phosphatase concentration of the plasma as a result of lack of activation are virtually diagnostic in cases of rickets. A deficiency of vitamin D affects chiefly the bones and teeth, in which calcium and phosphorus deposition is faulty; this syndrome is known as rickets. There are also general retarded growth, lack of vigor, and usually a loss of muscle tone.

Vitamin E (*Tocopherols*). During the period between 1920 and 1923, experimental work with rats kept under certain dietary regimens led to the discovery of **vitamin E.** It was found that young rats kept on certain diets that contained sufficient quantities of protein, fat, carbohydrate, and minerals as well as all vitamins known at that time (that is, A, B, C, and D) ceased to grow and lost the function of reproduction. Through a process of trial and elimination, it was demonstrated that the loss of the function of reproduction in the rat (that is, sterility in the male and degeneration of the fetus in the female) was caused by lack of another nutritional factor hitherto unrecognized. The name proposed for this vitamin was the next letter in the vitamin alphabet, E. Later, it was found that this new nutritional factor existed in three forms. For these individual vitamin E factors, Evans and his co-workers[11] proposed the term **tocopherol:** one to be designated as alpha-, the second as beta-, and the third as gamma-tocopherol. All these vitamin E factors are viscous oils at room temperature, soluble in fat solvents, and insoluble in water. They are resistant to heat in the absence of oxygen but are readily oxidized with complete loss of activity. These structures of the several tocopherols are as follows:

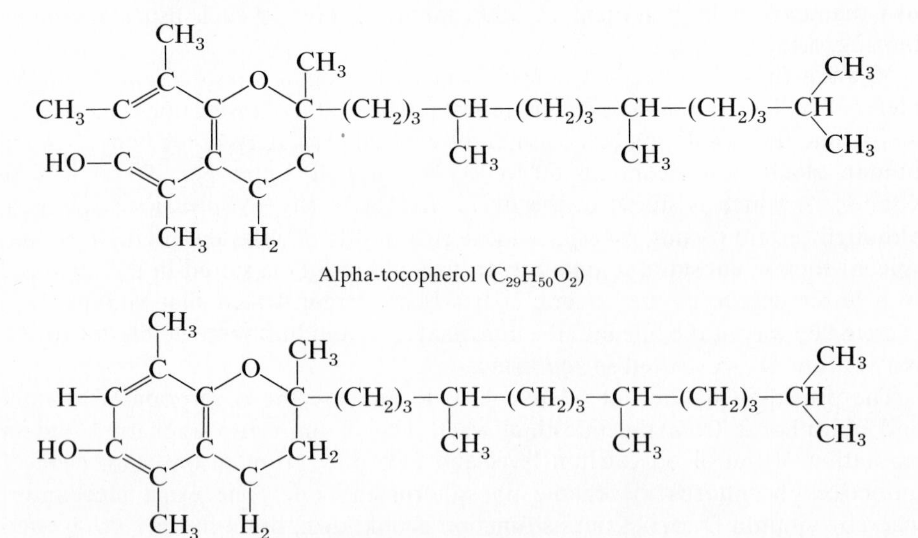

Alpha-tocopherol ($C_{29}H_{50}O_2$)

Beta-tocopherol ($C_{28}H_{48}O_2$)

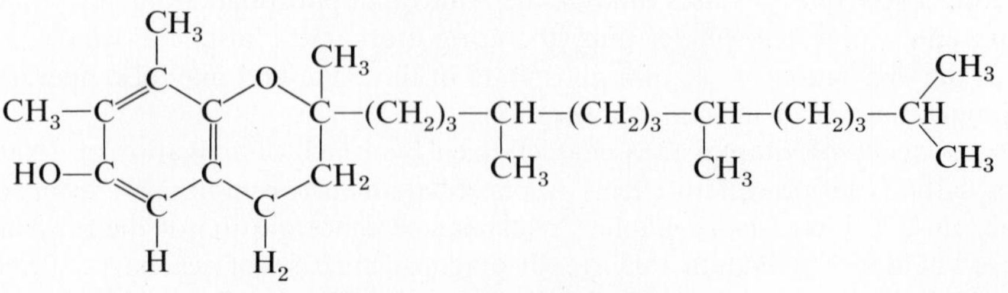

Gamma-tocopherol ($C_{28}H_{48}O_2$)

[11] H. M. Evans, O. H. Emerson, and G. A. Emerson, "The Isolation from Wheat Germ Oil of an Alcohol Having the Properties of Vitamin E." *J. Biol. Chem.*, **113**:319 (1936).

A fourth tocopherol has also been isolated and is referred to as delta-tocopherol, having a formula of $C_{27}H_{46}O_2$.

Vitamin E has been shown to be essential for normal reproduction in many animal species. However, its significance in human nutrition has not been definitely established. Vitamin E is a powerful antioxidant, and for this reason these compounds are used in foods to protect easily oxidized nutritional compounds, such as vitamin A and unsaturated fatty acids. In some animal species a lack of vitamin E produces muscular dystrophy. However, it has not been shown to benefit any type of muscular dystrophy in man. Vitamin E has been shown to protect the liver against damage from certain toxic agents, such as carbon tetrachloride and alloxan. Massive doses of vitamin E have been used in the treatment of cardiovascular disease, because it has been reported to maintain normal permeability of capillaries and to protect heart muscle against degeneration. However, such treatment and the beneficial effects of vitamin E in cardiovascular disease have been the subject of extensive controversy. The effect of vitamin E on oxygen consumption by various tissues suggests that the tocopherols may act as regulators of the metabolism of certain cells. Although considerable work has been done, little is known concerning the human requirements for vitamin E.

Good sources of vitamin E include meats, milk, eggs, fish, cereals, and leafy vegetables. Wheat-germ oil is particularly rich in the various tocopherols.

Vitamin K. Vitamin K, also known as the antihemorrhagic vitamin, exists in two forms, designated K_1 and K_2. Both vitamins are soluble in the common fat solvents (ether, acetone) but are insoluble in water and only slightly soluble in alcohol. They are very sensitive to alkalies and to various sources of light but are fairly resistant to heat. Vitamin K_1 is a yellowish oil and has the formula

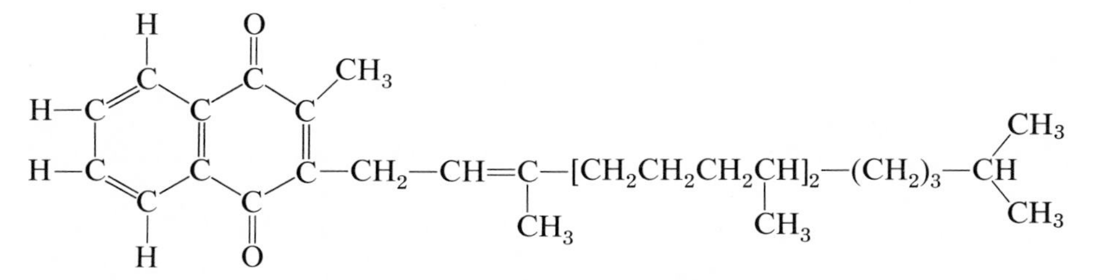

Vitamin K_1 ($C_{31}H_{46}O_2$) (phylloquinone)

Vitamin K_2 occurs as yellow crystals and has the formula

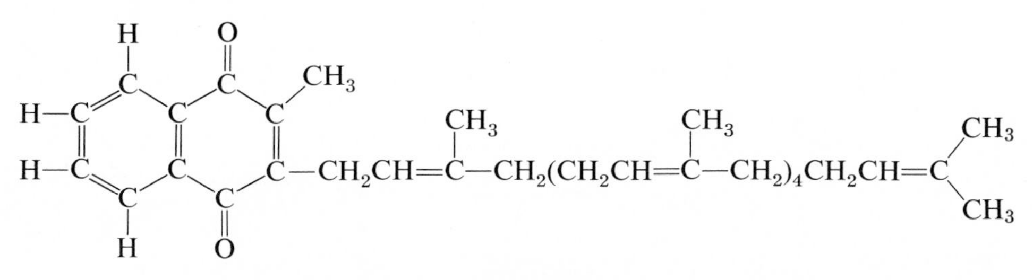

Vitamin K_2($C_{41}H_{56}O_2$) (flavinoquinone)

Synthetic vitamin K has been prepared and is known as vitamin K_3, or commercially as **menadione.** It is much more active than natural vitamin K_1 and is used as a standard of reference in studying K activity. Menadione is extensively used clinically in the form of a light-yellow crystalline powder. Its formula is

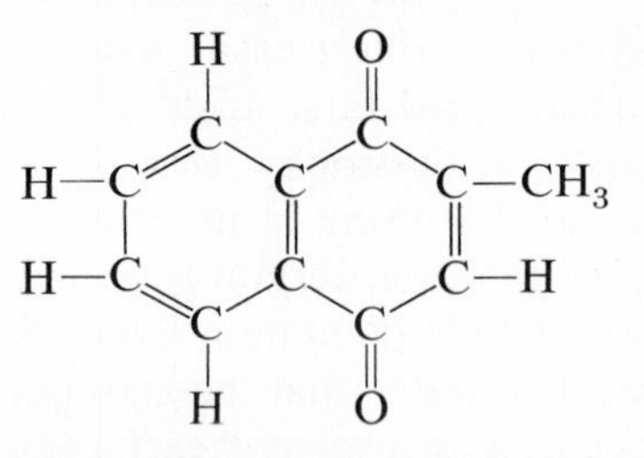

Vitamin K_3 (menadione: $C_{11}H_8O_2$)

Vitamin K is essential for the synthesis of prothrombin by the liver and therefore for normal blood clotting. Deficiency of vitamin K results in a lowering of the prothrombin concentration of the blood (hypoprothrombinemia), and the clotting time of blood is greatly prolonged. Vitamin K exerts its effect indirectly since adding vitamin K to blood in vitro does not shorten clotting time.

A deficiency of vitamin K is unlikely since the vitamin is fairly widely distributed in foods. Spinach, cabbage, and cauliflower are excellent sources of the vitamin. In addition, the intestinal bacteria synthesize considerable quantities of this vitamin.

The absorption of vitamin K from the intestinal tract depends on bile and moderate amounts of fat. Whenever there is a defect in fat absorption or an interference with normal bile flow, a deficiency of vitamin K in the body develops. Multiple hemorrhage is a symptom of vitamin K deficiency. Newborn infants usually are deficient in vitamin K because the intestinal tract is sterile at birth, and there are no bacteria to produce vitamin K. Thus a low prothrombin concentration usually appears, which will persist until the intestinal microorganisms become established and start synthesizing the vitamin. Vitamin K is commonly administered to the mother before parturition or to the newborn body in a small dose to prevent deficiencies from occurring.

Water-Soluble Vitamins. *Thiamine (Vitamin B_1).* **Thiamine** was first isolated from rice bran by Jansen and Donath[12] in 1926. Williams [13] first established its structure and later, with his co-workers, reported its synthesis. The vitamin is also known as **thiamine hydrochloride.** It is a colorless crystalline substance, very soluble in water and glycerin but insoluble in fat solvents (acetone, ether, chloroform). It is also comparatively stable in acid solutions and resistant to heat. The vitamin is very susceptible to both oxidation and reduction, and in neutral or alkaline solutions it is rapidly destroyed by heat. The structure of vitamin B_1 is

[12] B. C. P. Jansen and W. F. Donath, "The Isolation of the Anti-Beriberi Vitamin." *Chem. Weekblad,* **23**:201 (1926).

[13] R. R. Williams, "Structure of Vitamin B_1." *J. Am. Chem. Soc.,* **57**:229 (1935).

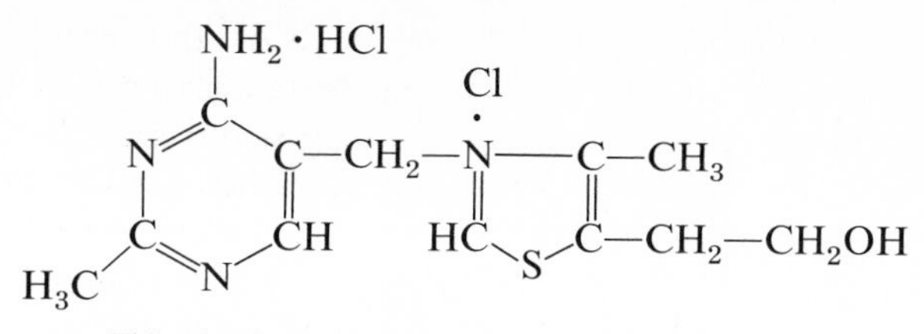

Thiamine hydrochloride (B_1) $C_{12}H_{17}ClN_4OS \cdot HCl$

The principal biological function of the vitamin is its use by the body in the formation of the coenzyme cocarboxylase; that is, cocarboxylase is the pyrophosphate ester of B_1. Cocarboxylase participates in all oxidative decarboxylations that lead to the formation of carbon dioxide. So far as is known, all animals except sheep and cattle need a dietary supply. Small amounts of free vitamin B_1 are found in the blood plasma and spinal fluid. Most of the cells of the body appear to have the ability to phosphorylate thiamine to form cocarboxylase. Thiamine is also essential for the synthesis of acetylcholine and for normal functioning of the nervous system.

The animal has no ability to store this vitamin and depends on a daily intake to supply its needs. Excess vitamin, that is, whatever is not needed by the body, is excreted in the urine. In cases of deficiencies in vitamin B_1, the oxidation of pyruvic acid by way of the citric acid cycle to carbon dioxide and water is prevented, resulting in the impairment of carbohydrate metabolism. Neurological involvement also occurs since lesions of peripheral nerve fibers develop, resulting in atrophy of the muscles and impairment of the sense of touch. If vitamin B_1 deficiency is prolonged, reflexes, particularly those related to the sense of position (kinesthesis), are impaired. The neurological syndrome is called **polyneuritis** and is characteristic of the disease **beriberi.** This disease falls into various clinical types, depending on whether the nervous tissue involvement occurs alone or in combination with edema, enlarged heart, and circulatory failure. These are as follows: (1) "dry" beriberi, in which the symptoms are restricted to the nervous system; (2) "wet" beriberi, in which the polyneuritis is accompanied by edema; (3) "cardiac" beriberi, in which the polyneuritis is associated with circulatory failure; and (4) "mixed" beriberi, in which any combination of the above may be found.

Thiamine is present in small concentrations in most plant and animal foods. Excellent sources of cereal grains, liver, heart, and kidney. Since thiamine is soluble in water and easily destroyed by heat if the solution is alkaline, improper cooking may destroy the vitamin B_1 content of foods. Today, many foods, such as flour, bread, and macaroni products, are enriched with thiamine, and it has been estimated that as much as 40 per cent of the daily thiamine requirement is supplied by these foods. The Food and Nutrition Board recommends a daily thiamine intake for adults of 0.5 mg for each 1000 cal and a minimum of 1 mg, even if the caloric intake falls below 2,000 cal.

Riboflavin (*Vitamin B_2*). **Riboflavin** has also been called **vitamin G,** *lactoflavin* when derived from milk, *ovaflavin* from eggs, and *hepatoflavin* from liver. It is required in the metabolic activity of all animals and is an essential constituent of the yellow and greenish fluorescent pigments in the tissues of plants and animals. Crystalline forms of this enzyme have been isolated from a large number of plant and animal products, such as liver, egg white, egg yolk, milk, kidney, and grasses. In 1935, two research groups[14,15] published methods for synthesizing riboflavin. It occurs as fine orange-yellow needles and is relatively stable under ordinary storage conditions. It is soluble in water, where it shows a yellow-green fluorescence, but is insoluble in fat solvents. The substance must be protected from light, because it is rapidly destroyed in the presence of alkalies. The formula for riboflavin is

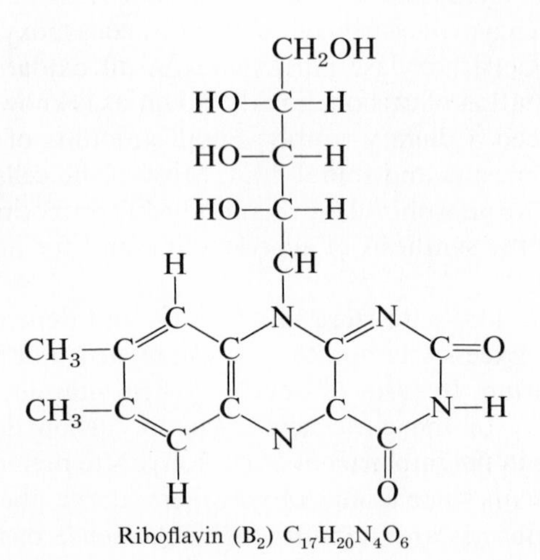

Riboflavin (B_2) $C_{17}H_{20}N_4O_6$

Riboflavin is an important component of all enzyme systems that regulate oxidation and thus plays an important role in general carbohydrate metabolism. For example, Warburg's yellow oxidation enzyme is a combination of riboflavin with a protein and phosphoric acid. The yellow enzymes are required along with coenzymes I and II in the breakdown of glucose to produce energy for body processes. The action of riboflavin precedes that of thiamine previously described. Riboflavin is also a constituent of cytochrome *c* reductase and a series of enzymes, such as D-amino acid oxidase and xanthine oxidase, that function in the final stages of protein metabolism. Most of the enzymes of which riboflavin is an essential constituent are capable of alternate oxidation and reduction reactions and therefore participate in cellular oxidation. They are especially important for respiration in tissues having poor circulation, such as the cornea. Because of the basic roles vitamin B_2 plays in the release of energy and in the assimilation of nutrients by the cells of the body, a deficiency is reflected in a wide variety of symptoms. The eyes, skin, nervous tissues, and blood are the areas of the body most commonly affected. Free riboflavin, for example, plays an important part

[14] R. Kuhn and K. Deinemund, "The Synthesis of Vitamin B_2." *Ber. Deut. Chem. Ges.,* **68:**1765 (1935).

[15] P. Karrer and B. Becker, "The Synthesis of Lactoflavins." *Helv. Chim. Acta,* **18:**1435 (1935).

in the visual mechanism in the retina since it is converted into a very sensitive photo compound by the action of light, and, as stated previously, it is also important in the respiration of such tissue as the cornea. Deficiencies of vitamin B_2 can cause corneal ulceration, cloudiness, cataracts, dimness of vision, and impairment of visual acuity. The skin of vitamin B_2-deficient animals shows a scaling, greasy dermatitis, and fissures at the angles of the mouth with local inflammation (angular stomatitis) may also occur. In experimental animals myelin degeneration of nerve fibers has been demonstrated, and symptoms resembling those of degeneration of the spinal cord have developed. Impairment of red blood cell formation with resultant anemia has also been demonstrated in experimental animals.

Nutritional studies have shown that riboflavin is an essential nutrient for man, and unless the diet contains adequate amounts, there are a retardation of growth and impairment of general health. Although present in all living cells, riboflavin is not stored in large amounts nor can the normal tissue levels be increased by the administration of large doses. The normal concentration of riboflavin in the blood plasma is about 0.5 μg per gram of blood and is quite constant. Most of the vitamin is excreted from the body in the free form by way of the kidney. Small amounts appear in the urine as the phosphoric acid ester. It has been shown that an inverse relationship exists between the amount of protein in the diet and the urinary excretion of riboflavin. Low-protein diets may lead to vitamin B_2 deficiency as a result of excessive excretion in the urine.

Riboflavin is generously distributed in most plant and animal tissues and excellent sources are liver, heart, eggs, chicken (dark meat), veal, beef, lamb, whole-wheat products, asparagus, peas, beets, and peanuts. The bacterial organisms within the intestinal tract may supply man with small amounts of riboflavin. However, it is safer and easier to plan the diet so that it supplies the entire requirement of the vitamin.

Although many factors may contribute to the daily requirement, it is generally accepted that mammalian needs are determined by body size and weight and metabolic activity, as well as the amount and type of food ingested. The Food and Nutrition Board has correlated its recommended riboflavin allowance with the recommended protein intake of the individual. Thus, if one multiplies the daily recommended protein intake by 0.025, the recommended riboflavin intake in milligrams can be determined. The recommended daily intake varies from 0.5 mg for infants to 2.5 for adults.

Niacin (*Nicotinic Acid*). The importance of **nicotinic acid** was established in 1937 when Elvehjem and his co-workers[16] at the University of Wisconsin made the discovery that nicotinic acid, a compound that was commonly found on the chemist's shelf, would cure the disease of black tongue in dogs. After this, it was also demonstrated that the same compound could cure human pellagra, a disease characterized by signs of dermatitis, diarrhea, stomatitis, and dementia. The tongue of pellagra sufferers is usually smooth, red, and painful, and there is a

[16] C. A. Elvehjem, "Relation of Nicotinic Acid and Nicotinic Acid Amide to Canine Black Tongue." *J. Am. Chem. Soc.*, **59**:1767 (1937).

burning sensation in the mouth. Recovery from the signs and symptoms of pellagra is quite spectacular once adequate amounts of niacin are administered.

Niacin was first synthesized in 1873 by Huber[17] and by Weidel.[18] It is a white crystalline compound, soluble in hot water and resistant to heat, weak acids, and alkalies. Niacin in the amide form, known as **niacinamide,** is much more soluble in water and has the same biological activity as niacin. In human beings, ingested niacin is readily converted to niacinamide. In fact, animal tissues contain the vitamin in the form of the amide, whereas plant tissues contain it in the form of the acid. The formulas of nicotinic acid and its amide, niacinamide, are as follows:

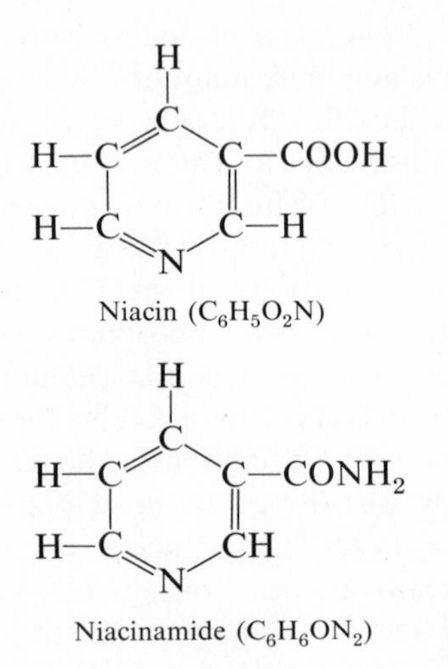

Niacinamide is a component of two coenzymes, present in all plant and animal cells, which are concerned with various dehydrogenation reactions (removal of hydrogen from the molecule) as well as the transport of hydrogen. These two coenzymes are nicotinamide-adenine dinucleotide, or NAD, and nicotinamide-adenine dinucleotide phosphate, or NADP:

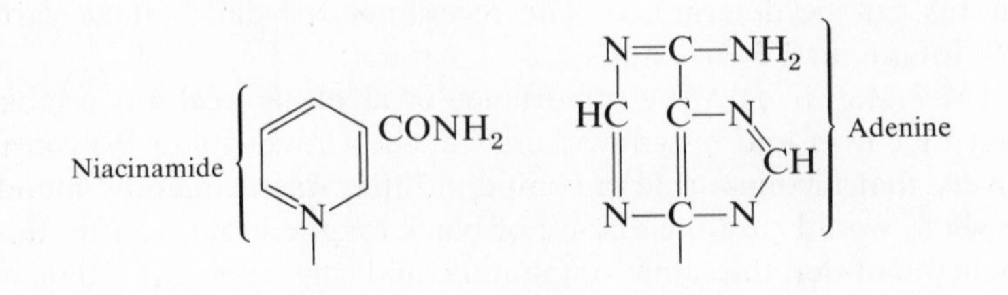

[17]C. Huber, "Voraufige Mitteilung. Einwirkung von Kaliumbichromat und Schwefelsäure and Nikotin." *Ber. Deut. Chem. Ges.,* **3**:849 (1870).

[18]H. Weidel, "Zür Kenntnis des Nicotins." *Ann. Chem.,* **165**:328 (1873).

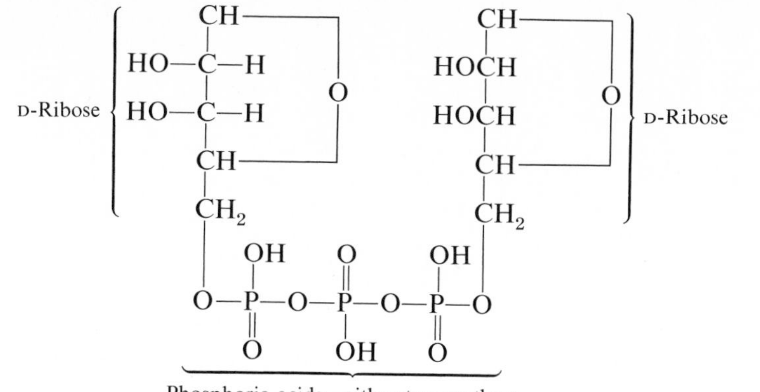

Phosphoric acids—either two or three
NAD (2 phosphoric acids), NADP (3 phosphoric acids)

NAD participates in the dehydrogenation of hexose phosphate and triose phosphate in the tissues and is especially concentrated in heart muscle and red blood cells. NADP is concerned with the dehydrogenation of glutamate, lactate, malate, and glyceraldehyde diphosphate.

The clinical use of nicotinic acid produces in many patients a considerable vasodilatation, so that such side effects as flushing of the skin, as well as some disturbance of pulse rate and intensity of heartbeat, are noted. These reactions are not manifested with the amide form of nicotinic acid, and for this reason, niacinamide has largely replaced the acid in clinical use.

Although it is certain that all living tissues require niacin, the minimum requirements for man are not known. In mammalian tissues niacin can be synthesized from the amino acid tryptophan. However, it has been shown that man and other animals require an external supply of the vitamin. Niacin-deficiency syndrome is produced when the diet is lacking in preformed niacin and in tryptophan. It has also been demonstrated that adequate amounts of tryptophan are as curative in human pellagra as is the administration of niacin.

Good sources of niacin are the following foods: meats, liver, fish, poultry, whole-grain breads and cereals, peas, beans, and nuts. Milk, although a poor source of niacin, is a good pellagra preventative because of its high content of tryptophan.

Pyridoxine (*Vitamin* B_6). P. György separated the nonthiamine part of the B complex into riboflavin and an unknown factor that he named **vitamin B_6.** The isolation of the vitamin in crystalline form was first accomplished by Keresztesy and Stevens, of Merck and Company, in 1938. Its chemical structure was determined, and later its complete synthesis was accomplished. It was found that the pure, natural vitamin and the synthetic form were identical in their chemical and physiological properties.

Pyridoxine hydrochloride, the form of the vitamin in commercial use, is a white, odorless powder with a salty taste. It is readily soluble in water and relatively heat-stable. The vitamin is light-sensitive. Its formula is

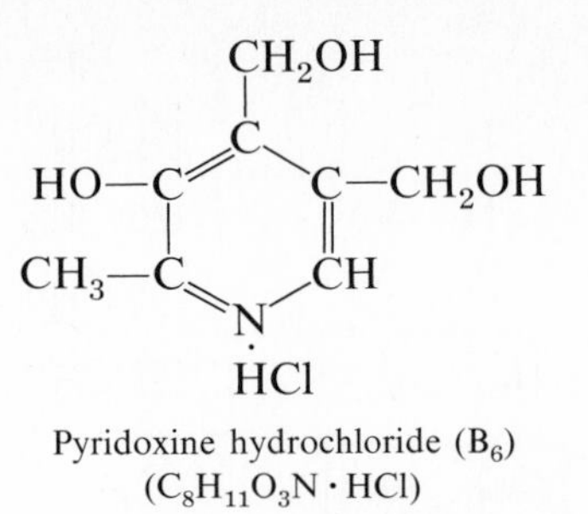

Pyridoxine hydrochloride (B_6)
($C_8H_{11}O_3N \cdot HCl$)

Vitamin B_6 appears to be linked with the metabolism of the essential unsaturated fatty acids, since it appears that the synthesis of arachidonic and hexonoic acids (unsaturated fatty acids) from linoleic and linolenic is dependent on this vitamin. It is also essential for normal amino acid metabolism, including the conversion of tryptophan to niacin, for its phosphorylated form (pyridoxal-5-phosphate) functions as the coenzyme for a number of enzymes involved in decarboxylation and transamination reactions.

The most common deficiency symptom in man is a dermatitis about the eyes, nose, and mouth. Cheilosis[19] and glossitis[20] have been reported, and some patients show signs of a peripheral neuritis, followed by motor impairment. The administration of pyridoxine in dosages as low as 5 mg per day results in the prompt relief of all symptoms.

The daily requirement for vitamin B_6 in human beings has been estimated to be 1 to 2 mg. However, no official standards have yet been set. Deficiencies of B_6 rarely occur in Western countries, since the vitamin is widely distributed in plant and animal tissue, and an ordinary mixed diet provides sufficient amounts. Excellent sources of B_6 are salmon, tomatoes, yellow corn, and spinach. Good sources are green beans, whole-grain cereals, liver, meat, asparagus, and peas.

Pantothenic Acid. **Pantothenic acid** is obtained as a pale-yellow viscous oil, predominantly acid in character, or as a calcium salt, which is a white crystalline solid. Pantothenic acid is readily soluble in water and stable to light and air under ordinary conditions. Its formula is

$$HOCH_2-\underset{CH_3}{\overset{CH_3}{\underset{|}{\overset{|}{C}}}}-CHOH-CO-NH-CH_2-CH_2-COOH$$

Pantothenic acid ($C_9H_{17}O_5N$)

Pantothenic acid is present in all animal tissues, with the liver and kidney containing the highest concentration of the vitamin. The heart, brain, pancreas, and lung contain much more pantothenic acid than the muscle tissues. It is a constituent of coenzyme A, the coenzyme necessary for acetylation reactions. The utilization of acetic acid by the cells is dependent on the activation of this acid

[19] Cheilosis: a condition marked by lesions on the lips and angles of the mouth.
[20] Glossitis: inflammation of the tongue.

by coenzyme A; that is, coenzyme A combines with acetic acid to form "active acetate," in which form acetic acid participates in a number of important metabolic processes. Acetic acid, which is a common product of fat, carbohydrate, and some amino acid metabolism, combines with oxalacetic acid to form citric acid, which initiates the Krebs cycle. "Active acetate" also combines with choline to form acetylcholine. The utilization of acetic acid in the formation of cholesterol, a substance necessary for the formation of steroid hormones, is catalyzed by pantothenic acid as coenzyme A. All these facts make it obvious that pantothenic acid is of fundamental importance in cellular metabolism. It has also been found that pantothenic acid is necessary to maintain the integrity of the adrenal cortex since a deficiency of the vitamin results in lesions of the cortical layer. Pantothenic acid in the form of coenzyme A is presumed to be involved in adrenal cortical function.

Pantothenic acid deficiency in human beings is practically unknown, since the vitamin occurs in almost everything consumed by man. The name of the vitamin actually means "derived from everywhere."

Excellent sources are liver, brain, pancreas, kidney, eggs, oatmeal, and peanuts. The daily requirement for pantothenic acid is about 10 to 15 mg. It is apparent that bacterial synthesis in the intestinal tract plus the vitamin contained even in poor diets provides amounts adequate for growth and maintenance of normal tissue function.

Folic Acid. **Folic acid,** which is a water-soluble, yellow crystalline compound, has also been known as vitamin M, vitamin B_c, and factor U.L. Folic acid occurs in nature in combination with glutamic acid, which, when ingested, is split by the digestive system into free glutamic acid and folic acid. The vitamin is unstable in heat in an acid medium and is easily destroyed by sunlight when in solution. The formula for folic acid is

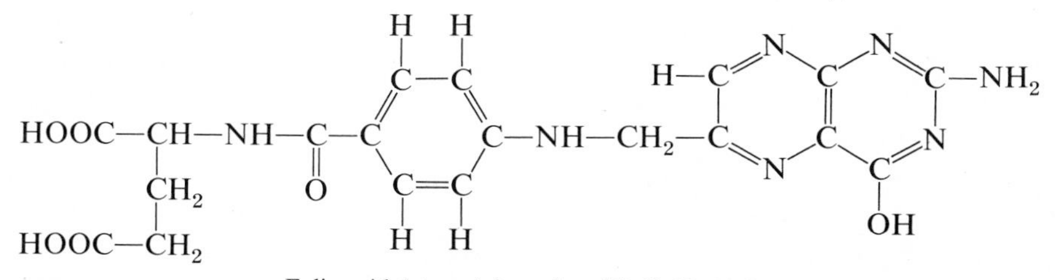

Folic acid (pteroylglutamic acid) $C_{19}H_{19}N_7O_6$

The most important activity of folic acid is the role it plays in the synthesis of purine and pyrimidine compounds, which are utilized for the formation of nucleoproteins. An adequate supply of nucleoproteins is needed for the normal development of mature red blood cells from the parent magaloblast. A deficiency of folic acid and/or nucleoprotein results in the faulty development of red blood cells, and a blood picture typical of a macrocytic anemia is characteristic. Thus, folic acid is effective in the treatment of this type of anemia without neurological involvement. Folic acid is a relatively new addition to the list of B-complex factors; the daily requirement in man has not yet been determined with accuracy, but it has been estimated to be from 0.5 to 1.0 mg. It appears that these amounts will maintain the normal metabolism of growing cells and tissues.

The richest dietary sources for man are mammalian liver and fresh leafy green vegetables. The cooking process destroys approximately 50 per cent of the vitamin content in food. Breakfast cereals prepared from wheat are also important dietary sources. The bacterial organisms within the intestinal tract are also able to synthesize folic acid and may provide adequate amounts for human needs. However, the quantity produced may vary under different environmental and dietary conditions.

Vitamin B_{12} (Cyanocobalamine). **Vitamin B_{12}** was first isolated from liver by Rickes and co-workers. Later it was demonstrated by a number of different scientists that this vitamin was effective in the treatment of pernicious anemia and was the antianemic factor associated with the liver. It was also demonstrated that B_{12} was the long-sought-after growth factor for animals. It is a water-soluble, red crystalline compound, unstable in acid and alkaline medi and also to light. Vitamin B_{12} is unique in that it is the only vitamin containing an essential mineral element, cobalt. The empirical formula of the vitamin is $C_{63}H_{84}O_{14}N_{14}PC_0$.

The absorption of vitamin B_{12} from the intestinal tract is dependent on hydrochloric acid and a substance secreted by the normal, healthy gastric mucosa known as the **intrinsic factor.** The basic defect in patients with pernicious anemia is usually the lack of intrinsic factor because of degenerative changes in the gastric mucosa. Consequently, a deficiency of vitamin B_{12} develops because of poor transfer of the vitamin across the mucosal membrane in the absence of the intrinsic factor. Pernicious anemia responds promptly to subcutaneous administration of B_{12}. Administration by way of the intestinal tract is not effective so long as the defect in the gastric mucosa remains.

Under normal conditions human beings depend on a supply of the vitamin from the foods ingested. The richest food sources of the vitamin are liver and kidney. Meats, eggs, and milk are fair sources, but plant tissues do not contain appreciable amounts of vitamin B_{12}. It has been suggested that 1 μg per day will meet the normal requirement for adults, since this amount is sufficient to maintain a patient subject to pernicious anemia in a normal condition.

Choline. **Choline** does not fall within the strict classification of a vitamin since it can be synthesized in the body by the methylation of ethanolamine. Methionine can function as a methyl donor, and ethanolamine is readily available from the decarboxylation of serine or the reduction of glycine. Thus, a choline deficiency can be produced in the animal only as a result of a combined deficiency of choline and methyl donors.

Choline is a colorless, viscous, strongly alkaline liquid with the formula

$$\begin{array}{c}(CH_3)_3\\ |\\ N{-}OH\\ |\\ CH_2\\ |\\ CH_2OH\end{array}$$

Choline ($C_5H_{15}O_2N$)

It is very soluble in water and alcohol but insoluble in ether. Choline readily forms salts (chloride, citrate, borate), which occur as white water-soluble crystals and are the form used commercially.

Choline is important in fat metabolism for it is a **lipotropic substance** (has the ability to decrease the fat content of the liver). As a constituent of the phospholipid lecithin, it is essential for normal transport of fat from the liver to the fat depots because it enhances phospholipid turnover. In the absence of choline, fat accumulates in the liver. Thus fatty infiltration of the liver can be reversed by the administration of choline, especially if the fatty deposition was originally caused by a choline deficiency. When fatty infiltration of the liver is due to hepatic injury, resulting in lowered capacity of the organ to metabolize the fats brought to it, some degree of lipotropic action can be achieved by the administration of choline.

In addition to its lipotropic action, choline is also important in tissue metabolism as a methyl donor and as a precursor of acetylcholine. Choline can serve as a methyl donor in methionine-deficient animals. The methyl groups furnished can be transferred to homocystine to form the essential amino acid methionine.

The synthesis of acetylcholine, which is the chemical transmitter of the impulse across synapses and neuromuscular junctions, depends on the acetylation of choline. This reaction, which is activated by the enzyme choline acetylase, has been extensively studied by Nachmansohn and his associates.

The human requirement for choline has been estimated to be from 1.5 to 3.0 g daily. This amount is readily supplied by the average diet. Additional amounts are supplied by the methionine content of the diet and by the synthesis of choline in the tissues.

Good food sources of choline are beef liver, egg yolk, peanuts, and wheat germ. Beef, milk, beans, and spinach are fair sources.

Vitamin C (Ascorbic Acid). Scurvy, the disease of sore gums, painful joints, and hemorrhages, was the first deficiency disease to be recognized as such. As early as 1720 it was observed that medicines gave no relief in this condition, but it could be cured by the addition of fresh vegetables to the diet or by the administration of fruit juices. In 1757, Lind showed experimentally in human subjects that oranges and lemons incorporated in the diet could prevent the onset of the disease.

The first definite isolation and identification of **vitamin C** were made by Waugh and King[21] and later confirmed by Svirbely and Szent-Györgyi.[22] It was shown by these research workers that the crystalline material isolated from lemons, oranges, and cabbages and effective against scurvy was a single substance. The substance was named hexuronic acid since it was a derivative of a hexose sugar. Elaboration of the structure and synthesis of vitamin C were carried out in other laboratories.[23-26]

[21] W. A. Waugh and C. G. King, "Isolation and Identification of Vitamin C." *J. Biol. Chem.*, **97**:320 (1932).

[22] J. L. Svirbely and A. Szent-Györgyi, "The Chemical Nature of Vitamin C." *Biochem. J.*, **27**:279 (1933).

[23] E. G. Cox and E. L. Hirst, "Hexuronic Acid as the Antiscorbutic Factor." *Nature*, **130**:888 (1932).

The chemical structure of **ascorbic acid** is

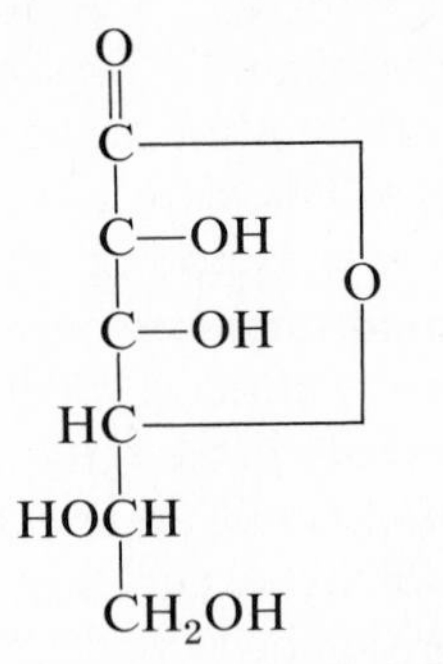

Ascorbic acid (vitamin C) $C_6H_8O_6$

The first oxidation product of ascorbic acid is dehydroascorbic acid, which has the same antiscorbutic potency of its precursor:

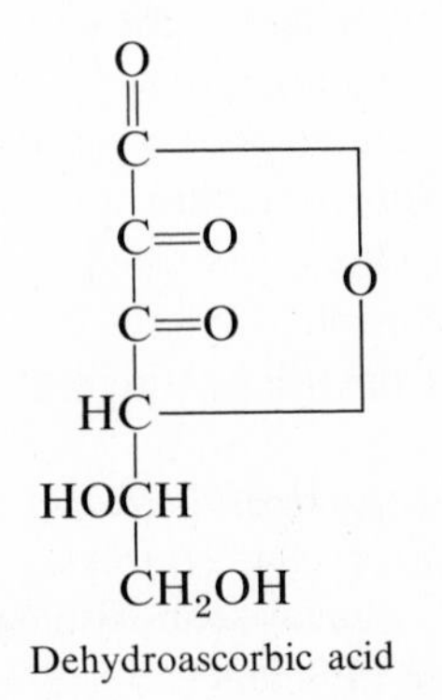

Dehydroascorbic acid

Vitamin C is the least stable of the vitamins and is readily oxidized by heat to diketogulonic acid, an inactive compound. The vitamin is fairly stable in acid solutions but is quite sensitive to an alkaline medium. It is markedly destroyed by cooking when the pH is alkaline. There are also losses in storage as a result of its sensitivity to oxidation.

Severe vitamin C deficiency produces scurvy. The symptoms of this deficiency are almost entirely confined to the supporting connective tissue structures of the body, such as bone, dentine, and cartilage. Scurvy is also characterized by a hemorrhagic syndrome, occurring at sites of greatest stress and where growth processes are most active. In the mouth the gums become swollen and spongy and bleed easily. The teeth become loose with extensive changes in tooth structure involving enamel, cementum, and most of all the dentine. During C deficiency, the bones cease growing and are easily fractured, for they are porotic and are

---

[24] P. Karrer and H. Solomon, "Vitamin C, Ascorbinsäure." *Biochem. Z.*, **258**:4 (1933).

[25] F. Micheel and K. Kraft, "Die Konstitution des Vitamins C." *Z. Physiol. Chem.*, **222**:235 (1933).

[26] T. Reichstein, A. Grussner, and R. Oppenauer, "Die Synthese der d-Ascorbinsäure." *Helv. Chim. Acta*, **16**:561 (1933).

lacking in strength. Poor wound healing is also characteristic of vitamin C deficiency.

From a metabolic viewpoint, vitamin C is essential since it is believed that this vitamin plays an important role in biological oxidation and reduction and in cellular respiration. Ascorbic acid can also serve as a hydrogen donor by being oxidized to the dehydroascorbic acid form. The metabolism of aromatic amino acids such as tyrosine appears to be involved with vitamin C, since the removal of hydrogen from tyrosine in its oxidation seems to depend on dehydroascorbic acid's acting as a hydrogen acceptor, which is in turn reduced to ascorbic acid. The subsequent transfer of hydrogen to oxygen regenerates the dehydroascorbic acid. A very important metabolic function of ascorbic acid is the role of the vitamin in the conversion of folic acid to the active folinic acid, an essential growth factor in some species of animals.

Ascorbic acid occurs in high concentrations in the adrenal glands. In the medulla it may function to prevent the oxidation of epinephrine. After stress or the administration of adrenocorticotropic hormone, the increased secretion of the cortical hormones is always associated with a rapid decrease in the amount of vitamin C in the gland. It has been suggested from these observations that the vitamin may play an important role in the reaction of the body to stress.

Vitamin C is readily absorbed from the intestinal tract and is distributed to all parts of the body and occurs in varying amounts in the tissues and the body fluids. Tissues that exhibit the highest metabolic activity, such as glandular tissue, usually have the highest concentration. A plasma level of 1 to 2 mg per 100 ml is considered to be a level of complete saturation, and when this level is exceeded, excretion of the vitamin readily occurs. Levels of 0.15 mg or below are invariably associated with clinical scurvy.

Since vitamin C is constantly being destroyed by the body by oxidation, the daily intake must equal the amount oxidized if a deficiency is to be avoided. The Food and Nutrition Board recommends a daily intake of 75 mg for adults except during pregnancy and lactation, when the intake should be increased to 100 to 150 mg per day. Adolescent girls and boys require 80 to 100 mg per day.

The best food sources of vitamin C are citrus fruits, tomatoes, green peppers, leafy green vegetables, raw cabbage, berries, and melons. The vitamin is easily destroyed by cooking because it is readily oxidized. Losses are also extensive during storage and processing of food, particularly where heat is involved.

The specific therapeutic use for vitamin C is in the treatment of scurvy. Vitamin C has also been widely employed for the treatment of such conditions as bleeding gums, gum infections, pyorrhea, hemorrhagic states, anemia, and undernutrition. It is usually administered by the oral route in the form of ascorbic acid tablets. In the dry state the vitamin is reasonably stable. The vitamin can also be given in solutions of the sodium salt by intramuscular or intravenous injection. Solutions of the vitamin deteriorate rapidly if exposed to air.

Normal Diet

The Food and Nutrition Board of the National Research Council has formulated nutrition standards designed to maintain good nutrition for healthy persons in the United States under ordinary conditions. These standards are recorded in table form and called "Recommended Dietary Allowances" (Table 13-1). The recommended allowance values of the table do not take into consideration prior losses in storage and cooking but are for nutrients in unprepared foods. In planning practical diets provision should be made for such losses.

Most workers in the field of nutrition have used in the formulation of diets the system of the "seven basic food groups." These food groups include the following: (1) breads and cereals, (2) dairy foods, (3) meats, (4) butter and margarine, (5) green, leafy, and yellow vegetables, (6) citrus fruits, tomatoes, and raw cabbage, (7) potatoes, vegetables, and fruits other than those listed above.

Recently, the Agricultural Research Service of the U.S. Department of Agriculture has developed a food plan based on four food groups. These are as follows: (1) **breads and cereals,** which are relied on to provide thiamine, niacin, iron, and low-quality protein; (2) **milk foods,** including milk, cheese, and ice cream; which are counted on to provide most of the calcium requirement, riboflavin; high-quality protein, and other nutrients; (3) **meats,** which provide high-quality protein, iron, thiamine, riboflavin, and niacin; and (4) **vegetables and fruits,** which are relied on to provide a good source of vitamins and minerals, especially vitamins A and C.

It has been suggested that the food intake from each of these groups in one day be as follows:

Bread–cereal group	4 or more servings
Milk group	
Children and teenagers	3 to 4 cups
Adults	2 or more cups
Pregnant women	4 or more cups
Meat group	2 or more servings
Vegetable group	4 or more servings, including at least a citrus fruit and a dark-green or deep-yellow vegetable every day

Carbohydrates (sugar) and fats (butter, margarine, oils) are usually taken in with the specific foods mentioned and will be eaten to satisfy the appetite. The caloric requirement will also be satisfied by the ingestion of these substances.

A good diet provides the necessary required level of not only vitamins and the energy principle (calories), but also protein, minerals, essential fatty acids, and water.

Since the diet provides a source of energy, the first and perhaps the most important characteristic of an adequate diet is the quantity of energy it provides. The caloric adequacy of a diet will depend on the total metabolism of the individual. As we have already seen, if the intake exceeds the expenditure, nutrients are converted into fat for storage, and weight will be gained; whereas if the nutrient intake is smaller than the energy output, a loss of body weight will occur. Under conditions of equilibrium, that is, intake approximately equal to output, the individual does not lose or gain weight and is said to be in physiological equilibrium. If growth is occurring, extra nutrients must be provided for the increase of protoplasmic materials, and in a pregnant or lactating woman, extra calories are necessary to compensate for the extra loss of energy under these circumstances. Table 13-1 summarizes the caloric requirements recommended by the Food and Nutrition Board for men, women, and children of every age group. It has been suggested by DuBois and Chambers[27] that the caloric requirement be provided by all three of the energy foods, that is, 40 per cent by carbohydrates, 45 per cent by fat, and 15 per cent by protein.

Today, suitable menus can be prepared with relative ease by anyone to meet the necessary caloric requirements of a particular person and also to maintain optimum health. Tables have been prepared to show not only how much is needed of a particular nutrient to maintain good nutrition (Table 13-1), but also what foods to select from and the approximate amounts to be used for the achievement of an adequate diet ("Food Groups," The Institute of Home Economics of the Agricultural Research Service). In addition, many tables are available listing various foods and their values. An excellent source of such tables is Clara Taylor's *Food Values in Shares and Weights* (New York: Macmillan, Inc., 1959). Other sources will be found listed in the references at the end of this chapter. Table 13-5 lists some common foods with their caloric values in weights.

Recommended allowances for carbohydrates and fat have not as yet been established. However, it may be said that if the intake is less than 3,000 cal per day, fat should account for at least 25 per cent of the total calories, and if the caloric intake is more than 3,000, 30 per cent. The proportion of carbohydrate in the diet is subject to wide variations, but in Europe and America carbohydrates account for 56 per cent of the total caloric intake. It is recommended that no less than 45 per cent nor more than 66 per cent be obtained from this source.

Problem of Weight and Its Control

Today medical and nutritional experts are alarmed at the number of overweight persons in the United States; approximately 20 per cent of the population is overweight. Studies of college freshmen have shown that 23 per cent of the male students and nearly 36 per cent of the female students are overweight. Statistics also suggest that at least 10 million young people under the age of 19 are overweight. Why do people become fat? This question concerns many research workers.

Meaning of Obesity

The terms **obesity** and **overweight** have come to be used interchangeably, but a distinction should be made between the two. The common interpretation of obesity is excessive deposition of fat, but more specifically obesity refers to the amount of fat stored relative to the lean body mass. Overweight, on the other hand, is determined by evaluation of an individual's weight with respect to certain norms presented in age–height–weight tables compiled by various insurance companies. Thus, individuals could be overweight by these standards without being fat. An example would be muscular persons whose total muscle mass is greater than the average person's. There are also instances of persons who are obese because the ratio of body fat to the lean mass is greater than normal, but the total body weight is normal or even below normal. However, since increases or decreases in a person's body weight are usually related to increases or decreases in his body fat, the two terms have been used interchangeably.

[27] E. F. DuBois and W. H. Chambers, *Calories in Medical Practice, Handbook of Nutrition.* (Chicago: American Medical Association, 1943).

Table 13-5. ***Caloric Value of an Average Serving of Some Foods from the Four Basic Food Groups***

Food	*Approximate Measure*	*Calories*
Bread–cereal group		
Cornflakes	1 cup	79
Shredded Wheat	1 large biscuit	100
Wheat flakes	1 cup	101
Cream of Wheat	$\frac{3}{4}$ cup	95
Puffed Wheat	1 cup	43
Puffed Rice	1 cup	55
Oatmeal	$\frac{2}{3}$ cup	99
Pancakes	1, 4-inch diameter	59
White bread	1 slice	63
Rye bread	1 slice	57
Parkerhouse rolls	1	114
Saltines	1	17
Oyster crackers	10	43
Zweibach	1 piece	31
Protein bread	1 slice	51
Raisin bread	1 slice	82
Matzoth	$6\frac{3}{4}$-inch square	159
Bagels	1 medium	150
Milk group		
Milk, whole	1 cup	166
Skim milk	1 cup	84
Buttermilk	1 cup	86
Chocolate milk	1 cup	184
Ice cream, plain	$3\frac{1}{2}$ oz	129
American cheese	1 slice	105
Swiss cheese	1 slice	101
Cottage cheese	2 tbsp	27
Cream cheese	2 tbsp	110
Sour cream	2 tbsp	62
Sweet cream	2 tbsp	62
Coffee cream	2 tbsp	30
Butter	1 pat, $\frac{1}{4}$-inch thick	50
Meat group		
Porterhouse steak	85 g	293
Roast round of beef, lean	85 g	197
Corned beef	45 g	100
Hamburger	1, 88 g	150
Lamb chop	1, 46 g	100
Veal chop	1, 60 g	110
Veal cutlet	1, 80 g	125
Ham	1 slice, 100 g	340
Bacon, broiled	2 slices, 16 g	97

Table 13-5. ***Caloric Value of an Average Serving of Some Foods from the Four Basic Food Groups***

Food	Approximate Measure	Calories
Fried chicken		
Breast	1	201
Drumstick	1	101
Thigh	1	177
Roast chicken	3 slices, 100 g	194
Roast duck	3 slices, 100 g	322
Turkey	1 slice, 51 g	130
Egg, whole, fresh	1	66
Fried flounder	100 g	293
Fried haddock	100 g	158
Salmon	$\frac{1}{2}$ cup	152
Sardines	4	100
Tuna, drained	$\frac{1}{2}$ cup	255
Clams, medium	6	100
Oysters, medium	5	63
Shrimp, medium	12	78
Liver	94 g	125
Frankfurter	1	124
Bologna	1 slice, 22 g	50
Salami	1 slice, 31 g	134
Vegetable–fruit group		
Vegetables		
Baked beans	$\frac{1}{2}$ cup	147
Lima beans	$\frac{1}{2}$ cup, cooked	76
Green beans	$\frac{1}{2}$ cup, cooked	14
Kidney beans	$\frac{1}{2}$ cup, cooked	134
Peas	$\frac{1}{2}$ cup	56
Corn	$\frac{1}{2}$ cup, cooked	70
Corn on the cob	1 ear	100
Potatoes, white	1 medium	98
Potatoes, sweet	1 medium	78
Asparagus	6 stalks	21
Cabbage	$\frac{1}{2}$ cup, cooked	20
Cauliflower	$\frac{1}{2}$ cup, cooked	15
Carrot, raw	1	20
Beets, medium	1	17
Mushrooms	$\frac{1}{3}$ cup	44
Spinach	$\frac{1}{2}$ cup	21
Sauerkraut	$\frac{1}{2}$ cup	23
Tomato, fresh	1 medium	32
Tomato, canned	$\frac{1}{2}$ cup	24
Turnips	$\frac{1}{2}$ cup	78
Celery	$\frac{1}{2}$ cup	14
Radishes, small	1	2
Peppers, medium	1	64

Table 13-5. ***Caloric Value of an Average Serving of Some Foods from the Four Basic Food Groups***

Food	*Approximate Measure*	*Calories*
Lettuce	3 large leaves	10
Onion	1 medium	30
Fruits		
Apples	1 medium	76
Apricots, dried	1	20
Banana	1 medium	100
Cantaloupe	$\frac{1}{2}$ melon	37
Cherries, raw	1	4
Grapefruit	$\frac{1}{2}$ medium	75
Grapes	100 g	55
Orange	1 medium	70
Pear	1 medium	95
Peaches	1 medium	46
Pineapple, canned	1 slice	43
Prunes, dried	1 medium	18
Plums	1 medium	25
Watermelon	1 slice, 357 g	100

Table 13-6. ***Approximate Caloric Value of Common Beverages***

Type	*Measure*	*Calories*
Milk, skim	1 cup	84
Milk, whole	1 cup	165
Buttermilk	1 cup	84
Water		0
Tea, no sugar		0
Coffee, no sugar		0
Orange juice, fresh	1 cup	108
Orange juice, canned	1 cup	135
Tomato juice	1 cup	50
Beer	12-oz bottle	173
Cola beverages	6-oz bottle	83
Ginger ale	6-oz bottle	63
Whiskey, gin, rum	1 jigger ($1\frac{1}{2}$ oz)	109
Cocktail	1 cocktail glass	155
Old-fashioned	1 glass	194
Cordials	1 cordial glass	67
Wine, dry	1 wineglass (3 oz)	80
Wine, sweet	1 wineglass (3 oz)	143

Fat is a major constituent of the normal body, comprising about 15 per cent of the total weight in young adults. A number of methods have been developed to determine indirectly the amount of fat in the body. In general, body volume–density measurements are determined either by weighing the body under water

(Archimedes' principle) or by direct volume displacement of water. Density is equal to mass divided by volume. By weighing the subject in air and measuring his volume, one can obtain his total density. From appropriate regression equations derived from studies on animals and human beings that utilize the differences in density of body components (fat, 0.90; cells, about 1.50; bone, 3.0; and water, 1.0), one can estimate body fatness to within 5 per cent in a given person. In addition, a skinfold technique has been developed for determining subcutaneous fat. Calipers designed to measure a double layer of skin plus subcutaneous tissue have been developed. Equations have been derived that make it possible to estimate total body fat from several skinfold measurements.

Causes of Obesity

Two general types of obesity are now recognized and are referred to as **metabolic obesity** and **regulatory obesity.** We have already seen that adipose tissue is not merely an inert storage area but is continually undergoing change. Active synthesis and degradation of fat together with continual interaction with carbohydrate and protein metabolism are always occurring. Oddly enough, it has been found that not all the fatty tissue takes part in this activity. In fact, in normal animals, only 1 per cent of the mass of each individual fat cells is involved. Thus 99 per cent of each fat cell is indeed impassive. It appears that in metabolic obesity, lipogenesis and other metabolic activities are abnormal. In working with fat mice, research workers have found that the amount of passive fat in fat cells is greater than 99 per cent, thus making the required amount of "active" fat abnormally low in contrast to that of normal mice. It has been suggested that perhaps this inadequate proportion of active fat leads in some way to overstimulation of the appetite, whereby the body can create enough fatty tissue to provide a normal amount of active fat. By the time this active amount is built up to the requisite level, the sum total of all the body fat, active and inactive, has ballooned the subject to obese proportions. How does one get shortchanged on the amount of active fat in the cell? This is where heredity and gene influence become the important determining factors. Experimental evidence also indicates that there are differences in the way the body uses food. In some persons it is converted into energy rapidly; in others the process is slower. In this regard, hormonal factors and enzymatic mechanisms appear to be implicated.

Regulatory obesity involves excessive fat deposition produced by overeating alone, which in itself can produce adaptive metabolic changes if continued over a long period of time. Such adaptive metabolic changes may alter the response of the appetite-regulating mechanism and aggravate the tendency toward the development of obesity. Tense persons often seek consolation in food. Another important factor in compulsive overeating is food habits. Many persons form habits of eating rich, calorie-laden snacks between meals. Still another highly important factor in obesity is lack of physical exercise.

Appetite Regulation

Appetite and hunger are controlled by a small area in the hypothalamus of the brain, which consists of two sets of bilateral nuclei (Fig. 13-1). These nerve cells form the **appetite center,** or **appestat.** When the nerve cells forming the lateral nuclei are stimulated, one wants to eat. Experimental animals from which these

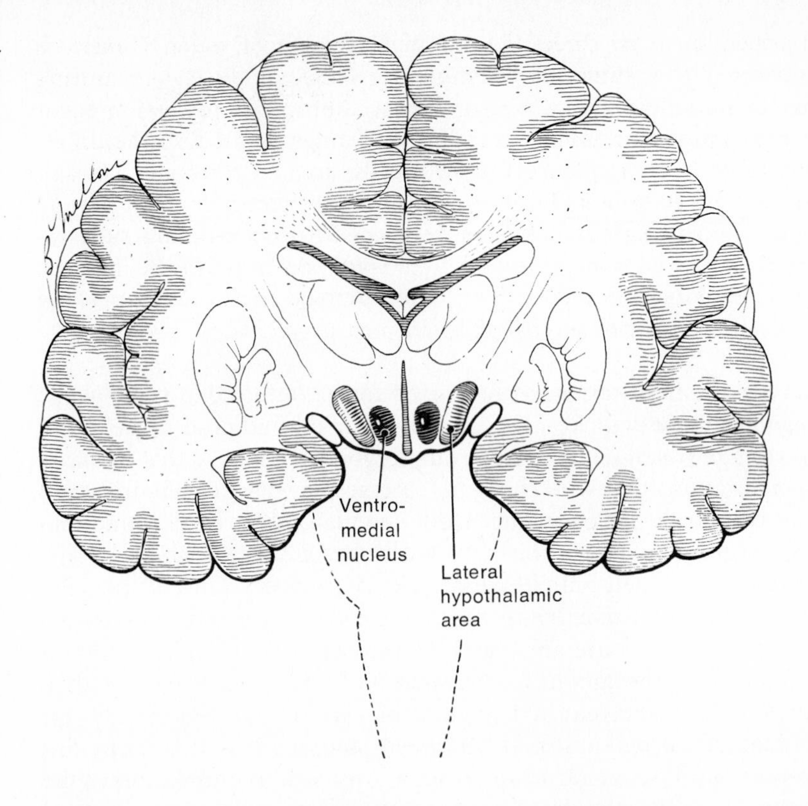

Figure 13-1. Hypothalamic nuclei regulating appetite and satiety.

lateral nuclei have been removed do not take food when it is made available nor do they seek food even when starved. The ventromedial nuclei act to inhibit food intake or act as a satiety center. When this area is ablated or lesions of this area are produced in experimental animals, such animals eat more or less continuously in the waking state and cannot distinguish between caloric and noncaloric materials. The appestat regulates how much one eats in somewhat the same way that a thermostat regulates temperature in a room. It continually and automatically matches average food intake with the energy needed.

The stimuli that activate the appetite centers have not been well established. Some of the theories that have been proposed are the following:

1. The concentration of blood glucose. Low glucose concentration stimulates the lateral nuclei and thus stimulates appetite and food intake. As the glucose concentration rises with food intake, the ventromedial nuclei in the hypothalamus becomes stimulated, and hunger is satisfied.
2. The temperature variation of the blood. A drop in temperature from normal stimulates food intake, and an increase in temperature acts to stimulate the satiety centers.
3. The proportionate amounts of active fat in fat cells to passive fat. This acts as a regulator.
4. The concentration of serum amino acids. This acts to regulate the appestat. It has also been suggested that instead of the lowering of some blood constituent's being the primary stimulus to the food intake center, an increase in some factor produced in the fasting state may be the causal agent.

The hypothalamic appetite center also receives nerve projections from the cortex, and thus the appestat may be consciously controlled. Thus, other factors come into play to stimulate the center, especially the emotions. The appestat appears to be peculiarly responsive to worry, tension, and other emotional stresses. Many people overeat because they are bored, frustrated, or discontented with their family relationships, their jobs, or their social relationships.

It must also be remembered, as Tepperman[28] has pointed out in his work on the etiological factors in obesity and leanness, that appestats vary in different individuals, who show differences in their levels of response to stimuli. These differences in threshold of response may be inherited along with all the other inherited characteristics of cells. The variation encountered in different persons in the voluntary control of appetite may be a reflection of this response sensitivity. In fact, the common experience of the difficulty encountered in voluntary control of food intake may be due to a constitutional defect at the level of the hypothalamic center.

No matter what the specific initiating mechanism, obesity is basically caused by an intake of excess calories above the metabolic requirements; that is, more calories are consumed in the diet than are expended by the body in work, exercise, heat loss, or other metabolic activities.

What Can Be Done

First, in order to lose weight the subject must have a desire or strong motivation to slim down. There are more reasons for maintaining a proper weight than just to look more attractive. Studies show that overweight shortens life and is an etiological factor in several diseases. Cardiovascular disease, diabetes, gallbladder disease, cirrhosis of the liver, certain forms of cancer, and arthritis are more common or more serious in obese persons than in those whose weight is normal. Obesity has also been considered a predisposing factor in such conditions as gout, varicose veins, high blood pressure, pulmonary emphysema, acute and chronic nephritis, and toxemia of pregnancy. The heart suffers, since it must pump blood through the extra blood vessels that develop in adipose tissue. It has been estimated that an additional two thirds of a mile of blood vessels is brought into play for every pound of fat added. The risks in major surgery and childbirth are also increased with obesity. Thus obesity not only detracts from an individual's physical appearance, but is also a predisposing and detrimental factor in many diseases and is known to shorten life.

Second, if one holds to the theory that obesity results from overnutrition, one might reason that the best antidote for this condition would simply be the reversal of the inequality, so that output exceeds intake. This is the assumption that underlies most prescribed programs of weight control. Generally speaking, this reversal can be accomplished by two means: (1) decreasing the caloric intake, or dieting; and (2) increasing the caloric expenditure by means of exercise or administration of drugs (notably hormones) that increase the metabolic rate. When loss energy is consumed in the form of food or more energy is expended in the form of work, or both, body fat diminishes.

[28] J. Tepperman, "Etiologic Factors in Obesity and Leanness." *Perspectives Biol. Med.*, **1**:296 (1958).

Diet. A large portion of the literature relating to obesity is concerned with the effects of various diets in reducing weight or adiposity. The relative merits of high-protein, high-fat, and high-carbohydrate proportions in the diet have been studied by a number of research workers. Most have found that, regarding isocaloric diets with constant water and salt content, weight loss in obese patients varied with the composition of the diet, being greatest with high fat and least with high carbohydrate intake. Pennington[29] concluded that restriction of carbohydrate alone seemed to make possible the treatment of obesity with a calorically unrestricted diet composed chiefly of protein and fat. The weight changes observed in the instances cited were not due to defective gut absorption or to fluctuations in body water, since the insensible weight loss and values of oxygen uptake throughout the day suggests that the metabolism was altered in response to changes in the diet, being increased by high fat and high protein intake and decreased by high carbohydrate intake. However, basic metabolic rate measurements were unchanged.

Some diets may be more satiating than others, more convenient to prepare, and more palatable than others. Any safe diet, however (most fad diets are not safe), faithfully followed will do the slimming job about as well as any other. A safe reducing diet must contain all the essential nutrients required by the subject and at least in minimal quantities, and it must supply fewer calories than the subject expends each day. In addition, since a good diet usually must be followed for many, many long months and eventually even be adopted as a lifelong practice, it should be reasonably compatible with the food pattern to which the subject is accustomed. Only a dietary program that results in permanent weight loss and lifetime control of weight will be a satisfactory one.

The caloric content of a good reducing diet will vary, of course, with the subject and his metabolic needs but normally should range between 1,200 and 1,800 cal. In a diet containing less than 1,200 cal it is rather difficult to supply the minimal amounts of all essential nutrients. When a diet contains less than 1,200 cal, for example, extremely low-calorie regimens (800 to 1,000 cal), a supplement of specific nutrients, such as vitamins and minerals, should be taken. Such supplements prevent any possible mineral or vitamin deficiency. In any case, supplementary additions to the diet should always be taken only on the advice of a competent physician.

Some research workers have recommended that total fasting with free access to water and supplements such as vitamins and minerals be used as an introduction to the treatment of obesity. Bloom[30] found that prolonged total fasting was well tolerated by obese subjects, and their re-education to proper eating habits was facilitated. Duncan and his associates, working at the Pennsylvania Hospital in Philadelphia, have tried such drastic methods on approximately 40 hopelessly obese patients. All were deprived of food for 10 days; only water, weak tea, coffee, and vitamins were ingested. Instead of being ravenously hungry, as they were

[29] A. W. Pennington, "Treatment of Obesity with Calorically Unrestricted Diets." *J. Clin. Nutrition,* **1**:343 (1953).

[30] W. L. Bloom, "Fasting as an Introduction to Treatment of Obesity." *Metab. Clin. Exptal.,* **8**:214 (1959).

on low-calorie diets, all reported a loss of appetite. After the initial fast (which the doctors advise should be carried out in the hospital), by utilizing intermittent 1- and 2-day fasts, the patient can gain a means of controlling what had appeared to be a hopeless situation.

High fat–high protein diets can also work quite well in a slimming program provided that the carbohydrates are rigidly limited and that the total fuel intake is less than the subject's expenditure. Calories do count, even if an individual cannot, does not, or will not count them. There is no mystery as to why an individual loses weight on a combination high fat–high protein diet. First, such diets appease hunger better than high-carbohydrate diets. Second, weight loss is accelerated at first because protein tends to counteract water retention and high fat dehydrates. At the point when a subject comes into caloric equilibrium, however, he will no longer lose weight on this diet or any diet unless the number of calories is reduced.

About 3,500 cal are the equivalent of 1 lb of body weight. Therefore, in order to lose 1 lb of body weight, 3,500 cal must be cut out of the diet. If one wanted to lose 2 lb a week, the diet would have to be adjusted so that 7,000 cal less were taken in per week, or 1,000 cal per day. A reduction of 500 cal per day would result in a weight loss of about 1 lb weekly. Ideally, the reducing individual should lose weight slowly but steadily, 1 to 2 lb per week, until the weight for his height and age is reached.

Food is not the only factor in a reducing campaign. What about exercise? Does it have any value? Generally exercise has been discounted as an effective means of dealing with obesity because of two misconceptions: (1) Exercise requires little caloric expenditure, and therefore, a lot of exercise gets rid of only a little fat; and (2) an increase in physical activity is always automatically followed by an increase in appetite, and the excess food intake more than makes up for the calories expended in the exercise.

Tables showing the measured energy cost of different types of physical exercise over the basal metabolic rate have been prepared by several investigators. Walking a mile, for example, uses up only 92 cal. More vigorous activities, however, such as swimming, rowing, and tennis, require an additional energy expenditure of 70 to 1,000 cal. Physical activity involving 500 to 600 cal per hour above the resting level can be endured by the average, middle-aged adult, not in training, for a period of 35 minutes without undue discomfort. A physically active person resists weight gain through expenditure of a greater amount of energy. In a sedentary person, weight gain is more rapid and pronounced. It has been estimated that a 3 per cent deficit in energy consumption can do the same thing as a 3 per cent excess in food intake—accumulate more than 100 lb of fat in 10 years.

Experimentally it has been shown that in normal, reasonably exercised human beings and animals, an increase in food intake does ordinarily follow an increase in physical activity. However, this is true only within a certain range, termed by Mayer[31] the *normal activity range.* In his experiments with animals he found

[31] J. Mayer, "Exercise and Weight Control," in: *Weight Control.* (Ames, Iowa: Iowa State College Press, 1955).

that those accustomed to a sedentary (caged) existence showed (1) no increase in food intake for brief durations of moderate exercise and consequent weight loss, (2) an increase in food intake that was linear with the amount of exercise for longer durations, and (3) a decrease in food intake, weight loss, and deterioration of appearance in cases where exercise was quite prolonged. In general, it may be said that if exercise is indulged in infrequently and strenuously, typical of the weekend athlete, the appetite is stimulated and the subject eats a lot more, thus gaining back any calories lost due to the energy consumed. With moderate and regular exercise the calories are used up without an increase in the appetite.

There is a great deal of evidence to indicate that physically active persons who subsequently become sedentary do not show a decrease in food intake proportional to their decrease in activity. Greene,[32] has studied more than 200 overweight adult patients in whom the beginnings of obesity could be traced directly to a sudden decrease in activity. Most of the studies relating overweight to activity in children show that obese children generally tend to be inactive or to engage in activities of a sedentary nature, avoiding unnecessary physical activity such as outdoor play and athletics.

In this respect age also plays an important role. Energy requirements decline progressively after the first years of adulthood, because of a decrease in basal metabolic rate as well as a lessened physical activity. According to the Food and Agriculture Organization of the United Nations,[33] the caloric expenditure at age 45 is 6 per cent less than at age 25; and at age 65, 21 per cent less. It has been proposed by the Food and Nutrition Board, National Research Council, that calorie allowances be reduced by 3 per cent per decade between ages of 30 and 50, by 7.5 per cent per decade from age 50 to 70, and by 10 per cent for the years from 70 to 80.

Recent research indicates that the spacing of meals, or the time between food intake, plays a role in the subsequent metabolic rate of the food and therefore influences the storage of calories as adipose tissue. Experiments with rats maintained on isocaloric diets, in which feeding was varied, demonstrated that animals forced to take their rations in one meal every 24 hours tend to deposit more fat as compared with their continuously fed littermates. In general, meal eating tends to favor lipogenesis as a result of an increased hexosephosphate shunt, which appears to be related to the rate of fat formation. It has also been suggested that the long-term practice of eating widely spaced meals, the large proportion of the daily calories being taken in the evening meal, upsets the appetite-regulatory mechanism and causes adaptive metabolic changes, leading to increased fat deposition. Eating small amounts of food (nibbling) through many feedings through the day might not only inhibit lipogenesis but also allow the appestat to control the appetite to the point where caloric intake could be matched with caloric output automatically, and the maintenance of a remarkably constant adult

[32] J. A. Greene, "A Clinical Study of the Etiology of Obesity." *Ann. Internal Med.,* **12:**1797 (1939).

[33] Food and Agriculture Organization of the United Nations, *Calorie Requirements* (Nutritional Studies 15) (Rome: FAO, 1957).

weight would be achieved. Janowitz and Hollander,[34] in their studies on precision of appetite regulation, found that most animals, including man, when fed ad libitum (to take food when wanted) are able to do just this.

Pharmaceutical Aids. Different chemical compounds have been employed as aids in weight reduction. These compounds act either to decrease appetite (anorexigenic) or to increase metabolism.

Sapoznik[35] reports the findings of a clinical study in which six different compounds were utilized to induce decreased food intake and permit weight reduction. It was found that Benzedrine or Dexedrine, alone or in combination with thyroid hormone, produced loss of weight but also produced a number of undesirable side effects. Best results were obtained with a compound consisting of a combination of D-amphetamine, phenobarbitol, lipotropic factors, vitamin B_{12}, and methyl cellulose.

Other appetite-inhibiting compounds found to be useful are L-amphetamine sulfate and D-amphetamine sulfate. Phenobarbitol is recommended for use with these compounds to reduce side effects such as insomnia and hyperirritability. Endocrine medication is not usually recommended unless an endocrine disorder is a factor in the development of the obese condition. The use of such preparations for reducing purposes is not justified since the fine balance that exists between the various endocrine glands may be upset by such use, leading to more serious difficulties than obesity. The compounds cited should be used only on specific medical advice.

Rules to Follow for Successful and Healthful Dieting. Unfortunately there is no magic formula or ready-made guaranteed diet for permanent loss of weight. However, by approaching the problem with intelligence and determination, one can in time reduce to a proper weight and maintain that weight throughout his lifetime without undue hardship. The following should always be taken into account:

1. **Know what your desirable weight for your height should be.** Tables are presented from which one may determine his desirable body weight.
2. **Know how many calories are necessary to maintain desirable body weight.** These data can be obtained from tables. Table 13-1 can be used for this purpose. To adapt calorie allowances for persons whose weight and height are different from those of the reference man and woman, the following formulas, adapted to 20° C, have been derived by the Food and Agriculture Organization of the United Nations.

$$\text{Calorie allowance for men: } 0.95(815 + 36.6W),$$
$$\text{Calorie allowance for women: } 0.95(580 + 31.1W).$$
$$W = \text{desirable body weight in kilograms.}$$

[34] H. D. Janowitz and F. Hollander, "The Time Factor in the Adjustment of Food Intake to Varied Caloric Requirement." *Ann. N.Y. Acad. Sci.* **63**:56 (1955).

[35] H. Sapoznik, "Clinical Study on Weight Reduction in Obesity." *Am. J. Digest Diseases,* **22**:159 (1955).

3. **Plan a diet.** The diet should contain the calories necessary to maintain desirable body weight. The caloric content of the reducing diet will also necessarily vary with the individual and activity. Sedentary persons require less than more active persons. Remember that 3500 cal are the equivalent of 1 lb of body weight. A desirable weight loss is from 1 to 2 lb weekly.

   In planning a diet or using one you see or hear about, make sure it includes all the basic food groups:

   Meat
   Milk
   Vegetable–fruit
   Bread–cereal

   It is important also to remember that you do not have to cut out certain foods you like; just cut down. High protein–high fat diets aid in weight loss; high-carbohydrate diets do not.
4. **Exercise.** Moderate and regular exercise increases energy output without increasing your appetite. It is also important to remember that regular exercise, by toning up the muscles and improving the breathing and circulation, contributes to general good health.
5. **Have motivation, patience, determination, and willpower.** Reducing is always a struggle and usually a long one. Normally, keeping one's weight at a desirable level is a lifetime job, and therefore it takes determination and patience. A strong motivating force, a reason for weight reduction, is needed. Usually the fact that a reduction in weight makes one more physically appealing is sufficient, but the motivation can be made stronger if one keeps in mind the health benefits and the possible increased years of life. It is a definite statistical fact that the obese do not live as long as the lean. Willpower is needed to break bad habits of eating, which in some instances have been acquired from our environment. It is also needed to prevent one from turning to sweets or fattening foods to lift the spirits. If one perseveres, however, and follows his plan to its successful completion so that weight is lost and the diet becomes a permanent habit, the rewards are more than worth the struggle.

References

Agricultural Research Service. A Report on Better Foods and Nutrition. Washington, D.C.: U.S. Dept. of Agriculture, 1970.

Allison, J. B., and W. H. Fitzpatrick. *Dietary Proteins in Health and Disease.* Springfield, Ill.: Thomas, 1960.

Burton, B. T., et al. *The Heinz Handbook of Nutrition.* New York: McGraw-Hill, 1959.

Clark, G. W. *A Vitamin Digest.* Springfield, Ill.: Thomas, 1951.

Committee on Dietetics of the Mayo Clinic, *Mayo Clinic Diet Manual.* Philadelphia: Saunders, 1961.

Davidson, S., A. P. Meiklejohn, and R. Passmore. *Human Nutrition and Dietetics.* Baltimore: Williams & Wilkins, 1963.

Food and Nutrition Board. *Recommended Dietary Allowances* (Publication 1146). Washington, D.C.: National Research Council, 1964.

Gerard, R. W. *Food for Life.* Chicago: U. of Chicago, 1952.

Goldsmith, G. *Nutritional Diagnosis.* Springfield, Ill.: Thomas, 1959.

Hoebel, B. G. "Feeding: Neural Control of Intake," in *Annual Review of Physiology,* Vol. 33. Palo Alto, Calif.: Annual Reviews, 1971, p. 533.

Mellanby, E. *A Story of Nutritional Research.* Baltimore: Williams & Wilkins, 1950.

Merrill, A. L., and B. R. Watt. *Energy Value of Foods* (Agricultural Handbook 74). Washington, D.C.: U.S. Department of Agriculture, 1955.
Nizel, A. E. *The Science of Nutrition and Its Application to Clinical Dentistry*. Philadelphia: Saunders, 1966.
Rinkel, H. J., et al. *Food Allergy*. Springfield, Ill.: Thomas, 1950.
Sebrell, W. H., and R. S. Harris. *The Vitamins,* Vols. 1, 2, 3. New York: Academic, 1971.
Sinclair, H. M. *Essential Fatty Acids*. New York: Academic, 1958.
Taylor, C. M. *Food Values in Shares and Weights*. New York: Macmillan, Inc., 1959.
——— et al. *Foundations of Nutrition,* 6th ed. New York: Macmillan, Inc., 1966.
Underwood, E. J. *Trace Elements in Human and Animal Nutrition*. New York: Academic, 1971.
Wasserman, R. H., and R. A. Corradino. "Metabolic Role of Vitamins A and D," in: *Annual Review of Biochemistry,* Vol. 40. Palo Alto, Calif.: Annual Reviews, 1971, p. 501.

14

Kidney Function

We have already seen that man's food consists of certain materials such as proteins, fats, carbohydrates, various salts, and water. In previous chapters we traced the food from the digestive tract into the blood and from the blood into the tissues. In their passage through the blood and tissues of the body these nutrients are metabolized into final products such as urea, carbonic acid, water, and various salts. Many of the proteins contain sulfur and also have phosphorus attached to them; some fats taken in as food also contain phosphorus. These elements ultimately undergo oxidation into phosphates and sulfates and leave the body in that form along with other salts.

Generally speaking, then, the waste products of animal metabolism are urea, carbon dioxide, salts, and water. These are eliminated from the body by one of three main routes: the lungs, the skin, or the kidney. The lungs serve as the channel for the elimination of the carbon dioxide and a considerable quantity of water; this has already been discussed in Chapter 10. Through the sweat glands in the skin a small quantity of salts, a little carbon dioxide, and a variable but significant quantity of water are eliminated. The kidneys eliminate almost all the urea, the greater portion of the salts, and a large amount of water. Since practically all the nitrogenous waste leaves the body by way of the kidneys, they are especially important as excretory organs. In addition to removing metabolic waste, the kidneys also perform important **homeostatic** functions; that is, it is essential to the normal functioning of the cells that the surrounding extracellular fluid be maintained relatively constant in composition. The extracellular fluid, or the internal environment, as it is sometimes called, is the medium in which the cells carry out their vital activities. Changes in the chemical composition of the extracellular fluid are reflected in changes in the intracellular fluid, which will in turn influence cell function. Thus, it is imperative that this internal environment be maintained relatively constant in chemical composition, a function that is delegated mainly to two pairs of organs. They are the lungs, which control the levels of oxygen and carbon dioxide, and the kidneys, which maintain the chemical composition of the body fluids at the proper concentrations. In this chapter we will be concerned with the kidneys and how they function in the elimination of waste products and the regulation of the internal environment.

Structure of the Kidney

Macroscopic Structure of the Kidney

The kidneys consist of two bean-shaped structures situated on the posterior abdominal wall opposite the last thoracic and the first three lumbar vertebrae (Fig. 14-1). The right kidney is usually situated lower than the left, a condition due to the presence of the liver on the right side. The kidneys are occasionally congenitally located quite low in the abdomen, a condition known as **ectopia** of the kidney. The shape of the kidneys is an elongated oval with a notch taken out of one side (Fig. 14-2). The region of the notch is called the **hilum,** and it is through this region that the blood vessels and nerves pass. Above this region the upper pole of the organ is surrounded by the adrenal gland, and below this region the ureter leaves the kidney. In position the kidney occupies a vertical position with the hilus facing medially. The surface of the kidney is relatively smooth and is covered with a white, thin, rather resistant membrane 0.1 to 0.2 mm in thickness, called the **fibrous capsule.** The whole organ is surrounded and kept in place by a fatty capsule consisting of grayish-yellow, pale, soft adipose tissue

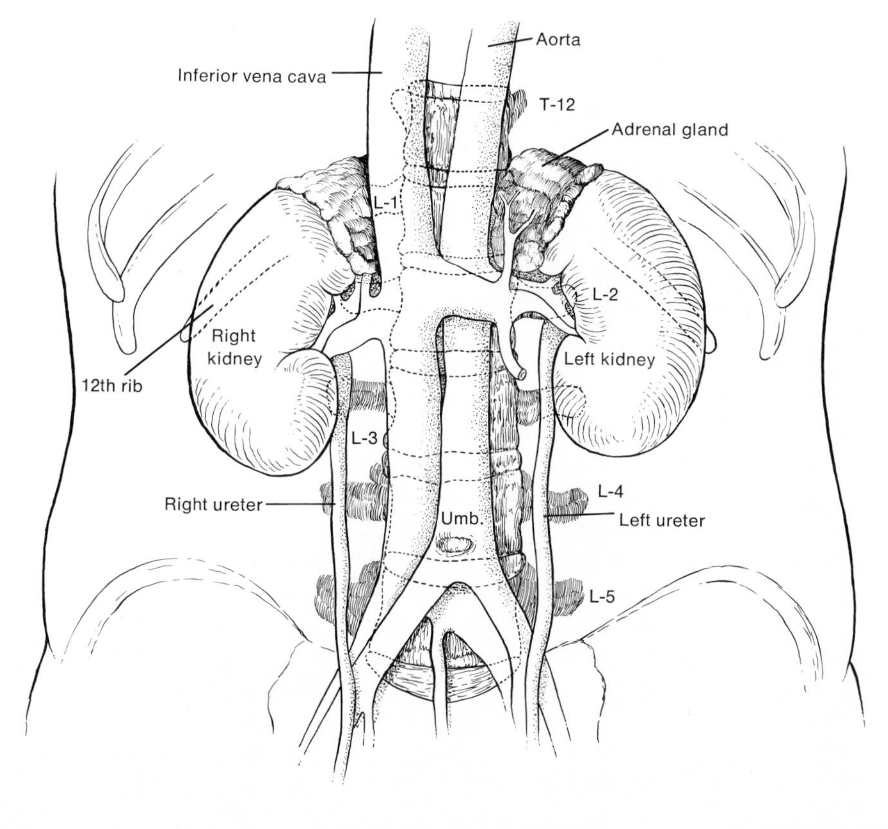

Figure 14-1. Diagram showing location of the kidneys. They are situated on the posterior abdominal wall opposite the last thoracic and the first three lumbar vertebrae.

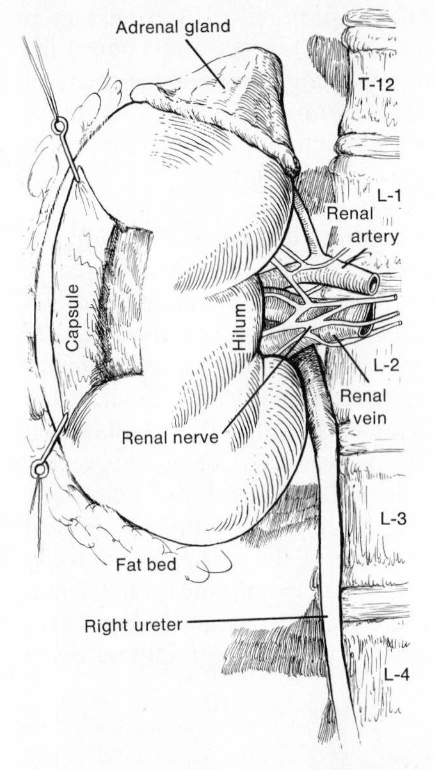

Figure 14-2. Anterior view of right kidney showing location of specific structures, such as adrenal gland, capsule, hilum, renal nerve, artery, and vein.

2 to 3 cm in thickness. In spite of its tendency to fix the kidney in place, the fatty capsule permits movement of the organ up and down and possibly also laterally.

The size and weight of the kidneys vary considerably in different persons. The average kidney measures approximately 5 inches in length, 3 inches in width, and 1 inch in thickness. Its total volume is about 130 to 150 cc; its total weight is about 170 g but may vary from 107 to 284 g.

If the kidney is sectioned longitudinally and the ureter followed inward, it opens into a large cavity called the **renal pelvis,** from which are given off a number of short, broad tubes, called **calyces,** projecting still further into the kidney substance (Fig. 14-3). The solid substance of the kidney can be differentiated into (1) an outer layer of soft granular material, reddish-brown in color, called the **cortex;** and (2) an inner portion, called the **medulla,** which is divided into cone-shaped **pyramids.** The cortex measures from 2 to 5 mm and extends from the fibrous capsule to the base of the pyramids, between which it dips down and extends toward the pelvis, forming a number of columns known as the **columns of Bertin.** The medullary substance surrounds the renal pelvis and is characterized by its hard consistency and deep red color. It is made up of eight to fifteen triangular bodies known as pyramids, which present a broad convex base facing toward the cortex and a blunt apex, the **papilla,** which projects into the calyces. These papillae contain the openings of from seven to fifty ducts known as the collecting tubules. The pyramids are separated from each other by a quantity of cortical substance, which creeps down their sides toward the pelvis.

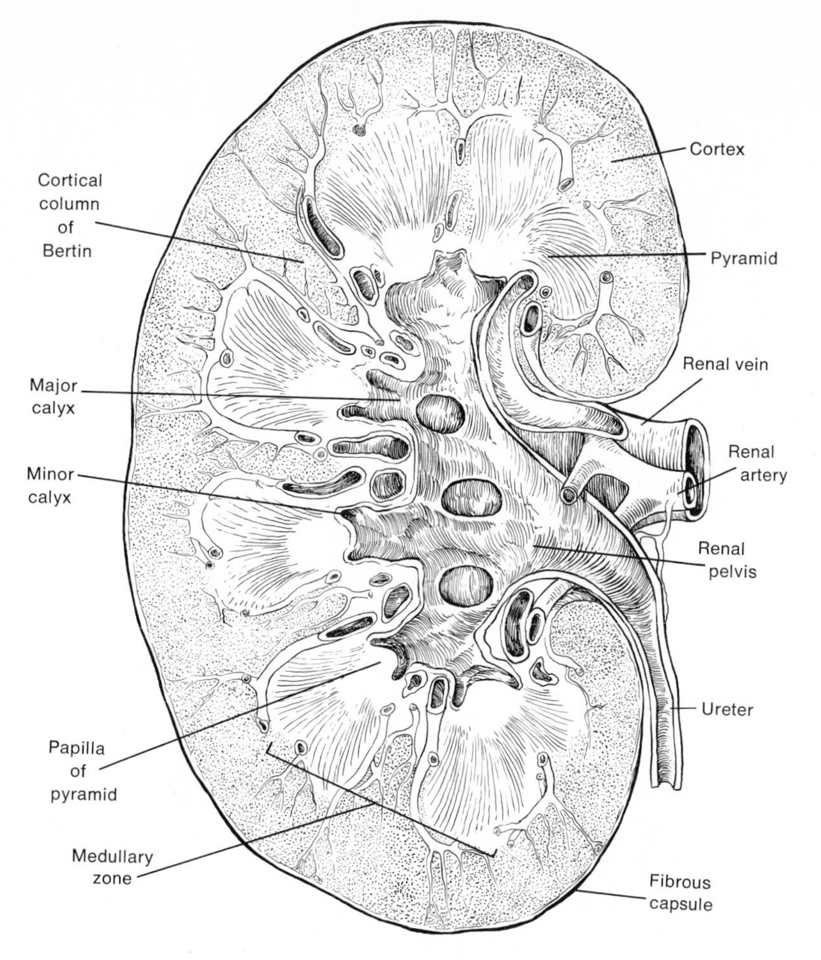

Figure 14-3. Longitudinal section of human kidney.

Microscopic Structure of the Kidney

In a microscopic section of the kidney there are six main elements to be seen consisting of (1) renal tubules; (2) the blood vessels—arteries, veins, and capillaries—that surround the tubules; (3) lymphatics; (4) nerves; (5) connective tissue; and (6) the renal, or malpighian, corpuscles.

Each tubule (Fig. 14-4) begins somewhere in the cortex, in a peculiar expanded structure called **Bowman's capsule,** from which point it takes a complicated course throughout the kidney. Coming away from Bowman's capsule the tube undergoes a series of twists, forming the portion of the tubule known as the **proximal convoluted tubule.** The last portion of the proximal convoluted tubule acquires a straight downward course, passing toward the renal pelvis, dipping down into the medulla for a certain distance, forming the **descending limb** of Henle's loop. Before reaching the renal pelvis it turns sharply, forming **Henle's loop,** and pursues a backward, nearly straight course parallel to its former one, until it finds itself back again in the cortex. This portion of the tubule is called the **ascending limb** of Henle's loop. Having reached some part or other of the cortex in a more or

Figure 14-4. Kidney tubule, its parts and its vascular relations.

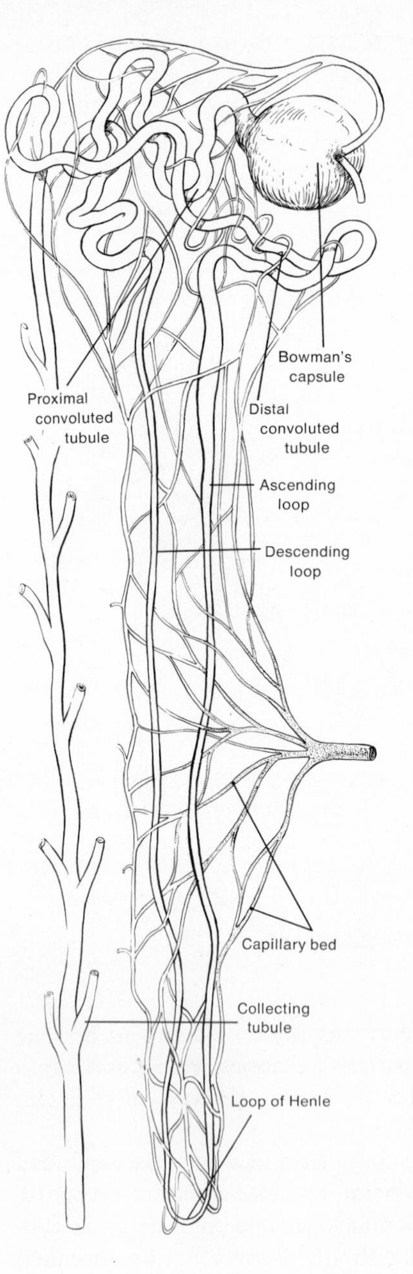

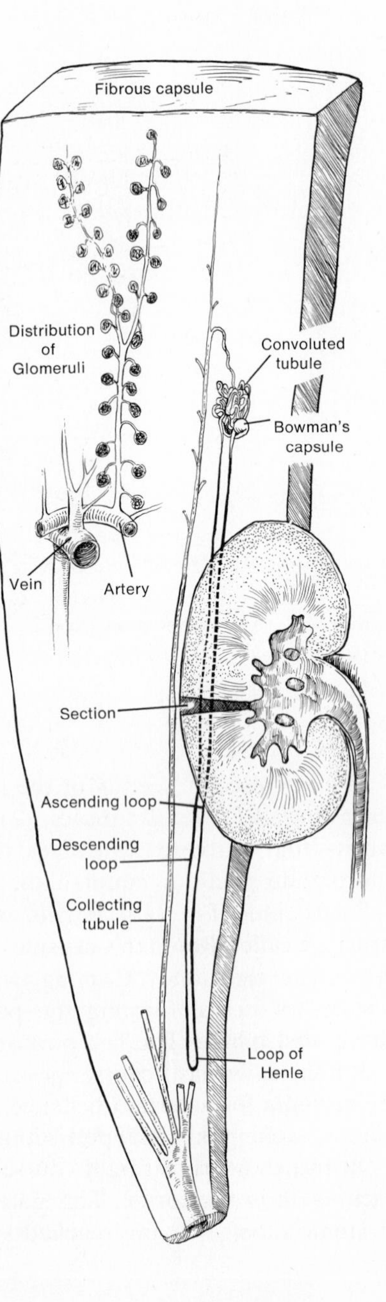

Figure 14-5. Relative position of the renal tubule and the course it takes through the kidney substance. The distribution of glomeruli through the cortex is shown also.

less straight line, the ascending limb undergoes a series of twists, or convolutions, almost identical with those of the initial convoluted portion and is called the **distal convoluted tubule.** The distal convoluted tubule then passes straight downward toward the medulla, becoming a **collecting tubule,** which passes toward the papilla, where it joins with similar collecting tubules, forming a large papillary duct measuring 0.2 to 0.3 mm in diameter and which emerges in the papilla. A collecting tubule may receive a number of short junctional tubules that connect several distal convoluted tubules with the collecting tubule. Thus, each renal tubule starting from a Bowman capsule becomes in succession a proximal convoluted tubule, a descending limb, a loop of Henle, an ascending limb, a distal convoluted tubule, a collecting tubule, and finally a papillary duct, which is the discharging tubule opening into the calyces. The loop of Henle, the lower part of the collecting tubule, and the papillary duct lie in the medulla and form part of the pyramids. The convoluted portions of the tubules, the upper portions of the collecting tubules, and the ascending and descending limbs, along with the malpighian corpuscles, lie in the cortex. The nephrons that have glomeruli in the outer portions of the renal cortex have relatively short thick loops of Henle. Some of these lie entirely within the cortex. The nephrons that have glomeruli in the inner layer of the cortex just above the medulla (juxtamedullary nephrons) have extremely long thin loops extending down to the tips of the papillae. Figure 14-5 illustrates the structure of the renal tubule and the course it takes through the kidney substance.

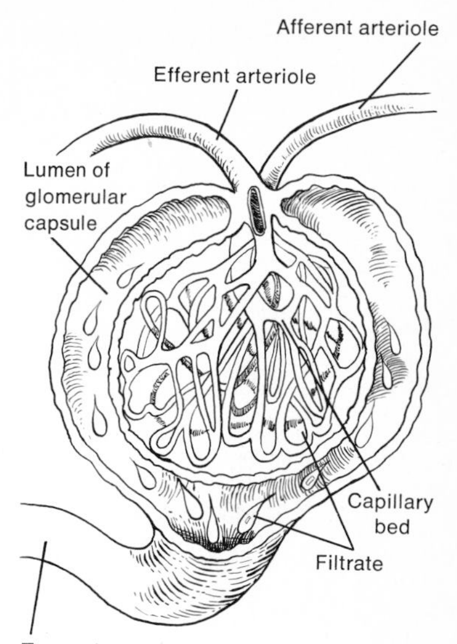

Figure 14-6. Structure of the malpighian corpuscle, showing the glomerulus embedded in Bowman's capsule.

The malpighian corpuscle (Fig. 14-6) is a structure about 0.2 mm in diameter, consisting of the upper expanded end of a tubule (Bowman's capsule) and a vascular tuft of blood vessels contained therein, known as a **glomerulus.** The glomeruli are confined to the cortex and are arranged in twelve to sixteen layers, which vary in direct proportion to the size of the cortex. Each glomerulus consists of a network of specialized capillaries arising from an arteriole called the **afferent arteriole** and terminating by uniting into an **efferent arteriole,** which carries blood away from the glomerulus. The efferent arteriole breaks up into a second capillary network surrounding the convoluted tubules, the descending and ascending limbs, and from which the blood eventually passes into the renal vein. The functional unit of the kidney, the **nephron,** consists of a malpighian (renal) corpuscle and a renal tubule, which is approximately 3 cm long. There are nearly a million of these units in each human kidney.

The kidney is richly supplied with nerves, mainly from the sympathetic, which arise from the tenth, eleventh, and twelfth thoracic segments of the spinal cord. Branches from the vagus supply the parasympathetic innervation. The renal nerves run in a plexus, the branches of which follow the blood vessels supplying these structures and also the renal corpuscles, the epithelial cells of the tubules, and the muscular tissue of the pelvis. Stimulation of the renal nerves produces vasoconstriction and a reduction in urinary output. There is no evidence that tubular activity is affected directly by such stimulation.

The various structures of the kidney, including the blood vessels, renal tubules, lymphatics, and nerves, are held together by a reticulum of delicate and coarse connective tissue fibers, which are continuous with the fibrous capsule.

Formation of Urine

The formation of urine and the regulation of the internal environment by the kidney are a composite of the following four processes: (1) the filtration of the blood through the renal corpuscle; (2) the selective reabsorption of materials by the renal tubule, which are required in maintaining the internal environment; (3) the excretion and secretion by the tubules of certain substances from the blood into the tubular lumen; and (4) the conservation of base for the body by exchange of hydrogen ions for sodium ions and the production of ammonia in metabolic acidosis. The anatomical unit of the kidney carrying out these functions is the nephron, and the end product of all these activities is urine. Urine is collected by the collecting tubule and carried to the renal pelvis. From the renal pelvis it is carried by way of the ureter to the bladder.

Filtration of Blood Plasma Through the Glomerulus

Bowman, in 1842, was the first to find the afferent and efferent arterioles of the glomerulus. He also discovered the capsule that bears his name and the fact that this capsule is an expansion of the upper end of the renal tubule. He guessed that the glomerular capsule was there to produce a filtrate of blood plasma. At about the same time Ludwig studied the peculiar arrangement of the glomerulus and concluded, because of the abrupt transition of the narrow afferent arteriole to the wide vascular bed of the capillaries and the issuing efferent arteriole with a diameter narrower than the afferent, that the renal corpuscle was designed for filtration. He also concluded, because of this peculiar arrangement, that the pressure in the glomerular capillaries is higher than that in the capillaries in other regions of the body and also higher than the pressure in the capsular space. Ludwig assumed, therefore, that, because of the higher pressure in the glomerular capillaries, water and simple solutes of low molecular weight are pressed out of the blood plasma through the walls of the capillaries of the glomerulus and through the membranous layer of Bowman's capsule. This was known as the *filtration theory of Ludwig*. Since then, the correctness of the filtration theory has been shown by the work of Richards, Walker, and their associates. They have found that the glomerular fluid consists essentially of plasma without protein and that it contains the same concentration of glucose, amino acids, chloride, and urea as blood plasma. They have also demonstrated by injecting into an experimental animal's circulation a polysaccharide of large molecular size (inulin) and finding that it appeared in the glomerular filtrate in the same concentration as the plasma, that the transference cannot be a diffusion process but is one of filtration. If it were a process of diffusion, the large molecules would pass across the membrane much more slowly than the small molecules. The only possible explanation from the facts is that the transference is one of filtration in which all molecules smaller than the pores of the filter pass through. Thus, the only difference between the fluid to be filtered (blood) and the filtrate is the absence in the latter of substances of molecular size too large to pass through the openings. Clearance studies with polysaccharides of different molecular sizes have shown that the pores in the glomerular membrane are not of equal size. All are permeable to molecules of molecular weight 5,000, 20 per cent pass molecules of 30,000, and a few allow molecules of molecular weight up to 70,000 to pass.

Since the transference process in the glomeruli is one of filtration, the amount of the filtrate must depend on the pressure and the rate of flow of the blood in the glomerular capillaries. Whenever there is an increased blood flow through the kidney with an increase in blood pressure, the flow of urine is increased. A decrease in the rate of blood flow through the kidney with a decrease in blood pressure is accompanied by a decrease in the flow of urine. When the general blood pressure falls to 40 mm Hg, there is a complete cessation of urine flow. Starling has shown that the proteins of the blood plasma exert an attraction force for water (osmotic pressure), which is equivalent to about 30 mm Hg. For the glomerulus to produce a protein-free filtrate, this pressure must be overcome, and a pressure of 40 mm is the least that is necessary to overcome that. Indirect measurements by Winton of glomerular pressure in the excised dog kidney give a value of 70 per cent of the arterial pressure, which would mean that approximately 75 mm Hg pressure is actually exerted on the glomerular capillaries. The other capillaries of the body usually have a pressure of only 25 to 30 mm Hg. The osmotic pressure of the plasma proteins is equivalent to 30 mm Hg and opposes this outflow pressure of 75 mm exerted by the arterial blood. The pressure of the interstitial fluid, which is about 10 mm Hg, together with the pressure within the tubular system, which is also approximately 10 mm Hg, also contributes to the opposition to the outflow pressure exerted by the arterial blood. This reduces the effective filtration pressure of 75 mm to a net of 25 mm Hg. Figure 14-7 illustrates the production of this net pressure of 25 mm Hg, which supplies the energy for filtration.

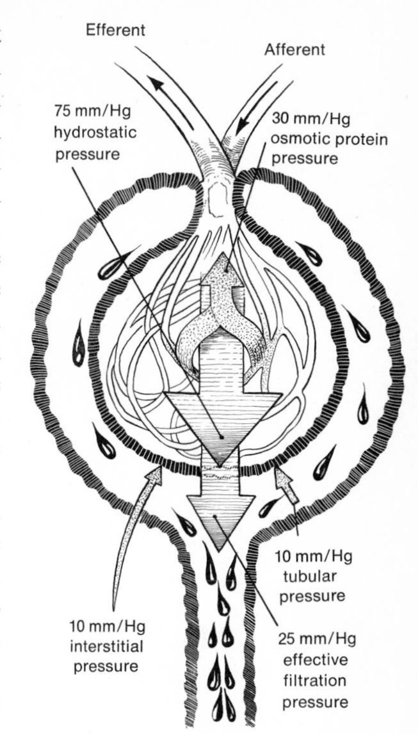

Figure 14-7. Production of an effective filtration pressure of 25 mm Hg within the glomerulus.

From what has already been discussed, it is apparent that the main function of the glomerulus is the filtration of blood plasma. This process involves the pressing out of water from the blood with its dissolved substances, such as glucose, chloride, amino acids, urea, sodium, potassium, and carbonate, through the wall of the blood capillaries and the epithelium of the capsule into the lumen of the tubule. The blood cells, droplets of fat, and very large molecules, such as proteins, are unable to pass through the pores of the glomerular membrane and are held back. The filtrate in the lumen of the renal tubule contains all the other constituents of the blood in about the same concentration as before filtration. It is also apparent that the quantity of filtrate produced by the glomerulus is determined by the net filtration pressure and by the amount of blood flowing through the kidneys.

Factors that could affect filtration would be

1. Changes in the normal arterial pressure. A drop in pressure would reduce the net glomerular filtration pressure, and urine formation would decrease or cease entirely if the pressure fell low enough. An increase in pressure would cause an increase in filtration and an increase in urine formation.
2. Gross changes in the vessels of the kidneys. Richards and his associates have shown that glomerular pressure and therefore glomerular filtration are also regulated by adjustment of the size of the lumen of the afferent and efferent vessels in relation to one another. For example, minute doses of epinephrine, which cause a constriction of blood vessels, cause a greater degree of constriction in the efferent than in the

afferent arteriole. This causes a decrease in the outflow of blood as compared with the inflow, resulting in an increase in the intraglomerular pressure and therefore in the flow of urine. Large doses of epinephrine, by constricting the afferent artery considerably and thereby decreasing the inflow of blood to the kidney to a great extent, lessen glomerular pressure and the flow of urine. The action of the sympathetic nervous system is the same as that of epinephrine. Fibers from the sympathetic nerves supply both the afferent and efferent arterioles and act to regulate their caliber and thus control intraglomerular pressure and urinary flow according to the needs of the organism for excretion.

Various substances that are excreted by the kidneys also aid in regulating glomerular pressure. Water, salt, urea, and certain drugs have a dilating effect on the afferent arteriole, more so than on the efferent arteriole. More blood flows into the glomerulus. As a result, these substances cause an increase in glomerular pressure and therefore an increase in filtration and urine formation. Other factors can also affect filtration: an increase in the interstitial pressure such as occurs in renal edema or an interstitial inflammatory process could reduce the net effective filtration pressure and thereby reduce urine formation. Obstruction of the arteriole pathway of the glomerulus could also produce the same results.

Glomerular Filtration Rate

A large volume of blood is filtered each minute by the glomeruli of both kidneys. In the normal adult from 750 ml to 1 liter of blood per minute flows through both kidneys. Thus, a volume of blood equal to the total blood volume of the body passes through the renal circulation in 4 or 5 minutes. At this volume of blood and a net filtration pressure of 25 mm Hg, approximately 125 ml of filtrate are forced through the glomerular membranes of the two kidneys per minute. The glomerular filtration rate is therefore about 125 ml per minute. Since the glomerular membrane is a simple filter, no work is done by the kidney in the filtration process. The energy for this action is supplied by the heart.

Tubular Function

As the glomerular filtrate passes from the capsular space into the lumen, it contains all the constituents of the blood except those that are of a molecular size too large to pass through the pores of the glomerular membrane (proteins, droplets of fat, blood cells) and in about the same concentration as before filtration. As a result, the filtrate contains many substances necessary for normal metabolism, such as water, glucose, amino acids, and chlorides. In addition, it contains waste substances to be eliminated from the body, such as urea, uric acid, creatinine, sulfates, phosphates, pigments, and other substances. The passage of the filtrate down the lumen of the tubules brings it into contact with an enormous cellular surface. There the filtrate becomes concentrated by two activities that appear to be the function of the epithelial cells of the renal tubule. First, the epithelial cells take certain substances out of the filtrate and pass them back into the blood by way of the capillaries surrounding the tubule. Second, the epithelial cells of the tubule take certain substances out of the blood and add them to the concentrated filtrate. Furthermore, under various conditions greater or smaller amounts of useful substances are retained in accordance with the need to maintain the constancy of the internal fluids of the body. Thus, the kidney tubule, by its highly selective reabsorption capacity and its ability to excrete waste substances from the body, modifies the glomerular filtrate, and the constancy of the internal environment is maintained. The waste products of protein metabolism are eliminated, and the acid–base balance of the body is supported. The end product of all this activity is **urine.**

The maintenance of the normal volume and composition of the blood is of fundamental importance to the body, because human cells are extremely sensitive to marked changes in certain constituents of their environment. Those substances that constitute the chief part of this internal environment and that the body needs in its economy are proteins, amino acids, fats, glucose, water, chloride, sodium, potassium, calcium, carbonate, and magnesium. The kidneys, as organs of homeostatic function and excretion, hold back entirely such substances as fats and proteins, which therefore do not, under normal circumstances, appear in the urine at all. Other substances that are contained in the plasma at a definite concentration and appear in the glomerular filtrate are reabsorbed almost completely by the tubule when their concentrations in the plasma are within normal range. These same substances are excreted by the kidneys when their normal plasma levels are exceeded. Cushny spoke of these substances as **threshold substances.** Glucose, amino acids, chloride, sodium, potassium, and creatine are found only in the urine in very slight concentration but appear in larger amounts when their normal value is exceeded in the blood.

Substances that are reabsorbed only slightly or not at all by the kidney tubule are referred to as **low-threshold substances.** Waste products such as urea, uric acid, creatinine, phosphate, and sulfates are examples of low-threshold substances.

Tubular Reabsorption

Richards and his associates have obtained samples of glomerular filtrate by introducing a slender, pointed tube (micropipette) into the intracapsular space and removing the glomerular fluid. Analyses of such samples have shown them to be filtrates of blood plasma. However, on comparison of glomerular filtrate with that of urine, the concentrations of substances in these filtrates are found to be very different. For example, chloride and sugar are found in large concentrations in glomerular filtrate but only in traces in urine obtained from the bladder. It is apparent that the composition of the tubular fluid must be altered considerably in its passage through the kidney tubule, and some substances that are passed through the glomeruli are reabsorbed in the tubules. The capacity of the tubular cells to absorb these various substances from the tubular filtrate is limited and has a different value for different materials. The maximum reabsorptive capacity of the tubular epithelium for a particular substance is referred to as the Tm (tubular maximum).

Reabsorption of the various constituents occurs in different areas of the tubule and may occur by active or passive mechanisms. Glucose, sodium salts, and part of the water and chloride are reabsorbed in the area of the proximal tubule. The ascending limb and the distal convoluted tubule absorb chloride, alkaline salts, and most of the remaining water.

The mechanism involved with the **reabsorption of glucose** is still not clearly understood. At one time it was thought that glucose transport was regulated by an enzymatic mechanism which involved the phosphorylation of glucose in the cells of the proximal convoluted tubule, the reaction being catalyzed by the enzyme **hexokinase.** The glucose was thought to be released to the blood through the peritubular capillaries, where the glucose phosphate complex is broken. Proof for this view was dependent on the fact that the drug phlorhizin was also thought to inhibit the enzymatic hexokinase phosphorylating mechanism, which results in a failure of the renal tubes to absorb glucose. Thus glucose was excreted in the urine, producing a "phlorhizin glycosuria."

It is probable that glucose is not phosphorylated in the course of its transport but is actively transported against a chemical gradient by an energy-requiring membrane **carrier system.** The carrier is thought to be an enzyme with a prosthetic group specific for the transported substance. The carrier substance is thought to be present in the luminal membrane of the tubular cells in fixed and limited amount. At the luminal surface the carrier combines reversibly with glucose present within the tubular fluid to form a complex which then migrates to the cytoplasmic surface of the tubular cells, where it uncouples delivering glucose to the cytoplasm. Energy is required from the splitting of ATP either to effect combination or to split the carrier complex. Phlorhizin is now thought to bind strongly to the membrane carrier, preventing the attachment of glucose and its entry into the cell. Anything that interferes with the chemical processes supplying adenosine triphosphate blocks glucose reabsorption also.

Normally all the glucose filtered by the glomeruli, which amounts to about 125 mg per 125 ml of filtrate formed per minute, is reabsorbed by the tubules. However, the capacity of the carrier system is limited. If the plasma concentration of glucose is raised from a normal level of approximately 100 mg per 100 ml to 200 mg per 100 ml, the glomerular filtrate, which is formed at the rate of 125 ml per minute, will contain 250 mg of glucose instead of 125 mg. Extra glucose presented to the tubule will be reabsorbed until the capacity of the carrier system is exceeded. This normally occurs when the glucose plasma concentration reaches about 280 mg per 100 ml, which would result in a glomerular filtrate concentration of 350 mg per 125 ml. The Tm for glucose reabsorption from the above example would be 350 mg per minute. Under normal circumstances when the glucose concentration of the blood reaches 180 mg per 100 ml, resulting in a glomerular filtrate glucose concentration of 225 mg per minute, glucose is presented to the tubule cells at a faster rate than they can absorb it, and it begins to appear in the urine, producing a glycosuria. This level of sugar is referred to as the **renal threshold** for glucose.

Wesson and his associates demonstrated that sodium, chloride, and bicarbonate are reabsorbed from the proximal convoluted tubule by an active process, whereas water is reabsorbed in this area passively. Adrenocortical hormone favors sodium reabsorption. Bicarbonate reabsorption appears to take place in two steps: one step involving a nonenzymatic mechanism and the second step involving an enzymatic component.[1] Concentrations of plasma pCO_2 influence the reabsorption

[1] W. B. Schwartz, A. Falbriard, and A. S. Relman, "An Analysis of Bicarbonate Reabsorption During Partial Inhibition of Carbonic Anhydrase." *J. Clin. Invest.*, **37**:744 (1958).

of bicarbonate, especially at normal or diminished pCO_2 limits. The enzyme carbonic anhydrase contributes a constant fixed amount to bicarbonate reabsorption at all plasma bicarbonate concentrations.[2]

Water reabsorption occurs in two different areas of the renal tubule. Approximately 80 per cent of water reabsorption occurs in the proximal tube; it is considered to occur by a passive mechanism and is termed **obligatory reabsorption.** That is, water is reabsorbed in the proximal tubule as a solvent for such solutes as glucose and sodium. The observations made by Windhager and his associates[3] clearly demonstrated this role of solute transfer in water transport. As these solutes are reabsorbed, the tubular filtrate tends to become hypotonic to the interstitial fluid surrounding the tubular epithelium. As a result more water passes from the lumen into the interstitial spaces to satisfy osmotic equilibrium. Obligatory reabsorption of water thus takes place purely in a passive fashion and occurs without regard for the water requirement of the body. By the time the tubular filtrate reaches the distal convoluted tubule, the original 125 ml of filtrate produced each minute is reduced in volume approximately 80 per cent, or to a volume of 25 ml.

In the distal convoluted tubule water is reabsorbed by an active process to satisfy the water needs of the body, and this is referred to as **facultative reabsorption.** In this case water must be transferred through the tubular membrane against an osmotic gradient, since osmotic equilibrium was reached in the proximal portion, and therefore requires an energy source. The mechanism that mediates facultative reabsorption is the action of a hormone produced by the posterior pituitary called the **antidiuretic hormone (ADH).** ADH in some way affects the permeability coefficient of the tubular epithelium. Some research workers observed that hyaluronic acid disappeared from the basement membrane of the distal tubule when ADH was administered to rats and proposed that ADH increases the permeability of the tubules to water by its effect on hyaluronidase activity. In the absence of ADH extreme diuresis occurs and up to 3000 ml of urine may be eliminated per day. Ingestion of large quantities of water suppresses the secretion of ADH, resulting in the elimination of excess water from the body.

Facultative reabsorption of water accounts for approximately 18 per cent of the water reabsorbed, so the volume of filtrate (25 ml) that reaches the distal tubule is further reduced by 23 ml. Thus from an original glomerular filtrate volume of 125 ml per minute, a final volume of 2 ml of urine per minute is produced.

Countercurrent Theory of Urine Concentration. As the tubular fluid passes through the loop of Henle the **osmotic concentration** (the osmolarity) is drastically altered and enters the distal tubule in a hypotonic state. The osmotic concentration refers to the number of particles dissolved in solution. The glomerular filtrate has a concentration close to that of plasma and although 80 per cent of the glomerular filtrate is reabsorbed in the proximal tube, the remaining 20 per cent

[2]F. C. Rector and D. W. Seldin, "In Vivo Kinetics of Renal Carbonic Anhydrase." *Federation Proc.,* **18:**125 (1959).

[3]E. E. Windhager, G. Whittembury, D. E. Oekn, H. J. Schatzmann, and A. K. Solomon, "Single Proximal Tubules of the Necturus Kidney." *Am. J. Physiol.,* **197:**313 (1959).

passing into the descending loop of Henle has essentially the same concentration; that is, the fluid is approximately isotonic. However, as this isotonic fluid passes through the descending limb, it becomes hypertonic due to the passive diffusion of sodium into the tube from the surrounding medullary tissue, which has had its sodium concentration increased by active transport of sodium out of the ascending limb into this area and also by the passive diffusion of water from the descending limb into the surrounding tissue. The situation is illustrated in Fig. 14-8. The hypertonic tubular fluid now entering the ascending limb has its sodium concentration progressively decreased by active transport of sodium from the ascending limb into the surrounding medullary tissue as described above without a compensatory shift in water because the ascending limb of Henle is impermeable to water. Thus, the tubular filtrate becomes hypotonic by the time it reaches the distal convoluted tubule. The antidiuretic hormone (ADH) specifically increases the permeability to water of the distal convoluted tubule. In the presence of ADH hypotonic fluid in the distal tubule becomes isotonic. The collecting ducts act as osmotic exchangers which permit the isotonic fluid which enters it from the distal convolution to come to equilibrium with the hypertonic condition of the medullary tissue. Water is given up by the isotonic tubular filtrate

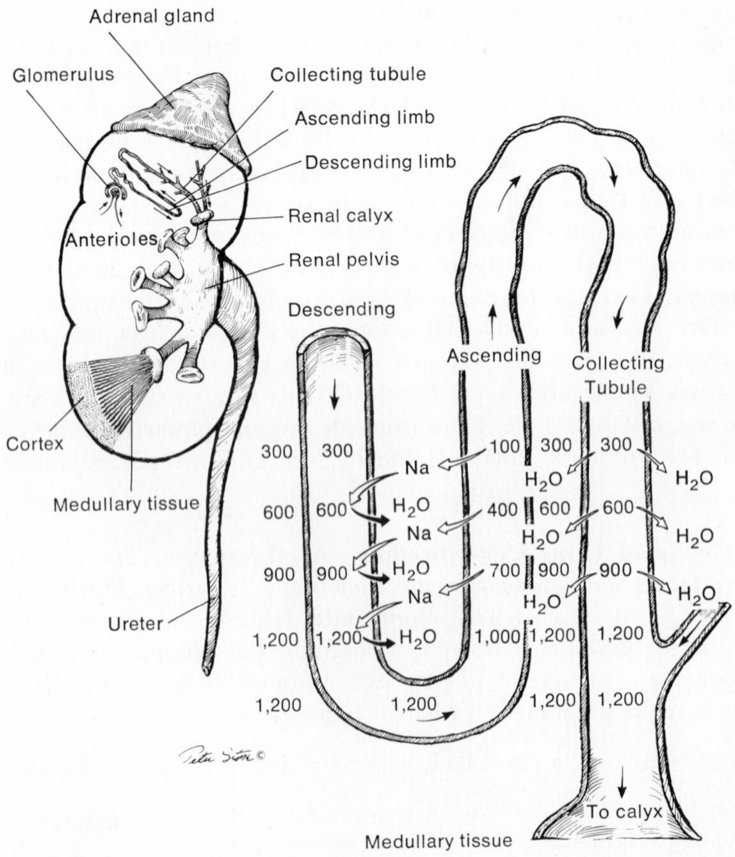

Figure 14-8. Mechanism of the counter-current theory of urine concentration. Flow in the descending limb is in the opposite direction of flow in the ascending limb. The ascending limb shows active transport of sodium from lumen to surrounding medullary tissue but is impermeable to water. This causes the medullary tissue to have a progressively higher concentration of sodium going from the convolutions to the bend of the loop of Henle. This provides an osmotic gradient that will remove water from the collecting duct as shown. The numbers indicate the concentration of sodium in milliosmoles. An osmole (Osm) is the number of particles in solution that will depress the freezing point of water to −1.86°C. One osmole is equal to 1,000 milliosmoles (mOsm).

to the hypertonic medullary tissue and the final urine attains an osmotic concentration equal to that of the medullary tissue.

Tubular Excretion and Secretion

As it passes down the renal tubule, the glomerular filtrate is modified and concentrated by the selective reabsorption of water and various threshold substances dissolved in it by the tubule epithelium. In addition, the filtrate can be modified when the tubule cells take substances from the blood in the peritubular capillaries and transfer them into the lumen of the tubule, an action more appropriately called **tubular excretion,** but more commonly referred to as **tubular secretion.** It has been demonstrated that the tubules in some animals are capable of excreting or secreting water, creatine, creatinine, magnesium, potassium sulfate, chloride, urea, uric acid, and certain dyes and drugs. Whether all of these can be excreted by the human kidney is not known, although undoubtedly the human renal tubule can excrete certain substances, especially creatinine, potassium, and hydrogen ion, since these substances appear in greater concentrations in the urine than can be accounted for on the basis of glomerular filtrate concentration.

The renal cells can produce certain substances such as ammonia and hippuric acid and secrete them into the tubular fluid. Ammonia formation occurs in the distal tubule cells, with glutamine being the most important source of the ammonia. In the presence of the enzyme glutaminase, which is especially concentrated in the distal tubule cells, glutamine is hydrolyzed into glutamic acid and ammonia. This reaction may be written

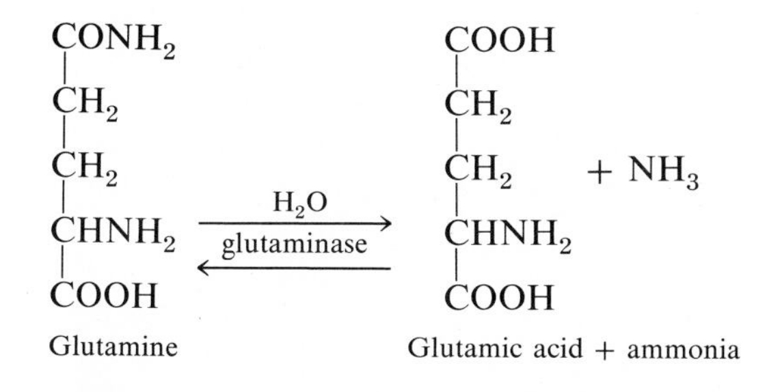

Ammonia secretion by the kidney is stimulated in cases of increased plasma hydrogen-ion concentration so that basic sodium ion may be conserved. In this way, together with its ability to excrete hydrogen ion and nonvolatile acids produced in metabolism, such as lactic, sulfuric, and phosphoric acids, the kidneys play an important part in acid–base regulation.

The ability of the kidney to exchange hydrogen ion for basic sodium ion is also an important device by which the kidney conserves base for the body, and the normally alkaline glomerular filtrate (pH 7.4) is converted to an acid urine of pH 6. This exchange of H^+ occurs in the distal tubule, where the acidifying cells exchange a hydrogen ion for a sodium ion in the glomerular filtrate. Alkaline sodium phosphate present in the glomerular filtrate gives up a sodium ion and takes on a hydrogen ion secreted by the tubular cells, becoming acid sodium phosphate. The sodium ion is taken up by the distal cells and combined with bicarbonate to produce sodium bicarbonate. The mechanism of the hydrogen-ion exchange may be illustrated as follows:

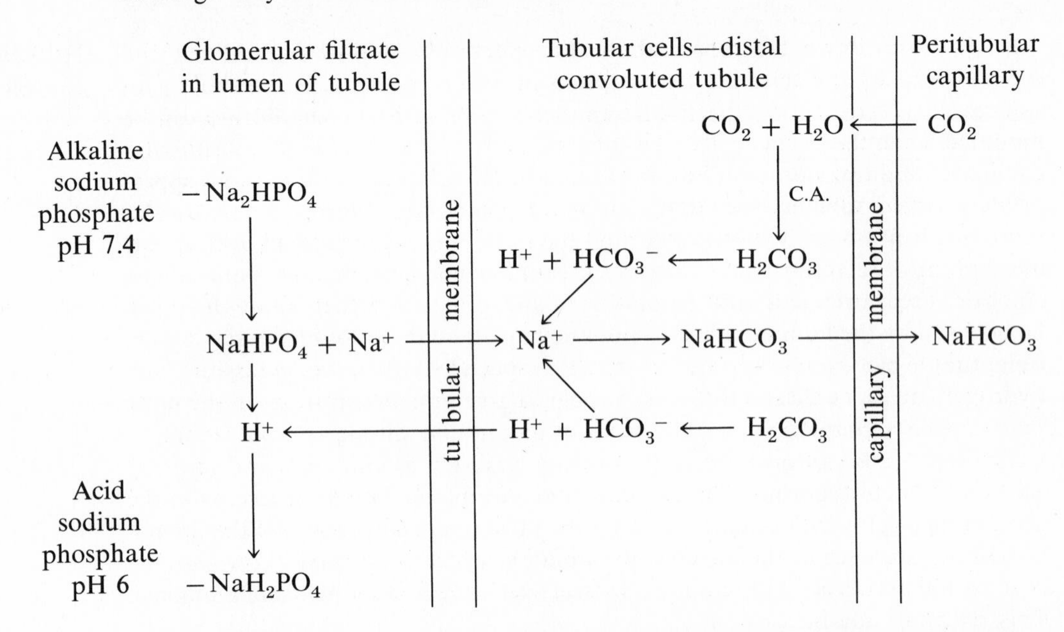

The renal tubules thus, by a considerable range of functional activity in modifying the glomerular filtrate by selective reabsorption, excretion, and secretion, produce urine. In so doing, the concentration of the normal constituents of blood is maintained at a constant level (homeostasis); nonvolatile products of metabolic activity are eliminated from the body, and acid–base equilibrium is supported. The amount of urine produced, as well as its composition and reaction, is adjusted to the needs of the body.

Volume and Characteristics of Urine

The quantity of urine eliminated per day in the normal adult varies from 600 to 2,500 ml and is influenced by such factors as water intake, environmental temperature, diet, physical condition, and mental state of the individual.

Since the skin is also responsible for eliminating large quantities of water from the body, we find a special relationship between the skin and the kidneys whereby the excretion of water from the skin in the form of sweat is stopped almost completely, and the cutaneous vessels are constricted. At the same time the blood vessels of the abdominal viscera and the kidneys are dilated. A full and rapid stream of blood is shunted through the glomeruli, resulting in an increased flow of urine. Conversely, when the body is exposed to warmth, the skin perspires freely and the cutaneous vessels are dilated, whereas the abdominal and renal vessels are constricted. The flow of blood through the glomeruli is decreased, and the amount of urine produced is scanty. Thus urine volume is lower in the summer and in warm climates.

When large quantities of fluid are ingested, practically all the water passes from the alimentary tract into the blood. The fluid so absorbed may influence the kidney in two ways: (1) excess water in the blood produces a local dilatation of the renal vessels, causing an increased volume of blood flow and thus increased filtration rate and urine formation; and (2) excess water suppresses the secretion of the antidiuretic hormone, resulting in the discharge of a greater quantity of a more dilute urine.

Certain substances taken in with water and food may increase the urine flow by acting either directly on the tubular epithelium as they are carried through the kidney in the blood or indirectly through some effect on the central nervous system. Such substances that increase urine flow are known as **diuretics.** Coffee, tea, alcoholic beverages, and nitrogenous end products have diuretic effects and will thus influence the over-all volume of urine output. Caffeine produces its diuretic effect by diminishing the output of ADH and by stimulating glomerular circulation. Low-threshold substances such as urea, sodium sulfate, and sodium phosphate resulting from protein metabolism, remaining in the tubules after selective reabsorption, retain water in the tubule to keep them in solution. After a meal, especially one high in proteins, an increased flow of urine is not unusual, for the products of digestion can act as diuretics.

In the same way high-threshold substances, such as glucose, sodium, chloride, and amino acids, when present in excess which cannot be completely reabsorbed, produce diuresis since they must be accompanied by water to keep them in solution. The diuresis that occurs in diabetes is a result of the excess glucose in the glomerular filtrate.

The powerful influence of emotions in promoting the production of urine is also well known. In extreme nervousness the discharge of urine in enormous quantity results from impulses originating in the brain and passing down to the kidney along the vasodilator fibers, leading to a dilatation of the renal vessels and an increased glomerular filtration rate.

In a normal subject the average rate of filtration through the glomeruli is approximately 125 ml per minute, with a urine formation of 1 to 2 ml per minute. The maximum urine flow does not exceed 20 ml per minute.

Normally urine is a yellow or amber color because of the presence of a pigment, urochrome, but this varies with the quantity voided and the concentration of solutes in the urine. The normal concentration of solids in the urine results in a specific gravity ranging from 1.003 to 1.030; the greater the concentration of solids, the higher the specific gravity; conversely, the lower the concentration, the lower the specific gravity. Water has a specific gravity of 1.000. In fever, because of concentration, that is, a larger amount of solids in comparison with water, the urine may be dark yellow or brown. Blood in the urine results in a reddish or red color. Coal-tar derivatives and bile pigments cause the urine to become greenish yellow or greenish brown. Certain compounds, such as methylene blue, when excreted in the urine color it a dirty green or blue, and various drugs, such as santonin, chrysophanic acid, and senna, cause the urine to become orange. Phenol derivatives produce a smoky-brown color in urine, and melanin and alkapton cause the urine to become black on standing. Table 14-1 lists some of the variations in the color of urine and their possible causes.

Normal urine, when freshly voided, is usually clear and acid in its reaction

Table 14-1. ***Some Variations in the Color of Urine and Their Possible Causes***

Color	*Cause of Coloration*	*Pathological Condition*
Nearly colorless	Dilution or diminution of normal pigments	Nervous conditions; hydruria, diabetes insipidus, granular kidney
Dark yellow to brown	Increase of normal, or occurrence of pathological, pigments; concentrated urine	Acute febrile diseases
Milky	Fat globules	Chyluria
	Pus corpuscles	Purulent diseases of the urinary tract
Orange	Excreted drugs	Santonin, chrysophanic acid
Red or reddish	Uroerythrin, uroporphyrin, coproporphyrin, hemoglobin, and myoglobin	Porphyrin, hemorrhages, hemoglobinuria, trauma
	Pigments of food (logwood, madder, bilberries, fuchsin)	
Brown to brown-black	Hematin	Small hemorrhages
	Methemoglobin	Methemoglobinuria
	Melanin	Melanotic sarcoma
	Hydroquinol and catechol	Phenol poisoning
Greenish yellow, greenish brown, approaching black	Bile pigments	Jaundice
Dirty green* or blue	A dark-blue scum on the surface, with a blue deposit, owing to an excess of indigo-forming substances	Cholera, typhus, seen especially when the urine is putrefying
Brown-yellow to red-brown, becoming blood red on addition of alkalies	Substances contained in senna, rhubarb, and chelidonium that are introduced into the system	

Courtesy Blakiston Division, McGraw-Hill Book Company, Inc., New York. From Hawk, Oser, and Summerson: *Practical Physiological Chemistry,* 1954.

*This dirty green or blue color also occurs after the use of methylene blue in the organism.

to litmus, having a pH of about 6. During digestion the urine may become alkaline because more bicarbonate is filtered. The increase in bicarbonate filtered is the result of an increase in the plasma level of bicarbonate produced during hydrochloric acid production. This may cause a precipitation of calcium phosphate, resulting in a turbid urine. This turbidity, of course, is without significance, and usually the urine may be made to clear by addition of a drop or two of acetic acid, which causes the phosphates to go into solution again. A persistent turbidity that cannot be removed by filtration may be due to the presence of pus or large numbers of bacteria, both of which are indicative of some pathology. The passage of cloudy urine is suggestive of the presence of blood, pus, or bacteria or of ammoniacal decomposition in the bladder. In strongly acid urine or when normal urine is chilled, uric acid salts are precipitated as a white, pink, or red sediment. When a chilled specimen is heated, the urates go into solution, and the turbidity disappears.

The odor of urine is normally aromatic but may be modified by various substances excreted in it. After the ingestion of asparagus, for example, urine takes on a characteristic odor due to the presence of methyl mercaptan, an inert substance left over from the metabolism of asparagus and excreted in the urine. In diabetes the urine has been described as having a sweetish odor, and in terminal conditions of the disease, because of the presence of acetone, the odor of this substance is quite noticeable. In cystitis the odor of ammonia may be present, and a fecal odor may be indicative of a heavy contamination with *Bacterium coli*. On standing, normal urine develops an ammoniacal odor because of the conversion of urea to ammonia, which also causes the urine to give an alkaline reaction.

Composition of Urine

Normal Constituents of Urine

Urine may be regarded as a complex substance composed of water and dissolved substances. Approximately one half of the dissolved substance is **urea** and is the main form in which nitrogen leaves the body. Urea is the chief end product of protein metabolism; 95 per cent of the total nitrogen of the urine comes from the urea, uric acid, creatinine, and ammonia. **Uric acid,** which is closely allied to urea, is an end product in the oxidation of purines in the body. It is always found in the urine of man, occurring in small but variable quantities. In the urine of some animals, such as birds and reptiles, uric acid replaces urea as the chief nitrogenous excretion. In addition to urea and uric acid, the urine contains small but variable quantities of **creatine** and **creatinine.** Creatinine is a hydrated form of creatine, which we have already discussed as a constituent of muscles. After being introduced into the alimentary tract or into the blood, creatine appears in the urine as creatinine. Creatine is present in the urine of children in much larger amounts than in the urine of adults. Normally there is very little **ammonia** in freshly voided urine.

Besides the above, all the natural-occurring **amino acids** have been found in urine in very small amounts. This very low concentration of amino acid is due to the high reabsorptive capacity the renal tubule has for these constituents.

The chief **inorganic salts** found in urine as sodium, potassium, calcium, and magnesium in the form of chlorides, phosphates, and sulfates. Sodium chloride is the most abundant and important inorganic constituent, composing approximately 44 per cent of the urine solids. A large part of the phosphoric acid seems to exist as acid sodium phosphate, the remainder as soluble calcium, magnesium, and potassium phosphates. Other inorganic salts present, but occurring in smaller quantity, are potassium and sodium sulfates and calcium chloride. The sulfates are derived mainly from protein because of the presence of the sulfur-containing amino acids, methionine and cystine. Vitamins, hormones, and enzymes can be detected in small quantities in normal urine. The enzymes amylase and pepsin

Table 14-2. ***Some of the Principal Urinary Constituents***

Constituent	*Grams per 24 Hours*
Water	600–2,500 ml, 600–2,500 g
Urea	33.0
Uric acid	0.6
Creatinine	1.0
Sulfate as H_2SO_4	2.0
Phosphate	1.7
Chlorine	7.0
Ammonia	0.7
Potassium	2.5
Sodium	6.0
Calcium	0.2
Magnesium	0.2
Glucose	trace

have been found in varying quantities, depending on diet. Rennin has also been found and, at times, trypsin. From this it appears that some of the enzymes of the digestive tract escape from the body by way of the urine. However, the quantity eliminated by this route is insignificant.

Glucose is found in normal urine in trace amounts and not more than 1 g is excreted in 24 hours. When more than this amount is found in the urine, glycosuria or diabetes is indicated. However, a transient glycosuria can result from emotional stress and after heavy ingestion of carbohydrate, so a true glycosuria or diabetes must be confirmed by blood studies. Table 14-2 lists some of the urinary constituents and the approximate amounts passed in 24 hours. The range of variation in healthy persons is considerable.

Abnormal Constituents of Urine

The kidneys are one of the most important avenues for the excretion of the products of metabolism. As such, the constituents of urine can be used as an index to the nature, effects, and progress of disease. Disease may be regarded as the manifestation of disturbances or loss of function of some part of the body, and since the functional mechanisms of the body are so closely coordinated and correlated that disturbance of one is reflected by a greater or lesser disturbance in the others, the functional efficiency of the kidney not only reflects the physiological efficiency, but also influences the functional mechanisms of the body as a whole. Kidney function may be fairly well estimated by a study of the characteristics and constituents of urine (urinalysis).

Proteins. The most common and important abnormal constituents of urine are **proteins,** giving rise to proteinuria, and **sugar,** giving rise to glycosuria.

The protein bodies that may be encountered in the urine under pathological conditions and that are of definite clinical significance are **serum albumin** and **serum globulin,** both derived from the blood. In addition **Bence-Jones protein** and **proteoses** may also be found; these proteins are not derived from the blood.

Normally, traces of protein (albumin, globulin) may be found in healthy urine, but it takes extremely delicate tests to detect these amounts. Normally not more than 30 to 200 mg of protein are excreted daily in the urine. **Proteinuria** refers to the presence of albumin and globulin in the urine in abnormal concentrations, and most clinical tests are designed to detect only abnormal amounts.

Two types of proteinuria are recognized: (1) accidental, sometimes called false or physiological proteinuria, and (2) renal, or true, proteinuria. **Accidental (physiological) proteinuria** is not looked upon as either an important or a grave sign. It may occur after severe exercise, after the ingestion of foods rich in protein, such as cheese and eggs, especially if the latter are eaten raw, and as a result of standing erect for long periods of time, which may temporarily impair renal circulation. Frequently accidental proteinuria occurs in children between the ages of 6 and 12 years and appears to be more common in girls. It is frequently associated with anemia and malnutrition (malnutrition proteinuria) and also posture, in which case it is referred to as **orthostatic proteinuria.**

Accidental albuminuria is most commonly due to (1) exposure to cold; (2) severe emotions; (3) inflammatory or destructive lesions of the genitourinary tract below the level of the kidney (postrenal), which would necessarily entail a greater flow of blood and a mobilization of white blood cells to the part and the formation of albuminous exudates; (4) the introduction of blood into the urine, as would occur during or shortly after the menstrual period; and (5) the administration of large volumes of antitoxin; the albuminuria in this case represents an excretion of the excess of serum protein from that contained in the antitoxin.

The proteinuria of renal origin **(true proteinuria)** can be distinguished from the accidental type by the finding of casts on microscopic examination of the urine and the absence of findings that would be indicative of inflammation or lesions outside the kidney, such as would be evident in inflammations of the bladder, vagina, or urethra.

Casts are definite evidence of renal lesions and are either epithelial casts, which are cylinders of more or less altered epithelial cells shed from the tubules, coarse granular casts, or waxy casts. In general, casts are composed of an albuminous material, degenerated tubule cells, red blood cells, and pus cells, and they represent a plugging of the renal tubule in the coarse of inflammation or hemorrhage of the kidney substance. Casts are always accompanied by proteinuria. The protein is predominantly albumin because the smaller albumin molecule passes more easily through the injured membrane than the larger globulin molecule.

Nephritis (Bright's disease), which represents a diffuse inflammation of the glomeruli and is always associated with such urinary changes as albuminuria, casts in the urine, hematuria, and reduced urine flow, may be classified into the following types.

1. **Glomerulonephritis,** characterized by inflammation of the glomeruli with secondary changes in the tubule; albuminuria is marked during the degenerative phases.
2. **Nephrosis,** or degenerative nephritis, in which degenerative changes in the tubules alone, or both in the tubules and in the glomeruli, are the chief pathological process. Nephrosis presents such signs of renal disease as albuminuria, blood in the urine (hematuria), edema, hypertension, and anuria.

3. **Renal arteriosclerosis,** or **nephrosclerosis,** characterized by pathological changes in the glomeruli as well as marked arteriosclerosis of the small arteries and veins of the kidney. As a result of the extreme narrowing or complete occlusion of the small arteries, the glomeruli do not receive blood and collapse. They are finally surrounded by connective tissue and are completely obliterated. The slow progress of arteriosclerosis in the kidney causes such patients to run a clinical course with a prominent symptom of chronic hypertension without renal insufficiency. The loss of protein in renal arteriosclerosis is generally less than that found in glomerulonephritis or nephrosis.

Proteinuria also occurs in some cases of pregnancy in which a symptom complex appears that closely resembles glomerulonephritis and is called the **kidney of pregnancy.** The amount of urine is diminished and contains albumin to a varying degree, sometimes as much as 20 to 30 g per liter. In some cases the urine contains blood or hemoglobin. Edema and an increase in blood pressure are also usually present. The blood pressure is characterized by sudden quick rises, which are accompanied by headaches and finally followed by convulsions. The condition at this stage is then known as **eclampsia.**

Poisoning with heavy metals such as mercury, arsenic, or bismuth can cause changes in the renal tubules in which the cells of the tubule become necrotic and slough off, blocking the lumen of the tubule. The glomeruli and vessels usually become congested with blood. The urine shows a definite albuminuria, which disappears when the patient recovers. Hematuria and casts are also prominent abnormal constituents.

In addition to albumin and globulin, the Bence-Jones protein may also appear as an abnormal constituent of urine. It can be detected by a color test in which 0.2 ml of a 1 per cent aqueous solution of Ninhydrin is added to 5 ml of urine. The urine is boiled for 1 minute. A deep reddish-purple color is indicative of Bence-Jones protein. This protein is rarely encountered in the urine but is commonly present in persons with multiple myeloma, and its presence in the urine is considered to be diagnostic of that disease.

Glucose. **Glucose** in the urine (glycosuria) is not always indicative of diabetes, for the glycosuria may be a transient one, which is usually encountered after ingestion of carbohydrates in excess and is known as **alimentary glycosuria.** Another condition is renal glycosuria, which is due to a disturbance in the permeability of renal epithelium to sugar, and glycosuria is also present when there is a sudden and excessive liberation of sugar from the glycogen stores through injury or derangement of the regulatory mechanism, such as injury to the nervous system or endocrine dysfunction. The glycosuria in these conditions is not accompanied by an excessive concentration of sugar in the blood (hyperglycemia).

An impairment of the ability of the body to utilize sugar, resulting in hyperglycemia, and its excretion in the urine is known as **diabetes mellitus.**

Ketone Bodies. When **ketone bodies** are present in excess, the condition is spoken of as **ketonuria.** Normally only 3 to 15 mg of ketones are excreted in a day; they include acetone, acetoacetic acid, and beta-hydroxybutyric acid. The amounts normally eliminated vary, of course, and since they are products of fat metabolism, they will be increased in the presence of a high-fat diet or after fasting when adipose tissue is destroyed. On the other hand, on a low-fat diet the ketone bodies are decreased in amount. The production of ketone bodies in excess is usually associated with metabolic disturbances. Thus, in uncontrolled diabetes, as a result of increased fat metabolism, ketonuria is prevalent. Ketonuria is also encountered in certain nervous and mental diseases, such as paresis and tabes, in pregnancy, after ether anesthesia, and in diseases accompanied by fever, in which case the ketonuria is the result of increased destruction of body fats as a consequence of the fever, which overstimulates the metabolic processes. The lack of sufficient nutrients taken in by such patients may also be a contributing factor. Among such febrile diseases are typhoid fever, scarlet fever, measles, septicemia, and acute rheumatic fever.

Indican. **Indican** in the urine in excessive concentrations is known as **indicanuria.** Indican is normally excreted in amounts of 4 to 20 mg per day, so traces may be encountered in normal urine. It is formed principally as a result of putrefaction of protein material in the intestinal tract and therefore is not related to renal dysfunction but to metabolic or pathological disturbances in other parts of the body.

Pigments. A number of **pigments** may be found in the urine under varying conditions and include compounds such as **urochromogen, bilirubin,** and **porphyrin.** The color of normal urine is due to the presence of **urochrome,** which is derived from hemoglobin through a series of reactions that may be described in the following way:

Hemoglobin → Hematin → Bilirubin → Urochromogen → Urochrome.

Urochromogen in the urine may be indicative of tuberculosis.

Bilirubin is the only pigment of the bile encountered in the urine, and its presence is always indicative of some interference with the biliary outflow. It can be detected only in freshly voided urine, because it is oxidized to biliverdin on standing. Urines containing bile are usually yellowish green to brown in color. **Bilirubinuria** may occur in such conditions as inflammation of the biliary ducts, gallstones, and in various diseases of the liver.

The presence of **porphyrin** in the urine is known as **porphyrinuria** and may be suspected when the urine voided takes on the color of sherry or port wine. It may occur in such conditions as cirrhosis, obstructive jaundice, Addison's disease, typhoid fever, and lead poisoning. The mechanism of its origin is not clearly understood. A congenital porphyrinuria has also been noted that begins in late childhood or middle age.

Melanin. **Melanin** may also be found as an abnormal constituent of urine and may be present in wasting disease, chronic malaria, and malignant pigmented (melanotic) tumors. Urines containing melanin gradually become black on standing exposed to air.

The Kidney and Blood Pressure

One of the main signs occurring in the various forms of nephritis is an increase in blood pressure, or hypertension. Thus, it appears that a relationship must exist between the nephritis and the hypertension. However, it has been demonstrated experimentally that hypertension can also be induced by impairing the supply of blood to the kidney. This can be demonstrated by compressing the renal artery or by inducing compression around the kidney by artificial means. From this it appears that the causative agent of the persistent hypertensive state that develops in nephritis is the reduction of the blood supply to the kidney. In the various forms of kidney disease it will be found that either the glomeruli or the arteries leading to the glomeruli are affected. The arteries can be affected either by narrowing due to vasoconstriction or by narrowing due to obstruction, as in renal arteriosclerosis. There may be a compression of the arterioles, as in tumors of the kidney, or a narrowing of the lumen of the glomeruli by an inflammatory process. In all these cases the increase in resistance produces a decrease in the flow of blood through the glomerular capillaries and a tendency for the pressure to drop in the glomerulus. Whenever there is a threatened fall in the intra-glomerular pressure, there is a general increase in body blood pressure, which may be a compensatory mechanism that the body makes use of to overcome such a fall in pressure. In order to understand the mechanism of hypertension produced by renal disorder, the regulation of normal blood pressure should be reviewed.

We have already noted that the normal pressure in the arterial system is dependent on the following factors: (1) cardiac action, which involves rate and force of beat and output of the heart per minute; (2) the total blood volume; (3) the nature of the walls of the large blood vessels; (4) the viscosity of the blood; and (5) the peripheral resistance. Of these factors the peripheral resistance is perhaps most important, because it is through the change in the lumen size of the small arterioles of the abdominal viscera, which constitute the peripheral resistance, that the influence in changing blood pressure exerted by the other factors can be counteracted. Thus, the diameter of the arterioles in the splanchnic region can vary according to the contraction of their muscular coats, and, depending on the number of small arteries affected, any change in blood pressure can be counteracted. Whenever there is an increase in blood pressure above normal, it must be assumed that there is a constriction of a great number of small arteries in the splanchnic region. The contraction of the muscular coats of these arterioles which causes narrowing of the lumen can be brought about in various ways: (1) by the inorganic constituents present in both the intra- and extracellular fluid, especially by the antagonistic action of potassium and calcium and of hydrogen and hydroxyl ions; (2) by a change in vasomotor impulses coming into the area by way of the sympathetic and parasympathetic fibers; and (3) by means of hormones such as the secretion of the posterior pituitary, which causes vasoconstriction, as well as epinephrine from the adrenal glands, which also causes vasoconstriction and an increase in blood pressure.

The mechanism of renal hypertension is believed to be a **renal pressor system,** which is brought into play whenever the proper maintenance of intraglomerular pressure is threatened, either by local vasoconstriction of the renal arteries or by a pathological narrowing of the lumen due to a disease process, or by an increased pressure in the capsular space, as in urinary obstruction from an enlarged prostate. After a reduction of blood flow through the kidney, the renal cortex, perhaps from the cells of the convoluted tubule, releases a proteolytic enzyme called **renin.** This enzyme, which is liberated into the blood, acts on a pseudoglobulin fraction of blood called **renin substrate,** or **hypertensinogen,** to produce a vasoconstrictor substance known as **angiotonin,** or **hypertensin.** The renin substrate is believed to be a product of the liver. Hypertensin causes generalized constriction of the splanchnic arterioles, and thus a compensatory increase in arterial pressure occurs. The force of the heartbeat is also increased by this substance, and filtration through the glomerulus is normal, probably because the increased blood pressure overcomes the effects of vasoconstriction, or narrowing of the lumen. Hypertensin produced under normal conditions is probably destroyed by another renal proteolytic enzyme called **hypertensinase.** Thus, the kidney produces not only a component (renin) responsible for the production of a vasopressor substance (hypertensin), but also an antipressor substance (hypertensinase).

Discharge of Urine

Urine is produced by the kidney tubules normally at a rate of 1 to 2 ml per minute but can reach a maximum of 20 ml. The urine collects in the pelvis of the kidney, where it is carried down to the bladder by way of the ureters. From the bladder the urine is eliminated from the body by way of the urethra.

The Ureters

The **ureters** (Fig. 14-9) are two large ducts consisting of an epithelial lining surrounded by a connective tissue layer, which in turn is surrounded and strengthened with smooth muscle fibers. The muscular fibers are arranged in three layers, an inner longitudinal, a thicker middle circular, and a thinner outer longitudinal layer. The pelvis of the kidney is actually the expanded upper end of a ureter and is lined by epithelium like that of the ureter, which is continued into the calyces and over the projecting papillae of the pyramids. The ureter proper, which is that portion that passes out of the kidney and terminates in the fundus of the bladder, is approximately 10 to 12 inches long and about $\frac{1}{2}$ inch in diameter. It terminates by an oblique opening, which serves as a valve, into the cavity of the bladder.

Parasympathetic and sympathetic nerves innervate the ureter. Parasympathetic branches come from the vagus and sacral spinal nerves. Sympathetic fibers come from the renal, the spermatic or ovarian, and the hypogastric plexuses. Sensory fibers pass from the ureters into the spinal cord by way of the eleventh and twelfth thoracic and the first lumbar spinal nerves.

Urine is secreted continuously by the kidney tubules and is collected into the renal pelvis. From here the fluid is carried along the ureters into the bladder, partly by pressure and gravity but mostly by peristaltic contractions of the muscular walls of the ureter. This peristaltic activity does not seem to depend on the mechanical stimulation produced by distention of the tube by urine or by chemical stimulation of the inner epithelium, for regularly recurring contractions can be observed in a ureter removed from the body. It appears that the rhythmically repeated contractions arise spontaneously in the muscular coat; that is, they are myogenic. The spontaneous waves of contraction, which pass in one direction

Figure 14-9. Structure of the ureters showing the ureter terminating in the fundus of the bladder.

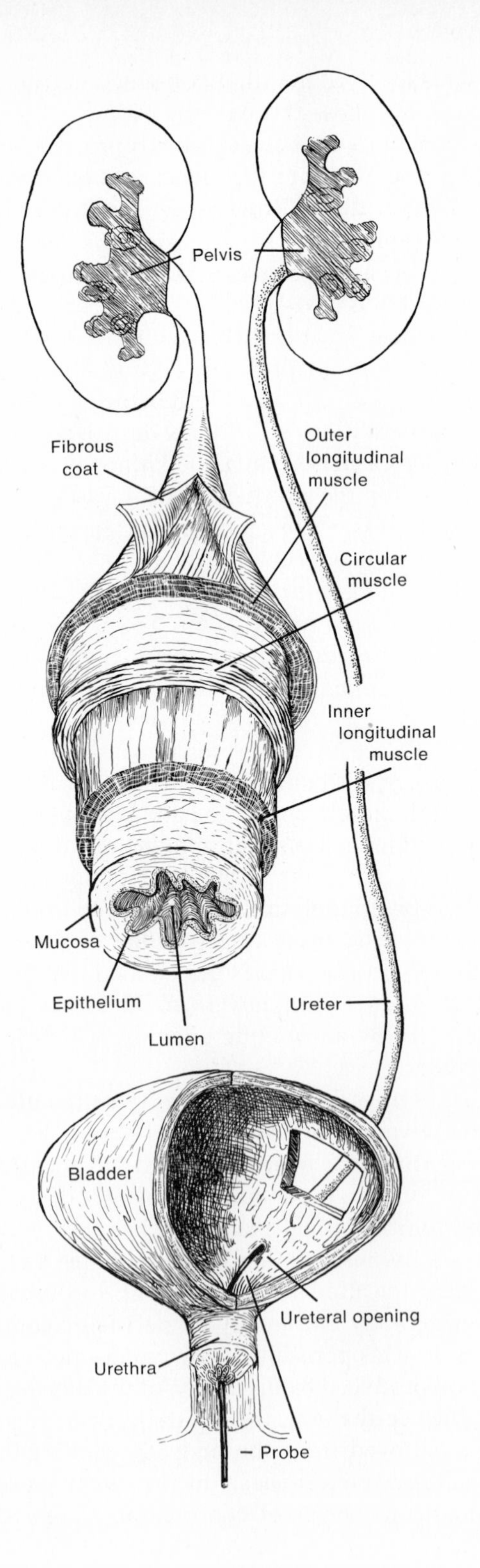

Figure 14-10. Structure of the bladder. A section has been removed from its anterior wall to show the trigone.

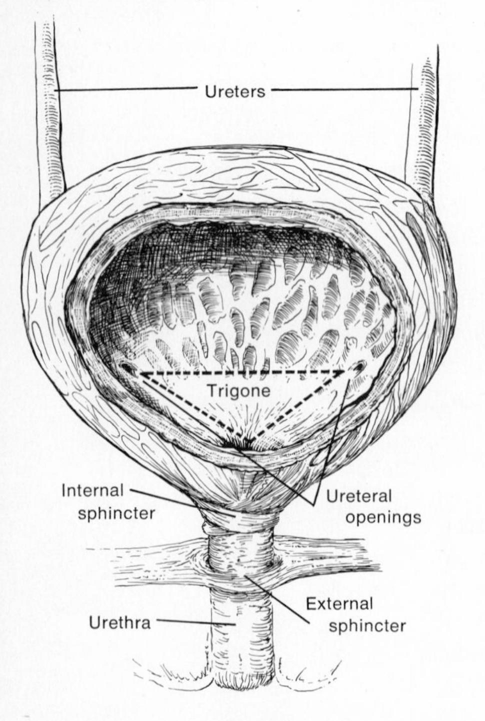

from the kidney to the bladder, vary in rate from one to five per minute (in human beings), depending on the rate of urine formation.

The Bladder

The **bladder,** a hollow muscular bag situated in the pelvic cavity behind the pubes, varies in shape depending on the volume of urine contained within it. When empty, it looks much like a deflated balloon, becoming more spherical when slightly distended. When the contents of the bladder increase in volume, the organ becomes more pear-shaped and can rise to a considerable height in the abdominal cavity.

The epithelium of the bladder resembles that of the ureter, but the appearance of the individual cells in sections of prepared bladder will vary greatly depending on whether the bladder was distended or contracted. Thus, it is said that the bladder is lined with **transitional epithelium.** The smooth muscular coat, the **detrusor muscle,** surrounding the epithelial lining consists of three layers: an inner longitudinal, a middle circular, and an outer longitudinal layer. At the base of the bladder the smooth muscle coat passes around the urethral opening in a series of loops to form the **internal sphincter.** This sphincter is normally in a state of contraction when the bladder is empty. In its posterior wall the bladder has three openings arranged in a triangular fashion forming the **trigone** of the bladder. The two ureters open into the lower part and form the base of the triangle, and the opening to the urethra lies in the median plane below and in front of the openings of the ureters, forming the vertex of the triangle (Fig. 14-10).

Parasympathetic fibers from the sacral spinal segments (second, third, and fourth pelvic nerves) pass through the hypogastric ganglia and innervate the detrusor muscle and the internal sphincter (Fig. 14-11). Parasympathetic stimulation causes contraction of the detrusor muscle and relaxation of the internal sphincter. Sympathetic fibers from the lumbar segment via the hypogastric nerves relax the detrusor muscle and increase tone of the internal sphincter. The **external sphincter** surrounding the urethra a short distance below the internal sphincter is composed of striated muscles and is innervated by the pudenal nerves, which are spinal somatic nerves, so that the regulation of this sphincter is under voluntary control. Sensory fibers accompany both sympathetic and parasympathetic paths. Since both systems pass through the hypogastric ganglion, section of nerves passing from the bladder to the ganglion is usually done for the relief of intractable bladder pain.

The urine is collected in the bladder; its regurgitation back into the ureters is prevented by the oblique entrance of the ureters through the bladder wall to form a flaplike valve. From time to time a considerable quantity of urine is discharged to the outside by way of the urethra, a process called **micturition** (Fig. 14-12).

Micturition

Micturition is a reflex act but under voluntary control. The center of micturition is found in the brain stem with a second center located in the sacral portion of the spinal cord. If the cord is transected between this area and the brain stem, the bladder will empty itself at regular intervals without voluntary control. This

Figure 14-11. *Nerve pathways of the bladder. The innervation of the detrusor muscle and internal and external sphincters is shown.*

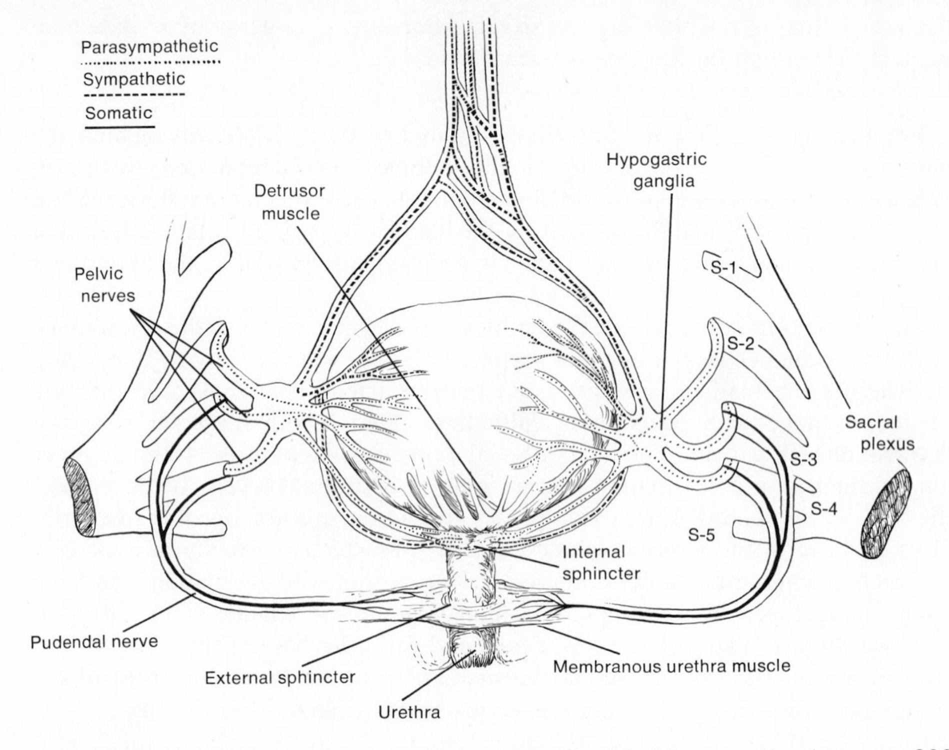

automatic emptying takes place usually when the bladder contains between 200 and 300 ml.

The empty bladder is contracted to a much smaller size than a bladder filled with urine. This is because of the great tone associated with the muscular wall when the bladder is empty. As the urine collects, the bladder becomes more and more distended, but the pressure within the bladder rises very little because the detrusor muscle relaxes as the bladder expands. It has been shown experimentally that the pressure is almost independent of the volume of the contents until the volume reaches 400 ml. The reflex action for micturition starts as a result of the excitation of sensory receptors in the bladder wall that is caused by the distention of the bladder as it fills. As the sensory impulses increase in number and frequency with greater distention and are transmitted through the spinal cord, the motor outflow to the bladder via the pelvic nerves increases also, causing (1) reflex contraction of the bladder muscle, (2) relaxation of the internal sphincter, and (3) a reduction in the number of impulses coming through the hypogastric (sympathetic) nerves, aiding in the relaxation of the internal sphincter. The passage of urine into the urethra at the beginning of micturition reflexly relaxes the external sphincter, which permits the discharge of urine to the outside. The desire to void is initiated when the volume of urine in the bladder reaches 200 to 300 ml. The external sphincter cannot be made to relax voluntarily, but it can be made to contract voluntarily. Thus, urination may be consciously deferred. Micturition can be produced voluntarily before the bladder wall is distended sufficiently to stimulate the reflex emptying by removal of the inhibition of the normal reflex

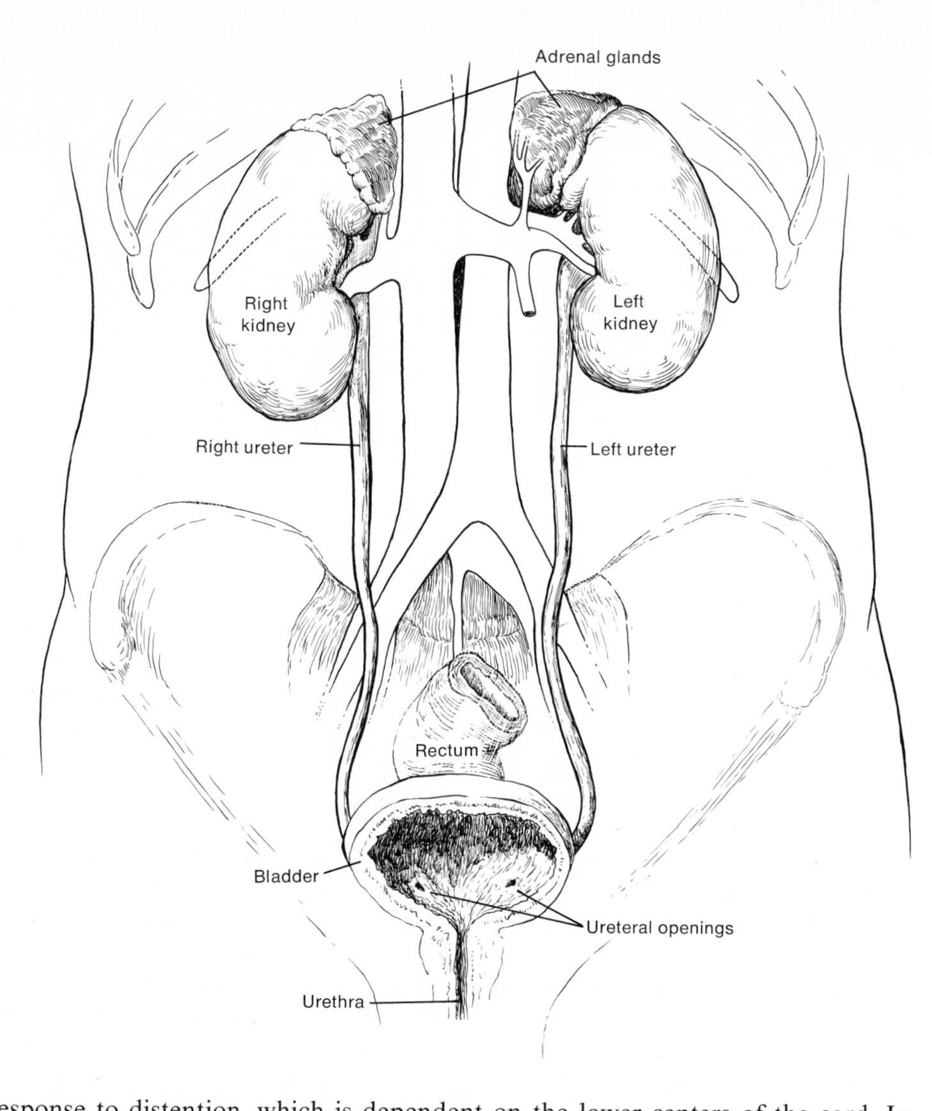

Figure 14-12. Urinary tract. The urine produced by the kidneys is carried to the bladder by way of the ureters. From time to time a considerable quantity of urine is discharged to the outside by way of the urethra.

response to distention, which is dependent on the lower centers of the cord. In other words, when we wish to micturate by a conscious effort of the will, what occurs is not a direct action of the will on the muscular walls of the bladder; rather, impulses arising in the cortex as a result of conscious effort descend down the cord and in a reflex manner throw into action the micturition center in the spinal cord. Voluntary inhibition of the detrusor muscle can also be mitigated by way of the cortex, resulting in a conscious ability to delay micturition. Thus, voluntary control, or the ability to start or delay micturition, may be looked upon as removal of inhibition or an inhibition of the normal reflex response to distention, the center of which is located in the lower parts of the spinal cord.

Premature involuntary micturition may occur in certain emotional conditions, in which the excitement produces an increase in the tone of the bladder muscle. As a result, a smaller volume of urine may exert the same effect on the bladder wall as the larger quantity, that is, the volume that would normally initiate reflex emptying. By the same mechanism, lesions of the cord may cause a loss of the inhibitory control of the brain on the normal reflex response and cause hypertonicity of the bladder muscle and frequent urination. Conversely, if the parasympathetic nerve supply to the smooth muscle of the bladder is cut, the bladder becomes dilated and flaccid. Urine accumulates, and the bladder distends more and more as the urine is retained. Because of the loss of its motor nerve supply, the bladder is unable to empty itself by contraction, but does so by continually dribbling through the urethra, the fullness of the bladder being sufficient to overcome the resistance at the neck of the urethra. The loss of voluntary control over micturition is referred to as **incontinence** and in the situation described above as **retention incontinence.** Incontinence in young infants is normal since the conscious ability to control micturition has not been developed as yet, and the infant voids whenever the bladder is sufficiently distended to arouse a reflex stimulus. The infant develops voluntary control of micturition by acquiring the capacity to inhibit the spinal reflexes.

Changes in Amount of Urine

Normally the volume of urine produced daily ranges from 600 to 2500 ml, and from what has already been said it is obvious that these figures may be influenced by numerous factors independent of disease. A diminution in urine flow is referred to as **oliguria,** and an increase is known as **polyuria.** Both of these are important when persistent. Persistent polyuria suggests diabetes, chronic glomerulonephritis, or, in cases of deposition of amyloid in the glomeruli, a condition more specifically known as amyloid kidney. Oliguria or decreased output when persistent suggests disease conditions that prevent the blood plasma from reaching the glomerular membrane in adequate amounts. This is the case in acute renal congestion, acute nephritis, and passive congestion of the kidneys resulting from circulatory conditions such as cardiac dysfunction or pressure against the kidneys due to pregnancy or intraabdominal growths.

Complete suppression of urine formation is referred to as **anuria.** It usually occurs whenever the blood plasma is completely prevented from reaching the filtering membrane. Inflammatory states involving all the glomeruli are instances in which anuria usually develops. It occurs also when the pressure within the intracapsular space increases to the point where there is no positive filtration pressure and the passage of fluid through the membrane stops. This can occur when both ureters are occluded or blocked. Reflex contraction of the small arteries and arterioles perfusing the kidney can also cause anuria.

Tests of Renal Function

To test the functional efficiency of the kidney, from which its structural condition can be analyzed, various clinical procedures have been devised. In some of these procedures the efficiency of renal excretion is tested by determining the degree to which substances normally excreted are retained in the blood. These are methods of blood chemistry, and the substances normally studied are urea and creatinine. In some procedures the rate at which the kidney excretes certain materials introduced into the mouth or into the circulation is tested. Materials usually used in such tests are urea, water, or an injected dye, such as phenolsulfonphthalein. The best tests for renal function are the renal clearance tests. In such tests a substance that is known to be filtered through the glomerulus, but neither reabsorbed nor excreted by the tubules, is injected into the blood, and the rate at which it is cleared from the plasma per unit time is determined. It is usually expressed as the amount of plasma that is completely cleared of some substance per minute.

Renal Clearance

A substance may clear from blood in three possible ways: (1) glomerular filtration, (2) filtration with tubular reabsorption, or (3) filtration and tubular secretion. The ideal substance for measuring glomerular filtration is one that is filtered by the glomerulus but neither reabsorbed or excreted by the tubules. Most substances used in renal clearance tests, such as urea, creatinine, and Diodrast, do not completely meet these requirements as they are either excreted or slightly reabsorbed by the kidney tubules. **Inulin,** a polysaccharide derived from artichokes, is one substance that is filtered but neither excreted nor reabsorbed by the kidney tubule. It is thus possible to use inulin to determine the volume of glomerular filtration and also to use it as a yardstick to determine tubular excretion and reabsorption of other constituents.

Clearance represents the volume of blood plasma that contains the amount of a substance excreted in the urine in 1 minute. Inulin has a clearance of 125; that is, 125 ml of plasma contains the inulin eliminated in 1 minute of renal activity. Since inulin is not reabsorbed or excreted through the kidney tubule, the clearance of this substance is equal to the glomerular filtration rate.

The clearance for any substance (the plasma volume) is calculated by multiplying the rate of urine flow by the concentration of the substance in the urine, divided by concentration of the substance in the plasma. The following equation is used:

$$C = \frac{UV}{P},$$

where C = clearance, or the quantity of plasma in milliliters that would be completely cleared of the substances in 1 minute or the volume of plasma that contained the substance excreted by the kidney during 1 minute; U = the concentration of the substance in urine; V = the volume of urine produced per minute; and P = the concentration of the substance in the plasma.

Example: The concentration of urinary inulin was found to be 3,125 mg per 100 ml. Blood inulin concentration was determined through analysis to be 50 mg per 100 ml; volume of urine excreted per minute, 2 ml (normally assumed from experimental evidence).

$$C = \frac{UV}{P}$$

$$= \frac{3125 \times 2}{50} = \frac{6250}{50} = 125.$$

Thus the clearance of inulin is 125, and since inulin is known not to be reabsorbed or excreted by the kidney tubule, the plasma volume that contained the amount of substances eliminated, that is, the clearance, equals the volume of glomerular filtrate per minute.

The proof of this can be analyzed in the following way. If every 100 ml of blood contained 50 mg of inulin, each milliliter of blood would contain 0.5 mg. With a glomerular filtration rate of 125 ml per minute, each 125 ml of glomerular filtrate would contain 62.5 mg of inulin (125×0.5). Since inulin is neither reabsorbed nor excreted by the kidney tubule, and urine formation is 2 ml per minute, this volume of urine would contain the same amount of inulin that was filtered through the glomerular capsule per minute (62.5 mg). Our urine sample would then contain 3,125 mg per 100 ml of urine (62.5×50).

In case of a substance that is not only filtered through the glomerulus, but also excreted by the tubules, for example Diodrast, the clearance increases with the tubular excretion. Diodrast gives a clearance value of 750 ml. If 125 ml is the glomerular filtration rate, 625 ml must be excreted by the tubules per minute. Since the kidney comes close to clearing the blood of this substance completely in a single passage, this value of 750 ml is approximately the actual amount of blood flowing through the kidneys per minute.

References

Bell, G. H., et al. *Textbook of Physiology and Biochemistry*. Baltimore: Williams & Wilkins, 1965.

Braun-Menendez, E., et al. *Renal Hypertension*. Springfield, Ill.: Thomas, 1946.

Dalton, A. S., and S. Haguenau. *Ultrastructure of the Kidney*. New York: Academic, 1966.

Deane, N. *Kidney and Electrolytes*. Englewood Cliffs, N.J.: Prentice-Hall, 1966.

Fisher, J. W. (ed.). *Kidney Hormones*. New York: Academic, 1971.

Giebisch, Gerhard. "Kidney, Water and Electrolyte Metabolism," in: *Annual Review Physiology*, Vol. 24. Palo Alto, Calif.: Annual Reviews, 1962, p. 357.

Guyton, A. C. *Textbook of Medical Physiology*. Philadelphia: Saunders, 1971.

Handley, C., and J. H. Moyer. *The Pharmacology and Clinical Use of Diuretics*. Springfield, Ill.: Thomas, 1959.

Hawk, P. B., et al. *Practical Physiological Chemistry*. New York: McGraw-Hill, 1954.

Hinman, F. *Hydrodynamics of Micturition*. Springfield, Ill.: Thomas, 1971.

Keitzer, W. A., and G. C. Huffman. *Urodynamics*. Springfield, Ill.: Thomas, 1971.

Lewis, A. G. *Kidney* (A Ciba Foundation Symposium). Boston: Little, Brown, 1954.

Lippman, R. W. *Urine and the Urinary Sediments*. Springfield, Ill.: Thomas, 1962.

Lotspeich, W. D. *Metabolic Aspects of Renal Function*. Springfield, Ill.: Thomas, 1959.

Maxwell, M. H., and C. R. Kleeman (eds.). *Clinical Disorders of Fluid and Electrolyte Metabolism*. New York: McGraw-Hill, 1972.

McManus, J. F. A. *Medical Diseases of the Kidney*. Philadelphia: Lea, 1950.

Orloff, J., and M. Burg. "Kidney," in: *Annual Review of Physiology*, Vol. 33. Palo Alto, Calif.: Annual Reviews, 1971, p. 83.

Smith, H. W. *The Kidney: Structure and Function in Health and Disease*. New York: Oxford U.P., 1951.

Of Water, Salt and Life. Milwaukee: Lakeside Laboratories, Inc., 1956.

Wirz, H. "Kidney, Water and Electrolyte Metabolism," in: *Annual Rewiew of Physiology*, Vol. 23. Palo Alto, Calif.: Annual Reviews, 1961, p. 23.

15

The Reproductive System

Most tissues of the body undergo, during life, considerable regeneration. The epithelial cells, which are shed from the surface of the body, are replaced by new cells from the germinating layers of the epidermis; old blood cells are replaced by new ones; worn-out muscles, or those that have atrophied from disease, are renewed by the acquisition of new, fresh muscle fibers; divided peripheral nerves grow together again; broken bones are united; new secreting cells take the place of the old ones that are cast off; in fact, almost all the individual constituents of the body that, within limits, are fixed more by bulk than anything else are capable of regeneration. In higher animals regeneration of whole organs and members never occurs, although it can be seen in the lower animals. For example, the limbs of a newt can be restored by growth, but not those of a man.

In higher animals the reproduction of the whole individual can be effected in no other way than by the process of sexual generation, through which the female reproductive cell, the **ovum,** under the influence of the male reproductive cell, the **spermatozoon,** develops into an adult individual. This chapter will concern itself not with the morphological problems connected with the series of changes by which the ovum becomes the adult being (embryology), but with the physiological and morphological aspects attendant on the fertilization of the ovum and on the nutrition and birth of the embryo.

The Female Organs of Generation

The female reproductive organs are anatomically divided into the internal and external organs. The **external organs** (Fig. 15-1) consist of the labia majora and minora, the clitoris, the hymen, the meatus urinarius, the vulvo-vaginal glands, and the mucous and sebaceous glands that are distributed in the mucous membrane covering the parts. The female external organs play a very minor role in

Figure 15-1. Female external genital organs.

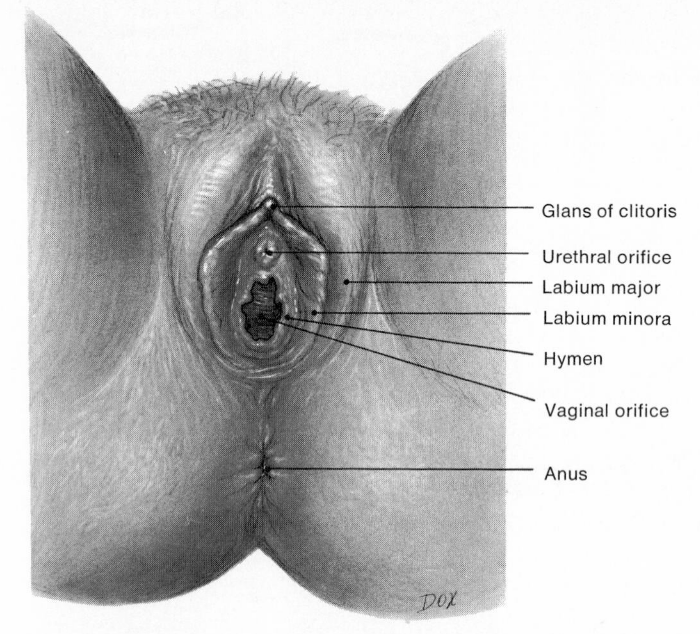

the function of reproduction and therefore will not be considered in great detail. The **hymen** is a thin fold of mucous membrane guarding the opening into the vagina and is a structure peculiar to the human race. It usually is perforated at the first coition. The **clitoris** is a small erectile organ homologous with the penis. This organ, however, is not traversed by the urethra but lies within the vaginal folds just above the urethral opening. The triangular space above the orifice of the vagina into which the female urethra opens is often called the **vestibule.** The clitoris, like the penis, contains numerous sensory nerve endings and undergoes erection during sexual excitement. It is sensitive to tactile stimulation and plays an important role in sexual satisfaction.

The **internal organs** consist of the vagina, uterus, fallopian tubes, and ovaries.

The **vagina** is a broad musculomembranous passage, about 4 to 6 inches long, directed obliquely upward and backward and extending from the hymen to the cervix of the uterus. Its walls contain both longitudinally and circularly arranged muscle fibers. Internally it is lined with a mucous coat surrounded by erectile tissue. The mucous membrane is thickly covered with sensitive papillae and mucous glands called **Bartholin's glands.** Bartholin's glands secrete a viscous fluid and are stimulated into activity during sexual play. The secretion serves to lubricate the penis and thus facilitates the sexual act. At the climax of sexual intercourse there are rhythmical contractions of the vaginal musculature and a diffuse pleasurable sensation, which constitutes the **orgasm** of the female. This corresponds to ejaculation in the male, but there is no production or expulsion of fluid in the female.

The **uterus** is a flattened muscular organ consisting of two parts: the corpus, or body, of the uterus and the cervix, or neck, which opens into the vagina (Fig. 15-2). The cervix is the lower constricted portion of the organ and extends from the corpus (fundus) to the vagina, into which it projects. It opens into the vagina by a transverse fissure, called the os uteri. The body, or fundus, is the upper expanded portion of the uterus, which is somewhat triangular in shape. The lower portion of the fundus is continuous with the cervix. The upper portion tapers off into two branches called the **cornua,** which lead into the fallopian tubes. The uterus is composed of three coats: (1) a serous layer, formed by the peritoneum; (2) a thick muscular layer; and (3) a still thicker layer, a mucous coat sometimes called the **endometrium.** The mucous coat is continuous with that lining the fallopian tubes and vagina. It is lined by ciliated columnar epithelium and contains numerous tubular glands, which open into the cavity of the uterus. The subepithelial mucosa, which is sometimes called the *uterine stroma,* contains a number of blood vessels and lymph spaces. The vessels are branches of the ovarian and uterine arteries and veins. The uterus is also supplied by nerves by way of the autonomic system.

The **fallopian tubes,** sometimes called the **oviducts,** or uterine tubes, are about 4 inches in length and extend from the cornua of the uterus to the ovaries, where they end in enlarged expansions, the margins of which are covered by long, slender processes. This expanded portion of the tube is known as the fimbriated end and opens into the abdominal cavity about an inch from the ovary, being connected to the ovary by only one of the slender processes. The tubes consist of an external serous coat, a middle muscular coat, containing both longitudinal and circular

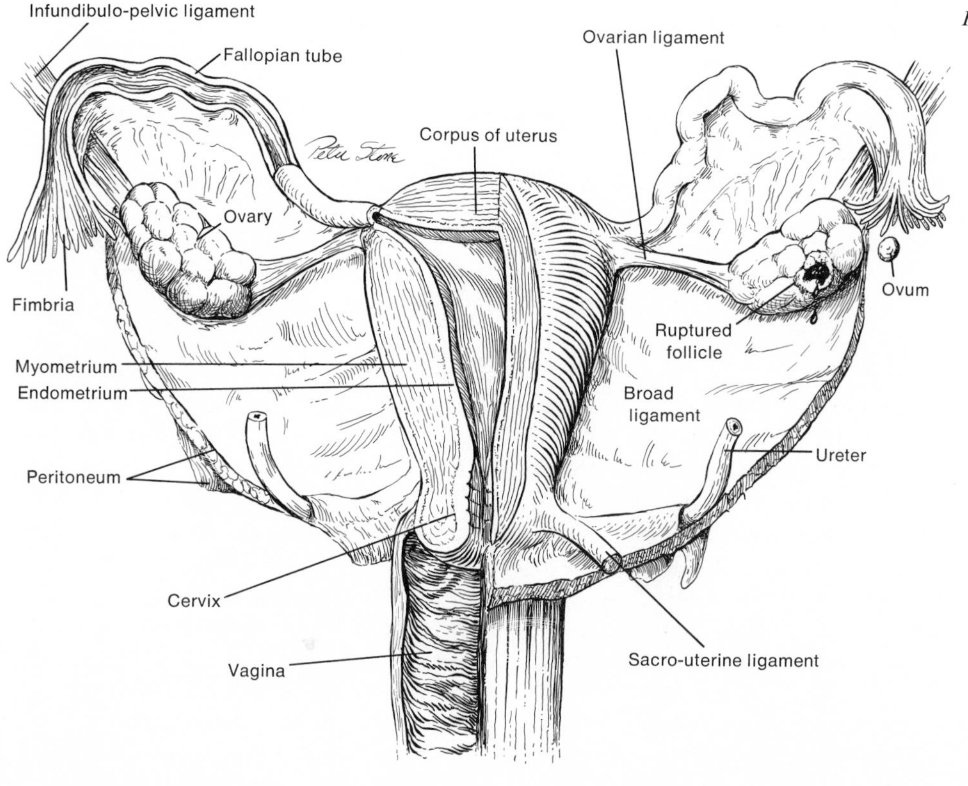

Figure 15-2. Uterus and associated organs.

fibers, and an internal mucous membrane, which is highly vascular and lined with ciliated columnar epithelium. When the ovum is ejected from the ovary, the slender processes of the fimbriated end of the fallopian tube literally catch the ovum and pass it on into the tube, where it is propelled by the ciliated columnar epithelium down into the uterus.

The two **ovaries,** which are flattened, ovoidal bodies weighing 2 to 3 g, are situated deep in the pelvic cavity, one on each side of the uterus (Fig. 15-2), and enclosed in the folds of the broad ligament. They are connected with the uterus by a ligament and with the fallopian tube by one of its fimbriae. Each ovary, which represents the female gonad, or organ in which the ova are produced, consists of a fibrous coat that encloses the **stroma** of the organ (Fig. 15-3). The stroma consists of fibrous connective tissue containing some smooth muscle cells, especially in the area of the attachment to the broad ligament. The surface of the ovary is covered by a layer of columnar cells, which constitute the **germinal epithelium.** Within, are a number of vesicles of various sizes, each with an ovum surrounded by an epithelium. These are called ovarian follicles. It has been estimated that there are approximately 400,000 immature follicles in both ovaries at birth; these are called **primary follicles.** These primary follicles are produced during fetal life from the germinal epithelium and contain the cells that become ova when the female matures. These cells represent the primitive oocyte and are

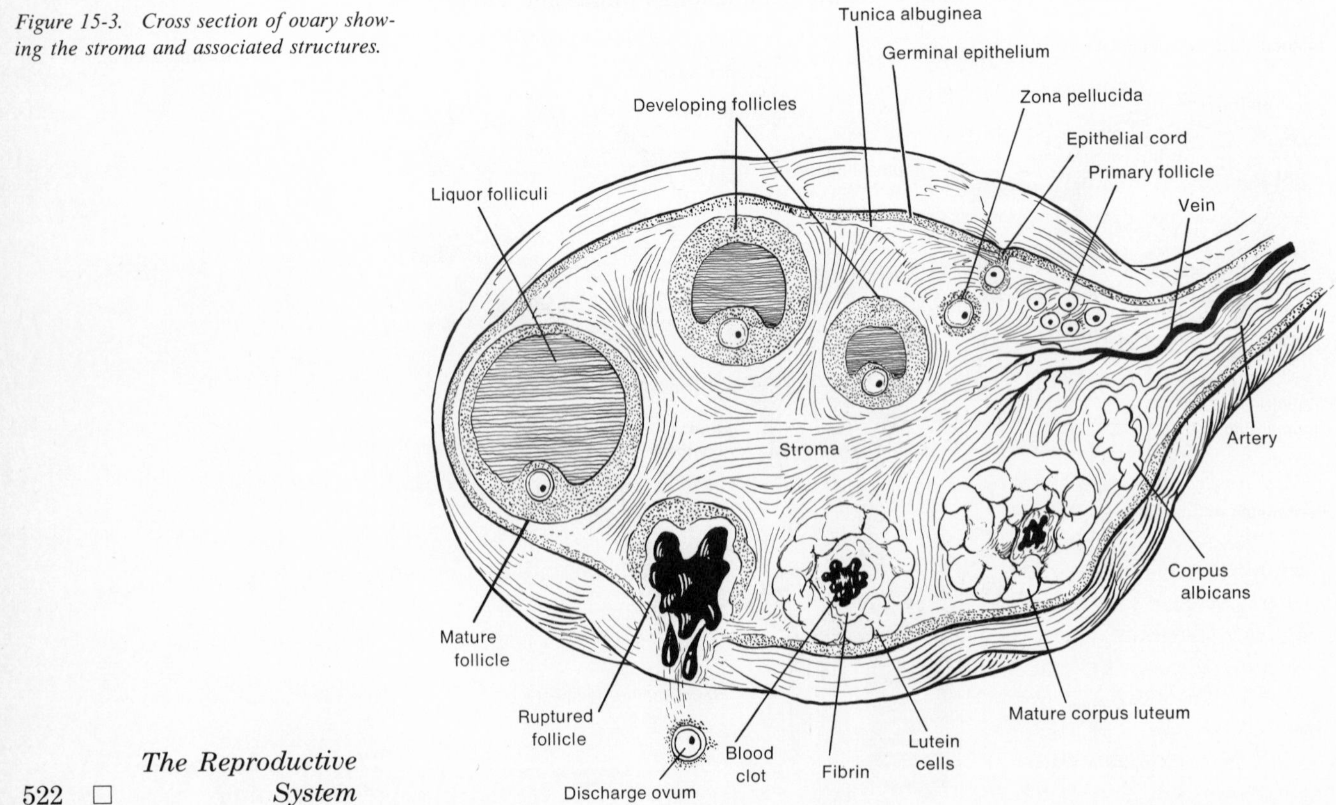

Figure 15-3. Cross section of ovary showing the stroma and associated structures.

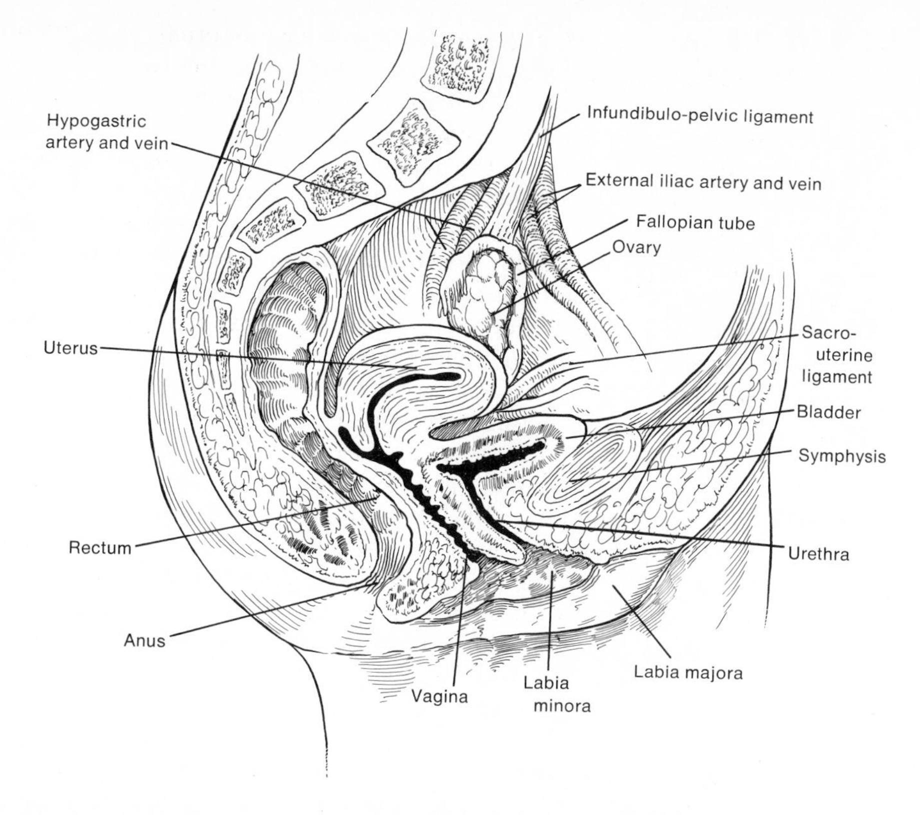

Figure 15-4. Midsagittal section of female pelvis showing the reproductive tract and its relation to other organs.

called **oogoniums.** Beginning at puberty, which is usually between the twelfth and fifteenth years of life, the primordial follicles mature, one approximately every 28 to 30 days. The oocyte within becomes a mature ovum, and the entire structure is known as a **graafian follicle.** Since the reproductive life of the female spans approximately 38 years (ages 12 to 50) and one ovum is developed per month during this time, it would appear that only 450 ova out of a potential 400,000 are ever released. This is another example where nature is lavish in providing a safety factor in important functions, in this case to ensure the reproductive capacity of the human species.

The entire reproductive tract just described may best be viewed by examining Fig. 15-4, which presents a midsagittal section of the reproductive organs as they appear in the female pelvis.

Ovarian Function

The ovaries are homologous with the testes of the male, since they arise from the same tissue and from the same site in the developing embryo, and, like the testes, they serve a dual function in producing sexual reproductive cells and elaborating a complex group of sex hormones.

Production of Mature Ova. The ovaries, as already mentioned, begin their major function of producing mature reproductive cells with the beginning of puberty. One mature ovum develops from a primordial follicle approximately every 28 to 30 days. Each primordial follicle consists of an **oocyte,** or primitive, nonmatured ovum, surrounded by a single layer of flattened epithelial cells (Fig. 15-3). Under the influence of *follicle-stimulating hormone* (*FSH*), which is secreted by the anterior pituitary gland, the development of the mature egg **(oogenesis)** and a mature graafian follicle begins. The oocyte continues to increase in size, fat globules appear in the cytoplasm, and a thin transparent structure develops between the follicular cells and the oocyte called the **zona pellucida.** The enlarged oocyte, now known as the **primary oocyte,** divides unequally, forming one large **secondary oocyte** and a small cell called the **first polar body.** The size difference is not due to nuclear size difference but to the amount of cytoplasm. The secondary oocyte undergoes another division to form a large mature egg cell, or ovum, and a second small polar body. The divisions of the oocyte thus result in one functional mature germ cell and two nonfunctional polar bodies. In immature germ cells the number of chromosomes characteristic of the species is present and is said to be **diploid.** Under these conditions chromosomes are represented in pairs. During the process of maturation the chromosome pairs are separated and distributed between daughter cells, with the result that in the mature sex cell only half the number of chromosomes characteristic of the species is present. The reduced number is called the **haploid** number of chromosomes. It is the halving of the chromosome number in mature sex cells that maintains constant the number of chromosomes characteristic of the species. Figure 15-5 schematically illustrates maturation, or gametogenesis (oogenesis in the female, spermatogenesis in the male).

During **meiosis,** The term applied to the several divisions of the oocyte described above, the structure surrounding the egg are also undergoing change. The primitive follicular cells proliferate, forming a layer of stratified cuboidal epithelial cells called the **membrana granulosa.** A cavity begins to develop between the cells of the membrana granulosa, which becomes filled with a clear fluid known as **liquor folliculi.** Extending into the cavity but entirely surrounded by a mass of cells from the membrana granulosa is the ovum. This entire structure is the graafian follicle (Fig. 15-6), which is surrounded by a connective tissue wall known as the **theca folliculi.** The theca consists of two distinct layers: the outer known as the **theca externa** and the inner as the **theca interna.**

The graafian follicle begins its development deep in the cortex of the ovary but completes its maturation at the surface, appearing as a blisterlike bulge on the surface of the ovary just prior to ovulation. Finally, as a result of the degeneration of its wall, it ruptures, discharging the ovum, the follicular fluid, and many of the follicular cells into the peritoneal cavity. This process is called **ovulation** and takes place approximately midway between menstrual periods, that is, 14 ± 2 days prior to the next expected menses. Siegler[1] has made the observation that ovulation may occur as early as the eighth day or as late as the twentieth. Although the ovum is released into the peritoneal cavity, the fimbriated end of the fallopian tube immediately grasps it, and the ovum is propelled along, by the ciliated epithelial cells lining its lumen, to the uterus. It has been estimated that it takes

[1] S. L. Siegler, *Fertility of Women.* (Philadelphia: Lippincott, 1944).

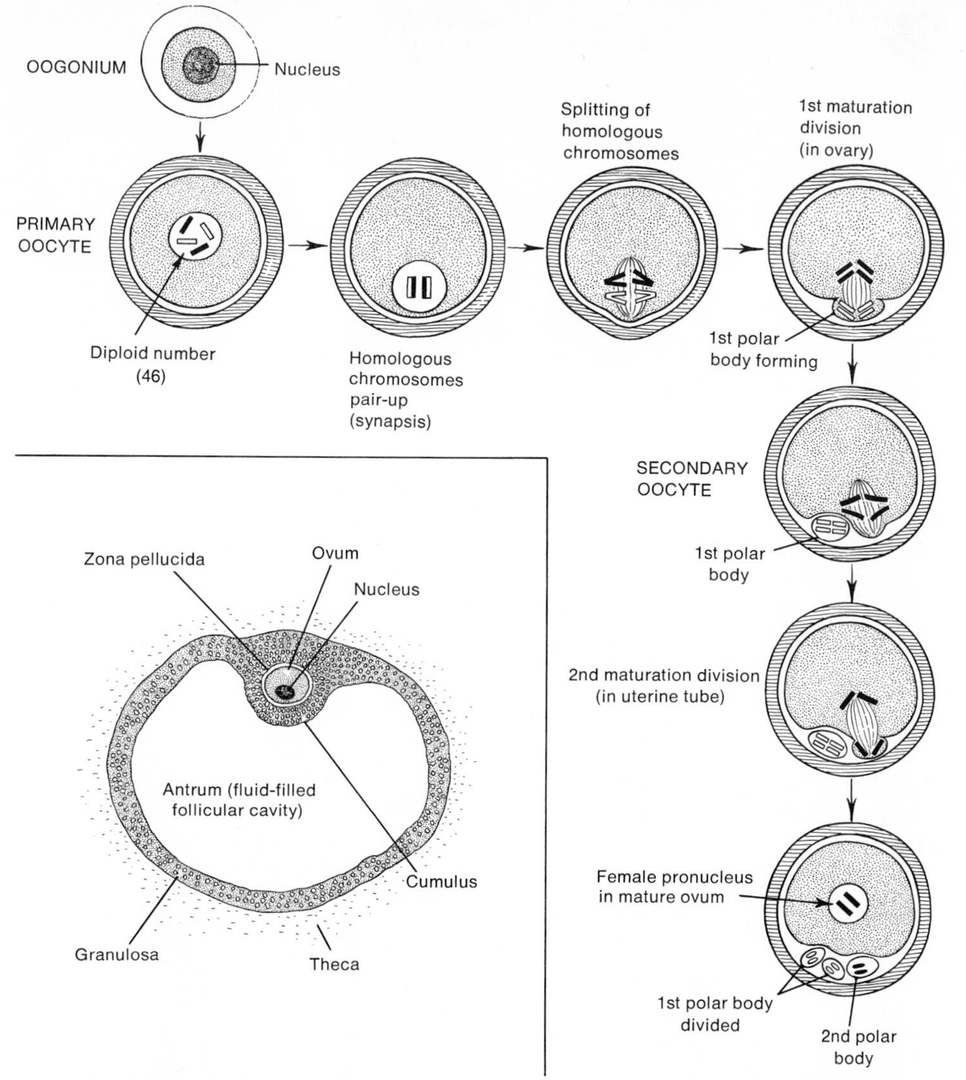

Figure 15-5. Oogenesis, showing the reduction in chromosomes to form a mature female ovum. For simplification only two pairs of homologues are included. Insert illustrates the mature ovum located within the graafian follicle.

the ovum approximately 72 hours to arrive at the uterus after ovulation. However, fertilization of the ovum usually occurs at the upper end of the fallopian tube and not in the uterus. If fertilization is to occur, experimental evidence indicates that it must take place within 24 hours after ovulation.

Formation of the Corpus Luteum. Immediately after ovulation a blood clot forms at the site of the ruptured follicle; it is often referred to as the **corpus hemorrhagicum** (Fig. 15-3). However, it should be mentioned that although this structure is definitely present for a number of lower animals, it is not readily found in the human female or monkey. In human beings, the ruptured site of the follicle is closed almost immediately with little bleeding. The remaining granulosal and

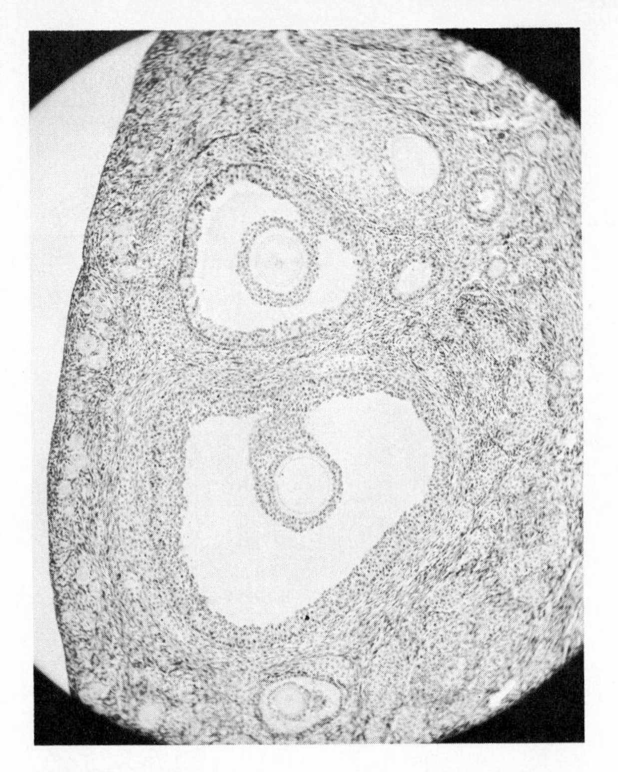

Figure 15-6. Section of mammalian ovary, showing mature graafian follicles with germ hill (cumulus oophorus) and large ovum. (Courtesy Ward's Natural Science Establishment, Inc., Rochester, N.Y.)

thecal cells undergo a series of changes, becoming large and glandular-appearing, containing many yellowish granules. These cells, now called **luteal cells** because of the yellow granules found within their cytoplasm, replace the blood clot at the site of the evacuation of the follicle and give rise to a definite glandlike spherical organ called the **corpus luteum,** which is partly embedded in the ovarian cortex. The corpus luteum is maintained for about 14 to 15 days, although its functional life, during which it elaborates progestational hormones such as progesterone, is about 7 to 11 days. With the onset of menstruation the corpus luteum undergoes rapid degeneration because of fatty infiltration, and many of the cells are removed. It now appears as a white fibrous structure and is aptly called the **corpus albicans.** This structure atrophies and after several months sinks deep within the stroma of the ovary as a tiny scar called the **corpus fibrosum.**

If fertilization of the ovum takes place and pregnancy occurs, the functional activity of the corpus luteum may be prolonged for 5 to 6 months. The corpus luteum of the pregnant female is known as the **corpus luteum verum** in contrast to that of the luteal body of the regular cycle, which is referred to as the **corpus luteum spurium.**

Endocrine Function of the Ovary. The ovaries, in addition to producing ova, also secrete two sex hormones. These are estradiol and progesterone.

Estradiol, also called the follicular hormone, is secreted by the maturing graafian follicle. Actual isolation and characterization of this female hormone (estrogen) from ovarian tissue were accomplished by MacCorquodale and co-workers.[2] It is a steroid, as are all the sex hormones, specifically referred to as 17β-estradiol, having the following formula:

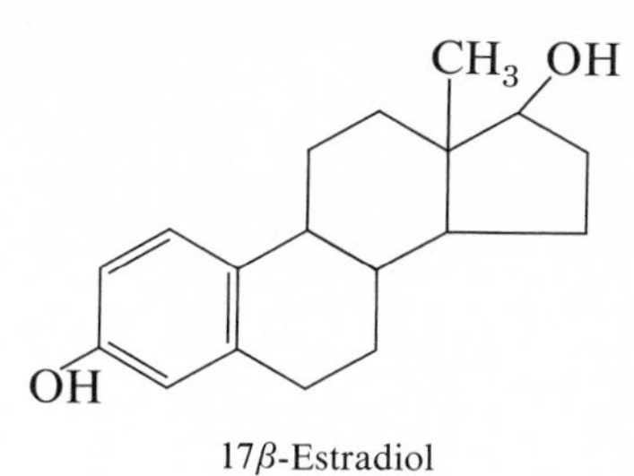

17β-Estradiol

Other estrogenic substances found in the various body fluids are actually derived from the metabolic degradation of the primary estrogen 17β-estradiol. The interrelationship that exists between the various estrogens in the human being is believed to be as follows:

$$17\beta\text{-Estradiol} \rightleftharpoons \text{Estrone} \longrightarrow \text{Estriol}$$

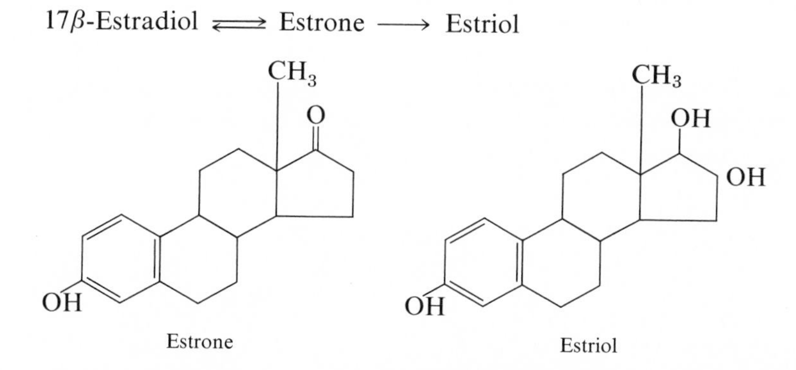

Estrone

Estriol

These naturally occurring estrogens have been isolated from various tissues of the body, including the ovary, testes, adrenals, and placenta. Large amounts are also found in the urine of pregnant women and mares. Strangely enough, the testes of stallions are the richest tissue source of female hormones (estrogens).

The estrogens bring about maturation of the secondary sex characteristics and ths sexual organs and also function in their maintenance. The secondary sex characteristics are the changes that occur at puberty in which certain structures typical of the sex begin to appear. For example, in the female the breasts develop; deposition of fat around the hips is more prevalent; the hair distribution on the body is somewhat different, and even its pattern of distribution is different, especially in the pubic area; the voice remains lighter and softer in tone. In addition to these effects, the entire female reproductive tract and the mammary glands are under the influence of the changing levels of the estrogens.

Complete removal of the ovaries (oophorectomy) before puberty causes girls

[2] S. MacCorquodale, S. A. Thayer, and E. A. Doisy, "The Isolation of the Principal Estrogenic Substance of Liquor Folliculi. *J. Biol. Chem.*, **115**:435 (1936).

to remain in the neutral or sexually undifferentiated state; that is, the female secondary sex characteristics fail to develop. The subcutaneous fat in its typical feminine distribution fails to appear; the sexual organs remain immature; the breasts do not develop; the sex instinct is suppressed; and menstruation does not occur. If oophorectomy is performed after puberty, menstruation ceases to occur; the body configuration becomes more masculine; there is a gradual atrophy of the sex organs and a suppression of the sexual drive. In addition, there may be some growth of hair in areas of the body typical for the male, that is, on the chest and face.

Thus, it may be concluded from such observations that the ovary is an organ of internal secretion and that secretions referred to as estrogens are responsible for the following: (1) the maturation of the female sexual organs and the cyclical changes associated with the structure of the mature uterus, and (2) the development of the breasts and the characteristic feminine proportions, the suppression of male characteristics, and the development of the sexual drive.

Progesterone is secreted by the corpus luteum and is therefore known also as the corpus luteum hormone. The biological and chemical characteristics of progesterone were not demonstrated until 1956, by Zander, although it had been isolated in 1952 by several workers. Progesterone is found in a variety of tissues of the body and has been isolated from blood, placental tissues, and, of course, corpora lutea. Like the other sex hormones, it is a sterol and differs only slightly from the estrogens. Its structural formula is

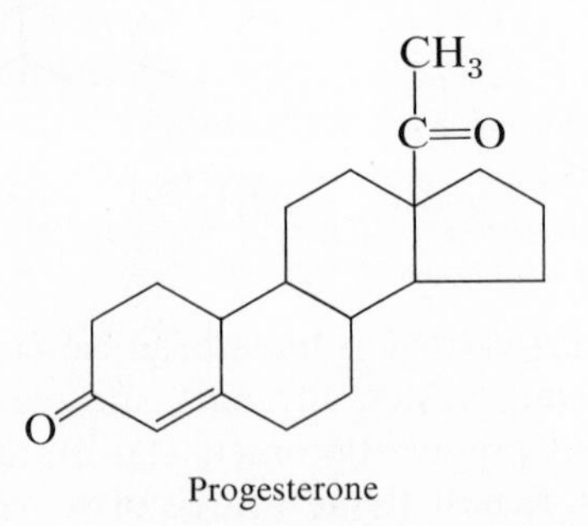

Progesterone

Unlike the estrogens, progesterone is not found in the urine; however, its main metabolic end product, **pregnanediol,** is:

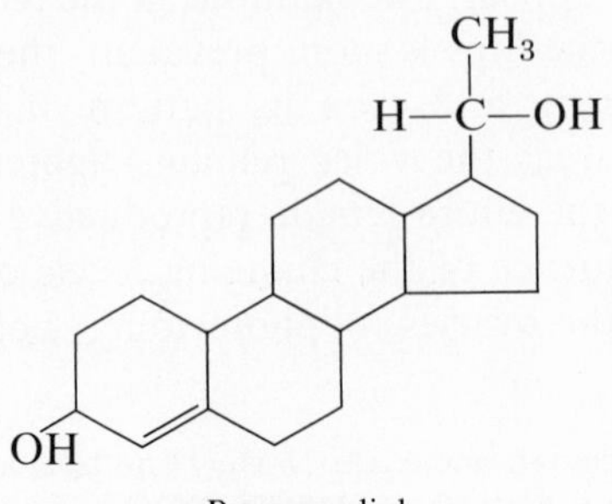

Pregnanediol

Pregnanediol concentrations in the urine have been used by clinicians as indicators of corpus luteum function.

The primary function of progesterone is to maintain the uterus and its implanted fertilized egg during the early months of pregnancy. Fraenkel[3] in 1903 was the first to demonstrate this function when he removed the corpora lutea from pregnant rabbits, resulting in the failure of the rabbit embryos to survive. Many years later Corner[4] confirmed this observation and also showed that the corpus luteum is responsible for aiding in the development of the uterus for the reception of the fertilized egg. Thus progesterone plays an important role in the sequence of events associated with the endometrium during the menstrual cycle and pregnancy. If the fertilization of the ovum does not occur, the corpus luteum disintegrates, and the production of the hormone ceases. In general, the actions of progesterone are dependent on the prior action of estrogen; that is, the effects of progesterone that may be described as differentiation are not displayed until after estrogens have accomplished their action on the reproductive tract, which may be described as a growth action. Thus an interaction exists between these two hormones of reproduction.

Changes in the Uterus Caused by Ovarian Activity—Menstruation

From **puberty** (when the ovaries become functional), which occurs at from 12 to 15 years, to the **menopause** (when the ovaries become nonfunctional), which occurs at from 45 to 50 years of age, the mucous membrane (endometrium) of the body of the uterus shows cyclical changes every 28 or 30 days. The cyclical changes in the histology of the endometrium are manifested by the regular monthly occurrence of a bloody discharge through the vaginal orifice, which persists from 3 to 7 days. This flow, known as the **menstrual flow,** or **menses,** consists of necrotic tissue as well as blood, and although varying in volume, amounts to approximately 8 oz (1 cup) for the entire discharge period. The histological picture of the menstrual discharge is so distinctive that it can almost always be distinguished from nonmenstrual bleeding. Another distinctive feature of menstrual blood is that it remains unclotted or clots imperfectly. The inability of the menstrual fluid to clot is believed to be due to a lack of fibrinogen and the absence of thrombin. Some research workers believe that a specific anticoagulant is present, which is provided by the activity of the anterior lobe of the pituitary gland.

Menstrual Cycle. The term **menarche** is applied to that point during puberty when the menstrual flow begins. The endometrium in the sexually mature female is that part of the uterus that reflects the actions of the ovarian hormones. The complicated cycle of events associated with this structure under the influence of estrogenic and progestative hormones that finally termines in the menses is referred to as the **menstrual cycle.** Usually four phases are observed during the typical 28-day menstrual cycle: these are (1) the proliferative, or follicular, phase;

[3] L. Fraenkel, "The Function of the Corpus Luteum." *Arch. Gynak.*, **68:**438 (1903).
[4] G. W. Corner, "Physiology of the Corpus Luteum." *Am. J. Physiol.*, **86:**74 (1928).

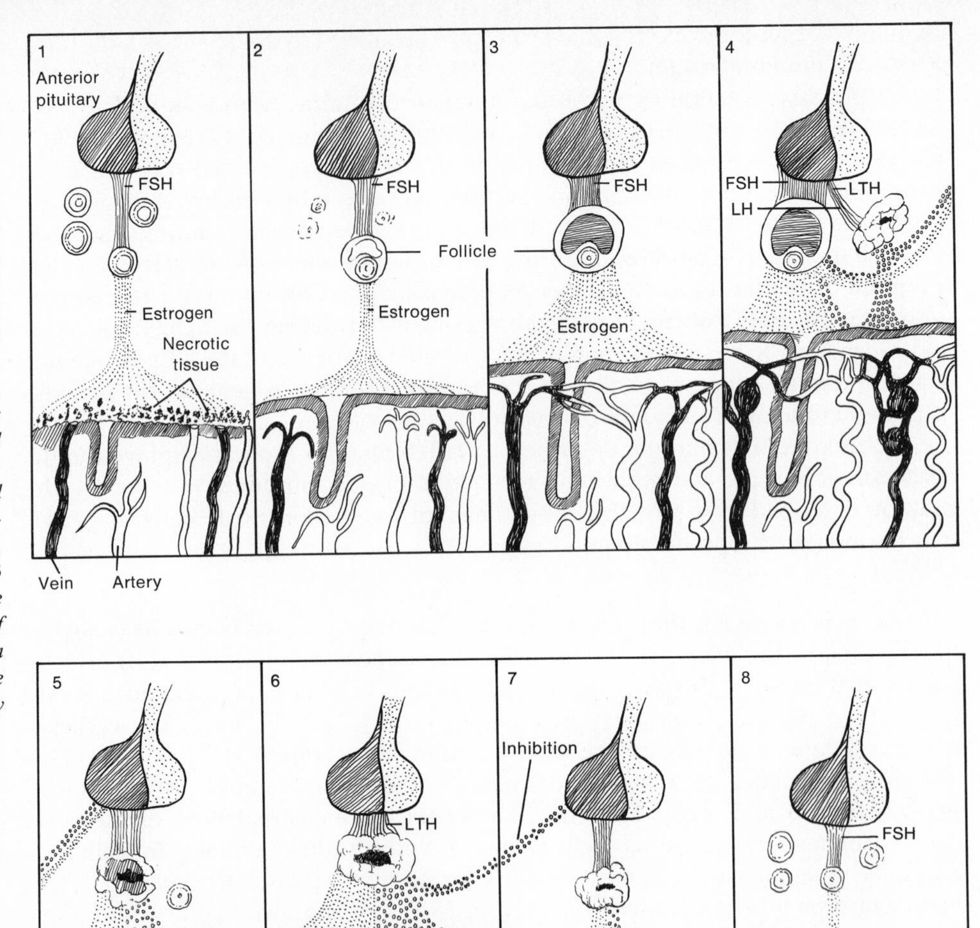

Figure 15-7. Ovarian–pituitary–endometrium relationship during the menstrual cycle. After menstruation the wall of the uterus is relatively thin, and the glands within are straight, short, and narrow (2). *During the development of the mature graafian follicle in the ovary, the endometrium, under stimulation of follicular hormone estrogen, undergoes a process of growth* (3, 4). *Ovulation occurs* (5) *always between the twelfth and sixteenth days and usually on the fourteenth day after the onset of the previous menstruation. As the ruptured follicle is transformed into a corpus luteum, the endometrium begins to poliferate under the influence of the remaining estrogen and the newly formed progesterone. The endometrial glands become quite enlarged and tortuous and undergo their maximal secretory activity* (6, 7). *This secretory, or luteal, phase of the menstrual cycle lasts about* 13 *to* 14 *days. If the ovum is not fertilized, the corpus disintegrates, and the decrease of both estrogen and progesterone causes a necrosis of the stromal cells, resulting in the destructive phase* (8, 1). *This phase usually persists for about* 4 *or* 5 *days.*

(2) the ovulatory phase; (3) the secretory, or luteal, phase; and (4) the destructive, or menstrual, phase.

Proliferative, or Follicular, Phase. After menstruation the wall of the uterus is relatively thin, and the glands contained within the endometrium are straight, short, and narrow. The length of the **proliferative phase** is variable but averages approximately 10 days. It has now been definitely proved that ovulation occurs nearly always between the twelfth and sixteenth days and usually on the fourteenth day after the onset of the previous menstruation. During the development of the mature graafian follicle in the ovary, the endometrium, under the stimulation of the follicular hormone estrin, undergoes a process of growth. By the tenth day the endometrium is approximately 0.5 cm in thickness, with an undulating surface and actively growing glands. The stromal cells within the endometrium are actively growing, and the entire structure appears quite dense.

Ovulatory Phase. The concentration of estrin in the blood reaches its greatest peak at the time of **ovulation.** During the 24- to 36-hour period after ovulation there is very little appreciable change in the endometrium. However, as the ruptured follicle is transformed into a corpus luteum, the endometrium continues to proliferate under the remaining estrogen and the stimulatory effect of the newly formed progesterone.

Secretory, or Luteal, Phase. During the **luteal phase** the endometrium is mainly under the influence of progesterone and the remaining concentration of estrogens. As a result of the combined hormonal action, the endometrium continues to increase in size and may reach a diameter of 5 mm or more. The endometrial glands become quite enlarged and tortuous and undergo their maximal secretory activity. This phase of the cycle lasts about 13 to 14 days. If the ovum is not fertilized, the corpus luteum disintegrates, and the concentration of progesterone also falls sharply. The decrease in the concentration of both estrogen and progesterone causes a cessation of secretion of the endometrial glands and a necrosis of the stromal cells, resulting in the destructive phase. If the ovum liberated during the ovulatory phase is fertilized, it becomes embedded in the highly developed endometrium, which then continues to develop into the maternal placental structures (decidua). These structures, that is, the placenta as well as the developing embryo, are able to produce sufficient quantities of progesterone to permit them to continue to develop within the uterus.

Destructive, or Menstrual, Phase. The **menstrual phase** is started when blood cells begin to pool in the intercellular spaces on the surface of the endometrial epithelium. Random areas of stromal cells and glandular tissue are sloughed off, which leads to the escape of blood and tissue into the cavity of the uterus. This phase usually persists for about 4 or 5 days.

Figure 15-7 summarizes the endometrial changes throughout the typical 28-day cycle.

Menstruation does not occur in mammals other than primates; however, a period somewhat analogous to menstruation does take place in other animals. This is spoken of as **estrus,** or **heat,** and is the period during which the female organism is receptive to the male.

Regulation of Ovarian Function and Menstruation. From what has already been said, it is obvious that the changes in the endometrium and other accessory sex organs are dependent on the action of the several hormones secreted by the ovary. What starts and maintains the cyclical activity known as the female sex cycle? The answer to this question lies in extraovarian influences. It is known that the anterior pituitary gland secretes three hormones that activate and influence ovarian function, and for this reason they are known as the **gonadotropic hormones.** They are commonly referred to as the **follicle-stimulating hormone (FSH),**

the **luteinizing hormone (LH** or **ICSH),** and the **lactogenic hormone (LTH).** These gonadotropins are produced in both sexes but have been named according to their actions in the female.

Experimental work has established that these hormones are essential at the time of puberty and throughout the reproductive life of the ovaries for continued graafian follicle development, ovulation, and the cyclical secretions of ovarian hormones. The interaction of the pituitary-ovarian axis appears to be established somewhere between the twelfth and fifteenth years of life. FSH is carried to the ovaries by way of the blood stream, where it stimulates the ovarian follicle to mature. As a result of the development of the theca interna, estrogen secretion is also stimulated, since experimentation has shown this to be the area of estrogen formation. As a result of estrogen secretion, dramatic changes occur in the secondary sexual characteristics. At the time of ovulation, which is initiated by the combined action of FSH and LH, the level of estrogen, which has reached its peak, suppresses the secretion of FSH. It appears that the output of FSH is controlled by the concentration of estrin in the blood, and, conversely, the level of estrin is regulated by the concentration of FSH. As the concentration of FSH falls, its stimulating effect on the graafian follicle is decreased, and the secretion of estrogen is decreased. As the concentration of estrogen falls, more FSH is secreted, resulting in the development of another follicle and more estrogen secretion (Fig. 15-8).

LH, in addition to its working with FSH to initiate ovulation, also stimulates the production of the corpus luteum. The third hormone, the lactogenic hormone, seems to be responsible for the secretory activity of the corpus luteum in the production of progesterone. Since progesterone is necessary for the further growth and maintenance of the endometrium, it is necessary for the corpus luteum to be sustained until the embryo has been able to establish itself. If the corpus luteum is not maintained for any reason, the endometrium sloughs off, and if the woman is pregnant, the implanted fertilized ovum is carried away with the debris. In such instances the woman is said to have spontaneously aborted. Removal of the pituitary gland in experimental animals causes such animals to abort if the removal is carried out during the early stages of pregnancy.

A reciprocal relationship exists between the levels of LH and LTH and progesterone similar to that which exists between FSH and estrogen. As the level of progesterone increases in the blood, the secretion of LH and LTH is suppressed. As a result, the corpus luteum begins to fade, and the concentration of progesterone decreases. If the egg is not fertilized, a supplemental supply of these hormones will not be forthcoming from the placental membranes, and the corpus luteum disintegrates to form the corpus albicans. The endometrium sloughs off as a result of the decreased progesterone level. The waning in the concentration of the progesterone stimulates the pituitary to secrete more LH and LTH. These hormones again exert their influence on the newly developing graafian follicle. Most of the available evidence indicates that it is this reciprocal relationship between the gonadotropins and the ovarian hormones that is responsible for the regulation of the ovary and the cyclical responses that are characteristic of the female sex cycle.

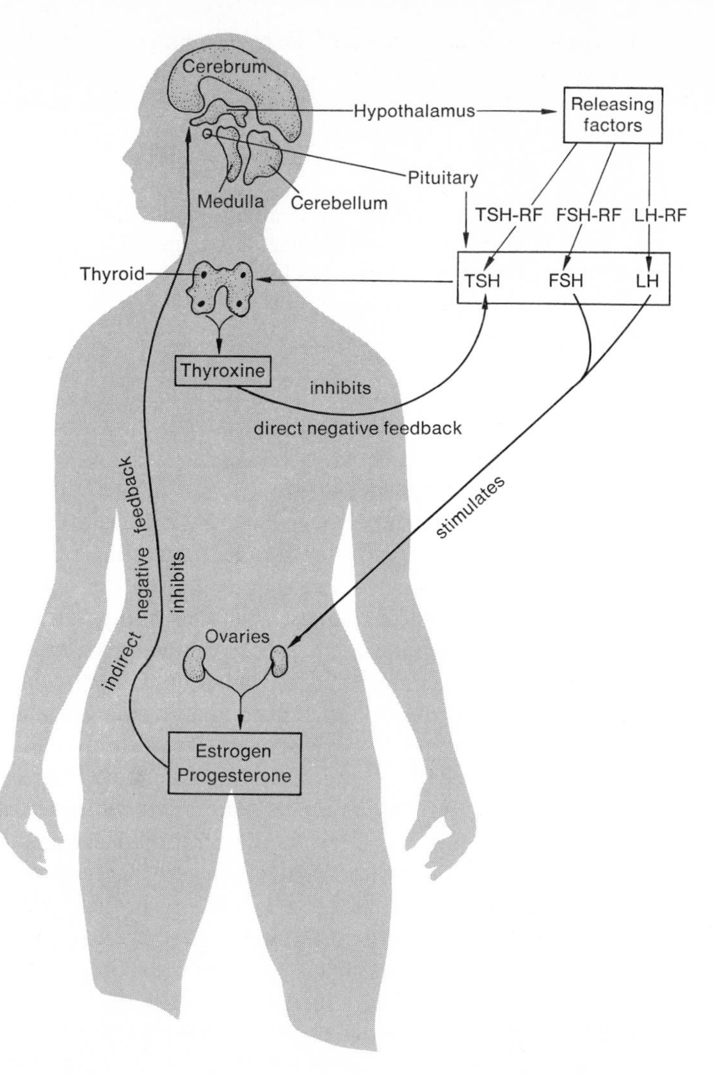

Figure 15-8. Influence of pituitary hormones on ovaries and how the level of estrogen influences the pituitary by negative indirect feedback.

The physiological mechanisms that regulate ovarian function and the phases of the menstrual cycle may be summarized as follows:

1. At puberty the ovary comes under the influence of the three gonadotropic hormones secreted by the anterior pituitary. The period from the beginning of increased growth of the follicle to the appearance of the menstrual flow is 28 to 30 days.
2. FSH stimulates the ovary to produce a mature graafian follicle and ovum.
3. Simultaneous with the changes in the ovary, corresponding changes take place in the uterus. The mature follicle secretes estrogen, which influences the development of the secondary sex characteristics and also stimulates the development of the endometrium. Increased concentrations of estrogen inhibit the production of FSH.
4. FSH and LH work together to initiate ovulation. Decreased concentrations of estrogen stimulate FSH secretion.
5. LH stimulates the production of the corpus luteum.

6. The luteal cells are stimulated to secrete progesterone by LTH. Increased concentrations of progesterone inhibit the production of LH and LTH.
7. Progesterone aids in the development of the endometrium and maintains it during the early stages of pregnancy.
8. Decreasing concentrations of progesterone stimulate the pituitary to secrete more LH and LTH.
9. The fertilized egg, on being implanted in the uterine wall, develops the placenta. The placenta secretes hormones similar to the gonadotropic hormones, which enable it to maintain itself after the corpus luteum has faded. If the corpus luteum fades before the placental hormones can be secreted, the uterine wall sloughs off, causing the embryo to abort.

Disturbances of Menstruation. Deviations from the normal course of menstruation are not uncommon. The most important and most common are the menorrhagias and metrorrhagias. **Menorrhagia** is a condition characterized by an abnormal loss of blood during the monthly period. It is usually caused by endocrine or ovarian functional disturbances and, in some instances, by disturbances in the physiological contractility of the uterus. **Metrorrhagias** are irregular hemorrhages in which the menstrual cycle may or may not be preserved. The hemorrhage may appear before the menstrual period, in the middle of the interval, or after the menstrual period. Sometimes the hemorrhage may persist throughout the entire menstrual cycle. The cause may be hormonal, or the hemorrhage may be due to inflammation of the endometrium or to disturbances of the uterine musculature. Tumors and polyps can also be a causative factor.

Amenorrhea is the term used to denote the absence of menstruation. The most common cause of amenorrhea is pregnancy. Conception results in the maintenance of the corpus luteum. As a result, the endometrium is still under the influence of progesterone and does not disintegrate. In fact, amenorrhea is one of the first and most reliable symptoms a woman has to indicate that conception has taken place. Amenorrhea may also be indicative of nonfunctional ovaries. In such instances, though, the menstrual cycle would not be established, and since the hormones elaborated by the ovaries control the development of the sexual organs and the secondary sex characteristics, other symptoms would be present to confirm the cause. Another cause of amenorrhea in women with normally functioning ovaries and who also show normal fertility may be the fact that the ovarian hormone concentrations do not fall low enough to permit sloughing off of the endometrium.

Menopause. The term *menopause* is used to denote that period in the female reproductive cycle when menstrual flow ceases. It can occur at any time after the menarche, but on the average the onset is at around 47 to 48 years. However, there are instances of menopause occurring as early as 36 years and instances of normal menstruation to the age of 60. The termination of the genital cyclical function in the woman does not occur abruptly but involves a period of several years; that is, menses may be irregular at the beginning of menopause, gradually decreasing in occurrence and volume until there is no menstrual flow at all. The entire period involved with the change is often referred to as the "change of life," or **climacteric;** the actual cessation of menstruation, as the **menopause.** Recurrence of the menses after the menopause is not possible, since once the menstrual cycle has ceased, the internal sex organs atrophy, and the function of the ovaries in producing sex hormones is not possible.

The climacteric is part of the natural development of the woman, and although in itself it is not pathological, the period is characterized by certain disturbances, both mental and physical. Among the symptoms that are most commonly associated with this period are "hot flushes," which take the form of a sudden rush of blood to the head, with a feeling of dizziness, heat, and perspiration. In addition to hot flushes, palpitation, irritability, anxiety and depression, loss of appetite, insomnia, and headache may develop, so that the woman feels quite ill. Since the ovaries undergo progressive degeneration and there is no maturation of follicles, little, if any, estrogen is produced. The sexual organs and the secondary sexual characteristics undergo marked alterations. The external features of the woman may take on a more intersexual aspect. There is a tendency for the skin to become drier and less elastic, so that the folds of the skin deepen and wrinkles appear. In some instances the head hair and pubic hair may decrease, whereas there may be an increased growth of hair on the face. There is a tendency for the breasts to become smaller, and the vaginal secretions diminish so that the vulva may be dry and bothersome itching may be present. The symptoms of the menopause are probably precipitated by estrogen deficiency. However, in the absence of estrin inhibition, FSH is secreted in large quantities, and the concentration of this hormone in the blood and urine becomes abnormally high. Whether the vasomotor symptoms of menopause are related to the excess of the follicle-stimulating hormone or to the absence primarily of estrogenic hormones is not definitely known.

Mammary Glands

The mammary glands, or breasts, of infancy and childhood are similar in form and size in both sexes. At puberty the female mammary glands become large, firm, and rotund, with the nipples rather centrally located. The adult female breast consists of masses, or lobules, of lemon-yellow fat, which constitute the bulk of the structure. Supporting connective tissue courses throughout the fat, and many small, glistening orangeish-colored bodies about 1 mm or less in diameter are present. These are the mammary gland alveoli and are composed of two types of cells: columnar cells, with a dense oval nucleus located near the base, and flat, oval-shaped cells. The columnar cells are secretory and produce milk, and the flattened, ovoidal cells are regenerative and serve to replace the columnar cells when the latter are destroyed or shed. Ducts, or tubules, connect the alveoli with the surface of the nipple (Fig. 15-9). The secretion of appreciable amounts of milk is dependent on the development of the alveolar system within the mammary gland. The alveolar system as well as the duct system is influenced by estradiol, progesterone, and lactogenic hormone. The activities of the ovary, uterus, and mammary glands are interrelated, and the essential stimulus for the development and secretory activity of the gland is hormonal. Estradiol causes duct development and an increase in the supporting connective tissue structures, and progesterone is responsible for the growth of the alveoli and lobules. In fact,

Figure 15-9. Cross section of mammary gland, showing the development of the alveolar system under different conditions. (1) *Nonpregnant woman,* (2) *middle pregnancy, and* (3) *during lactation.*

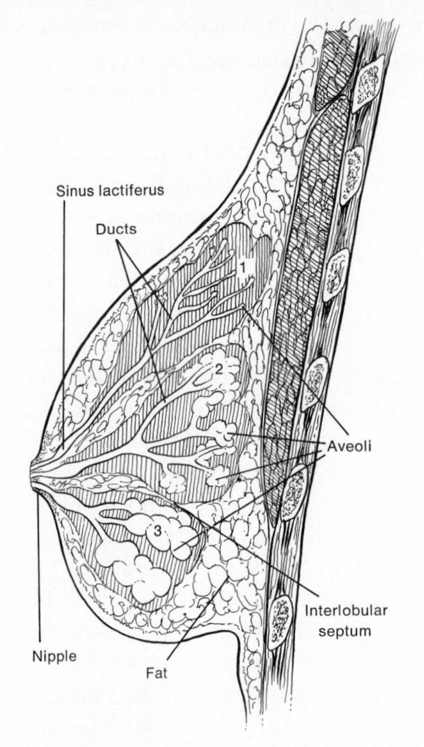

for the complete development of the breasts, progesterone is of prime importance. Finally, lactogenic hormone from the pituitary, also called **prolactin,** or **luteotropic hormone,** initiates the secretory activity of the mammary gland (lactation). It is believed that this hormone has a direct action on the gland.

Before puberty the breasts lie close to the thoracic wall and have a compact structure. Between the menarche and the climacterium the breasts show variations in their structure, which are related to the stages of the menstrual cycle and the levels of both ovarian and gonadotropic hormones in the blood. During pregnancy, especially after the fifth month, the breasts become fuller and show a definite increase in size above that normally shown during the normal menstrual cycle. This is because of the extensive increase in the number of cells in the lobules as well as an increase in the number of lobules themselves. And there is an additional engorgement of the vascular tissue. After the normal reproductive function of the female ovary has ceased (menopause), the breasts show a general shrinkage due to the contraction and extensive atrophy of the connective tissue, as well as the lactiferous ducts and alveoli. The breasts sag with age and lie closer to the chest wall.

Female Eunuchoidism

The term *eunuchoidism,* as generally used, applies to males as well as females in whom the gonads have been removed or destroyed by disease or, if present, are not functioning normally to produce the sex hormones. If the ovaries of the normal female are removed before puberty because of some disease or have been rendered inactive by X-ray therapy, the child never develops the secondary sexual characteristics, and the sexual organs remain immature. Growth continues at a normal rate but does not show the usual spurt associated with puberty, and the individual ends up with the disproportionate skeletal configuration typical of eunuchs; that is, the length of the lower limbs is much greater than that of the trunk. In the normal person the distance from the floor to the pubis is practically the same as the distance from the pubis to the head. Removal of the ovaries after puberty will also cause changes; that is, the breasts usually atrophy, growth of hair on the face and body may occur, the body configuration of the female becomes more masculine and sexual drive may decrease. It is obvious that the hormones secreted by the ovaries control the development of the secondary sexual characteristics and the sexual organs and that the degree of development depends on the activity of these glands.

The administration of estrogens is highly satisfactory in the treatment of eunuchoidism, as it will not only cause development of the secondary sexual characteristics and reproductive organs but also usually will result in a psychological uplift for such patients.

The Male Generative Organs

The male genital organs (Fig. 15-10) include the penis, the prostate and Cowper's glands, the testes, the vasa deferentia, and the seminal vesicles. The gonads in the male are called the testes, and they arise embryonically from the same tissue site as their female counterpart, the ovaries. The factors involved in the transformation of the undifferentiated gonad into a testis or ovary are not fully known. The most important factor appears to be the chromosomal pattern of the zygote. Of the 46 chromosomes present in all the somatic cells and primitive germ cells, two are concerned with the determination of sex. These are the **X and Y chromosomes;** female cells contain two X chromosomes, and those of the male

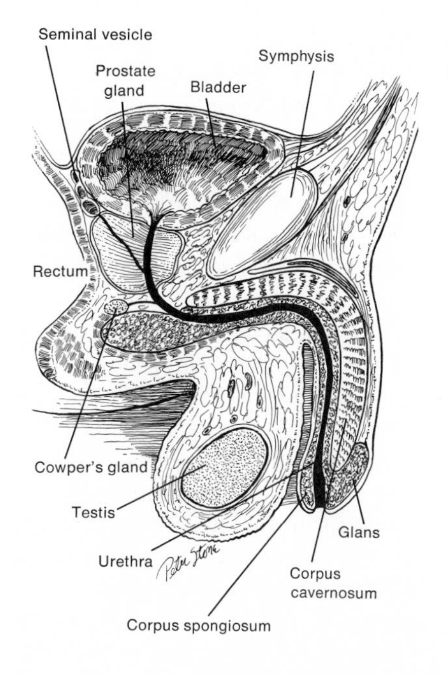

Figure 15-10. Midsagittal section of male reproductive organs.

contain an X and a Y chromosome. The genetic composition of these chromosomes appears to exert the sex influence. However, the ultimate sex of the individual is thought to be determined not only by the sex-influencing genes in the sex chromosomes, but also by the balance that exists between these genes and the sex-influencing genes known to be present in certain other chromosomes of the cell. The development of certain forms of intersexuality is believed to be due to a disturbance of this balance. The manner in which sex-determining genes cause development of the undifferentiated gonad into a testis is unknown.

The Testes

The testes, or testicles, are two small, flattened, oval-shaped glands which are situated in a musculomembranous pouch, called the **scrotum,** and are suspended by the spermatic cords. These glands are not developed within the scrotal sac; as has already been discussed, the undifferentiated gonad appears during the fifth or sixth week of embryonic life as a thickening of the peritoneal epithelium. By the seventh week the specific characteristics of the differentiated gonad (testicle or ovary) can be recognized. During the seventh to ninth month of fetal life, the testes descend from the peritoneal cavity, through a narrow canal connecting the peritoneal cavity with the scrotal sac, called the **inguinal canal,** to come to rest in the location they will occupy throughout life. The inguinal canal closes just prior to birth. Incomplete closure of this canal at birth predisposes the individual to subsequent inguinal hernia formation. Very little growth or development of the testes occurs during the first 10 years of life. With the approach of puberty, from the eleventh to fifteenth years, there is a gradual acquisition of adult proportions. Each adult testis measures approximately 4 to 5 cm in length and 2.5 to 3 cm in width, with an average weight of 25 g.

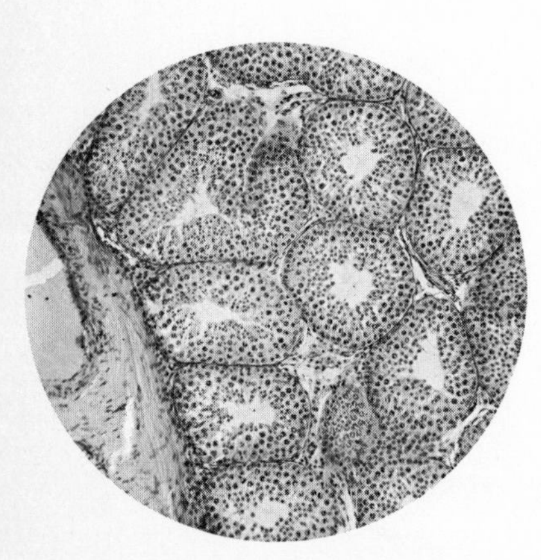

Figure 15-11. Human testis, showing capsule tubules and interstitial tissue. (Courtesy General Biological Supply House, Chicago.)

The testis (Fig. 15-11) has an outer fibrous coat, the **tunica albuginea.** The substance of the gland is composed of a number of pyramid-shaped **lobules,** which are situated with their bases toward the surface. Each lobule is composed of several **seminiferous tubules** between which are the **interstitial cells of Leydig.** There are estimated to be 800 or more tubules in the testis. The seminiferous tubules unite and form a plexus of canals known as the **rete testis,** which end in the upper part of the testis in a series of ducts called the **vasa efferentia,** or efferent ducts. These pass through the tunica albuginea and form convoluted masses lying on top of the testis called the **epididymis.** A single duct leads from the epididymis, the **ductus deferens,** to join the urethra (Fig. 15-12).

Testicular Function

The testes, like the ovaries, have a dual function of producing reproductive cells and sex hormone.

Production of Mature Sperm. The production of mature sperm cells by the seminiferous tubules is referred to as **spermatogenesis** and is a process that begins at puberty and continues without interruption throughout the life of the male. Spermatozoa are formed from the primitive parent cells, called **spermatogonia,** which form the basilar layers of the seminiferous tubules (Fig. 15-13). Spermatogonia undergo extensive mitotic activity to give rise to smaller spermatogonia, which are pushed closer toward the lumen of the tubule. These pass through a period of growth in which each spermatogonium undergoes considerable enlarge-

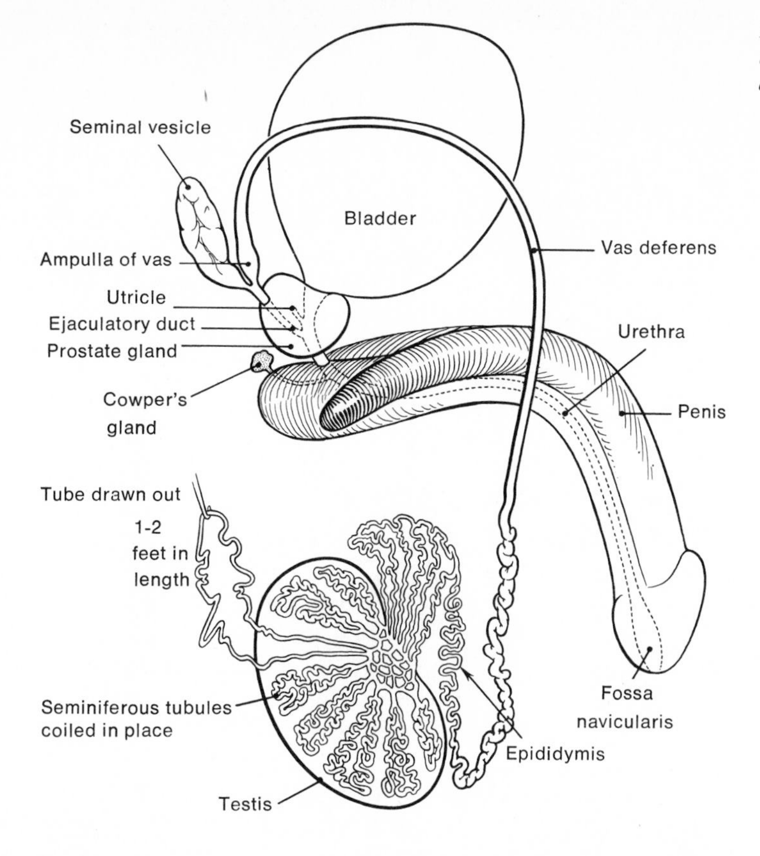

Figure 15-12. Testis, showing its duct system and the relation of the ducts to the accessory sex glands and penis.

ment, becoming a **primary spermatocyte.** Each primary spermatocyte divides, forming two **secondary spermatocytes.** Previous to this division, the chromosome number characteristic of the species is present. Each secondary spermatocyte containing half of the characteristic chromosome number undergoes division to produce two cells, called **spermatids.** Spermatids are attached to **Sertoli's cells,** which appear to act as sources of nourishment for the developing sperm cells. Each spermatid, by a process of differentiation without any division, becomes a mature **sperm cell,** or **spermatozoon.** Beginning with a primary spermatocyte, two cell divisions have occurred, resulting in the formation of four functional sperm cells, each with the haploid number of chromosomes. The two separate mitotic divisions involved in this process constitute reduction division, or meiosis (Fig. 15-14). In most mammalian species the process of mature sperm cell production requires about 30 days. Since the sex chromosomes in the male are X and Y chromosomes, half the spermatozoa contain the X chromosome and half the Y chromosome.

Although there are differences in shape of mature sperm cells among different species of animals, they all share in common basic structural features such as a head, neck, middle piece, and long tail. In man the spermatozoon (Fig. 15-15) measures 55 to 65 μm in length. The **head** is oval or elliptical in shape and consists

Figure 15-13. *Enlarged section of a portion of a seminiferous tubule, showing spermatogonia at different stages of development. Top of illustration indicates lumen, into which are discharged mature sperm cells.*

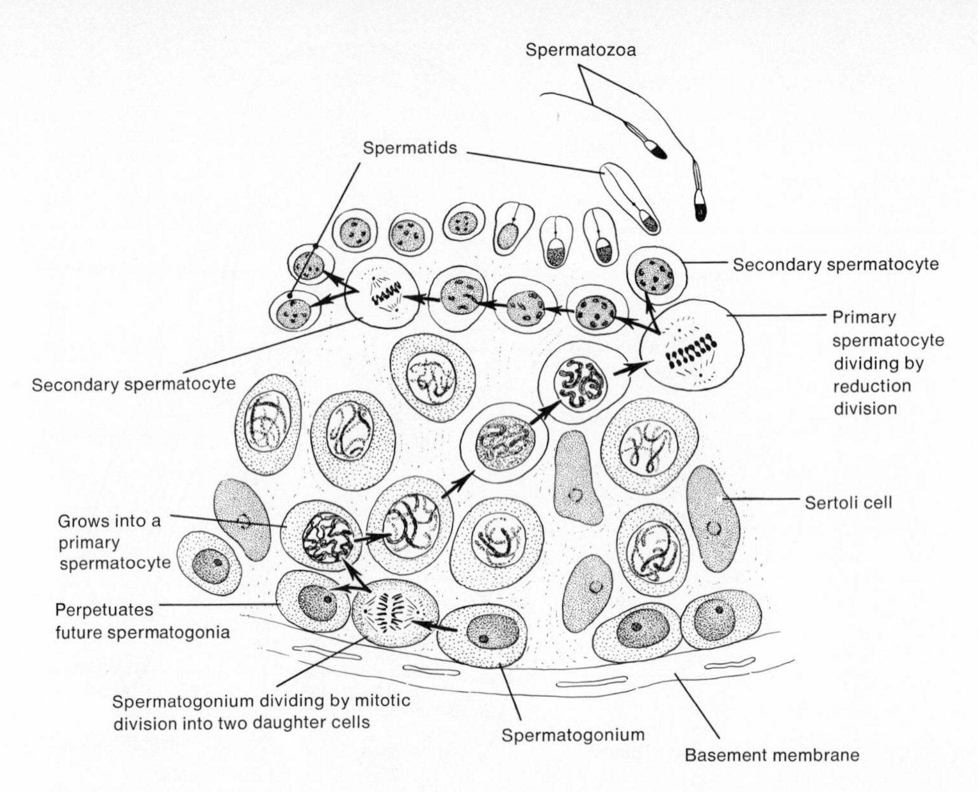

mostly of nuclear material with little cytoplasm. The **neck** is a rather delicate structure, believed to be derived from the centriole of the undifferentiated sperm cell. It connects the head with the middle piece, or body, of the spermatozoon. The **middle piece** is cylindrical or spindle-shaped and contains many fine fibrils grouped into a fascicle called the **axial filament.** A spiraled structure called **Jensen's spiral body** surrounds the axial filament. Enclosing the entire middle piece is a sheath of cytoplasm in which are suspended minute particles called **microsomes.** The **tail** is the longest structure of the mature sperm and appears to be a continuation of the axial filament of the middle piece, which tapers off, finally terminating in man in an end piece as either a single fibril or several fibrils in tassellike form. Under the ordinary light microscope only the head and a long, whiplike tail can be distinguished.

The nuclear material that makes up most of the head of the sperm cell carries the chromatin material, which consists approximately of 43 per cent DNA. At fertilization it is this material that contributes to the zygote the paternal genetic characteristics. The small amount of cytoplasm surrounding the nuclear material of the head, which is called the **acrosome,** is believed to be concerned with the release of enzymes necessary for fertilization. The middle piece is the area of metabolic activity wherein the energy released in the breakdown of glycogen is transferred to high-energy phosphate compounds. The tail is the means by which the spermatozoon propels itself through its surrounding medium; that is, it is the locomotor organ of the cell. However, spermatozoa are nonmotile while in the

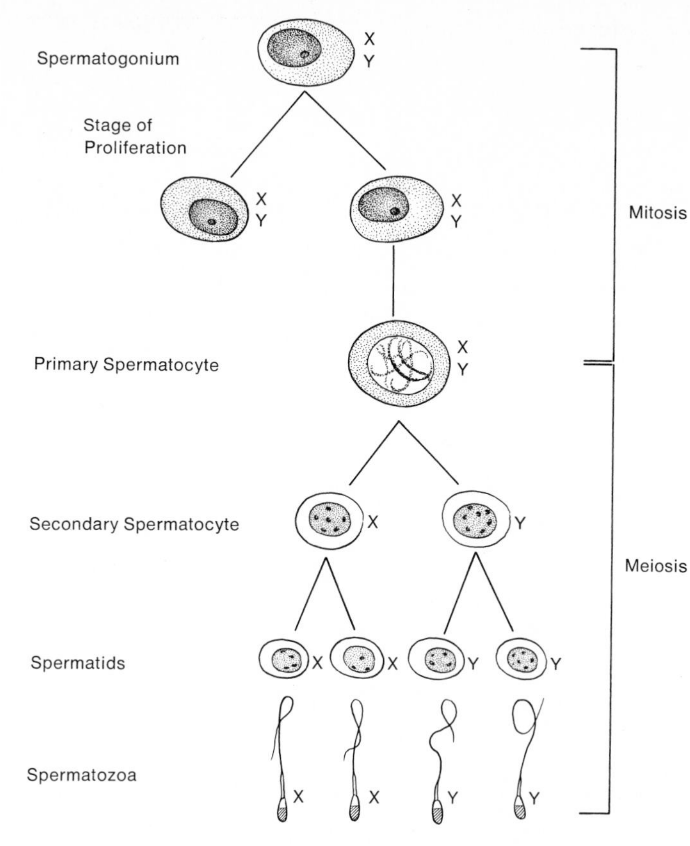

Figure 15-14. Maturation of male gametes (spermatogenesis). Spermatogonia at the same time renew themselves to maintain their numbers to produce a new generation of germ cells, the primary spermatocyte. The primary spermatocyte divides into two smaller, secondary spermatocytes. They in turn divide, each one giving rise to two smaller cells, spermatids, which undergo metamorphosis to become spermatozoa.

Figure 15-15. Mature sperm cell (spermatozoon).

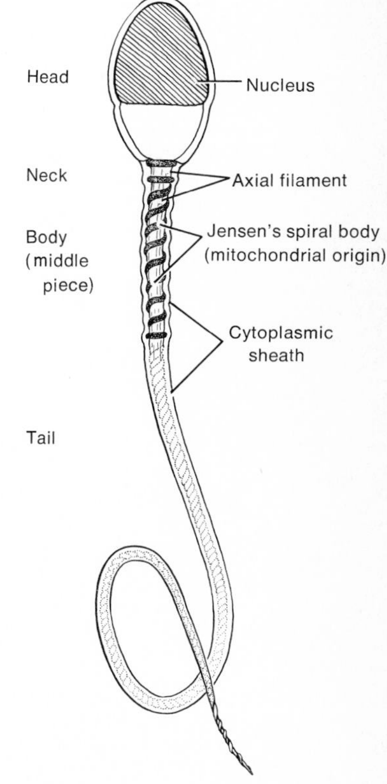

seminiferous tubules. As sperm cells mature, they become detached from Sertoli's cells and are transported to the epididymis, where they are stored. Ciliary action of the rete tubule cells and pressure are believed responsible for the transmission of these forms to the epididymis. The exposure to air and the addition of alkaline secretions from the prostate, seminal vesicles, and Cowper's gland that occur at ejaculation stimulate sperm motility. In the absence of the emission of spermatozoa from the epididymis, the sperm cells disintegrate and are reabsorbed by the tubules. In addition to being stored in the epididymis, sperm cells are probably also stored in the vas deferens, which ascend in the posterior part of the spermatic cord, becoming the spermatic canal, where it runs to the base of the bladder and becomes enlarged, sacculated, and narrowed, forming the **ampulla,** which receives the duct of the seminal vesicle. From here the **ejaculatory duct** conducts the sperm through the substance of the prostate gland to the penile urethra (Fig. 15-16). The seminal vesicles, the prostate, and Cowper's glands produce an alkaline secretion that serves as a suspending medium for the spermatozoa. The **seminal fluid** is thus a complex secretion, being composed of the spermatozoa produced in the testes and the secretions produced by the above-listed structures. It has a thick, whitish, striated appearance, and if examined microscopically, it is seen to contain innumerable motile spermatozoa.

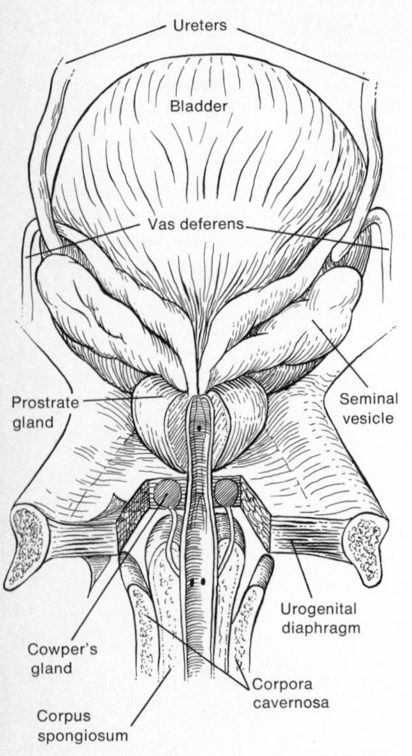

Figure 15-16. Posterior view of bladder, showing the relation of the vas deferens to the accessory sex glands and the opening of the ejaculatory duct into the penile urethra.

At the time of ejaculation, the chief factor involved is the propulsive force generated by the contraction of the perineal muscles. The average volume of an ejaculation is approximately 4 ml, and in man the average sperm count is 120,000,000 sperm per milliliter. It can be seen that in the total ejaculation there are more than a half billion sperm; since usually only one sperm fertilizes one ovum, this is another example of nature's providing an overabundance in order to protect a vital function, in this case to assure the continuation of the species. If the sperm count falls below 50 million per milliliter, good fertility is not assured. In addition, if abnormal forms of sperm cells exceed 25 per cent of the sperm population, the man will, in all probability, be sterile. Abnormal forms of spermatozoa are usually encountered in normal seminal fluid and involve only differences in size and shape of the head.

Spermatogenesis is affected adversely by both too high and too low temperatures. In human beings, as in most mammals, normal spermatogenesis cannot occur at the temperature prevailing in the body cavity but can occur at the temperature prevailing in the scrotum. Although mature spermatozoa are capable of surviving very low temperatures, the process of spermatogenesis is inhibited by exposure to cold. The scrotum, therefore, appears to be a structure serving a dual thermoregulatory function. It permits the testes to maintain a temperature 3 to 4 degrees below body temperature, which appears to be the optimal temperature for sperm production, and also helps to protect the testes against too low environmental temperature. It is the mobility exhibited by the scrotal sac that allows for this thermoregulatory control, whereby an optimal temperature can be maintained for spermatogenesis. The scrotal sac is suspended primarily by the **cremasteric fascia,** which is composed of a double layer of areolar and elastic tissue enclosing a definite but thin layer of striated muscle. It is principally the cremasteric fascia that affords proper retraction of the testicles and also their descent from the body in response to temperature changes. When the temperature drops or when there is danger of trauma, the muscles contract, causing the testes to be brought tight up against the body (retraction) where they can have the benefit of the body heat and protection of the body cavity. When the temperature increases, the muscle relaxes, allowing the scrotal sac to fall away from the body cavity and the body heat and thus permitting the testicles to maintain an optimal temperature for spermatogenesis. It has been suggested that certain types of sterility in man due to spermatic arrest may be caused by the insulating effects of tight-fitting jockey-type shorts and to occupations that expose the man to very high temperature for prolonged periods of time. In cases of **cryptorchidism** (Fig. 15-17*A, B*), where the testes have failed to descend into the scrotum, no viable sperm are produced because of the temperature. Such men are sterile but are quite capable of sexual intercourse, and they develop all the male secondary sexual characteristics since testosterone is produced by such testes in adequate quantities.

Endocrine Function of the Testes. In addition to producing spermatozoa, the testes also produce the male sex hormone. The fact that the testes exert an endocrine influence was first demonstrated by Berthold in 1849 when he carried out transplantation experiments with excised testes in roosters. However, it was

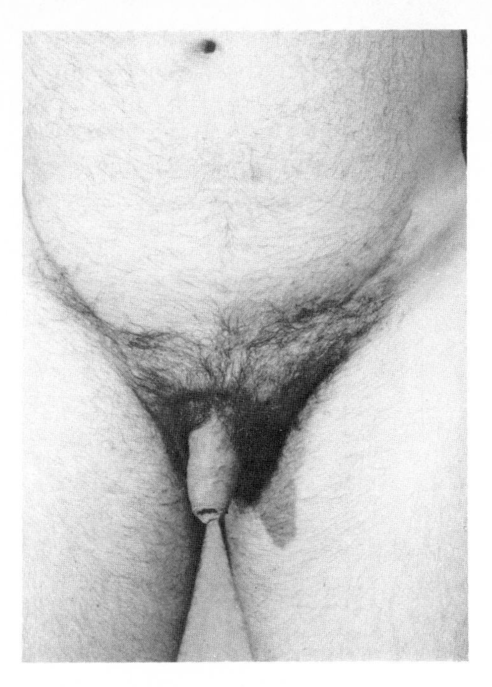

(A)

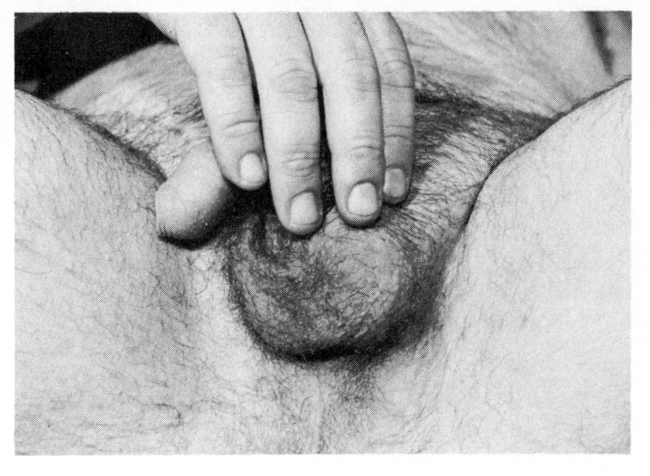

(B)

Figure 15-17. Cryptorchidism. In order to exclude the many cases of pseudocryptorchidism in which the testes are merely retracted from the scrotum, the patient should be examined in the upright (A) *as well as in the recumbent position* (B) *to lessen the possibility of a mistaken diagnosis.* (*Courtesy Armed Forces Institute of Pathology, Washington, D.C., picture M-51354-5-7.*)

Brown-Séquard, a French physician, who, in 1889, was the first to use testicular extract as a therapeutic product. After a series of experiments on himself, in which he took a series of doses of testicular juice, he found an increased capacity for both physical and mental work and a general sense of well-being. From his experiments he concluded that the degeneration and impotence associated with old age in men were due to the absence of testicular secretion on the nerve centers and that this condition could be corrected to a great degree by the administration of testicular extracts. Thus, the legend of the rejuvenating power of testicular extract had its birth with Brown-Séquard's experiments. It is interesting to note that Brown-Séquard used aqueous extracts. Since the estrogens and androgens are steroids, they are insoluble in salt solutions, and it is unlikely that Brown-Séquard had any hormonal component in his extract.

McGee in 1927 isolated a relatively pure active principle from bulls' testes, which, acting in a quantitative way, was able to relieve the symptoms of castration. Later, extracts of male urine were found to be effective substitutes for the testicular principle, thus demonstrating its urinary excretion. The work of Budenandt and his co-workers[5] led to the identification and preparation of pure crystalline male hormone. Since the secretion of the testes stimulates the development of the male reproductive organs and the male secondary sexual characteristics, it is known as an **androgenic hormone.**

[5] A. Budenandt, "The Chemical Investigation of the Male Sex Hormone." *Z. Angew. Chem.*, **44**:905 (1931).

Five principal androgens are secreted in the adult male body. The most active of these is **testosterone,** secreted only by the testes of adult men. Another male hormone, **androstene-3,17-dione,** is also secreted by the testes. The other three are secreted by the adrenal tissue and are known as **11β-hydroxyandrostene, dehydroepiandrosterone,** and **adrenosterone.** However, it is the testicular androgen that is most important in the physiology of the male reproductive system. Testosterone has the following structure:

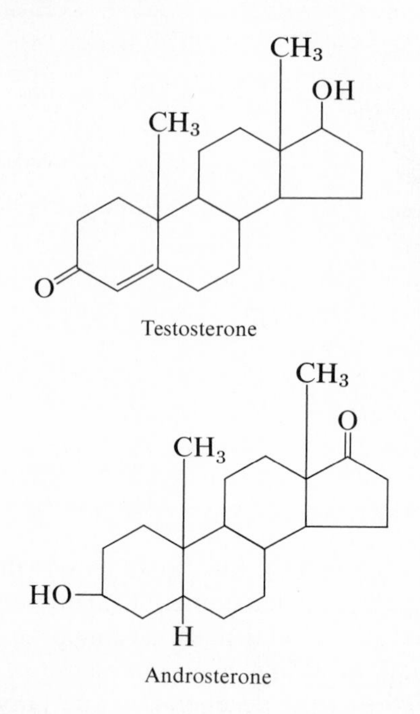

The androgens are metabolized and excreted in the urine as less active androgens, mainly **androsterone.**

It is believed that the male sex hormone is probably formed by the **interstitial cells of Leydig.** These cells appear singly or in groups between the seminiferous tubules. They usually contain a single nucleus, and the nucleoplasm is poor in chromatin. Many mitochondria are present in the cytoplasm, and variable amounts of fat granules are present. Pigment granules appear in the cytoplasm at about the age of 21.

The androgen secreted by the testes functions to bring about the development and maintenance of the male accessory sex organs and genitalia, such as the vas deferens, epididymis, prostate gland, and penis. In addition, it is responsible for the development of the male secondary sexual characteristics. These include muscular development, skeletal structural characteristics of narrow pelvis and broader pectoral girdle, deep voice, facial hair, chest hair, and masculine distribution of pubic hair. In the female the hair covering the upper border of the pubic region is usually horizontal, whereas in the male it extends up toward the umbilicus along the midline. The pattern of distribution of the hair on the head is also different in the two sexes and is probably determined by the male hormone. In mature men the hairline on the forehead is usually marked by a recession over the lateral frontal region on each side. These are referred to as the **calvities.** In boys, girls, and women the hairline on the forehead forms a continuous curve.

It has been definitely established that the testes, in addition to secreting androgens, secrete a hormone having estrogenic properties. Small amounts of estradiol have been isolated from the human testis by several research workers. The site of its formation is believed to be the germinal epithelium of the seminiferous tubules. However, it has also been postulated that Sertoli's cells, as well as the interstitial cells of Leydig, are probable sites of estrogen production. Testicular tumors, in which Sertoli's cells usually dominate, have a feminizing effect on the male.

Regulation of Testicular Function. Adolescence in both male and female is apparently initiated by the action of the anterior pituitary gonadotropic hormones. We have already seen in our discussion of ovarian regulation that three such hormones act on the ovaries, the follicle-stimulating hormone (FSH), the luteinizing hormone (LH), and the lactogenic hormone (LTH). These substances have also been identified and their concentrations determined in human male pituitary glands. The gonadotropins first become demonstrable in boys at the age of 13 years, but in girls they appear earlier, approximately at the age of 11. The **follicle-stimulating hormone** acts primarily, in the male, on the epithelium of the seminiferous tubules to stimulate spermatogenesis, and the **luteinizing hormone** stimulates the interstitial cells of Leydig to produce the male hormones (androgens). The latter hormone is therefore called the **interstitial cell-stimulating hormone (ICSH).** As male hormone is elaborated, it stimulates the development of the reproductive system and the male secondary sexual characteristics. The growth spurt occurring at this time, as well as the development of the masculine type of musculature, is also due to the male hormone. Interestingly enough, it is believed that a similar growth spurt in girls is also due to androgens secreted by the adrenals rather than to ovarian estrogen. The **luteotropic hormone,** so named because it maintains the activity of the corpus luteum in the female and also known as lactogenic hormone and prolactin, also exists in the anterior pituitary of males. It is believed to function in males in the same manner as LH and thus aids in the elaboration of male hormone. Experimental evidence also suggests that it may act directly on the male accessory organs along with the male hormone to stimulate their growth and development.

A reciprocal relationship exists between the gonadotropic hormones and the testicular hormones in much the same way that it exists in the female between ovarian and gonadotropic hormones. The male hormone secreted by the testes exerts an inhibiting effect on the anterior pituitary. A decreased output of testosterone or a fall in its normal concentration in the circulation has a stimulating effect. In general, clinical evidence has established beyond doubt that a reciprocal relationship exists between pituitary and testicular secretory activity. Whether there is, in the male, anything similar to a cyclical rhythm of activity, as is apparent in the female because of the reciprocity between gonad and pituitary, is not easily

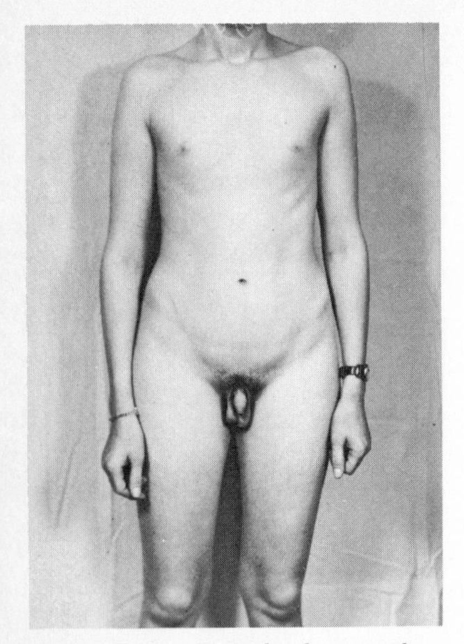

Figure 15-18. Testicular hypogonadism. Note the feminine contour of the body, the mammary glands, and the absence of body hair. Age of patient, 25 years. (Courtesy Armed Forces Institute of Pathology, Washington, D.C., picture 54-663.)

demonstrated. Production of sperm appears to vary rhythmically, being highest at the peak of FSH concentration and low when it falls off; similarly, the production of testosterone would increase and decrease.

Disturbances of Testicular Function—Hypogonadism and Hypergonadism. The term **hypogonadism** is used to designate the conditions of total loss of hormone activity of the testes, as occurs in castrated or eunuchoid men, and insufficient activity of the testes due to a primary functional deficiency of the pituitary. Withdrawal of the gonadal influence probably affects all the tissues of the body, but predominantly genital, osseous, adipose, and dermal tissue. Castration before puberty or hypogonadism is usually followed by an excessive longitudinal growth of bones (Fig. 15-18). Sex hormones stimulate the fusion of the epiphysis; in their absence there is a delayed fusion, producing a definite and characteristic disproportionate growth of long bones. The hypogonadal person is usually tall, with the lower body measurement (floor to pubis) being greater than the upper one (pubis to top of head). The head and face are usually of the long, narrow, type with delicate features. Castration, or hypogonadism occurring in the adult after the epiphyses have closed, is without effect on the skeletal system.

The distribution of adipose tissue appears to be influenced by androgens. Fat is more commonly stored in the abdominal region above the umbilicus in males, and in females it usually accumulates below the umbilicus, around the hips, in the mammary glands, and in the axillary region. Hypogonadism in the male usually causes an accumulation of fat in the areas typical of the female; thus in the male castrate, fat frequently accumulates in the mammary region, around the hips, and below the umbilicus.

Early hypogonadism leads to a significant retardation of the accessory sexual organs (Fig. 15-19). The penis, prostate, seminal vesicles, and vas deferens do not develop in such persons but remain infantile. Late hypogonadism leads to a very minor regression of the genital organs; however, a reduction in sexual drive and potency is more characteristic of this condition. The hair on the body and pubic region becomes sparse, and there is a tendency to feminization of the pubic contour. Such men are sterile because of a lack of sperm (Fig. 15-20).

The skin is also influenced by androgens. In castrated and eunuchoid men the skin is velvety in character, soft, and sallow. After the administration of male hormone, the skin becomes firmer and ruddier and has a darker color—characteristics typical of the normal male. Castrated men have little or no ability to tan because of the subnormal levels of melanin in the skin in comparison with those of normal men. The skin also becomes prematurely wrinkled in the hypogonadal person.

The administration of androgens to the hypogonadal man restores the masculine sexual characteristics and also stimulates the growth of the accessory sex organs.

Hypergonadism refers to overactivity of the gonads, which leads to excessive development of the genitalia, secondary sexual characteristics, and the body as a whole. Such men would have unusually broad shoulders, narrow hips, extremely muscular bodies, inordinately large sexual organs, and excessive body hair. However, it has never been established that such a primary hypergonadism is a clinical entity since such men would not consider the hyperactivity an undesirable abnormality and thus would not seek corrective treatment. In most instances the manifestation of hypergonadism is a hypergenitalism (enlarged genital organs), and this is usually found in association with tumors of the pineal, adrenal, pituitary, or gonads.

Hypergonadism is found frequently in boys and girls, with excessive development of the genitalia, secondary sexual characteristics, and body structure being common to all. Usually such persons end up being short because of premature closure of the epiphyses. The sexual drive usually parallels the sexual development, so that boys experience ejaculations, practice masturbation, and perform cohabitation. In girls the genitalia are overdeveloped, as are the mammae, and menstruation is common.

The hypergonadism or hypergenitalism may be either the result of hereditary factors or the result of an increased secretory activity of the gonads due to local lesions or to lesions of other endocrine glands, which may act directly on the bodily structures or through the gonads.

Hypergonadism developing in the postpubertal normal men is not attended by clinically recognizable manifestations, except perhaps by an increase in the excretion in the urine of 17-ketosteroids and a possible depression of spermatogenesis. Usually testicular or adrenal tumors are removed surgically before such studies of the effects of increased androgen secretion on the adult male may be made.

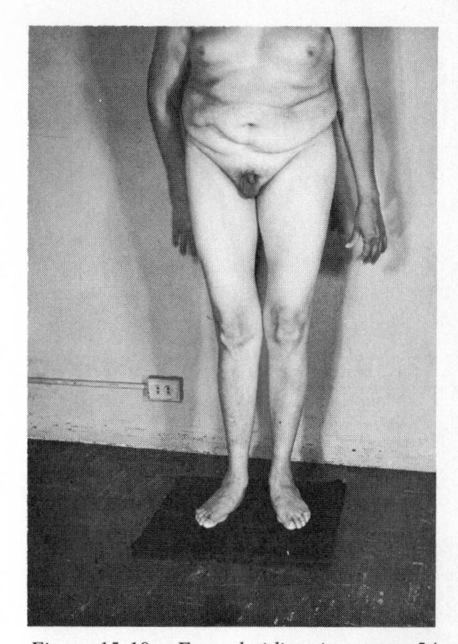

Figure 15-19. Eunuchoidism in a man 54 years old. Note the obesity of female distribution, the enlarged breasts, and the absence of body hair. (*Courtesy Armed Forces Institute of Pathology, Washington, D.C., picture 61-9682* [*Br 20200*].)

From what has already been discussed it becomes obvious that the androgenic function of the testes can range from low activity to high activity. It is most probable that normal testicular function occurs between the two extremes, so that normal masculine characteristics also fall between the extremes; at one extreme the male approaches the feminine, and at the other he is most masculine and virile.

Hermaphroditism

Since the male and female gonads develop from the same embryonic site and type of tissue, it is not impossible for the embryonic gonad to develop into a gland containing both testicular and ovarian components. Such a gonad is referred to as an **ovotestis,** and the condition is termed **true hermaphroditism.** However, testes and ovaries are not always combined in one organ but may exist separately on two sides of the body. In human hermaphroditism the sexual characters are usually intermediate between the male and female. The external genitalia include a penislike organ with the urethra opening under its base. This condition is known as **hypospadias.** A small vaginal opening is usually present and also a uterus. A scrotal sac may or may not be present and may or may not contain a testis or ovotestis. In most instances the testes remain within the abdominal cavity, resulting in a cryptorchid condition. The sex interests of most hermaphrodites are usually those of a male; that is, they are oriented toward females and have usually been reared as males. The cause of human hermaphroditism is unknown. However, since the embryo is possessed of a sexually indifferent, or sexually neutral, protoplasm, containing structures necessary for the formation of either set of sexual characters depending on differentiation by genetic factors, the condition probably is caused by a defect in differentiation.

Studies on hermaphroditism, together with those on the genetic factors associated with sex, have led to the view expressed by Biedl[6] that there is no such thing as a pure male or female but that each contains the elements of the other sex. It has been postulated that the secondary sexual characteristics develop in a masculine or feminine direction depending on which of the sex hormones dominate.

Coitus

The Male Copulatory Organ

We have already seen that the testes function to produce the male sex cells, the sperm. The duct system, consisting of the epididymis and ductus deferens, conducts the sperm cells to the ejaculatory organ, the penis. The accessory male reproductive glands, consisting of the seminal vesicles, prostate, and Cowper's gland, pour their secretions into the duct system, whereby the spermatozoa become suspended and motile. The **penis,** besides serving to conduct the urine to the exterior through the channel of the urethra, also has the function of introducing the semen into the genital passages of the female, whereby fertilization of the female reproductive cell may take place. In order to accomplish the latter function, the penis, which is usually flaccid, must become firm, or erect, to be effectively introduced into the vagina.

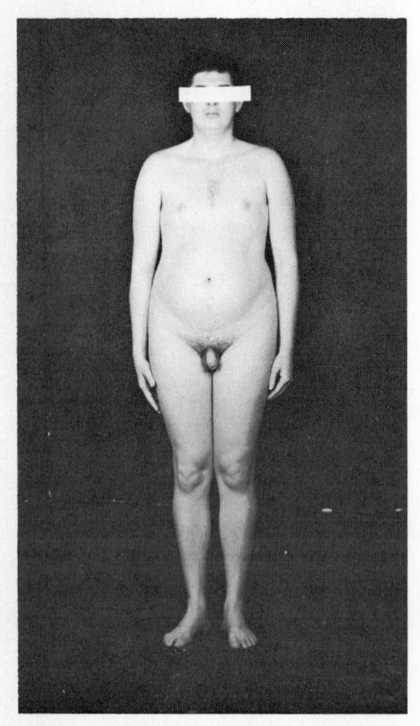

Figure 15-20. Hypogonadism in an 18-year-old boy. Although the penis is well developed, the testes are but $1\frac{1}{2}$ cm long. Note the feminine contour of the body and the scanty development of body hair. (Courtesy Armed Forces Institute of Pathology, Washington, D.C., picture PJ 4348.)

The penis (Fig. 15-10) contains three distinct tracts of erectile tissue: the two **corpora cavernosa** and a smaller **corpus spongiosum,** which surrounds the urethra (Fig. 15-21). At the distal end the corpus spongiosum becomes enlarged, forming the **glans penis,** which is richly supplied with sensory receptors, making it especially sensitive to external stimulation. The tissue of the three corpora is composed of large venous sinuses, which are devoid of blood when the penis is flaccid but become dilated and engorged with blood during the erection of the penis. Each of the corpora is enclosed in a fibrous capsule, called the **tunica albuginea,** and all three are bound together by a connective tissue sheath called **Buck's fascia,** or the **fascia penis.** The large, deep dorsal vein lies between the corpora cavernosa and the fascial sheath. The arterial supply runs through the tissue of the corpora, with some of the arteries branching into the intertrabecular spaces of the corpora cavernosa, forming dilated vessels. In most instances the arteries open into the venous sinuses through capillaries. Figure 15-22 shows a cross section of the penis with its component structures.

The genitourinary structures receive a supply of autonomic fibers through the pelvic ganglia (Figs. 15-23 and 15-24). The parasympathetic fibers are derived from branches of the second, third, and possibly the fourth sacral spinal nerves. These branches, after passing through the sacral foramen as the **nervi erigentes,** enter the pelvic nerve plexus and then follow the course of the blood vessels to the genitourinary structures. The sympathetic fibers are derived from the twelfth thoracic and upper lumbar segments of the spinal cord. These fibers descend to the presacral area to form a distinct nerve plexus, called the superior hypogastric plexus, from which branches pass on to the pelvic organs. Nerves emanating from

[6] A. Biedl, *The Internal Secretory Organs,* translated by Linda Forster. (Baltimore: William Wood & Company, 1913).

the pelvic autonomic plexuses apparently innervate the smooth muscle surrounding the cavernous spaces and the arterioles within the penis. Thus they control changes in the blood capacity by which the venous sinuses are filled, causing erection.

Mechanism of Erection and Ejaculation

The erection of the penis is brought about mainly by the reflexly produced dilatation of its blood vessels. The glans penis is most liberally supplied with sensory fibers and receptors in which nerve impulses may be initiated by mechanical stimulation of the skin of the penis. The impulses are transmitted to the spinal cord, where they are then passed by way of parasympathetic fibers to the penile arteries and arterioles. The arterioles dilate, causing the venous sinuses in the corpora to become engorged with blood. First, the base of the organ increases in size, and then swelling extends throughout the cavernous structures and eventually reaches the glans. The swelling of the corpora squeezes the veins located between these structures and the inelastic connective tissue sheath, the fascia penis, thereby reducing the outward flow of blood. In addition, the contraction of the ischiocavernosus (erector penis) and bulbocavernosus muscles also occludes the veins and aids in arresting the outward flow of blood. The result of these actions is that more blood can enter the dilated venous spaces of the penis than can leave. The corpora expand and become rigid, remaining in that condition, which facilitates intromission of the penis into the vaginal tract, until the arteries and arterioles once again constrict. Erection is due to three factors: dilatation of the pudendal artery and dorsal artery of the penis, contraction of penile muscle fibers, and compression of veins between the corpora cavernosa and the surrounding fibrous capsule. Stimulation of the second, third, and fourth sacral spinal nerves initiates erection. If these nerves are severed, erection is abolished. Stimulation of the sympathetic fibers from the twelfth thoracic and upper lumbar segments of the spinal cord is known to constrict the arteries and arterioles of the penis and make it flaccid.

Figure 15-21. Penis, showing its structure.

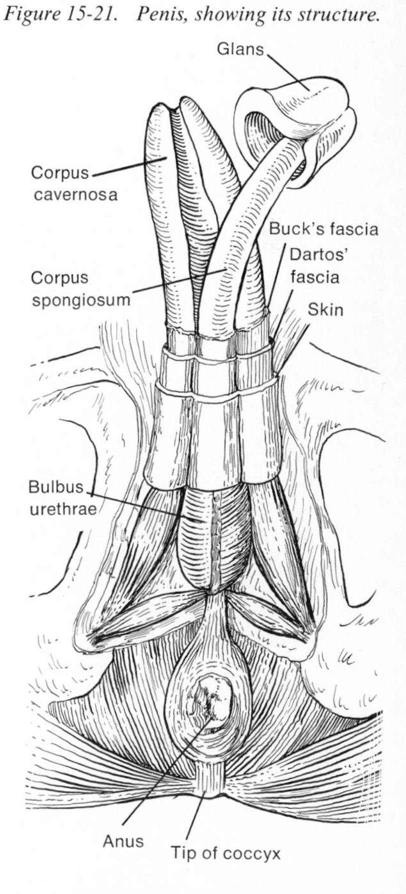

Coitus, or **copulation,** which is the act of union whereby the male deposits sperm in the vaginal tract of the female, is dependent on successful erection of the penis and ejaculation. Copulation is attended in both sexes by a high degree of mental excitement. The friction that is set up between the male and female organs during coition causes a discharge of motor impulses in both sexes, the uterus undergoing a series of contractions and at the same time the accessory sexual glands in the woman, Bartholin's glands, discharging a secretion that is added to the semen. In men the act is culminated by the ejaculation, or expulsion, of seminal fluid from the urethra by contraction of the skeletal bulbocavernosus muscle. In both sexes the culmination of coitus is accomplished not only by the action described above, but also by a paroxysm of pleasurable sensation largely contributed by sensory elements in the glans penis and glans clitoris. The culmination of coitus in this paroxysm of sensation, accompanied in the man by the ejaculation of semen, is known as the **sexual orgasm.**

Ejaculation really consists of two distinct actions. The first is the delivery of semen into the urethra from the sperm ducts by contraction of the smooth muscle of the internal genital organs. This phase is usually referred to as "emission."

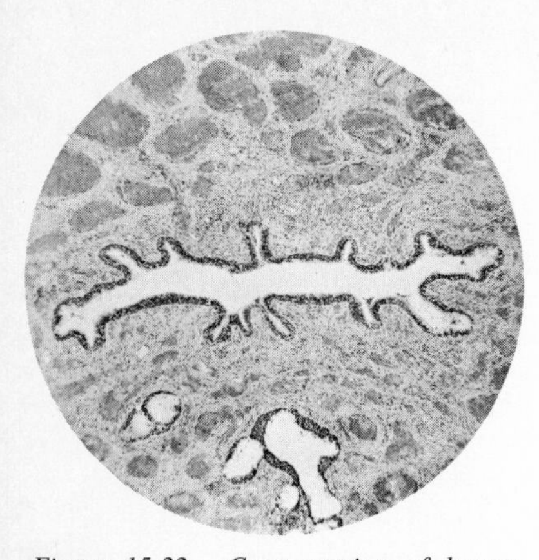

Figure 15-22. Cross section of human penis, showing urethra and corpora cavernosa. (Courtesy General Biological Supply House, Chicago.)

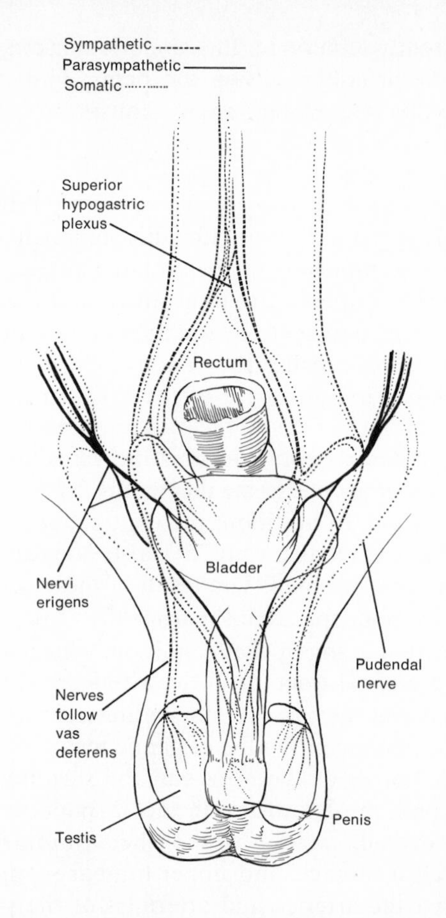

Figure 15-23. Innervation of the male genitourinary structures.

It is during this phase that the accessory male glands (prostate, seminal vesicles, Cowper's glands) add their fluid to the semen. The second phase is the actual ejaculation, whereby the seminal fluid is expulsed from the urethra with considerable force by contraction of the bulbocavernosus muscle. The process is basically a reflex phenomenon with its center located in the lumbosacral region of the spinal cord. Sensory impulses arise in the receptors located in the glans and are transmitted by the internal pudendal nerves to the spinal cord. Motor impulses leave the upper lumbar segments and travel back over parasympathetic fibers to the muscles, causing emission and ejaculation.

Although erection and orgasm are basically reflex actions, it is known that they can also be induced voluntarily by stimuli conveyed from the cortex of the brain, that is, by sexual thoughts or emotion. Experimentally, erection and ejaculation have been induced by electrical stimulation of the cervical spinal cord and the pons and by stimulation of definite regions of the cerebral cortex. It is not unusual that erection and ejaculation do occur when these areas are stimulated, as it has been noted that hanging and decapitation in man are sometimes followed by these reactions.

Experimental evidence also indicates that the higher nerve centers exercise an inhibitory influence over the sexual processes. It has been shown that it is easier to induce erection of the penis by mechanical stimulation of the glans in animals whose spinal cord has been severed than in a normal animal. In general, then, although erection and ejaculation are basically reflex phenomena occuring at the spinal level, influences from the higher centers can alter the response so that stimuli arising elsewhere in the body can exert their effect. Erection and ejaculation can occur without tactile stimulation of the penis. Sensual thoughts or activities (petting, stimulation of other erotic areas of the body) may lead to orgasm in some highly responsive persons in the absence of coition or tactile stimulation of the penis. On the other hand, orgasm may be retarded or prevented by inhibitory influences of the higher centers being brought into play by emotional situations despite the presence of the usual stimulus.

Reproduction

Fertilization

Fertilization of the mature ovum normally occurs in the fallopian tube and is completed when the nucleus of the sperm cell fuses with the nucleus of the ovum. Having been deposited in the vaginal tract of the female, the spermatozoa find their way into the fallopian tubes and here come in contact with the ovum. The passage is probably effected mainly by the swimming activity of the spermatozoa. Normal spermatozoa are known to swim in a more or less straight line. In certain cases of sterility, where there are large numbers of morphologically defective sperm cells, the only defect is their loss of ability to move in a straight line; instead they move or swim in circles. They are unable to reach the egg to initiate the activity on which fertilization depends. Viable sperm can generally be found in the cervical canal for as long as 24 to 48 hours. However, there is only a very short time period during the menstrual cycle when fertilization can occur. Normally ovulation takes place approximately 14 $\pm$ 2 days before the onset of the next expected menses. The ovum normally has a fertilizing life that has been estimated to be approximately 18 to 24 hours, and thus it would be exptected that fertilization could occur only in the fallopian tubes, since it has been estimated that it takes the ovum approximately 72 hours to arrive at the uterus after ovulation, and the viability life of the egg cell limits the time period in which an effective pregnancy may be initiated. Some authorities place the time limit of conception closer to 36 than to 24 hours. If the egg is not fertilized in the fallopian tubes, it passes on into the uterus, where it still can be fertilized, but if it is not, it undergoes degeneration and is eliminated from the tract with the menstrual flow.

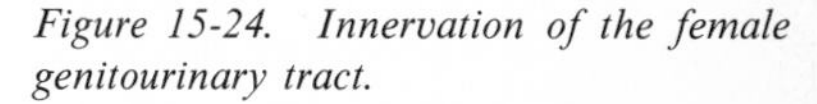

Figure 15-24. Innervation of the female genitourinary tract.

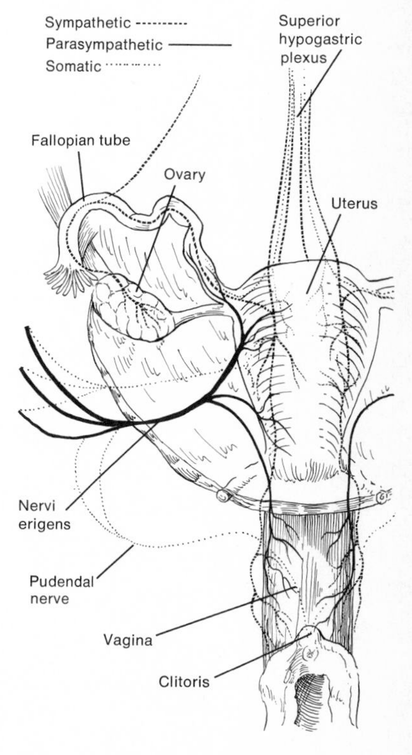

Since sperm cells are generally considered to be capable of fertilization for approximately 18 to 24 hours but may remain viable in the female genital tract up to 3 days, the length of time in which the ovum, with a fertilizable life of only 18 to 24 hours, may be penetrated by a sperm is extended to about 4 days. If deposition of sperm occurs several days before ovulation, pregnancy can occur, or it can also occur if sperm are deposited 1 or 2 days after ovulation. It has been estimated that there is a period of about 4 days in the middle of the menstrual cycle during which fertilization is most likely to be successful. In some instances when conception takes place before or after this 4-day estimated period, there may be unknown variables in fertilizing potential of ova and/or sperm cells that make it difficult to correlate the estimated time of ovulation with the time of coitus.

Penetration of the ovum, of course, is necessary before fertilization can occur. We have already seen that considerable numbers of spermatozoa are necessary in order that one spermatozoon may penetrate the egg cell membrane. A count of fewer than 50,000,000 spermatozoa in 1 ml of semen is usually associated with sterility. This great number of spermatozoa is needed to ensure penetration of the ovum by a single sperm cell because the egg cell membrane must be liquefied before penetration can occur. Sperm cell suspensions contain **hyaluronidase,** an enzyme secreted by the spermatozoa. Its function is to liquefy the cell membrane so that the development of the egg can proceed, but no less than approximately 60,000,000 sperm cells are needed to produce a high enough concentration of this enzyme so that one of the spermatozoa is able to pierce the egg cell barrier. As soon as one sperm penetrates into the ovum, the membrane thickens, forming the **fertilization membrane** and thus preventing the entry of more than one sperm cell. The hereditary function of the spermatozoon now replaces its developmental function by transmitting the paternal characters to the developing embryo by fusing its nuclear material with that of the ovum. When the two nuclei are fused, fertilization is complete, and the structure is now referred to as a **zygote.** Fertilization merges immediately into the division of the zygote (fertilized cell) into two daughter cells by **mitosis.** The chromosomes from the two nuclei are combined on the mitotic spindle, restoring the diploid number of chromosomes. It must be remembered that each of the nuclei of the ovum and spermatozoon contained only half the number of chromosomes characteristic of the species and other cells of the body. By their union in the fertilization process, a cell is formed whose nucleus has the normal sum. As a result of the action of the spermatozoon on the ovum, the latter, instead of disintegrating as will occur when fertilization fails, is stimulated to great metabolic activity accompanied by morphological changes; it enlarges and develops into an embryo. The woman is then said to be pregnant.

Diagnosis and Tests of Pregnancy

In most instances the pregnant woman makes her own diagnosis of pregnancy before consulting a physician. However, in very early stages of pregnancy and occasionally in more advanced stages, diagnosis with certainty is not always as easy as it would appear. usually the pregnant woman's symptoms, certain physical signs, and specific tests supply the criteria needed for establishing pregnancy with certainty. The most significant symptoms are (1) cessation of menstrual flow (amenorrhea), (2) changes in the breasts, (3) nausea and vomiting, and (4) urinary frequency.

The first evidence of pregnancy is a missed menstrual cycle. However, this sign cannot always be relied on, since some women may be amenorrheic for reasons other than pregnancy. Some women may bleed during pregnancy from other causes and may therefore simulate menstruation. Such causes may be extrauterine pregnancy, lesions of the vagina and cervix, and bleeding that may follow implantation of the fertilized ovum. However, conception without cessation of menstrual flow is not common.

Very early in pregnancy the woman may notice an unusual fullness of the breasts, with the nipples exhibiting an unusual tenderness. The breasts also undergo a considerable increase in size during the period of pregnancy; the nipples enlarge, and their pigmentation increases. In some women a turbid watery secretion, known as the **colostrum,** is expressible from the nipples, which changes to milk as pregnancy approaches its end. If the hypertrophy of the breasts subsides during pregnancy, this is suggestive of the death of the developing embryo.

Nausea and vomiting are symptoms present in about two thirds of all pregnant women. They usually begin about a week after the first missed menstrual period and are more pronounced in the morning. For this reason this condition is often referred to as "morning sickness." Morning sickness lasts for about 4 to 6 weeks in the average pregnant woman. The cause of morning sickness is not known, but a relationship appears to exist between the woman's emotional make-up and the severity and duration of the vomiting.

Urinary frequency is common between the twelfth and sixteenth weeks of pregnancy and also reappears in late pregnancy. The early urinary frequency is believed to be due to the vascular engorgement of the bladder mucosa, modifying the tonal qualities of the bladder wall. In late pregnancy the urinary frequency develops as a result of the enlargement of the developing embryo and its encroachment on the bladder, decreasing the capacity of this organ.

In addition to the above-described symptoms, several physical signs associated with the generative organs are of diagnostic importance. During pregnancy the vagina shows three principal changes: vascularity, softening, and increased secretion. The increased vascularity of the vaginal wall changes the color of its usual pink mucosa to a bluish or cyanotic color. This change in color may appear as early as 2 weeks after the first missed menses and is referred to as **Chadwick's sign.** Since there is a great variation in the time of appearance of Chadwick's sign, it is not considered a reliable indicator for excluding pregnancy in the very early stages. Another physical sign that is one of the most reliable and consistent for the early diagnosis of pregnancy is **Hegar's sign.** This is noted as a softening of the isthmus or junction between the cervix and body of the uterus, which results in a greater mobility between the two areas. The change may be so pronounced that the body of the uterus and cervix may feel like separate unconnected structures on bimanual pelvic examination by the physician.

Pregnancy tests are usually performed when confirmation of pregnancy is needed. Most of the tests are carried out with the patient's urine and are based on the fact that the placenta produces gonadotropic hormones, which are excreted in the urine. If the urine contains such hormones and it is injected into an immature female test animal, the ovaries will undergo maturation. The test animals most commonly used are the mouse, rat, rabbit, South African clawed toad, and the common American laboratory frog, *Rana pipiens.* The first satisfactory pregnancy test was devised by Aschheim and Zondek in 1928. Most of the other tests are modifications of the original, and all are frequently referred to as A–Z tests.

In the **Aschheim-Zondek test,** or **mouse test,** 3- to 4-week-old immature mice are used. A total of six evenly spaced injections of about 0.4 ml of urine from a woman thought to be pregnant are administered for a 3-day period. Three to four days later the mice are sacrificed and the ovaries examined. If the ovaries show blood spots or hemorrhagic follicles, the test is considered positive.

The **rat test** similarly uses 3- to 4-week-old immature females. Two milliliters of test urine are injected intraperitoneally, and the test animal is sacrificed 24 hours later and the ovaries examined. Hyperemia of the immature ovaries constitutes a positive test.

The **rabbit test,** also referred to as the **Friedman test,** uses mature female rabbits that have been isolated from males for at least 6 weeks. The rabbits, usually two, are injected with 10 ml of test urine. One rabbit is examined 24 hours later, the second in 2 days. A positive test is indicated when fresh corpora lutea appear in the ovaries. The test is accurate in more than 98 per cent of cases but is used infrequently because of the greater cost of the test animals.

The **toad test** involves the injection of test urine into the female toad. A positive test is indicated if the toad is induced to deposit eggs approximately 12 hours later. The **frog test** involves the use of male frogs. Urine is injected into the dorsal lymph sacs of the test animal. One to two hours later the frog's urine is aspirated from the cloaca with a pipette and examined under the microscope for spermatozoa. The presence of sperm cells is indicative of a positive test.

Most of these tests are more than 95 per cent accurate when carried out 2 weeks after the first missed menses. The test urine is collected in a clean bottle by the patient and should be the first urine voided in the morning. The patient should also make sure that she has taken no drugs or alcoholic beverages for at least 12 hours before collection of the urine, since these substances, being excreted in the urine, are also toxic to the test animals.

Development of the Ovum in Pregnancy

After fertilization, the ovum is still a minute cell surrounded by a structureless, transparent membrane, called the **zona pellucida,** or **vitelline membrane.** Within this is the yolk, or **vitellus,** which consists of a granular, semifluid mass having suspended in it a fused male and female pronucleus. When the ovum is fertilized, the corpus luteum persists instead of degenerating at the end of the menstrual cycle, and menstruation does not occur. It is not known with certainty, but there is good evidence that the presence of a fertilized egg and later the embryo in the uterus causes a neurally transmitted signal to go from the uterus to the pituitary gland, which responds by releasing lactogenic hormone. This hormone is responsible for the maintenance of the corpus luteum and therefore the endometrium. We have already seen that removal of the corpus luteum early in pregnancy before implantation invariably leads to abortion. The human female can tolerate removal of the corpus luteum after the first third to half of pregnancy because the developing placental membranes secrete progesterone and estrogen, which are essential for maintenance of the proper uterine environment during pregnancy.

The fertilized ovum immediately undergoes, by mitotic activity, division into two daughter cells, a process commonly called **cleavage,** or **segmentation.** The amount of yolk stored in the ovum as food material for the growing embryo will cause variations in the way cleavage occurs in different groups of animals. However, all fertilized eggs will pass through the same stages, as schematically illustrated in Fig. 15-25, which represents the various stages of embryogeny of the starfish. In the ova of higher mammals, including man, the amount of stored food material is exceedingly small because the embryo at a very early stage in its development will draw upon the uterine circulation of the mother for its nutrition. We find that the pattern assumed by the daughter cells during cleavage of such eggs are different from that found in Fig. 15-25. This variation in pattern will be discussed following the general description of the various stages of cleavage. The two daughter cells formed by cleavage of the fertilized ovum divide, forming four cells, and these four into eight, and so on indefinitely until an agglomerate mass of cells results resembling a mulberry; at this stage the mass of cells is called a **morula.** The morula continues to divide, and the cells arrange themselves about the interior of the zona pellucida and form the blastodermic membrane. Fluids begin to accumulate between the peripheral and central mass of cells, giving rise to the cavity of the **blastocyst;** the structure is now said to be in the **blastula stage.** The developing ovum (blastocyst) at this stage reaches the uterus 4 days or so after ovulation, when the endometrium is in the premenstrual stage, and sinks into it about the seventh day by digesting away the superficial layers. This process is known as **implantation,** or **nidation** (see Fig. 15-27). It is believed that the membrane of the cleaving ovum secretes a proteolytic enzyme, which facilitates implantation in the thick vascular endometrium of the uterus, which has been prepared for such reception by the synergistic action of estrogen and progesterone. However, very little, if anything, is known about the changes that occur within the uterine lumen that may satisfy the conditions necessary for the embryo to continue its growth, make attachment to the uterine epithelium, and implant.

The cavity within the blastocyst is known as the **blastocoele.** As cell division

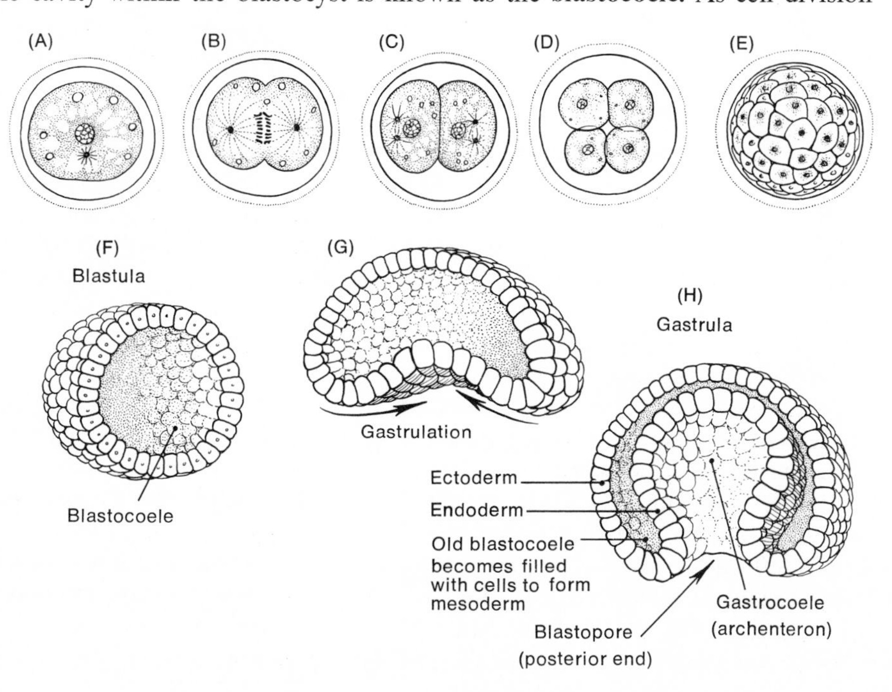

Figure 15-25. Stages of segmentation (cleavage). The fertilized egg (A) *undergoes mitotic activity* (B) *to form two daughter cells* (C). D. *Four-cell state.* E. *Morula stage.* F. *Hollow-ball stage (blastula).* G. *The wall of the blastula begins to pucker in on one side much as a soft rubber ball might be dented by pressure at one point. The process is referred to as invagination, which marks the process of gastrulation.* H. *Gastrula stage. The compressed side extends inward until it almost completely fills the original cavity. The embryo has the form of a double-walled sac with an opening at one end called the blastopore.*

continues, now more rapidly at animal than at vegetative pole, invagination of the cells occurs near the vegetative pole. This process is known as **gastrulation** and results in the gradual obliteration of the blastocoele, the embryo now becoming an oblong ball of cells whose new cavity, formed by the outer cells pushing in, is known as the **gastrocoele,** or **archenteron.** The entire structure is now known as the **gastrula.** The gastrula opens to the outside by an opening called the blastopore, and its wall now consists of two layers of cells, the outer **ectoderm** and the inner **endoderm.** Between the ectoderm and endoderm, cells proliferate to occupy what is left of the old blastocoele. This layer of cells forms the third germ layer, the mesoderm. The mesoderm cells separate into two layers, an outer layer near the ectoderm called the **somatic mesoderm** and an inner layer lying close to the endoderm called the **splanchnic mesoderm.** The space that is formed between the somatic and splanchnic layers becomes the **body cavity,** or **coelom.** With the beginning of the formation of the body cavity, **organogeny,** or differentiation of the various organs, begins.

In higher mammals it is at the blastula stage that the restraining zona pellucida disintegrates and the rapid growth of the embryo in size first begins. The blastocoele accumulates more and more fluid within itself, thereby expanding the outer layer of the blastoderm into a voluminous membrane through which the yolkless embryo can draw food from the uterine circulation of the mother. During this development an internal mass of cells is established at one area which is destined to be concerned primarily with the formation of the embryonic germ layers from which will develop all the organs (Fig. 15-26). Shortly after the blastocoele has become enlarged, some of the cells of the inner cell mass become detached and rapidly increase in number, forming a second complete layer inside the enlarged blastular cavity. This cell layer forms the entoderm, and the internal cavity bounded by it is known as the archenteron or primitive gut (Fig. 15-26*B*). The mass of cells from which the entodermal cells became detached become more regularly arranged and is known as the **embryonic disk** (Fig. 15-26*C*). Shortly after the establishment of the entoderm the embryonic disk begins to differentiate and shows an increased rate of cell proliferation at one of its margins. This marks the beginning of mesoderm formation. The newly formed cells spread out in a radial manner but do not converge upon one another as did the entodermal cells. However, they do extend far beyond the region of the embryonic disk (Fig. 15-26*D, E*). The more lateral portions of the mesoderm split into two layers, the somatic mesoderm and the splanchnic mesoderm. The cavity formed between the two layers is the coelom. The original aggregation of cells which is left covering the outside of the embryo is the ectoderm.

The organs developing from the various embryonic germ layers are as follows: From the ectoderm arise the integument, the nervous system, and the lining of the mouth and anus. From the entoderm arise the lining of the digestive tract, except the mouth and anus, and the lining of the various structures that arise from the digestive tract, such as the lungs, liver, and pancreas. The bulk of the body's organs are derived from the mesodermal layer. These include the skeleton, muscles, heart and blood vessels, kidneys, gonads, and connective tissues.

After implantation, the extensive development described above continues for

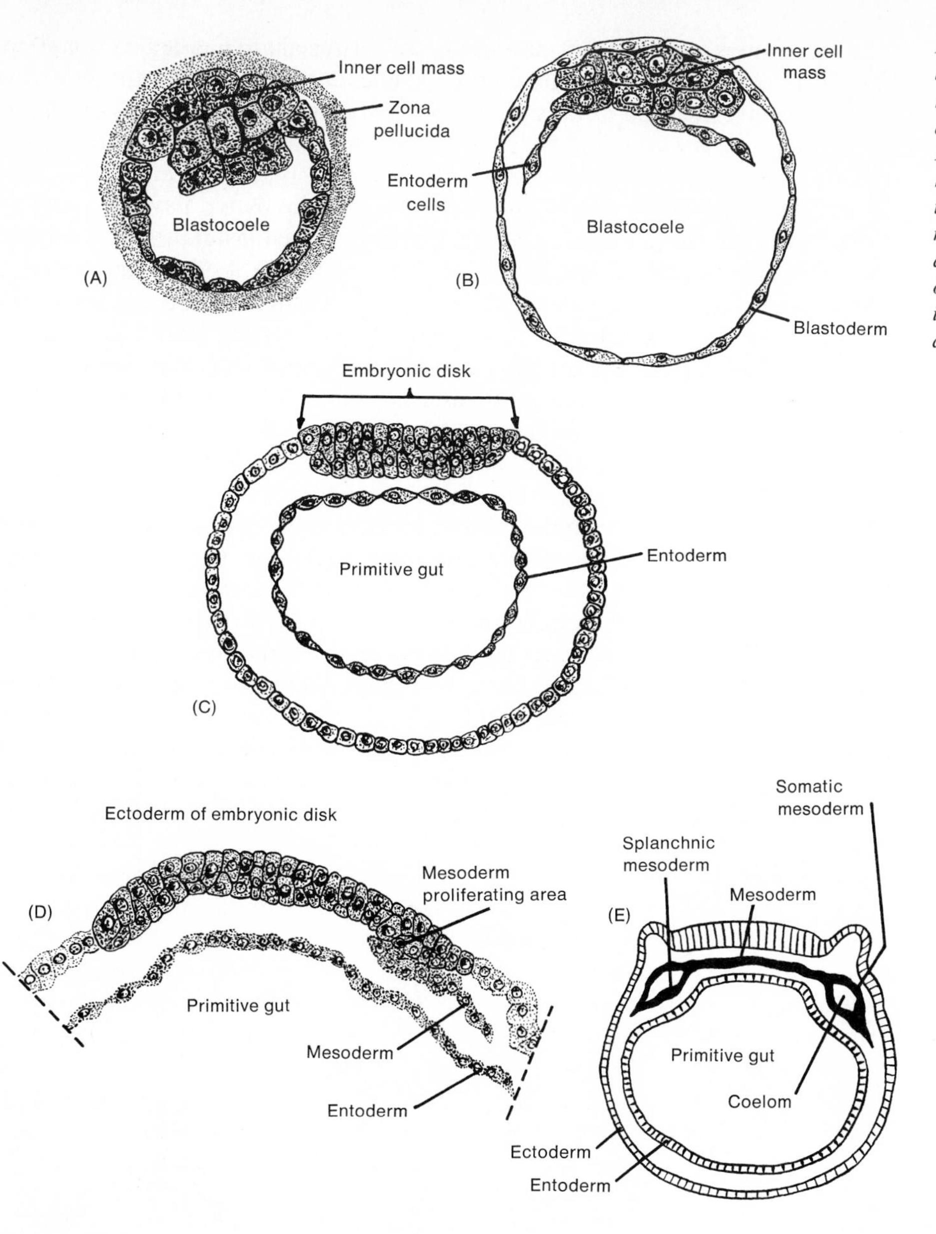

Figure 15-26. Development of a mammalian-form blastocyst (A-B) *to the development of the three primary germ layers (ectoderm, entoderm, and mesoderm).* B. *Blastocyst, showing the first appearance.* C. *Subsequent rapid extension of the entoderm to form the primitive gut.* D. *Formation of mesoderm by a local differentiation in an area of the embryonic disk.* E. *Cross section of embryo in primitive streak stage to show interrelations of the primary germ layers and formation of the coelom.*

approximately 280 days, that is, slightly in excess of 9 months. During all this time the developing embryo must be nourished by some means since the mammalian ovum is small and does not contain a sufficient supply of nutriment. By special modifications of the uterine mucosa and a part of the developing ovum, the placenta is formed, and a transmission of nutriment from the mother to the embryo is made possible.

The Placenta

After the ovum is fertilized and cleavage of the fertilized egg has begun, correlative changes also take place in the uterus. The endometrium continues to grow, becoming congested and quite thick. This new growth does not degenerate and hemorrhage but remains and may be looked upon as a new, temporary lining to the uterus. Since this layer of the endometrium is cast off at parturition, it is often called the **decidua.** It is into this layer that the ovum, having undergone development changes, extending probably as far as the blastula stage, is received and embedded, the decidua closing in over it (Fig. 15-27). As the ovum continues to increase in size, it bulges into the cavity of the uterus, carrying with it the portion of the decidua that has closed over it. The portion of the decidua that covers the projecting ovum is called the **decidua reflexa,** and that portion which intervenes between the ovum and the uterine wall is frequently spoken of as the **decidua basalis.** In the decidua basalis there is a great dilatation of the maternal blood vessels, and small, delicate, fingerlike outgrowths of the outer wall of the blastocyst, the **chorionic villi,** grow toward these hypertrophied blood vessels. The villi then undergo a rapid increase in size and vascularity because they are invaded by mesoderm, which forms fetal blood vessels. At the same time, owing to obliteration of small arteries in the decidua, many cells die, leaving spaces called follicles, which fill with maternal blood. The villi and follicles thus grow simultaneously and finally become blended with each other and are no longer separate structures but become the placenta. However, there is no direct connection between the maternal and fetal circulations, so blood cannot pass from mother to embryo, or vice versa. The transfer of nutritive material from the mother necessary for the growth and development of the fetus takes place through the walls of the villi. The placenta, in addition to being the means whereby the fetus

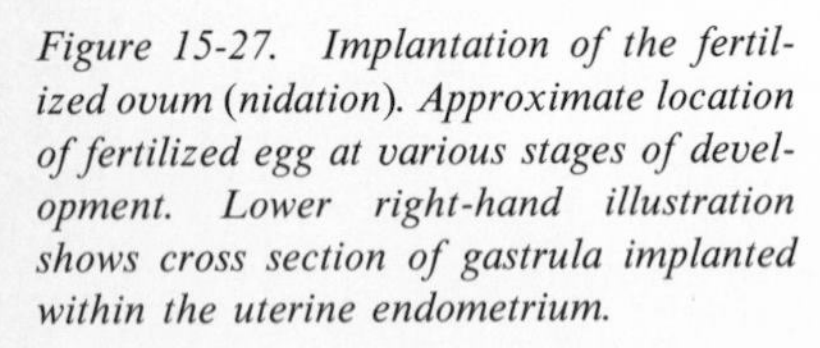

Figure 15-27. Implantation of the fertilized ovum (nidation). Approximate location of fertilized egg at various stages of development. Lower right-hand illustration shows cross section of gastrula implanted within the uterine endometrium.

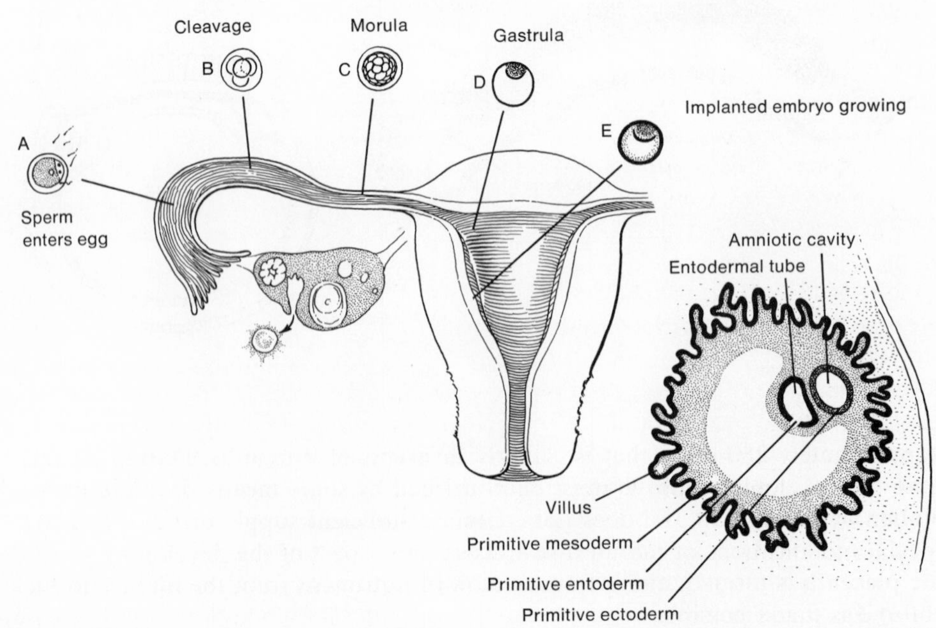

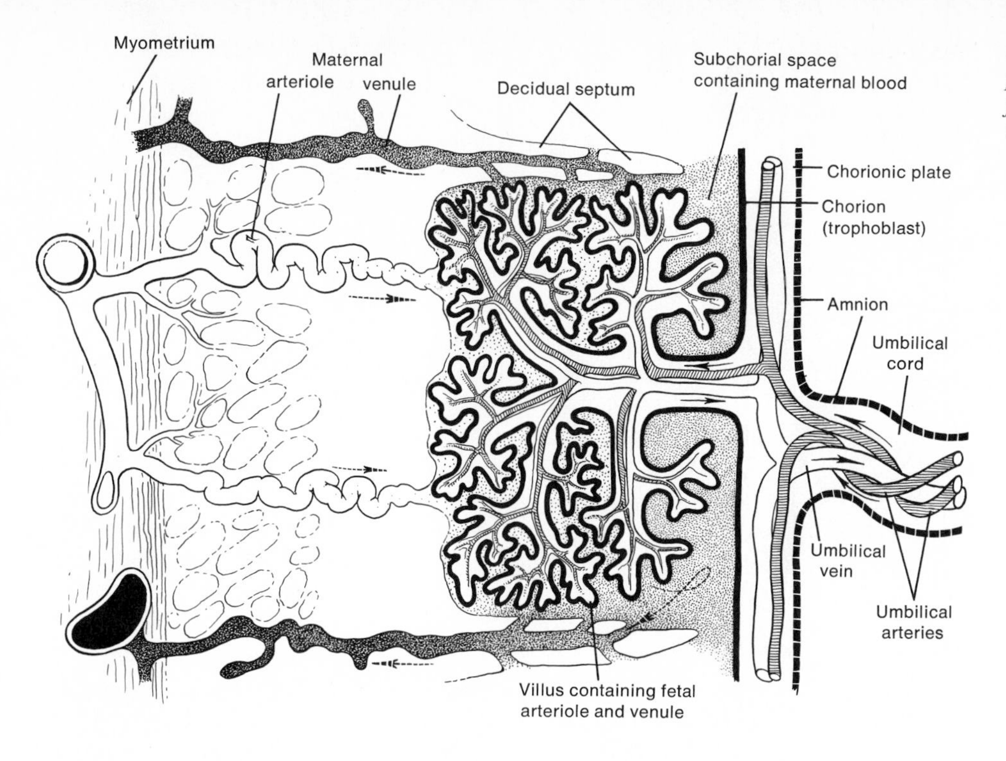

Figure 15-28. Portion of a fully formed placenta showing the relation between the fetal circulation and the maternal circulation.

is nourished, also produces hormones, which assist in maintaining the endometrium. Figure 15-28 shows a portion of a fully formed placenta.

As the embryo develops, it increases in size and eventually occupies the entire uterine cavity. The embryo is attached to the placenta by the **umbilical cord** and is itself surrounded by a double membrane consisting of the **amnion** and **chorion,** often called the "bag of membranes" (Fig. 15-29). The space between the embryo and the amnion is filled with **amniotic fluid,** which bathes the developing embryo. This fluid is an excretory product of the amniotic epithelium and functions to provide space for embryonic growth and to distribute the pressure due to uterine contractions evenly over the embryo. It is freely exchanged with the maternal fluids and contains wastes, such as urea and bile salts, derived from the fetus.

Parturition

In the human being the period involved in the development of the fetus, usually referred to as the **gestation period,** generally lasts from 275 to 280 days. It is customary to expect parturition at about 280 days from the last menstruation. **Parturition,** also referred to as **labor,** confinement, delivery, or childbirth, is the process by which the viable fetus is expelled from the uterus through the vagina into the external world. If the fetus is expelled before the stage of development when it is able to continue an extrauterine existence, the term **abortion,** or **miscarriage,** is applied. Parturition, or labor, is **premature** when it occurs after the child is viable but before the full term.

Three factors always involved in labor are (1) the birth passages, (2) the fetus and its membranes, and (3) the muscular contractions that accomplish the deliv-

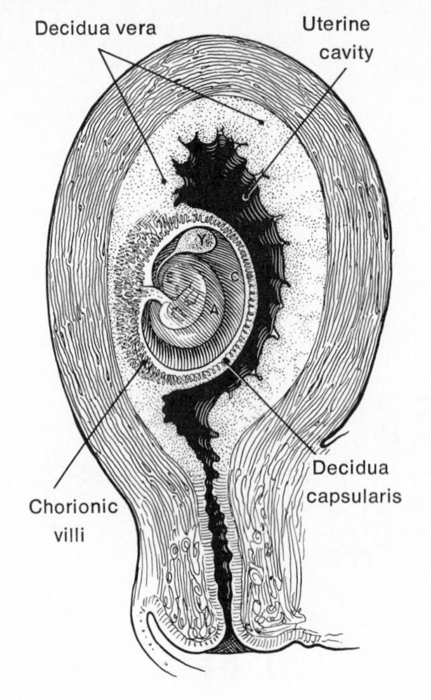

Figure 15-29. Early fetal development. The embryo is attached to the placenta by the umbilical cord. A, *Amnion;* C, *chorion;* Y, *yolk sac.*

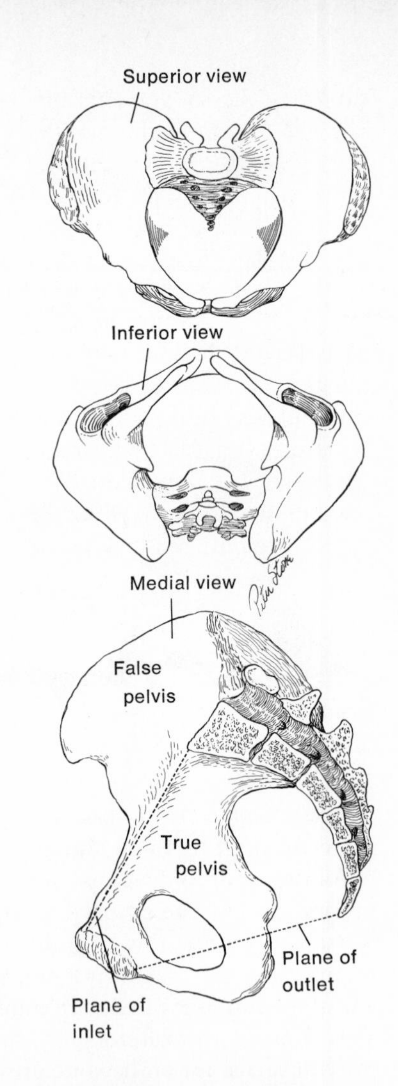

Figure 30. The female pelvis, showing its true pelvis portion, which makes up part of the birth canal.

ery. The **birth canal** is the passageway through which the fetus must pass from the uterine cavity into the external environment. It consists essentially of the cervix, the true pelvis, and the soft parts that line the pelvic cavity and close its outlet. The true pelvis (Fig. 15-30) consists of an inlet, cavity, and outlet. The **inlet,** or **brim,** is formed in front by the upper border of the symphysis pubis, behind by the sacral promontory and alae, and on each side by the iliopectineal line. The inlet is normally the narrowest portion of the birth canal, and the passage of the head of the fetus through this portion indicates that it can pass through and out of the birth canal. The **pelvic cavity** is the space between the inlet and the outlet and is formed by the walls of the pelvis. It is approximately cylindrical and is normally spacious. The **outlet** is formed posteriorly by the tip of the coccyx, laterally by the sacrosciatic ligaments and ischial tuberosities, and anteriorly by the lower margin of the pubic arch. The soft tissue almost completely closing the pelvic outlet, called the **pelvic floor,** is the chief factor determining and directing the movements of the fetus when, under the driving forces of labor, it seeks exit from the genital tract.

The fetus together with the placenta and membranes makes its exit through these passages. The fetus is composed of head and trunk with the attached extremities. The head, relatively large and incompressible, presents the greatest obstacle to delivery. The body normally offers little difficulty to passage through the canal if the head can pass through. At the end of pregnancy the fetus is enclosed in the amniotic sac and enveloped in fluid. It is normally in the attitude of **universal flexion,** with the head and limbs flexed on the trunk. This flexed position is caused by two factors: (1) the growing embryo develops stronger flexor muscles first, and (2) the intermittent contractions of the uterus tend to force the fetus into the least space necessary. The head of the fetus is the portion normally in advance or lowest in the birth canal, and the term **cephalic presentation** is used to denote this position (Fig. 15-31). Cephalic presentations are of three main types: occiput, brow, and face, depending on the degree of flexion or extension

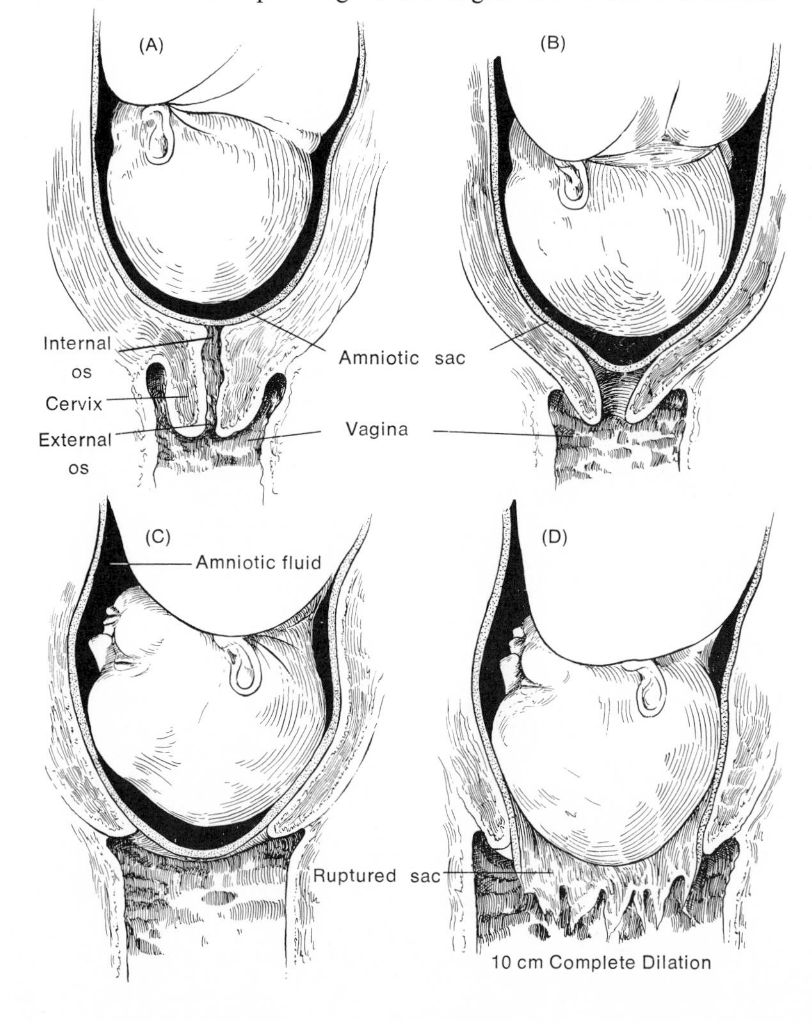

Figure 15-31. First stage of labor. A. *Head in low uterine position, amniotic sac unruptured, cervical canal closed.* B. *Beginning of dilatation, internal os dilated.* C. *Amniotic sac protruding through partly dilated cervix.* D. *Amniotic sac ruptured, almost complete dilatation of cervix.*

of the fetal head on the thorax. If the head is well flexed on the chest, the occiput, or crown of the head, is in advance, and this is called a **vertex presentation.** If the head is not completely flexed and is somewhat extended, the brow, or bregma, may be the presenting part, and this is therefore called a **brow,** or **bregma, presentation.** More commonly, extension of the head is complete, with the face presenting and the chin becoming the presenting part. This is called a **face presentation.** In approximately 96 per cent of full-term pregnancies, cephalic presentation occurs, and in nearly all cases it is a vertex presentation.

When the fetus is presented transversely instead of in its normal longitudinal position, the shoulder or arm may be the presenting part. Shoulder presentation, fortunately, is very infrequent. When the buttocks or one or both feet become the presenting part, this is referred to as a **breech presentation.** Approximately 4 per cent of full-term pregnancies result in breech presentations.

The expulsion of the fetus is brought about partly by involuntary rhythmical contractions of the uterus itself and partly by pressure exerted by the voluntary contraction of the abdominal muscles, similar to that described in defecation. The contractions of the uterus are the first to appear, and their first effect is to bring about a dilatation of the cervix; second, the fetus is propelled along the birth canal. It is not until the later stages of labor, while the fetus is passing into the vagina, that the abdominal muscles are brought into play.

Stages of Labor. Labor is normally divided into three stages. The first stage, called the **stage of dilatation,** covers that period from the onset of labor to complete dilatation of the cervix (Figs. 15-31 and 15-32). The second stage, also called the **stage of expulsion,** covers that period from complete cervical dilatation to the birth of the child (Fig. 15-33). The third stage, also known as the **placental stage,** covers that period from the birth of the child to the expulsion of the placenta and membranes (Fig. 15-34).

First Stage of Labor. At the end of pregnancy the uterine muscles begin to contract rhythmically. The saclike pregnant uterus has been irritable throughout pregnancy, and there have been irregular, intermittent, painless contractions. However, at the onset of labor the contractions become regular, stronger, more prolonged, and painful. When labor begins, the patient usually experiences

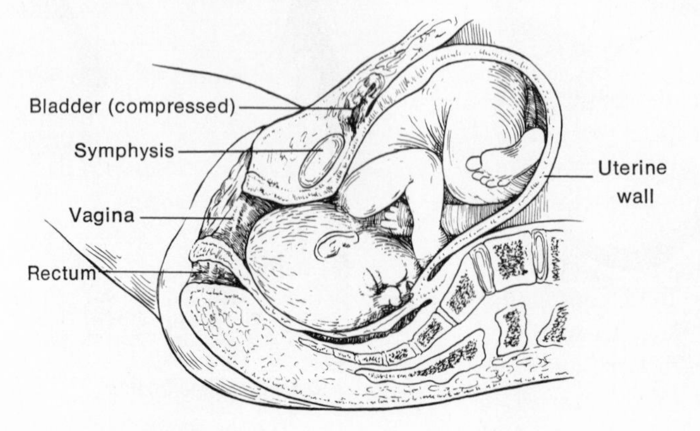

Figure 15-32. End of first stage of labor. Complete dilatation of cervical canal is produced, and the amniotic fluid has escaped from the ruptured amnion.

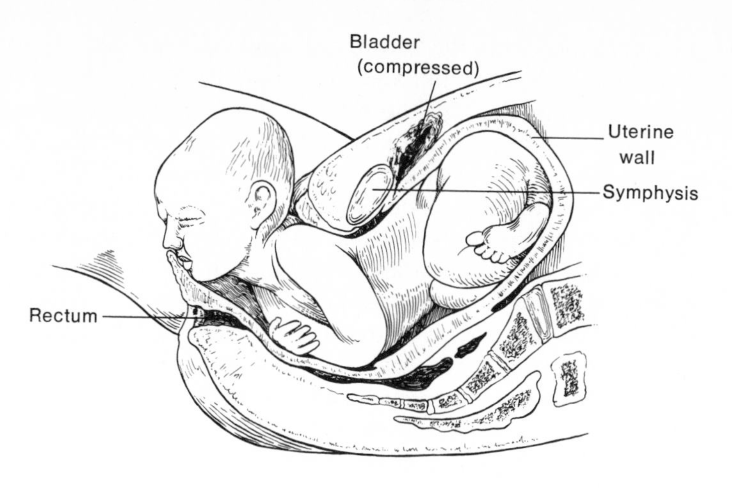

Figure 15-33. Start of the second stage of labor.

recurrent pains, which may be 15 minutes apart and last for only 10 to 15 seconds. As labor progresses, the pains become more frequent and longer lasting. At the end of the first stage, they usually occur about 3 minutes apart and last from 35 to 40 seconds.

The cause of the onset of labor is not known; however, it is believed that labor is hormonally initiated and controlled. The strong contractions initiated at this time exert considerable pressure on the amniotic sac, which causes it to bulge into the internal os. This opening is the weakest point of the cervical canal, and therefore it dilates as the contractions recur. Dilatation of the internal os and cervical canal renders this funnel-shaped depression a part of the cavity of the lower uterine segment. Dilatation continues until the external os is about 10 cm in diameter, at which time the process is considered complete. The amniotic sac, having fulfilled its chief function in labor of exerting hydrostatic pressure on the cervical canal to produce dilatation, frequently ruptures, releasing the amniotic fluid. This marks the end of the first stage.

Second Stage of Labor. During the second stage the contractions may be more frequent, approximately 4 to 5 minutes apart and lasting up to 50 to 60 seconds. At the beginning of the second stage the cervix and upper vagina are dilated, the membranes are ruptured, and the head of the fetus begins to descend through the vaginal tract. The force required to push the fetus through the vaginal tract is supplied by the strong contractions of the uterus in combination with the voluntary contractions of the abdominal muscles. The latter are effected by the active voluntary efforts of the mother unless the reflex has been inhibited by saddle block anesthesia or by heavy sedation. The bearing-down efforts of the abdominal muscles, as in defecation, added to the force of the uterine contractions aid in the expulsion of the fetus to the outside. This terminates the second stage. The newborn infant is held head down in one hand by the delivering physician. Its neck is stroked downward toward the mouth to expel any mucus or amniotic fluid from the pharynx. This material may be further aspirated from the mouth and nostrils with a rubber syringe. The infant, then being held at a lower level than the maternal abdomen to allow drainage of blood back into it from the placental

Figure 15-34. Third stage of labor. (A) *Area of placental attachment prior to birth, showing placenta starting to separate from uterine wall.* (B) *Uterine wall contracting to expel placenta.*

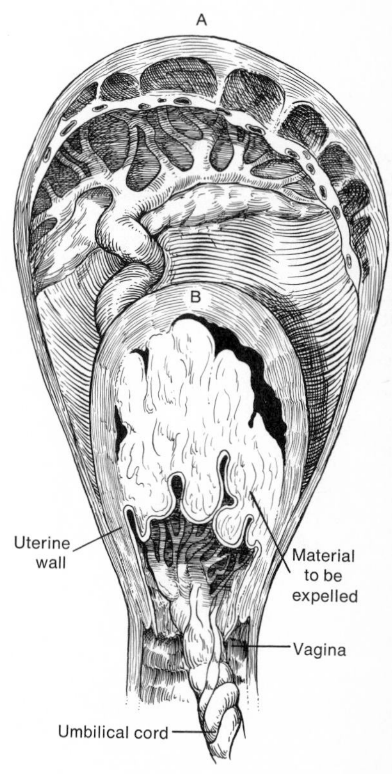

vessels, is now detached from the mother by cutting the umbilical cord about 3 inches from the umbilicus. Two clamps are placed on the cord before the severance is made, and the cut is made between the two clamps.

Third Stage of Labor. Immediately after the birth of the child the uterus slowly contracts, decreasing in size. This action results not only in taking up the large slack left by the expulsion of the fetus, but also in a buckling of the placenta and a detachment of the decidua underlying it. As the uterus continues to contract and retract, the placenta and its membranes are expelled into the passive lower uterine segment or upper vagina. As the placenta slides into the lower vagina, the physician can aid in its delivery by grasping the partially extruded organ and guiding it into a sterile basin held by an assistant.

The average blood loss in the third stage is about 200 ml. The contraction of the uterine muscles serves to constrict the blood vessels that have been ruptured during labor, thus reducing hemorrhage to a minimum. In addition, blood loss can be minimized by the routine use of an **oxytocic,** such as Pitocin. Pitocin is a drug usually injected into the mother intramuscularly immediately after the birth of the infant; it helps the uterine muscles regain and maintain a good tone.

Duration of Labor. The average duration of labor in women who are being delivered for the first time (primigravidas) ranges from about 11 to 13 hours. In patients who have had deliveries before (parous patients), the delivery time (labor) ranges from 6 to 8 hours. In primigravidas the median duration of the expulsion of the child during the second stage is 50 minutes. In women who have borne children before, the median duration of the second stage is 20 minutes. In general, labors vary widely in duration, depending on the strength and frequency of uterine contractions, the experience of the patient, the size of her pelvis, and the size and position of the fetus. Nature is usually capable of accomplishing the delivery of fetuses weighing up to 9 lb without very great difficulty if the pelvis, uterine wall, and fetal position are all normal.

Fetal Circulation

The fetal circulatory system is fully developed about the middle of intrauterine life (Fig. 15-35). Blood is oxygenated in the **placenta** and then passes along the **umbilical vein** to the **ductus venosus.** Part of the umbilical blood, however, enters into the hepatic circulation by the portal vein, from which it returns to the ductus venosus by the hepatic veins. Thus blood passing from the fetal liver also drains into the ductus venosus to join the oxygenated blood coming from the placenta, whereby the mixed blood is carried into the **inferior vena cava** and on into the **right atrium.** Most of the blood is directed through the **foramen ovale** into the left atrium. Some blood passing into the right ventricle passes through the pulmonary circulation (lungs) without picking up oxygen and is returned to the left atrium. From the left atrium the blood passes into the left ventricle, from which it is driven into the aorta and thus throughout the fetal tissues. Only a small quantity of the contents of the right ventricle finds its way into the lungs. The blood descending as fetal venous blood from the head and upper limbs by the **superior vena cava** also falls into the right ventricle, from which it is discharged into the aorta by way of the **ductus arteriosus.** This duct transfers blood directly

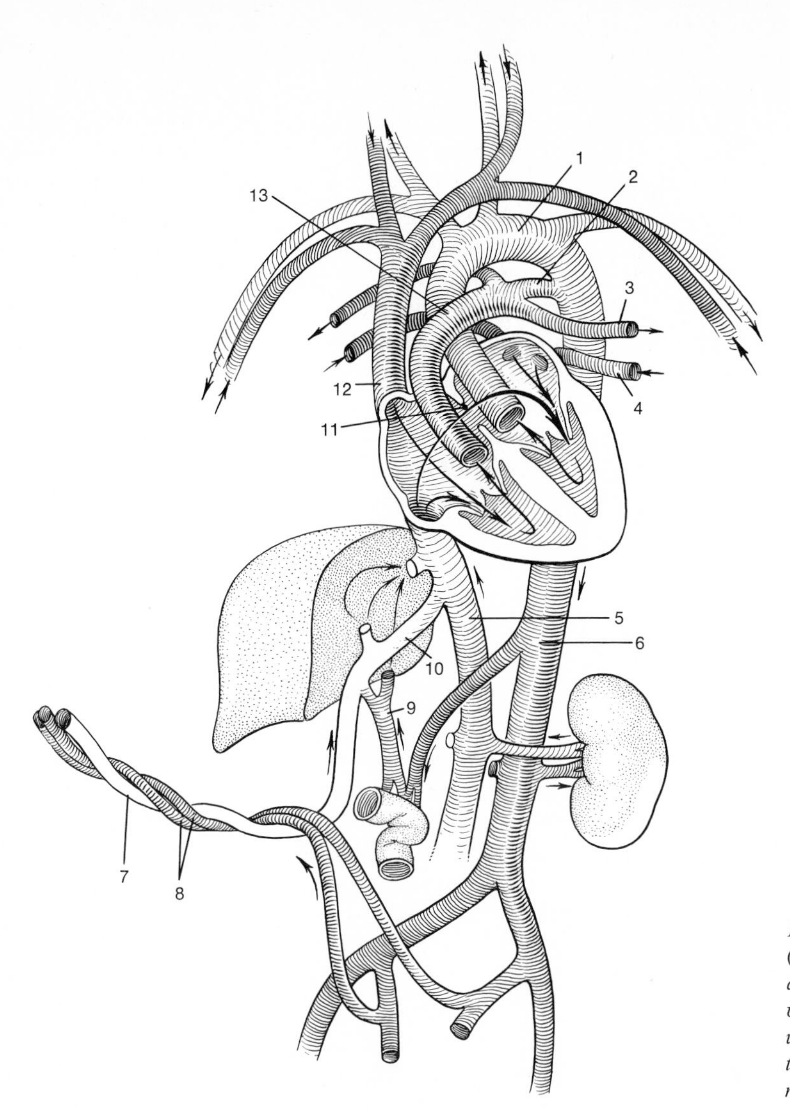

Figure 15-35. Fetal circulation. (1) *Aorta,* (2) *ductus arteriosus,* (3) *left pulmonary artery,* (4) *left pulmonary vein,* (5) *inferior vena cava,* (6) *aorta,* (7) *umbilical vein,* (8) *umbilical arteries,* (9) *portal vein,* (10) *ductus venosus,* (11) *foramen ovale,* (12) *superior vena cava,* (13) *pulmonary artery.*

from the **pulmonary artery** into the aorta, below the arch, where it flows partly to the lower trunks and limbs but chiefly to the placenta, by way of the umbilical arteries, to be oxygenated. From the placenta the blood is again shunted by way of the umbilical vein to the right atrium and, though mixed with a small amount of blood coming from the liver, is on the whole arterial blood. It is this blood which, passing into the left ventricle by way of the foramen ovale and left atrium, is sent by the left ventricle through the arch of the aorta into the carotid and subclavian arteries. Thus, the head of the fetus is provided with blood comparatively rich in oxygen.

The blood descending from the head and upper limbs by the superior vena cava is distinctly venous and, as we have seen, passes into the right ventricle, from which it is driven into the descending aorta by way of the ductus arteriosus. This blood, together with some of the blood passing from the left ventricle into the aortic arch, falls into the umbilical arteries and so reaches the placenta. The fetal circulation is so arranged that while most of the venous blood is driven by the right ventricle back to the placenta to be oxygenated, the arterial (although still mixed) blood is driven by the left ventricle to the cerebral structures, which have more need of oxygen than the other tissues.

During the whole time of intrauterine life the amount of oxygen in the blood is sufficient to prevent any inspiratory impulses from being originated in the respiratory center, which has established itself in the medulla. This center remains dormant, the oxygen supply never falling so low as to stimulate it into activity. As soon as the child is born and the intercourse between the maternal and umbilical blood is interrupted by separation of the placenta or by clamping of the umbilical cord, the fetal blood ceases to be oxygenated by the maternal blood. The concentration of oxygen falls in the fetal blood, and when it reaches a certain point, the respiratory center is stimulated, an impulse of inspiration is generated, and the child breathes for the first time. The action of the respiratory center may be assisted by sensory impulses reaching the center along various sensory nerves, such as those started by the exposure of the body to air or to cold. A retarded first breath may be hurried on by slapping the infant on the buttock. These sensory impulses, though, are subordinated and are not essential.

When the first breath is taken, with free access to the atmosphere the lungs become filled with air. With the reduction of resistance in the pulmonary circulation as a result of the expansion of the thorax, a larger supply of blood passes into the pulmonary artery instead of into the ductus arteriosus. In several months the ductus arteriosus becomes obliterated and becomes the nonfunctional **ligamentum arteriosum.** Corresponding to the greater flow of blood into the pulmonary artery, a greater quantity of blood returns from the pulmonary veins into the left atrium. The greater pressure in the left atrium opposes the flow of blood from the right atrium through the foramen ovale. Any return of blood from the now active left atrium into the right atrium is prevented by the **valve of Eustachius,** which has been growing up in the left atrium over the foramen ovale. During the first year, adhesion of this valve takes place, and the separation of the two atria becomes complete. The detached umbilical arteries becomes the **umbilical ligaments,** and the obliterated umbilical vein, because of its round shape, is called the **teres ligament.**

References

Bacci, G. *Sex Determination*. Elmsford, N.Y.: Pergamon, 1965.
Balinski, B. I. *Embryology*. Philadelphia: Saunders, 1970.
Bishop, P. M. F. *Chemistry of the Sex Hormones*. Springfield, Ill.: Thomas, 1962.
Calkins, L. A. *Normal Labor*. Springfield, Ill.: Thomas, 1955.
Dorfman, R. I., and F. Ungar. *Metabolism of Steroid Hormones*. New York: Academic, 1965.
Engle, E. T. *Studies on Testis and Ovary, Eggs and Sperm*. Springfield, Ill.: Thomas, 1952.
Hadek, R. *Mammalian Fertilization*. New York: Academic, 1969.
Hoffman, J. W. *Gynecology and Obstetrics*. Philadelphia: Saunders, 1962.
Jackson, H. *Antifertility Compounds in the Male and Female*. New York: Thomas, 1966.

Johnson, A. D., W. R. Gomez, and N. L. Vandemark (eds.). *The Testis,* Vols. 1, 2, 3. New York: Academic, 1971.

Jones, H. W., and W. W. Scott. *Hermaphroditism, Genital Anomalies and Related Endocrine Disorders.* Baltimore: Williams & Wilkins, 1958.

McKerns, K. W. (ed.). *The Sex Steroids.* New York: Appleton, 1971.

Metz, C. B., and A. Monroy (eds.). *Fertilization,* Vols. 1 and 2. New York: Academic, 1969.

Miller, N. F. *Human Parturition.* Baltimore: Williams & Wilkins, 1960.

Moore, C. R. *Embryonic Sex Hormones and Sexual Differentiation.* Springfield, Ill.: Thomas, 1948.

Parkes, A. S. (ed.). *Marshall's Physiology of Reproduction,* Vol. 3. London: Longmans, 1966.

Pincus, G. *The Control of Fertility.* New York: Academic, 1965.

Velardo, J. T. *Endocrinology of Reproduction.* New York: Oxford U. P., 1958.

Williams, A., and A. H. Reddi. "Actions of Vertebrate Sex Hormones," in: *Annual Review of Physiology,* Vol. 33. Palo Alto, Calif.: Annual Reviews, 1971, p. 31.

Wolstenholme, G. E. W., and M. O'Connor. *Preimplantation Stages of Pregnancy.* London: Churchill, 1965.

16

The Endocrine Glands

As has already been pointed out, the complex animal organization and all its functions are coordinated and regulated by the nervous system. To this system, especially in the more complex organism, there must be added a special integrating mechanism of chemical regulation. These chemical regulators are the **hormones,** the specific products of the ductless glands or the glands of internal secretion, called **endocrine glands.** These glands, although present on invertebrate forms of animal life, are especially characteristic of vertebrates. The endocrine glands play an especially important part in the organism in coordinating and regulating the physiological processes of the whole animal. We have already studied the interrelationships between the different parts of the digestive glands and the way that hormones such as gastrin and secretin help regulate and harmonize the digestive processes.

Although the endocrine glands, which are widely distributed throughout the animal body, do not constitute an organ system, they are usually grouped together and referred to as the *endocrine system.* These glands differ from the other glands of the body (exocrine glands) in that the former secrete their products directly into the blood stream for transmission to various tissues rather than in specialized ducts. The hormones secreted by the endocrine glands exert profound effects on certain tissues and, like the vitamins, are needed in relatively small amounts to produce their effect. Hormones, unlike vitamins and enzymes, are not used to initiate reactions in order to bring about an appropriate response in the organism, but act to regulate.

The tissue on which a specific hormone exerts its effect is called the **target tissue;** in addition, a state of balance normally exists between the various glands, and a reciprocal interaction among them is often demonstrable. The interrelationships of the endocrine organs are rather complex since one and the same organ may at the same time stimulate simultaneously or consecutively many organs, which in their turn may produce a series of stimulations and inhibitions. Where one endocrine organ can stimulate another, the organ that is stimulated may have a contrastimulating or inhibiting effect on the stimulator. This we have seen demonstrated in our discussion of the gonads and their effect on the pituitary gland, that is, the relationship between estrogen and FSH and progesterone and LH. If one decreases, the other increases, and vice versa. This concept in regulating secretion is the idea of **negative feedback.**

Mechanism of Hormone Action on Target Tissue

Hormones never initiate a reaction; they regulate cell reactions by acceleration or deceleration of specific cellular processes. Before hormones can carry out their activity, they must bind to a specific tissue or tissue component whereby they may then alter the rate of membrane transport of a substance or alter the rate of enzyme activity.

Many hormones are known to facilitate or inhibit the transport of substances through the cell membrane. For example, insulin increases the rate at which glucose enters the cell as well as the total amount entering. The increase in cell permeability to glucose is actually in both directions, but once in the cell, glucose is rapidly phosphorylated. Since glucose-6-phosphate is not able to get out of the cell, the glucose is effectively trapped inside and the net effect is an increased glucose uptake.

The specific mechanism by which insulin gets glucose into the cell is unknown. It would appear that some sort of loose chemical union with glucose is part of the transfer process. There is evidence that insulin also becomes loosely fixed or bound to the cell membrane. It may be that insulin is providing a surface configuration for which the glucose molecule has an affinity, where the normal contours of the cell membrane's outer surface represent a barrier to its entry into the cell (Fig. 16-1). It has also been postulated that insulin reduces the activity of a cellular inhibitor of glucose that is normally present. Other hormones inhibit glucose transport. The transport of other organic metabolites as well as water and inorganic ions is also influenced by hormones.

Hormones may also influence the rates at which specific cellular processes proceed by changing the amount or activity of enzyme(s) or other specific proteins.

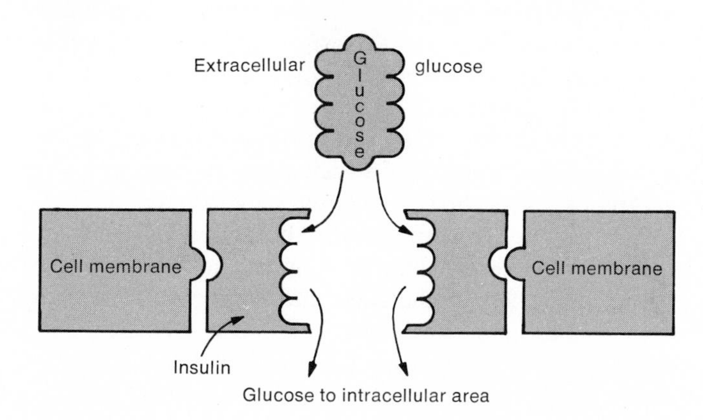

Figure 16-1. The molecular configuration of extracellular glucose prevents it from becoming attached to the knobed contours of the cell membrane's outer surface, providing a barrier to its entry into the cell. Insulin facilitates transport by providing a surface for its attachment.

A major effect of certain hormones is to stimulate production of more enzyme by the cell. They may do this by changing DNA-directed messenger RNA synthesis, perhaps by regulating which area of DNA is to be active or by decreasing intrastrand bonds of DNA. Hormones may also affect protein synthesis at the ribosome level, where they may act by controlling the polymerization of polypeptides (see discussion of protein synthesis, p. 412). The synthesis of new enzyme molecules requires hours or days. The hormone may exert a slow effect, or it may exert a rapid effect by the activation of enzyme molecules already present.

An enzyme's activity in many instances depends upon the presence of other molecules or ions (cofactors). The ability of such cofactors to increase or decrease an enzyme's activity is accomplished by combining with the molecules so as to change its shape. This effect is referred to as the *allosteric effect* of the molecule. Certain hormones exert an allosteric effect on enzymes by binding to one part of the molecule, changing the shape, and thus changing the action, of the molecule. It is highly likely that some of the effects of certain hormones is to change the amount or availability of such cofactors. It is thought by some that adrenocorticotropic hormone (ACTH) and thyroid-stimulating hormone (TSH) do this.

Hormonal Mediators

The concept and discovery that many hormones exert their effects on target cells by first causing the release of an intracellular hormonal mediator(s) is perhaps the most significant advance in basic endocrinology in recent years. This substance, first described in 1957 by Rall et al.[1] is cyclic 3′, 5′-adenosine monophosphate (cyclic AMP). Cyclic AMP is formed in the cell from adenosine triphosphate (ATP). The reaction is catalyzed by the enzyme adenyl cyclase. It is believed that hormones that act by stimulating the formation of cyclic AMP do so by activating the enzyme adenyl cyclase. It is, then, the cyclic AMP and not the original hormone that causes the effects inside the cell, giving to this substance the function of an intracellular hormonal mediator. Cyclic AMP once formed is rapidly broken down within the cell to adenosine monophosphate (AMP). This reaction is catalyzed by the enzyme phosphodiesterase. In fact, the same hormones may stimulate cyclic AMP formation by inhibiting the activity of phosphodiesterase. Figure 16-2 illustrates the mechanism whereby hormones may exert their effect on cell function through their stimulation of cyclic AMP synthesis. Since adenyl cyclase is found in the cell membrane, it is believed that the stimulating hormone acts on the membrane of the target cell with a receptor site, which shows a high degree of specificity for that particular hormone before the stimulation of adenyl cyclase is possible.

An example of such an action is demonstrated by parathyroid hormone (PTH) and vasopressin, both of which activate adenyl cyclase in the kidney, but the receptor sites for the two hormones are different. PTH acts primarily in the cortex of the kidney, consistent with the action of PTH on calcium and phosphate transport in the proximal portion of the tubule; vasopression acts primarily in the medulla. The cyclic AMP mechanism has been implicated for such a large

[1] T. W. Rall, E. W. Sutherland, and J. Berthet, "The Relationship of Epinephrine and Glucagon to Liver Phosphorylase." *J. Biol Chem.*, **224**:463 (1957).

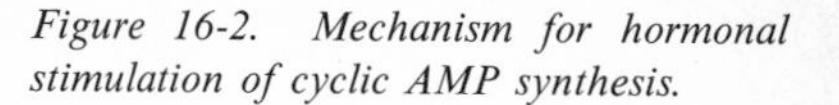
Figure 16-2. Mechanism for hormonal stimulation of cyclic AMP synthesis.

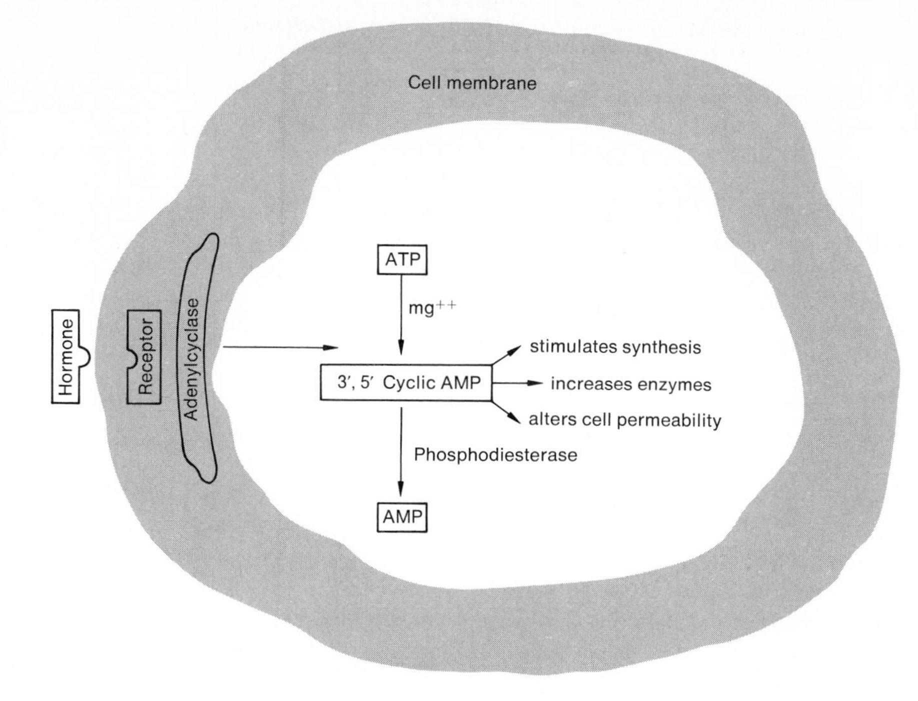

number of hormones [adrenocorticotropin (ACTH), corticotropin, thyrotropin (TSH), thyroxine, antidiuretic hormone (ADH), insulin, glucagon, serotonin, epinephrine, estradiol, parathyroid hormone (PTH), melanocyte-stimulating hormone (MSH), and vasopressin] that it may now be considered a universal mediator of hormone activity.

Another group of intracellular substances which probably are also hormonal mediators are the substances called **prostaglandins.** They are found in many tissues and have very powerful and varied effects. Where cyclic AMP acts as a mediator of hormone stimulation, in many instances, the prostaglandins act as mediators of hormonal inhibition—most probably through their interfering with the function of adenyl cyclase, blocking the formation of cyclic AMP. The prostaglandins may also be true hormones, circulating in the blood and acting in a way unrelated to intracellular mediation. In addition, the nervous system, especially the midbrain and the sympathetic nervous system, influence the endocrine glands; in turn it is influenced by the hormones secreted by these glands. In fact, the ties that the endocrine system has with the nervous system are so extensive they are now recognized as being of such importance that a new field of study, neuroendocrinology, has developed.

Some of the more important endocrine glands are the following: (1) the pituitary, or hypophysis; (2) thyroid; (3) parathyroids; (4) thymus; (5) adrenals; (6) pancreas; and (7) gonads. The structure and function of each of these glands will be discussed. Table 16-1 summarizes all the endocrine glands and their secretions.

Table 16-1. ***Summary of Hormones and Their Sites of Formation***

Gland	*Hormone(s)*
Pituitary (anterior)	STH, TSH, ACTH, FSH, LH, LTH
Pituitary (posterior)	ADH, oxytocin
Thryoid	Thyroxine
Parathyroid	PTH
Thymus	Lymphocyte-stimulating factor (LSF), thymus hormone (questionable)
Adrenal (cortex)	Hydrocortisone (cortisol)—main one in man, aldosterone, androgens, corticosterone—little in humans, main one in rat
Adrenal (medulla)	Catecholamines—epinephrine, norepinephrine
Pancreas	Insulin, glucagon
Kidneys	Renin-angiotensin, erythropoietin, medullin (PGE_2)
Gonads	
Male	Testosterone
Female	Estrogen, progesterone
Placenta	Gonadatropins, progesterone
Gastrointestinal tract	See Chapter 11
Pineal gland	Melatonin, serotonin (questionable)
Blood	Bradykinin, kallidin

The Pituitary Gland

The human **pituitary,** located in the brain just behind the optic chiasma (Fig. 16-3), is a reddish-gray oval structure about 10 mm in diameter. The average weight of the gland in a man is 0.5 to 0.6 g. In a woman the gland is usually larger, weighing from 0.6 to 0.7 g. The pituitary gland is formed embryologically from two sources: (1) part from a neural source, developing from a downgrowth of the floor of the thalamus, the infundibulum; and (2) a portion that develops from the ectoderm of the primitive oral cavity. That portion of the gland formed by the neural component is referred to as the **neurohypophysis,** and the term **adenohypophysis** is applied to that part formed from the primitive oral cavity. Generally the gland is divided into **anterior, intermediate,** and **posterior lobes** (Fig. 16-4). The adenohypophysis includes the anterior lobe and the infundibular stalk, which attachès the gland to the hypothalamus of the brain.

The **anterior lobe,** also known as the **pars distalis,** comprises the largest portion of the gland, forming 70 per cent of the total weight of the gland. The blood supply of the pituitary (Fig. 16-5) is independent of that of the brain proper and is derived from the hypophysial arteries, which branch from the internal carotid. The circulation of the anterior lobe and hypophysial stalk is separate from that of the posterior lobe. In other words, all the blood to the anterior pituitary comes through portal vessels from the hypothalamus, the **hypothalamopituitary portal system.** It has been clearly established that the blood flows downward from the hypothalamus to the pituitary, and it is this blood supply arrangement that gives to the hypothalamus its control over the anterior pituitary. More recently, additional vessels, which arise from the posterior lobe, have been shown also to perfuse the anterior lobe. The hypothalamus is now known to contain releasing or inhib-

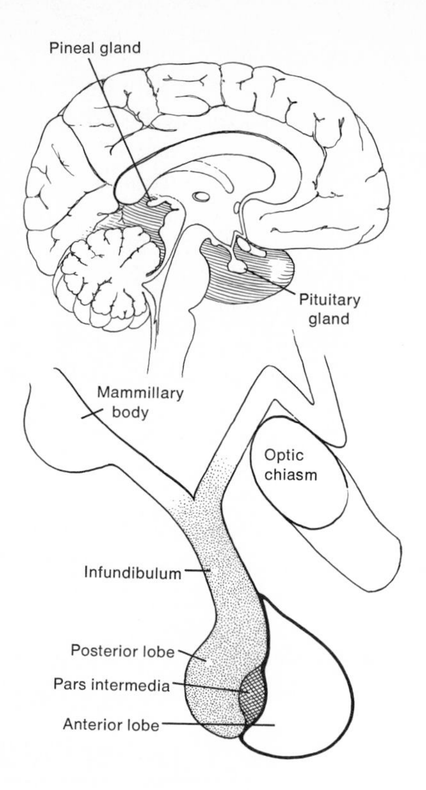

Figure 16-3. Pituitary gland showing its location and the three lobes associated with the gland.

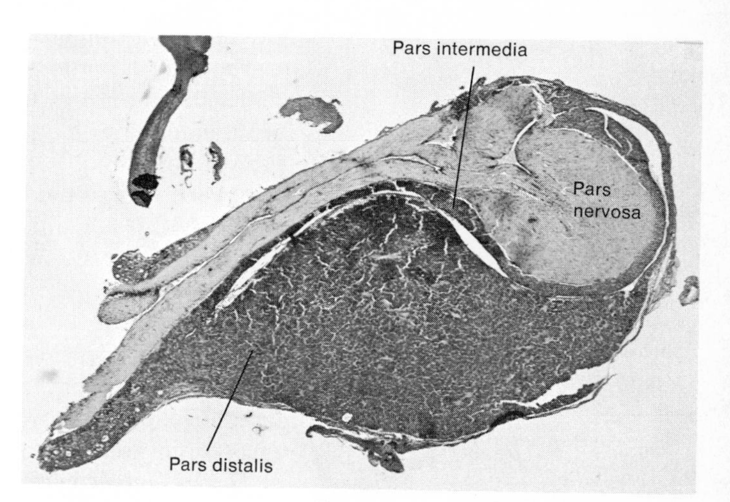

Figure 16-4. Mammalian pituitary gland (hypophysis). Sagittal section of entire body, showing glandular, intermediate, and nervous portions. (Courtesy Ward's Natural Science Establishment, Inc., Rochester, N.Y.)

iting factors for all the hormones of the anterior pituitary. Most of these factors arise in the lower part of the hypothalamus in an area referred to as the **median eminence.** Neurons originating in higher centers in the hypothalamus or in other areas of the brain send nerve fibers into the median eminence. These nerve fibers secrete small polypeptide hormones which are the actual releasing or inhibiting factors. These are transmitted by way of the portal system to the anterior pituitary to control release of anterior pituitary hormones. By changing the rate of secretion of these releasing or inhibiting factors the brain and hypothalamus can increase or decrease the secretion of hormones produced by the cells of the anterior pituitary.

The human hypothalamus contains **somatotropin-releasing factor (STHRF), thyrotropin-releasing factor (TRF), corticotropin-releasing factor (CRF), luteinizing hormone-releasing factor (LH-RF), follicle-stimulating hormone-releasing factor (FSH-RF),** and **melanocyte-stimulating hormone-inhibiting factor (MSH-IF).** In other species one can find **prolactin-inhibiting factor (PIF)** and **MSH-releasing factor (MSH-RF).** It is thought that the releasing factors may act by depolarizing the membrane of the anterior pituitary cell and their effects may be mediated by cyclic AMP.

A large proportion of the anterior lobe is practically devoid of nerve fibers; however, there has been much interest in the innervation of this gland, since disturbances in the central nervous system give rise to changes in the secretion rate of the anterior pituitary.

Hormones of the Anterior Pituitary

The anterior pituitary secretes hormones that mainly influence other endocrine glands and best illustrates the reciprocal relationship that exists within this system. Such hormones are referred to as **tropins** and are carried by the blood to other target glands, where they aid in the maintenance of the glands as well as stimulating them to produce their own hormones. In addition to the tropic hormones, the anterior pituitary secretes other hormones. Six hormones are known to be secreted by this part of the pituitary and include the growth hormone, the lactogenic hormone, the follicle-stimulating hormone, the luteinizing hormone, the thyrotropic hormone, and the adrenocorticotropic hormone.

The **growth hormone,** also known as **somatotropin (STH),** is a protein with a molecular weight of about 49,000 and has been prepared in a highly purified crystalline state. It appears that the amino acid tyrosine is necessary for the growth-promoting activity of this hormone. Growth hormone functions to stimulate the growth of the long bones at the epiphyses as well as growth of soft tissue. It also exerts an anabolic effect on nitrogen metabolism, thereby causing a considerable retention of nitrogen. Carbohydrate and lipid metabolism are similarly influenced. Elevation of glycogen stores of skeletal and cardiac muscle is effected by STH. Removal of the pituitary **(hypophysectomy)** in young animals results in a complete cessation of growth. Replacement of the growth hormone with purified preparations of pituitary extracts by injection will restore growth. The injection of growth hormone into normal animals produces an excessive increase in growth.

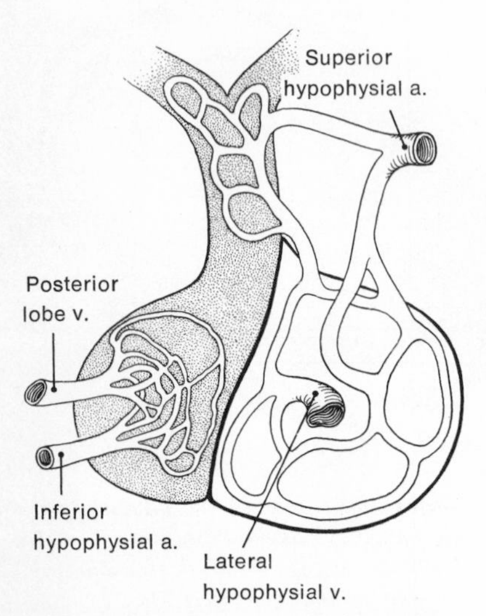

Figure 16-5. Blood supply of the pituitary. The circulation of the anterior lobe and hypophysial stalk is separate from that of the posterior lobe.

The **lactogenic hormone,** which stimulates growth of the mammary glands; the **follicle-stimulating hormone (FSH),** which stimulates the graafian follicle in the ovaries to mature and produce estrogen and stimulates spermatogenesis in the male; and the **luteinizing hormone (LH),** which causes ovulation and development of the corpus luteum and liberation of progesterone in the female and stimulates testosterone production by the testis in the male, have been discussed in Chapter 15.

The **thyrotoropic hormone,** also known as **thyroid-stimulating hormone** or **TSH,** functions to maintain the thyroid gland and stimulates it to produce the hormone thyroxine. Excessive liberation of this hormone causes excessive production of the thyroid hormone and thus produces all the symptoms of excessive thyroid activity (hyperthyroidism). On the other hand, a deficiency of pituitary function causes the thyroid to atrophy. A fine balance exists between the production of thyrotropic hormone and the concentration of thyroid hormone. Thyroxine inhibits the production of TSH; the reciprocal relationship existing between the target organ and the tropic hormone is again demonstrated.

The **adrenocorticotropic hormone,** more commonly referred to as **ACTH,** is a protein of low molecular weight. A polypeptide fraction containing approximately eight amino acids and retaining the activity of the raw hormone has been isolated. The administration of ACTH stimulates the secretion of adrenal cortical steroids by the adrenal cortex. It has also been shown to cause an increased excretion of nitrogen, potassium, and phosphorus, but it causes retention of sodium and chloride. The injection of ACTH to normal human beings has also been shown to cause an elevation of blood glucose and an increased excretion of uric acid.

Other tropic hormones elaborated by the anterior pituitary have been described, such as parathyrotropic hormone and a pancreatropic hormone. However, since they have not been isolated in purified form, their existence is not established. Levin and Farber[2] believe that a pituitary hormone referred to as **adipokinin** is an active factor in the mobilization of depot fat to the liver. In Table 16-2 the hormones secreted by the pituitary gland and their functions are summarized.

The **intermediate lobe,** often referred to as the **pars intermedia,** secretes a hormone known as the **melanocyte-stimulating hormone (MSH).** It has also been known as **intermedin.** Little is known of the chemistry of MSH, except that it is a protein that acts in the dispersion of pigment granules in the melanocytes, which are important in the darkening of the skin. MSH may also affect the functions of the central nervous system, where it causes an increased excitability.

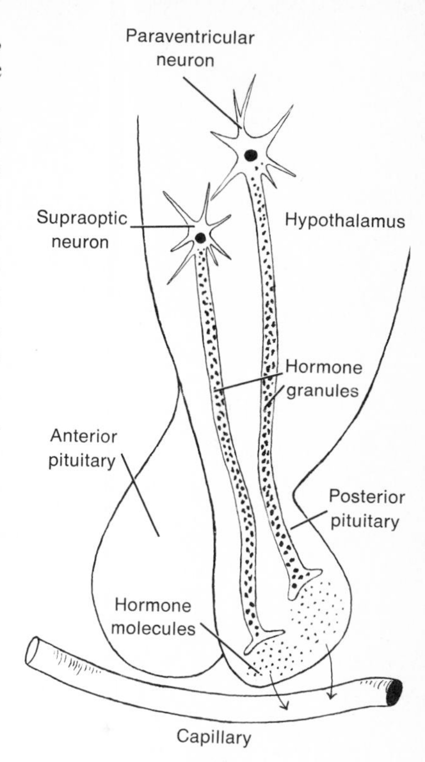

Figure 16-6. Synthesis and storage of posterior pituitary hormone.

The posterior pituitary and its hypothalamic connections are together called the **neurohypophysis.** It is known to secrete at least two active substances: a vasopressor-antidiuretic principle called **vasopressin,** which has little to do with maintaining blood pressure but is of prime importance in stimulating facultative resorption by the kidney tubule. It is preferably called **antidiuretic hormone (ADH).** Vasopressin acts to elevate blood pressure by constricting the peripheral arterioles; however, this effect is seen only at dose levels rarely reached physiologically. **Oxytocin,** the other active substance secreted by the neurohypophysis, stimulates smooth muscle contraction particularly of the uterus.

It is believed that oxytocin and vasopressin (ADH) are formed in the hypothalamus with the posterior pituitary acting largely as a storage site. There is evidence that oxytocin originates in the **paraventricular nuclei** of the hypothalamus and ADH in the **supraoptic nuclei.** From here the hormones in the form of small granules travel down the neurons to the posterior pituitary, where they are then stored or released to the blood stream (Fig. 16-6). Vasopressin has been used in cases of surgical shock in conjunction with other drugs to elevate the blood pressure. Oxytocin stimulates smooth muscle contraction, particularly of the uterus. It is therefore used in obstetrics to prevent postpartum hemorrhage resulting from an overly relaxed uterus. Oxytocin also acts on the postpartum mammary gland, causing ejection of milk.

Control of Pituitary Hormone Secretion

The controlling mechanism of hormone secretion from the pituitary gland is one of **negative feedback,** which may be either direct or indirect. The classic type is of the direct negative feedback, where, for example, thyrotropic-stimulating hormone (TSH) stimulates the secretion of thyroxine from the thyroid, which in turn shuts off the secretion of TSH by acting directly on the pituitary. The feedback may be indirect, whereby the pituitary hormone stimulates the secretion by another hormone, which then inhibits the secretion of the pituitary hormone by inhibiting the output of the hypothalamic-releasing factor. For example, the secretion of gonadtropin from the anterior pituitary stimulates the secretion of gonadal steroids, which exert their feedback effects to inhibit the secretion of gonadotropins not on the anterior pituitary directly but on the hypothalamus to

[2] L. Levin and R. K. Farber, *Recent Progress in Hormone Research,* **7**:399 (1952).

Table 16-2. *Hormones Secreted by the Pituitary Gland and Their Functions*

Hormones	*Principal Actions*
ADENOHYPOPHYSIS (ANTERIOR [PARS DISTALIS] AND INTERMEDIATE LOBES)	
Anterior lobe	
Somatotropin (STH, growth hormone)	Controls growth of bone and muscle; anabolic effect on nitrogen metabolism; carbohydrate and fat metabolism; elevates glycogen stores of skeletal and cardiac muscle
Thyrotropic hormone (thyrotropin, thyroid-stimulating hormone, TSH)	Controls the rate of iodine uptake by thyroid tissue and influences the synthesis of thyroxine from diiodotyrosine
Adrenocorticotropic hormone (ACTH, adrenocorticotropin)	Stimulates the secretion of cortical hormones by the adrenal cortex
Prolactin (lactogenic hormone, luteotropin)	Controls proliferation of the mammary gland and initiation of milk secretion; prolongs the functional life of the corpus luteum, the secretion of progesterone
Gonadotropin interstitial cell-stimulating hormone or luteinizing hormone (ICSH, LH)	*Ovary:* controls formation of corpora lutea, secretion of progesterone; probably acts in conjunction with FSH *Testes:* stimulates the interstitial cells of Leydig, promoting the production of androgen
Follicle-stimulating hormone (FSH)	*Ovary:* controls growth of ovarian follicles; functions with LH to cause estrogen secretion and ovulation *Testes:* has possible action on seminiferous tubules to promote spermatogenesis
Pars intermedia	
Melanocyte-stimulating hormone (MSH)	Controls dispersion of pigment granules in the melanophores; darkening of the skin
NEUROHYPOPHYSIS (POSTERIOR LOBE)	
Vasopressin (antidiuretic hormone)	Elevates blood pressure through action on arterioles; promotes resorption of water by kidney tubules
Oxytocin	Affects postpartum mammary gland, causing ejection of milk; promotes contraction of uterine muscle; possible action in parturition and in sperm transport in female tract

inhibit the secretion of gonadotropin-releasing factor. Figure 16-7 illustrates these two types of negative feedback.

Abnormalities of Pituitary Function

Most endocrine disorders are best considered from the viewpoint of either an increased (hyper) or decreased (hypo) functional activity of the gland. Disorders of the anterior pituitary gland in childhood are best represented by the **pituitary dwarf** and, in rarer instances, the **pituitary giant. Cushing's disease** can also occur in childhood as a result of an increased secretion of adrenocorticotropic hormone from the anterior pituitary.

Excess production of growth hormone, known as **hyperpituitarism,** during childhood or adolescence results in **gigantism.** Normally, growth hormone is active only up to the time of maturity. Pituitary growth hormone in excess increases the length of bone and in some instances its width in excess of normal growth before closure of the epiphyses. In some instances a suppression of gonadotropic hormones occurs. Since the testes and adrenals in the male, or adrenal glands in the female, do not develop as in normal puberty, epiphyseal closure, which is dependent on the male hormone, fails to take place. The end result is that the individual reaches a height of 7 to 8 ft or more (Fig. 16-8). In addition, there are associated metabolic changes attributed to a generalized pituitary hyperfunction.

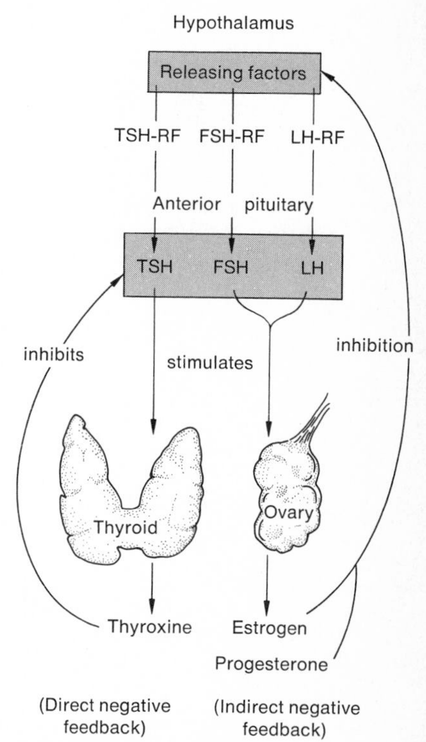

Figure 16-7. Control of pituitary and thyroid hormone secretion by negative feedback. Two types of negative feedback, direct and indirect, are illustrated.

Should the hyperfunction occur near the end of puberty or after epiphyseal closure, there is no further increase in linear growth but an increase in the width of bone. This results in an unusual thickening of the hands and feet, prominence of the jaw, enlargement of the nose, and thickening of the skin. **Acromegaly** is the term applied to this condition (Fig. 16-9). Usually hyperactivity of the gland is associated with a tumor of this structure.

Excess production of ACTH as a result of a basophilic cell tumor of the anterior pituitary gland is known as Cushing's disease. The excessive quantities of adrenocorticotropic hormone stimulate the adrenal cortex to produce steroids, which are the actual cause of the rapidly increasing adiposity of the face, which gives a characteristic roundness to the face, sometimes described as "pig-eye" or "moonface." In addition to the roundness of the face, there is usually an increased adiposity at the nape of the neck and trunk similar to the fat distribution in the buffalo; hence the term "buffalo obesity" has been applied. Purplish striations may be present on the arms, the sides of the hips, the legs, or the abdomen, and there is usually an increased growth of hair (hirsutism) on the face (Fig. 16-10). High blood pressure is usually present, and there is also a tendency to depletion of potassium and to alkalosis. Cushing's disease is not always due to tumors of the anterior pituitary; it has been demonstrated that tumors of the adrenal cortex, which secreted excessive corticoids, could also produce the condition.

Hypopituitarism, or deficiency of the anterior lobe of the pituitary, may occur as a result of injury or atrophy of the gland or as a result of certain types of tumors. Hypopituitarism before puberty manifests itself chiefly by a retarded growth rate (Fig. 16-11); if it begins very early in life, dwarfism will result (Fig. 16-12). Midgets are examples of such deficiencies. Adult hypopituitarism may be classified according to whether there is a selective deficiency of various anterior

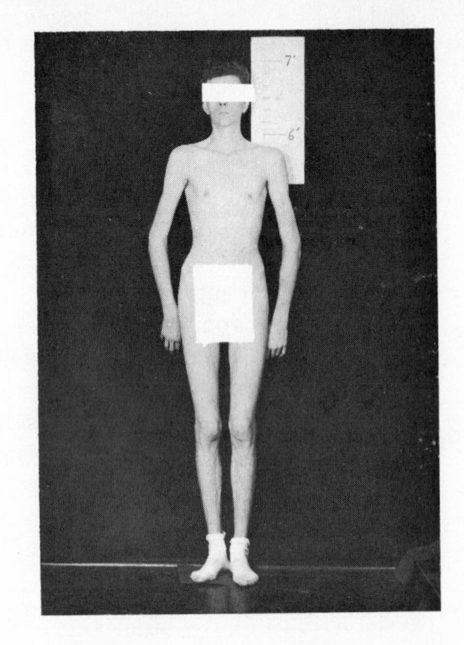

Figure 16-8. Pituitary giant. (Courtesy Armed Forces Institute of Pathology, Washington, D.C., picture Ol 8196.)

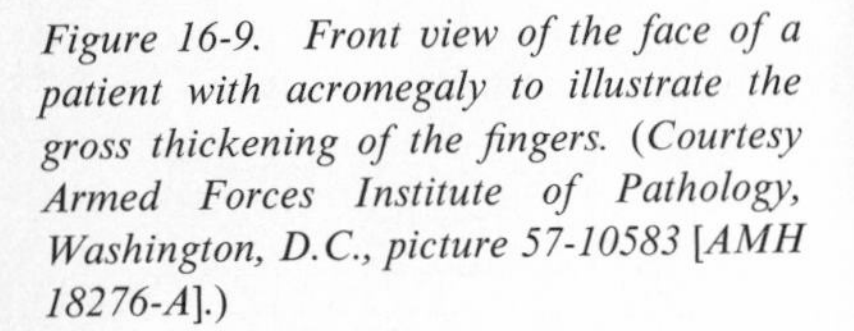

Figure 16-9. Front view of the face of a patient with acromegaly to illustrate the gross thickening of the fingers. (Courtesy Armed Forces Institute of Pathology, Washington, D.C., picture 57-10583 [AMH 18276-A].)

pituitary hormones or whether all the anterior pituitary hormones appear to be deficient. The latter condition is referred to as **panhypopituitarism.** In the selective type, **eunuchoidism** in the male or female may result as a consequence of a deficiency of gonadotropic hormones. Lack of thyrotropic-stimulating hormone produces symptoms similar to those of hypothyroidism, the condition being referred to as **pituitary myxedema** (see pp. 583–584). If all anterior pituitary function is decreased (panhypopituitarism), the clinical picture of myxedema does not seem to occur. This form usually occurs as a result of the destruction of the gland because of hemorrhage, atrophy, tuberculosis, syphilis, irradiation, fibrosis, or cystic degeneration. It was first described by Simmonds in 1914 and is usually referred to as **Simmond's disease.** The disease is characterized by extreme emaciation of the body because of a profound decline in the function of the endocrine system and of the metabolic processes that this system controls.

The Thyroid Gland

The human **thyroid** (Fig. 16-13) is a yellowish or amber-red lobulated, shield-shaped organ located in the neck region near the junction of the larynx and trachea. The gland consists of two lateral lobes that lie on either side of the trachea and are connected usually by an isthmus that crosses ventral to the second and third tracheal rings. The thyroid is invested by two layers of connective tissue, the outer layer forming a fibrous capsule that is loosely connected to the inner elastic layer, from which trabeculae extend within the gland, dividing it into ill-defined lobules. The size and general shape of the thyroid fluctuate with age, reproductive state, diet, season of the year, and geographical position, but it weighs on an average from 16 to 23 g in the adult. It is larger in the female than in the male.

The blood supply to the thyroid is exceptionally rich, and probably more blood flows through this gland in proportion to its size than through any other organ of the body (Fig. 16-14). The unit of structure of the thyroid gland is an alveolus or follicle consisting of a round or oval closed space lined with a single layer of low, cuboidal epithelium. The follicles (Fig. 16-15) vary in size and shape; some are large enough to be seen with the naked eye when the gland is sectioned. The interfollicular areas are occupied by a richly anastomosing system of blood and lymph vessels, which are supported in a network of connective tissue.

The closed cavities of the follicles are filled with **colloid,** a viscid, homogeneous translucent fluid that coagulates on fixation. It is believed that the colloids serve as a store for the hormone thyroxine, which is the secretory product of this gland. The amount of colloid in the follicles varies with the physiological condition of the gland, being increased during periods when there is no great demand for the hormone and reduced when there is an excessive utilization of the hormone.

In contrast to the pituitary gland, the thyroid gland has only three hormones. These are the iodine-containing amino acids **thyroxine** and **triiodothyronine;** they are present in the gland incorporated in the protein colloidal material known as **thyroglobulin.** The presence of triiodothyronine in the thyroid gland was discovered by Gross and Leblond in 1951 with the aid of radioactive iodine and two-dimensional chromatography. Previous to this it was believed that thyroxine was the only active principle secreted by the thyroid. The new active principle was found to be qualitatively similar to thyroxine in its biological actions but much more potent. The thyroid gland also produces **thyrocalcitonin,** a hormone that lowers serum calcium.

The synthesis of thyroxin and triiodothyronine by the thyroid is believed to

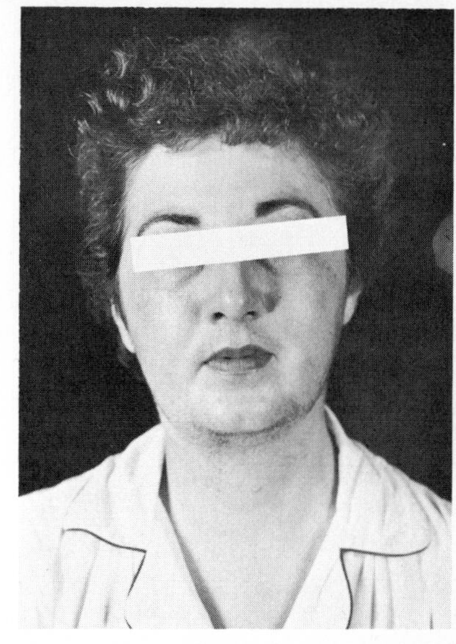

Figure 16-10. Adrenal cortical hyperfunction with Cushing's syndrome. (Courtesy Armed Forces Institute of Pathology, Washington, D.C., picture 56-20699 [CAMH 17221-A].)

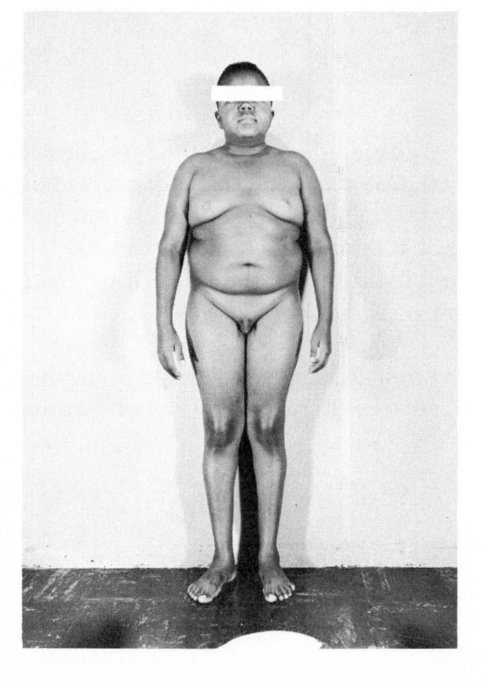

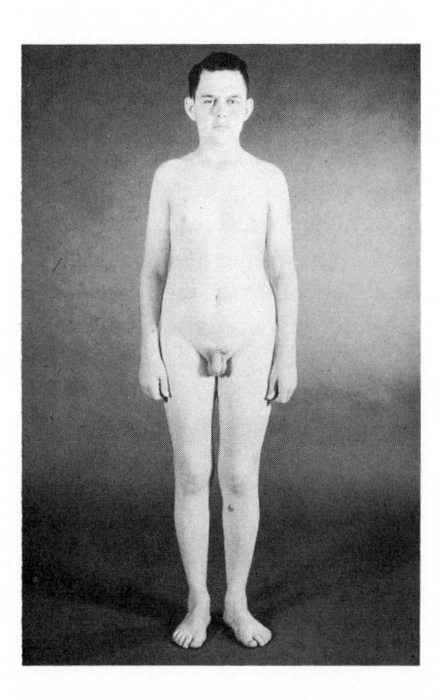

Figure 16-11. Delayed puberty due to anterior pituitary dysfunction (sex infantilism). Patient is 22 years old and has the growth and development of a 12-year-old. (Courtesy Armed Forces Institute of Pathology, Washington, D.C., picture 61-12477.)

Figure 16-12. Pituitary dwarfism with infantilism and obesity. Aged 20 years, the patient looks like a small fat boy of 8 years. (Courtesy Armed Forces Institute of Pathology, Washington, D.C., picture 58-10730 [Br 17883].)

Figure 16-13. *Position of the thyroid gland.*

Figure 16-14. *Blood supply of the thyroid.*

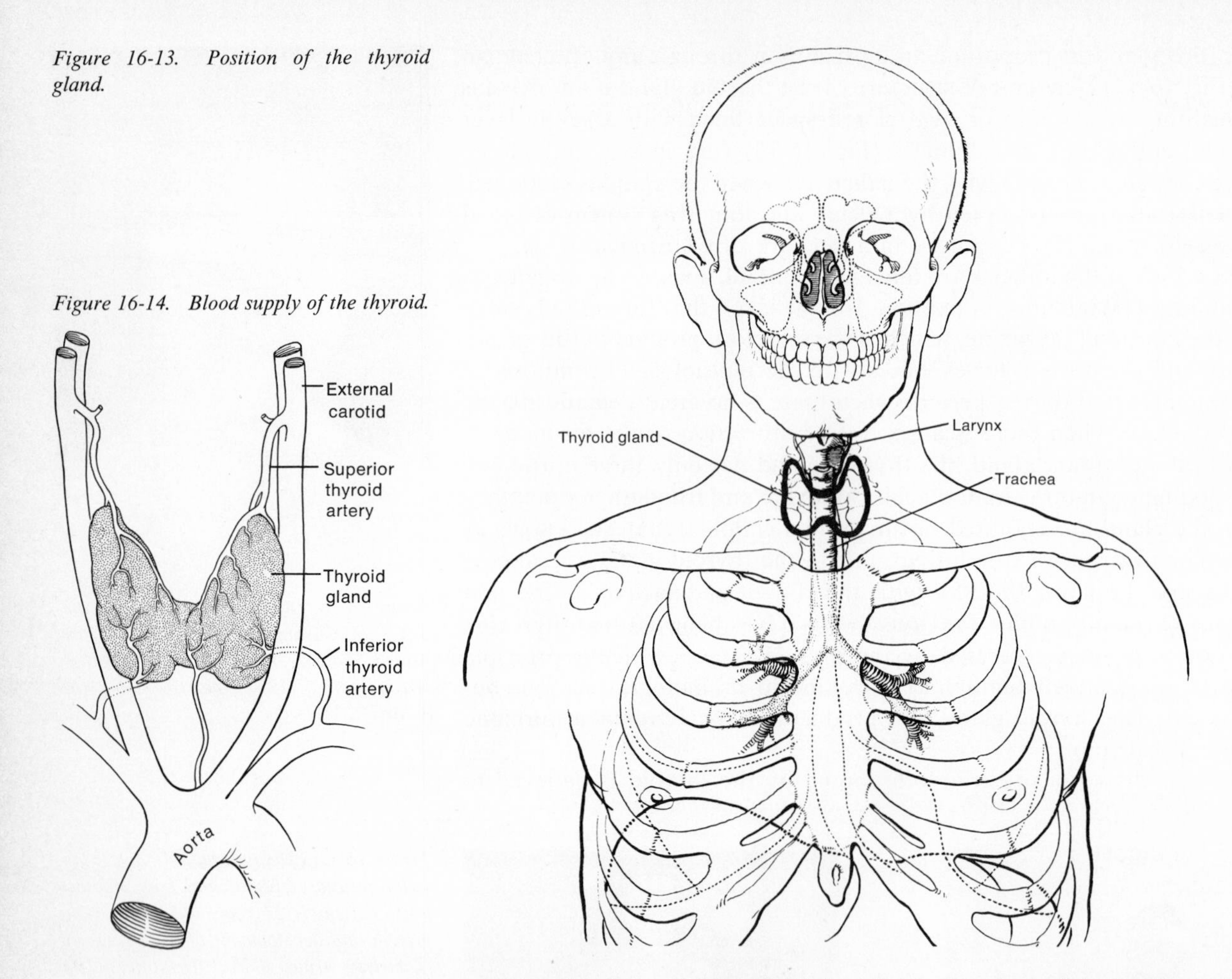

occur in the following way: (1) the gland absorbs iodide ion from the blood; (2) the iodide ion is oxidized to elemental iodine; (3) in the presence of elemental iodine, the amino acid tyrosine is iodinated for form diiodotryosine; experimental evidence indicates that diiodotyrosine is the direct precursor of thyroxine; (4) the coupling of two diiodotyrosine molecules results in the formation of thyroxine; and (5) it is believed that the deiodination of thyroxine gives rise to triiodothyronine.

Thyroxine is abbreviated T_4 because of the four iodine atoms on the molecule. In like manner, triiodothyronine is called T_3 because of the presence of only three iodine atoms:

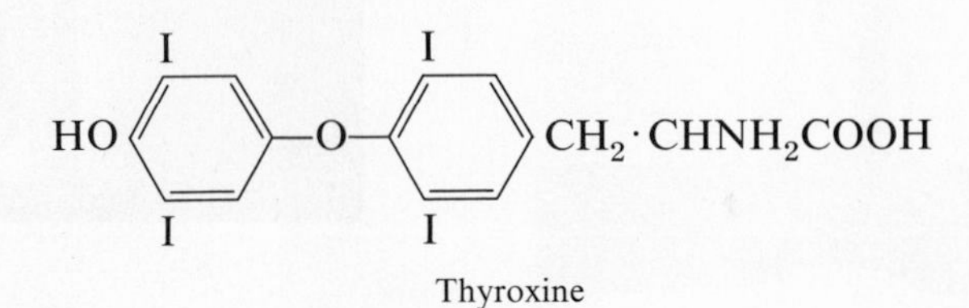

Thyroxine

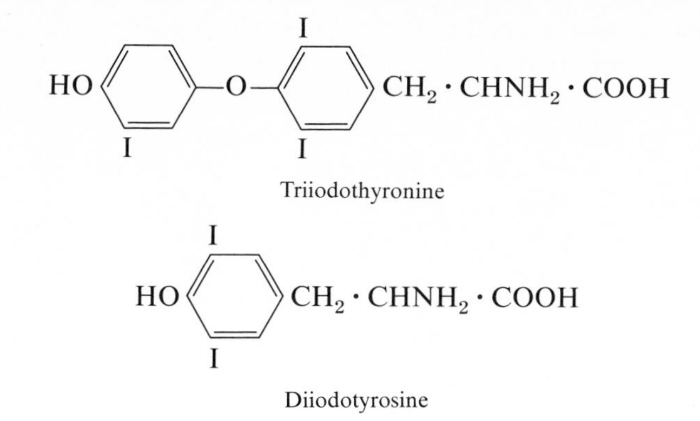

As we have seen, the active principles of the thyroid are bound up within the gland in the colloidal material, thyroglobulin, a protein with a molecular weight of approximately 675,000. The hormones gain access to the circulation from the gland by hydrolysis of thyroglobulin; the high molecular weight of thyroglobulin makes it impossible for the hormones to pass into the circulation through the follicular epithelium in this form. Furthermore, it has been demonstrated that the hydrolytic activity within the thyroid can be increased or decreased, depending on the level of thyrotropic hormone. Once the hormones get into the blood, they are bound to the plasma protein and circulated to all the tissues of the body. Under normal conditions the gland secretes about 0.35 mg of hormone per day.

Function of Thyroid Hormones

Knowledge of the action of thyroid hormone has been derived from the study of patients and laboratory animals that have been deprived of thyroid tissue or subjected to an excess of the hormone.

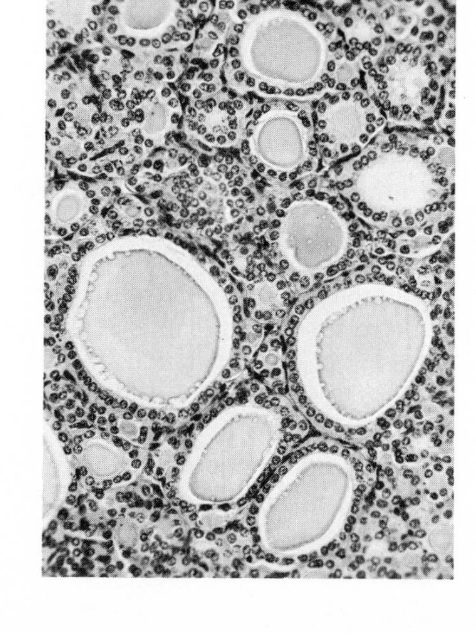

Figure 16-15. Section through thyroid gland showing cuboidal cells, resting on reticular tissue, lining the follicles. (*Courtesy Ward's Natural Science Establishment, Inc., Rochester, N.Y.*)

The primary action of thyroxine and triiodothyronine is on the metabolic rate; this is sometimes referrred to as **calorigenic action.** These hormones directly increase the rate of oxidation of foodstuffs within the cells of the tissues. Since there is no change in the respiratory quotient, it is assumed that all three nutrients utilized by the body are affected equally. The calorigenic action of thyroid is essential for normal growth and development, for in man and other vertebrate animals, it is impossible to attain adult form and dimensions in the absence of thyroid hormone.

The thyroid hormone accelerates the rate of absorption of sugar and depletes the liver of glycogen, possibly because of the increase rate of oxidation of sugar by body tissues. It is also essential to the normal development of bone. Epiphyseal fusion is delayed, and the epiphyses show radiological changes when the hormone is not present in normal quantities before puberty. Since in stimulating the general metabolism there is an increased demand for vitamins in general, any excess of the hormone may result in signs of vitamin deficiency. The thyroid is known to be necessary for the conversion of carotene to vitamin A.

Thyrocalcitonin (TCT) is secreted by the thyroid gland of many mammalian species. In birds, however, TCT is found in the ultimobranchial gland, which is separate from the thyroid. Embryologically, a similar gland begins in mammals as an outpouching of the fifth pharyngeal pouch but soon becomes mixed with the thyroid gland. The function of thyrocalcitonin is to lower the concentration of serum calcium. In conjunction with parathyroid hormone (PTH) and vitamin D, which keep the level of serum calcium high, the maintenance of the normal plasma concentration of calcium of approximately 10 ± 1 mg per ml is maintained. Any fluctuation from this concentration will upset many functions of the body (see Chapter 12, section on Mineral Metabolism).

Many normal functions of the organism are dependent on thyroid secretion. These include water and mineral metabolism and the normal function of the nervous system, muscular systems, and circulatory system. All these functions are disturbed with either a deficiency or excess of the hormone.

The thyroid gland also functions in harmony with other endocrine glands. Normal thyroid function is necessary for involution of the thymus. Excess thyroid secretion leads to parathyroid hypertrophy. The gonads are also affected by hypofunction or hyperfunction of the thyroid gland. Of particular importance is the inhibitory effect thyroid extract has on the cells of the thyroid gland itself. Additionally there is the reciprocal relationship that exists between the thyroid and pituitary. We have already seen that the activity of thyroid is regulated by the thyrotropic hormone of the anterior pituitary. Conversely, the rate of secretion of thyrotropic hormone is regulated primarily by the concentration of thyroid hormone in the circulation. However, it is also known that the anterior pituitary can also be stimulated by the hypothalamus via the hypophyseal-portal circulation under conditions of stress.

Thyrotropic-stimulating hormone (TSH) from the pituitary is the main regulator of thyroid hormone secretion. TSH stimulates thyroid hormone secretion quickly and over the long term and also stimulates the growth of the thyroid gland. The effect of TSH on the thyroid is uncertain, but it may well be on the thyroid cell membrane to activate adenyl cyclase with the resulting increase in cyclic AMP. Cyclic AMP is probably the mediator of the effects [increase in thyroid hormone(s) secretion]. TSH secretion is regulated in turn by the plasma levels of thyroxine (see Fig. 16-7). TSH hormone from the pituitary is also increased in response to cold and other stimuli. These responses are mediated by the central nervous system, which stimulates the secretion of releasing factors from the hypothalamus.

Abnormalities of the Thyroid Gland

Abnormalities associated with the thyroid gland are due either to a deficiency of iodine, the essential element of the thyroid hormone, or to an insufficient production of the hormone itself, or to an overactivity of the gland with the liberation of an excessive amount.

A deficiency of iodine in the diet as well as certain other disorders of the thyroid is frequently accompanied by an enlargement of the gland, which is designated as a goiter. **Simple,** or **endemic, goiter** is a deficiency disease caused by a lack of a sufficient amount of iodine in the diet. In this condition the thyroid is enlarged, apparently because of an increase in the number of cells (hyperplasia) as a compensatory reaction to the lack of iodine. There is neither inflammation nor malignancy, and the condition is not associated with toxic features. Simple goiter is endemic, occurring in areas in which the available supply of iodine in the drinking water and in the soil is low. In the United States the goitrous areas are the Pacific Northwest, the Great Plains, the basin of the St. Lawrence River, and the Great Lakes.

In simple goiter there is no clinical syndrome, for there is no hyposecretion or hypersecretion, the gland being capable of elaborating a normal concentration of hormone. However, the gland may become so enlarged as to compress the trachea and other structures located in the neck region. Under these conditions surgical removal of the goiter is necessary, but generally in the early stages of its development the administration of iodine can induce a regression of the glandular enlargèment. The value of iodine as a prophylactic against goiter is now generally accepted, and its use is advocated in all goitrous areas. The most feasible method of administration has been the addition of sodium or potassium iodide to table salt.

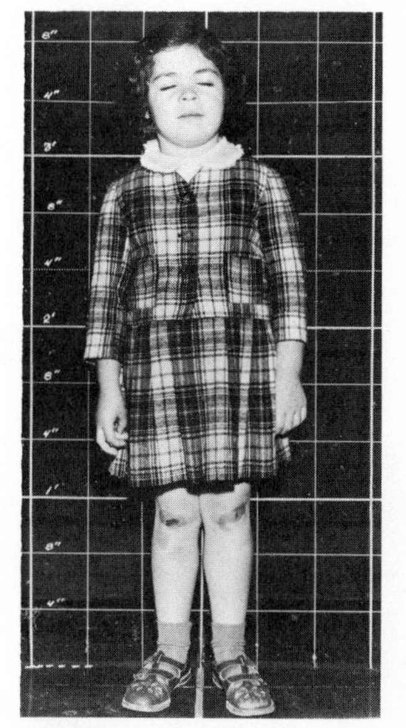

Figure 16-16. Childhood hypothyroidism (juvenile myxedema). (Courtesy Armed Forces Institute of Pathology, Washington, D.C., picture F 12059-A.)

A deficiency of thyroid hormone, referred to as hypothyroidism, produces a number of symptoms depending on the degree of deficiency and the age at which it occurs. Deficiency or lack of thyroid hormone in very early childhood leads to **cretinism,** a condition characterized by a low basal metabolic rate, cessation of mental and physical development, slow heart rate, poor appetite, and constipation. The face is usually puffy, with a characteristic apathetic expression, and the skin is dry, coarse, and pale yellow in color. Cretinism follows the incomplete development or congenital absence of the thyroid gland, but it can be successfully treated if desiccated thyroid is administered sufficiently early in life.

Thyroid insufficiency occurring later in childhood results in **childhood hypothyroidism,** or **juvenile myxedema** (Fig. 16-16). This condition differs from cretinism in appearing later in life. The severity of the symptoms depends primarily on the degree of thyroid activity and secondarily on the age at which the deficiency occurred. In general, the most characteristic symptoms are as follows: the child

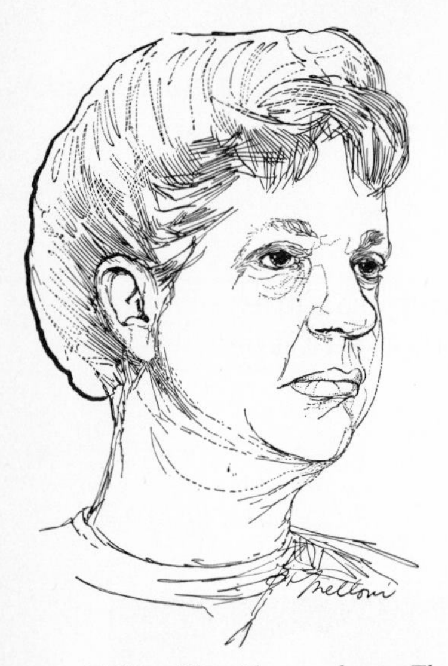

Figure 16-17. Thyroid myxedema. The face is quite expressionless, puffy, and pallid.

Figure 16-18. Exophthalmic goiter (toxic goiter). The eyes are prominent and staring, and characteristic protrusion of the eyeballs (exophthalmos) is common. (Courtesy Armed Forces Institute of Pathology, Washington, D.C., picture 60-10307 [AMH 23440-A].)

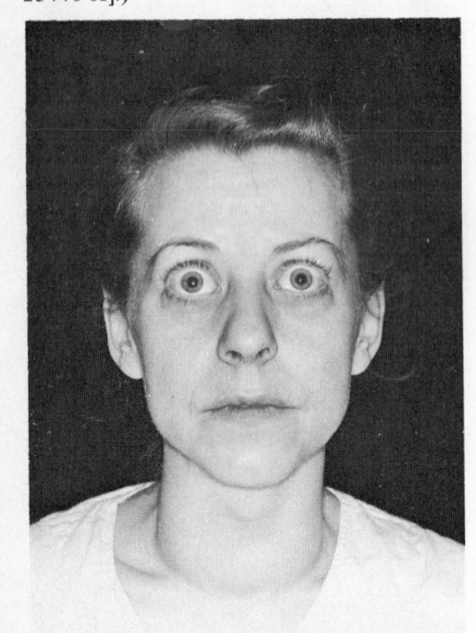

shows a tendency to be stout, short, and squatty; the head is proportionately larger than the normal for the age of the child and is attached to the trunk by a short neck; the skin is dry, and there is puffiness or bloating about the eyes and face, which usually has a dull, stupid look; the mouth is usually open and drooling and the tongue thick and protruding.

Treatment of childhood myxedema with desiccated thyroid gives remarkable results and is considered the one triumph of organotherapy. The outcome, however, depends on the degree of thyroid deficiency, the time of inception of treatment, and the regularity with which it is continued.

In the adult, hypothyroidism gives rise to the condition known as myxedema (Fig. 16-17). It usually results from atrophy of the thyroid gland; the cause of the atrophy is unknown. If the degree of deficiency is mild, a state of hypothyroidism without myxedema may exist, and this is characterized chiefly by a lowered metabolic rate of variable degree.

In true myxedema the face is quite expressionless, puffy, and pallid. Both the mental and physical processes slow down. The skin becomes dry and thick, and the hair and teeth tend to be lost. Sometimes obesity and an undue sensitivity to cold are other signs. The administration of adequate amounts of thyroid hormone usually results in a dramatic relief of all the symptoms within 10 days of the start of the treatment. With adequate continuous therapy, symptoms do not return.

Excessive thyroid function, known as **hyperthyroidism,** results from enlargement of the gland, with overproduction and excessive release of thyroid hormone. The most common form of hyperthyroidism, resulting from a diffuse increase in thyroid tissue, is known as **exophthalmic goiter, Graves' disease,** or **Basedow's disease.** Sometimes the condition is referred to as **thyrotoxicosis,** or **toxic goiter.**

The most characteristic symptoms of hyperthyroidism, in addition to goiter, are nervousness, irritability, purposeless movements, fatigue, loss of weight, increased heart rate, elevated metabolic rate, emotional instability, and increased body temperature with excessive sweating. The face of such a patient is often flushed, and the facial expression is one of anxiety. The eyes are prominent and staring, and characteristic protrusion of the eyeballs (exophthalmos) is common but may be absent (Fig. 16-18). It is due to the increase in the retro-orbital fat.

Treatment of toxic goiter usually involves the use of antithyroid drugs, such as propylthiouracil and methylthiouracil. These compounds interfere with the processes involved in the synthesis of thyroxine and triiodothyronine by the thyroid gland. Surgical removal of the thyroid gland is often resorted to, especially after treatment with antithyroid drugs. If the hyperthyroidism is due to malignant growth of the thyroid, only surgical intervention can be considered. Iodine, in the form of Lugol's solution, is considered to be the best and safest form of treatment of mild hyperthyroidism. It probably acts by depressing the activity of the gland, through decreasing the output of pituitary thyrotropic hormone. Because iodine becomes concentrated in the gland, radioactive iodine has recently come into use as a form of treatment. The internal radiations destroy the tissue and induce remission of the disease.

Clinical Tests of Thyroid Function

Protein-bound iodine (PBI) determination is used primarily as a test for thyroid function. The thyroid gland utilizes inorganic iodine and protein to form thyroxine. Since iodine exists in the blood in both an organic (protein-bound) and an inorganic (iodide) form, all PBI procedures require separation of the inorganic iodine from the protein-bound form. This is accomplished by preliminary precipitation and washing of the proteins of the serum sample. The precipitated and washed proteins containing the organic iodine are then digested in order to release the iodine for colorimetric analysis. For healthy adults, serum PBI concentrations are 3.5 to 8.0 mg per 100 ml.

In addition to the above method, the basal metabolism rate (BMR) and the radioactive iodine uptake (RAI) are also used for thyroid-function tests. It is the opinion of many clinicians specializing in thyroid diseases that none of these tests is infallible and therefore should be used to complement one another rather than to replace each other.

The Parathyroids

The **parathyroids** in man are yellow-brown, ovoid bodies and are probably the smallest of the known endocrine organs. There are usually four parathyroids, two on each side, situated on and more or less intimately connected with the posterior surface of the thyroid gland (Fig. 16-19). However, their number and position are extremely variable. The combined weight of the four glands varies from 0.05 to 0.3 g. The upper pair of parathyroids, usually designated the **superior,** are fairly

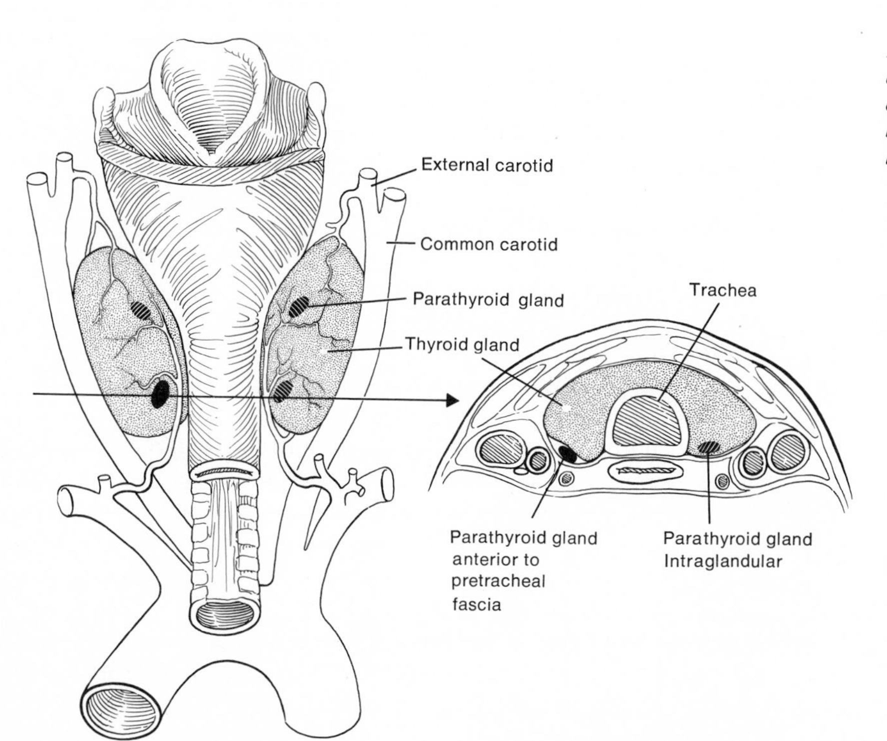

Figure 16-19. Parathyroid glands. There are usually four glands. The upper pair are designated as superior and are usually embedded into the thyroid substance. The lower pair are referred to as inferior.

constant in position at approximately one third of the distance from the top to the lower tip of the thyroid lobes. In man they are usually embedded in the thyroid substance and are separated from it by a connective tissue capsule. The lower pair, referred to as the **inferior,** are usually found at the level of the inferior extremity of the lateral lobe of the thyroid but may descend into the thorax along with the thymus. Although four glands are normally found, there may be as few as two or as many as twelve in the tissue throughout the neck area from above the thyroid down into the chest cavity.

Histologically, the parathyroids show the same general structure in all vertebrates. The bulk of the cells are round with a scant cytoplasm and a deeply staining nucleus. These are arranged in cords between which are numerous vessels with little connective tissue. A second type of cell, shaped similarly to the first, with a finely granular cytoplasm, which stains with eosin, is also present but not as numerous. A delicate fibrous capsule surrounds each gland, from which fine branches penetrate into the substance of the gland, dividing it into lobules.

It is interesting to note that early attempts to treat goiter by removal of the thyroid resulted in unexpected effects as a result of the unwitting removal of the parathyroids. Raynard, in 1835, was probably the first to recognize and record these unexpected results in experimental thyroidectomy. Credit for the discovery of the parathyroid glands is usually given to Sandstrom (1880). However, his work attracted very little attention among physiologists, and the glands were forgotten until Gley (1891) rediscovered them in some experimental work, in which he determined the effects of their extirpation. He called attention to the location of the inferior parathyroids of the rabbit some distance caudal to the thyroid lobes and to the fact that total thyroidectomy could be performed on this species without disturbing one or more of the parathyroids. It became possible to account for the discrepancies encountered in experimental removal of the thyroid and ascribe separate physiological roles in the thyroid and parathyroid glands.

Parathyroid Hormone

It is now recognized that the parathyroids produce a hormone that plays a vital role in the metabolism of calcium and phosphorus. **Parathyroid hormone,** sometimes referred to as **parathormone (PTH),** was first prepared in crude extract form by a number of early workers. Of these methods, Hanson's is now utilized in preparing the extract for commercial use. Collip's studies (1924) first demonstrated the activity of the parathyroids and formed the basis for much of our present knowledge of the action of parathormone.

Although much progress has been made in purifying extracts, no hormone has been obtained in pure form. Available evidence indicates that the parathyroid hormone is of protein nature with a molecular weight of approximately 20,000 to 80,000. The extracts are acted on and rendered ineffective by proteolytic enzymes and hydrochloric acid and therefore cannot be administered orally.

Function of Parathormone. The action of parathyroid hormone is not yet clearly defined. Experimental injections of human beings and animals with parathyroid extracts tend to give results that conflict with each other. These results may arise from artifacts in the gland extracts or from instability of their active constituents, which makes the action of injected material different from that of the endogenous hormone. The primary function of the parathyroid gland appears to be the maintenance of the proper calcium–phosphorous ratio in the blood and tissues. It is able to do this through its influence on the absorption of calcium from the intestinal tract, the deposition of calcium in bone and its mobilization therefrom, and the excretion of calcium ion in the urine and possibly the feces. The most important activity of the gland, however, is to promote the mobilization of calcium from the bone.

The secretory activity of the parathyroid gland apparently is controlled directly by the level of ionized calcium in the circulation, for whenever serum calcium changes, the activity of the parathyroid gland changes. This in turn affects the rate at which phosphorus is lost in the urine and the rate of calcium resorption by bone. The effect of the parathyroid gland in calcium and phosphorus metabolism is discussed more thoroughly in Chapter 12.

The action of the hormone is considered to be on two sites. The first is the kidneys, where it promotes the renal excretion of phosphate by decreasing tubular resorption, whereby a low serum phosphorus is produced. A decrease in the concentration of phosphate ion allows larger amounts of calcium ion to exist in solution, and thereby the mobilization of the ion from bone is favored. The second site of action is probably directly on bone tissue to cause its dissolution through increased osteoclastic activity.

Abnormalities of Parathyroid Function

The multiple nature and wide distribution of the parathyroid glands are such that abnormalities associated with disease or injury are, fortunately, not common. However, cases of parathyroid insufficiency, hypoparathyroidism, and excessive secretion, hyperparathyroidism, are known.

Hypoparathyroidism is caused most commonly by the accidental removal or injury of the glands as a result of operation on the thyroid. Degenerative diseases of the gland are also encountered. The symptoms associated with this condition are muscle weakness, tetany, and irritability. **Tetany** is the most characteristic symptom and is characterized by an abnormally increased sensitivity of the nervous system to external stimuli which causes painful spasms of the musculature to occur (Fig. 16-20). In the presence of tetany, prompt injection of calcium salts, either intramuscularly or intravenously, is required. Generally hypoparathyroidism is treated by the administration of calcium lactate in doses of 1 to 2 teaspoonfuls three times a day dissolved in warm or hot solution.

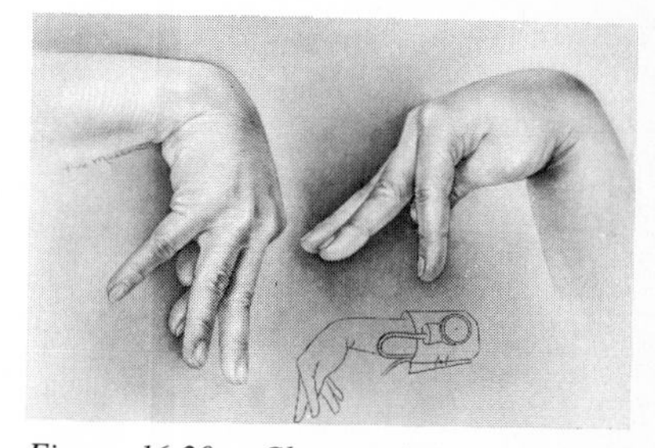

Figure 16-20. Characteristic attitude of the hand in tetany, called main d'accoucheur, *caused by muscle hypertonia. It may be elicited by arrest of the circulation in the forearm in a patient with parathyroid tetany.*

An increase in parathyroid hormone production, **hyperparathyroidism,** is usually due to a tumor of the gland, a condition known as **von Recklinghausen's disease.**

Hyperparathyroidism causes an excessive secretion of calcium. As a result, polyuria usually occurs and may be the first and sometimes the only symptom of an increased activity of the parathyroids. Kidney stones may form, and the decalcification of the bones that usually occurs causes pain and deformities as well as spontaneous fractures. There may be softening and bending of the long bones or ribs. Occasionally there is considerable weakness of the muscles with decreased response to stimulation.

The treatment of hyperparathyroidism is usually the surgical removal of the parathyroid tumor. After the operation, the parathyroid deficiency that usually occurs requires the administration of calcium and vitamin D.

The Thymus

The **thymus** (Fig. 16-21) is composed of two elongated, flask-shaped lobes, which occupy a region just above the heart where the chest cavity narrows at the root of the neck. The two glands are connected to one another by areolar connective tissue, which also invests each gland, forming a distinct capsule. Septa of similar connective tissue divide the organ into a number of irregular, more or less cylindrical anastomosing lobules. Finer radiating trabeculae enter into the interior of each lobule. Each lobule consists of two kinds of tissue, **medullary** on the inside and **cortical** on the outside.

The medullary substance contains structures known as **Hassall's corpuscles,** which are nests of concentrically arranged, flattened epithelial cells. In addition, the medullary core contains epithelial reticular cells and some lymphocytes.

The cortex of each lobule consists of a framework of reticular connective tissue, which is identical with or closely allied to adenoid tissue. The meshes of the cortex are crowded with lymphocytes similar to those found in the tonsils, other lymph glands, and spleen.

The thymus is a prominent organ at birth but reaches its greatest size at puberty, after which it undergoes a gradual regression. During regression the thymic tissue is usually replaced by fat, so that the adult thymus is composed largely of fat and connective tissue.

The exact function of the thymus is unknown. Because the thymus attains its greatest relative size during childhood, gradually involuting after sexual maturity and being eventually replaced by fat and connective tissue, it is obvious that its functions are in some way associated with events in early life.

Many functions and interpretations of the functions of the thymus gland have been offered. The studies of Park and McClure (1919) of the effects of thymectomy are classic in this field. Animals lacking the gland seem to be abnormally sensitive to environmental stresses and particularly to malnutrition or suboptimal vitamin intake. In the literature two correlations stand out with rather impressive consist-

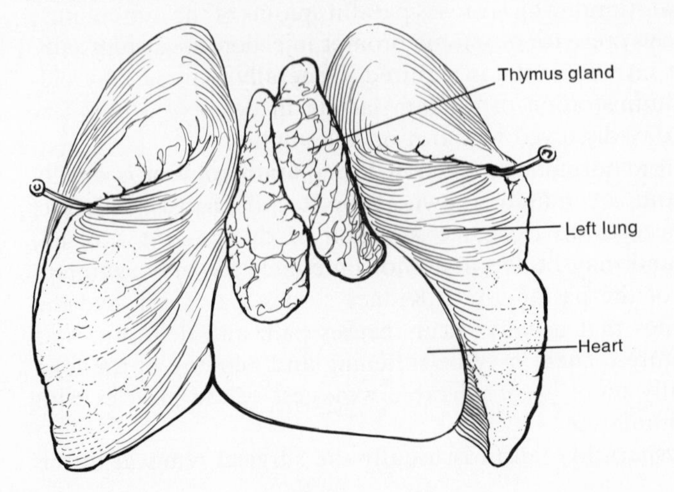

Figure 16-21. Thymus gland of a 12-year-old child.

ency: thymus enlargement after castration and a similar hypertrophy under conditions in which diminution of adrenal cortical function occurs. After treatment with androgenic, estrogenic, and adrenal cortical hormones, the thymus involutes. Endocrine relationships other than those just referred to are unknown.

Gudernatsch (1914), who has extensively reviewed the literature on the effect of thymus feeding on metamorphosis and sexual development, has suggested that a thymic hormone is involved. Asher (1936) has reported on extensive observations of the influence of aqueous thymus extract in producing significant acceleration of growth and development, suggesting the possibility of the presence of a hormone affecting growth. Hornedo (1952) has proposed a theory that the thymus gland should be regarded as the real regulator or main defense factor in the over-all defense mechanism of the body. He suggests that the thymus in the embryo and the young child secretes steroid and other hormones, which probably act directly or indirectly as trigger hormones or precursors of other gland hormones. He further suggests that this action on the hormone production of other glands leads to the establishment of a conditioned reflex, which is brought into play whenever the body is subjected to stress. As the individual grows older, this conditioned reflex secretory activity of the other glands becomes fixed and almost automatic, needing very small amounts of thymus hormones to set it off. In consequence, because of the reduced production of needed hormones, the gland gradually decreases in size until it becomes vestigial in the adult. It must be pointed out, however, that Hornedo has no experimental evidence to substantiate his theory.

The thymus is definitely considered to be most important in the production of lymphocytes. Comsa and Bezssonoff[3,4] have reported on the preparation and bioassay of a thymus hormone and have presented a theory that the lymphocytes act as carriers of the hormone. Metcalf[5] has also described a lymphocytosis-stimulating factor (LSF) found in the thymus, which he believes is involved in the maintenance of normal lymphocytes. Perhaps it could be hypothesized that the leukocytosis (increase in white blood cells) that usually follows initial stress is partly due to the lymphocytosis-stimulating factor of Metcalf.

The thymus gland has been implicated in many of the pathological conditions to which the human body is subject. Recently it has been associated with leukemia in the mouse, probably as a bridge between the leukemia stimulus and the disease. Myasthenia gravis, a neuromuscular disease, has been associated with thymic tumors. There is also the possibility that the apparent rejuvenation in health of a pregnant woman during the second half of the period of gestation may be due to the great hormone-producing activity of the thymus gland of the embryo. In general, there is a lack of information about the thymus, and more intensive research work on the function of this gland in health and disease is needed.

[3] J. Comsa and N. A. Bezssonoff, "Preparation and Bioassay of a Purified Thymus Extract." *Acta Endocrinol.,* **29**:257 (1959).

[4] J. Comsa and N. A. Bezssonoff, "Origin of the Supposed Thymus Hormone." *Acta Endocrinol.,* **30**:621 (1959).

[5] D. Metcalf, "Lymphocytosis Stimulating Factor in Thymus." *Brit. J. Cancer,* **10**:442 (1956).

The Adrenals

The **adrenals,** also referred to as the **suprarenal glands,** are two pyramid-shaped structures lying, one on each side of the body, close to the upper pole of the kidney (Fig. 16-22). The two glands usually differ somewhat in size, and there is also a difference in size between the glands in men and women. Between the ages of 16 and 20 the female adrenals are larger than the male. Generally, each gland measures from 1 to 2 inches. Each gland consists of two functionally distinct parts: a central portion, called the **medulla,** and a surrounding zone of tissue, called the **cortex,** which comprises eight to nine times the volume of the former (Figs. 16-23 and 16-24).

In the human being and in other mammals the medullary tissue consists of irregularly shaped cells measuring approximately 16 to 36 μm and containing a markedly basophilic protoplasm. It has a different origin embryologically from cortical tissue, because the medulla develops from a group of cells, split off from the neural crest of the embryo, to which the ganglion cells of the sympathetic nervous system also owe their origin. The nerve supply of medullary tissue is more abundant than that of any other tissue. These nerves are chiefly secretory nerves.

The cortex of the adrenal gland arises from a mass of cells in close relation to those from which the gonads are developed. The elements composing this tissue are rounded cells of a smaller average size than the medullary cells and containing a protoplasm abundant in lipoid globules.

Function of the Adrenals

The two parts of the adrenal gland not only differ in origin, but their functions are also in no way related. The only function of the adrenal medulla is the

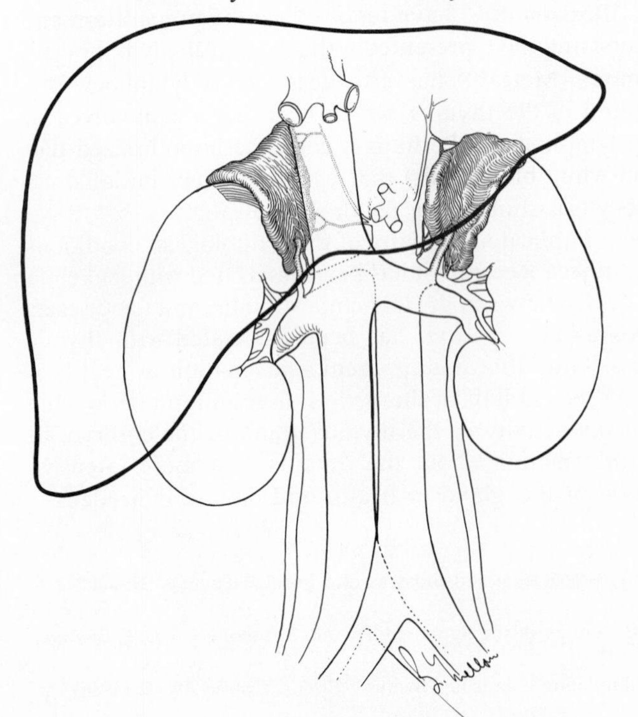

Figure 16-22. Adrenal glands showing their location, one on each side of the body, close to the upper pole of the kidney.

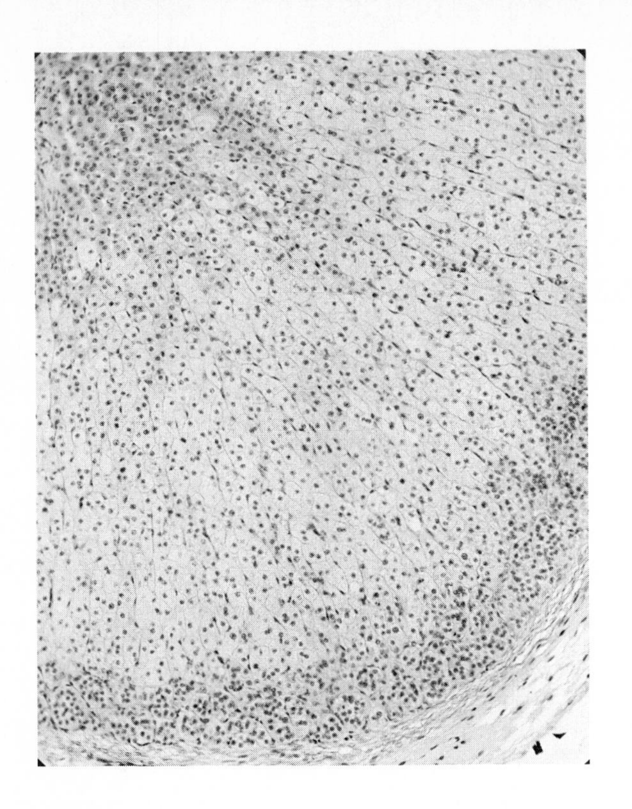

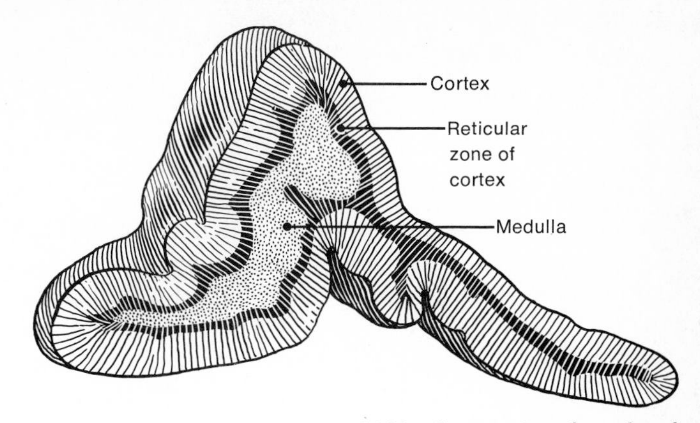

Figure 16-23. Cross section of an adrenal gland showing the medulla and cortex.

Figure 16-24. Section of mammalian adrenal gland, showing details of cortex and medulla. (Courtesy Ward's Natural Science Establishment, Inc., Rochester, N.Y.)

secretion of **epinephrine.** Its formation in the body depends on the amino acid tyrosine or phenylalanine, and it has the structural formula

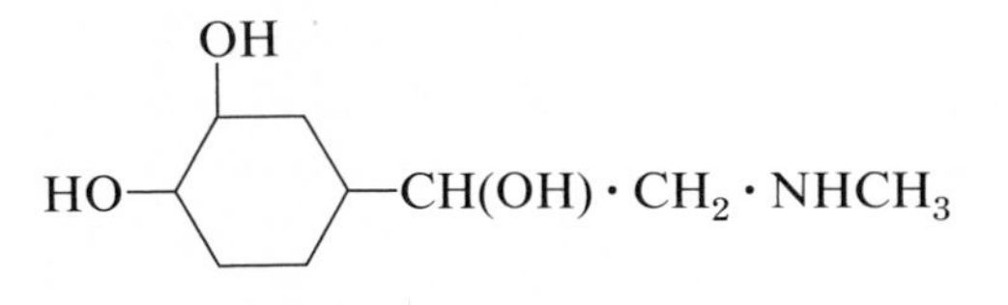

It is a nonalkaloid base, very slightly soluble in water, less soluble in alcohol, and insoluble in ether and other fat solvents.

Epinephrine has a cardiovascular, circulatory action, exerting a constricting action on the arteries, veins, and capillaries and also on the lymph vessels. However, the vasomotor effect is not the same in all areas. For example, the blood vessels of the brain and muscles react only slightly to epinephrine, and those of the adrenal, thyroid, and placental vessels not at all. In order for epinephrine to produce its effect it must be in a certain minimum concentration. If the concentration is below this minimum, there is no response, or even an opposite effect can result. Epinephrine normally produces a vasoconstriction, which gives rise to a hypertensive state, but very small concentrations can provoke vasodilatation and hypotension. This varying response to epinephrine in different vascular areas, as well as in different dosages, produces an increased blood volume in some organs and a reduced blood volume in others.

In general, epinephrine has the same effect on the organism as stimulation of the sympathetic nerves. However, epinephrine does not stimulate the sweat glands; sympathetic stimulation does. The rise in blood pressure after the release or injection of epinephrine is primarily the result of increased cardiac output. The action of epinephrine on the heart is, in general, to increase and accelerate contraction. This is effected by activation of both the SA and AV nodes, acceleration of conduction, and increase of irritability of the ventricular musculature. The coronary vessels of the heart and the skeletal muscles are relaxed (dilated) by epinephrine. It will also produce constriction of the small blood vessels chiefly of the splanchnic region and of the skin. As a result of the vasoconstriction and increased power of the heart, a general increase in blood pressure is induced.

Usually intravenous injection of only 0.005 mg will cause a measurable rise in blood pressure. Injection of 0.01 mg will cause a rise in blood pressure of 10 to 30 mm Hg beginning 10 to 20 seconds after injection, reaching its maximum in about 45 seconds and subsiding in 2 minutes. In normal subjects 0.04 to 0.05 mg will give a rise in blood pressure of 100 mm Hg. The degree of blood pressure attained and the subsequent blood pressure curve are also to some degree dependent on the condition of the autonomic nervous system. In those perons who are parasympathetic-sensitive (vagotonic), an initial fall in blood pressure may occur. The increased pressure in the ventricle and aorta reflexly stimulates the vagus. This vagus stimulation suffices to suppress the direct action of the epinephrine by slowing the heart and thus modifies the blood pressure.

Other effects produced by epinephrine are inhibition of the intestinal musculature and contraction of the cardiac, pyloric, ileocecal, and anal sphincters. Of practical importance is the dilating action of epinephrine on the smooth muscles of the bronchioles, which has been of great benefit in relieving spasm of the bronchioles in asthma. Epinephrine causes contraction of the sphincter musculature of the bladder and also the muscles of the trigonium. The urethra and ureters are also affected similarly, but no action can be demonstrated on the bladder itself. Epinephrine dilates the pupil of the eye and contracts the smooth muscle of the skin, so that after subcutaneous injection, gooseflesh is often observed. Epinephrine acts on all glands innervated by the sympathetic, except the sweat glands, and exerts a favorable effect on the contraction of skeletal muscle.

Epinephrine increases the blood sugar as a result of its effect on the liver; it enhances the transformation of liver glycogen into glucose. It also appears that epinephrine increases glycogen formation from sugar in the liver and accelerates the breakdown of glycogen into lactic acid in the muscles. The effect of epinephrine on the blood is to increase the number of red blood cells and platelets. A very frequent finding is the absolute increase in lymphocytes. These changes are brought about in part by the compression of the spleen resulting from the contraction of the smooth muscles fibers in the spleen capsule and in part perhaps from an increased formation and discharge of such cells from their sites of origin.

Most of the effects listed form the basis of the general pharmacology of epinephrine, and these same effects can also be produced by stimulation of the sympathetic nervous system. It becomes obvious that epinephrine mimics the effect of sympathetic excitation, and for this reason it is often referred to as a **sympathomimetic** compound. This relationship does not seem too unusual when we recall that the medullary portion of the adrenal has a common origin with sympathetic nerve cells.

One of the most puzzling and enigmatic actions of epinephrine, however, is that in certain areas it acts as a vasoconstrictor, whereas in other areas it acts as a vasodilator. To explain this dual action of epinephrine, several theories have been suggested although none is substantiated by experimental proof. One theory suggests that epinephrine stimulates not only the sympathetic but also the parasympathetic and that differences in concentration differentiate the type of stimulation. Minute doses of epinephrine may affect only the parasympathetic vasodilators. Under these circumstances it is assumed that the threshold of irritability of the parasympathetic fibers to epinephrine stimulation is lower than that of the sympathetic system.

Another possible explanation for the inverse reaction of epinephrine assumes two different types of sympathetic fibers; that is, the vasodilators that react to minimal doses of epinephrine, referred to above, are not parasympathetic fibers but are of a sympathetic nature (sympathetic vasodilators), and under various conditions epinephrine may produce in the "receptive substance" of the effector organ various opposing metabolic processes. The chemical nature of this receptive substance and the way it reacts with epinephrine are still unknown.

The secretion of the adrenal medulla is under control of the sympathetic nervous system, and in emergency situations or exposure to stress, such as injury, excessive muscular exercise, infection, hemorrhage, cold, fever, burn, nervous shock, and anoxia, epinephrine is discharged into the blood. The increase of epinephrine, produced by impulses passing over the splanchnic nerve into the medulla, helps the organism to re-establish the physiological equilibria that have been disturbed. Prolonged stimulation can cause exhaustion of the epinephrine stores. The importance of epinephrine secretion in conditions of emergency has been definitely established. Nevertheless, careful study has shown that the adrenal medulla, unlike the adrenal cortex, is not indispensable. Removal of the medullary areas of both adrenals in experimental animals causes no significant disturbances in any of the main functions of the organism if the adrenal cortex remains intact. The blood pressure, heart action, blood sugar level, and sympathetic responses, with the release of **sympathin** at sympathetic nerve endings, remain approximately the same. However, such animals are less adaptable to emergency situations, which provoke epinephrine secretion in the normal animal.

Sympathin is the chemical substance responsible for the transmission of nerve impulses through most sympathetic neuroeffector junctions. It resembles epinephrine in its action but is not identical with it. There are two kinds of sympathin, each causing a specific type of reaction. Inhibitory sympathin, referred to as **sympathin I,** is liberated at the neuroeffector junctions of the intestine, coronary vessels, and so on, causing dilatation. The other, **sympathin E,** is excitatory, causing constriction of blood vessels, increased heart rate, contraction of smooth muscle,

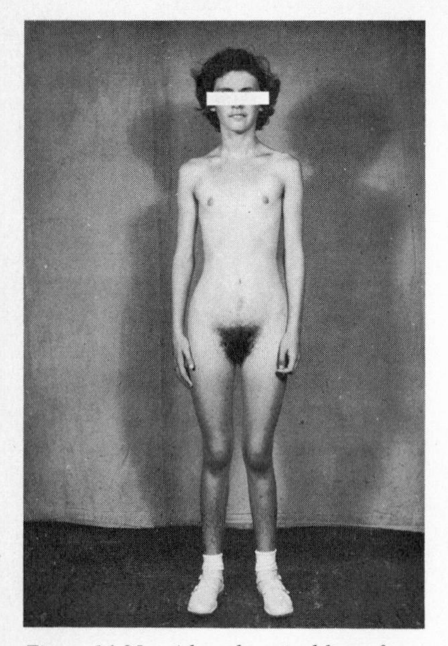

Figure 16-25. Adrenal cortical hyperfunction with premature puberty and virilism (pubic hair since 18 months of age). (Courtesy Armed Forces Institute of Pathology, Washington, D.C., picture 53-10936 [Br 9942-1].)

and so on. Epinephrine, as we have seen, causes both inhibitory and excitatory effects, according to the particular organ on which its action is exerted. Today, the terms *epinephrine* and *norepinephrine* (Arterenol) have replaced the terms *sympathin I* and *sympathin E.* Norepinephrine is the excitor sympathin in the sense that it is the over-all vasoconstrictor, except in the vessels of the skin. It raises the blood pressure through vasoconstriction without directly affecting cardiac output. Epinephrine is the over-all vasodilator, except that it constricts the skin vessels. It raises the blood pressure by increasing the heart rate and output through direct stimulating action on the heart muscle.

The **adrenal cortex** is necessary for life. We have seen that excision of the adrenal medulla is not fatal. However, complete removal of both adrenals leads to death within a short period of time. Approximately 28 steroid hormones have been separated from cortical extracts, but only six to seven have shown to have physiological activity. These are generally called **corticoids** and their effects, **corticoid effects.** Some of the steroids have a masculinizing effect; others have feminizing or progestational activity. Some are active in water and mineral metabolism—regulating the distribution of water, sodium, potassium, and chloride in the body, for example, aldosterone. The adrenal cortex secreting **glycocorticoids** also exerts an effect on carbohydrate metabolism. It increases the blood sugar and liver glycogen concentrations and stimulates gluconeogenesis from protein. Fat catabolism is also increased, resulting in the production of ketone bodies. Other corticoids aid the organism to resist many physical and chemical agents and increase resistance to infection. The cortical extract **cortisone** is now used therapeutically in certain disease states. When given in large doses, it has proved in certain instances to be successful in relieving symptoms of rheumatoid arthritis and improving rheumatic fever patients. Secretion of the adrenal cortex also functions importantly in interrelationships with other endocrine glands and tissues.

Abnormal Function of the Adrenals

The medullary portion of the adrenal gland is not often involved in disease, and no deficiency state has ever been recognized. However, hyperactivity of the adrenal medulla does occur as a result of certain tumors effecting the chromaffin cells. The symptom of excessive medullary activity is hypertension, which may eventually lead to complications such as coronary insufficiency, ventricular fibrillation, and pulmonary edema.

Degeneration of the adrenal cortex resulting in **hypoadrenocorticalism** causes **Addison's disease.** The symptoms of Addison's disease are pigmentation of the skin, loss of appetite, loss of weight, weakness, low blood pressure, and some anemia. There is a loss of sodium because of failure of the renal tubules to reabsorb this ion; at the same time, chloride and bicarbonate are lost, and potassium is retained. As a result of the changes in the ion concentration of the circulation, water is lost from the blood and the tissue spaces with severe dehydration and hemoconcentration.

The treatment of Addison's disease consists of the administration of cortisone orally and the implantation of deoxycorticosterone acetate (DOCA) pellets. DOCA may be used in buccal tablets, which are absorbed through the mucous membranes of the mouth, but the implantation of pellets relieves the patient from the bother of daily buccal therapy.

Hyperadrenocorticalism may be caused by cortical hyperplasia as a result of increased production of adrencorticotropic hormone from the pituitary or by tumors of the cortex. Since the adrenal cortex secretes both male and female hormones, among many others, certain tumors of the adrenals cause the production of increased amounts of these steroid hormones and give rise to physical abnormalities referred to as the **adrenogenital syndrome.** In the young male with adrenogenital syndrome there is precocious development of the penis and pubic hair, along with advanced bone age and early and increased growth rate. A girl born with the adrenogenital syndrome may show an enlarged clitoris and may be mistaken for a male; thus she may be raised as a boy (Fig. 16-25). In the adult female with adrenogenital syndrome there is repression of female characteristics with the assumption of male characteristics (female virilism) (Fig. 16-26). When it occurs in the male, there is excessive masculinization. Feminization of the male due to adrenocortical tumors is considered to be rare.

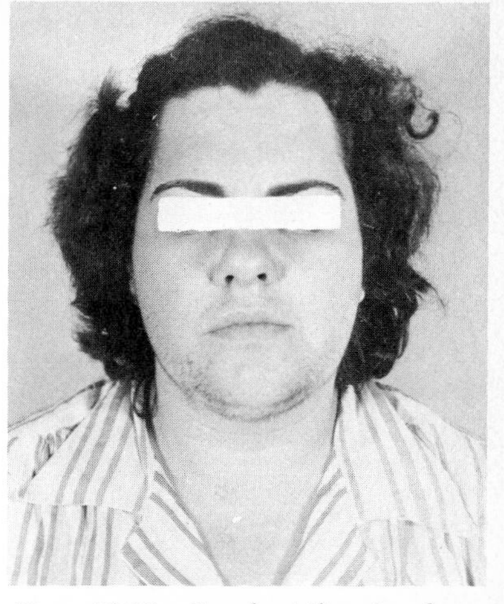

Figure 16-26. Female virilism in adrenogenital syndrome. (*Courtesy Armed Forces Institute of Pathology, Washington, D.C., picture Br 5216-3.*)

Other symptoms of hyperadrenocorticalism include retention of sodium and water, resulting in increased blood volume, hypertension, and edema; increased gluconeogenesis with hyperglycemia and glycosuria; potassium depletion; and hirsutism.

The Pancreas

The **pancreas** serves a dual function in man. The bulk of the gland produces an external secretion concerned with digestion, and certain isolated groups of cells, the **islets of Langerhans,** produce at least one and probably two hormones.

The inlet tissue is easily differentiated from the exocrine tissue and accounts for approximately 1 to 2 per cent of the weight of the pancreas. The islets consist of discrete groups of cells and are widely scattered throughout the pancreas, numbering up to two million. There are two main types of cells in the islets. The **alpha cells** are large cells with large spherical nuclei and granular cytoplasm. The granular material is thought to be **glucagon,** or the **hyperglycemic factor** of the pancreas. The **beta cells** are small cells, lying mainly at the center of the islets nearer the rich capillary supply. They contain large numbers of acidophil granules during the inactive phase of the gland. These granules are believed to be the hormone **insulin.**

Insulin is a protein with a molecular weight of approximately 34,500. It has been isolated from the pancreas and was first prepared in crystalline form by Abel in 1926. The hormone is inactivated by proteolytic enzymes, and for this reason it cannot be administered orally. The structure of insulin, as determined by Sanger, includes a relatively high content of leucine, tyrosine, glutamic acid, and cystine and a low content of methionine, phenylalanine, and proline.

The part that insulin plays in carbohydrate metabolism has already been discussed in Chapter 12. In general, it promotes the removal of glucose from the blood by four available routes of disposal. After the injection of insulin the following effects are observed:

1. An increase in the conversion of glucose to glycogen in the liver.
2. An increase in the conversion of glucose to glycogen in the muscle.

3. An increase in the oxidation of glucose by the tissue.
4. An increase in the rate of conversion of carbohydrate to fat and an increase in the rate of protein synthesis.

Excessive amounts of insulin in the blood cause a rapid fall of the blood sugar level, resulting in a **hypoglycemic** condition. This is characterized by symptoms of weakness, faintness, tremors, and convulsions. If adequate amounts of glucose are administered, the symptoms are relieved.

In reviewing the fate of carbohydrate in the body, which is discussed in Chapter 12, we find that, regardless of the eventual metabolic end of glucose, it must first be converted to glucose-6-phosphate as follows:

$$\text{Glucose} + \text{ATP} \xrightarrow{\text{hexokinase}} \text{Glucose-6-phosphate} + \text{ADP}$$

This reaction precedes all subsequent reactions involving glucose, and it is generally regarded to be the rate-determining step in the utilization of glucose. It is only through this reaction that glucose can be converted via various other intermediates into muscle glycogen, liver glycogen, or fat. Glycogen and fat can later be mobilized and oxidized to provide energy, or glucose can be directly utilized as a source of energy. Glucose can be released from glucose-6-phosphate only in the liver, and thus only hepatic glycogen provides an immediate source of blood sugar. The reason for this is that the reaction

$$\text{Glucose-6-phosphate} \xrightarrow{\text{phosphatase}} \text{Glucose} + \text{PO}_4$$

requires an enzyme which is a specific glucose-6-phosphatase and is found only in the liver. Against this brief background the gross effect of insulin may now be considered.

Mechanism of Insulin Action

Three theories of insulin action have been proposed. One theory proposes that insulin indirectly affects hexokinase activity by counteracting the inhibitory effects of pituitary and adrenocortical hormones on the enzyme. The experimental evidence substantiating this theory was accumulated from a series of studies carried out by Cori (1945–1946), Price et al. (1945), and Colowick et al. (1947). They found that the hexokinase activity in muscle of normal rats can be inhibited by pituitary extract and that the inhibition can be counteracted by the addition of insulin. Insulin alone has no significant action on the enzyme hexokinase.

A second theory proposes that insulin promotes the utilization of glucose by promoting the passage of glucose from the extracellular fluid compartment into certain cell systems by increasing the permeability of cell membranes to glucose. In this theory the actions of insulin are divorced from the complex enzymatic events of carbohydrate metabolism.

The third theory proposes that insulin acts in the oxidative phosphorylating reactions that generate high-energy phosphate in the form of ATP. The proponents of this theory believe that a defect in oxidative phosphorylation could account for most of the observed defects in metabolism occurring in the diabetic person.

These three theories represent the current thought on the mechanism of insulin action. It is obvious that the exact mechanism of action of this hormone remains to be defined.

The secretion of insulin is not under nervous or hormonal control. Experimental evidence indicates that the concentration of glucose in the blood flowing through the pancreas regulates the rate at which insulin is released into the blood stream.

Glucagon, the other possible hormone of the pancreas, was first discovered in pancreatic extracts, and most workers experienced difficulty in separating it from insulin. There is considerable evidence that it is formed by the alpha cells, although it can also be obtained from certain portions of the gastric and intestinal mucosa. The principal action of glucagon is to stimulate the conversion of glycogen to glucose in the liver. Additionally, glucagon has been shown to inhibit gastric motility and to relieve hunger. A rise in blood insulin and a fall in blood glucose stimulate the secretion and release of this hormone.

Abnormalities of Pancreatic Function

Hyperinsulinism, denoting an excessive amount of insulin in the blood, commonly results from improper insulin medication and occasionally from tumors of the pancreatic islets. The resulting symptoms are caused by the fall in blood sugar, a condition referred to as **hypoglycemia.** One of the first symptoms may be blurring of vision and an inability to focus the eyes. A feeling of drowsiness and yawning usually occurs, and the hypoglycemic person may become excited or perspire or appear to be under the influence of alcohol. Convulsive seizures may occur, and finally the patient will go into coma and die.

The symptoms of hypoglycemia are relieved almost immediately after the intravenous injection of dextrose unless organic changes in the brain have already occurred. Hypoglycemia deprives the brain of glucose, on which it is almost exclusively dependent for its energy source. When the hypoglycemia is due to an organic cause, particularly pancreatic tumor, surgical removal is required.

A deficiency of insulin, known as **hypoinsulinism,** occurs when the pancreas is removed surgically or when there is a decreased secretory activity or complete inactivation of the islet tissue. Under these conditions the peripheral utilization of glucose decreases, but the breakdown of glycogen to glucose as well as the conversion of fat and protein to glycogen is accelerated. As a result, the level of blood glucose rapidly rises, producing a condition known as **hyperglycemia,** which results in a series of symptoms referred to as **diabetes mellitus.** Diabetes is considered to be a metabolic disease characterized by a disturbance in carbohydrate metabolism and by secondary disturbances in the metabolism of protein and fat. The hyperglycemia that results always gives rise to a **glycosuria,** since the kidney's capacity for reabsorbing glucose is exceeded; hyperglycemia and glycosuria are considered to be definite evidence clinically of diabetes mellitus. The unreabsorbed glucose in the renal tubules acts as a diuretic, so that polyuria develops, which consequently leads to **dehydration.** Further, with glucose being unavailable for the production of energy, large amounts of fat are mobilized and oxidized, resulting in the production of ketone bodies in quantities that exceed the capacity of the tissue to oxidize them. As a result, ketones such as beta-hydroxybutyric acid and acetoacetic acid accumulate in the blood (ketosis), leading to an **acidosis.** If the diabetic state is allowed to persist, the excretion of nitrogen is increased. The loss of protein contributes to the decrease in body weight. The increased acidity of the blood stimulates the respiratory center, resulting in hyperventilation; the terminal stage of diabetic acidosis is characterized by coma. The exact mechanism of coma production in diabetes is not known. Some observers believe that the depression of cerebral metabolism is due to the increased acidity of the body fluids and the toxic action of ketone bodies on the brain cells.

The administration of insulin to diabetic patients rapidly restores the ability to oxidize carbohydrates and to form glycogen. The blood sugar falls to normal levels. The production of ketone bodies from increased fat metabolism is retarded, and those present in the extracellular fluid are soon oxidized and the acidosis is relieved. The dehydration may be relieved by the intravenous infusion of large volumes of isotonic saline solution.

The Intestinal Glands and Gonads

The intestinal glands are endocrine glands, owing to their secretion of gastrin, secretin, and cholecystokinin. The physiological significance of these hormones has been studied together with the digestive system in Chapter 11. The gonads are endocrine glands of great importance, owing to the secretion of testosterone by the testes and of estrogen and progesterone by the ovaries. For the hormonal function of the gonads the student is referred to Chapter 15.

References

Axelsson, J. "Catecholamine Functions," in: *Annual Review of Physiology,* Vol. 33. Palo Alto, Calif.: Annual Reviews, 1971, p. 1.

Bittar, E. E. (ed.). *The Biological Basis of Medicine,* Vol. 2: *The Hormones.* New York: Academic, 1968.

Blum, J. J. (ed.). *Biogenic Amines as Physiological Regulators.* Englewood Cliffs, N.J.: Prentice-Hall, 1970.

Briggs, M. H. (ed.). *Advances in Steroid Biochemistry and Pharmacology.* New York: Academic, 1970.

Fallis, B. D., and R. D. Ashworth. *A Textbook of Human Histology.* Boston: Little, Brown, 1970.

Foà, P. P. *The Action of Hormones.* Springfield, Ill.: Thomas, 1971.

Ramwell, P., and J. E. Shaw. "Prostaglandins." *Ann. N.Y. Acad. Sci.,* **180** (1971).

Root, A. *Human Pituitary Growth Hormone.* Springfield, Ill.: Thomas, 1971.

Sawin, C. T. *The Hormones, Endocrine Physiology.* Boston: Little, Brown, 1970.

Turner, C. D., and J. T. Bagnara. *General Endocrinology.* Philadelphia: Saunders, 1971.

Yates, F. E., S. M. Russell, and J. W. Maran. "Brain-Adenohypophysial Communication in Mammals," in: *Annual Review of Physiology.* Palo Alto, Calif.: Annual Reviews, 1971, pp. 393–444.

Index

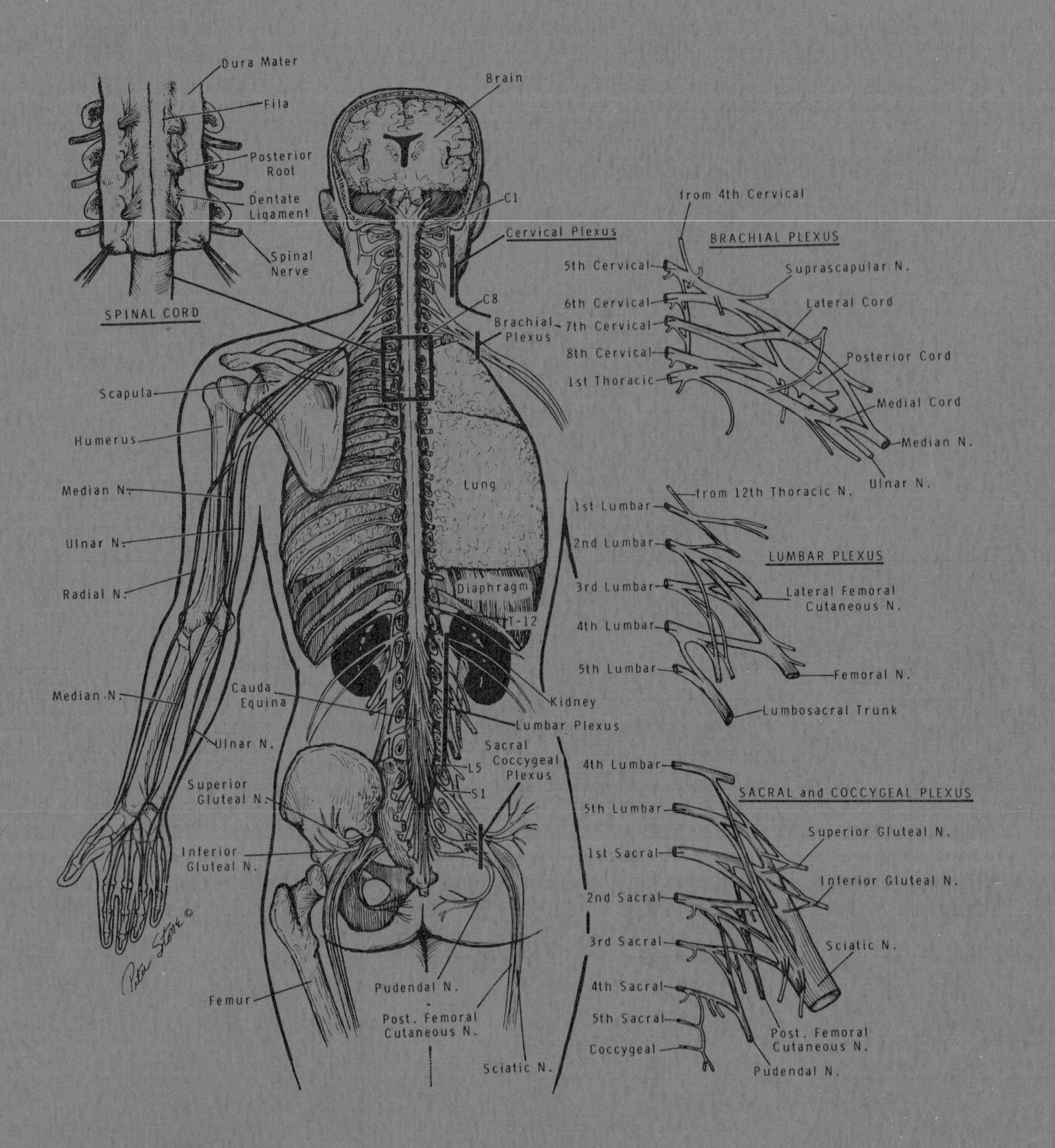
Dura Mater
Fila
Posterior Root
Dentate Ligament
Spinal Nerve
SPINAL CORD
Brain
C1
Cervical Plexus
C8
Brachial Plexus
Scapula
Humerus
Median N.
Ulnar N.
Radial N.
Lung
Diaphragm
T-12
Median N.
Cauda Equina
Kidney
Lumbar Plexus
Ulnar N.
Sacral Coccygeal Plexus
L5
S1
Superior Gluteal N.
Inferior Gluteal N.
Pudendal N.
Post. Femoral Cutaneous N.
Femur
Sciatic N.
from 4th Cervical
BRACHIAL PLEXUS
5th Cervical
6th Cervical
7th Cervical
8th Cervical
1st Thoracic
Suprascapular N.
Lateral Cord
Posterior Cord
Medial Cord
Median N.
Ulnar N.
from 12th Thoracic N.
1st Lumbar
2nd Lumbar
3rd Lumbar
4th Lumbar
5th Lumbar
LUMBAR PLEXUS
Lateral Femoral Cutaneous N.
Femoral N.
Lumbosacral Trunk
SACRAL and COCCYGEAL PLEXUS
4th Lumbar
5th Lumbar
1st Sacral
2nd Sacral
3rd Sacral
4th Sacral
5th Sacral
Coccygeal
Superior Gluteal N.
Inferior Gluteal N.
Sciatic N.
Post. Femoral Cutaneous N.
Pudendal N.